PIPELINE
RULES OF THUMB
HANDBOOK

PIPELINE
RULES OF THUMB
HANDBOOK

A Manual of Quick, Accurate Solutions to Everyday Pipeline Engineering Problems

E. W. McAllister, Editor

AMSTERDAM • BOSTON • HEIDELBERG • LONDON
NEW YORK • OXFORD • PARIS • SAN DIEGO
SAN FRANCISCO • SINGAPORE • SYDNEY • TOKYO
Gulf Professional Publishing is an imprint of Elsevier

Gulf Professional Publishing is an imprint of Elsevier.
225 Wyman Street, Waltham, MA 02451, USA
The Boulevard, Langford Lane, Kidlington, Oxford OX5 1GB, UK

Seventh edition 2009

Library of Congress Cataloging-in-Publication Data
A catalog record for this book is available from the Library of Congress.

British Library Cataloguing-in-Publication Data
A catalogue record for this book is available from the British Library.

ISBN: 978-0-12-387693-5

For information on all Gulf Professional Publishing publications
Visit our Web site at www.books.elsevier.com

Transferred to Digital Printing in 2014

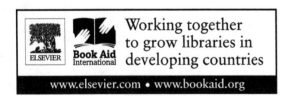

Contents

As an added bonus to this edition, Elsevier has created online interactive engineering widgets to accompany selected material found in this book. To make the selected table or graph interactive, please scan the code provided next to the table or graph, and you will be sent to an online companion website of widgets. You will need internet access in order to link to the widgets.

2: Construction 43

3: Pipe Design...................... 93

4: Electrical Design 125

Floodlighting Concepts 143

5: Hydrostatic Testing 149

The Benefits and Limitations of Hydrostatic Testing 150

Appendix A 167

Appendix B 169

6: Pipeline Drying 173

Pipeline Dewatering, Cleaning, and Drying 174

13: Liquids—Hydraulics..... 413

Marine Hose Data 414

Pressure Loss through Valves and Fittings 431

14: Pumps 471

Pulsation Control for Reciprocating Pumps................................. 487

Key Centrifugal Pump Parameters and How They Impact Your Applications—Part 1 500

Key Centrifugal Pump Parameters and How They Impact Your Applications—Part 2 506

1: General Information

1

Basic Formulas

1. Rate of Return Formulas:

$$S = P(1+i)^n$$

a. Single payment compound amount (SPCA). The $(1+i)^n$ factor is referred to as the compound amount of $1.00.

b. Single payment present worth (SPPW):

$$P = S\left[\frac{1}{(1+i)^n}\right].$$

The factor $[1/(1+i)^n]$ is referred to as the present worth of $1.00.

c. Uniform series compound amount (USCA):

$$S = R\left[\frac{(1+i)^n - 1}{i}\right]$$

The factor $= \left[\frac{(1+i)^n - 1}{i}\right]$

is referred to as the compound amount of $1.00 per period.

d. Sinking fund deposit (SFD):

$$R = S\left[\frac{i}{(1+i)^n - 1}\right].$$

The factor $= \left[\frac{i}{(1+i)^n - 1}\right]$

is referred to as the uniform series, which amounts to $1.00.

e. Capital recovery (CR):

$$R = S\left[\frac{i}{(1+i)^n - 1}\right] = P\left[\frac{i(1+i)^n}{(1+i)^n - 1}\right].$$

The factor $= \left[\frac{i(1+i)^n}{(1+i)^n - 1}\right]$

is referred to as the uniform series that $1.00 will purchase.

f. Uniform series present worth (USPW):

$$P = R\left[\frac{(1+i)^n - 1}{i(1+i)^n}\right]$$

The factor $[((1+i)^n - 1)/i(1+i)^n]$ is referred to as the present worth of $1.00 per period.

Where

$P =$ A present sum of money,
$S =$ A sum of money at a specified future date,
$R =$ A uniform series of equal end-of-period payments,
$n =$ The number of interest periods,
$i =$ The interest rate earned at the end of each period.

Mathematics—areas

Form		Method of Finding Area
Triangle		Base x 1/2 perpendicular height $$\sqrt{s(s-a)(s-b)(s-c)}$$ s = 1/2 sum of the three sides a,b,c
Trapezium		Sum of area of the two triangles
Trapezoid		1/2 sum of parallel sides x perpendicular height
Parallelogram		Base x perpendicular height
Reg. polygon		1/2 sum of sides x inside radius
Circle		πr^2 = 0.78540 x $diam^2$. = 0.07958 x $circumference^2$
Sector of a circle		$\dfrac{\pi r^2 A^\circ}{360}$ = $0.0087266 r^2 A^\circ$ = arc x 1/2 radius
Segment of a circle		$r^2 \left(\dfrac{\pi A^\circ}{180} - \sin A^\circ \right)$
Circle of same area as a square		Diameter = side x 1.12838
Square of same area as a circle		Side = diameter x 0.88623
Ellipse		Long diameter x short diameter x 0.78540
Parabola		Base x 2/3 perpendicular height
Irregular plane surface		The larger the value of n, the greater the accuracy of approximation (Simpson's Rule). $$A = \frac{h}{3}\Big[(y_o + y_n) + 4(y_1 + y_3 + \ldots + y_{n-1}) + 2(y_2 + y_4 + \ldots + y_{n-2})\Big]$$

Mathematics—surfaces and volumes

Method of Finding Surfaces and Volumes of Solids
S = lateral or convex surface   V = Volume

Parallelopiped

S = perimeter P perp. to sides x lat. length, l

V = area of base x perpendicular height, h

V = area of section A perp. to sides x lat. length, l

Prism-right, oblique, regular or irregular

S = perimeter P perp. to sides x lat length l

V = area of base x perp. height, h

V = area of section A perp. to sides x lat. length, l

Cylinder-right, oblique, circular or elliptic, etc.

S = perimeter of base x perp. height

V = area of base x perp. height

V = area of section A perp. to sides x lat. length, l

Frustum of any prism or cylinder

V = area of base x perpendicular distance h, from base to center of gravity of opposite face. For cylinder, $\frac{1}{2}A(l_1+l_2)$

Pyramid or Cone, right or regular

S = perimeter of base x 1/2 slant height, l

V = area of base x 1/3 perp. height, h

Pyramid or Cone-right, oblique, regular or irregular

V = area of base x 1/3 perp. height, h

V = 1/3 vol. of prism or cylinder of same base and perp height

V = 1/2 vol. of hemisphere of same base and perp. height

Frustum of pyramid or cone, right and regular, parallel ends

S = (sum of perimeter of base and top) x 1/2 slant height, l

$$V = \frac{h}{3}\left(A_1 + A_2 + \sqrt{A_1 x A_2}\right)$$

Where A_1 and A_2 are the areas of the bases

Frustum of any pyramid or cone, parallel ends

$$V = \frac{h}{3}\left(A_1 + A_2 + \sqrt{A_1 x A_2}\right)$$

Where A_1 and A_2 are the areas of the bases

Sphere

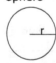

$S = 4\pi r^2$

$V = 4/3\pi r^3$

Spherical Sector

$V = 2/3\pi r^2 b$

$S = 1/2\pi r(4b+c)$

Spherical Zone

$S = 2\pi rb$

$V = 1/24\pi b(3a^2+3c^2+4b^2)$

Rules of exponents

$$a^n \times a^m = a^{n+m},$$

$$a^n / a^m = a^{n-m},$$

$$(a^n)^m = a^{nm},$$

$$(ab)^n = a^n b^n,$$

$$(a/b)^n = a^n / b^n,$$

$$a^{n/m} = \left(a^{1/m}\right)^n.$$

Recommended drill sizes for self-tapping screws

Self-Tapping Screw Size			Major Thread Diameter			Minor Thread Diameter			For Heavy Metals		For Light Metals	
No.	Threads/In	OD	Max	Mean	Min	Max	Mean	Min	Drill Size		Drill Size	
2	32	0.086	0.088	0.0850	0.082	0.064	0.0620	0.0060	49	0.0730	49	0.0730
4	24	0.112	0.114	0.1110	0.108	0.086	0.0840	0.082	41	0.0960	41	0.0960
5	20	0.125	0.130	0.1265	0.123	0.094	0.0920	0.090	36	0.1065	36	0.1065
6	20	0.138	0.139	0.1355	0.132	0.104	0.1015	0.099	32	0.1160	32	0.1160
7	19	0.151	0.154	0.1505	0.147	0.115	0.1120	0.109	30	0.1285	30	0.1285
8	18	0.164	0.166	0.1625	0.159	0.122	0.1190	0.116	28	0.1405	29	0.1360
10	16	0.190	0.189	0.1855	0.182	0.141	0.1380	0.135	20	0.1610	21	0.1590
12	14	0.216	0.215	0.2115	0.208	0.164	0.1605	0.157	13	0.1850	14	0.1820
1/4	14	0.250	0.246	0.2415	0.237	0.192	0.1885	0.185	3	0.2130	4	0.2090
5/16	12	0.313	0.315	0.3105	0.306	0.244	0.2400	0.236	I	0.2720	H	0.2660
3/8	12	0.375	0.380	0.3755	0.371	0.309	0.3040	0.299	R	0.3390	Q	0.3320

Determine pulley speed

Speed of Driven Pulley Required:

Diameter and speed of driving pulley and diameter of driven pulley are known.

$D_1 =$ Diameter of driving pulley 15 in,
$\text{RPM}_1 = 180$ (driving pulley speed),
$d_2 =$ Diameter of driven pulley 9 in,
$\text{RPM}_2 =$ Speed of driven pulley,

$$\text{RPM}_2 = \frac{15 \times 180}{9} = 300 \text{ RPM}.$$

Diameter of Driven Pulley Required:

Diameter and speed of driving pulley and speed of driven pulley are known.

$D_1 =$ Diameter of driving pulley 24 in,
$\text{RPM}_1 = 100$ (driving pulley speed),

$$d_2 = \frac{24 \times 100}{600} = 4 \text{ in},$$

$\text{RPM}_2 =$ Speed of driven pulley $= 600$.

Diameter of Driving Pulley Required:

$D_1 =$ in,
$d_2 = 36$ in,
$\text{RPM}_2 = 150,$
$\text{RPM}_1 = 600,$

$$D_1 = \frac{36 \times 150}{600} = 9 \text{ in}.$$

Speed of Driving Pulley Required:

$D_2 = 4$ in,
$\text{RPM}_2 = 800$,
$D_1 = 26$ in,

$$\text{RPM}_1 = \frac{4 \times 800}{26} = 123.$$

Speed of Driven Pulley in Compound Drive Required:

$\text{RPM}_A = 260$,
$\text{RPM}_D = 720$ (required),

$$\frac{260}{720} = {}^{13}\!/_{36} \text{ required speed ratio.}$$

Resolve $^{13}\!/_{36}$ into two factors $\dfrac{1 \times 13}{2 \times 18}$.

Multiply by trial numbers 12 and 1:

$$\frac{(1 \times 12) \times (13 \times 1)}{(2 \times 12) \times (18 \times 1)} = \frac{12 \times 13}{24 \times 18}.$$

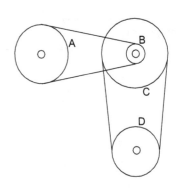

The values 12 and 13 in the numerator represent the diameter of driven pulleys B and D, and the values 24 and 18 in the denominator represent the diameter of the driving pulleys A and B.

Calculate volume in horizontal storage tank with ellipsoidal or hemispherical heads

Total volume = volume in 2 heads + volume in cylinder.
Total volume = $1/6 m K_1 D^3 + \frac{1}{4} m D^2 L$.

$K_1 = 2b/D$,
$Z_e = H_1/D$,
$Z_c = H_1/D$,

Partial volume = $1/6 m K_1 D^3 \times [f(Z_e)] + 1/4\, m D^2 L \times [f(Z_c)]$,

$$F(Z_c) = \left(\frac{\alpha - \sin \alpha \times \cos \alpha}{\pi} \right) \text{ horizontal cylinder}$$
coefficient,

$$F(Z_e) = \left(\frac{H_1}{D} \right)^2 \times \left(-3 + \frac{2H_1}{D} \right) \text{ ellipsodial coefficient,}$$

$$\alpha = 2 \times A \tan \left(\frac{H}{\sqrt{\left(2 \times H_1 \times \dfrac{D}{2} \right) - H_1^2}} \right) \text{ where } \alpha \text{ is in radius}$$

For elliptical 2:1 heads, $b = 1/4D$, $K_1 = \frac{1}{2}$.

Example: Find total volume.

$L = 50$ ft,
$D = 20$ ft,
$H_1 = 6$ ft,
$b = 4$ ft.

Total volume = $1/6\pi K_1 D^3 + \frac{1}{4}\pi D^2 L$

$K_1 = 2b/D = (2 \times 4)/20 = 8/20 = 0.4$.

Total volume = $0.1667 \times 3.1416 \times 0.4 \times 8000 + 0.25 \times 3.1416 \times 400 \times 50 = 17,383.86$ cu ft.

ASTM standard reinforcing bars

Soft Metric Size	Nom Diam (mm)	Area (mm²)	Weight Factors		Imperial Size	Nom Diam (in)	Area (in²)	Weight Factors	
			kg/m	kg/ft				kg/m	kg/ft
10	9.5	71	0.560	0.171	3	0.375	0.11	0.376	1.234
13	12.7	129	0.994	0.303	4	0.500	0.20	0.668	2.192
16	15.9	199	1.552	0.473	5	0.625	0.31	1.043	3.422
19	19.1	284	2.235	0.681	6	0.750	0.44	1.502	4.928
22	22.2	387	3.042	0.927	7	0.875	0.60	2.044	6.706
25	25.4	510	3.973	1.211	8	1.000	0.79	2.670	8.760
29	28.7	645	5.060	1.542	9	1.128	1.00	3.400	11.155
32	32.3	819	6.404	1.952	10	1.270	1.27	4.303	14.117
36	35.8	1006	7.907	2.410	11	1.410	1.56	5.313	17.431
43	43.0	1452	11.384	3.470	14	1.693	2.25	7.650	25.098
57	57.3	2581	20.239	6.169	18	2.257	4.0	13.600	44.619

Pressure rating for carbon steel flanges

Temperature (°F)	Flange Class						
	150	300	400	600	900	1500	2500
<100	285	740	990	1480	2220	3705	6170
200	260	675	900	1350	2025	3375	5625
300	230	655	875	1315	1970	3280	5470

CABLES AND ROPES

Estimating strength of cable

Rule

1. Change line diameter to eighths
2. Square the numerator
3. Divide by the denominator
4. Read the answer in tons

Example. Estimate the strength of ½-in. steel cable:

$$\text{Diameter} = \frac{1}{2} = \frac{4}{8}$$

$$\frac{4^2}{8} = \frac{16}{8} = 2$$

The approximate strength of ½-in steel cable is 2 tons.

Find the working strength of Manila rope

The working strength of Manila rope is approximately $900 \times (\text{diameter})^2$:

$$W = 900d^2,$$

where d is expressed in inches and W is given in pounds.

Example. What is the working strength of a ¾ in Manila rope?

The maximum recommended pull is

$$W = 900 \times \frac{3}{4} \times \frac{3}{4} = 506 \text{ lb.}$$

Example. Find the maximum working pull for a 1½ in Manila rope.

$$W = 900 \times \frac{3}{2} \times \frac{3}{2} = 2025 \text{ lb.}$$

For rope diameters greater than 2 in, a factor lower than 900 should be used. In working with heavier rigging it is advisable to refer to accepted handbooks to find safe working strengths.

How large should drums and sheaves be for various types of wire rope?

Type of Wire Rope	6 × 19	6 × 37	8 × 19	5 × 28	6 × 25	18 × 7	6 × 7
For best wear	45	27	31	36	45	51	72
Good practice	30	18	21	24	30	34	42
Critical	16	14	14	16	16	18	28

The diameter of sheaves or drums should preferably fall within the table[a] given above for most efficient utilization of the wire rope.

Example. What size should the hoisting drum on a dragline be, if the wire rope is 6 × 19 construction, ¾ in in diameter?

From the table, good practice calls for 30 diameters, which in this instance would be 22½ in. Loads, speeds, bends, and service conditions will also affect the life of wire rope, so it is better to stay somewhere between the "good practice" and "best wear" factors in the table.

[a] Construction Methods and Machinery, by F. H. Kellogg, Prentice-Hall, Inc., 1954.

Find advantages of block and tackle, taking into account pull out friction

The efficiency of various sheaves differs. For one with roller bearings the efficiency has been estimated at 96.2%. For plain bearing sheaves a commonly used figure is 91.7%. The following formula will give close results:

$$MA = \frac{W}{w} = E\frac{1 - E^n}{1 - E},$$

where

MA = Mechanical advantage,
W = Total weight to be lifted by the assembly,
w = Maximum line pull at the hoist,
n = Number of working parts in the tackle,
E = Efficiency of individual sheaves.

It is assumed that the line leaving the upper block goes directly to the hoist without additional direction change (requiring a snatch block).

Example. Find the mechanical advantage of a four-part block and tackle using upper and lower blocks having journal bearings, which have an efficiency of 91.7%.

$$MA = 0.917\frac{1 - 0.917^4}{1 - 0.917} = 0.917,$$

$$\left(\frac{1 - 0.707^4}{1 - 0.917}\right) = 0.917\left(\frac{0.293}{0.083}\right),$$

$$MA = 3.25.$$

If the load weighed 3250 lb, what pull would be required on the lead line?

$$\frac{W}{w} = MA,$$

$$\frac{3250}{w} = 3.25,$$

$$w = 1000 \text{ lb}.$$

Safe loads for wire rope

Safe Loading Factors For Wire Rope and Chains

Dia. In.	SINGLE PART LINE Wire Rope (Tons)	SINGLE PART LINE Alloy Chain (Tons)	SINGLE PART LINE Wrought Chain (Tons)	SINGLE PART-CHOKER HITCH Wire Rope (Tons)	SINGLE PART-CHOKER HITCH Alloy Chain (Tons)	SINGLE PART-CHOKER HITCH Wrought Chain (Tons)	TWO PARTS 60° Wire Rope (Tons)	TWO PARTS 60° Alloy Chain (Tons)	TWO PARTS 60° Wrought Chain (Tons)	TWO PARTS 30° Wire Rope (Tons)	TWO PARTS 30° Alloy Chain (Tons)	TWO PARTS 30° Wrought Chain (Tons)
¼	.5	....	....	.35	....	....	.6	....	....	.35	....	....
9/32	.5	1.3	....	....	1	....	....	1.8	....	....	1	....
⅜	1	2.3	.8	.75	1.8	.63	1.4	3.1	1.1	.8	1.8	.64
7/16	1	3	1.1	....	2.4	.9	....	4.2	1.5	....	2.4	.9
½	1.8	4	1.4	1.4	3.2	1.1	2.4	5.6	1.9	1.4	3.2	1.1
9/16	1.8	5	....	....	4	....	....	7	....	....	4	....
⅝	2.8	6	2.2	2.1	4.8	1.8	3.7	8.2	3	2.2	4.8	1.8
¾	3.9	7.8	3.2	2.8	6.2	2.6	5.2	10.7	4.4	3	6.2	2.6
⅞	5.2	10.4	4.3	3.7	8.3	3.4	7	14.4	6	4	8.3	3.4
1	6.6	13.8	5.7	4.6	10.8	4.6	9	18.7	7.8	5.2	10.8	4.6
1⅛	6.6	17.8	6.7	....	14.2	5.4	....	24.7	9.3	....	14.2	5.4
1¼	9.7	23.	8.3	6.6	18.2	6.6	13	31.9	11.5	7.5	18.4	6.6
1⅜	9.7	....	10	....	....	8	....	....	13.9	....	....	8
1½	13.2	....	12	9.2	....	9.6	17.3	....	16.6	10	....	9.6
1⅝	13.2	....	14	....	....	11.2	....	....	19.4	....	....	11.2
1¾	17	....	16	12.1	....	12.8	21.5	....	22.2	12.4	....	12.8
2	20.6	....	21	15.4	....	16.8	26.7	....	29.1	15.4	....	16.8

There are various kinds of wire rope and chain available, and where it is possible strength factors should be reviewed before using any particular kind. Tables here may be used on construction jobs. They allow for a safety factor of four as based on mild plow steels, etc. Where the wire rope is made of improved plow steel, etc., it will have a greater safety factor.

Stress in guy wires

Guys are wire ropes or strands used to hold a vertical structure in position against an overturning force. The most common types of guyed structures are stacks, derricks, masts for draglines, reversible tramways, and radio transmission towers.

As a general rule, stresses in guys from temperature changes are neglected, but in structures such as radio masts this is an important feature and must be subject to special analysis. The number of guys used for any particular installation is contingent on several variable factors such as type of structure, space available, contour of the ground, etc., and is not a part of this discussion.

It is desirable to space guys uniformly whenever possible. This equalizes the pull, P, on each guy insofar as possible, particularly against forces that change in direction, as when a derrick boom swings in its circle.

It is also desirable to equalize the erection tensions on the guys. When no external force is acting on the structure, the tension in each guy should be the same. A tension indicator is sometimes used to determine the tension in guys. If this instrument is not available, the tension can be very closely approximated by measuring the deflection at the center of the span from the chord drawn from the guy anchorage to the point of support on the structure. A good average figure to use for erection tension of guys is 20% of the maximum working tension of the guy.

This discussion outlines the method for determining the stresses in guys. One of the first considerations is the location of the guy anchorages. The anchorages should be so located that the angle α, between the horizontal plane and the guy line, is the same for all guys (to equalize erection tensions). Angle α, in good practice, seldom exceeds 45°, with 30° being commonly used. The tension in the guys decreases as angle α becomes smaller. The direct load on the structure is also less with a smaller value of α.

To find the maximum extra tension, T, that will be applied to any single guy by the force, F, first determine the pull, P, which is the amount required along the guys, in the same vertical plane as the force to resist the horizontal component

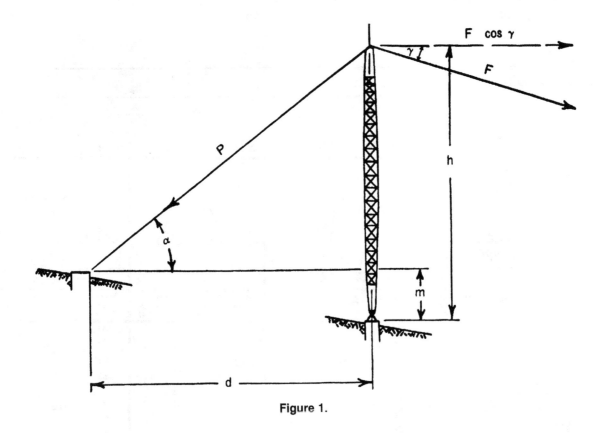

Figure 1.

of the force. This pull is entirely independent of the number of guys. Assume that the following are known:

F = The total resultant external force acting on the structure,
G = The angle between the horizontal plane and the force F,
h = The height of the structure,
d = The horizontal distance from structure to guy anchorage,
m = The vertical height of anchorage above or below the base of the structure,

The horizontal component of the force, $F = F \cos \gamma$.

α = The angle whose tangent is $(h \pm m) \div d$.

m is added if the anchorage is below the base of the structure and subtracted if it is above.

$$P = F \cos \gamma \div \cos \alpha.$$

As $\cos \alpha$ is always less than 1, P is always greater than $F \cos \gamma$, the horizontal component of force F.

It must be remembered that P represents the total pull acting along the guys at an angle, α, with the horizontal, and in the same vertical plane as the force, F.

If only one guy were used, P would represent the extra tension, T. In practice, however, a number of guys are always used, and, therefore, the pull on any one guy will not be equal to P. The following table gives factors for any number of guys from 3 to 15, equally spaced about a central structure. To find the maximum extra tension, T, that will be applied to any single guy by the force, F, capable of rotating 360° around a vertical axis, it is only necessary to multiply the value of P, as determined above, by the factor for the number of guys used. It must be clearly understood in using this table that the guys are uniformly spaced and under equal tension when no load is acting on the structure.

Example. A derrick mast 90 ft high is supported by nine equally spaced guys anchored at a horizontal distance of 170 ft from the mast, and the elevations of the guy anchorages are 10 ft below the base of the mast. The load on the structure is equivalent to a force of 10,000 lb, acting on an angle of 10° below the horizontal. What is the maximum pull on any single cable?

From Figure 1:

h = 90 ft,
d = 170 ft,

Table 1

No. of Guys	Factors[*]
3	1.15
4	1.00
5	0.90
6	0.75
7	0.65
8	0.55
9	0.50
10	0.45
11	0.40
12	0.37
13	0.35
14	0.32
15	0.30

[*]These factors are for average conditions. If the guys are erected under accurately measured tensions of not less than 20% of the working load, the factors for five or more guys may be reduced by 10%. If the erecting tensions are low or not accurately equalized, the factors for five or more guys should be increased 10%.

m = 10 ft,
g = 10°00′,
F = 10,000 lb,

$$\tan \alpha = \frac{90 + 10}{170} = 100 = 0.588,$$

$$\alpha = 30°28′,$$

$$P = \frac{F \cos \gamma}{\cos \alpha} = \frac{10,000 \times 0.985}{0.862} = 11,427 \text{ lb}.$$

From Table 1, $T = 11{,}427 \times 0.50 = 5714$ lb.

If erection tension is 10% of total working tension, 5714 is 90% of total working tension. Therefore, working tension = $(5714 \times 100)/90 = 6349$ lb.

Strength and weight of popular wire rope

The following tables give the breaking strength for wire rope of popular construction made of improved plow steel.

Size	6 × 19	
	Breaking strength	Weight
1/4	5480	0.10
5/16	8520	0.16
3/8	12,200	0.23
7/16	16,540	0.31
1/2	21,400	0.40
9/16	27,000	0.51
5/8	33,400	0.63
1 3/4	47,600	0.90
7/8	64,400	1.23
1	83,600	1.60
1 1/8	105,200	2.03
1 1/4	129,200	2.50
1 3/8	155,400	3.03
1 1/2	184,000	3.60
1 5/8	214,000	4.23
1 3/4	248,000	4.90
1 7/8	282,000	5.63
2	320,000	6.40

Conversion factors for wire rope of other construction

To apply the above table to wire rope of other construction, multiply by the following factors:

Wire Rope Construction	6 × 19	6 × 29	6 × 37	18 × 7
Strength factors	1.00	0.96	0.95	0.92
Weight factors	1.00	0.97	0.97	1.08

Example. Find the breaking strength of 6 × 29 improved plow steel wire rope 2 in in diameter.

Strength $= 320,000 \times 0.96 = 307,000$ lb.

The weight can be found the same way.

Measuring the diameter of wire rope

Correct way

Incorrect way

Wire rope: field troubles and their causes

All wire rope will eventually deteriorate in operation or have to be removed simply by virtue of the loads and reversals of load applied in normal service. There are, however, many conditions of service or inadvertent abuse that will materially shorten the normal life of a wire rope of proper construction although it is properly applied. The following field troubles and their causes give some of the field conditions and practices that result in the premature replacement of wire rope (it should be borne in mind that in all cases the contributory cause of removal may be one or more of these practices or conditions):

Wire-Rope Trouble	**Cause**
a. Rope broken (all strands)	Overload resulting from severe impact, kinking, damage, localized wear, weakening of one or more strands, or rust-bound condition and loss of elasticity.
b. One or more whole strands parted	Overloading, kinking, divider interference, localized wear, or rust-bound condition. Fatigue, excessive speed, slipping, or running too loosely. Concentration of vibration at dead sheave or dead-end anchor.
c. Excessive corrosion	Lack of lubrication. Exposure to salt spray, corrosive gases, alkaline water, acid water, mud, or dirt. Period of inactivity without adequate protection.
d. Rope damage in hauling to the well or location	Rolling reel over obstructions or dropping from car, truck, or platform. The use of chains for lashing, or the use of lever against rope instead of flange. Nailing through rope to flange.
e. Damage by improper socketing	Improper seizing, which allows slack from one or more strands to work back into rope; improper method of socketing or poor workmanship in socketing, frequently shown by rope being untwisted at socket, loose or drawn.
f. Kinks, dog legs, and other distorted places	Kinking the rope and pulling out the loops such as in improper coiling or unreeling. Improper winding on the drum. Improper tie-down. Open-drum reels having longitudinal spokes too widely spaced. Divider interference. The addition of improperly spaced cleats to increase the drum diameter. Stressing while rope is over small sheave or obstacle.
g. Damage by hooking back slack too tightly to girt	Operation of walking beam causing a bending action on wires at clamp and resulting in fatigue and cracking of wires, frequently before rope goes down into hole.
h. Damage or failure on a fishing job	Rope improperly used on a fishing job, resulting in damage or failure as a result of the nature of the work.
i. Lengthening of lay and reduction of diameter	Frequently produced by some type of overloading, such as an overload resulting in a collapse of the fiber core in swabbing lines. This may also occur in cable-tool lines as a result of concentrated pulsating or surging forces, which may contribute to fiber-core collapse.
j. Premature breakage of wires	Caused by frictional heat developed by pressure and slippage, regardless of drilling depth.
k. Excessive wear in spots	Kinks or bends in rope due to improper handling during installation or service. Divider interference; also, wear against casing or hard shells or abrasive formations in a crooked hole. Too infrequent cut-offs on working end.
l. Spliced rope	A splice is never as good as a continuous piece of rope, and slack is liable to work back and cause irregular wear.
m. Abrasion and broken wires in a straight line. Drawn or loosened strands. Rapid fatigue breaks.	Injury due to slipping rope through clamps.
n. Reduction in tensile strength or damage to rope	Excessive heat due to careless exposure to fire or torch.
o. Distortion of wire rope	Damage due to improperly attached clamps or wire-rope clips.
p. High strands	Slipping through clamps, improper seizing, improper socketing or splicing, kinks, dog legs, and core popping.
q. Wear by abrasion	Lack of lubrication. Slipping clamp unduly. Sandy or gritty working conditions. Rubbing against stationary object or abrasive surface. Faulty alignment. Undersized grooves and sheaves.
r. Fatigue breaks in wire	Excessive vibration due to poor drilling conditions, i.e. high speed, rope slipping, concentration of vibration at dead sheave or dead-end anchor, undersized grooves and sheaves, and improper selection of rope construction. Prolonged bending action over spudder sheaves, such as that due to hard drilling.
s. Spiraling or curling	Allowing rope to drag or rub over pipe, sill, or any object during installation or operation. It is recommended that a block with sheave diameter 16 times the nominal wire-rope diameter, or larger, be used during installation of the line.

	Wire-Rope Trouble	**Cause**
t.	Excessive flattening or crushing	Heavy overload, loose winding on drum, or cross winding. Too infrequent cut-offs on working end of cable-tool lines. Improper cutoff and moving program for cable-tool lines.
u.	Bird-caging or core-popping	Sudden unloading of line such as hitting fluid with excessive speed. Improper drilling motion or jar action. Use of sheaves of too small diameter or passing line around sharp bend.
v.	Whipping off of rope	Running too loose.
w.	Cutting in on drum	Loose winding on drum. Improper cutoff and moving program for rotary drilling lines. Improper or worn drum grooving or line turn-back late.

Capacity of drums

The capacity of wire line drums may be figured from the following formula:

$$M = (A + B) \times A \times C \times K.$$

Rope Size	K	Rope Size	K
$\frac{3}{8}$	1.86	1	0.262
7/16	1.37	$1\frac{1}{8}$	0.207
½	1.05	1¼	0.167
9/16	0.828	$1\frac{5}{8}$	0.138
$\frac{5}{8}$	0.672	1½	0.116
$\frac{3}{4}$	0.465	$1\frac{5}{8}$	0.099
$\frac{7}{8}$	0.342	$1\frac{3}{4}$	0.085

where

M = Rope capacity of drum, inches,
A = Depth of flange, inches,
B = Diameter of drum, inches,
C = Width of the drum between flanges, inches,
K = Constant depending on rope size shown below.

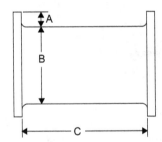

Figure 1. Capacity of wire line drums.

BELTS AND SHAFTS

Determine length of a V-belt

Rule: To find the nominal length of a V-belt, lay the belt on a table and place within it a pair of circular objects of the same diameter (flanges, tin cans, or whatever may be at hand). Pull them apart until the belt is fully extended without stretching. Then measure the shortest distance between the two circles, in inches. The belt size is then twice this figure, plus 5.14 times the diameter of the circles. This value, for any standard belt, should be a whole number of inches, which is the belt size.

V-belts are made in four standard sections, classified as A, B, C, and D; the widths (at the widest part) are ½, ⅝, ⅞, and 1⅛ in, respectively. The complete designation of the belt is the letter showing the width, followed by the length in inches; thus, an A26 belt is ½-in. wide and 26 in. long on the inside edge. The pitch length of the belt is measured along a median section and corresponds to the length that runs on the pulley diameter, which determines the actual speed ratio—about half of the depth of the groove. Pitch lengths for A, B, C, and D belts are greater than their nominal lengths by 1.3, 1.8, 2.9, and 3.3 in, respectively.

Calculate stress in shaft key

The shear and compressive stresses in a key are calculated using the following equations:

$$Ss = \frac{2T}{d \times W \times L}, \quad Sc = \frac{2T}{d \times h_1 \times L}$$

Ss = Shear stress in psi,
Sc = Compressive stress in psi,
T = Shaft torque pounds, inches,
d = Shaft diameter, inches (for taper shafts, use average diameter),
w = Width key, inches,
L = Effective length of key, inches,
h_1 = Height of key in the shaft or hub that bears against the keyway, inches,
$h_1 = h_2$ for square keys. For designs where unequal portions of the key are in the hub or shaft, h_1 is the minimum portion.

Key material is usually AISI 1018 or AISI 1045 with the following allowable stresses:

Material	Heat Treatment	Allowable Stresses—psi	
		Shear	Compressive
AISI 1018	None	7500	15,000
AISI 1045	225–300 Bhn	15,000	30,000

Example. Determine key stresses for these conditions: 300 HP @ 600 RPM; 3″ diameter shaft, ¾ × ¾ key, 4″ key engagement length.

$$T = \text{Torque} = \frac{\text{HP} \times 63,000}{\text{RPM}} = \frac{300 \times 63,000}{600}$$

$$= 31.500 \text{ in.}, -\text{lbs}$$

$$Ss = \frac{2T}{d \times W \times L} = \frac{2 \times 31,500}{3 \times 3/4 \times 4} = 7000 \text{ psi},$$

$$Sc = \frac{2T}{d \times h_1 \times L} = \frac{2 \times 31,500}{3 \times 3/8 \times 4} = 14,000 \text{ psi}.$$

The AISI 1018 key should be used since it is within allowable stresses. *Note:* If shaft had been $2\frac{3}{4}''$ diameter (4″ long hub), the key would be $\frac{5}{8} \times \frac{5}{8}$ and $Ss = 9200$ psi, $Sc = 18,400$ psi, and a heat-treated key, AISI 1045, would have been required.

Reprinted with permission: The Falk Corporation.

Calculate V-belt length using simple equation

Rule.

$$L = 1.57(D + d) + 2C,$$

where

L = Belt length, inches,
D = Diameter of larger sheave, inches,
d = Diameter of smaller sheave, inches,
C = Distance between sheave centers, inches.

Estimate the horsepower that can be transmitted by a shaft

1. Where there are no stresses due to bending, weight of the shaft, pulleys, gears, or sprockets, use

$$HP = \frac{D^3 N}{50},$$

where

   D = diameter of shaft, inches,
   N = revolutions per minute.

2. For heavy duty service use

$$HP = \frac{D^3 N}{125}.$$

Example. What horsepower can be transmitted to an atmospheric cooling coil by a 2 in shaft turning at 1800 revolutions per minute?

$$HP = \frac{(2)^3(1800)}{125} = 115.2 \text{ horsepower.}$$

MISCELLANEOUS

How to estimate length of material contained in roll

Where material of uniform thickness, like belting, is in a roll, the total length may be obtained by the following rule:

Measure the diameter of the hole in the center, and of the outside of the roll, both measurements in inches; count the number of turns; multiply the sum of the two measured diameters by the number of turns, and multiply this product by 0.13; the result is the total length of the material in feet.

Example. A roll of belting contains 24 turns. The diameter of the hole is 2 in, and that of the outside of the roll is 13 in.

$$(2 + 13) \times 24 \times 0.13 = 46.8$$

The roll contains 46.8 ft of belting.

Note: The rule can even be applied to materials as thin as pipeline felt; counting the turns is not as difficult as it might appear without a trial.

Convenient antifreeze chart for winterizing cooling systems

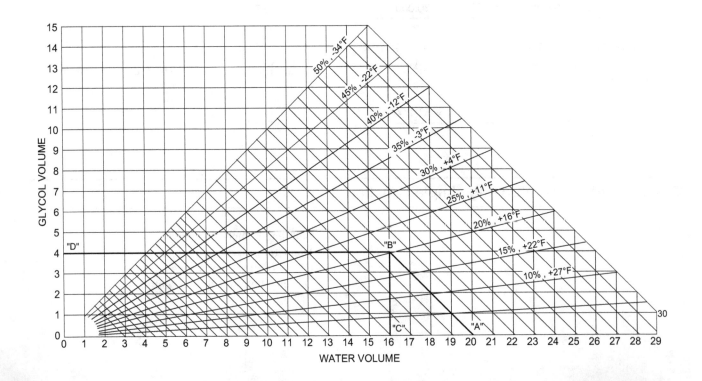

This plot of water volume versus glycol volume at various conditions of temperature and percent of glycol in the system makes winterizing field engines relatively easy.

Example. Determine the amount of glycol to be used in a 200 gallon system for protection to 16 °F.

To use the chart, first find the total system capacity, 200 gallons, at point A on the water volume axis. Point A also can represent 2, 20, 2000, etc. Proceed along the 45° line from point A to point B at the intersection with the 16 °F line. Then, move from point B to point C and read 160 on the water volume axis. This is the amount of water the system should contain. Move from point B to point D on the glycol axis and read 40. This is the amount of glycol that must be added to the system.

How to determine glycol requirements to bring a system to a desired temperature protection level

Solve the equation below to find the number of gallons of aqueous solution that must be removed and replaced with glycol.

$$d = \frac{D}{B},$$

where

d = Gallons of aqueous solution to be removed from the system and replaced with an equal number of gallons of glycol,
D = Total gallons of glycol required for protection at desired temperature less gallons of glycol in original system,
B = 1.0 − fraction of glycol in system.

Example. Find the number of gallons of existing aqueous solution that must be removed and replaced with glycol in a 1200 gallon system to give protection to −10 °F. Glycol in the system determined by a hydrometer test is 28% (0.28 fraction). The published figure for −10 °F protection is 519 gallons of glycol in the 1200 gallon system.

Then:

Gallons of glycol in system = 1200 × 0.28 = 336 gallons,
D = 519 − 336 = 183 gallons difference,
B = 1.0 − 0.28 = 0.72.

Thus $d = \dfrac{183}{0.72} = 254$ gallons.

Therefore, 254 gallons of existing aqueous solution need to be removed and replaced with 254 gallons of glycol for the desired protection to −10 °F.

Weight in pounds of round steel shafting

Diameter of Shaft (in)	Length in Feet																	Weight per Inch
	1	2	3	4	5	6	7	8	9	10	12	14	16	18	20	22	24	
15/16	2	5	7	9	12	14	16	19	21	23	28	33	37	42	47	51	56	0.195
1 3/16	4	8	11	15	19	23	26	30	34	38	45	53	60	68	76	83	91	0.314
1 7/16	6	11	17	22	28	33	39	44	50	55	66	77	88	99	110	121	133	0.460
1 11/16	8	15	23	30	38	46	53	61	68	76	91	107	122	137	152	167	183	0.634
1 15/16	10	20	30	40	50	60	70	80	90	100	120	140	161	181	201	221	241	0.835
2 3/16	13	26	38	51	64	77	90	102	115	128	153	179	205	230	256	281	307	1.06
2 7/16	16	32	48	63	79	95	111	127	143	159	190	222	254	286	317	349	381	1.32
2 11/16	19	39	58	77	97	116	135	154	174	193	232	270	309	348	386	425	463	1.61
2 15/16	23	46	69	92	115	138	161	184	208	231	277	323	369	415	461	507	553	1.92
3 3/16	27	54	81	109	136	163	190	217	244	272	326	380	435	489	543	598	652	2.26
3 7/16	32	63	95	126	158	189	221	253	284	316	379	442	505	568	631	695	758	2.63
3 11/16	36	73	109	146	182	218	255	291	328	364	437	510	582	655	728	801	874	3.03
3 15/16	41	83	124	166	207	248	290	331	373	414	497	580	662	745	828	911	994	3.45
4 7/16	53	105	158	210	263	315	368	421	473	526	631	736	841	946	1052	1157	1262	4.38
4 15/16	65	130	195	260	326	391	456	521	586	651	781	911	1041	1172	1302	1432	1562	5.43
5	67	134	200	267	334	401	467	534	601	668	801	935	1068	1202	1335	1469	1602	5.56
5 7/16	79	158	237	316	395	474	553	632	711	790	947	1105	1263	1421	1579	1737	1894	6.58
5 1/2	81	162	242	323	404	485	565	646	727	808	969	1131	1292	1454	1615	1777	1938	6.73
5 15/16	94	188	282	377	471	565	659	753	847	941	1130	1318	1506	1695	1883	2071	2259	7.85
6	96	192	288	385	481	577	673	769	865	961	1154	1346	1538	1731	1923	2115	2307	8.01
6 1/2	113	226	338	451	564	677	790	903	1015	1128	1354	1579	1805	2031	2256	2482	2708	9.40
7	131	262	393	524	654	785	916	1047	1178	1309	1507	1832	2094	2355	2677	2878	3140	10.90
7 1/2	150	300	451	601	751	901	1051	1202	1352	1502	1803	2103	2103	2704	3004	3305	3605	12.52
8	171	342	513	684	855	1024	1196	1367	1538	1709	2051	2051	2303	3076	3418	3760	4102	14.24

Properties of shafting

Diameter of Shaft (in)	Area of Shaft	Weight per Inch	Weight per Foot	Section Modulus		Moment of Inertia		Radius of Gyration
				Bending	Torsion	Bending	Torsion	
15/16	0.6903	0.195	2.34	0.0809	0.1618	0.0379	0.0758	0.2343
1 3/16	1.1075	0.314	3.76	0.1644	0.3288	0.0976	0.1952	0.2969
1 7/16	1.6230	0.460	5.52	0.2916	0.5832	0.2096	0.4192	0.3594
1 11/16	2.2365	0.634	7.61	0.4718	0.9435	0.3981	0.7961	0.4219
1 15/16	2.9483	0.835	10.03	0.7140	1.4281	0.6971	1.3835	0.4844
2 3/16	3.7583	1.06	12.78	1.0276	2.0553	1.1240	2.2480	0.5469
2 7/16	4.6664	1.32	15.86	1.4218	2.8436	1.7328	3.4656	0.6094
2 11/16	5.6727	1.61	19.29	1.9057	3.8113	2.5607	5.1215	0.6719
2 15/16	6.7771	1.92	23.04	2.4885	4.9770	3.6549	7.3099	0.7344
3 3/16	7.9798	2.26	27.12	3.1794	6.3589	5.0672	10.1345	0.7969
3 7/16	9.2806	2.63	31.56	3.9878	7.9755	6.8539	13.7079	0.8594
3 11/16	10.680	3.03	36.31	4.9226	9.8452	9.0761	18.1521	0.9219
3 15/16	12.177	3.45	41.40	5.9932	11.9865	11.7992	23.5984	0.9844
4 7/16	15.466	4.38	52.58	8.5786	17.1571	19.0337	38.0674	1.1094
4 15/16	19.147	5.43	65.10	11.8174	23.6348	29.1742	58.3483	1.2344
5	19.635	5.56	66.76	12.2718	24.5437	30.6796	61.3592	1.2500
5 7/16	23.221	6.58	78.96	15.7833	31.5666	42.9108	85.8217	1.3594
5 1/2	23.758	6.73	80.78	16.3338	32.6677	44.9180	89.8361	1.3750
5 15/16	27.688	7.85	94.14	20.5499	41.0999	61.0077	122.0153	1.4844
6	28.274	8.01	96.12	21.2058	42.4115	63.6172	127.2345	1.5000
6 1/2	33.183	9.40	112.82	26.9612	53.9225	87.6240	175.2481	1.6250
7	38.485	10.90	130.84	33.6739	67.3479	117.8588	235.7176	1.7500
7 1/2	44.179	12.52	150.21	41.4175	82.8350	155.3155	310.6311	1.8750
8	50.265	14.24	170.90	50.2655	100.5310	201.0619	402.1239	2.0000

Steel shaft sizes calculated for strength

Base formula: $d = \sqrt{\dfrac{C(HP)}{N}}$

Base Fiber Stress = 60,000 lb sq in.

Diameter of Shaft (in) $\frac{HP}{N}$		C = 50 Very Little Bending Fac. Saf. 9.4	C = 75 Average Conditions of Bending Fac. Saf. 14.1	C = 100 Very Severe Conditions of Bending Fac. Saf. 18.8
Fractional	Decimal			
1/16	0.0625	0.000004883	0.000003256	0.000002492
1/8	0.125	0.00003905	0.00002604	0.00001953
3/16	0.1875	0.0001318	0.00008788	0.00006592
1/4	0.25	0.0003124	0.0002083	0.0001563
5/16	0.3125	0.0006105	0.0004070	0.0003053
3/8	0.375	0.001054	0.0007030	0.0005272
7/16	0.4375	0.001675	0.001116	0.0008375
1/2	0.5	0.002500	0.001666	0.001250
9/16	0.5625	0.003558	0.002372	0.001779
5/8	0.625	0.004883	0.003256	0.002442
11/16	0.6875	0.006500	0.004333	0.003250
3/4	0.75	0.008439	0.005626	0.004220
13/16	0.8125	0.01074	0.007155	0.005367
7/8	0.875	0.01340	0.008932	0.006699
15/16	0.9375	0.01647	0.01098	0.008235

Diameter of Shaft (in) $\frac{HP}{N}$		C = 50 Very Little Bending Fac. Saf. 9.4	C = 75 Average Conditions of Bending Fac. Saf. 14.1	C = 100 Very Severe Conditions of Bending Fac. Saf. 18.8
Fractional	Decimal			
1	1.0	0.02000	0.01333	0.01000
1 1/16	1.0625	0.02396	0.01598	0.01199
1 1/8	1.125	0.02846	0.01898	0.01423
1 3/16	1.1875	0.03353	0.02236	0.01677
1 1/4	1.25	0.03905	0.02604	0.01953
15/16	1.3125	0.04519	0.03019	0.02259
1 3/8	1.375	0.5199	0.03465	0.02599
1 7/16	1.4375	0.05940	0.03960	0.02970
1 1/2	1.5	0.06750	0.04500	0.03375
1 9/16	1.5625	0.07628	0.05086	0.03815
1 5/8	1.625	0.08578	0.05719	0.04289
1 11/16	1.6875	0.09524	0.06409	0.04807
1 3/4	1.75	0.1072	0.07144	0.05358
1 13/16	1.8125	0.1191	0.07939	0.05955
1 7/8	1.875	0.1318	0.08788	0.06592

Diameter of Shaft(in) $\frac{HP}{N}$		C = 50 Very Little Bending Fac. Saf. 9.4	C = 75 Average Conditions of Bending Fac. Saf. 14.1	C = 100 Very Severe Conditions of Bending Fac. Saf. 18.8
Fractional	Decimal			
1 15/16	1.9375	0.1456	0.09707	0.07281
2	2.0	0.1600	0.1066	0.07998
2 1/16	2.0625	0.1755	0.1169	0.08774
2 1/8	2.125	0.1918	0.1278	0.09592
2 3/16	2.1875	0.2095	0.1397	0.1048
2 1/4	2.25	0.2278	0.1519	0.1140
2 5/16	2.3125	0.2474	0.1649	0.1237
2 3/8	2.375	0.2678	0.1785	0.1339
2 7/16	2.4375	0.2897	0.1932	0.1449
2 1/2	2.5	0.3124	0.2083	0.1563
2 9/16	2.5625	0.3363	0.2243	0.1682
2 7/8	2.625	0.3616	0.2411	0.1808
2 11/16	2.6875	0.3884	0.2589	0.1942
2 3/4	2.75	0.4159	0.2772	0.2079
2 13/16	2.8125	0.4449	0.2966	0.2225
2 7/8	2.875	0.4754	0.3170	0.2378
2 15/16	2.9375	0.5074	0.3383	0.2537
3	3.0	0.5399	0.3683	0.2700
3 1/16	3.0625	0.5742	0.3836	0.2871
3 1/8	3.125	0.6105	0.4070	0.3053
3 3/16	3.1875	0.6479	0.4319	0.3240
3 1/4	3.25	0.6866	0.4577	0.3434
3 5/16	3.3125	0.7266	0.4844	0.3634
3 3/8	3.375	0.7685	0.5123	0.3842
3 7/16	3.4375	0.8127	0.5418	0.4061
3 1/2	3.5	0.8567	0.5718	0.4288
3 9/16	3.5625	0.9038	0.6026	0.4520

Diameter of Shaft(in) $\frac{HP}{N}$		C = 50 Very Little Bending Fac. Saf. 9.4	C = 75 Average Conditions of Bending Fac. Saf. 14.1	C = 100 Very Severe Conditions of Bending Fac. Saf. 18.8
Fractional	Decimal			
3 5/8	3.625	0.9526	0.6351	0.4763
3 11/16	3.6875	1.003	0.6688	0.5017
3 3/4	3.75	1.054	0.7030	0.5272
3 13/16	3.8125	1.108	0.7382	0.5538
3 7/8	3.875	1.163	0.7759	0.5820
3 15/16	3.9375	1.222	0.8143	0.6108
4	4.0	1.280	0.8535	0.6401
4 1/16	4.0625	1.340	0.8933	0.6701
4 1/8	4.125	1.404	0.9356	0.7018
4 3/16	4.1875	1.469	0.9792	0.7345
4 1/4	4.25	1.536	1.023	0.7678
4 5/16	4.3125	1.603	1.069	0.8019
4 3/8	4.375	1.675	1.116	0.8375
4 7/16	4.4375	1.748	1.165	0.8742
4 1/2	4.5	1.823	1.214	0.9111
4 9/16	4.5625	1.900	1.266	0.9497
4 5/8	4.625	1.978	1.319	0.9893
4 11/16	4.6875	2.059	1.373	1.030
4 3/4	4.75	2.143	1.429	1.072
4 13/16	4.8125	2.228	1.486	1.114
4 7/8	4.875	2.316	1.544	1.158
4 15/16	4.9375	2.407	1.604	1.203
5	5.0	2.500	1.666	1.250

Haven-Swett "Treatise on Leather Belting," American Leather Belting Association.

Tap drills and clearance drills for machine screws

Screw Size	Coarse Thread		Fine Thread		Clearance Drill
	TPI	Drill	TPI	Drill	
4	40	43	48	42	32
5	40	38	44	37	30
6	32	36	40	33	27
8	32	29	36	29	18
10	24	25	32	21	9
12	24	16	28	14	2
14	20	10	24	7	D
1/4	20	7	28	3	F
5/16	18	F	24	1	P
3/8	16	5/16	24	Q	W
7/16	14	U	20	25/64	29/64
1/2	13	27/64	20	29/64	33/64
9/16	12	31/64	18	33/64	37/64
5/8	11	17/32	18	37/64	41/64
3/4	10	21/32	16	11/16	49/64

Common nails

Size	Length	Gauge No.	Head Diameter	Approx. No./lb
2d	1.00	15	$^{11}/_{64}$	845
3d	1.25	14	$^{13}/_{64}$	540
4d	1.50	12.5	$^1/_4$	290
5d	1.75	12.5	$^1/_4$	250
6d	2.00	11.5	$^{17}/_{64}$	265
7d	2.25	11.5	$^{17}/_{64}$	250
8d	2.50	10.25	$^9/_{32}$	200
9d	2.75	10.25	$^9/_{32}$	90

Size	Length	Gauge No.	Head Diameter	Approx. No./lb
10d	3.00	9	$^5/_{16}$	65
12d	3.25	9	$^5/_{16}$	60
16d	3.50	8	$^{11}/_{32}$	45
20d	4.00	6	$^{11}/_{32}$	30
30d	4.50	5	$^7/_{16}$	20
40d	5.00	4	$^{15}/_{32}$	17
50d	5.50	3	$^1/_2$	13
60d	6.00	2	$^{17}/_{32}$	10

Drill sizes for pipe taps

Tap Size (in)	TPI	Drill Diameter (in)
$^1/_8$	27	$^{11}/_{32}$
$^1/_4$	18	$^7/_{16}$
$^3/_8$	18	$^{37}/_{64}$
$^1/_2$	14	$^{23}/_{32}$
$^3/_4$	14	$^{59}/_{64}$
1	11 ½	$1\,^5/_{32}$
1 ¼	11 ½	$1\,^1/_2$
1 ½	11 ½	$1\,^{49}/_{64}$

Tap Size (in)	TPI	Drill Diameter (in)
2	11 ½	$2\,^3/_{16}$
2 ½	8	$2\,^9/_{16}$
3	8	$3\,^3/_{16}$
3 ½	8	$3\,^{11}/_{16}$
4	8	$4\,^3/_{16}$
4 ½	8	$4\,^3/_4$
5	8	$5\,^5/_{16}$
6	8	$6\,^5/_{16}$

Carbon steel—color and approximate temperature

Color	Temperature (°F)
Black red	990
Dark blood red	1050
Dark cherry red	1175
Medium cherry red	1250
Light cherry, scaling	1550

Color	Temperature (°F)
Salmon, free scaling	1650
Light salmon	1725
Yellow	1825
Light yellow	1975
White	2220

Bolting dimensions for flanges

Nom. Pipe Size	150 ANSI RF WN			300 ANSI RF WN			400 ANSI RF WN			Notes
	Quan.	Stud Diam.	Stud Length	Quan.	Stud Diam.	Stud Length	Quan.	Stud Diam.	Stud Length	
2	4	⅝	3½	8	⅝	3½	See Note 1			Notes:
4	8	⅝	4	8	¾	4¼	8	⅞	5½	1. For 3½″ and smaller, use ANSI 600 flange
6	8	¾	4	8	¾	4½	12	⅞	6	
8	8	¾	4½	12	¾	5	12	1	6¾	2. For 2½″ and smaller, use ANSI 1500 flang
10	12	⅞	5	16	⅞	5½	16	1⅛	7½	
12	12	⅞	5	16	1	6½	16	1¼	8	Std. gasket thickness = 1/16″ to 1/8″
14	12	1	5½	20	1⅛	6¾	20	1¼	8¼	
16	16	1	5½	20	1⅛	7	20	1⅜	8¾	If insulating kit is installed, increase stud length
18	16	1⅛	6	24	1¼	7½	24	1⅜	9	
20	20	1⅛	6½	24	1¼	7¾	24	1½	9¾	If orifice plate is used, increase stud length
24	20	1¼	7	24	1¼	8¼	24	1¾	10¾	
26	24	1¼	8½	28	1½	9¼	28	1¾	11¾	
28	28	1¼	8¾	28	1⅝	10¼	28	1¾	11¾	
30	28	1¼	9	28	1⅝	10¾	28	1⅞	12½	
32	28	1½	10	28	1¾	11½	28	2	13¼	
34	32	1½	10¼	28	1⅞	12¼	28	2	14	
36	32	1½	10¾	32	2	13	32	2	14¼	

Nom. Pipe Size	600 ANSI RF WN			900 ANSI RF WN			1500 ANSI RF WN		
	Quan.	Stud Diam.	Stud Length	Quan.	Stud Diam.	Stud Length	Quan.	Stud Diam.	Stud Length
2	8	⅝	4¼	See Note 2			8	⅞	6¼
3	8	¾	5	8	⅞	5¾	8	1⅛	7
4	8	⅞	5¾	8	1⅛	6¾	8	3¾	7¾
6	12	1	6¾	12	1⅛	7¾	12	1⅜	10¼
8	12	1⅛	7¾	12	1⅜	8¾	12	1⅝	11½
10	16	1¼	8½	16	1⅜	8¾	12	1⅞	13½
12	20	1¼	8¾	20	1⅜	10	16	2	15
14	20	1⅜	9¼	20	1½	10¾	16	2¼	16¼
16	20	1½	10	20	1⅝	11¼	16	2½	17¾
18	20	1⅝	10¾	20	1⅞	12¾	16	2¾	19½
20	24	1⅝	11½	20	2	13½	16	3	21¼
24	24	1⅞	13	20	2½	17¼	16	3½	24¼
26	28	1⅞	13½	20	2¾	17¾			
28	28	2	14	20	3	18½			
30	28	2	14¼	20	3	19			
32	28	2¼	15	20	3¼	20¼			
34	28	2¼	15¼	20	3½	20¼			
36	28	2¼	16	20	3½	12¾			

Flange Bolt Tightening Sequence

Flange Bolt Torque Sequence

Sequential Order	Rotational Order
1-2	1
3-4	2
5-6	3
7-8	4
	5
	6
	7
	8

Sequential Order	Rotational Order
1-2	1
3-4	5
5-6	9
7-8	3
9-10	7
11-12	11
	2
	6
	10
	4
	8

Sequential Order	Rotational Order	
1-2	1	2
3-4	9	10
5-6	5	6
7-8	13	14
9-10	3	4
11-12	11	12
13-14	7	8
	15	16
	2	
	10	
	6	

Sequential Order	Rotational Order	
1-2	1	2
3-4	13	14
5-6	5	6
7-8	17	18
9-10	9	10
11-12	3	4
13-14	15	16
15-16	7	8
17-18	19	20
19-20	11	12

Sequential Order	Rotational Order	
1-2	1	2
3-4	9	10
5-6	17	18
7-8	5	6
9-10	13	14
11-12	21	22
13-14	3	4
15-16	11	12
17-18	19	20
19-20	7	8
21-22	15	16
23-24	23	24

Sequential Order	Rotational Order	
1-2	1	2
3-4	17	18
5-6	9	10
7-8	25	26
9-10	5	6
11-12	11	12
13-14	13	14
15-16	29	30
17-18	3	4
19-20	19	20
21-22	11	12
23-24	27	28
25-26	7	8
27-28	23	24
29-30	15	16
31-32	31	32

used with permissino of Sypris Technologies, Inc, Tube Turns Division, Louisville, KY

Steel fitting dimensions

90° Long radius elbow · 90° Reducing long radius elbow · 45° Long radius elbow · 90° Short radius elbow · Cap · Lap joint stub end · 180° Long radius return · 180° Short radius return · 90° 3R elbow · 45° 3R elbow

Nom. Pipe Size	Pipe O.D. (In.)	Light Wall	SCH 20	SCH 30	STD	SCH 40	SCH 60	X STG	SCH 80	SCH 100	SCH 120	SCH 140	SCH 160	XX STG	A	B	D	E	F ANSI	F MSS	G	O	K	S	V	X	Y	Nom. Pipe Size
1/2	.84				.109	.109		.147	.147				.188	.294	1.50	.62		1.00	3.00	2.00	1.38	3.00	1.88					1/2
3/4	1.05				.113	.113		.154	.154				.219	.308	1.12	.44		1.00	3.00	2.00	1.69	2.25	1.69					3/4
1	1.32				.133	.133		.179	.179				.250	.358	1.50	.88	1.00	1.50	4.00	2.00	2.00	3.00	2.19	2.00	1.62			1
1 1/4	1.66				.140	.140		.191	.191				.250	.382	1.88	1.00	1.25	1.50	4.00	2.00	2.50	3.75	2.75	2.50	2.06			1 1/4
1 1/2	1.90				.145	.145		.200	.200				.281	.400	2.25	1.12	1.50	1.50	4.00	2.00	2.88	4.50	3.25	3.00	2.44			1 1/2
2	2.38				.154	.154		.218	.218				.344	.436	3.00	1.38	2.00	1.50	6.00	2.50	3.62	6.00	4.19	4.00	3.19	6.00	2.50	2
2 1/2	2.88	.188			.203	.203		.276	.276				.375	.552	3.75	1.75	2.50	1.50	6.00	2.50	4.12	7.50	5.19	5.00	3.94			2 1/2
3	3.50	.188			.216	.216		.300	.300				.438	.600	4.50	2.00	3.00	2.00	6.00	2.50	5.00	9.00	6.25	6.00	4.75	9.00	3.75	3
3 1/2	4.00				.226	.226		.318	.318						5.25	2.25	3.50	2.50	6.00	3.00	5.50	10.50	7.25	7.00	5.50			3 1/2
4	4.50	.188			.237	.237		.337	.337		.438		.531	.674	6.00	2.50	4.00	2.50	6.00	3.00	6.19	12.00	8.25	8.00	6.25	12.00	5.00	4
5	5.56				.258	.258		.375	.375		.500		.625	.750	7.50	3.12	5.00	3.00	8.00	3.00	7.31	15.00	10.31	10.00	7.75			5
6	6.62	.219			.280	.280		.432	.432		.562		.719	.864	9.00	3.75	6.00	3.50	8.00	3.50	8.50	18.00	12.31	12.00	9.31	18.00	7.50	6
8	8.62	.219	.250	.277	.322	.322	.406	.500	.500	.594	.719	.812	.906	.875	12.00	5.00	8.00	4.00	8.00	4.00	10.62	24.00	16.31	16.00	12.31	24.00	10.00	8
10	10.75	.219	.250	.307	.365	.365	.500	.500	.594	.719	.844	1.000	1.125	1.000	15.00	6.25	10.00	5.00	10.00	5.00	12.75	30.00	20.38	20.00	15.38	30.00	12.50	10
12	12.75	.250	.250	.330	.375	.406	.562	.500	.688	.844	1.000	1.125	1.312	1.000	18.00	7.50	12.00	6.00	10.00	6.00	15.00	36.00	24.38	24.00	18.38	36.00	15.00	12
14	14.00	.250	.312	.375	.375	.438	.594	.500	.750	.938	1.094	1.250	1.406		21.00	8.75	14.00	6.50	12.00	6.00	16.25	42.00	28.00	28.00	21.00	42.00	17.50	14
16	16.00	.250	.312	.375	.375	.500	.656	.500	.844	1.031	1.219	1.438	1.594		24.00	10.00	16.00	7.00	12.00	6.00	18.50	48.00	32.00	32.00	24.00	48.00	19.88	16
18	18.00	.250	.312	.438	.375	.562	.750	.500	.938	1.156	1.375	1.562	1.781		27.00	11.25	18.00	8.00	12.00	6.00	21.00	54.00	36.00	36.00	27.00	54.00	22.38	18
20	20.00	.250	.375	.500	.375	.594	.812	.500	1.031	1.281	1.500	1.750	1.969		30.00	12.50	20.00	9.00	12.00	6.00	23.00	60.00	40.00	40.00	30.00	60.00	24.88	20
24	24.00	.250	.375	.562	.375	.688	.969	.500	1.219	1.531	1.812	2.062	2.344		36.00	15.00	24.00	10.50	12.00	6.00	27.25	72.00	48.00	48.00	36.00	72.00	29.81	24
30	30.00	.312	.500	.625	.375			.500							45.00	18.50	30.00	10.50				90.00	60.00	60.00	45.00	90.00	37.25	30
36	36.00	.312	.500	.625	.375	.750		.500							54.00	22.25	36.00	10.50						72.00	54.00	108.00	44.75	36
42	42.00				.375			.500							63.00	26.00	42.00	12.00								126.00	52.19	42
48	48.00				.375			.500							72.00	29.88	48.00	13.50								144.00	59.69	48

 Straight tee   Reducing tee   Concentric reducer   Eccentric reducer

Nom. Pipe Size Run	Outlet	C	M	H
3/4	3/4	1.12		
	1/2	1.12	1.12	1.50
1	1	1.50		
	3/4	1.50	1.50	2.00
	1/2	1.50	1.50	2.00
1 1/4	1 1/4	1.88		
	1	1.88	1.88	2.00
	3/4	1.88	1.88	2.00
	1/2	1.88	1.88	2.00
1 1/2	1 1/2	2.25		
	1 1/4	2.25	2.25	2.50
	1	2.25	2.25	2.50
	3/4	2.25	2.25	2.50
	1/2	2.25	2.25	2.50
2	2	2.50		
	1 1/2	2.50	2.38	3.00
	1 1/4	2.50	2.25	3.00
	1	2.50	2.00	3.00
	3/4	2.50	1.75	3.00
2 1/2	2 1/2	3.00		
	2	3.00	2.75	3.50
	1 1/2	3.00	2.62	3.50
	1 1/4	3.00	2.50	3.50
	1	3.00	2.25	3.50
3	3	3.38		
	2 1/2	3.38	3.25	3.50
	2	3.38	3.00	3.50
	1 1/2	3.38	2.88	3.50
	1 1/4	3.38	2.75	3.50

Nom. Pipe Size Run	Outlet	C	M	H
3 1/2	3 1/2	3.75		
	3	3.75	3.62	4.00
	2 1/2	3.75	3.50	4.00
	2	3.75	3.25	4.00
	1 1/2	3.75	3.12	4.00
4	4	4.12		
	3 1/2	4.12	4.00	4.00
	3	4.12	3.88	4.00
	2 1/2	4.12	3.75	4.00
	2	4.12	3.50	4.00
	1 1/2	4.12	3.38	4.00
5	5	4.88		
	4	4.88	4.62	5.00
	3 1/2	4.88	4.50	5.00
	3	4.88	4.38	5.00
	2 1/2	4.88	4.25	5.00
	2	4.88	4.12	5.00
6	6	5.62		
	5	5.62	5.38	5.50
	4	5.62	5.12	5.50
	3 1/2	5.62	5.00	5.50
	3	5.62	4.88	5.50
	2 1/2	5.62	4.75	5.50
8	8	7.00		
	6	7.00	6.62	6.00
	5	7.00	6.38	6.00
	4	7.00	6.12	6.00
	3 1/2	7.00	6.00	6.00
10	10	8.50		
	8	8.50	8.00	7.00
	6	8.50	7.62	7.00
	5	8.50	7.50	7.00
	4	8.50	7.25	7.00

Nom. Pipe Size Run	Outlet	C	M	H
12	12	10.00		
	10	10.00	9.50	8.00
	8	10.00	9.00	8.00
	6	10.00	8.62	8.00
	5	10.00	8.50	8.00
14	14	11.00	11.00	
	12	11.00	10.62	13.00
	10	11.00	10.12	13.00
	8	11.00	9.75	13.00
	6	11.00	9.38	13.00
16	16	12.00		
	14	12.00	12.00	14.00
	12	12.00	11.62	14.00
	10	12.00	11.12	14.00
	8	12.00	10.75	14.00
	6	12.00	10.38	
18	18	13.50		
	16	13.50	13.00	15.00
	14	13.50	13.00	15.00
	12	13.50	12.62	15.00
	10	13.50	12.12	15.00
	8	13.50	11.75	
20	20	15.00		
	18	15.00	14.50	20.00
	16	15.00	14.00	20.00
	14	15.00	14.00	20.00
	12	15.00	13.62	20.00
	10	15.00	13.12	
	8	15.00	12.75	

Nom. Pipe Size Run	Outlet	C	M	H
24	24	17.00		
	22	17.00	17.00	20.00
	20	17.00	17.00	20.00
	18	17.00	16.50	20.00
	16	17.00	16.00	20.00
	14	17.00	16.00	
	12	17.00	15.62	
	10	17.00	15.12	
30	30	22.00		
	24	22.00	21.00	24.00
	20	22.00	20.00	24.00
	18	22.00	19.50	
	16	22.00	19.00	
	14	22.00	19.00	
36	36	26.50		
	30	26.50	25.00	24.00
	24	26.50	24.00	24.00
	20	26.50	23.00	
	18	26.50	22.50	
	16	26.50	22.00	
42	42	30.00		
	36	30.00	28.00	24.00
	30	30.00	28.00	24.00
	24	30.00	26.00	
	20	30.00	26.00	
48	48	35.00		
	42	35.00	32.00	28.00
	36	35.00	31.00	
	30	35.00	30.00	
	24	35.00	29.00	

NOTES:
Fittings are to ASTM A234 WPB with dimensions to ANSI B16.9 except that short radius elbows and returns are to ANSI B16.28 and 3R elbows are to MSS-SP75.

Other types, sizes and thicknesses of fittings on application.

ANSI forged steel flanges

Weld neck flange   Slip-on flange   Socket type   Threaded flange   Lap joint flange   Bind flange

CLASS 150 FLANGES

Nom. Pipe Size	O	C†	Y† Weld Neck	Y† Slip-on Thrd. Socket	Y† Lap Joint	Bolt Circle	No. and Size of Holes
1/2	3.50	.44	1.88	.62	.62	2.38	4-0.62
3/4	3.88	.50	2.06	.62	.62	2.75	4-0.62
1	4.25	.56	2.19	.69	.69	3.12	4-0.62
1 1/4	4.62	.62	2.25	.81	.81	3.50	4-0.62
1 1/2	5.00	.69	2.44	.88	.88	3.88	4-0.62
2	6.00	.75	2.50	1.00	1.00	4.75	4-0.75
2 1/2	7.00	.88	2.75	1.12	1.12	5.50	4-0.75
3	7.50	.94	2.75	1.19	1.19	6.00	4-0.75
3 1/2	8.50	.94	2.81	1.25	1.25	7.00	8-0.75
4	9.00	.94	3.00	1.31	1.31	7.50	8-0.75
5	10.00	.94	3.50	1.44	1.44	8.50	8-0.88
6	11.00	1.00	3.50	1.56	1.56	9.50	8-0.88
8	13.50	1.12	4.00	1.75	1.75	11.75	8-0.88
10	16.00	1.19	4.00	1.94	1.94	14.25	12-1.00
12	19.00	1.25	4.50	2.19	2.19	17.00	12-1.00
14	21.00	1.38	5.00	2.25	USE SLIP-ON DIMS.	18.75	12-1.12
16	23.50	1.44	5.00	2.50		21.25	16-1.12
18	25.00	1.56	5.50	2.69		22.75	16-1.25
20	27.50	1.69	5.69	2.88		25.00	20-1.25
24	32.00	1.88	6.00	3.25		29.50	20-1.38
30	38.75	2.94	5.38			36.00	28-1.38
36	46.00	3.56	6.19			42.75	32-1.62
42	53.00	3.81	6.75			49.50	36-1.62
48	59.50	4.25	7.56			56.00	44-1.62
54	66.25	4.75	8.50			62.75	44-1.88
60	73.00	5.19	9.44			69.25	52-1.88

CLASS 300 FLANGES

Nom. Pipe Size	O	C†	Y† Weld Neck	Y† Slip-on Thrd. Socket	Y† Lap Joint	Bolt Circle	No. and Size of Holes
1/2	3.75	.56	2.06	.88	.88	2.62	4-0.62
3/4	4.62	.62	2.25	1.00	1.00	3.25	4-0.75
1	4.88	.69	2.44	1.06	1.06	3.50	4-0.75
1 1/4	5.25	.75	2.56	1.06	1.06	3.88	4-0.75
1 1/2	6.12	.81	2.69	1.19	1.19	4.50	4-0.88
2	6.50	.88	2.75	1.31	1.31	5.00	8-0.75
2 1/2	7.50	1.00	3.00	1.50	1.50	5.88	8-0.88
3	8.25	1.12	3.12	1.69	1.69	6.62	8-0.88
3 1/2	9.00	1.19	3.19	1.75	1.75	7.25	8-0.88
4	10.00	1.25	3.38	1.88	1.88	7.88	8-0.88
5	11.00	1.38	3.88	2.06	2.00	9.25	8-0.88
6	12.50	1.44	3.88	2.06	2.06	10.62	12-0.88
8	15.00	1.62	4.38	2.44	2.44	13.00	12-1.00
10	17.50	1.88	4.62	2.62		15.25	16-1.12
12	20.50	2.00	5.12	2.88		17.75	16-1.25
14	23.00	2.12	5.62	3.00	USE SLIP-ON DIMS.	20.25	20-1.25
16	25.50	2.25	5.75	3.25		22.50	20-1.38
18	28.00	2.38	6.25	3.50		24.75	24-1.38
20	30.50	2.50	6.38	3.75		27.00	24-1.38
24	36.00	2.75	6.62	4:19		32.00	24-1.62
30	43.00	3.62	8.25			39.25	28-1.88
36	50.00	4.12	9.50			46.00	32-2.12
42	50.75	4.69	7.88			47.50	32-1.75
48	57.75	5.25	8.81			54.00	32-2.00
54	65.25	6.00	9.94			61.00	28-2.38
60	71.25	6.44	10.75			67.00	32-2.38

CLASS 400 FLANGES

Nom. Pipe Size	O	C†	Y† Weld Neck	Y† Slip-on Thrd. Socket	Y† Lap Joint	Bolt Circle	No. and Size of Holes
1/2	3.75	.56	2.06	.88	.88	2.62	4-0.62
3/4	4.62	.62	2.25	1.00	1.00	3.25	4-0.75
1	4.88	.69	2.44	1.06	1.06	3.50	4-0.75
1 1/4	5.25	.81	2.62	1.12	1.12	3.88	4-0.75
1 1/2	6.12	.88	2.75	1.25	1.25	4.50	4-0.88
2	6.50	1.00	2.88	1.44	1.44	5.00	8-0.75
2 1/2	7.50	1.12	3.12	1.62	1.62	5.88	8-0.88
3	8.25	1.25	3.25	1.81	1.81	6.62	8-0.88
3 1/2	9.00	1.38	3.38	1.94	1.94	7.25	8-1.00
4	10.00	1.38	3.50	2.00	2.00	7.88	8-1.00
5	11.00	1.50	4.00	2.12	2.12	9.25	8-1.00
6	12.50	1.62	4.06	2.25	2.25	10.62	12-1.00
8	15.00	1.88	4.62	2.69	2.69	13.00	12-1.12
10	17.50	2.12	4.88	2.88		15.25	16-1.25
12	20.50	2.25	5.38	3.12		17.75	16-1.38
14	23.00	2.38	5.88	3.31	USE SLIP-ON DIMS.	20.25	20-1.38
16	25.50	2.50	6.00	3.69		22.50	20-1.50
18	28.00	2.62	6.50	3.88		24.75	24-1.50
20	30.50	2.75	6.62	4.00		27.00	24-1.62
24	36.00	3.00	6.88	4.50		32.00	24-1.88
30	43.00	4.00	8.62			39.25	28-2.12
36	50.00	4.50	9.88			50.00	32-2.12
42	52.00	5.25	8.81			48.25	32-2.00
48	59.50	6.00	10.12			55.25	28-2.38
54	67.00	6.69	11.38			62.25	28-2.62
60	74.25	7.31	12.56			69.00	32-2.88

CLASS 600 FLANGES

Nom. Pipe Size	O	C†	Y† Weld Neck	Y† Slip-on Thrd. Socket	Y† Lap Joint	Bolt Circle	No. and Size of Holes
1/2	3.75	.56	2.06	.88	.88	2.62	4-0.62
3/4	4.62	.62	2.25	1.00	1.00	3.25	4-0.75
1	4.88	.69	2.44	1.06	1.06	3.50	4-0.75
1 1/4	5.25	.81	2.62	1.12	1.12	3.88	4-0.75
1 1/2	6.12	.88	2.75	1.25	1.25	4.50	4-0.88
2	6.50	1.00	2.88	1.44	1.44	5.00	8-0.75
2 1/2	7.50	1.12	3.12	1.62	1.62	5.88	8-0.88
3	8.25	1.25	3.25	1.81	1.81	6.62	8-0.88
3 1/2	9.00	1.38	3.38	1.94	1.94	7.25	8-1.00
4	10.75	1.50	4.00	2.12	2.12	8.50	8-1.00
5	13.00	1.75	4.50	2.38	2.38	10.50	8-1.12
6	14.00	1.88	4.62	2.62	2.62	11.50	12-1.12
8	16.50	2.19	5.25	3.00	3.00	13.75	12-1.25
10	20.00	2.50	6.00	3.38		17.00	16-1.38
12	22.00	2.62	6.12	3.62		19.25	20-1.38
14	23.75	2.75	6.50	3.69	USE SLIP-ON DIMS.	20.75	20-1.50
16	27.00	3.00	7.00	4.19		23.75	20-1.62
18	29.25	3.25	7.25	4.62		25.75	20-1.75
20	32.00	3.50	7.50	5.00		28.50	24-1.75
24	37.00	4.00	8.00	5.50		33.00	24-2.00
30	44.50	4.50	9.75			40.25	28-2.12
36	51.75	4.88	11.12			47.00	28-2.62
42	55.25	6.62	11.00			50.50	28-2.62
48	62.75	7.44	12.44			57.50	32-2.88
54	70.00	8.25	13.75			64.25	32-3.12
60	78.50	9.19	15.31			71.75	28-3.62

CLASS 900 FLANGES

Nom. Pipe Size	O	C†	Y† Weld Neck	Y† Slip-on Thrd. Socket	Y† Lap Joint	Bolt Circle	No. and Size of Holes
1/2	4.75	.88	2.38	1.25	1.25	3.25	4-0.88
3/4	5.12	1.00	2.75	1.38	1.38	3.50	4-0.88
1	5.88	1.12	2.88	1.62	1.62	4.00	4-1.00
1 1/4	6.25	1.12	2.88	1.62	1.62	4.38	4-1.00
1 1/2	7.00	1.25	3.25	1.75	1.75	4.88	4-1.12
2	8.50	1.50	4.00	2.25	2.25	6.50	8-1.00
2 1/2	9.62	1.62	4.12	2.50	2.50	7.50	8-1.12
3	9.50	1.50	4.00	2.12	2.12	7.50	8-1.00
4	11.50	1.75	4.50	2.75	2.75	9.25	8-1.25
5	13.75	2.00	5.00	3.12	3.12	11.00	8-1.38
6	15.00	2.19	5.50	3.38	3.38	12.50	12-1.25
8	18.50	2.50	6.38	4.00		15.50	12-1.50
10	21.50	2.75	7.25	4.25		18.50	16-1.50
12	24.00	3.12	7.88	4.62		21.00	20-1.50
14	25.25	3.38	8.38	5.12	USE SLIP-ON DIMS.	20.75	20-1.50
16	27.75	3.50	8.50	5.25		23.75	20-1.62
18	31.00	4.00	9.00	6.00		25.75	20-1.75
20	33.75	4.25	9.75	6.25		28.50	24-1.75
24	41.00	5.50	11.50	8.00		33.00	24-2.00
30	48.50	5.88	12.25			40.25	28-2.12
36	57.50	6.75	14.25			47.00	28-2.62
42	61.50	8.12	14.62			50.50	28-2.62
48	70.25	9.19	16.50			57.50	32-2.88

CLASS 1500 FLANGES

Nom. Pipe Size	O	C†	Y† Weld Neck	Y† Slip-on Thrd. Socket	Y† Lap Joint	Bolt Circle	No. and Size of Holes
1/2	4.75	.88	2.38	1.25	1.25	3.25	4-0.88
3/4	5.12	1.00	2.75	1.38	1.38	3.50	4-0.88
1	5.88	1.12	2.88	1.62	1.62	4.00	4-1.00
1 1/4	6.25	1.12	2.88	1.62	1.62	4.38	4-1.00
1 1/2	7.00	1.25	3.25	1.75	1.75	4.88	4-1.12
2	8.50	1.50	4.00	2.25	2.25	6.50	8-1.00
2 1/2	9.62	1.62	4.12	2.50	2.50	7.50	8-1.12
3	10.50	1.88	4.62	2.88	2.88	8.00	8-1.25
4	12.25	2.12	4.88	3.56	3.56	9.50	8-1.38
5	14.75	2.88	6.12	4.12	4.12	11.50	8-1.62
6	15.50	3.25	6.75	4.69	4.69	12.50	12-1.50
8	19.00	3.62	8.38	5.62	5.62	15.50	12-1.75
10	23.00	4.25	10.00	6.25		19.00	12-2.00
12	26.50	4.88	11.12	7.12		22.50	16-2.12
14	29.50	5.25	11.75			25.00	16-2.38
16	32.50	5.75	12.25			27.75	16-2.62
18	36.00	6.38	12.88			30.50	16-2.88
20	38.75	7.00	14.00			32.75	16-3.12
24	46.00	8.00	16.00			39.00	16-3.62

CLASS 2500 FLANGES

Nom. Pipe Size	O	C†	Y† Weld Neck	Y† Slip-on Thrd. Socket	Y† Lap Joint	Bolt Circle	No. and Size of Holes
1/2	5.25	1.19	2.88	1.56	1.56	3.50	4-0.88
3/4	5.50	1.25	3.12	1.69	1.69	3.75	4-0.88
1	6.25	1.38	3.50	1.88	1.88	4.25	4-1.00
1 1/4	7.25	1.50	3.75	2.06	2.06	5.12	4-1.12
1 1/2	8.00	1.75	4.38	2.38	2.38	5.75	4-1.25
2	9.25	2.00	5.00	2.75	2.75	6.75	8-1.12
2 1/2	10.50	2.25	5.62	3.12	3.12	7.75	8-1.25
3	12.00	2.62	6.62	3.62	3.62	9.00	8-1.38
4	14.00	3.00	7.50	4.25	4.25	10.75	8-1.62
5	16.50	3.62	9.00	5.12	5.12	12.75	8-1.88
6	19.00	4.25	10.75	6.00	6.00	14.50	8-2.12
8	21.75	5.00	12.50	7.00	7.00	17.25	12-2.12
10	26.50	6.50	16.50	9.00	9.00	21.25	12-2.62
12	30.00	7.25	18.25	10.00	10.00	24.38	12-2.88

NOTES:

Flange material is ASTM A105 unless specified otherwise.

Flange dimensions are to ANSI B16.5 for sizes 24" and smaller. Sizes over 24" are to MSS-SP44.

Always specify bore or matching pipe wall when ordering.

Other types, sizes and facings on application.

† Includes .06 raised face in Class 150 and Class 300 standards. Does **NOT** include .25 raised face in Class 400 and heavier standards.

Trench shoring–minimum requirements

Trench jacks may be used in lieu of, or in combination with, cross braces. Shoring is not required in solid rock, hard shale, or hard slag. Where desirable, steel sheet piling and bracing of equal strength may be substituted for wood.

Depth of Trench Feet	Kind or Condition of Earth	Uprights Minimum Dimensions Inches	Uprights Maximum Spacing Feet	Stringers Minimum Dimensions Inches	Stringers Maximum Spacing Feet	Cross Braces — Width of Trench — Up to 3 ft Inches	3 to 6 ft Inches	6 to 9 ft Inches	9 to 12 ft Inches	12 to 15 ft Inches	Maximum Spacing Vertical Feet	Maximum Spacing Horizontal Feet
5 to 10	Hard, compact	3 × 4 or 2 × 6	6	—	—	2 × 6	4 × 4	4 × 6	6 × 6	6 × 8	4	6
	Likely to crack	3 × 4 or 2 × 6	3	4 × 6	4	2 × 6	4 × 4	4 × 6	6 × 6	6 × 8	4	6
	Soft, sandy, or filled	3 × 4 or 2 × 6	Close sheeting	4 × 6	4	4 × 4	4 × 6	6 × 6	6 × 8	8 × 8	4	6
	Hydrostatic pressure	3 × 4 or 2 × 6	Close sheeting	4 × 6	4	4 × 4	4 × 6	6 × 6	6 × 8	8 × 8	4	6
10 to 15	Hard	3 × 4 or 2 × 6	4	4 × 6	4	4 × 4	4 × 6	6 × 6	6 × 8	8 × 8	4	6
	Likely to crack	3 × 4 or 2 × 6	2	4 × 4	4	4 × 6	6 × 6	6 × 6	6 × 8	8 × 8	4	6
	Soft, sandy, or filled	3 × 4 or 2 × 6	Close sheeting	4 × 6	4	4 × 6	6 × 6	6 × 8	8 × 8	8 × 10	4	6
	Hydrostatic pressure	3 × 6	Close sheeting	8 × 10	4	4 × 6	6 × 6	6 × 8	8 × 8	8 × 10	4	6
15 to 20	All kinds or conditions	3 × 6	Close sheeting	4 × 12	4	4 × 12	6 × 8	8 × 8	8 × 8	10 × 10	4	6
Over 20	All kinds or conditions	3 × 6	Close sheeting	6 × 8	4	4 × 12	6 × 8	8 × 10	10 × 10	10 × 12	4	6

Reuniting separated mercury in thermometers

The largest single cause for failure of precision thermometers is separated mercury columns. This can occur in transit or in use. The mercury may be reunited by cooling the thermometer in a solution of solid CO_2 (dry ice) and alcohol so that the mercury column retreats slowly into the bulb. Do not cool the stem or mercury column. Keep the bulb in the solution until the main column, as well as the separated portion, retreats into the bulb. Remove and swing the thermometer in a short arc, forcing all of the mercury into the bulb.

Most mercury thermometers can be reunited using this method regardless of range (with the exception of deep immersion thermometers) provided only the bulb is immersed in the CO_2.

Caution: Do not touch the bulb until it has warmed sufficiently for the mercury to emerge from the bulb into the capillary.

Never subject the stem or mercury column to the CO_2 solution, as it will freeze the mercury column in the capillary and may cause the bulb to fracture.

Typical wire resistance

(Stranded Copper Conductors at 59 °F).

Wire Size AWG	Resistance (ohm/ft)
0000	0.00005
000	0.00006
00	0.00008
0	0.00010
1	0.00012
2	0.00016
4	0.00025
6	0.00039
8	0.00063
10	0.00098
12	0.00160

Dimensions of hex nuts and hex jam nuts

Nominal size	F			G		H			J		
	Width Across Flats			Width Across Corners		Thickness Hex Nuts			Thickness Hex Jam Nuts		
	Basic	Max	Min	Max	Min	Basic	Max	Min	Basic	Max	Min
¼	⁷⁄₁₆	0.438	0.428	0.505	0.488	⁷⁄₃₂	0.263	0.212	⁵⁄₃₂	0.163	0.150
⁵⁄₁₆	½	0.500	0.489	0.557	0.577	¹⁷⁄₆₄	0.273	0.258	³⁄₁₆	0.195	0.180
³⁄₈	⁹⁄₁₆	0.562	0.551	0.650	0.628	²¹⁄₆₄	0.337	0.320	⁷⁄₂₃	0.227	0.210
⁷⁄₁₆	¹¹⁄₁₆	0.688	0.675	0.794	0.768	³⁄₈	0.385	0.365	¼	0.260	0.240
½	¾	0.750	0.736	0.855	0.840	⁷⁄₁₆	0.448	0.427	⁵⁄₁₆	0.323	0.302
⁹⁄₁₆	⁷⁄₈	0.875	0.861	1.010	0.982	³¹⁄₆₄	0.496	0.473	⁵⁄₁₆	0.324	0.301
⁵⁄₈	¹⁵⁄₁₆	0.938	0.922	1.083	1.051	³⁵⁄₆₄	0.559	0.535	³⁄₈	0.387	0.363
¾	1 ⅛	1.125	1.088	1.299	1.240	⁴¹⁄₆₄	0.665	0.617	²⁷⁄₆₄	0.446	0.398
⅞	1 ⁵⁄₁₆	1.312	1.269	1.516	1.447	³⁄₁₄	0.776	0.724	³¹⁄₆₄	0.510	0.458
1	1½	1.500	1.450	1.732	1.653	⁵⁵⁄₆₄	0.887	0.831	³⁵⁄₆₄	0.575	0.519
1⅛	1¹¹⁄₁₆	1.688	1.631	1.949	1.859	³¹⁄₃₂	0.999	0.939	³⁹⁄₆₄	0.639	0.579
1¼	1 ⅞	1.875	1.812	2.165	2.066	1 ¹⁄₁₆	1.094	1.030	³²⁄₃₂	0.751	0.687
1⅜	2 ¹⁄₁₆	2.062	1.994	2.382	2.273	1 ¹¹⁄₆₄	1.206	1.138	²⁵⁄₃₂	0.815	0.747
1½	2¼	2.250	2.175	2.598	2.480	1 ⁹⁄₅₂	1.317	1.245	²⁷⁄₃₂	0.880	0.808

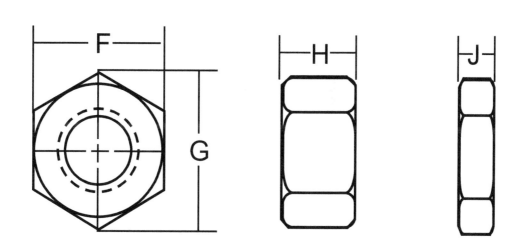

Reprinted with permission—Barnhill Bolt Co., Inc. Albuquerque, NM.

Color codes for locating underground utilities

Approximate angle of repose for sloping sides of excavations

This information may be used by anyone desiring to do so. Where excavation activities are regulated by a regulatory body, the user should consult and use the regulations promulgated by the regulatory body.

Note: Clays, SIMS, loams, or non-hermorporus soils require shoring and bracing. The presence of groundwater requires special treatment.

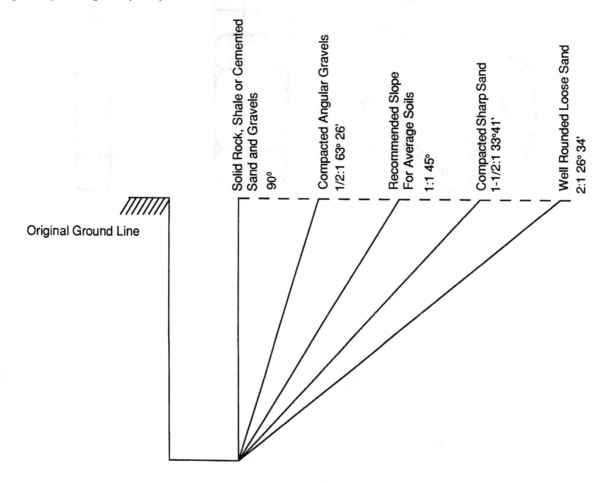

Wind chill chart

Wind Speed		Cooling Power of Wind Expressed as "Equivalent Chill Temperature," °F																				
Knots	MPH	Equivalent Chill Temperature																				
Calm	Calm	40	35	30	25	20	15	10	5	0	-5	-10	-15	-20	-25	-30	-35	-40	-45	-50	-55	-60
3–6	5	35	30	25	20	15	10	5	0	-5	-10	-15	-20	-25	-30	-35	-40	-45	-50	-55	-65	-70
7–10	10	30	20	15	10	5	0	-10	-15	-20	-25	-35	-40	-45	-50	-60	-65	-70	-75	-80	-90	-95
11–15	15	25	15	10	0	-5	-10	-20	-25	-30	-40	-45	-50	-60	-65	-70	-80	-85	-90	-100	-105	-110
16–19	20	20	10	5	0	-10	-15	-25	-30	-35	-45	-50	-60	-65	-75	-80	-85	-95	-100	-110	-115	-120
20–23	25	15	10	0	-5	-15	-20	-30	-35	-45	-50	-60	-65	-75	-80	-90	-95	-105	-110	-120	-125	-135
24–28	30	10	5	0	-10	-20	-25	-30	-40	-50	-55	-65	-70	-80	-85	-95	-100	-110	-115	-125	-130	-140
29–32	35	10	5	-5	-10	-20	-30	-35	-40	-50	-60	-65	-75	-80	-90	-100	-105	-115	-120	-130	-135	-145
33–36	40	10	0	-5	-10	-20	-30	-35	-45	-55	-60	-70	-75	-85	-95	-100	-110	-115	-125	-130	-140	-150

Winds above 40 °F have little additional effect

Little danger

Increasing danger (Flesh may freeze within 1 min)

Great danger (Flesh may freeze within 30 s)

PIPELINE PIGGING

Sizing plates

Often the debris-removal operation, after completion of construction, is combined with gauging to detect dents and buckles. This operation will prove that the pipeline has a circular hole from end to end. Typically an aluminum disc with a diameter of 95% of the nominal inside diameter of the pipe is attached to the front of a pig and inspected for marks at the end of the run. The pig could also be equipped with a transmitter to facilitate tracking the location of the pig. If the pig hangs, this will facilitate easy location of the pig to locate the dent/buckle.

When constructing off-shore pipelines, the most likely place for a buckle to occur during the lay operation is in the sag bend just before the pipe touches the bottom. A gauging pig can be placed inside the pipe and pulled along the pipe. If the lay barge moves forward and the pig encounters a buckle or dent, the pull line will become taut. This indicates that it will be necessary to pick up and replace the dented section of pipe.

Caliper pigging

Caliper pigs are used to measure pipe internal geometry. Typically they have an array of levers mounted in one of the cups. The levers are connected to a recording device in the body. As the pig travels through the pipeline, the deflections of the levers are recorded. The results can show up details such as girth-weld penetration, pipe ovality, and dents. The body is normally compact, about 60% of the internal diameter, which, combined with flexible cups, allows the pig to pass constrictions up to 15% of bore.

Caliper pigs can be used to gauge the pipeline. The ability to pass constrictions such as a dent or buckle means that the pig can be used to prove that the line is clear with minimum risk of jamming. This is particularly useful on subsea pipelines and long landlines, where it would be difficult and expensive to locate a stuck pig. The results of a caliper pig run also form a baseline record for comparison with future similar surveys, as discussed further below.

Cleaning after construction

After construction, the pipeline bore typically contains dirt, rust, and millscale; for several reasons it is normal to clean these off. The most obvious of these reasons is to prevent contamination of the product. Gas feeding into the domestic grid, for example, must not be contaminated with particulate matter, since it could block the jets in the burners downstream. A similar argument applies to most product lines, in that the fluid is devalued by contamination.

A second reason for cleaning the pipeline after construction is to allow effective use of corrosion inhibitors during commissioning and operation. If product fluid contains corrosive components such as hydrogen sulfide and carbon dioxide, or the pipeline has to be left full of water for some time before it can be commissioned, one way of protecting against corrosive attack is by introducing inhibitors into the pipeline. These are, however, less effective where the steel surface is already corroded or covered with millscale, because the inhibitors do not come into intimate contact with the surface they intended to protect.

Thirdly, the flow efficiency is improved by having a clean line and keeping it clean. This applies particularly to longer pipelines where the effect is more noticeable.

Therefore, most pipelines will be required to be clean commissioning. Increasingly, operators are specifying that the pipe should be sand blasted, coated with inhibitor, and have its ends capped after traction in order to minimize the post-construction cleaning operation. A typical cleaning operation would consist of sending through a train of pigs driven by water. The pigs would have wire brushes and would permit some bypass flow of the water so that the rust and millscale dislodged by the brushing would be flushed out in front of the pigs and kept in suspension by the turbulent flow. The pipeline would then be flushed and swept out by batching pigs until the particulate matter in the flow has reduced to acceptable levels.

Following brushing, the longer the pipeline the longer it will take to flush and sweep out the particles to an acceptable level. Gel slugs are used to pick up the debris into suspension, cleaning the pipeline more efficiently. Gels are specially formulated viscous liquids that will wet the pipe surface and pick up and hold particles in suspension. A slug of gel would be contained between two batching pigs and would be followed by a slug of solvent to remove any traces of gel left behind.

Flooding for hydrotest

In order to demonstrate the strength and integrity of the pipeline, it is filled with water and pressure tested. The air must be removed so that the line can be pressurized efficiently because, if pockets of air remain, these will be compressed and absorb energy. It will also take longer to bring the line up to pressure and will be more hazardous in the event of a rupture during the test. It is therefore necessary to ensure that the line is properly flooded and all of the air is displaced.

A batching pig driven ahead of the water forms an efficient interface. Without a pig, in downhill portions of the line, the water will run down underneath the air, trapping pockets at the high points. Even with a pig, in mountainous terrain with steep downhill slopes, the weight of water behind the pig can cause it to accelerate away, leaving a low pressure zone at the hill crest. This would cause dissolved air to come out of the solution and form an air lock. A pig with a high pressure drop across it would be required to prevent this.

Alternatives to using a pig include flushing out the air or installing vents at high point. For a long or large-diameter pipeline, achieving sufficient flushing velocity becomes impractical. Installing vents reduces the pipeline integrity and should be avoided. So for flooding a pipeline, pigging is normally the best solution.

Dewatering and drying

After the hydrotest has been completed, the water is generally displaced by the product or by nitrogen. The same arguments apply to dewatering as to flooding. A pig is used to provide an interface between the hydrotest water and the displacing medium so that the water is swept out of all low points. A bidirectional batching pig may be used during hydrotest and left in the line during hydrotest and then reversed to dewater the line.

In some cases, it may be necessary to dry the pipeline. This is particularly so for gas pipelines where traces of water may combine with the gas to form hydrates. Drying is also required for chemical pipelines such as ethylene and propylene pipelines, since water will contaminate the material and make it unusable. After dewatering the pipe walls will be damp and some water may remain trapped in valves and dead legs. This problem is solved by designing dead legs to be self-draining and installing drains on valves.

Refer to Chapter 6, Pipeline Drying, for a detailed discussing of pipeline drying methods.

Reprinted with permission—Gulf Publishing Co. "Pipeline Pigging Technology 2nd Edition".

Pig Trap Design

Pipeline Pig Traps

The primary purpose of a pig trap is to provide a means for safely and without flow interruption inserting and launching or receiving and retrieving a pig from a pipeline.

Notes:

For sphere launchers utilizing an automatic retractable flap system, a larger oversize may be required. Consult the manufacturer. The length of the barrel will need to be sized for the number of spheres to be handled.

The reducer between the major barrel and the minor barrel should be eccentric for horizontal traps (with the bottom of the major and minor barrel at the same level) and concentric for vertical traps.

The launcher should be designed for the same working pressure as the pipeline and to the same design code. This will usually be ASME B 31.4 for pipelines transporting liquids and ASME B 31.8 for pipelines transporting natural and other gas. It may be necessary to design to ASME B 31.3 for pipelines located within the boundaries of a processing plant. If the pipeline is subject to regulation by a governmental agency, it will be necessary to conform to the requirements of the regulation. In the US, the regulations for

Figure 1. Typical pig launcher.

Table 1
Typical Launcher Dimensions (All Dimensions Given in Inches)

Pipeline Diameter	Bypass Line	Kicker Line	Balance Line	Drain Line	Major Barrel Diameter	Major Barrel "A"	Minor Barrel "B"	Major Barrel "A"	Minor Barrel "B"
						Launcher		Receiver	
4	3	2	2	2	8	110	20	110	110
6	4	2	2	2	10	110	59	110	110
8	4–6	4	2	2	12	161	59	154	154
10	6	4	2	2	14	169	59	169	169
12	6–8	4	2	2	16	169	59	169	169
14	6–10	4	4	2	16	189	59	189	189
16	8–12	6	4	4	20	200	59	200	200
18	10–12	8	4	4	22	200	59	200	200
20	10–16	8	4	4	24	200	59	200	200
24	12–18	8	4	4	28	224	59	224	224
28	16–20	10	4	4	32	228	59	228	228
30	16–24	10	4	4	36	236	59	236	236
32	16–24	10	4	4	36	260	59	260	260
36	18–28	10	4	4	40	260	59	201	260
38	20–28	12	4	4	42	260	59	216	260
40	20–32	12	4	4	44	260	59	216	260
42	20–36	16	4	4	46	260	59	252	260
48	24–36	18	4	4	52	260	59	260	260
56	32–40	20	4	4	60	260	59	260	260

pipelines transporting hazardous liquids can be found in 49 CFR Part 195 and the regulations for transporting natural and other gases can be found in 49 CFR Part 192. Other design codes may apply depending on the location. If any electrical apparatus is attached to the trap, e.g. a pig sig, the device must be compatible with the product being handled. For example, devices rated for Group D would be suitable for crude oil. Systems handling ethylene will require devices suitable for Group C materials. Consult NFPA 70 for complete information on the group classification.

Two types of pigs (maintenance pigs and intelligent pigs) are typically used in pipeline operations. The trap shown here is designed for handling intelligent pigs that are used for performing in-line inspection.

barrel. The barrel can be extended to accommodate intelligent pigs by inserting a flanged extension.

A balance line is used to enable filling and pressurizing the barrel on both sides of the pig at the same time. This will prevent the pig from possibly moving forward and damaging the pig trap valve or from moving backward and losing the seal in the reducer. Refer to Table 1 for balance line sizing.

A vent valve should be provided near the trap valve for a launcher and near the closure for a receiver. For liquid systems, a 2″ vent is normally adequate. For high pressure gas systems, consideration should be given to a blow down line equipped with a valve that can be used for depressurization of the trap. For systems handling other materials such as

Figure 2. Typical pig receiver.

The trap should be designed to accommodate the length of the longest pig (plus 10%) anticipated to be used. This is usually an intelligent pig. The major barrel is normally two pipe sizes larger than the size of the pipeline for pipelines with a diameter less than 20 in and for pipelines larger than 20 in in diameter, the oversize should be 4 in. A quick opening closure is welded to the end of the major barrel to allow for the quick insertion and removal of pigs. The closure must be fitted with a device to prevent opening the closure until the pressure in the barrel has been relieved. The closure must have a pressure and temperature rating equal to the pipe to which it is attached. The closure should be designed in accordance with the ASME Boiler and Pressure Vessel Code, section VIII, Pressure Vessels, Division 1. If intelligent pigs will be used infrequently, e.g. every 5–10 years, the major barrel may be shortened to facilitate handling maintenance type pigs. If the barrel is to be shortened, a flange should be installed between the hinged closure and the major

ethylene, consideration will need to be given to the reduced temperature as a result of throttling a high pressure to a low pressure. For high pressure gas systems, consideration needs to be given to the thrust developed during the blow down operation.

The pig trap and bypass valves should be motor operated for 10 in and larger valves. The pig trap and kicker valves should be tight shutoff valves to allow for complete isolation of the trap from the main line. Refer to Figure 3 for valve identification.

The closure should be hinged or held by a built-in davit to prevent dropping and to facilitate handling during operation. The "O" ring seal should be carefully examined every time the closure is opened. If the pipeline is in HVL service, the O ring should be replaced every time the closure is opened.

For horizontal traps, the reducer between the major and minor barrels should be an eccentric type with the flat side down so that the bottom of both barrels will be at the same level.

Figure 3. Typical launcher configuration.

Flange insulating kits should be used to isolate the pipeline system from station/terminal system for cathodic protection purposes.

Estimate volume of on-shore oil spill

An irregular shape can be converted into a series of rectangles that approximate the area of the irregular shape. There will be about the same amount of spill area out side the rectangle as there is dry area inside the rectangle. This can be done by stretching a steel tape along the ground outside the spill area. The area can then be quickly estimated by multiplying the length of the sides. In Figure 1, the following area is determined:

Figure 1.

Area A $= 70' \times 20' = 1400$ sq ft,
Area B $= 60' \times 10' = 600$ sq ft,
Area C $= 35' \times 20' = 700$ sq ft,
Total $= 2700$ sq ft.

The more rectangles used, the more accurate the estimate becomes.

The next task is to estimate the average depth of oil in each of the areas. The oil will vary from very shallow at the edge to whatever depth the terrain is at the lowest point. This can be determined by "gauging" with a stick if it is shallow or accessible. If the pool is wider, you can heave a large stone into the pool to confirm depth. A good estimate can usually be made by observing the slope of the ground around the pool and assuming that the slope continues under the surface of the oil.

If you estimate that the deepest point in area A is 20″ and area A has three boundaries of "shore," divide the depth figure by 3 to obtain average depth. If it has two "shore" boundaries, like area B, divide the depth by 2 to obtain average area depth.

The irregular-shaped area with unseen bottom has now been reduced to a familiar shape. The volume of free oil in area A is

Area "A" $= 70' \times 20' = 1400$ sq ft,
Average depth $= 20''/3 = 7''$ or 0.6 ft,
Area "A" volume $= 1400$ sq ft $\times 0.6$ ft $= 840$ cu ft.

The total volume will be the sum of volumes for areas A, B, and C.

Next, convert 840 cu ft to barrels. Each cubic foot is equivalent to 0.178 bbls.

Area "A" volume $= 840$ cu ft and therefore 840×0.178
$= 150$ bbls.

Determining how much additional oil has penetrated into the soil can be accurately measured by taking a core sample of the oil-covered soil; however, the following rule should suffice for estimates of oil spilled:

For penetration allowance in normal sand or soil, add 5% to the total volume for every foot of average depth.

In the case of area A, the average depth was 0.6 ft; therefore, $0.6 \times 5\% = 3\%$ to be added. 150 bbls $\times 1.03 =$ 154.5 bbls total volume spilled in area A.

- Do not add a penetration allowance to areas with slopes that allow a reasonable flow rate.
- Add an allowance for slow-flowing areas.
- Reduce allowance by half if the area is wet from rain.

If more precise determination is required, drive a clear plastic tube, about 2″ or larger in diameter, to a depth of 6″ in the uncontaminated soil adjacent to the spill. Twist and remove with soil core. Seal the bottom of the tube with plastic and tape. Pour free oil into the tube to the depth of the oil in the pool, mark the level, and let it set for 1 h. Measure how much the oil level has dropped. Observe how deep the oil has penetrated. Retain the model to observe increased penetration with time.

Walk-around method

If the pool of oil is roughly circular, you can estimate its area by pacing around the pool and counting your paces. Walk as close to the pool edge as possible. Try to make your paces 3 ft, or 1 yard long. If you counted 700 paces, the circumference is 700×3 or 2100 ft. The next step is to guess how much smaller the actual pool is, compared to the circle you walked. If you were pretty close, deduct 10%.

$2100 \times 0.9 = 1890$ ft adjusted circumference.

The diameter of a circle is related to the circumference by the following equation:

$C = \pi D$

where

$\pi = 3.14159$,
$D =$ diameter,
$C =$ circumference,

$D = 1890/\pi = 601.6$ ft,
The radius of the pool is $D/2$ or 300.80 ft,

The area of the pool $= \pi r^2$,
$A = 3.14159 \times 301 \times 301$,
$A = 284,253$ sq ft.

Now you can estimate the average depth by guessing the maximum depth. Assume the depth from the exposed slope to be 12″ at the deepest part, divide by 4 (four sloping sides) to estimate an average depth of 3″ or 0.25 ft.

The volume is

$V = 284,253 \times 0.25 = 71,063$ cu ft,

Volume of oil $= 71,063 \times 0.178 = 12,649$ bbls.

The average depth was 3″ and therefore, we need to add about 1% for penetration or $1.01 \times 12,660$ bbls $= 12,775.7$ bbls.

Average diameters

You can also estimate the area of an oval-shaped pool by pacing off (3 ft per step) the width of the short diameter and the long diameter and averaging the diameters.

Pace off the short diameter, but stop short to allow for the irregular shape. Repeat the procedure for the long diameter. Add the diameters together and divide by 2 to get the average diameter.

Short diameter $= 75$ paces $= 75 \times 3 = 225$ ft,
Long diameter $= 120$ paces $= 120 \times 3 = 360$ ft,
Average diameter $= (225 + 360)/2 = 292$ ft,
Radius $= 292/2 = 146$ ft,

$A = \pi r^2 = 3.14159 \times 146 \times 146 = 66{,}966$ sq ft,
Average depth $= 3''$ or 0.25 ft,
Volume $= 66{,}966 \times 0.25 = 16{,}741.5$ cu ft,
Volume $= 2981.79$ bbls.

Figure 2. Average diameters.

Estimating spill volume on water

When conditions permit, direct measurements of spill parameters are preferred over visual estimates.

A rough estimate of spill volume can be generated from observations of the oil slick's size and thickness. Figure 3 and Table 1 relate the appearance, thickness, and the light conditions. For example, slick thickness greater than 0.08 in cannot be determined by appearance alone.

Since oil slick spreading is influenced by the spill volume as well as physical forces, stopping the spill at its source is critical in controlling the spread of a slick on water. The more conservative the first estimate of the spill volume, the better the chances that response forces will arrive at the spill site prepared with adequate and appropriate equipment. It is preferable to overrespond early rather than underrespond and risk unpreparedness. To underrespond will impede the effectiveness of spill control and cleanup efforts. A slow or poorly prepared initial response can incur more operational costs and increase the risk of damage to marine and shoreline resources and environments. Therefore, properly planning the initial response is critical in a spill situation.

SHADED AREA INDICATES THE RANGE OF OIL SLICK OBSERVATIONS
FOR WHICH THICKNESS AND AREA COVERED CAN BE DETERMINED BY APPEARANCE.
ANY VALUE BELOW THE SHADED AREA WOULD NOT BE VISIBLE, AND ANY VALUE
ABOVE THE SHADED AREA WOULD BE DARK BROWN OR BLACK.

Figure 3.

Table 1
Estimating Spill Volume by Color and Coverage Area

Width × Length	Sq ft	Silvery Sheen	Trace of Color (Yellow, Bronze Violet)	Bright Bands of Color (Purple, Blue to Green)	Colors Turning Dull (Brick Red, Turquoise, Pale Yellow)
		Gals	Gals	Gals	Gals
100 × 500	50,000	0.1	0.2	0.4	1.2
100 × 1000	100,000	0.2	0.4	0.7	2.5
100 × 2000	200,000	0.4	0.7	1.5	4.9
200 × 2000	400,000	0.7	1.5	3.0	9.9
500 × 1000	500,000	0.9	1.9	3.7	12.3
200 × 5000	1,000,000	1.9	3.7	7.5	24.7
500 × 5000	2,500,000	4.7	9.4	18.7	61.7
500 × 10,000	5,000,000	9.4	18.7	37.4	123.4

Fluid Power Formulas

Heat equivalent of fluid power: BTU/H = PSI × GPM × 1½

GPM = gallons per minute,
PSI = pounds per square inch.

Hydraulic cylinder piston travel speed: S = CIM ÷ A

S = piston travel speed, inches per minute,
CIM = oil flow into cylinder, cubic inches per minute.

Force for piercing or shearing sheet metal: $F = P \times T \times$ PSI

F = force required, pounds,
P = perimeter around area to be sheared, inches,
T = sheet thickness, inches,
PSI = shear strength rating of the material, pounds per square inch.

Side load on pump or motor shaft: F = (HP × 63,024) ÷ (RPM × R)

F = side load, pounds, against shaft,
R = pitch radius of sheave on pump shaft, inches,
HP = driving power applied to shaft.

Effective force of a cylinder working at an angle to direction of the load travel: $F = T \times \sin A$

T = total cylinder force, pounds,
F = effective force, pounds,
A = least angle between cylinder axis and load direction, degrees.

Relationship between displacement and torque of a hydraulic motor: $T = D \div$ PSI ÷ 24π

T = torque in foot pounds,
D = displacement in cubic inches per revolution,
PSI = pressure difference across the motor.

Thrust or force of a cylinder: $T = A \div$ PSI

T = thrust, pounds,
A = effective area of cylinder, square inches,
PSI = pounds per square inch gauge pressure.

Recommended maximum oil velocity in hydraulic lines

Pump suction lines: 2–4 feet per second (fps),
Pressure lines to 500 psi: 10 to 15 fps,
Pressure lines to 3000 psi: 15 to 20 fps,
Pressure lines over 3000 psi: 25 fps.

2: Construction

43

PROJECT SCOPING DATA

Project scoping data worksheet for major facilities

Project Name

_____

Location_____

Latitude_____

Longitude_____

☐ Offshore   ☐ Onshore   ☐ Arctic

☐ Marsh   Site elevation, ft _____

Project description _____

_____

_____

_____

1.0 Type of contract required

1.1 Describe the type of contract that the project will be based on.

1.2 What is the completion date?

1.3 What type of permits will be required and who will obtain them?

1.4 Will an environmental impact statement be required?

1.5 Describe currency requirements if in a foreign country.

1.6 If the project is located in a foreign country, does the country have regulations/codes that will need to be complied with?

2.0 General engineering

2.1 Does the client have engineering standards and specifications, or will it be necessary to develop these, or can the Engineering Contractor's specifications and standards be used?

2.2 Will in-country engineering concerns, such as design institutes, be required to provide engineering services?

2.3 If in-country engineering facilities are required, how much assistance will be required?

2.4 Describe any existing above-ground or underground obstructions at the job site.

2.5 Obtain copies of any applicable local codes dealing with air and/or water pollution, sanitary systems, electrical systems, and structures. Obtain copies of any regulations dealing with waste disposal.

2.6 Are soil bearing data available?

2.7 If the project involves a pipeline that transports heat-sensitive fluid, are any soil thermal conductivity data available?

2.8 Describe type of soil—sandy, rocky, caliche, etc.

2.9 Does the site drain naturally? If fill material is required, what is the availability? Is sand-asphalt mix available?

3.0 Drafting

3.1 Will the drawings need to be sealed by a professional engineer?

3.2 Determine drawing forms to be used and type of drawings required—single line or double line.

4.0 Structural and architectural

4.1 Will all exposed steel need to be galvanized or have other protective coating?

4.2 Is sand, rock, cement, and ready-mixed concrete available locally? Price?

4.3 Requirements for building design:
Steel
Concrete/brick
Cooling and heating requirements
Other requirements

5.0 Piping

5.1 How will storm water, waste water, and sewage be disposed of? Will an oil-water separator be required?

5.2 Will connections to any existing pipelines be required?

5.3 Are there any facilities outside the job site that must be given consideration in the development of a plot plan?

5.4 What is the pressure rating for any existing manifolds, suction lines, or outgoing pipelines?

5.5 Are there any special requirements for isolating the facility?

5.6 Will scraper traps be required? Incoming, outgoing, or both?

5.7 How many relief headers are required?

5.8 If system relief valves are required, where will they relieve to?

5.9 Will a fire water system be required? Is a source of water available? If a source is available, what is the capacity of the source?

5.10 Is the water fresh, brackish, or seawater?

6.0 Electrical

6.1 Is power available from the local power grid? If yes, specify voltage, frequency, phase, and system capacity.

6.2 If power generation will be required in connection with the project, will it be a requirement to connect to a local grid?

6.3 If local power is available, define the approximate location where the power will enter the job site.

6.4 Will the local power entity furnish and install any substation equipment that will be required?

6.5 If power is to be supplied from a local power grid, attempt to determine the short-circuit capacity of the system.

6.6 Will across-the-line starting of large electric motors be permitted?

6.7 If across-the-line starting is not permitted, what is the maximum horsepower that will be permitted for across-the-line starting?

7.0 Instrumentation equipment

7.1 What type of control panel will be required?

7.2 Define electrical classification that will be required for a control house if a control house is required.

7.3 Are local communications facilities available? Describe.

8.0 Equipment

8.1 Have any long-delivery items been placed on order? List items and delivery schedule.

8.2 Who will supply general equipment specifications?

8.3 Will in-country purchasing be required?

8.4 List design criteria for equipment sizing.

8.5 Define spare equipment requirements. If equipment is purchased outside the country where the project is located, consideration should be given to transit time from the country of origin to the job site.

8.6 Define the design life for the facility.

8.7 Will oil storage tanks be required? Will secondary seals be required? Will tank mixers be required?

9.0 Material purchasing

9.1 If a preferred vendor list is available, obtain a copy of the list.

9.2 If such a list is not available, will it be a requirement to develop a list for the client's approval?

9.3 Determine the shipping address.

9.4 If material is shipped to a foreign location, who will handle receipt of the material and clearance with local customs?

9.5 Is the location near an established seaport? If not, determine the nearest seaport and logistics for moving material from the seaport to the job site. This should include any weight limitations, load width restrictions, and requirements for moving heavy equipment at night during periods of light road traffic.

9.6 Where will material be stored pending installation?

9.7 What procedure will be used for turning material over to the contractor?

10.0 Special considerations

10.1 Is heavy equipment available locally and does it have the capacity to lift the heaviest piece of material?

10.2 Is an adequate supply of sufficiently skilled local labor available?

10.3 If the project is in a foreign location, is there a limit on the use of expatriates?

10.4 What is the procedure for obtaining a work permit?

10.5 What is the availability of contractors that may be working in the area?

10.6 Will a construction camp be required?

10.7 If a construction camp is required, what is the availability of a local catering service?

11.0 Environmental information

11.1 Ambient Temp. ($^\circ$F) Max._____Min._____

11.2 Relative humidity
Wet bulb ($^\circ$F) Max._____Min._____
Dry bulb ($^\circ$F) Max._____Min._____

11.3 Water temp. surface ($^\circ$F) Max._____Min._____

11.4 Water temp. bottom ($^\circ$F) Max._____Min._____

11.5 Potential for icing Y or N

11.6 Rainfall (in/h) Max._____Min.___

11.7 Prevailing wind direction

11.8 Velocity Max._____Min.___

11.9 Weather window (time of year)_____

12.0 Fluid characteristics

12.1 Crude oil

12.2 Sediment and water, %_____

12.3 Free water_____

12.4 Salt content (#/1000 bbls)_____

12.5 API gravity _____ @ _____ $^\circ$F and _____ @ _____ $^\circ$F (2 required)

12.6 Viscosity _____ cP @ _____ $^\circ$F and _____ cP @ _____ $^\circ$F (2 required)

12.7 Oil pour point _____ $^\circ$F

12.8 Reid vapor pressure _____

12.9 Produced water: oil & grease (ppm)___

12.10 Total suspended solids (mg/kg oil)_____

12.11 Natural Gas

12.12 MOL % inerts_____

12.13 MOL % CO_2_____

12.14 ppm H_2S or grains/100 SCF _____

12.15 Water content_____#/MMSCF or dew point _____ $^\circ$F

12.16 Max. heating value (BTU/SCF)_____

12.17 Min heating value (BTU/SCF)_____

12.18 NGL content _____ gal/MCF or dew point_____ $^\circ$F

13.0 Offshore Environmental Information

Parameter	1 Year	5 Years	25 Years	100 Years
Waves				
Direction				
Significant wave height, ft				
Maximum wave height, ft				
Crest elevation, ft				
Significant max. wave period, s				
Peak period, s				
Tides				
Astronomical tide, ft				
Storm surge, ft				
Wind @ 30 ft elevation above MLW				
1-min average, kt				
1-h average, kt				
2-s gust, kt				
Current				
Surface speed, kt				
3 ft from bottom, kt				
Wave force coefficients				
C_D with marine growth				
C_D with no marine growth				
C_M				
Marine growth to 150 ft, inches				
Sea temperature, °F				
Maximum surface				
Minimum surface				
Maximum bottom				
Minimum bottom				
Air temperature, °F				
Maximum				
Minimum				
Relative humidity				
Wet bulb, °F				
Dry bulb, °F				
Notes/Comments				

RIGHT-OF-WAY

How to determine the crop acreage included in a right-of-way strip

Multiply the width of the strip in feet by the length in rods; divide this by 2640 to obtain the acreage. If the ends of the strip are not parallel, use the length of the center line of the right-of-way.

Example. A right-of-way 35 ft wide crosses a cultivated field for a length of 14 rods; how many acres of crop were destroyed?

$35 \times 14/2640 = 0.18$ acres, or almost ⅛ acre.

Example. A right-of-way 50 ft wide crosses a field for 330 rods. How many acres of crop were destroyed?

$50 \times 330/2640 = 6.25$ acres.
The rule is exact, not an approximation.

Clearing and grading right-of-way: labor/equipment considerations

To estimate labor crew and equipment spread for clearing and grading operations, the following items, as they may apply to a given project, should be given consideration:

1. Removal of trees, brush, and stumps.
2. Grubbing and removal of stumps that are in the way of the ditch.
3. Disposal of all debris, including method of disposal and length of haul.
4. Clearing area spoil a sufficient distance from the ditch line so that the spoil-bank from the ditching operations will not fall in any foreign material that might become mixed with the excavated spoil.
5. Cutting of merchantable timber into standard lengths and stacked along the right-of-way for disposition by others if specifically required by the right-of-way agreement.
6. Providing temporary walks, passageways, fences, or other structures so as not to interfere with traffic.
7. Providing sufficient and proper lighting where required.
8. Providing guards where required.
9. Preserving all trees, shrubs, hedges, and lawns where required.
10. Grading irregularities where required.
11. Preserving topsoil for replacement, through all cultivated or improved fields and pastures, to its original position.
12. Proper grading of the terrain so as to allow passage of loaded trucks and equipment hauling materials and so ditching operations can be properly performed.
13. Protecting and preserving existing drainage facilities.
14. Protecting any existing structures or pipelines.
15. Protecting any telephone or utility lines and keeping them in service.
16. Cutting through fences and hedges where required and replacing these when necessary.
17. Installing gates and fences where required.

Estimating man hours for removing trees

Net Man Hours—Each

Average Tree Diameter in Inches	Softwood Trees		Hardwood Trees	
	Open Area	Congested Area	Open Area	Congested Area
Cross-Cut Saws				
4	1.49	1.86	1.88	2.35
6	2.26	2.83	2.82	3.53
8	3.24	4.02	4.00	4.96
10	4.10	5.08	5.00	6.20
12	4.98	6.18	6.00	7.44
14	6.39	7.86	7.70	9.47
16	7.39	9.09	8.80	10.82
18	8.32	10.23	9.90	12.18
20	10.58	12.91	12.60	15.37
24	12.71	15.51	15.12	18.45
30	17.09	20.85	20.10	24.52
36	20.50	25.01	24.12	29.43

Net Man Hours—Each (continued)

Average Tree Diameter in Inches	Softwood Trees		Hardwood Trees	
	Open Area	Congested Area	Open Area	Congested Area
Chain Saws				
4	0.37	0.46	0.47	0.59
6	0.57	0.71	0.71	0.89
8	0.81	1.00	1.00	1.24
10	1.03	1.28	1.25	1.55
12	1.25	1.55	1.50	1.86
14	1.60	1.97	1.93	2.37
16	1.85	2.28	2.20	2.71
18	2.08	2.56	2.48	3.05
20	2.65	3.23	3.15	3.84
24	3.18	3.88	3.78	4.61
30	4.28	5.22	5.03	6.14
36	5.13	6.26	6.03	7.36

Man hours include ax trimming, cutting down with cross-cut saws or chain saws, and cutting into 4-ft lengths for the tree diameter sizes as listed previously. Man hours are average for various heights of trees.

Man hours do not include hauling, piling, and burning of trees or branches, or the removal of stumps.

Estimating man hours for removing tree stumps

Net Man Hours—Each

Item	Laborer	Oper. Engr.	Powder Man	Total
Grub & Removal by Hand				
8″–12″ diameter	6.00	–	–	6.00
14″–18″ diameter	7.50	–	–	7.50
20″–24″ diameter	9.00	–	–	9.00
26″–36″ diameter	11.20	–	–	11.20

Net Man Hours—Each (continued)

Item	Laborer	Oper. Engr.	Powder Man	Total
Blast & Pull with Tractor				
8″–12″ diameter	0.83	0.11	1.50	2.44
14″–18″ diameter	1.05	0.23	2.33	3.61
20″–24″ diameter	1.50	0.30	3.38	5.18
26″–36″ diameter	2.11	0.42	4.76	7.29

Man hours include excavating and removing by hand or blasting and removing with cables and tractors. Man hours do not include burning or removal from premises.

Clearing and grading right-of-way

Equipment Spread

Equipment Description	Number of Units for											
	50 Linear ft Width				80 Linear ft Width				100 Linear ft Width			
	L	M	MH	H	L	M	MH	H	L	M	MH	H
D8 Tractor W/Dozer	1	1	1	1	1	1	2	2	1	2	3	3
D7 Tractor W/Dozer	0	0	1	1	1	1	1	1	1	1	2	2
Truck—2½ Ton Dump	2	2	2	2	2	2	2	2	2	2	3	3
Truck—Pick-up	1	1	2	2	1	2	3	3	2	3	4	4
Ripper or Brushrake	1	1	1	1	1	1	1	1	1	2	2	2

Above equipment spread should be ample for clearing and grading 1 mile of right-of-way per 10-hour day for the width and conditions outlined. Haul trucks are based on a round-trip haul of 2 miles. If brush and trees are to be burned on site, omit the above dump trucks. Small tools such as saws, axes, etc., must be added as required for the individual job.

Code description

L = Light—light brush and grass, no trees.
M = Medium—considerable brush of larger size.
MH = Medium Heavy—large brush and small trees.
H = Heavy—much small brush, many small trees, and occasional large trees.

Labor Crew

Personnel Description	Number of Men for											
	50 Linear ft Width				80 Linear ft Width				100 Linear ft Width			
	L	M	MH	H	L	M	MH	H	L	M	MH	H
Foreman	1	1	2	2	1	2	3	3	2	3	4	4
Operator	1	1	2	2	1	2	3	3	2	4	5	5
Mechanic	1	1	1	1	1	1	1	1	1	1	1	1
Swamper	1	1	2	2	1	2	3	3	2	4	5	5
Truck Driver	2	2	2	2	2	2	2	2	2	2	3	3
Laborer	10	15	20	30	15	25	30	40	20	35	40	50
Total Crew	16	21	29	39	21	34	42	52	29	49	58	68

Above total crew should be ample for clearing and grading 1 mile of right-of-way per 10-hour day for the width and conditions outlined. Crew spread includes cutting, stacking or piling, loading, and hauling a round-trip distance of 2 miles. If burning is necessary or permitted, substitute fire tenders for dump truck drivers. See Clearing and Grading equipment spread for number of dump trucks.

Code description

L = Light—light brush and grass, no trees.
M = Medium—considerable brush of larger size.

MH = Medium Heavy—large brush and small trees.
 H = Heavy—much small brush, many small trees, and occasional large trees.

Source

Page, J. S., *Cost Estimating Man-Hour Manual for Pipelines and Marine Structures*, Gulf Publishing Co., Houston, Texas, 1977.

DITCHING

How many cubic yards of excavation in a mile of ditch?

Multiply the width in inches by the depth in inches by 1.36; the answer is cubic yards per mile.

Example. How many cubic yards of excavation in a mile of 12-in ditch 30-in deep?

$12 \times 30 \times 1.36 = 490$ cubic yards per mile.

The rule is correct within about 1/5 of 1%. Actually, the errors in depth and width are much greater than this. To get the cubic yards per 1000 ft, as in computing rock ditching, use 0.257 instead of 1.36.

Shrinkage and expansion of excavated and compacted soil

Ever notice how the spoil from a ditch occupies a greater volume than the ditch itself? There's a reason. Excavate sand and it expands about 10%. Ordinary soil expands about 25% and clay expands about 40%.

Here's a summary of the bulk you can expect from excavated soil

Type soil	Undisturbed	Excavated	Compacted
Sand	1.00	1.11	0.95
Ordinary earth	1.00	1.25	0.90
Clay	1.00	1.43	0.90

Example. Find the volume of loose spoil from a pipeline ditch 42 in wide, 60 in deep. The excavation is through clay.

Volume of undisturbed clay soil:

$3.5 \times 5.0 = 17.5$ cu ft per lineal foot of ditch.

Volume of the spoil will be 143% of the undisturbed volume.

Volume of spoil $= 1.43 \times 17.5 = 25 +$ cu ft per lineal ft.

Through mechanical compaction, this volume can be reduced to 15.75 cu ft.

Ditching and trenching: labor/equipment considerations

In determining the labor crew and equipment spread for ditching and trenching operations, the following should be given consideration should they apply to the particular project:

1. Ditching or trenching for buried pipelines should be in accordance with the following table of minimum width and coverage for all soil formations.
2. In rock, cut ditches at least 6 in wider.

3. If dirt-filled benches are used, ditch should be excavated deeper to obtain proper coverage.
4. Trench should be excavated to greater depth when required for proper installation of the pipe where the topography of the country warrants same.
5. Repair any damage to and maintain existing natural or other drainage facilities.
6. Do not open ditch too far in advance of pipelay crew.
7. Obtain permits for blasting.
8. When blasting, use extreme caution and protection.
9. Clean up blasted rock to prevent damage to coated pipe.

Nominal Pipe Size Inches	Minimum Width Inches	Normal Minimum Coverage Inches
4	22	30
6	26	30
8	26	30
10	26	30
12	30	30
14	32	30
16	36	30
18	38	30
20	40	30
24	44	30
30	50	30
36	52	36
42	58	36

CONCRETE WORK

How to approximate sacks of cement needed to fill a form

To obtain a close estimate of the number of sacks of cement that will be required, first determine the volume (cubic feet) to be filled in the form. Divide the volume by 4.86 to approximate the number of sacks of cement needed.

Example. How many 94-lb sacks of cement will be required to fill a form for a concrete base 10 ft by 10 ft if it is to be 6 in thick?

$$\frac{10 \times 10 \times 0.5}{4.86} = 10.3, \text{ or } 11 \text{ sacks needed.}$$

What you should know about mixing and finishing concrete

To determine the proper mix, divide the constant 44 by the sum of the parts of cement, the parts of sand, and the parts of gravel to determine the number of bags of Portland cement required. Multiply the number of bags of cement as determined above by parts of sand and the constant 0.035 to calculate the number of cubic yards of sand needed. To determine cubic yards of gravel needed, multiply bags of cement by parts of gravel × 0.035.

Example. Calculate the quantities of cement, sand, and gravel required for 1 cubic yard of 1:2:4 concrete.

$$C = \frac{44}{1+2+4}$$

$$= 6.28 \text{ bags of cement}$$

$$S = 6.28 \times 2 \times 0.035$$

$$= 0.44 \text{ cubic yard of sand}$$

$$G = 6.28 \times 4 \times 0.035$$

$$= 0.88 \text{ cubic yard of gravel}$$

To increase or decrease the slump of concrete, add or subtract 1 gallon of water per cubic yard of mix and subtract or add 20 lb of aggregate to maintain yield.

To adjust from no air to air-entrained concrete and maintain strength, reduce water ¼ gallon per sack of cement and reduce sand 10 lb per sack of cement for each 1% of entrained air.

PIPE LAYING

How to determine the degrees of bend in a pipe that must fit a ditch calling for a bend in both horizontal and vertical planes

Rule. To find the number of degrees in the combination bend, square the side bend and the sag or overbend; add them together and extract the square root. The answer will be the number of degrees necessary to make the pipe fit the ditch.

Example. Determine the bend to make in a pipe whose ditch has a 3° overbend and a 4° side bend. Let $X =$ the unknown angle:

$$X^2 = 3^2 + 4^2$$
$$X^2 = 9 + 16$$
$$X^2 = 25$$
$$X = 5°$$

Example. Determine the bend to make in a pipe whose ditch has a 9° sag and a 12° side bend. Let $X =$ the unknown angle:

$$X^2 = 9^2 + 12^2$$
$$X^2 = 225$$
$$X = 15°$$

How to bend a pipe to fit a ditch—sags, overbends, and combination bends

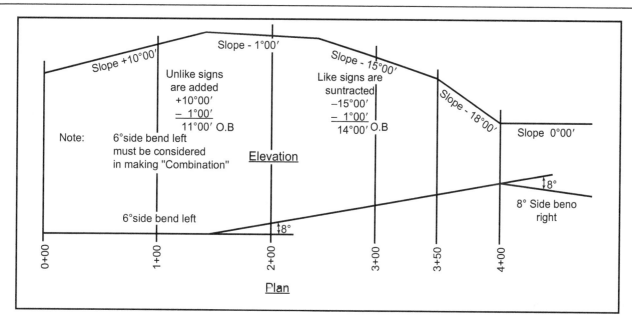

To make straight sags or overbends fit the ditch (see drawing), add angles of unlike signs and subtract those of like signs.

Example. (Sta. 2 + 40)

Slope − 1500′
Slope − 100′
(sub.)   1400′ overbend

In cases involving either a sag or overbend, in addition to a side bend, the rule becomes: *Make the combination bend equal to the largest angle plus $\frac{1}{3}$ of the smallest.*

Example. (Sta. 1 + 40)

Slope + 1000′
Slope − 100′
(add)   1100′ overbend (1)
        600′ overbend (2)

Combination bend $= 11° + (\frac{1}{3} \times 6°) = 13°00′$ overbend left.

Note: This rule gives an error of approximately 1° for a maximum bend of 18.5°.

Pipe bending computations made with a hand-held calculator

Maximum code radii for pipe cold bends require a printer for calculations

Cold bending pipe is subject to provisions of Liquid Petroleum Transportation Piping Systems, B31.4, and Gas Transmission and Distribution Piping Systems, B31.8. These codes stipulate a minimum bending radius for each size pipe in addition to requirements for thinning, flattening, etc.

The minimum bending radius for cold field bends is specified in ASME B31.4 (Liquid Lines) and ASME B31.8 (Gas Pipelines)

ASME B31.4

Nominal Pipe Size	Minimum Radius in Pipe Diameters
≤12	18 D
14	21
16	24
18	27
≥20	30

Nominal Pipe Size	Deflection of longitudinal Axis—degrees	Minimum Radius in Pipe Diameters
<12	Refer to 841.2.3 (a)(4) in ASME B31.8	18 D
12	3.2	18
14	2.7	21
16	2.4	24
18	2.1	27
≥20	1.9	30

A 6 ft tangent is preferred; however, some bending machines may accommodate a shorter tangent.

The following equations address the geometry of fabricating offset bends, sloping scraper traps, connections, and direction changes common to all pipelines. Minimum code radii are calculated for each size of oil or gas line. The vertical distance between below- and above-ground center lines is matched to a design distance with given tangents (5 ft is a convenient tangent for pipe bending machines). The bend angle, overall horizontal distance, and total length of pipe can be calculated for each configuration.

(The code states "longitudinal axis of the pipe shall not be deflected more than 1½ degrees in any length along the pipe axis equal to the diameter of the pipe.")

Note: This distance can be readily determined by inspection.

Figure 1. Typical offset bend.

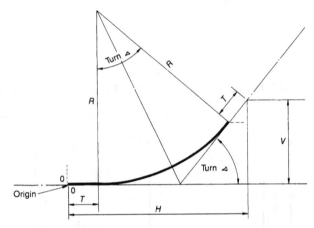

Figure 2. A conventional direction change.

Note: Aboveground pipe ends are modified as shown for traps not level. Face to face of valve included in *LRT* and *LLT*.

Figure 3. Receiver and launcher scraper trap connections, where the traps are not level.

Formulas for reverse bends

$$V = 2R(1 - \cos \angle 1) + 2T \sin \angle 1$$

Figure 4. A crossing under a foreign line, where offset bends are used.

$$H = 2R \sin \angle 1 + 2T \cos \angle 1 + 2T$$

$$L = 2R \frac{\angle 1}{57.2958} + 4T$$

Formulas for scraper trap connections

Receiver.

$VR = V - R(1 - \cos \text{Trap } \angle)T \sin \text{Trap } \angle$
$HR = H - T + R \sin \text{Trap } \angle + T \cos \text{Trap } \angle$

$$LL = L - R \frac{\text{Trap } \angle}{57.2958}$$

Launcher.

$VL = V - R(1 - \cos \text{Trap } \angle) + T \sin \text{Trap } \angle$
$HL = H - T + R \sin \text{Trap } \angle + T \cos \text{Trap } \angle$

$$LL = L - R \frac{\text{Trap } \angle}{57.2958}$$

Receiver trap.

$VRT = VR - LRT \sin \text{Trap } \angle$
$HRT = HR + LRT \cos \text{Trap } \angle$

Launcher trap.

$VLT = VL - LLT \sin \text{Trap } \angle$
$HLT = HL + LLT \cos \text{Trap } \angle$

Formulas for direction change

$$V = R(1 - \cos \text{Turn } \angle) + T \sin \text{Turn } \angle$$

$$H = R(\sin \text{Turn } \angle) + T(1 + \cos \text{Turn } \angle)$$

$$L = 2T + R \frac{\text{Turn } \angle}{57.2958}$$

Assume a trial angle, say 20° for a small pipe. Let $T = 0.5$ ft.

For noncode work: user may employ a radius that is suitable in his judgment. Caution: Be aware that short radii for large pipes, say 8 in and larger, soon get into a hot bend category.

Example 1. Offset bend

$12\frac{3}{4}$ -in oil line. Let $V = 8$ ft, $T = 5$ ft. Smaller diameters have larger bend $\angle$. Try 25.4°. Minimum radius for 12″ is $18 \times 12.750″ = 19.125$ ft. sin 25.4 = 0.428935 cos 25.4 = 0.903335 $V = 2R(1 - \cos \angle 1) + 2T$ sin $\angle 1$ $V = 2 \times 19.125 (1 - 0.903335) + 2 \times 5 \times 0.428935$ $V = 7.9868$ ft.

Example 2. Direction change

24-in gas line, say 30° turn. Let $R = 80$ ft, $T = 5$ ft. $V = R(1 - \cos \text{Turn } \angle) + T \sin \text{Turn } \angle$ cos 30 = 0.866025, sin 30 = 0.5000 $V = 60(1 - 0.866025) + 5 \times 0.5000$ $V = 10.53850$ ft.

Example 3. Crossing

A 14-in oil line crosses under $10\frac{3}{4}$ in foreign line. Take 2 ft, 0 in min. clearance. Center distance = 3.0313 ft. Min. height of bend = 2.513 ft = "V". Take 10° as trial $\angle$, $T = 5$ ft. $V = 2R(1 - \cos \angle 1) + 2T$ sin $\angle 1$ sin 10° = 0.173648, cos 10° = 0.984808 $V = 2 \times 24.5$ $(1 - 0.984808) + 2 \times 5 \times 0.173648$ $V = 2.4809$ ft. Increment angle to 10.1° sin 10.1° = 0.175367, cos 10.1° = 0.984503 $V = 2 \times 24.5(1 - 0.984503) + 2 \times 5 \times 0.175367$ $V = 2.513$ ft.

Note: This is a possible solution. However, this may be an alternative solution. Builders will take advantage of sag bends where possible, depending on site conditions, space available, pipe size, etc.

Calculate maximum bend on cold pipe

How much of a bend can a cold straight pipe take without overstressing? In other words, in laying straight pipe cross-country, what is the smallest radius of curvature possible within the allowable stress?

The following is an analytical solution to these questions.

The deflection equation for a length of straight pipe used as a cantilever beam with a concentrated end load is

$$\Delta y \frac{Pl^3}{3EI} \tag{1}$$

The maximum stress in the beam is

$$s = \frac{M}{Z} \tag{2}$$

The maximum bending moment is

$$M = Pl \tag{3}$$

If the allowable stress is S_A, rewriting Equation (2) gives

$$S_A = \frac{M}{Z} \tag{4}$$

$$S_A = \frac{Pl}{Z} \tag{5}$$

$$Pl = ZS_A \tag{6}$$

Rewriting Equation (1):

$$\Delta y = \frac{(Pl)l^2}{3EI} = \frac{ZS_A l^2}{3EI} \tag{7}$$

Taking the first and second derivatives of Equation (7):

$$\frac{d\Delta y}{dl} = \frac{2lZS_A}{3EI} \tag{8}$$

$$\frac{d^2\Delta y}{dl^2} = \frac{2ZS_A}{3EI} \tag{9}$$

The radius of curvature is given by the equation

$$R = \frac{\left[1 - (d\Delta y/dl)^2\right]^{3/2}}{d^2\Delta y/dl^2} \tag{10}$$

Substituting in Equation (10):

$$R = \frac{\left[1 + (2lZS_A/3EI)^2\right]^{3/2}}{2ZS_A/3EI} \tag{11}$$

$$\lim_{l \to 0} R = \frac{1^{8/2}}{2ZS_A/3EI} = \frac{3EI}{2ZS_A} \tag{12}$$

Notation

Δy	Deflection, inches
l	Length, inches
P	Load, pounds
E	Modulus of elasticity, pounds per square inch
I	Cross-sectional moment of inertia, inch^4
S	Stress, pounds per square inch
S_A	Allowable stress, pounds per square inch from Code for Pressure Piping
M	Bending moment, inch-pounds
Z	Section modulus, inch^3
R	Radius of curvature, inches

Example

What is the smallest radius of curvature that can be used without overstressing straight 24-in Schedule 20 seamless steel pipe ASTM 53? The pipe is at 70 °F and is to be used for oil outside of refinery limits.

$$R = \frac{3EI}{2ZS_A}$$

$E = 29 \times 10^6$ lb/in^2
$I = 1943$ in^4
$Z = 162$ in^3
$S_A = 25,500$ psi (from Code for Pressure Piping)

$$R = \frac{3 \times 20 \times 10^6 \times 1.943}{2 \times 162 \times 25,500} = 20,460 \text{ in or } 1705 \text{ ft}$$

The section modulus $= Z = 2I/D$, and may be simplified to $Z = I/r$ and substituted for Z in Equation (12).

$I =$ Moment of inertia
$r =$ Distance from center pipe
$D =$ Outside diameter of pipe

Equation (12) then becomes

$$R = 3Er/2S_A \tag{13}$$

$$R = (21.75 \times 10^6/S_A) \times D \tag{14}$$

S_A, the allowable stress, is determined by the design factor permitted by the design code used. Equation (14) may be further simplified by using the following table.

Example

What is the minimum radius of curvature that can be used without overstressing 42″ API 5LX X-65 line pipe at 72°F used in type C construction (gas line)? The design factor from ANSI B31.8 is 0.5; therefore, use the column headed $F = 0.5$.

$R = 669 \times 42″ = 28.098″$ or 2.342 ft

For liquid lines designed in accordance with ANSI B31.4, use the column headed $F = 0.72$.

Minimum Radius of Curvature, R				
Yield Stress (psi)	$F = 0.72$	$F = 0.6$	$F = 0.5$	$F = 0.4$
35,000	863 D	1036 D	1243 D	1553 D
42,000	719 D	863 D	1036 D	1295 D
46,000	657 D	788 D	946 D	1182 D
52,000	581 D	697 D	836 D	1046 D
56,000	539 D	647 D	777 D	971 D
60,000	503 D	604 D	725 D	906 D
65,000	465 D	558 D	669 D	836 D
70,000	431 D	518 D	621 D	776 D
80,000	377 D	453 D	544 D	680 D

Determine length of a pipe bend

For smooth bends (other than wrinkle bends) use the following formula:

$$L = R \times D \times .017 + T$$

where

L = length of bend
R = radius of bend (center line of pipe)
D = number of degrees in the bend
T = total length of the two tangents at either end

Example. Find the length of a piece of pipe required to make a 90° bend with a radius of 60 in. There should be a 4-in tangent at either end.

$L = ?$
$R = 60$
$D = 90$
$T = 8$

$$L = 60 \times 90 \times 0.01745 + 8$$
$$= 94.23 + 8$$
$$= 102.23 \text{ in}$$

Length of pipe in arc subtended by any angle

Application

To find the length of pipe subtended by an angle of 42°-35 ft with a radius of 24 in

Assuming $R - 1$ in.

Angle of 42° subtends an arc = 0.7330″

Angle of 35 ft subtends an arc = $\dfrac{0.01018″}{0.7432″}$

DEGREES						MINUTES			
Degrees	Length Of Arc	Degrees	Length Of Arc	Degrees	Length Of Arc	Minutes	Length Of Arc	Minutes	Length Of Arc
1	.0175	31	.5411	61	1.0647	1	.000291	31	.009018
2	.0349	32	.5585	62	1.0821	2	.000582	32	.009308
3	.0524	33	.5760	63	1.0996	3	.000873	33	.009599
4	.0698	34	.5934	64	1.1170	4	.001164	34	.009890
5	.0873	35	.6109	65	1.1345	5	.001454	35	.010181
6	.1047	36	.6283	66	1.1519	6	.001745	36	.010472
7	.1222	37	.6458	67	1.1694	7	.002036	37	.010763
8	.1396	38	.6632	68	1.1868	8	.002327	38	.011054
9	.1571	39	.6807	69	1.2043	9	.002618	39	.011345
10	.1745	40	.6981	70	1.2217	10	.002909	40	.011636
11	.1920	41	.7156	71	1.2392	11	.003200	41	.011926
12	.2094	42	.7330	72	1.2566	12	.003491	42	.012217
13	.2269	43	.7505	73	1.2741	13	.003782	43	.012508
14	.2443	44	.7679	74	1.2915	14	.004072	44	.012799
15	.2618	45	.7854	75	1.3090	15	.004363	45	.013090
16	.2793	46	.8029	76	1.3265	16	.004654	46	.013381
17	.2967	47	.8203	77	1.3439	17	.004945	47	.013672
18	.3142	48	.8378	78	1.3614	18	.005236	48	.013963
19	.3316	49	.8552	79	1.3788	19	.005527	49	.014254
20	.3491	50	.8727	80	1.3963	20	.005818	50	.014544
21	.3665	51	.8901	81	1.4137	21	.006109	51	.014835
22	.3840	52	.9076	82	1.4312	22	.006400	52	.015126
23	.4014	53	.9250	83	1.4486	23	.006690	53	.015417
24	.4189	54	.9425	84	1.4661	24	.006981	54	.015708
25	.4363	55	.9599	85	1.4835	25	.007272	55	.015999
26	.4538	56	.9774	86	1.5010	26	.007563	56	.016290
27	.4712	57	.9948	87	1.5184	27	.007854	57	.016581
28	.4887	58	1.0123	88	1.5359	28	.008145	58	.016872
29	.5061	59	1.0297	89	1.5533	29	.008436	59	.017162
30	.5236	60	1.0472	90	1.5708	30	.008727	60	.017453

The arc subtended for any angle is directly proportional to the radius:

$$\frac{1\ \text{in}}{24\ \text{in}} = \frac{0.7432\ \text{in}}{\text{Arc}}$$

$$\text{Arc} = 0.7432 \times 24 = 17^{13}/_{16}''.$$

Average pipelay table—underground

Linear feet of pipelay per 10 h day in trench in level, rock-free terrain

Nominal Pipe Size	Pipe Wall Thickness in Inches					
	0.000–0.250	0.251–0.375	0.376–0.500	0.501–0.625	0.626–0.750	0.756–1.000
4	10,800	10,500	10,300	—	—	—
6	9600	9400	9210	—	—	—
8	9120	8900	8720	8550	—	—
10	8900	8760	8580	8410	—	—
12	8600	8400	8230	8070	7910	—
14	8400	8230	8160	8000	7840	—
16	8000	7840	7680	7530	7380	7230
18	7800	7640	7500	7350	7200	7050
20	7400	7250	7100	6960	6820	6680
24	—	6600	6470	6340	6210	6080
30	—	6000	5880	5760	5640	5530
36	—	—	5400	5300	5200	5090
42	—	—	4800	4700	4600	4500

Productivity will vary where different types of terrain or rock are encountered.
The above footage is based on installing double random joints of pipe.

Average pipelay table—on supports

Linear Feet of Pipelay Per 10-hour Day on Waist-High Supports on Level Ground						
Nominal Pipe Size	Pipe Wall Thickness in Inches					
	0.000–0.250	0.251–0.375	0.376–0.500	0.501–0.625	0.626–0.750	0.756–1.000
4	11,320	11,100	10,880	–	–	–
6	10,080	9880	9680	–	–	–
8	9560	9370	9190	9010	–	–
10	9390	9200	9020	8840	–	–
12	8800	8620	8500	8330	8160	–
14	8700	8570	8400	8250	8090	–
16	8040	7880	7730	7580	7430	7280
18	7800	7650	7500	7350	7200	7060
20	7550	7400	7250	7100	6960	6820
24	–	6720	6590	6460	6330	6200
30	–	6120	6000	5880	5770	5660
36	–	–	5440	5350	5240	5140
42	–	–	4840	4740	4650	4560

Productivity will vary with contour of terrain and accessibility.
Above footage is based on installing double random joints of pipe.

Allowable pipe span between supports

When it is desired to run pipelines on supports above the ground, the engineer must often calculate the allowable pipe span between supports from a structural viewpoint. The formula for the span length can be obtained by equating the maximum bending moment caused by the loading to the resisting moment of the pipe. This formula is

$$L\left(\frac{SI}{we}\right)^{0.5}$$

where

   L = span length, ft
   c = radius to outside of pipe, inches
   I = moment of inertia of pipe, inches^4
   S = maximum allowable fiber stress of material, pounds per inch^2
   w = uniform loading per unit length of pipe, pounds per foot

The above formula applies to pipes above 2 in in diameter. Reference 1 gives calculated pipe span data, as well as estimated pipe loading for different pipe diameters, taking into account the fluid load, dead load, and the horizontal component of the wind loading.

Figure 1. Allowable pipe span between supports.

The design of the pipe span can be quickened up with Figure 1.

Example. Find the allowable span length for a 6-in continuous steel pipe strung between supports. The estimated loading (sum of pipe, fluid, dead, and wind loads) is 36 lb/ft. The allowable fiber stress for steel can be taken at 20,000 lb/in^2.

From any suitable handbook, c for a 6-in pipe is 6.625/2 = 3.3175 in, I is 28.14 in.[4] Now calculate $I/c = 28.14/3.3175 = 8.5$ in.[3] Connect $S = 20,000$ lb/in^2 on the S scale with 8.5 on the I/c scale to the intersection with the reference line. With the reference line as a pivot, connect the pivot point with $W = 36$. At the intersection with the L scale, obtain a pipe span of 68 ft.

Check the amount of deflection to make sure that it does not exceed your piping practices or applicable design code.

Source

Kuong, J. F., *Petroleum Refiner 39*, No. 10, 196 (1960).

Reference

1. Sweeney, R. J., *Chem. Eng., 63*, No. 3 199 (1956).

How engineers make pipe fit the ditch

By R. J. Brown

The five-men crews of engineers that keep track of the pipe and calculate side bends, overbends, sags, and combination bends can make money—or lose money—for any contractor crossing rough terrain. If they function smoothly, the bending crew and pipe gang have an easy time and the lowering-in can be a breeze. But when they make mistakes, the welded-up pipe may not fit the ditch, and this spells trouble for the contractor.

Bending crew

As a rule, there are two rod men, two chairmen, and one instrument man in the measurement crew. Their duties are to measure each joint of pipe, number it, and stake its position in the line. In addition, they calculate degrees of bends, etc. Here is how they perform these duties:

Measuring and numbering joints

(See Figure 1) First, the crew measures all joints of pipe and numbers them from 0 to 100. (When 100 is reached, the numbering starts over.) Next, they stake the position of each joint of pipe in the line. They use stakes that are numbered according to the pipe, with the number of the joint behind on front of the stake and the joint ahead on back of the stake.

Stakes are set back

Because the setup gang will move the pipe back half a joint's length and the pipe gang back another half, the engineers set the stakes from one to two joints' length behind the corresponding numbered pipe.

Notes on Figure 1

Note that joint 25 was measured 39.9 ft, and the gap between stakes 24 and 25 and 25 and 26 was set at 29.9 ft to allow a 10-ft lap for tie-in. This leaves a pup long enough to be carried ahead and worked into the line. If the pups are distributed over sections of bends, there is a possibility of saving a cut in a full joint. If the stakes catch up with the pipe, one joint is carried ahead of the stringing.

Calculating bends

The method used to determine the size of overbend and sags is illustrated in Figure 2. Note that at 0 + 00 the instrument is set up on the ditch bottom at a height of 5 ft. The rod is set on the ditch bottom ahead of the radius of overbend. (Care should be taken to set neither the transit nor the rod on the arc of the change in slope.) The transit is moved behind the radius of overbend at 1 + 00, and the height of the instrument is measured to be 4.90 ft.

The rod is set just ahead of the change in radius of overbend and is read at 4.90 ft, establishing a slope of −1°. The magnitude of bend can now be determined.

Add unlike angles

If the slope is positive and the tangent of the slope ahead is decreasing, an overbend is established. If the slope is positive

Figure 1.

ELEVATION

Figure 2.

Figure 3.

Figure 4.

Figure 5.

and the tangent ahead is greater than the positive slope, a sag is noted. The reverse holds true for a change in negative slope: if the tangent ahead is increasing, the transition is an overbend; if it is decreasing, the transition is a sag. The unlike algebraic signs are added and like signs are subtracted. Figure 2, station $2+00$, shows a positive $10°$ slope approaching the overbend and a negative $1°$ slope leaving it. Since the signs are unlike, the sum of the two angles is $11°$, and since the slope is positive and decreasing, the notation overbend is used.

Side bend angles

Side bends are determined by "eyeballing" a P.I. in the ditch and setting the instrument up over it. A back sight is taken on the ditch or preceding P.I., whichever is more easily read. The scope is flipped and a foresight is taken on the ditch ahead. Referring to Figure 2 at station $1+00$, the deflection angle is $6°$ left; station $4+00$ shows a deflection of $8°$ right.

Combination bends

At points where side bends and vertical bends occur, the angles are combined to make one angle rolled. The method of calculating is a rule of thumb: *the larger angle plus one-third the smaller angle.*

This rule is accurate enough to give satisfactory results in actual practice. The actual derivation of this rule comes from an approximation of the following:

$$\sin(\text{combination angle}) = \sqrt{\sin^2 A + \sin^2 B}.$$

Angles A and B are the horizontal and vertical angles. Figure 3 shows a plot of the $\frac{1}{3}$ rule and of the equation. From the plot it can be seen that for the maximum bend of 18.5, an error of slightly over $1°$ is introduced. The maximum error evidenced on the graph is due to the side and vertical bends being equal.

Marking pipe

The final step of the measurement crew is marking the joints to be bent. After the bend is established, its position between the pipe stakes is then transferred to the pipe joint number matching the pipe stakes number. The bend size, the centerline of the bend, and arrows indicating the direction of construction are marked on the joint (see Figure 5). If a bend falls too close to the end of the pipe, the joint would have to be moved so as to leave enough room for the bend. A pup could be inserted immediately behind the joint in question to give better bend positioning.

PIPE LOWERING

How to lower an existing pipeline that is still in service

Lowering a loaded pipeline is a low-cost alternative for constructing new facilities

Marshall D. Cromwell, Senior Project Engineer, PT Caltex Pacific Indonesia, Jakarta, Indonesia

Lowering an existing line is a dirty little job, but it can have big cost benefits. The line can be lowered while remaining in service with no lost production, and the cost of lowering an existing pipeline section is relatively cheap.

No expenses are incurred for new pipe, valves, stopples, and fittings. Construction is much faster than cutting and relocating a pipeline section.

Construction of new highways, buildings, airport runways, and other facilities is often planned at locations where above-ground pipelines are present. Relocating such lines can be extremely expensive in terms of downtime and new pipeline materials.

Long-delivery items, such as hot tap equipment, valves, and fittings, can add to the pipeline construction time, delaying an already tight schedule for the new facility. The alternative is to lower the existing lines, with adequate protection, so as to eliminate the obstruction.

Figure 1. Road crossing profiles.

The following information describes how to calculate an optimum profile for smooth transitions, while keeping the pipe stresses within allowable limits, and how to physically lower the pipeline in the field to match this profile.

Precise engineering

There is more to lowering an existing line than just digging a hole and letting gravity take over. Engineering and design of a new profile must be precise so that the line is not dropped too suddenly, causing a buckle or rupture (the larger the pipeline, the more critical the profile).

For instance, in Saudi Arabia, an extensive highway modernization program resulted in scores of large-diameter

Table 1
Profile Elevations

Station	X, ft	Deflection, ft	Top of Pipe Existing Elevation	Top of Pipe New Elevation
1 + 25	0	0.00	From field survey	Existing elevation minus deflection
1 + 50	25	0.07		
1 + 75	50	0.29		
2 + 00	75	0.65		
2 + 25	100	1.15		
2 + 50	125	1.80		
2 + 75	150	2.59	(Another column can be	
3 + 00	175	3.50	added for trench-bottom	
3 + 25	200	4.41	elevations if desired.)	
3 + 50	225	5.20		
3 + 75	250	5.85		
4 + 00	275	6.35		
4 + 25	300	6.71	Note: Depending on the	
4 + 50	325	6.93	configuration of the	
4 + 75	350	7.00	original pipeline, the left	
5 + 00	375	7.00	profile does not have to be	
5 + 25	400	7.00	identical to the right	
			profile. Transition lengths,	
			radii, and deflections may	
			be different.	
5 + 50	425	6.93		
5 + 75	450	6.71		
etc.	etc.	etc.		

After elevations are received from the field, end-points (Sta 1 + 25 and 8 + 75) should be checked by use of equations (7) and (4) if the pipeline is at the uphill slope.

Elevations

Station	Feet
1 + 00	7.25
1 + 25	7.50
1 + 50	8.00
1 + 50	7.43 New

$$A = \sqrt{(125-100)^2 + (7.50-7.25)^2} = 25.00125 \text{ ft}$$

$$B = \sqrt{(150-125)^2 + (7.50-7.43)^2} = 25.00010 \text{ ft}$$

$$C = \sqrt{(150-100)^2 + (7.43-7.25)^2} = 50.00032 \text{ ft}$$

$$R = 1,947.5 \text{ ft} = 23,370 - \text{in.}$$

$$S = 22,340 \text{ psi}$$

No good—must come up with a new profile, with wider transition length, and check again by trial-and-error.

Figure 2. Elevation coordinates.

1. $R = \dfrac{x^2 + y^2}{2y}$

2. $x = \sqrt{2Ry - y^2}$

3. $y = R - \sqrt{R^2 - x^2}$

4. $R = \dfrac{EC^1}{S}$ or $S = \dfrac{EC^1}{R}$

5. $y = \dfrac{EC^1}{S} - \sqrt{\left(\dfrac{EC^1}{S}\right)^2 - x^2}$

6. $x = \sqrt{2y\dfrac{EC^1}{S} - y^2}$

7. $R = \dfrac{A\,\text{SIN}\left[90 - \text{COS}^{-1}\dfrac{B^2 + C^2 - A^2}{2BC}\right]}{\text{SIN}\left[2\,\text{COS}^{-1}\dfrac{B^2 + C^2 - A^2}{2BC}\right]}$

(R = Feet here)

Note that formulas are independent of w.t., I, and cross-sectional area (cancelled out).

R = Radius of bend (in.)
x = Horizontal distance (in.)
y = Deflection or drop from the horizontal (in.)
S = Bending stress, psi

C¹ = One half diameter of pipeline (in.)
E = Young's modulus of elasticity (29 × 10⁶ psi for steel)
A, B & C = Dimensions of triangle (ft.)

P/L profile

3 points on a circle

● = Inflection points
x₁ = 25′, x₂ = 50 , etc.

Figure 3. Formulas for changes in slope.

line pipelines being lowered. A few line sections ruptured during the pioneer stages due to the fact that each situation was not properly evaluated in terms of a critical engineering problem.

Excavation begins. Ten-foot soil plugs remained at 50-ft centers to act as temporary pipe supports for the 36-in.

The key to the problem's solution is to provide a smooth profile with constant radius bends and to include a detailed list of instructions with the design package. Careful monitoring by survey equipment (levels) will assure smooth transitions in the field.

The following approach has been used effectively on numerous occasions in Saudi Arabia and Indonesia and can be recommended. Design steps to be included are as follows:

- Determine how much soil cover there should be over the top of the pipeline to assure protection from construction equipment and wheel loads (see API-RP-1102). Four feet of cover is adequate for pipelines 12 in and smaller. Larger pipelines may require additional protection by way of more cover, pipe sleeves, or concrete slabs.
- Determine allowable bending stress on pipeline, based on grade of pipe. Normally, 75% of allowable stress is acceptable for cross-country pipelines (ANSI B31.4, Section 419.6). However, 10,000-psi stress is very conservative and often preferred. If the pipeline is completely restrained (anchors on each end and no expansion loops), thermal stresses must also be considered.
- Once the total required drop of pipeline is established, calculate the minimum bending radius based on allowable bending stress. Then calculate the overall transition length from one end to the other.
- Obtain elevations on top of the pipeline every 25 ft (Table 1), marking station numbers and elevations with paint. If the line is on supports, also include grade elevations at each station.
- Starting at each end of transition, calculate the drop at 25-ft stations. Draw up a profile and fit it on the existing profile. Add stations and new elevations (top of pipe and trench bottom) to the profile drawing.
- Check the transition points at the beginning and end of the new profiles. If the pipeline was going uphill, and suddenly it needs to go downhill, the end points may be critical in terms of stress.

Since any three points not on the same line form an arc of constant radius, the stress may be approximated by Equations (7) and (4) (Figure 3). The end point should become the center of three points for calculation. If overstressed, widen the transition length and check again.

Example 1. A new village was being built in the jungle, but access was obstructed by an above-ground, 36-in pipeline. The village head requested that the line be buried or relocated so that bicycles, becaks, and water buffalo would have easier access.

It was determined that 4 ft of cover was required to protect the line from excessive wheel loads (people-packed buses and trucks). This means the line must be lowered 7 ft (Figure 1). The pipe is Grade B, allowable bending stress is 10,000 psi, and crossing width is 50 ft. To calculate the minimum bending radius, total transition length, and deflections at 25-ft intervals, use:

$$R = EC'/S = (29,000,000)(18)/10,000 = 52,200 \text{ in} = 4350 \text{ ft (min)}$$

Total transition length $= 4(X) + 50$ ft.
$Y = 7.0$ ft$/2 = 3.5$ ft.
$X = \sqrt{2RY - Y^2}$.
$X = \sqrt{(2)(4,350)(3.5) - (3.5)^2} = 175$ ft.

Total transition length $= 750$ ft.

It is suggested that field personnel should start the survey (elevations) 500 ft left of the crossing centerline and stop at station $10 + 00$.

Deflections at 25-ft intervals are calculated from Equation (3) (Figure 3), and a table is prepared accordingly (Table 1). Note that calculations for Y are required for 175 ft only. Since the curve is then inverted (Sta $3 + 00$ to Sta $4 + 75$), results are subtracted from 7 ft, working backwards, to arrive at true deflection or drop. For example, deflection at Sta $4 + 50 = 7.0 - 0.07 = 6.93$, etc.

Example 2. Given the following elevations, is pipeline being dropped too suddenly, and will it become overstressed between stations $1 + 00$ and $1 + 50$ (Figure 2)?

Construction

The design package to the field should include a drawing of new profiles, stations, tables of new pipeline elevations, trench elevations, and a very detailed construction sequence. The following may be used as a guide for construction.

- Starting at one end of transition, dig 40 ft of trench under the line. Leave 10 ft of soil for pipe support. Dig another 40 ft, repeat until the opposite end is reached. When complete, 10-ft soil plugs are supporting the line at 50 ft centers.
- Clean the line between soil supports, take ultrasonic test (UT) readings, and install a sleeve where required. Add primer and tape wrap between supports, maintaining station numbers and elevation points on the line.
- Add 12 in of sand or soil/sand premix to the bottom of the trench and compact to proper trench bottom elevation (profile elevation minus pipeline diameter).
- Install 6-in × 6-in wooden blocks every 50 ft (between soil plugs) to support the line.
- Remove soil plugs and complete line cleaning, UT readings, sleeve installations, tape wrapping, and trench preparation.
- Check elevations of new trench bottom and adjust as required. Check tape wrap with a holiday detector if desired.
- Starting at one end, use six sidebooms to raise approximately 250 ft of line off its supports. Remove 6 in (or less) of blocks on outside supports and 12 in of blocks on all inner supports. Lower line onto supports.

Hand cleaning. The 30-in. and 36-in. crude lines were cleaned by hand to expose corroded areas and prepare for tape wrapping.

Lowering lines. The two lines, full of crude oil, are gradually lowered with the use of six sidebooms.

Gradual lowering. A full sweep of the total transition length is made by lowering each line 6 in. or less at a time. The procedure is repeated until the lines finally reach desired profile elevations.

- Move convoy of sidebooms 250 ft, lift the line, and remove 6 in block at beginning and end supports and 12 in of blocks at inner supports.
- Repeat until total transition length is lowered 12 in, except for several end supports, which are lowered less than 6 in to match required deflections. Then come back and do it again, repeating several times until the pipeline is completely lowered. Note that each repeat becomes shorter. Check final pipeline elevations if desired.
- Check tape wrap for possible damage, add 12 in of sand or premix around pipe, backfill, and compact.

Surprisingly, this operation goes very quickly but must be well coordinated. One person should signal the sideboom operators to lower the pipeline concurrently, so as to eliminate excessive bends and sags. The pipeline is heavy when full of liquid, and if not handled carefully, it could easily rupture. Pressure should be reduced, and extreme caution exercised during the actual lowering process.

Strain calculations

A new profile means the pipeline has been stretched a certain amount, assuming the line was horizontal originally. (If the pipeline was convex, lowering it will put the steel into compression.) Fortunately, steel is very forgiving either way, and the added strain is generally not a problem. It can be checked, however, by use of Young's Modulus of Elasticity (E = stress over strain).

The new pipe length can be calculated from the following formula:

Arc Length = $\pi R\theta/180$, where $\theta = \sin^{-1} X/R$.

A comparison of the new pipe length with the original pipe length will result in a figure for strain as a result of the new profile. Knowing the strain, additional longitudinal stress can be easily calculated.

In Example 1, $\theta = \sin^{-1} 175$ ft/4350 ft = 2.3056°.

Arc Length (AL) = $\pi(4350$ ft$)(2.3056°)/180 = 175.047$ ft.

Length of new line = 4(AL) + 50 = 750.19 ft.

So the pipeline was actually stretched 0.19 ft or 2.28 in.

Strain = $\Delta L/L$ = (750.19 − 750.00)750.00 = 0.000253.

E = stress/strain, so longitudinal stress = $E \times$ Strain or LS = $29 \times 10^6 \times 0.000253 = 7347$ psi.

This is well within the 35,000-psi yield strength of Grade B pipe.

Source

Pipe Line Industry, July 1986.

WELDING

When should steel be preheated before welding?

From the chemistry of the steel, determine the carbon equivalent:

Carbon equivalent = $C + \dfrac{Mn}{4}$.

If it exceeds 0.58, the steel may be crack sensitive and should be preheated before welding in ambient temperatures below 40 °F.

Example. If steel pipe having a carbon content of 0.25 and manganese content of 0.70 is to be welded in springtime temperatures, ranging from 40 to 80 °F, is preheat necessary?

Carbon equivalent = $0.25 + \dfrac{0.70}{4} = 0.25 + 0.175 = 0.45$.

It is not necessary to preheat this particular steel before welding it.

But for another example:

Carbon = 0.20

Manganese = 1.60

Carbon equivalent = $0.20 + \dfrac{1.60}{4} = 0.60$.

This steel should be preheated, particularly for early morning welding.

Why does preheating prevent cracking? It slows the cooling rate and reduces the amount of austenite retained as the weld cools. This prevents microcracking. Other alloying elements and pipe wall thickness may also influence when joints of high strength pipe should be preheated.

Welding and brazing temperatures

Carbon Steel Welding	2700–2790 °F
Stainless Steel Welding	2490–2730 °F
Cast Iron Welding	1920–2500 °F
Copper Welding and Brazing	1980 °F
Brazing Copper-Silicon with Phosphor-Bronze	1850–1900 °F
Brazing Naval Bronze with Manganese Bronze	1600–1700 °F
Silver Solder	1175–1600 °F
Low Temperature Brazing	1175–1530 °F
Soft Solder	200–730 °F
Wrought Iron	2700–2750 °F

Reprinted with permission—Tube Turns, Inc.

Mechanical properties of pipe welding rods

When welded and tested in accordance with appropriate
AWS or MIL specifications[1]

Electrode	AWS Class	Tensile psi	Yield psi	%Elong. in 2″	Tensile psi	Yield psi	%Elong. in 2″
Fleetweld 5P+	E6010	62–86,000[1]	50–75,000[1]	22–28[1]	67–78,000	51–67,000	30–34
Fleetweld 5P	E6010	62–75,000[1]	50–64,000[1]	22–30[1]	60–69,000	46–56,000	28–36
Shield-Arc 85	E7010-A1	70–78,000[1]	57–71,000[1]	22–26[1]	70–83,000	57–72,000	22–28
Shield-Arc 85P	E7010-A1	70–78,000[1]	57–63,000[1]	22–27[1]	70–77,000	57–68,000	22–25
Shield-Arc HYP	E7010-G	70–82,000[1]	60–71,000[1]	22–28[1]	72–81,000	60–72,000	25–29
Shield-Arc 70+	E8010-G	80–86,000[1]	67–74,000[1]	19–29[1]	76–80,000	68–72,000	23–29
Jetweld LH-70	E7018[4]	72–86,000	60–76,000	22–30	65–74,000	55–60,000	24–34
Jetweld LH-75MR	E8018-B2	72–87,000	60–74,000	22–30	70–82,000	70–82,000	29–32
Jetweld LH-90MR	E8018-C1	97–107,000	84–97,000	17–24	80–105,000[2]	67–93,000[2]	19–23[2]
Jet-LH8018-C1MR	E8018-C3	80–95,000	67–81,000	19–25	80–86,000	67–75,000	19–31
JEt-LH8018-C3MR	H8018-C3	80–90,000	68–80,000	24–30	75–84,000	66–73,000	24–32
Jetweld LH-100M1MR	MIL10018-M1	95–101,000	82–91,000	20–27	93–96,000[5]	80–90,000[5]	20–28[5]
Jetweld LH-110MMR	E11018-M	110–123,000	98–109,000	20–24	110–120,000[3]	95–107,000[3]	20–25[3]

[1]Aged 48 h @ 220 °F.
[2]Stress relieved @ 1275 °F.
[3]Stress relieved @ 1025 °F.
[4]Imprinted identification 7018-1 to indicate meeting minimum CVN impact requirement of 20 ft-lbs at −50 °F.
[5]Stress relieved @ 1125 °F.

[1]Mechanical properties obtained with each type electrode also depend upon the chemistry of the pipe, the welding procedures, and the rate of cooling. Because these factors vary between actual applications and testing conditions, the mechanical properties may also vary.

Reprinted with permission—Lincoln Electric Co.

Lens shade selector

Operation	Shade No.	Operation	Shade No.
Soldering	2	Nonferrous	
Torch Brazing	3 or 4	Gas Tungsten-Arc Welding	
Oxygen Cutting		Gas Metal-Arc Welding	
Up to 1 in	3 or 4	$1/16$, $3/32$, $1/8$, $5/32$ in electrodes	11
1–6 in	4 or 5	Ferrous	
6 in and over	5 or 6	Gas Tungsten-Arc Welding	
Gas Welding		Gas Metal-Arc Welding	
Up to $1/8$ in	4 or 5	$1/16$, $3/32$, $1/8$, $5/32$ in electrodes	12
$1/8$ –½ in	5 or 6	Shielded Metal-Arc Welding	
½ in and over	6 or 8	$3/16$, $7/32$, ¼ in electrodes	12
Shielded Metal-Arc Welding		$5/16$, ⅜ in electrodes	14
$1/16$, $3/32$, $1/8$, $5/32$ in electrodes	10	Atomic Hydrogen Welding	10–14
		Carbon-Arc Welding	14

Reprinted with permission—Tube Turns, Inc.

PIPELINE WELDING

Meeting today's quality requirements for manual vertical down techniques

Pipelines are inspected more critically than ever before, and today's radiographic equipment and techniques produce clearer radiographs with greater sensitivity than in the past. Although codes have not changed drastically, interpretation standards have been upgraded. The combination of more rigorous inspection, better testing methods, and high acceptability standards often approaches an attitude requiring zero defects.

This poses some serious problems because the job of welding cross-country pipelines under typical conditions has always been an extreme challenge requiring specialized and highly developed skills. Now that the demands are greater, even the best welding operators are having trouble. Rejectable defects usually require cutting out the entire weld. This is expensive and can cost competent pipeline welders their jobs.

The purpose of this bulletin is to discuss some of the more common reasons given for rejecting a pipeweld and to suggest what may be done to correct some of these conditions.

Change in attitude

Any discussion on this subject should begin by accepting the fact that the standards are higher than they were a few years ago. Therefore, it will take a change in the attitude of the welders and everyone else involved if this new level of quality and workmanship is to be met. In turn, a commitment to new methods, equipment, and theory will be required by all concerned.

Importance of joint preparation

It must be recognized that all joint preparation details (Figure 6) are critical and any variation could directly contribute to rejected welds.

Without good cooperation from those who prepare the pipe edges and the line-up crew, the welder has very little chance of meeting today's rigid inspection requirements. Too often the attitude exists that variations in fit-up and joint preparation are permissible and that the welder can compensate for them. This attitude cannot be tolerated. It puts too much responsibility on the welder and inevitably leads to rejects.

Recommended procedures for properly cleaning pipe

For lowest cost and highest quality, careful attention to the cleaning of pipe joint surfaces is critically important.

Courtesy of Pipeline & Gas Industry.

Today's pipe welders face pipe that may be covered with a variety of coatings; these include primers, epoxy, tar, paper, varnish, rust, scale, or moisture.

While joint cleanliness is important in all welding, it is especially so in the root pass of pipe. Even a thin film of contaminants, which may be difficult to see with the naked eye, may be present. Failure to recognize and properly clean joints can result in a hollow bead or other welding defects. Follow these instructions to minimize costly defects:

1. Remove all moisture and condensations of any type prior to welding. The joint must be completely dry.
2. Clean BOTH ends of the pipe INSIDE AND OUT to remove traces of paint, rust, scale, oil, epoxy, or organic materials. The area to be cleaned should extend at least 1 in (25 mm) from the end of the bevel (per ANSI/AWS D10.11-80 Page 1) on both the INSIDE and OUTSIDE surfaces.
3. A recommended method for cleaning in the field as described above is the use of a heavy-duty straight shaft grinder with a rubber expanding wheel and a carbide coated sleeve. The small shaft and reduced overall weight allows easy access to the inside and outside surfaces of the pipe.

Internal undercut

See Figures 1–5 for X-rays of the various defects that can cause rejects. Of these, one of the most common and troublesome is internal undercut (undercut on the inside of the pipe). This is understandable because it occurs on the "blind side" and the operator cannot see it happening. Consequently, he cannot immediately do anything to correct it. To make matters worse, he seldom gets a chance to see the X-ray negative or the actual weld itself. All of this makes it difficult for him to correct the situation.

Undercut may be the direct result of poor joint preparation or fit-up. If the joint does not conform to the details specified in the established procedures (see Figure 6), every reasonable effort should be made to correct it before starting to weld.

Internal undercut will tend to occur if

1. the root face (land) is too small
2. the root opening is too large
3. a severe high-low condition exists
4. the current is too high.

When any undesirable variation occurs in the joint preparation, the normal reaction is to compensate for it by

Figure 1. Radiograph of internal undercut. This defect may be intermittent or continuous (Figure 2) and either on one side or both sides of the weld centerline.

Figure 2. Radiograph of internal undercut and the lack of penetration that tends to occur at a stop and start.

Figure 3. Radiograph of lack of penetration on the stringer bead appears as a single, straight, dark line.

Figure 4. Radiograph of unfilled wagon tracks. Distinguishing between this defect and internal undercut required skill and experience.

Figure 5. Radiograph of porosity.

juggling the root spacing. Within limits, this can be fairly effective. For example:

Condition	Change in Root Spacing
Land too small	Decrease root spacing
Land too large	Increase root spacing
High-low condition	Decrease root spacing
Bevel too small	Increase root spacing

A skillful line-up crew can be helpful in juggling the root spacing to bring about the most favorable results, but this has limitations. And it is at best a poor substitute for a uniform and consistent joint preparation.

When radiographs indicate poor workmanship, best results can be obtained by showing the radiographs to the welder so that they can understand what is being done wrong and correct it.

Pass	Electrode[1] in. (mm)	Current[2] Amps DC(+)[s]	Alternate stringer Bead Procedure Amps DC(−)[s]
Stringer Bead	1/8 (3.2)[3] or 5/32 (4.0)[3]	90 − 120 140 − 165[4]	85 − 115 115 − 140
Hot Pass	5/32 (4.0)	160 − 185	−
Fills and Stripper	3/16 (4.8)	160 − 190	−
Cap Pass	3/16 (4.8)	150 − 180	−

(1) Fleetweld 5P+, Shield-Arc[c] HYP or 70+ suggested depending upon the pipe steel requirements and contractor specifications.

(2) The ideal current should be selected from within the range shown. In general, the best quality will be obtained by operating at the lower end of the range. Hot pass currents up to 200 amps are often used but cause early coating breakdown and large stub losses.

(3) Use 1/8" (3.2 mm) diameter electrode for stringer beads when the wall thickness is .250" (6.35 mm) or less and when the gap is too small to permit use of the 5/32" (4.0 mm) size [1/8"(3.2 mm) diameter electrode should also be used on other passes when the wall thickness is .250" (6.35 mm) or less; however, 5/32" (4.0 mm) diameter can also be used on fill and cap passes on .250" (6.35 mm) wall].

(4) Weld stringer bead at 23-25 arc volts and 10-16 in/min arc speed. Measure arc voltage between the pipe and electrode holder.

(5) Polarity—DC (−) (negative polarity) should be used for stringer bead welding when burn-through, internal undercut and hollow bead defects are a problem. These problems generally occur on thin wall pipe steels containing over .1% silicon. Lower currents are used with DC(−) which helps to reduce these problems. Travel speeds with DC(−) will be equal to travel speeds with DC(+).

Hot pass and all other passes should be run DC(+) (positive polarity).

Welding DC(−) (negative polarity) on the stringer bead will not be harmful to either mechanical or metallurgical properties.

General Guidelines for Power Source Adjustment (SA- and SAE Machines)
In general, the "Current Range" or "Current Control" switch (whether a tap or continuous control), should always be set as low as possible to get the current wanted, and the "Fine Current Adjustment" or "Job Selector" control (which controls open circut voltage) should be set as high as possible. It is usually better to set this control (for OCV) at the mid-point or higher for best arc stability and fewest popouts.

When using a Lincoln SA-250 engine driven power source, it will normally be necessary to set the "Current Range Selector" tap switch one position lower than used on a Lincoln SA-200 power source to get the same operating characteristics.

Figure 6. Recommended joint preparation and typical procedures—operate within these tolerances to help ensure good-quality welds.

Internal chamfer

It is important to remove any burr or overhang on the inside edge of the pipe after the root face has been machined (or ground). However, this clean-up operation should never produce an internal chamfer. A condition such as this is almost certain to result in "internal undercut" (Figure 7). See Recommended Procedures for Cleaning Pipe on the previous page.

Undercut vs. welding current

Excessive current produces internal undercut. Further, the incidence of internal undercut is more prevalent in the 4–6–8 o'clock portions of the pipe. In this bottom area it is possible to use a considerably higher current than is necessary without getting burn-through, and this results in overheating and undercutting. Knowing this, it would appear to be a simple matter to specify an ideal current that would not give undercut. However, this is extremely difficult because of the many variables of the many interrelated factors.

A recommended current range is given in Figure 6, but it is necessarily rather broad to cover all reasonable variations. From a practical standpoint, the correct current should be determined by starting within the recommended range and then "fine tuning" to get precisely what "feels good" and produces the required results. This final current value will vary slightly from one operator to another and with different pipe materials, diameters, wall thickness, etc.

Because of inaccuracies in ammeters, the effect of arc voltage on current, the inability to accurately read a bobbing ammeter, etc., it is impractical to hold to an arbitrary current value. For the accomplished stringer bead welder, the selection of the ideal current is not too difficult—but he will be doing it primarily by "feel and touch."

Keyhole size vs. current and undercut

To get a good inside bead it is highly desirable to maintain a small, visible keyhole at all times. If the keyhole is allowed to

Figure 7. An internal chamfer will tend to leave an unfilled area on one or both sides of the stringer bead. On the X-ray this will be interpreted as internal undercut.

get too big, it will result in internal undercut and/or windows (Figure 8).

Assuming that the joint preparation is correct, the keyhole size is a function of current, electrode angle, and pressure. The current should be "fine-tuned" to produce a small keyhole when the electrode is dragged lightly using the normal electrode angle. If minor variations occur in the keyhole size (for any reason), the electrode angle and pressure can be manipulated to maintain the proper keyhole size (Figure 9).

In general, a keyhole that is about $1/8$" (3.2 mm) in length is ideal; if it becomes 5/32" (4.0 mm) or longer, undercut and windows are imminent.

Frequently, a good inside bead is obtained without having a visible keyhole. For example, at a tight spot, the keyhole may disappear and the arc goes "inside the pipe." With the proper manipulative skill, this condition is tolerable and a good inside bead will be obtained. However, when it is done in this manner, the welder is largely dependent on the sound of the arc inside the pipe. A small, visible keyhole is easier to work with and is a much more controllable condition.

Significance of arc voltage

It is recommended that meters be used since a fairly good correlation can be established between the arc voltage and the keyhole size. For example, under controlled conditions, an ideal keyhole size is consistently accompanied by an arc

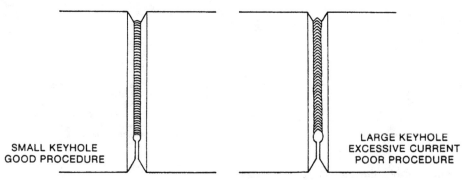

SMALL KEYHOLE
GOOD PROCEDURE

LARGE KEYHOLE
EXCESSIVE CURRENT
POOR PROCEDURE

Figure 8. Maintaining a small keyhole will help assure a good inside bead. The larger keyhole at the right is almost certain to give internal undercut and/or windows.

ONE OF THE PRIME OBJECTIVES OF THE STRINGER BEAD CREW SHOULD BE TO PROVIDE A GOOD, STOUT STRINGER WITH AS MUCH THROAT AS POSSIBLE.

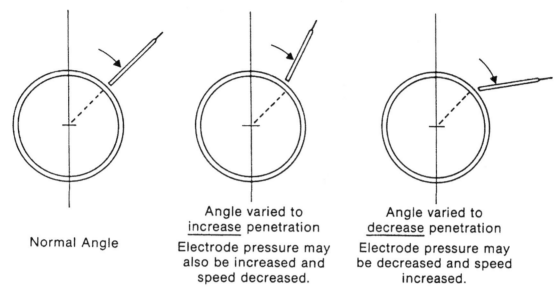

Normal Angle

Angle varied to
increase penetration
Electrode pressure may
also be increased and
speed decreased.

Angle varied to
decrease penetration
Electrode pressure may
be decreased and speed
increased.

Figure 9. The penetration and keyhole size can be effectively controlled by varying the electrode angle as shown—if the joint preparation is uniform.

voltage of no more than 25 volts. When keyhole size is increased by increasing current, increasing the gap, changing the electrode angle, etc., the arc voltage increases to 26–28 volts. This is reasonable because the arc voltage reflects arc length. At the higher arc voltages, internal undercut and wagon tracks may occur.

The same arc voltage numbers quoted above may not occur in all instances. However, it is readily apparent that there is a dependable correlation between keyhole size and arc voltage. A maximum tolerable arc voltage can be determined by experimentation.

It should be further noted that this arc voltage is determined while welding and it should not be confused with the open circuit voltage. Also, it can be affected by the welding technique; changes in electrode angles and drag techniques should not be allowed to cloud the issue.

Varying current/varying gap

Greater root penetration naturally occurs at the top and bottom portions of the pipe, and the least penetration tends to occur at the sides. This being the case, it would be desirable to alter the root spacing accordingly, but this obviously is not practical. It should also be noted that a condition which permits maximum spacing on top and bottom and minimum spacing on the sides is not tolerable.

Assuming a uniform gap all the way around, the ideal condition would be to vary the current as the weld progresses from 12 to 6 o'clock to compensate for the changes in penetration due to position. Instead, penetration changes are

usually controlled by the manipulative skills of the welder (see Figure 9). In an extreme case, a second pair of hands may be required to adjust the current up or down on a signal from the welder. This requires a good communication system and a well-coordinated effort to avoid overshooting or undershooting the current. In some cases, this becomes a survival technique to make up for other undesirable variations.

Undercut—real or imaginary?

If visual inspection were practical for determining internal undercut, it would be easy to accurately determine its presence and severity. However, this is not the case. Radiography is the judge of internal conditions, and this has led to many questionable interpretations. If internal undercut is present, it will show up on the film as a very narrow, dark line immediately adjacent to the stringer bead—on one or both sides. The darkness of the line will vary with the depth of the undercut (Figures 1 and 2).

Proper identification of undercut is sometimes difficult because its location is directly in line with the "wagon track" area (Figure 4). Distinguishing one from the other may be difficult.

Correct interpretation of internal undercut may be further complicated by the fact that a "stout" inside bead will stand out clearly (lighter on the film) against the adjacent area, which is thinner and therefore darker. The thinner, darker area will be further darkened by the presence of any fairly deep, widely spaced surface ripples in the cap pass. The net

Figure 10. Any deep surface ripples located directly above a point adjacent to the inside bead will contribute to a dark shading of the film at this area. This has been misinterpreted by some as internal undercut.

Figure 11. "Wagon tracks" are located on either side of the stringer bead.

effect is to produce a dark shading in the area where internal undercut might occur.

Although this darker area is considerably wider than any undercut, it has been mistaken for undercut and resulted in cutting out good welds. Normally, a "stout" inside bead is considered good, but in this instance, the stouter the bead, the darker the adjacent area would appear (Figure 10).

Wagon tracks

It is not possible to completely eliminate the sidewall undercut produced by the stringer bead. This condition is generally referred to as "wagon tracks" (Figures 4 and 11).

It is the responsibility of the stringer bead crew to minimize wagon tracks and the responsibility of the hot pass crew to burn them out. This is normally done with 5/32″ (4.0 mm) electrode at about 175–180 amps. The secret in melting out wagon tracks lies primarily in the skill of the hot pass crew and the penetration factor of the electrode. The majority of the hot pass men use a rapid stitching technique that exposes the wagon track momentarily and permits them to direct the arc into the undercut area and melt them out.

Condition	Results
Bevel too small (Figure 12)	Increases depth of W.T.
Root spacing (gap) too small	Increases depth of W.T.
Current and/or speed too high	Increases depth of W.T.

The depth of the wagon tracks may be affected by the following:

In extremes, it may be necessary to use a grinder to open up the sidewalls to minimize deep "wagon tracks" or, if they do occur, to grind the stringer to eliminate the high peaked center (Figure 13). In all cases, a 5/32″ (4.0 mm) thickness disc grinder should be used to grind root beads on all pipe from 12″ (304.8 mm) up irrespective of wall thickness (a 1/8″ (3.2 mm) disc grinder will grind the center and roll the sides over on the wagon track unless side pressure is applied). This will make it easier to melt them out.

If the stringer bead tends to wander to one side or the other, it will leave a deep undercut on the shallow side. This

Figure 12. The depth of the wagon tracks will vary inversely with the bevel angle.

Before grinding   After grinding

A narrow bevel or tight gap may require grinding as per dotted line.

Figure 13. A disc grinder, if used with restraint, can be helpful in correcting the condition shown above.

The electrode should be centered in the joint and not changed unless the bead wanders off to one side or the other. Changing the angle will bring it back.

Figure 14. The correct angle can only be determined by observing the shape and location of the bead. If the bead drifts to one side for any reason, it will leave a deep undercut on the opposite sidewall. Tilting the electrode just a few degrees toward the undercut side will straighten the bead up. This change must take place rapidly if the correct bead placement is to be maintained.

condition should be corrected immediately by changing the electrode angle (Figure 14).

The hollow bead

The stringer bead defect shown in this sketch is known by several names including the hollow bead, vermicular porosity, and wormhole porosity. Its length varies from a fraction of an inch to several inches. Radiography exposes the presence of the problem clearly.

Vermicular porosity occurs most readily when welding high silicon pipe—generally above 0.10% silicon. It is aggravated by excessive travel speeds and high currents. Welding the stringer bead with a 5/32″ (4.0 mm) electrode at 130–165 amps DC(+) and 12 to 14 in/min (0.3–0.4 m/min) travel speed minimizes its occurrence. DC(−) (negative polarity) should be used for stringer bead welding when burn-through, internal undercut, and hollow bead defects are a problem. These problems generally occur on thin wall pipe and on pipe steels containing over 0.1% silicon. Lower currents can be used with DC(−), which helps to reduce these problems. Travel speed with DC(−) will be equal to travel speeds with DC(+).

Hot pass and all other passes should be run DC(+) (positive polarity).

Welding DC(−) (negative polarity) on the stringer bead will not be harmful to either mechanical or metallurgical properties.

Hollow bead or wormhole porosity may also be caused by poor joint preparation and/or cleanliness. See Recommended Procedures for Properly Cleaning Pipe.

Filler, stripper, and cap

In general, there is little fault found with the fillers, strippers, and caps. Most of this welding today is being done with 3/16″ (4.8 mm) Shield-Arc® HYP or 70+ at 160–180 amps, and the results have been excellent.

A reasonably competent crew of firing line welders armed with FW5P+, HYP, or 70+ can do a very fine job of finishing the weld if the stringer and hot pass crew have been doing their work properly.

The size and consistency of the final weld will have its influence on the X-ray radiograph. Thus, the firing line welders should be encouraged to produce a cap pass that is as uniform as possible with neatly stitched close ripples and as much reinforcement as required.

Welding cracks

Since the advent of the higher tensile pipe steels (5LX 52, 60, 65, 70 etc.), it has been necessary to exercise better procedural control to eliminate the possibility of weld and heat-affected zone (HAZ) cracks. To do this effectively, all of the following factors must be controlled.

1. Joint preparation and root spacing must be as specified in the approved procedure. (See Figure 6.)
2. Hi-Low conditions must be held to a minimum.
3. Weld 5LX70 with Shield-Arc® 70+. Weld 5LX65 and lower with Shield-Arc HYP or Fleetweld 5P+. Fleetweld 5P+ is recommended for all stringer beads when lower hardness is a consideration. Lower stringer bead hardness will result in improved resistance to HAZ cracking.
4. The need for preheat varies considerably between applications. Cracking tendencies increase with higher carbon and alloy content, thicker pipe wall and diameter (faster quench rate), and lower ambient temperature. Preheat cold pipe to at least 70 °F (21 °C). Preheat to as much as 300 °F (149 °C) may be required to slow the quench rate and prevent cracking when welding high-strength, large-diameter, or heavy-wall pipe. Specific preheat requirements must be determined for each situation based on these considerations.
5. Ideally the line-up clamp should not be removed nor should any movement of the pipe take place until the stringer bead is 100% completed. Any movement of the pipe could cause a partially completed stringer bead to be overstressed and to crack. The tendency for such cracking varies with the chemistry of the pipe, its diameter and wall thickness, and its temperature. Under favorable conditions, it may be possible to remove the line-up clamp with as little as 60% of the stringer bead completed, but this should be done only when it has been clearly demonstrated that this practice does not cause cracks to occur.
6. After removal of the line-up clamp, the pipe must be gently and carefully lowered on skids.
7. Use only enough current on the stringer to get a good inside bead and travel slowly to get the maximum weld cross-section.
8. Restrict lack of penetration on the inside bead at tie-ins to ¼″ (6.4 mm) or less. Use a disc grinder to improve this situation on starts and stops only.
9. Remove slag from each bead by power wire brushing. Grinding should not be necessary except possibly to clean up a lumpy start or humped up center or perhaps to improve a crater condition. Note: Grinding of the stringer bead should be done with a ⁵/₃₂″ (4.0 mm) disc. Excessive grinding can be detrimental.
10. Weld stringer beads with two or more persons welding on opposite sides to equalize stress. Use three welders on 20–30″ (508–762 mm) pipe and four welders on larger pipe.

Recommended pass sequence

Start

Stop

11. Start the hot pass immediately after completion of the stringer—always within 5 min. At least two hot pass welders should be used on each joint, and to put this pass in as soon as possible after the stringer bead it may require a total of four hot pass welders leap-frogging each other to keep up.
12. Minimize the wagon tracks—this area is highly vulnerable to cracking until the hot pass is completed (Figure 15).

Figure 15. Cracks tend to occur at the area indicated on one side or the other of the stringer bead. This could eventually propagate up through the weld. A properly controlled procedure and good workmanship can eliminate this condition.

The Lincoln Electric Company strongly recommends for weldments intended for sour gas, sour crude, or other critical service applications that the customer verifies that both the level of hardness and variation in hardness of the weld, HAZ, and base plate are within acceptable limits.

Reprinted with permission—Lincoln Electric Co.

How many welds will the average welder make per hour?

Where stringer beads have been run, the average welder can complete about 140 in of weld per hour on ordinary ¼-in wall line pipe. To find the average number of welds per hour, divide the circumference of the pipe into 140.

$N = \dfrac{140}{C}$, where circumference $(C) =$ pipe diameter in inches times 3.14

Example. How many welds per hour will a welder complete on ¼-in wall $10\frac{3}{4}$ -in line pipe, if the stringer beads have been run?

$$N = \frac{140}{(10.75)(3.14)}$$

$$N = \frac{140}{(33.76)}$$

$N = 4.15$ welds per hour or about 42 welds per 10-h day.

On large diameter pipe where wall thickness has been increased, welders average about 100 in of completed weld per hour.

Example. How many welds per hour will a welder complete on 30-in line pipe if the stringer beads have been run?

$$N = \frac{100}{C}$$

$$N = \frac{100}{(30)(3.14)} = \frac{100}{94.2} = 1.1 \text{ welds per hour or 11}$$
welds per 10-h day

How much welding rod is required for a mile of schedule 40 pipeline?

Rule. For 40-ft joints of schedule 40 pipe, multiply the nominal pipe diameter by 22½; the answer is pounds of electrode per mile of pipeline.

Example. 6-in pipe:
6 × 22½ = 135 lb per mile.

Example. 10-in pipe, 30-ft joints:
10 × 22½ = 225 lb per mile.

For 40-ft joints:

$225 \times \dfrac{40}{30} = 300$ lb per mile for 30-ft joints.

The rule is based on standard weight (schedule 40) pipe and the usual number of passes, which varies from two for small pipe up to four for 16 in; larger pipe is usually thinner wall. The rule includes an allowance for wastage.

How many pounds of electrodes are required per weld on line pipe?

Divide the nominal pipe size by 2 and multiply the result by one-fourth pound.

Pounds of electrode $= \dfrac{N}{2} \times 0.25$.

Example. How many pounds of electrode will be used per weld on 10-in line pipe?

Pounds of electrode $= \dfrac{10}{2} \times 0.25 = 1.25$ lb per weld.

Example. How many pounds of electrode will be used per weld on 24-in line pipe?

Solution. Pounds of electrode $= 12 \times 0.25$ or 3 lb per weld.

Welding criteria permit safe and effective pipeline repair

Battelle Laboratories studies show weld sleeves and deposited weld metal can be applied without removing defective lines from service

J. F. Kiefner, Senior Research Engineer, Battelle Houston Operations

Research conducted by the American Gas Association (AGA)[1–5] and others[6–8] showed that pipeline defects can be repaired without removing the lines from service. Two repair methods evaluated were full-encirclement sleeves and direct deposition of weld metal on defects.[*] To ensure minimum risk when using these methods, criteria for repairing pipelines in service have been formulated (Table 1).

The criteria in the table are intended as a repair guide for lines that operate at 40% of their specified minimum yield strength (SMYS) or more. The criteria may by unnecessarily restrictive for lines that operate at or below 40% of SMYS.

Note that defects in pressurized pipelines, especially pneumatically pressurized lines, can cause sudden catastrophic ruptures. When the procedures are carefully planned, risks are minimized. Nevertheless, repair errors could reduce the safety margin of these criteria. Consequently, users of Table 1 are urged to exercise caution when performing repairs.

Choice of repair method

The repair method options included in Table 1 are as follows:

- Type A sleeve
- Type B sleeve
- Deposited weld metal
- Removal
- Removal by hot-tapping.

Type A sleeve repair consists of placing a reinforcing band concentric with the pipe while leaving the band ends unsealed. The sleeve strengthens a defective area by restraining bulging that would otherwise occur when the weakened pipe shell is pressurized. Such a sleeve may also carry a small portion of the hoop stress, but in no case can it become the sole pressure-containing element since its ends are not sealed. As a result, it cannot be used to repair a leaking defect. If a gap exists over the defect, filling of the gap is required. Two classes of Type A sleeve exist: no filler and with filler.

Type B sleeve repair requires a pressure-tight reinforcing concentric band. Its ends are sealed to the carrier pipe, and it can contain pressure. It can be used to repair leaking defects. In addition, a Type B sleeve can also be used as a Type A sleeve. Three classes of Type B sleeves exist: no filler, with filler, and pressurized.

Deposited weld metal completely fills and eliminates clearly visible defects such as corrosion. Access to an entire defect is required so that weld metal can penetrate and bond to sound metal. It may be possible to enlarge pits, laps, or undercut by grinding a groove wide enough to permit access. Deposited weld metal can be selected when requirements to prevent burn-through are met. These requirements are discussed later.

Removal of a damaged pipe section is another way of eliminating a defect. The line must be taken out of service, purged, removed, and replaced with a sound tie-in piece. Removal is often a poor economic choice in terms of wasted product and interrupted service. Thus, removal is recommended in Table 1 with the understanding that it may not be practical in certain instances.

Removal by hot-tapping takes advantage of a widely accepted method for making branch connections on live pipelines. The coupon removed from the line must contain the entire defect. Limitations of this technique will be discussed later.

Types of defects

Classes of defects listed in Table 1 are as follows:

- Manufacturing defects—cracks, undercut, lack of fusion, and lack of penetration in seam welds; laps, pits, cracks, and rolled-in slugs in the pipe body; hard spots.
- Environmentally caused defects—selective corrosion and hydrogen stress cracking in electric resistance welded or flash welded seams; general pipe corrosion; pitting corrosion; stress corrosion cracks; hydrogen stress cracking in hard spots.
- Defects caused by outside forces including dents and gouges.
- Construction defects.

Submerged arc-welded (SAW) seam defects, both straight seam and spiral, include undercut, incomplete fusion, incomplete penetration, and cracks. Repair methods recommended for these defects are shown in Table 1. No filler is required with sleeves, because after the weld reinforcement is ground flush, no gap between the pipe and sleeve should exist. Removal by hot-tapping is not recommended because the tap would involve cutting through the seam, a practice considered unacceptable by many companies. Repair by deposited weld metal is applicable only for undercuts and is subject to special requirements given in footnotes of Table 1.

Type B pressurized sleeves used on nonleaking defects can be pressurized only by using a tapping nipple. Since other means of sleeve repair are entirely adequate when used as directed, the intentional pressurization of a sleeve when no leak or near leak is present is not necessary.

Inside or interior defects require special consideration since they are not readily visible. Removal may be the best alternative unless one can be reasonably certain of the extent of the defects. Hot-tapping is not recommended because of the uncertainty of the extent of an inside or interior defect.

Electric resistance welded (ERW) or flash welded (FW) seam defects include upturned fiber imperfections, incomplete fusion, penetrators, cold welds, and cracks. These can only be repaired with pressurized sleeves and removal since the welds are susceptible to brittle fracture or low-resistance ductile fracture initiation. The value of restraint of bulging from either type of nonpressurized sleeve is uncertain. Hence they are not recommended for these defects. Hot-tapping also is not recommended because of the involvement of the seam weld. It is not recommended that weld metal be deposited because of possible low ductility. Required grinding before such welding would involve creating or enlarging a defect in a potentially low-toughness material. Only removal or use of a Type B pressurized sleeve that stress-relieves the defect is recommended.

Other seam defects include lap welds and furnace butt welds. These must be dealt with on a case-by-case basis. Such welds usually appear in older or smaller and lower pressure pipelines where fracture resistance requirements are less stringent. On the other hand, some of these older materials can be quite susceptible to low-toughness fracture behavior. It is probably best to treat these materials with the same caution as ERW and FW seams as they are being operated at stress levels exceeding 40% of SMYS.

Laps, pits, seams, cracks, and rolled-in plugs should be repaired subject to these limitations: special requirements of I.D. and interior defects and leaks must be observed; hot-tapping is acceptable on nonleaking, O.D. defects as long as the entire defect is removed with the coupon; deposited weld metal may be used to repair pits or laps if they can be entirely exposed by grinding.

Hard spots created in the plate by accidental quenching on the run-out table may become flat spots in the pipe since they do not yield when the plate is formed to pipe. Such hard spots do not fail spontaneously unless they contain quench cracks or unless they undergo hydrogen stress cracking. Such spots should be repaired if they are not cracked. Type A or B sleeves without filler would not be acceptable since they would not restrain flat spots from bulging.

Hard spots can be remedied by several methods if they occur as outside nonleaking defects. Sleeve repair methods with filler not only provide strengthening but shield the hard spot from the environment, which may cause it to become cracked.[a] Hard spots at the I.D. surface should not be repaired with Type A sleeves since they may be susceptible to cracking in certain types of product service. Such a sleeve would not prevent a leak if the resulting crack grew through the wall.

Selective corrosion and hydrogen stress cracking in ERW and FW seams are subject to low toughness fracture behavior. It is recommended that they be removed or repaired with pressurized Type B sleeves to relieve stress.

General or widespread corrosion is that which covers too wide an area to be repairable by means of hot-tapping or deposited weld metal. Any other repair means are acceptable if they cover the critically affected area. When internal corrosion is present, its extent must be reasonably well known and further corrosion must be prevented. Removal may sometimes be the only suitable choice.

Pitting corrosion may involve isolated pits or groups of pits. Generally, isolated single pits will not require repair if further corrosion can be halted. ASME guidelines can be used to determine whether or not a corroded region requires repair. Groups of interacting pits can substantially lower remaining strength. If strength has been lowered, the pipe should be repaired. Any of the methods of Table 1 are suitable if the noted precautions are observed.

Stress-corrosion cracks occur in clusters and often cover large areas. It is difficult to repair these cracks by hot-tapping or deposited weld metal. Repair methods for stress-corrosion cracks are confined to those which can strengthen the entire affected area.

Hydrogen stress cracking appears in hard spots attacked by hydrogen emitted from bacteria external to the pipe and from cathodic protection. Internally the hard spots can be attacked by certain types of products—especially those containing hydrogen sulfide. Because flatness often occurs near hard spots, nonfilled sleeves are not recommended. Deposited weld metal is not recommended because grinding of a hydrogen stress crack in a hard—and usually brittle—spot while the pipe is under pressure is not safe. Interior hydrogen stress cracking should not be repaired by a Type A sleeve since such a sleeve cannot prevent leaks if the crack grows through the wall.

Dents and gouges and a combination gouge-in-dent result from external encroachment by mechanical excavating equipment, other kinds of equipment, or rocks. Plain dents are usually innocuous unless they are quite large[b]—2% or

---

[a] One method of protecting hard spots from hydrogen stress cracking that so far, has proven adequate, is using a concentric band of sheet metal spaced away from the pipe by rubber seals. The annular space is filled with coal tar to exclude ground water, and the metal band (which is coated) shields the hard spot from hydrogen generating cathodic protection current.

[b] Present requirements of federal regulations Part 192 dictate removal of dents extending over 2% of the pipe diameter.

Table 1
Criteria for Selecting an Appropriate Repair Method (X Indicates an Acceptable Repair)

Repair Method	Type of Defect								
	Manufacturing Defect				Environmentally Caused Defect in ERW or FW, Seam Weld in Body, SAW Seam or Girth Weld				
	In Seam Weld		In the Body of the Pipe						
	SAW Undercut, Incomplete Fusion, Incomplete Penetration, Crack	ERW and FW Upturned Fiber Imperfection, Incomplete Fusion, Penetrator, Cold Weld Crack	Lap, Pit, Seam, Crack Rolled-In Slug	Hard Spot Exceeding $R_c > 35$ and 2 in in Extent or More	Selective Corrosion, Hydrogen Stress Cracking in Weld Zone	General Corrosion	Pitting Corrosion	Stress-Corrosion Cracking	Hydrogen Stress Cracking in Hard Spot
Nonleaking O.D. defect[a]									
Type A Sleeve, No Filler	X	...	X	...	...	X	X	X	...
Type A Sleeve, with Filler	...	...	...	X[g]	...	X	X	...	X
Type B Sleeve, No Filler	X	...	X	...	...	X	X	X	...
Type B Sleeve, with Filler	...	...	...	X	...	X	X	...	X
Type B Sleeve, Pressurized	X[b]	X[b]	X[b]	X[b]	X[b]	X[b]	X[b]	X[b]	X[b]
Deposited Weld Metal (h)	X[c]	...	X[c]	...	...	...	X[c]	...	...
Removal	X	X	X	X	X	X	X	X	X
Removal by Hot Tap	...	...	X[d]	X[d]	...	...	X[d]	...	X[d]
Nonleaking interior or I.D. defect[a]									
Type A Sleeve, No Filler	X	...	X	...	...	X	X	...	...
Type B Sleeve, No Filler	X	...	X	X	...	X	X	...	X
Type B Sleeve, Pressurized	X[b]	X[b]	X[b]	X[b]	X[b]	X[b]	X[b]	...	X[b]
Removal	X	X	X	X	X	X	X	...	X
Leaking defects									
Type B Sleeve Pressurized	X	X	X	X	X	X	X	X	X
Removal	X	X	X	X	X	X	X	X	X

Corrosion or other defect that is known to exceed 80% of the wall thickness in depth should be treated as though it were a leak.
Pressurization to be accomplished upon completion of repair by drilling hole through carrier pipe through a small tapping nipple.
Only for undercut, laps, and pits where the defect can be safely enlarged to a weldable groove and subject to remaining wall thickness rules.
Only if the branch coupon entirely contains and removes defect.
Only with a humped sleeve or if a flat sleeve is used. Girth weld must be ground flush. Do not use if extreme high-low condition exists.

Defect Caused by Outside Force					Construction Defect
In Seam or Other Weld		In Body of Pipe			In Girth Weld
Plain Dent or Gouge-in-Dent, SAW Seam or Girth Weld	Plain Dent, or Gouge-in-Dent ERW or FW Seam	Plain Dent Greater than 2% or Pipe Diameter	Gouge (Also Including Gouges in SAW Seams or Girth Welds)	Gouge-in-Dent	Undercut, Incomplete, Fusion, Incomplete Penetration, Crack
...	...	...	...	...	...
X	...	X	...	X	...
...	...	...	X	...	X(e)
X	...	X	...	X	...
X(b)	X(b)	X(b)	X(b)	X(b)	X(b,e)
...	...	...	X(b)	...	X(c)
X	X	X	X	X	X
...	...	...	X(d)	...	...
...	...	...	...	...	...
...	...	...	...	...	X(e)
...	...	...	...	...	X(b,e)
...	...	...	...	...	X
X	X	X	X	X	X(e)
X	X	X	X	X	X

One method of protecting hard spots from hydrogen stress cracking, which so far has proven adequate, is that of using a concentric band of sheet metal spaced away from the pipe by rubber seals. The annular space is filled with coal tar to exclude ground water, and the metal band (which is itself coated) shields the hard spot from hydrogen-generating cathodic protection current. The role of this band is merely to shield the pipe from the cathodic protection and not to strengthen the pipe.
Use of low hydrogen electrodes is recommended.

more of the diameter—or unless they involve a seam or girth weld. When the dent is large or welds are included in the dent or gouge-in-dent, the repair should prevent its outward movement. Sleeves with fillers or pressurized sleeves are required. If the included weld is an ERW or FW seam, the defect should be removed or the stress relieved with a pressurized sleeve.

Gouges without dents in the body of the pipe or in SAW seams or girth welds may be repaired by any means except deposited weld metal. Gouges-in-dents may be repaired only by means that prevent outward movement of the dent. Sleeves without fillers are not acceptable.

When dents are involved, hot-tapping is not recommended, since it may not remove the dent entirely. For any gouge or gouge-in-dent, repair by deposited weld metal is not recommended, since concealed cracks may exist. When a dent is present, weld metal may not have sufficient ductility to withstand the severe strains that accompany outward dent movement.

Construction defects in girth weld include undercut, incomplete fusion, incomplete penetration, and cracks. Deposited weld metal may be one of the best ways to repair undercut or other externally connected defects in girth welds that can be ground for access. Sleeves, if used, should have a special shape such as a central hump to avoid interference with the girth weld reinforcement. Only sleeves with welded ends are recommended. Such sleeves tend to strengthen the defective girth joint, whereas those with nonweld ends do not.

Burn-through

Burn-through into a live pipeline would defeat the purpose of a repair, would probably cause the line to be shut down, and might create a serious safety hazard to the repair crew.

An AGA study revealed that the following parameters control burn-through: remaining wall-thickness, welding heat

Table 2
Limitations on Remaining Wall-Thickness for Repairing Without a Burn-Through

Values are the minimum recommended thicknesses in inches with natural gas as the pressurizing medium at the pressures and flows shown				
Maximum welding voltage, 20 volts				
Maximum welding current, 100 amps				
Pressure, psia	Gas Flow Rate, ft/s			
	0	5	10	20
15	0.320	–	–	–
500	0.300	0.270	0.240	0.205
900	0.280	0.235	0.190	0.150

input, and the nature and state of the pressurization media in the pipe. Consequently, it is possible to prescribe safe limits for avoiding burn-through while depositing weld metal on a defect (Table 2).

Table 2 shows a relationship between I.D. surface temperature during welding and remaining wall thickness for various pressure and flow rates of natural gas. These values were taken from AGA curves. The table is based on the following: heat input characteristics of the $3/32$-in or $1/8$-in low hydrogen electrode, 80–100 amps welding current at 20 volts DC, and an average electrode travel speed of 4–5 in/min.

Energy input should not exceed these limits. Low hydrogen electrodes are recommended for reasons explained later. Under these conditions, repairs can safely be made in the flat, horizontal, or overhead positions provided flow or pressure is controlled. Analytic results predict that the I.D. wall temperature will not exceed 2000 °F and a burn-through is highly unlikely when repairs are made in this manner.

Values shown in Table 2 are believed to be quite conservative. Repairs were made without burn-throughs on 0.180-in remaining wall with air at ambient pressure and no flow inside the pipe. This conservatism provides an extra safety margin and allows linear extrapolations for values between those shown in the table. The minimum wall-thickness of 0.150 in is established on the basis of the experimental results. This thickness is also recommended by the British Gas Corp. Within these limits, repairs may be made by deposited weld metal (Table 1).

Underbead cracking

Underbead cracking can be minimized by making repairs with low hydrogen electrodes in a manner that avoids hard weld HAZs.

One way to help assure that extremely hard weld zones are not formed during repair welding is to limit repairs to well-known carbon equivalent ranges. Unfortunately, chemistries of specific samples needing repairs will seldom be known. Low hydrogen electrodes prevent hydrogen from being present in the welding atmosphere, but they do not prevent formation of hard weld zones when adverse chemistries are present. Making crack-free weldments requires careful control of heat input to avoid rapid quenching or post-repair treatments to assure that extreme hardness does not remain.

In the case of sleeves, the critical area in which creaking most often occurs is the fillet-weld at the ends. At least two procedures for making these welds, with a minimum risk of cracking, are available.[1,4,5,8] In the case of deposited weld metal repairs, heat input must be kept low to avoid burn-through during the first or second passes. During later passes, however, higher heat inputs can be used to soften the resulting repair metal microstructure and the HAZ of base metal. A high heat input final pass can be made with a non-fusing tungsten electrode as suggested by the British Gas Corp.[8] An alternative procedure is to make an extra pass with high heat input using a conventional electrode. This pass can be ground off, since its purpose is merely to soften the heat-affected microstructure of previous passes.

(Based on a paper, "Criteria for Repairing Pipelines in Service Using Sleeves and Deposited Weld Metal" presented by the author at the AGA Transmission Conference at Montreal, Quebec, May 8–10, 1978.)

Source

Pipe Line Industry, January 1980.

References

1. Kiefner, J. F., and Duffy, A. R., *A Study of Two Methods for Repairing Defects in Line Pipe*, Pipeline Research Committee of the American Gas Association, 1974, Catalog No. L22275.
2. Kiefner, J. F., Duffy, A. R., Bunn, J. S., and Hanna, L. E., "Feasibility and Methods of Repairing Corroded Line Pipe," Materials Protection and Performance, Vol. III, No. 10, Oct., 1972.
3. Kiefner, J. F., "Corroded Pipe: Strength and Repair Methods," *Fifth Symposium on Line Pipe Research*, Houston, Texas, 1974, American Gas Association, Catalog No. L30174.
4. Kiefner, J. F., "Repair of Line Pipe Defects by Full-Encirclement Sleeves," *Welding Journal*, Vol. 56, No. 6, June, 1977.
5. Kiefner, J. F., Whitacre, G. R., and Eiber, R. J., "Further Studies of Two Methods for Repairing Defects in Line Pipe," to Pipeline Research Committee of the American Gas Association, NG-18 Report No. 112, March 2, 1978.

Terminology[9–11]

Defect parameters

Defect: A crack, pit, crevice, gouge, dent, metallurgical anomaly, or combination of two or more of the above that is known or suspected to reduce the effective pipe strength level to less than 100% of its specified minimum yield strength (SMYS).

O.D. defect: A defect emanating at and extending radially inward from the outside surface but not entirely through the wall of the pipe.

I.D. defect: A defect emanating at and extending radially outward from the inside surface but not entirely through the wall of the pipe.

Interior defect: A defect emanating in the interior of the pipe wall but not of sufficient radial extent to be connected with either the inside or the outside surface.

Leaking O.D. defect: A defect that was initially an O.D. defect but which has grown through the wall to become a leak.

Leaking I.D. defect: A defect that was initially an I.D. defect but which has grown through the wall to become a leak.

Superficial defect: A lap, crevice, pit, group of pits, metallurgical anomaly, or plain dent (i.e., without scratches, gouges, or cracks) that is of insufficient extent to reduce the effective strength level of the pipe below 100% of SMYS.

Kinds of defects

*Definitions marked by an asterisk are from API Bulletin 5T1 *Nondestructive Testing Terminology*, Third Edition, April, 1972.

Not every conceivable kind of defect in pipe is covered by this list. The list is limited to those that are likely to be encountered in an in-service pipeline.

Defects originating from pipe manufacture

a. Defects not necessarily in the seam weld (primarily in the body of the pipe)
  * Lap: Fold of metal that has been rolled or otherwise worked against the surface of rolled metal but has not fused into sound metal.
  * Pit: A depression resulting from the removal of foreign material rolled into the surface during manufacture.
  * Rolled-in slugs: A foreign metallic body rolled into the metal surface, usually not fused.
  * Seam: Crevice in rolled metal that has been more or less closed by rolling or other work but has not been fused into sound metal.
  * Hard spot: An area in the pipe with a hardness level considerably higher than that of surrounding metal; usually caused by localized quenching.
  * Crack: A stress-induced separation of the metal that, without any other influence, is insufficient in extent to cause complete rupture of the material.

b. Defects in the seam weld
  * Incomplete fusion: Lack of complete coalescence of some portion of the metal in a weld joint.
  * Incomplete penetration: A condition where weld metal does not continue through the full thickness of the joint.
  * Undercut: Undercutting on submerged-arc-welded pipe is the reduction in thickness of the pipe wall adjacent to the weld where it is fused to the surface of the pipe.
  * Weld area crack: A crack that occurs in the weld deposit, the fusion line, or the HAZ. (Crack: A stress-induced separation of the metal that, without any other influence, is sufficient in extent to cause complete rupture of the material.)
  * Upturned fiber imperfection: Metal separation, resulting from imperfections at the edge of the plate or skelp, parallel to the surface, which turn toward the I.D. or O.D. pipe surface when the edges are upset during welding.
  * Penetrator: A localized spot of incomplete fusion.
  * Cold weld: A metallurgically inexact term generally indicating a lack of adequate weld bonding strength of the abutting edges, due to insufficient heat and/or pressure. A cold weld may or may not have separation in the weld line. Other more definitive terms should be used whenever possible.

Defects originating from external or internal environmental degeneration of the pipe

a. Seam weld defects
  Selective corrosion: Preferential corrosion in the fusion line of an electric resistance-welded or flash-welded longitudinal seam.
  Hydrogen stress cracking: Environmentally stimulated cracking of the weld metal or HAZ of the longitudinal seam.

b. Defects not in the body of the pipe (possibly in the seam weld—but not specifically because of the seam weld).

6. Cassie, B. A., "The Welding of Hot-Tap Connections to High Pressure Gas Pipelines," J. W. Jones Memorial Lecture, Pipeline Industries Guild, British Gas Corp., October, 1974.

7. Morel, R. D., *Welded Repairs on API 5LX-52 Pipe*, 13th Annual Petroleum Mechanical Engineering Conference, Denver, Colorado, September 24, 1958.

8. Christian, J. R., and Cassie, B. A., "Analysis of a Live Welded Repair on An Artificial Defect," ERS.C.96, October, 1975.

9. Smith, R. B., and Eiber, R. J., "Field Failure Survey and Investigation," *Fourth Symposium on Line Pipe Research*, American Gas Association, 1969, Catalog No. L30075.

10. Eiber, R. J., "Field Failure Investigations," *Fifth Symposium on Line Pipe Research*, American Gas Association, 1974, Catalog No. L30174.

11. Wright, R. R., "Proper Inspection Methods Minimize Pipeline Failures," *Oil and Gas Journal*, May 23, 1977, pp. 51–56.

Cross-country pipeline—vertical down electrode consumption, pounds of electrode per joint*

Wall Thickness	5/16"			3/8"			1/2"			5/8"			3/4"		
Elec. Dia.	5/32"	3/16"	Total	5/32"	3/16"	Total	5/32"	3/16"	Total	5/32"	3/16"	Total	5/32"	3/16"	Total
Nom. Pipe Dia., in			lb			lb			lb			lb			lb
6	0.34	0.54	0.88	0.34	0.86	1.2									
8	0.45	0.71	1.2	0.45	1.1	1.6	0.45	2.2	2.7						
10	0.56	0.88	1.4	0.56	1.4	2.0	0.56	2.7	3.3						
12	0.67	1.0	1.7	0.67	1.7	2.4	0.67	3.2	3.9						
14	0.73	1.1	1.8	0.73	1.8	2.5	0.73	3.5	4.2	0.73	5.7	0.73	8.3	9.0	
16	0.84	1.3	2.1	0.84	2.1	2.9	0.84	4.0	4.8	0.84	6.5	7.3	0.84	9.4	10.2
18	0.94	1.5	2.4	0.94	2.3	3.2	0.94	4.5	5.4	0.94	7.3	8.2	0.94	10.6	11.5
20	1.10	1.6	2.7	1.10	2.6	3.7	1.1	5.0	62.71	1.1	8.1	9.2	1.1	11.8	12.9
22	1.20	1.8	3.0	1.20	2.9	4.1	1.2	5.5	6.7	1.2	8.9	10.1	1.2	13.0	14.2
24	1.30	2.0	31.3	1.30	3.1	4.4	1.3	6.0	7.3	1.3	9.7	11.0	1.3	14.2	15.5
26	1.40	2.1	3.5	1.40	3.4	4.8	1.4	6.5	7.9	1.4	10.5	11.9	1.4	15.3	16.7
28	1.50	2.3	3.8	1.50	3.7	5.2	1.5	7.0	8.5	1.5	11.3	12.8	1.5	16.5	18.0
30	1.60	2.5	4.1	1.60	3.9	5.5	1.6	7.5	95.51	1.6	13.71	13.7	1.6	17.7	19.3
32	1.70	2.6	4.3	1.70	4.2	5.9	1.7	8.0	9.7	1.7	13.0	14.7	1.7	18.9	20.6
34	1.80	2.8	4.6	1.80	4.4	6.2	1.8	8.6	10.4	1.8	13.8	15.6	1.8	20.1	21.9
36	1.90	2.9	4.8	1.90	4.7	6.6	1.9	4.61	11.0	1.9	14.6	16.5	1.9	21.3	23.2
38	2.00	3.1	5.1	2.00	2.0	7.0	2.0	9.6	11.6	2.0	22.4	22.4	2.0	22.4	24.4
40	2.10	3.3	5.4	2.10	5.42	7.3	2.1	42.1	123.62	2.1	16.2	18.3	2.1	23.6	25.7
42				2.20	5.5	7.7	2.2	10.6	12.8	2.2	17.0	19.2	2.2	24.8	27.0
44				2.30	5.7	8.0	2.3	11.1	13.4	2.3	17.8	20.1	2.3	26.0	28.3
46				2.40	6.0	8.4	2.4	11.6	14.0	2.4	18.6	21.0	2.4	27.2	29.6
48				2.50	6.3	8.8	2.5	12.1	14.6	2.5	19.4	21.9	2.5	28.3	30.8

* "Electrode required." Figures in the table include 4-in stub lengths. These figures will vary with different stub loss practices. Quantities required for the 5/32" size will vary based on travel speeds of the stringer bead and hot pass. Slow travel speeds may increase these quantities by up to 50%. Reprinted courtesy of the Lincoln Electric Company.

Guidelines for a successful directional crossing bid package

The Directional Crossing Contractors Association (DCCA) has been addressing the issue of what information should be made available to contractors and engineers so that future projects proceed as planned. Crossings of rivers and other obstacles using directional drilling techniques are increasingly being utilized around the world. As in any construction project, it is necessary for the contractor to have as much information as possible to prepare a competitive and comprehensive proposal and to be able to successfully install the crossing. Better preconstruction information also allows the work to be undertaken more safely and with less environmental disturbance.

Development and uses

Originally used in the 1970s, directional crossings are a marriage of conventional road boring and directional drilling of oil wells. The method is now the preferred method of construction. Crossings have been installed for pipelines carrying oil, natural gas, petrochemicals, water, sewerage and other products. Ducts have been installed to carry electric and fiber-optic cables. Besides crossing under rivers and waterways, installations have been made crossing under highways, railroads, airport runways, shore approaches, islands, areas congested with buildings, pipeline corridors, and future water channels.

Technology limits

The longest crossing to date has been about 11,000 ft. Pipe diameters of up to 48 in have been installed. Although directional drilling was originally used primarily in the U.S. Gulf Coast through alluvial soils, more and more crossings are being undertaken through gravel, cobble, glacial till, and hard rock.

Advantages

Directional crossings have the least environmental impact of any alternate method. The technology also offers maximum depth of cover under the obstacle, thereby affording maximum protection and minimizing maintenance costs. River traffic is not interrupted, because most of the work is confined to either bank. Directional crossings have a predictable and short construction schedule. Perhaps most significantly, directional crossings are in many cases less expensive than other methods. Bundles of multiple pipes can be pulled into a common bore.

Technique

Pilot Hole—A pilot hole is drilled beginning at a prescribed angle from horizontal and continues under and across the obstacle along a design profile made up of straight tangents and long radius arcs. A schematic of the technique is shown in Figure 1. Concurrent to drilling a pilot hole, the contractor may elect to run a larger diameter "wash pipe" that will encase the pilot drill string. The wash pipe acts as a conductor casing providing rigidity to the smaller diameter pilot drill string and will also save the drilled hole should it be necessary to retract the pilot string for bit changes. The directional control is brought about by a small bend in the drill string just behind the cutting head. The pilot drill string is not rotated except to orient the bend. If the bend is oriented to the right, the drill path then proceeds in a smooth radius bend to the right. The drill path is monitored by an electronic package housed in the pilot drill string near the cutting head. The electronic package detects the relation of the drill string to the earth's magnetic field and its inclination. These data are transmitted back to the surface where calculations are made as to the location of the cutting head. Surface location of the drill head also can be used where there is reasonable access.

Pre-ream—Once the pilot hole is complete, the hole must be enlarged to a suitable diameter for the product pipeline. For instance, if the pipeline to be installed is 36-in diameter, the hole may be enlarged to 48-in diameter or larger. This is accomplished by "pre-reaming" the hole to successively larger diameters. Generally, the reamer is attached to the drill string on the bank opposite the drilling rig and pulled back into the pilot hole. Joints of drill pipe are added as the reamer makes its way back to the drilling rig. Large quantities of slurry are pumped into the hole to maintain the integrity of the hole and to flush out cuttings.

Pullback—Once the drilled hole is enlarged, the product pipeline can be pulled through it. The pipeline is prefabricated on the bank opposite the drilling rig. A reamer is attached to the drill string, and then connected to the pipeline pullhead via a swivel. The swivel prevents any translation of the reamer's rotation into the pipeline string, allowing for a smooth pull into the drilled hole. The drilling rig then begins the pullback operation, rotating and pulling on the drill string and once again circulating high volumes of drilling slurry. The pullback continues until the reamer and pipeline break ground at the drilling rig.

I.D.1. Pilot hole

Wash pipe

Wash pipe bit

Pilot string bit

Survey tool

Drill profile

Pilot string

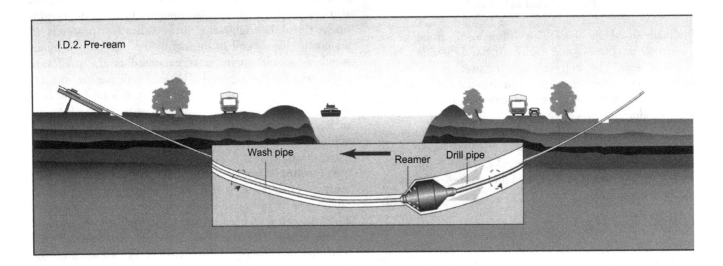

I.D.2. Pre-ream

Wash pipe

Reamer

Drill pipe

I.D.3. Pullback

Drill pipe

Reamer swivel

Product pipeline

Figure 1. Technique.

1. Rig unit
2. Control Cab/Power unit
3. Drill pipe
4. Water pump
5. Slurry mixing tank
6. Cutting separation equipment
7. Slurry pump
8. Bentonite storage
9. Power generations
10. Spares storage
11. Site office
12. Site office
13. Entry point slurry containment pit
14. Cuttings settlement pit

Figure 2. Rig side work space.

Layout and design

Access—Heavy equipment is required on both sides of the crossing. To minimize cost, access to either side of the crossing should be provided with the least distance from an improved road. Often the pipeline right-of-way is used for access. All access agreements should be provided by the owner. It is not practical to negotiate such agreements during the bid process.

Work space

Rig Side—The rig spread requires a minimum 100-ft-wide by 150-ft-long area. This area should extend from the entry point away from the crossing, although the entry point should be at least 10 ft inside the prescribed area. Since many components of the rig spread have no predetermined position, the rig site can be made up of smaller irregular areas. Operations are facilitated if the area is level, hardstanding, and clear of overhead obstructions. The drilling operation requires large volumes of water for the mixing of the drilling slurry. A nearby source of water is necessary (Figure 2).

Pipe Side—Strong consideration should be given to provide a sufficient length of work space to fabricate the product pipeline into one string. The width will be as necessary for normal pipeline construction, although a work space of 100-ft-wide by 150-ft-long should be provided at the exit point itself. The length will assure that during the pullback the pipe can be installed in one uninterrupted operation. Tie-ins of successive strings during the pullback operation increase the risk considerably, because the pullback should be continuous (Figure 3).

Profile survey

Once the work locations have been chosen, the area should be surveyed and detailed drawings prepared. The eventual accuracy of the drill profile and alignment is dependent on the accuracy of the survey information.

Profile design parameters

Depth of Cover—Once the crossing profile has been taken and the geotechnical investigation complete, a determination

Figure 3. Pipe side work space.

1. Cuttings settlement pit
2. Exit point slurry containment pit
3. Pipeline rollers
4. Product pipeline
5. Construction equipment
6. Drill pipe
7. Spares storage

of the depth of cover under the crossing is made. Factors considered may include flow characteristics of the river, the depth of scour from periodic flooding, future channel widening/deepening, and whether pipeline or cable crossings already exist at the location. It is normally recommended that the minimum depth of cover be 20 ft under the lowest section of the crossing. While 20 ft is a recommended depth of cover on a river crossing, crossings of other obstacles may have differing requirements.

Penetration Angles and Radius of Curvature—An entry angle between 8° and 20° can be used for most crossings. It is preferable that straight tangent sections are drilled before the introduction of a long radius curve. The radius of the curve is determined by the bending characteristic of the product pipeline, increasing with the diameter. A general "rule-of-thumb" for the radius of curvature is 100 ft/1-in diameter for steel line pipe. The curve usually brings the profile to the elevation providing the design cover of the pipeline under the river. Long horizontal runs can be made at this elevation before curving up toward the exit point. Exit angle should be kept between 5° and 12° to facilitate handling of the product pipeline during pullback.

Calculate minimum radius of curvature:

$$R = \frac{EC}{FS - \frac{PD}{4t}}$$

$$C = \frac{D}{12} \times 0.5$$

$$R = \frac{30,000,000 \times 0.58}{0.9 \times 35,000 - \frac{1440 \times 14}{4 \times 0.5}} = 816.99$$

$R =$ Minimum radius, ft
$E =$ Modulus of elasticity, 30,000,000
$C =$ One half of pipe OD in ft
$F =$ Design factor
$S =$ Pipe SMYS, psi
$P =$ Design pressure , psig
$D =$ Pipe outside diameter, inches
$t =$ Pipe wall thickness, inches

Example.
$D = 14$ in.
$t = 0.500$ in.

$F = 0.9$

$S = 35{,}000 \text{ psi}$

$E = 30{,}000{,}000$

$C = \frac{D}{12} \times 0.5 = 0.58$

$R = \dfrac{30{,}000{,}000 \times 0.58}{0.9 \times 35{,}000 - \dfrac{1440 \times 14}{4 \times 0.5}} = 816.99 - \text{ use } 817 \text{ ft}$

Drill Survey—Most downhole survey tools are electronic devices that give a magnetic azimuth (for "right/left" control) and inclination (for "up/down" control). Surface locators can also be used in conjunction with the downhole electronic package.

Accuracy—The accuracy of the drill profile is largely dependent on variations in the earth's magnetic field. For instance, large steel structures (bridges, pilings, other pipelines, etc.) and electric power transmission lines affect magnetic field readings. However, a reasonable drill target at the pilot hole exit location is 10 ft left or right, and −10 ft to +30 ft in length.

As-Built Drawings—Normally, survey calculations are conducted every 30 ft during pilot hole operations. As-built drawings that are based on these calculations should be provided by the contractor. Alternate methods such as gyroscoping, ground penetrating radar, or "intelligent" pigs may also be used to determine the as-built position.

Geotechnical investigation

Number of Borings—The number of exploration holes is a function of the proposed crossing length and the complexity of the strata. If the crossing is about 1000 ft, a borehole made on each side of the crossing may suffice. If an examination of these borings indicates that conditions are likely to be homogeneous on both sides, it may not be necessary to conduct further sampling. If the report indicates anomalies discontinuity in the strata or the presence of rock or large concentrations of gravel, it is advisable to make additional borings to better define the strata. Longer crossings (especially large-diameter pipelines) that indicate gravel, cobble, boulders, or rock should have samples taken about 600–800 ft apart unless significant anomalies are identified that might necessitate more borings. All borings should be located on the crossing profile along with their surface elevations being properly identified. If possible, the borings should be conducted at least 25 ft off of the proposed centerline. The bore holes should be grouted upon completion. This will help prevent the loss of drilling slurry during the crossing installation.

Depth of Borings—All borings should be made to a minimum depth of 40 ft below the lowest point in the crossing or 20 ft below the proposed depth of the crossing, whichever is greater. In some instances, it may be beneficial to the owner and the contractor to install the crossing at a greater depth than the owner requires for his permit. It is suggested that all borings be through the same elevation to better determine the consistency of the underlying material and note any patterns that may be present.

Standard Classification of Soils—A qualified technician or geologist should classify the material in accordance with the Unified Soil Classification System and ASTM Designations D-2487 and D-2488. It is beneficial to have a copy of the field-drilling log completed by the field technician or driller. These logs include visual classifications of materials, as well as the driller's interpretation of the subsurface conditions between samples.

Standard Penetration Test (SPT)—In order to better define the density of granular materials, the geotechnical engineer generally uses the Standard Penetration Test (SPT) in general accordance with ASTM Specifications D-1586. This is a field test that involves driving a 2-in split spoon sampler into the soil by dropping a hammer of a specific weight (usually 140 lb) a specified distance (usually 30 in) to determine the number of blows necessary to drive the sampler 12 in. In very dense soils, the field technician may note the number of blows required to drive the sampler less than the required 12 in (i.e., 50 blows for 3 in). The number obtained is the standard penetration resistance value (N) and is used to estimate the in situ relative density of cohesionless soils. Some geotechnical firms will conduct these penetration tests in cohesive materials and rock, and to a lesser extent, the consistency of cohesive soils and the hardness of rock can be determined.

Thinwalled "Shelby" Tube Sampling—Most geotechnical firms prefer to use a Thin-walled Tube Sampling method for obtaining samples of cohesive materials. These tests are conducted in general accordance with ASTM Specification D-1587. This test is similar to the Standard Penetration test except the sample is collected by hydraulically pushing a thin-walled seamless steel tube with a sharp cutting edge into the ground. The hydraulic pressure required to collect the sample is noted on the field log. This produces a relatively undisturbed sample that can be further analyzed in the laboratory. These samples can be field tested with hand-held penetrometers, but more accurate readings of density and consistency can be obtained by performing unconfined compressive strength tests where the results are noted in tons per square foot. Generally, for directional drilling contractors a standard penetration test using the split spoon sampler described earlier will suffice in both materials.

Sieve Analysis of Granular Materials—A sieve analysis is a mechanical test of granular materials performed on samples collected in the field during the standard penetration test with the split spoon sampler. The split spoon samples are taken to the laboratory and processed through a series of screens. The sample provides a percentage analysis of the granular material by size and weight. It is one of the most important tests undertaken.

Rock Information—If rock is encountered during the soil investigation borings, it is important to determine the type, the relative hardness, and the unconfined compressive strength. This information is typically collected by the geotechnical drilling firm by core drilling with a diamond bit core barrel. The typical core sample recovered with this process has a 2-in diameter. The type of rock is classified by a geologist. The geologist should provide the Rock Quality Designation (RQD), which rates the quality of the rock based on the length of core retrieved in relation to the total length of the core. The hardness of the rock (Mohs' Scale of Hardness) is determined by comparing the rock to ten materials of known hardness. The compressive strength is determined by accurately measuring the core and then compressing the core to failure. This information pertaining to the underlying rock formation is imperative to determine the type of downhole equipment required and the penetration rates that can be expected.

Pipe material selection

Wall Thickness—D/t "Rule-of-Thumb"—The following table provides generalized recommendations for the selection of steel pipe wall thicknesses relative to pipe diameter. These recommendations are meant to be used only as a starting point in the design. It is recommended that in the final design, specific stresses be calculated and compared with allowable limits.

(For high-density polyethylene (HDPE) pipe, a standard dimension ratio of D/t, SDR, of 11 or less is recommended and the pipe manufacturer should be consulted.)

Diameter (D)	Wall Thickness (t)
6″ and smaller	0.250″
6–12″	0.375″
12–30″	0.500″
>30″	$D/t < 50$

Stress Analysis—In finalizing the design, the stresses imposed during construction and in-service must be calculated and checked to be within allowable limits for the grade of material. The stresses at each stage must be considered acting individually and in combination. Stresses result due to

spanning between rollers prior to pullback, the hydrostatic testing pressures, pulling forces during installation, radius of curvature as the pipe enters the ground, the drilling profile curvature, external pressures in the drilled hole, and the working pressure.

1. Pre-installation
   a. Hoop and longitudinal stresses resulting from hydrostatic testing are calculated.
   b. Using the known distance between rollers as the free spanning distance, the maximum hogging and sagging moments can be calculated. Considering the greater of these two moments, the maximum spanning stress is calculated. Note: During hydrostatic testing the pipeline will be full of water; therefore, the additional weight of water must be included in these calculations.
2. Installation
   a. The spanning stresses calculated in stage 1.b. also apply in this installation phase.
   b. The theoretical pulling force must be determined in order to provide the stresses that will result. An assumed downhole friction factor of 1.0 is recommended to provide conservative results and to include the effect of the pipeline being pulled around a curve. The maximum predicted pulling force should then be used in calculating the resulting longitudinal stress.
   c. Allowing for a 10% drilling tolerance leads to the use of a radius of curvature 90% of the design radius when calculating the longitudinal curvature stresses.
   d. External pressure from static head in the drilled hole and/or overburden pressures must be considered. It is recommended that the static head resulting from the maximum envisaged drilling fluid density should be used with a factor of safety of 1.5 to provide conservative estimations of resulting hoop and longitudinal stresses.
3. Post-installation
   a. The longitudinal curvature stresses calculated for stage 2.c. above are used again here.
   b. External pressure stresses from 2.d. apply.
   c. Hoop and longitudinal stresses resulting from the final hydrostatic test are calculated.
4. In-service
   a. Curvature—see 2.c.
   b. External pressure—see 2.d.

The maximum working pressure of the pipeline is used in calculating longitudinal and hoop stresses that will be imposed during service.

Allowable Stresses—Having determined the individual and combined stresses at each stage of construction and those for the in-service condition, they must be compared with allowable limits.

1. ASME B31.8—1992, Table A842.22 provides the following limits
   - Maximum allowable longitudinal stress: 80% SMYS.
   - Maximum allowable hoop stress: 72% SMYS.
   - Maximum allowable combined stress: 90% SMYS.
   (Where SMYS is the Specified Minimum Yield Strength of the pipe material.)
2. Regulatory bodies may impose additional limits to those specified above—owner companies should identify any such further constraints and ensure the adequacy of the design.

Pipeline coating

Coatings are applied to provide a corrosion barrier and an abrasion barrier. Directional crossings generally encounter varying materials and often can be exposed to extra abrasion during the pullback. An outer abrasion-resistant overcoat is often warranted. To facilitate the pullback of the pipeline, the coating should bond well to the pipe to resist soil stresses and have a smooth, hard surface to reduce friction and maintain the corrosion barrier. As in any pipeline construction, the recommended external coating system should be compatible with any specifications for the field joint coating or any internal coating.

The recommended pipe coating is mill-applied fusion bonded epoxy (FBE). The recommended minimum thickness is 20 mils.

Joint Coating—The coating application of the weld area is the most critical field operation to maintain a smooth abrasion-resistant pipe string. It is recommended that the girth weld be coated with FBE powder utilizing the induction heating coil and powder application machine to a minimum dry film thickness of 25 mils. As an alternate, two component catalyzed liquid epoxy may be applied to the girth weld area to a minimum dry film thickness of 25 mils using a paintbrush or roller. Tape should never be used for joint coating on the pullback portion of a directional crossing.

Coating Repair—It is recommended that small coating damaged areas be repaired with a polymeric melt stick patching material. Coating Repair—It is recommended that small coating damaged areas be repaired with a polymeric melt stick patching material. Holidays larger than 1 in in diameter should be repaired utilizing the two-component catalyzed liquid epoxy applied with a paintbrush or roller. Tapes should never be used for repair of coating damaged areas on the pullback portion of a directional crossing. Shrink sleeves designed for directional drilling may be used to repair holidays in the coating.

Abrasion—Resistant Overcoat—As an extra abrasion-resistant barrier for crossings that may encounter stones. Abrasion-Resistant Overcoat—As an extra abrasion-resistant barrier for crossings that may encounter stones, boulders, or solid rock, it is recommended the FBE-coated pipeline be overcoated with an epoxy-based polymer concrete. The material should be applied at a mill or with a portable yard coating machine to a minimum thickness of 40 mils. Girth weld and coating damaged areas should be field coated using an epoxy-based polymer concrete compatible with the overcoat material. The patch material should be applied so that the material tapers uniformly and feathers into the original coating. Stability of the pipeline in drilled crossings is not normally a concern, so a Portland cement-type concrete coating is not recommended.

Drilling slurry—containment, recycling, and disposal

The directional crossing process requires the use of large volumes of slurry that provide the following functions. The directional crossing process requires the use of large volumes of slurry that provide the following functions:

1. Hydraulic cutting with a jet.
2. Provide energy to the drill motor.
3. Lubricate the cutting head.
4. Transport drill cuttings to the surface.
5. Stabilize the hole against collapse.
6. Guard against loss of slurry into surrounding formations.

Slurry Composition—The slurries most commonly used are bentonite-based. Bentonite is a naturally occurring Wyoming clay known for its hydrophilic characteristics. Often polymer extenders are also added to enhance certain characteristics. Material Safety Data Sheets (MSDS) are readily available from suppliers and can be presented to regulatory/disposal authorities.

Containment—The slurry is pumped downhole and circulates back to the surface and collected in "return pits." These pits typically have a volume of at least 500 cu. ft. Depending on the nature of the project, the slurry is pumped from the return pits to a "settling and containment pit." These pits vary in size depending on pumping rates and contain the slurry for recycling or disposal.

Recycling Slurry—Slurry that has been circulated downhole and collected in the containment pit is then passed through machinery that separates the cuttings from the slurry. This process involves a series of shaking sieves and various size hydroclones.

Slurry and Cuttings Disposal—Significant amounts of slurry are normally disposed of at the end of a project. Economics for disposal is extremely site specific. This slurry can be disposed of by:

1. Use at another drilling location.
2. Spread onto raw land for water retention improvement.
3. Evacuate to a dumpsite.

If working in an area of contaminated ground. If working in an area of contaminated ground, the slurry should be tested for contamination and disposed of in a manner that meets governmental requirements.

Cost mitigation for the owner

With prebid planning and research. With prebid planning and research, the owner can realize significant savings in slurry disposal. It is in the owner's interest to define and specify all disposal issues. In particular:

1. Define an approved disposal site as part of the project specifications.
2. Because it is difficult to estimate disposal quantities, disposal should be a separate bid item as either "cost plus" or on "unit rates."
3. Inadvertent returns are not uncommon and difficult to predict. The issue should be fairly represented to permitting bodies prior to construction. Contingency plans for containment and disposal of inadvertent returns should be priced as a separate bid item and agreed prior to construction.

Conditions of the contract

Always utilize a written contract to maximize communication and minimize controversy. Always utilize a written contract to maximize communication and minimize controversy. A contract should be used to anticipate what the parties intend to do if a problem occurs on the job. The contract should be readable and understandable.

Bid package

A proposal presented by a contractor to the owner is an offer by the contractor that becomes a binding contract if accepted by the owner. A proposal presented by a contractor to the owner is an offer by the contractor that becomes a binding contract if accepted by the owner. The parties, price, and performance must be specified. Define the project to be undertaken by detailing the scope of work and incorporate all plans and specifications from the bid package.

Differing ground conditions and walkaway provision

Owners should accept the responsibility of performing an adequate geotechnical investigation. Owners should accept the responsibility of performing an adequate geotechnical investigation. Despite adequate testing of ground conditions, unknown, unusual, and unexpected ground conditions may be encountered. The contract should provide solutions when the project encounters differing ground conditions. The walkaway provision in the contract should entitle the contractor to stop work and walk away from the job without the owner having the right to take over the contractor's equipment. The contractor should be entitled to receive compensation for demobilization, lost profits, and work performed prior to walking away. If the project is completed, the contractor should be paid on a cost-plus basis. Assumption of risk of unforeseen ground conditions by the contractor affects the bid price.

Environmental concerns

Before the project begins. Before the project begins, address environmental concerns because owners and contractors are included as potentially responsible parties when environmental damages and clean-up costs are assessed. Federal, state, and local laws must be evaluated, and licensing, permitting, and other regulations must be followed. Directional crossings that damage soil or water may cause liability.

1. Turbidity of Water and Inadvertent Returns—As these events are difficult to predict and work stoppage may occur, the contract should offer a mechanism to mutually address and mitigate the problem. Liabilities are generally shared by both the contractor and owner and many times can be insured.
2. Slurry Disposal—Comply with the regulations of the area regarding slurry disposal. Slurry disposal should be referred to in the contract and bid as a separate line item on a cost plus or unit price basis.

Allocation of risk of loss

Evaluate and allocate risks of loss that may occur during the project. Evaluate and allocate risks of loss that may occur during the project. Owners should share the risk of loss rather than shifting all the losses through the indemnification to the contractor, because the bid price is directly affected by contingent losses. Insurance may provide coverage by third parties for losses from differing ground conditions or environmental losses.

Dispute resolution

Provide for dispute resolution in the event of controversy by including mediation or arbitration provisions in the contract, 1940 Provide for dispute resolution in the event of controversy by including mediation or arbitration provisions in the contract. Disputes should be resolved in the following order: (1) negotiation, (2) nonbinding mediation through a third party, (3) landing arbitration, and lastly, (4) litigation. Determine whom should be parties to the resolution, what law will be used, and where the dispute will be resolved.

A Contract Law Seminar notebook with recommended model provisions is available from the office of the DCCA.

This section is provided courtesy of:
Directional Crossing Contractors Association
One Galleria Tower, Suite 1940
13555 Noel Road, Lock Box 39
Dallas, TX 75240-6613
TEL 972-386-9545 *FAX 972-386-9547
http://www.dcca.org

3: Pipe Design

Steel pipe design

The maximum allowable design pressure stress will depend upon the intended service for the pipeline.

Pipelines to be used for transporting liquid petroleum are covered by ANSI/ASME B31.4—"Liquid Petroleum Transportation Piping Systems." Pipelines used for transporting gas are covered by ANSI/ASME B31.8—"Gas Transmission and Distribution Systems."

Pipelines that must be operated in compliance with the Federal Pipeline Safety Regulations will also need to comply with the applicable parts of these regulations: Part 192 for gas transportation systems and Part 195 for liquid transportation systems.

Gas pipelines—ANSI/ASME B31.8

The maximum allowable pressure is calculated by the following equation:

$$P = (2St/D) \times F \times E \times T$$

where

P = Design pressure, lb/in^2,
S = Specified minimum yield strength, lb/in^2 (see Table 1),
t = Nominal wall thickness, in,
D = Nominal outside diameter, in,
F = Design factor (see Table 2),
E = Longitudinal joint factor (see Table 3),
T = Temperature derating factor (see Table 4).

Class location definitions may be obtained from p. 192.111 of Part 192 of the Federal Pipeline Safety Regulations. A typical calculation is as follows:

Pipe: 16″ OD × 0.250″ wt API 5LX X52 ERW
Location: Class 1, therefore $F = 0.72$ (see Table 2)
Temperature: 90 °F, Temp. factor $T = 1$ (see Table 4)
Joint Factor: $E = 1.0$ (see Table 3)

$$P = (2 \times 52,000 \times 0.250/16.0) \times 0.72 \times 1 \times 1$$

$$P = 1170 \text{ lb/in}^2 \text{ gauge}$$

Table 1
Specified Minimum Yield Strength for Steel and Iron Pipe Commonly Used in Piping System

Specification	Grade	Type[1]	SMYS, psi
API 5L	A25..	BW, ERW, S	25,000
API 5L	A...	ERW, FW, S, DSA	30,000
API 5L	B...	ERW, FW, S, DSA	35,000
API 5LS [Note (2)]	A...	ERW, DSA	30,000
API 5LS	B...	ERW, DSA	35,000
API 5LS	X42...	ERW, DSA	42,000
API 5LS	X46...	ERW, DSA	46,000
API 5LS	X52...	ERW, DSA	52,000
API 5LS	X56...	ERW, DSA	56,000
API 5LS	X60...	ERW, DSA	60,000
API 5LS	X65...	ERW, DSA	65,000
API 5LS	X70...	ERW, DSA	70,000
API 5LX [Note (2)]	X42...	ERW, FW, S, DSA	42,000
API 5LX	X46...	ERW, FW, S, DSA	46,000
API 5LX	X52...	ERW, FW, S, DSA	52,000
API 5LX	X56...	ERW, FW, S, DSA	56,000
API 5LX	X60...	ERW, FW, S, DSA	60,000

Table 1
Specified Minimum Yield Strength for Steel and Iron Pipe Commonly Used in Piping System (continued)

Specification	Grade	Type[1]	SMYS, psi
API 5LX	X65...	ERW, FW, S, DSA	65,000
API 5LX	X70...	ERW, FW, S, DSA	70,000
ASTM A53	Open Hrth. Bas. Oxy., Elec. Furn.............................	BW	25,000
ASTM A53	Bessemer..	BW	30,000
ASTM A53	A..	ERW, S	30,000
ASTM A53	B..	ERW, S	35,000
ASTM A106	A..	S	30,000
ASTM A106	B..	S	35,000
ASTM A106	C..	S	40,000
ASTM A134	—...	EFW	[Note (3)]
ASTM A135	A..	ERW	30,000
ASTM A135	B..	ERW	35,000
ASTM A139	A..	ERW	30,000
ASTM A139	B..	ERW	35,000
ASTM A333	1..	S, ERW	30,000
ASTM A333	3..	S, ERW	35,000
ASTM A333	4..	S	35,000
ASTM A333	6..	S, ERW	35,000
ASTM A333	7..	S, ERW	35,000
ASTM A333	8..	S, ERW	75,000

Reproduced from ANSI/ASME Code B31-8-1982, Appendix D. Reprinted courtesy of The American Society of Mechanical Engineers.

Table 2
Values of Design Factor F

Construction Type (See 841.151)	Design Factor F
Type A	0.72
Type B	0.60
Type C	0.50
Type D	0.40

Reproduced from ANSI/ASME Code B31-8-1982, Table 841.1A. Reprinted courtesy of The American Society of Mechanical Engineers.

Table 4
Temperature Derating Factor T for Steel Pipe

Temperature, °F	Temperature Derating Factor T
250 or less	1.000
300	0.967
350	0.933
400	0.900
450	0.867

Note: For intermediate temperatures, interpolate for derating factor. Reproduced from ANSI/ASME Code B31-8-1982, Table 841.1C. Reprinted courtesy of The American Society of Mechanical Engineers.

Table 3
Longitudinal Joint Factor E

Spec. Number	Pipe Class	E Factor
ASTM A53	Seamless	1.00
	Electric resistance welded	1.00
	Furnace welded	0.60
ASTM A106	Seamless	1.00
ASTM A134	Electric fusion arc welded	0.80
ASTM A135	Electric resistance welded	1.00
ASTM A139	Electric fusion welded	0.80
ASTM A211	Spiral welded steel pipe	0.80
ASTM A381	Double submerged arc welded	1.00
ASTM A671	Electric fusion welded	1.00*
ASTM A672	Electric fusion welded	1.00*
API 5L	Seamless	1.00
	Electric resistance welded	1.00
	Electric flash welded	1.00
	Submerged arc welded	1.00
	Furnace butt welded	0.60
API 5LX	Seamless	1.00
	Electric resistance welded	1.00
	Electric flash welded	1.00
	Submerged arc welded	1.00
API 5LS	Electric resistance welded	1.00
	Submerged welded	1.00

Note: Definitions for the various classes of welded pipe are given in 804.243.
*Includes Classes 12, 22, 32, 42, and 52 only.
Reproduced from ANSI/ASME Code B31-8-1982, Table 841.1B. Reprinted courtesy of The American Society of Mechanical Engineers.

Liquid pipelines—ANSI/ASME B31.4

The internal design pressure is determined by using the following formula:

$$P = (2St/D) \times E \times F$$

where P = Internal design pressure, lb/in^2 gauge,
  S = Specified minimum yield strength, lb/in^2 (see Table 5),
  t = Nominal wall thickness, in,
  D = Nominal outside diameter of the pipe, in,
  E = Weld joint factor (see Table 6),
  F = Design factor of 0.72.

Note: Refer to p. 195.106 of Part 195 Federal Pipeline Safety Regulations for design factors to be used on offshore risers and platform piping and cold worked pipe.

A typical calculation of the internal design pressure is as follows:

Pipe: 26″ OD × 0.3125″ wt API 5LX X52 ERW
Weld joint factor $E = 1.0$ (see Table 6)
Design factor $F = 0.72$

$$P = (2 \times 52,000 \times 0.3125/26) \times 1 \times 0.72$$

$$P = 900 \text{ lb/in}^2 \text{ in gauge}$$

Table 5
Tabulation of Examples of Allowable Stresses for Reference Use in Piping Systems

Allowable stress values (*S*) shown in this table are equal to 0.72 × E (weld joint factor) × specified minimum yield strength of the pipe. Allowable stress values shown are for new pipe of known specification. Allowable stress values for new pipe of unknown specification, ASTM A 120 specification or used (reclaimed) pipe shall be determined in accordance with 402.3.1.
For some code computations, particularly with regard to branch connections [see 404.3.1 (d) (3)] and expansion, flexibility, structural attachments, supports, and restraints (Chapter II, Part 5), the weld joint factor *E* need not be considered.
For specified minimum yield strength of other grades in approved specifications, refer to that particular specification.
Allowable stress value for cold worked pipe subsequently heated to 600 °F (300 °C) or higher (welding excepted) shall be 75% of value listed in table.
Definitions for the various types of pipe are given in 400.2.
(Metrics stress levels are given in MPa [1 Megapascal = 1 million pascals])

Specification	Grade	Specified Min Yield Strength psi (MPa)	Notes	(E) Weld Joint Factor	(S) Allowable Stress Value −20 °F to 250 °F (−30 °C–120 °C) psi (MPa)
Seamless					
API 5L	A25	25,000 (172)	(1)	1.00	18,000 (124)
API 5L, ASTM A53, ASTM A106	A	30,000 (207)	(1) (2)	1.00	21,600 (149)
API 5L, ASTM A53, ASTM A106	B	35,000 (241)	(1) (2)	1.00	25,200 (174)
ASTM A106	C	40,000 (278)	(1) (2)	1.00	28,800 (199)
ASTM A524	I	35,000 (241)	(1)	1.00	25,200 (174)
ASTM A524	II	30,000 (207)	(1)	1.00	21,600 (149)
API 5LU	U80	80,000 (551)	(1) (4)	1.00	57,600 (397)
API 5LU	U100	100,000 (689)	(1) (4)	1.00	72,000 (496)
API 5LX	X42	42,000 (289)	(1) (2) (4)	1.00	30,250 (208)
API 5LX	X46	46,000 (317)	(1) (2) (4)	1.00	33,100 (228)
API 5LX	X52	52,000 (358)	(1) (2) (4)	1.00	37,450 (258)
API 5LX	X56	56,000 (386)	(1) (4)	1.00	40,300 (278)
API 5LX	X60	60,000 (413)	(1) (4)	1.00	43,200 (298)
API 5LX	X65	65,000 (448)	(1) (4)	1.00	46,800 (323)
API 5LX	X70	70,000 (482)	(1) (4)	1.00	50,400 (347)
Furnace Welded-Butt Welded					
ASTM A53		25,000 (172)	(1) (2)	0.60	10,800 (74)
API 5L Class I & Class II	A25	25,000 (172)	(1) (2) (3)	0.60	10,800 (74)
API 5L (Bessemer),		30,000 (207)	(1) (2) (5)	0.60	12,950 (89)
ASTM A53 (Bessemer)					
Furnace Welded-Lap Welded					
API 5L Class I		25,000 (172)	(1) (2) (6)	0.80	14,400 (99)
API 5L Class II		28,000 (193)	(1) (2) (6)	0.80	16,150 (111)

Table 5
Tabulation of Examples of Allowable Stresses for Reference Use in Piping Systems (continued)

Specification	Grade	Specified Min Yield Strength psi (MPa)	Notes	(E) Weld Joint Factor	(S) Allowable Stress Value −20 °F to 250 °F (−30 °C−120 °C) psi (MPa)
API 5L (Bessemer)		30,000 (207)	(1) (2) (6)	0.80	17,300 (119)
API 5L electric		25,000 (172)	(1) (2) (6)	0.80	14,400 (99)
Furnace					
Electric Resistance Welded and Electric Flash Welded					
API 5L	A25	25,000 (172)	(1) (7)	1.00	18,000 (124)
API 5L, ASTM A53, ASTM A135	A	30,000 (207)	(2)	0.85	18,360 (127)
API 5L, API 5LS, ASTM A53, ASTM A135	A	30,000 (207)	(1)	1.00	21.600 (149)
API 5L, ASTM A53, ASTM A135	B	35,000 (241)	(2)	0.85	21,420 (148)
API 5L, API 5LS, ASTM A53, ASTM A135	B	35,000 (241)	(1)	1.00	25,200 (174)
API 5LS, API 5LX	X42	42,000 (289)	(1) (2) (4)	1.00	30,250 (208)
API 5LS, API 5LX	X46	46,000 (317)	(1) (2) (4)	1.00	33,100 (228)
API 5LS, API 5LX	X52	52,000 (358)	(1) (2) (4)	1.00	37,450 (258)
API 5LS, API 5LX	X56	56,000 (386)	(1) (4)	1.00	40,300 (279)
API 5LS, API 5LX	X60	60,000 (413)	(1) (4)	1.00	43,200 (297)
API 5LS, API 5LX	X65	65,000 (448)	(1) (4)	1.00	46,800 (323)
API 5LS, API 5LX	X70	70,000 (482)	(1) (4)	1.00	50,400 (347)
API 5LU	U80	80,000 (551)	(1) (4)	1.00	57,600 (397)
API 5LU	U100	100,000 (689)	(1) (4)	1.00	72,000 (496)
Electric Fusion Welded					
ASTM A 134	—	—		0.80	—
ASTM A 139	A	30,000 (207)	(1) (2)	0.80	17,300 (119)
ASTM A 139	B	35,000 (241)	(1) (2)	0.80	20,150 (139)
ASTM A 155	—	—	(2) (8)	0.90	—
ASTM A 155	—	—	(1) (8)	1.00	—
Submerged Arc Welded					
API 5L, API 5LS	A	30,000 (207)	(1)	1.00	21,600 (149)
API 5L, API 5LS	B	35,000 (241)	(1)	1.00	25,200 (174)
API 5LS, API 5LX	X42	42,000 (289)	(1) (2) (4)	1.00	30,250 (208)
API 5LS, API 5LX	X46	46,000 (317)	(1) (2) (4)	1.00	33,100 (228)
API 5LS, API 5LX	X52	52,000 (358)	(1) (2) (4)	1.00	37,450 (258)
API 5LS, API 5LX	X56	56,000 (386)	(1) (4)	1.00	40,300 (278)
API 5LS, API 5LX	X60	60,000 (413)	(1) (4)	1.00	43,200 (298)
API 5LS, API 5LX	X65	65,000 (448)	(1) (4)	1.00	46,800 (323)
API 5LS, API 5LX	X70	70,000 (482)	(1) (4)	1.00	50,400 (347)
API 5LU	U80	80,000 (551)	(1) (4)	1.00	57,600 (397)
API 5LU	U100	100,000 (689)	(1) (4)	1.00	72,000 (496)
ASTM A 381	Y35	35,000 (241)	(1) (2)	1.00	25,200 (174)
ASTM A 381	Y42	42,000 (290)	(1) (2)	1.00	30,250 (209)
ASTM A 381	Y46	46,000 (317)	(1) (2)	1.00	33,100 (228)
ASTM A 381	Y48	48,000 (331)	(1) (2)	1.00	34,550 (238)
ASTM A 381	Y50	50,000 (345)	(1)	1.00	36,000 (248)
ASTM A 381	Y52	52,000 (358)	(1)	1.00	37,450 (258)
ASTM A 381	Y60	60,000 (413)	(1)	1.00	43,200 (298)
ASTM A 381	Y65	65,000 (448)	(1)	1.00	46,800 (323)

Notes: (1) Weld joint factor E (see Table 402.4.3) and allowable stress value are applicable to pipe manufactured after 1958; (2) Weld joint factor E (see Table 402.4.3) and allowable stress value are applicable to pipe manufactured before 1959; (3) Class II produced under API 5L 23rd Edition, 1968, or earlier has a specified minimum yield strength of 28,000 psi (193 MPa); (4) Other grades provided for in API 5LS, API 5LU, and API 5LX not precluded; (5) Manufacture was discontinued and process deleted from API 5L in 1969; (6) Manufacture was discontinued and process deleted from API 5L in 1962; (7) A25 is not produced in electric flash weld; (8) See applicable plate specification for yield point and refer to 402.3.1 for calculation of (S). Reproduced from ANSI/ASME Code B31-4-1979, Table 402.3.1 (a). Reprinted courtesy of The American Society of Mechanical Engineers.

Table 6
Weld Joint Factor

Specification Number	Pipe Type (1)	Weld Joint Factor E	
		Pipe Mfrd. Before 1959	Pipe Mfrd. After 1958
ASTM A53	Seamless	1.00	1.00
	Electric-resistance-welded	0.85 (2)	1.00
	Furnace lap-welded	0.80	0.80
	Furnace butt-welded	0.60	0.60
ASTM A106	Seamless	1.00	1.00
ASTM A134	Electric-fusion (arc)-welded single or double pass	0.80	0.80
ASTM A135	Electric-resistance-welded	0.85 (2)	1.00
ASTM A139	Electric-fusion-welded single or double pass	0.80	0.80
ASTM A155	Electric-fusion-welded	0.90	1.00
ASTM A381	Electric-fusion-welded, double submerged arc-welded	—	1.00
API 5L	Seamless	1.00	1.00
	Electric-resistance-welded	0.85 (2)	1.00
	Electric-flash-welded	0.85 (2)	1.00
	Electric-induction-welded	—	1.00
	Submerged arc-welded	—	1.00
	Furnace lap-welded	0.80	0.80 (3)
	Furnace butt-welded	0.60	0.60
API 5LS	Electric-resistance-welded	—	1.00
	Submerged arc-welded	—	1.00
API 5LX	Seamless	1.00	1.00
	Electric-resistance-welded	1.00	1.00
	Electric-flash-welded	1.00	1.00
	Electric-induction-welded	—	1.00
	Submerged arc-welded	1.00	1.00
API 5LU	Seamless	—	1.00
	Electric-resistance-welded	—	1.00
	Electric-flash-welded	—	1.00
	Electric-induction-welded	—	1.00
	Submerged arc-welded	—	1.00
Known	Known	(4)	(5)
Unknown	Seamless	1.00 (6)	1.00 (6)
Unknown	Electric-resistance or flash-welded	0.85 (6)	1.00 (6)
Unknown	Electric-fusion-welded	0.80 (6)	0.80 (6)
Unknown	Furnace lap-welded or over NPS 4	0.80 (7)	0.80 (7)
Unknown	Furnace butt-welded or NPS 4 and smaller	0.60 (8)	0.60 (8)

Notes: (1) Definitions for the various pipe types (weld joints) are given in 400.2; (2) A weld joint factor of 1.0 may be used for electric-resistance-welded or electric-flash-welded pipe manufactured prior to 1959 where (a) pipe furnished under this classification has been subjected to supplemental tests and/or heat treatments as agreed to by the supplier and the purchaser, and such supplemental tests and/or heat treatment demonstrate the strength characteristics of the weld to be equal to the minimum tensile strength specified for the pipe, or (b) pipe has been tested as required for a new pipeline in accordance with 437.4.1; (3) Manufacture was discontinued and process deleted from API 5L in 1962; (4) Factors shown above for pipe manufactured before 1959 apply for new or used (reclaimed) pipe, if pipe specification and pipe type are known and it is known that the pipe was manufactured before 1959, or not known whether it was manufactured after 1958; (5) Factors shown above for pipe manufactured after 1958 apply for new or used (reclaimed) pipe if pipe specification and pipe type are known and it is known that the pipe was manufactured after 1958; (6) Factor applies for new or used pipe of unknown specification and ASTM A120 if type of weld joint is known; (7) Factor applies for new or used pipe of unknown specification and ASTM A120 if type of weld joint is known to be furnace lap-welded, or for pipe over NPS 4 if type of joint is unknown; (8) Factor applies for new or used pipe of unknown specification and ASTM A120 if type of weld joint is known to be furnace butt-welded, or for pipe NPS 4 and smaller if type of joint is unknown.
Reproduced from ANSI/ASME Code B31-4-1979, Table 402.4.3. Reprinted courtesy of The American Society of Mechanical Engineers.

Properties of pipe

D	t	d	W	A_o	V_m	A_f	I	Z
4.500	0.125	4.250	5.840	1.178097	92.644911	0.098516	4.12	1.83
	0.156	4.188	7.240		89.961575	0.095662	5.03	2.24
	0.172	4.156	7.950		88.592057	0.094206	5.49	2.44
	0.188	4.124	8.660		87.233042	0.092761	5.93	2.64
	0.219	4.062	10.010		84.629845	0.089993	6.77	3.01
STD	0.237	4.026	10.790		83.136406	0.088405	7.56	3.36
6.625	0.188	6.249	12.93	1.734421	200.292532	0.212985	19.7	6.0
	0.219	6.187	14.99		196.337808	0.208779	22.6	6.8
	0.250	6.125	17.02		192.422518	0.204616	25.5	7.7
	0.280	6.065	18.98		188.671072	0.200627	28.1	8.5
STD	0.312	6.001	21.04		184.710235	0.196415	30.9	9.3
	0.375	5.875	25.03		177.035128	0.188254	36.1	10.9
8.625	0.188	8.249	16.94	2.258020	349.016785	0.371133	44.4	10.3
	0.219	8.187	19.66		343.790038	0.365575	51.1	11.9
	0.250	8.125	22.36		338.602723	0.360059	57.7	13.4
	0.277	8.071	24.70		334.116868	0.355289	63.4	14.7
	0.312	8.001	27.70		328.346391	0.349153	70.5	16.3
	0.322	7.981	28.56		326.706916	0.347410	72.5	16.8
STD	0.344	7.937	30.43		323.114514	0.343590	76.9	17.8
	0.375	7.875	33.05		318.086203	0.338243	82.9	19.2
10.750	0.219	10.312	24.63	2.814343	545.418061	0.579980	87.0	16.2
	0.250	10.250	28.04		538.879221	0.573027	113.7	21.2
	0.279	10.192	31.20		532.797939	0.566560	125.9	23.4
	0.307	10.136	34.24		526.959102	0.560352	137.5	25.6
STD	0.365	10.020	40.49		514.966704	0.547599	160.8	29.9
	0.500	9.750	54.74		487.587921	0.518486	212.0	39.4
12.750	0.219	12.312	29.31	3.337942	777.500935	0.826770	169.3	26.6
	0.250	12.250	33.38		769.690071	0.818464	191.9	30.1
	0.281	12.188	37.43		761.918639	0.810200	214.1	33.6
	0.312	12.126	41.45		754.186639	0.801978	236.0	37.0
	0.375	12.000	49.57		738.594720	0.785398	279.0	43.8
STD	0.438	11.874	57.60		723.165661	0.768991	321.0	50.4
	0.500	11.750	65.42		708.140511	0.753014	363.0	56.7
14.000	0.219	13.562	32.24	3.665191	943.389822	1.003171	226.0	32.3
	0.250	13.500	36.72		934.783943	0.994020	255.0	36.5
	0.281	13.438	41.18		926.217495	0.984910	285.1	40.7
	0.312	13.376	45.62		917.690481	0.975843	315.0	45.0
	0.344	13.312	50.18		908.929763	0.966527	344.3	49.2
	0.375	13.250	54.58		900.482886	0.957545	373.0	53.3
STD	0.438	13.124	63.45		883.438151	0.939420	429.0	61.4
	0.500	13.000	72.10		866.822970	0.921752	484.0	69.1

(table continued on next page)

D	t	d	W	A_o	V_m	A_f	I	Z
16.000	0.219	15.562	36.92	4.188790	1242.151387	1.320864	338.0	42.3
	0.250	15.500	42.06		1232.273483	1.310360	384.0	48.0
	0.281	15.438	47.18		1222.435011	1.299899	429.0	53.6
	0.312	15.376	52.28		1212.635972	1.289479	474.0	59.3
	0.344	15.312	57.53		1202.562197	1.278766	519.0	64.8
STD	0.375	15.250	62.59		1192.843296	1.268432	562.0	70.3
	0.438	15.124	72.81		1173.213479	1.247558	649.0	81.1
	0.500	15.000	82.78		1154.054250	1.227185	732.0	91.5
18.000	0.250	17.500	47.40	4.712389	1570.796063	1.670335	549.0	61.0
	0.281	17.438	53.18		1559.685567	1.658520	614.2	68.2
	0.312	17.376	58.95		1548.614504	1.646747	679.0	75.5
	0.344	17.312	64.88		1537.227671	1.634639	744.0	82.7
	0.375	17.250	70.60		1526.236746	1.622952	807.0	89.6
STD	0.438	17.124	82.16		1504.021848	1.599329	932.0	103.6
	0.500	17.000	93.46		1482.318570	1.576250	1053.0	117.0
20.000	0.250	19.500	52.74	5.235988	1950.351683	2.073942	756.6	75.7
	0.281	19.438	59.19		1937.969163	2.060775	846.5	84.6
	0.312	19.376	65.61		1925.626075	2.047650	935.5	93.5
	0.344	19.312	72.22		1912.926185	2.034145	1026.5	102.6
	0.375	19.250	78.61		1900.663236	2.021105	1113.8	111.4
STD	0.438	19.124	91.52		1875.863256	1.994733	1288.6	128.9
	0.500	19.000	104.14		1851.615930	1.968950	1457.2	145.7
24.000	0.250	23.500	63.42	6.283185	2832.562043	3.012056	1315.7	109.6
	0.281	23.438	71.19		2817.635474	2.996184	1473.1	122.8
	0.312	23.376	78.94		2802.748338	2.980354	1629.2	135.8
	0.344	23.312	86.92		2787.422334	2.964056	1789.1	149.1
	0.375	23.250	94.63		2772.615336	2.948311	1942.8	161.9
STD	0.438	23.124	110.23		2742.645193	2.916442	2251.3	187.6
	0.500	23.000	125.51		2713.309770	2.885247	2550.0	212.5
26.000	0.250	25.500	676	6.806784	3335.216783	3.546564	1676.8	129.0
	0.281	25.438	77.20		3319.018190	3.529339	1878.0	144.5
	0.312	25.376	85.61		3302.859030	3.512155	2077.7	159.8
	0.344	25.312	94.27		3286.219968	3.494462	2282.3	175.6
	0.375	25.250	102.64		3270.140946	3.477364	2479.1	190.7
STD	0.438	25.124	119.59		3237.585722	3.442746	2874.5	221.1
	0.500	25.000	136.19		3205.706250	3.408846	3257.8	250.6
30.000	0.250	29.500	79.44	7.853982	4463.625383	4.746477	2585.8	172.4
	0.281	29.438	89.20		4444.882741	4.726547	2897.5	193.2
	0.312	29.376	98.94		4426.179533	4.706659	3207.1	213.8
	0.344	29.312	108.97		4406.914357	4.686173	3524.7	235.0
	0.375	29.250	118.66		4388.291286	4.666370	3830.4	255.4
STD	0.438	29.124	138.30		4350.565898	4.626254	4445.7	296.4
	0.500	29.000	157.55		4313.598330	4.586943	5043.5	336.2

(table continued on next page)

D	t	d	W	A_o	V_m	A_f	I	Z
32.000	0.250	31.500	84.78	8.377580	5089.379243	5.411884	3143.2	196.4
	0.281	31.438	95.20		5069.364577	5.390601	3522.7	220.2
	0.312	31.376	105.60		5049.389344	5.369360	3899.9	243.7
	0.344	31.312	116.32		5028.811111	5.347478	4287.0	267.9
STD	0.375	31.250	126.67		5008.916016	5.326322	4659.7	291.2
	0.438	31.124	147.66		4968.605547	5.283457	5410.3	338.1
	0.500	31.000	168.23		4929.093930	5.241442	6140.2	383.8
34.000	0.250	33.500	90.11	8.901179	5756.166143	6.120924	3775.3	236.0
	0.281	33.438	101.19		5734.879453	6.098289	4231.9	264.5
	0.312	33.376	112.25		5713.632196	6.075695	4685.9	292.9
	0.344	33.312	123.65		5691.740905	6.052416	5151.8	322.0
	0.375	33.250	134.67		5670.573786	6.029908	5600.7	350.0
	0.406	33.188	145.67		5649.446099	6.007442	6047.1	377.9
	0.438	33.124	157.00		5627.678235	5.984294	6505.2	406.6
	0.469	33.062	167.95		5606.630686	5.961913	6946.5	434.2
	0.500	33.000	178.89		5585.622570	5.939574	7385.4	461.6
36.000	0.250	35.500	95.45	9.424778	6463.986083	6.873597	4487.0	280.4
	0.281	35.438	107.20		6441.427369	6.849609	5030.4	314.4
	0.312	35.376	118.92		6418.908087	6.825663	5570.9	348.2
	0.344	35.312	131.00		6395.703739	6.800988	6125.9	382.9
	0.375	35.250	142.68		6373.264596	6.777127	6660.6	416.3
	0.406	35.188	154.34		6350.864885	6.753308	7192.6	449.5
	0.438	35.124	166.35		6327.783964	6.728764	7738.7	483.7
	0.469	35.062	117.97		6305.464391	6.705031	8265.0	516.6
	0.500	35.000	189.57		6283.184250	6.681339	8788.5	549.3
42.000	0.344	41.312	153.04	10.995574	8753.790482	9.308503	9767.7	610.5
	0.375	41.250	166.71		8727.535266	9.280584	10624.3	664.0
	0.406	41.188	180.35		8701.319482	9.252707	11477.1	717.3
	0.438	41.124	194.42		8674.299389	9.223974	12353.3	772.1
	0.469	41.062	208.03		8648.163743	9.196183	13198.3	824.9
	0.500	41.000	221.61		8622.067530	9.168433	14039.4	877.5
48.000	0.344	47.312	175.08	12.566371	11481.174585	12.208716	14625.3	914.1
	0.375	47.250	190.74		11451.103296	12.176739	15912.4	994.5
	0.406	47.188	206.37		11421.071439	12.144805	17194.3	1074.6
	0.438	47.124	222.49		11390.112175	12.111883	18512.4	1157.0
	0.469	47.062	238.08		11360.160456	12.080034	19784.2	1236.5
	0.500	47.000	253.60		11330.248170	12.048226	21050.9	1315.7

D = Outside diameter, ins.; t = Wall thickness, ins.; d = Inside diameter, ins.; W = Pipe weight #/ft.; $I = 0.0491 (D^4 - d^4)$; Z = Section modulus, in^3; A_o = Outside surface area sq ft/ft; V_m = Displacement, bbls/mile; A_f = Flow area, sq ft; I = Moment of Inertia, in^4; $Z = 0.0982 (D^4 - d^4)/D$; Gals/ft = $d^2 \times 0.0408$; Gals/mile = $d^2 \times 215.4240$; Bbls/mile = $d^2 \times 5.1291$.

Length of pipe in bends

Radius of Pipe Bends		90° Bends	180° Bends	270° Bends	360° Bends	540° Bends
Inches	Feet					
1		1½"	3"	4¾"	6¼"	9½"
2		3	6¼	9½	12½	18¾
3	¼	4¾	9½	14¼	18¾	28¼
4		6¼	12½	18¾	25¼	37¾
5		7¾	15¾	23½	31½	47¼
6	½	9½	18¾	28¼	37¾	56½
7		11	22	33	44	66
8		12½	25¼	37¾	50¼	75½
9	¾	14¼	28¼	42½	56½	84¾
10		15¾	31½	47¼	62¾	94¼
11		17¼	34¼	51¾	69	103¾
12	1	18¾	37¾	56½	75½	113
12.5		19¾	39¼	59	78½	117¾
14		22	44	66	88	132
15	1¼	23½	47	70¾	94¼	141¼
16		25¼	50¼	75½	100½	150¾
17.5		27½	55	82½	110	165
18	1½	28¼	56½	84¾	113	169¾
20		31½	62¾	94¼	125¾	188½
21	1¾	33	66	99	132	198
24	2	37¾	75½	113	150¾	226¼
25		39¼	78½	117¾	157	235½
30	2½	47¼	94¼	141¼	188½	282¾
32		50¼	100½	150¾	201	301½
36	3	56½	113	169½	226¼	339¼
40		62¾	125¾	188½	251¼	377
48	4	75½	150¾	226¼	301½	452½
50		78½	157	235½	314¼	471¼
56		88	176	264	351¾	527¾
60	5	94¼	188½	282¾	377	565½
64		100½	201	301½	402	603¼
70		110	220	329¾	439¾	659¾
72	6	113	226¼	339¼	452½	678½
80		125¾	251¼	377	502¾	754
84	7	132	263¾	395¾	527¾	791½
90	7½	141¼	282¾	424	565½	848¼
96	8	150¾	301½	452½	603	904¾
100		157	314¼	471¼	628¼	942½
108	9	169½	339¼	509	678½	1017¾
120	10	188½	377	565½	754	1131
132	11	207¼	414¾	622	829½	1244
144	12	226¼	452½	678½	904¾	1357¼
156	13	245	490	735¼	980¼	1470¼
168	14	263¾	527¾	791½	1055½	1583½
180	15	282¾	565½	848¼	1131	1696½
192	16	301½	603	904¾	1206¼	1809½
204	17	320½	640¾	961¼	1281¾	1922½
216	18	339¼	678½	1017¾	1357¼	2035¾
228	19	358	716¼	1074½	1432½	2148¾
240	20	377	754	1131	1508	2262

To find the length of pipe in a bend having a radius not given above, add together the length of pipe in bends whose combined radii equal the required radius.

Example: Find length of pipe in 90° bend of 5′ 9″ radius.
Length of pipe in 90° bend of 5′ radius = 94¼″
Length of pipe in 90° bend of 9″ radius = 14¼″

Then, length of pipe in 90° bend of 5′ 9″ radius = 108½″

Calculation of pipe bends

Example No. 1—Given A, B, C, R

$D = B - C$
$E = R - A$
$F = \sqrt{D^2 + E^2}$
$\dfrac{E}{F} = \sin \angle G$
$H = \sqrt{F^2 - R^2}$

$\dfrac{R}{F} = \sin \angle P$
$\angle S = \angle P - \angle G$
$\angle K = 90° - \angle S$
$\angle L = \frac{1}{2} \angle S$
$M = \tan \angle L \times R$
$N = H + M$
$O = B - C - M$

Example No. 2—Given A, B, C, R

$D = B - C$
$E = A - R$
$F = \sqrt{D^2 + E^2}$
$\dfrac{E}{F} = \sin \angle G$
$H = \sqrt{F^2 - R^2}$

$\dfrac{R}{F} = \sin \angle P$
$\angle S = \angle P + \angle G$
$\angle K = 90° - \angle S$
$\angle L = \frac{1}{2} \angle S$
$M = \tan \angle L \times R$
$N = H + M$
$O = B - C - M$

Example No. 3—Given A, B, C, R

$D = B - C$
$E = \sqrt{A^2 + D^2}$
$\dfrac{A}{E} = \sin \angle F$

$\angle G = \frac{1}{2} \angle F$
$H = \tan \angle G \times R$
$P = C - H$

Example No. 4—Given A, B, C, R

$D = B - C$
$E = \sqrt{A^2 + D^2}$

$\dfrac{A}{E} = \sin \angle F$
$\angle G = \frac{1}{2} \angle F$

$H = \tan \angle G \times R$
$P = C - H$

Example No. 5—Given A, R

$B = 2R - A$ $C = \sqrt{(2R)^2 - B^2}$ $\dfrac{C}{2R} = \sin \angle D$

Example No. 6—Given A, B

$C = \frac{1}{2} B$
$D = \frac{1}{2} A$
$E = \sqrt{C^2 + D^2}$

$\dfrac{D}{E} = \sin \angle F$
$G = \frac{1}{2} E$

$\angle H = 90° - \angle F$
$R = \dfrac{A^2 + B^2}{4A}$
$L = 2F$

Example No. 7—Given A, B, C, D, R

$E = D - A - B$
$F = R - C$
$G = R + F$
$H = \sqrt{E^2 + G^2}$

$G/H = \sin \angle K$
$L = \frac{1}{2} H$
$M = \sqrt{L^2 - R^2}$

$M/L = \sin \angle N$
$\angle O = 90° - \angle K - \angle N$
$\angle P = \frac{1}{2} \angle O$
$S = \tan \angle P \times R$

*Example No. 8—***Given A, B, C, D, R**

$$E = D - A - B$$
$$F = \sqrt{E^2 + C^2}$$
$$\frac{C}{F} = \sin \angle G$$

$$\angle H = \tfrac{1}{2} \angle G$$
$$K = \tan \angle H \times R$$
$$L = A - K$$
$$P = B - K$$
$$N = F - 2K$$

*Example No. 9—***Given A, B, C, D, R**

$$E = D - A - B$$
$$F = 2R - C$$
$$G = \sqrt{E^2 + F^2}$$
$$\frac{F}{E} = \tan \angle H$$
$$K = \tfrac{1}{2} G$$

$$M = \sqrt{K^2 - R^2}$$
$$\frac{M}{K} = \sin \angle N$$
$$\angle P = 90° - \angle H - \angle N$$
$$\angle O = \tfrac{1}{2} \angle P$$
$$S = \tan \angle O \times R$$

*Example No. 10—***Given A, B, C, D, R**

$$E = D - A - B$$
$$F = \sqrt{C^2 + E^2}$$
$$\frac{C}{F} = \sin \angle G$$

$$\angle H = \tfrac{1}{2} \angle G$$
$$K = \tan \angle H \times R$$
$$M = A - K$$
$$N = B - K$$
$$P = F - 2K$$

*Example No. 11—***Given R and 45° Angles**

$$A = 3.414 \times R$$
$$B = 2.828 \times R$$
$$C = 0.828 \times R$$

$$T = \text{Tangent}$$
Length of pipe in bend =
$$9.425 \times R + 2T$$

*Example No. 12—***Given A, R**

$$C = A - 2R$$
$$D = \sqrt{(2R)^2 - C^2}$$
$$E = D - R$$
$$F = 2E$$

$$P = 2D$$
$$\angle H = 90° + \angle G$$

$$C/2R = \sin \angle G$$
$$\angle H = 90° + \angle G$$
$$\angle K = 180° + 2 \angle G$$

*Example No. 13—***Given A, B, R**

$$C = \tfrac{1}{2} B$$
$$D = R + C$$
$$E = A - 2R$$
$$F = \sqrt{D^2 + E^2}$$

$$E/F = \sin \angle G$$
$$H = \tfrac{1}{2} F$$
$$K = \sqrt{H^2 - R^2}$$
$$K/H = \sin \angle L$$

$$\angle M = \angle G - \angle L$$
$$\angle N = 180° + 2 \angle M$$
$$\angle O = 90° + \angle M$$
$$P = 2D$$

Spacing of pipe supports

For chart on deflection of horizontal pipe lines, see next page

When a horizontal pipeline is supported at intermediate points, sagging of the pipe occurs between these supports, the amount of sag being dependent upon the weight of the pipe, fluid, insulation, and valves or fittings that may be included in the line. If the pipeline is installed with no downward pitch, pockets will be formed in each span, in which case condensation may collect if the line is transporting steam. In order to eliminate these pockets, the line must be pitched downward so that the outlet of each span is lower than the maximum sag.

Crane has conducted tests to determine the deflection of horizontal standard pipelines filled with water, in pipe sizes 3/4 to 4″ inclusive, the results of which have indicated that for pipes larger than 2″ and with supports having center to center dimensions greater than 10 ft, the resultant deflection is less than that determined by the use of the formula for a uniformly loaded pipe fixed at both ends. For pipe sizes 2″ and smaller, the test deflection was in excess of that determined by the formula for pipe having fixed ends and approached, for the shorter spans, the deflection as determined by the use of the formula for pipelines having unrestrained ends.

The deflection of horizontal standard pipelines filled with water, for varying spans, based upon the results obtained from tests for sizes 2″ and smaller and upon the formula for fixed ends for the larger sizes of pipe is described previously. The deflection values given on the chart are twice those obtained from test or calculation, to compensate for any variables including weight of insulation, etc.

The formula given below indicates the vertical distance that the span must be pitched so that the outlet is lower than the maximum sag of the pipe.

$$h = \frac{144^2 y}{36S^2 - y^2}$$

where

h = Difference in elevation of span ends, inches,
S = Length of one span, feet,
y = Deflection of one span, inches.

By eliminating the inconsequential term "$-y^2$" from the denominator, the formula reduces to:

$$h = 4y$$

The pitch of pipe spans, called the Average Gradient, is a ratio between the drop in elevation and the length of the span. This is expressed as so many inches in a certain number of feet.

Reprinted with Permission—Crane Company.

$$\text{Average gradient} = \frac{4y}{S}$$

The dotted lines as shown on the chart on the opposite page are plotted from the above formula and indicate average gradients of 1″ in 10′, 1″ in 15′, 1″ in 20′, 1″ in 30′, and 1″ in 40′.

Example. What is the maximum distance between supports for a 4″ standard pipeline assuming a pitch or average gradient of 1″ in 30 ft?

Using the chart on the opposite page, find the point where the diagonal dotted line for an average gradient of 1″ in 30 ft intersects the diagonal solid line for 4″ pipe. From this point, proceed downward to the bottom line where the maximum span is noted to be approximately 22 ft.

Code for Pressure Piping: The Code for Pressure Piping, ASA B 31.1, makes the following statements relative to installations within the scope of the code:

"**605 (g)** Supports shall be spaced so as to prevent excessive sag, bending and shear stresses in the piping, with special consideration given to those piping sections where flanges, valves, etc., impose concentrated loads. Where calculations are not made, suggested maximum spacing of hangers or supports for carbon steel piping operating at 750 °F and lower are given in Table 21a (*see the table below*)."

"Where greater distance between supports, concentrated loads, higher temperatures, or vibration considerations are involved, special consideration should be given to effects of bending and shear stresses."

"**623 (b)** The design and spacing of supports shall be checked to assure that the sum of the longitudinal stresses due to weight, pressure, and other sustained external loading does not exceed the allowable stress (S value) in the hot condition."

Suggested maximum spacing between pipe supports for straight runs of standard wall and heavier pipe (at maximum operating temperature of 750 °F)

Nominal Pipe Size Inches	Maximum Span Feet	Nominal Pipe Size Inches	Maximum Span Feet
1	7	8	19
1½	9	10	22
2	10	12	23
2½	11	14	25
3	12	16	27
3½	13	18	28
4	14	20	30
5	16	24	32
6	17	—	—

Note: The values in the table do not apply where there are concentrated loads between supports such as flanges, valves, etc.

Spacing is based on a combined bending and shear stress of 1500 psi when pipe is filled with water and the pitch of the line is such that a sag of 0.1 in between supports is permissible.

Spacing of pipe supports

Deflection of horizontal pipe lines
(Based on standard pipe filled with water)

Reprinted with permission—Crane Company.

American standard taper pipe threads (NPT)

$E_0 = D - (0.050D + 1.1) p$

$E_1 = E_0 + 0.0625 L_1$

$L_2 = (0.80D + 6.8) p$

p = Pitch

Depth of thread = $0.80p$

Total Taper ¾ - in.per foot

Tolerance on product

One turn large or small from notch on plug gauge or face of ring gauge.

Notch flush with face of fitting. If chamfered, notch flush with bottom of chamfer.

Flush by hand

Dimensions, in Inches

Nominal pipe size	D Outside diameter of pipe	E₀ Pitch diameter at end of external thread	E₁† Pitch diameter at end of internal thread	L₁¶ Normal engagement by hand between external and internal threads	L₂§ Length of effective thread	p Pitch of thread	Depth of thread	Number of threads Per Inch
⅛	0.405	0.36351	0.37476	0.180	0.2639	0.03704	0.02963	27
¼	0.540	0.47739	0.48989	0.200	0.4018	0.05556	0.04444	18
⅜	0.675	0.61201	0.62701	0.240	0.4078	0.05556	0.04444	18
½	0.840	0.75843	0.77843	0.320	0.5337	0.07143	0.05714	14
¾	1.050	0.96768	0.98887	0.339	0.5457	0.07143	0.05714	14
1	1.315	1.21363	1.23863	0.400	0.6828	0.08696	0.06957	11½
1¼	1.660	1.55713	1.58338	0.420	0.7068	0.08696	0.06957	11½
1½	1.900	1.79609	1.82234	0.420	0.7235	0.08696	0.06957	11½
2	2.375	2.26902	2.29627	0.436	0.7565	0.08696	0.06957	11½
2½	2.875	2.71953	2.76216	0.682	1.1375	0.12500	0.10000	8
3	3.500	3.34062	3.38850	0.766	1.2000	0.12500	0.10000	8
3½	4.000	3.83750	3.88881	0.821	1.2500	0.12500	0.10000	8
4	4.500	4.33438	4.38712	0.844	1.3000	0.12500	0.10000	8
5	5.563	5.39073	5.44929	0.937	1.4063	0.12500	0.10000	8
6	6.625	6.44609	6.50597	0.958	1.5125	0.12500	0.10000	8
8	8.625	8.43359	8.50003	1.063	1.7125	0.12500	0.10000	8
10	10.750	10.54531	10.62094	1.210	1.9250	0.12500	0.10000	8
12	12.750	12.53281	12.61781	1.360	2.1250	0.12500	0.10000	8
14 O.D.	14.000	13.77500	13.87262	1.562	2.2500	0.12500	0.10000	8
16 O.D.	16.000	15.76250	15.87575	1.812	2.4500	0.12500	0.10000	8
18 O.D.	18.000	17.75000	17.87500	2.000	2.6500	0.12500	0.10000	8
20 O.D.	20.000	19.73750	19.87031	2.125	2.8500	0.12500	0.10000	8
24 O.D.	24.000	23.71250	23.86094	2.375	3.2500	0.12500	0.10000	8

†Also pitch diameter at gauging notch.

§Also length of plug gauge.

¶Also length of ring gauge, and length from gauging notch to small end of plug gauge.

British standard taper pipe threads

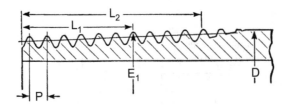

Whitworth 55°
Form of Thread

Total taper
¾-inch per ft

This information is taken
from the british engineering
standards association
report no. 21-1938.

Dimensions, in Inches

Pipe size Nominal Bore	D Approximate outside diameter of pipe	E₁ Pitch diameter at end of internal thread	L₁ Normal engagement by hand between external and internal threads	L₂ Length of effective thread	p Pitch of threads	Depth of threads	Number of threads Per Inch
⅛	¹³/₃₂	0.3601	0.1563	0.2545	0.03571	0.0229	28
¼	¹⁷/₃₂	0.4843	0.2367	0.3814	0.05263	0.0337	19
⅜	¹¹/₁₆	0.6223	0.2500	0.3947	0.05263	0.0337	19
½	²⁷/₃₂	0.7793	0.3214	0.5178	0.07143	0.0457	14
¾	1¹/₁₆	0.9953	0.3750	0.5714	0.07143	0.0457	14
1	1¹¹/₃₂	1.2508	0.4091	0.6591	0.09091	0.0582	11
1¼	1¹¹/₁₆	1.5918	0.5000	0.7500	0.09091	0.0582	11
1½	1²⁹/₃₂	1.8238	0.5000	0.7500	0.09091	0.0582	11
2	2⅜	2.2888	0.6250	0.9204	0.09091	0.0582	11
2½	3	2.9018	0.6875	1.0511	0.09091	0.0582	11
3	3½	3.4018	0.8125	1.1761	0.09091	0.0582	11
3½	4	3.8918	0.8750	1.2386	0.09091	0.0582	11
4	4½	4.3918	1.0000	1.4091	0.09091	0.0582	11
5	5½	5.3918	1.1250	1.5795	0.09091	0.0582	11
6	6½	6.3918	1.1250	1.5795	0.09091	0.0582	11
7	7½	7.3860	1.3750	1.9250	0.10000	0.0640	10
8	8½	8.3860	1.5000	2.0500	0.10000	0.0640	10
9	9½	9.3860	1.5000	2.0500	0.10000	0.0640	10
10	10½	10.3860	1.6250	2.1750	0.10000	0.0640	10
11	11½	11.3700	1.6250	2.3125	0.1250	0.0800	8
12	12½	12.3700	1.6250	2.3125	0.1250	0.0800	8

Normal engagement between male and female threads to make tight joints

Normal engagement

The normal amount of engagement to make a tight joint for various types of screwed material is given in the table. These dimensions have been established from tests made under practical working conditions.

The normal engagement specified for American Standard Pipe Thread and API Line Pipe Thread joints is based on parts being threaded to the American Standard for Pipe Threads or the API Standard for Line Pipe Threads.

In order to obtain the thread engagements listed in the table, it is necessary to vary the torque or power applied according to the size, metal, and weight of material used. For example, it requires considerably less power to make up a screwed joint using a light bronze valve than a high pressure steel valve.

Reprinted with Permission—Crane Company.

American standard and
API line pipe threads

Dimensions, in Inches			
Dimensions Given Do Not Allow for Variations in Tapping or Threading			
Size	A	Size	A
1/8	1/4	2½	15/16
1/4	3/8	3	1
3/8	3/8	3½	1 $1/16$
1/2	1/2	4	1 $1/8$
3/4	9/16	5	1¼
1	11/16	6	1 $5/16$
1 ¼	11/16	8	1 $7/16$
1 ½	11/16	10	1 $5/8$
2	3/4	12	1 $3/4$

Calculate pipe weight, contents, velocity

Pipeline engineers, accountants, and system operators are deeply involved with data concerning line pipe such as:

How much does it weigh (lb/ft, tons/mi)? What are the corresponding metric units?
How much will it contain (bbl/mi, gal/100 ft, cu m/km)?
What are velocities at various feed rates (ft/sec, mi/hr, m/hr)?
Results for the WEIGHT category are calculated by an API formula and given as: Pounds (mass) per ft; kg per m; short tons (2000 lb) per mile; metric tons per km.
The content results are: bbl/mi, gal (U.S.)/100 ft, cu ft/mi, L/m, cu m/km.
Flowrates can be input to give: b/h or b/d; gal (U.S.)/min; or gal/hr or gal/d; cu ft/min, /h, /d; cu m/min, /h, /d.
Velocity results: mph; fps; km/h; m/s.

Basic data calculated by this program can be used for determining the following criteria:

- Weight: Cost of construction, hauling and stringing, investments.
- Content: Amount of water for hydrostatic test, line fill, oil stocks. Amount of one grade to be displaced for another to reach a station, shut-off valve, or terminal. When a segment must be drained, barrels or gallons to be removed.
- Velocity: Time for interface or scraper to traverse distance between two points. Time to accumulate or withdraw a given quantity of fluid. Locating head of column of a grade of oil or a scraper from given station after a certain time. Determining expected time of arrival (ETA) of interface or scraper at a given point.

Formulas

Pipe weight LBM/FT (plain end) is calculated by formula used by API:

$$LBM/FT = 10.68(Dt - t^2) \quad D = \text{Pipe OD},$$

$$t = \text{pipe wall thickness(all in inches)}$$

Weights in other units are found by multiplying by conversion factors:

$$BBL/MI = 5.129131(D - 2t)^2$$

Basic velocity relation:

$$\frac{BBL}{HR} \div \frac{BBL}{MI} = \frac{MI}{HR}$$

This is converted to other units.

Example 1. Calculate Pounds Mass per foot.

$D = 12.750$ in

$t = 0.375$ in

Pounds/foot $= 10.68(12.750 \times 0.375 - 0.375^2)$

$\qquad\qquad\quad = 49.56$ pounds per foot

Example 2. Calculate Barrels per mile (use pipe data from example 1).

BBL/MILE $= 5.129131(12.750 - 2 \times 0.375)^2$

$\qquad\qquad\quad = 738.59$ bbls/mile

Formulas and constants of value in solving problems relating to tubular goods

Area of Metal in Tubular Section:

Area in square inches $= 3.1416(D - t)t$

where D = Outside diameter, in,
$\quad\quad\ t$ = Wall thickness, in.

Capacity

$B_{100} = 0.1237A = 0.0972D^2$

where $\quad B_{100}$ = Barrels per 100 ft,
$\quad\quad\quad\ A$ = Internal area of pipe, in^2
$\quad\quad\quad\ D$ = Internal diameter, in

and $F_B = \dfrac{808.5}{A} = \dfrac{1029.4}{D^2}$

where $\quad F_B$ = Number of feet filled by one barrel,
$\quad\quad\quad\ A$ = Internal area of pipe, in^2,
$\quad\quad\quad\ D$ = Internal diameter, in.

Complete tables giving capacities of different types of tubular goods are given in chapters dealing with these types.
Steel constants[a] (applicable to tubular goods)

One cubic inch $= 0.2833$ lb

One cubic foot $= 489.542$ lb

Specific gravity $= 7.851$

Approximate coefficient of expansion for commercial casing, drill pipe, and tubing is

Coefficient $= 6.9$

$\qquad\qquad \times 10^{-6}$ per $°F$, over range from 0 to 400 $°F$.

Formula for expansion of pipe due to change of temperature is as follows:

$L_t = L_o(1 + 0.0000069t)$

where

$\qquad L_o$ = Length at atmospheric temperature,
$\qquad L_t$ = Length after change of temperature t.

Modulus : Section Modulus $\dfrac{\pi}{64}\left(D^4 - D_1^4\right)$

Polar Modulus $\dfrac{\pi}{32}\left(D^4 - D_1^4\right)$

where

$\qquad D$ = Outside diameter,
$\qquad D_1$ = Inside diameter,
$\qquad \pi = 3.1416$.

Relation between twisting effort, torsional strain, and stress of cylindrical shaft or tube

$$\frac{q}{r} = \frac{T}{J} = \frac{C\theta}{L}$$

---

[a] Engineering data Spang-Chalfant Division of the National Supply Co.

How to calculate the contraction or expansion of a pipeline

Steel pipe contracts or expands approximately 0.8 in per 100 °F temperature change for each 100 ft of pipe. For construction in the United States, slack is not normally needed in welded lines for contraction or expansion unless abnormal variations are encountered. However, slack is often provided near road crossings where the pipe may have to be lowered at some future time.

$$\text{Contraction} = 0.8 \times \frac{\text{length of pipe(feet)}}{100}$$
$$\times \frac{\text{temp change(°F)}}{100}$$

Example. Calculate the amount a 1000-ft section of pipeline would contract if laid at a temperature of 100 °F and cooled to a low of 0 °F during winter operation (assuming the line were free of soil or other resisting loads).

$$\text{Contraction} = 0.8 \times \frac{1000}{100} \times \frac{100}{100} = 0.8 \times 10 \times 1 = 8 \text{ in.}$$

Large temperature changes may arise on exposed line or in lines downstream from compressor stations.

When slack is desired in a line, the amount of sag or rise needed may be estimated quickly. To provide for 0.8 in of movement longitudinally along the pipe, a sag of 21 in would be needed in a 100-ft loop. For a 150-ft loop, 31 in are necessary. Such slack put into the line at time of construction might allow for lowering the pipe under roadbeds without putting excessive stress upon the pipe.

Location of overbends, sags, and side bends should be considered when laying slack loops. When loops are placed, parts of the line should be backfilled to serve as a tie-down. Loops should be lowered during the coolest part of the day when the line is the shortest. After lowering in, sags should rest on the bottom of the ditch; overbends should "ride high." Side bends should rest on the bottom of the ditch and against the outside wall. The line should be lowered so that all sections of the pipe are in compression.

Estimate weight of pipe in metric tons per kilometer

To estimate the weight of pipe in metric tons per kilometer, multiply the nominal diameter by the number of sixteenths of an inch in wall thickness.

Example. Find the weight of pipe in metric tons per kilometer for 20-in diameter pipe, wall thickness 1/4-in.

$4 \times 20 = 80$ tons(metric) per kilometer.

Actual answer from pipe tables is 79 metric tons per kilometer.

This rule of thumb is based on a density of 490 lb. per cubic foot for steel. For larger diameter thin wall pipe, this approximation gives an answer usually about 1% low. The accompanying table provides a comparison between actual weights in metric tons per kilometer, as compared to that calculated by this rule of thumb.

Pipe Dia., in	Wall Thickness, in	Actual Tons, metric	R.O.T. Tons, metric
10	3/16	31.5	30
12	5/16	62	60
16	7/16	109	112
20	1/4	79	80
20	5/16	99	100
20	7/16	137	140
24	5/16	118	120
24	1/2	189	192
28	5/16	139	140
30	5/16	148	150
32	7/16	221	224
36	1/4	143	144
36	1/2	285	288

How to find pipe weight from outside diameter and wall thickness

Weight (pounds per foot) $= (Dt - t^2) \times 10.68$

where

D = outside diameter, in,
t = wall thickness, in.

Example. Outside diameter = 4.500″, Wall thickness = 0.250″

$$W = (4.5 \times 2.5 - 2.5^2) \times 10.68$$
$$= (1.125 - .625) \times 10.68$$
$$= 1.0625 \times 10.68 = 11.35 \text{ lb per ft}$$

The above equation is based on a density of 490 lb. per cubic foot for the steel. High-yield-point, thin-wall pipe may run slightly heavier than indicated.

What is the maximum allowable length of unsupported line pipe?

For Schedule 40 pipe, the maximum span between supports is given by

$$S = 66\sqrt{D}$$

where S is the span length in feet and D is the actual outside diameter of the pipe in inches. For pipe smaller than 12 in, the nominal size may be used, and

$$S = 7\sqrt{N}$$

where N is the nominal outside diameter in inches.

Example. How wide can a ditch be spanned with 4-in line pipe without the use of intermediate supports?

$$S = 66\sqrt{4.5} = 6.6 \times 2.12 = 13.99 (14 \text{ ft})$$

$$S = 7\sqrt{4} = 7 \times 2 = 14 \text{ ft.}$$

Note. The above rule is approximate, but conservative. It does not assume that the pipe is held fixed at either end; this condition can make longer spans safe. Also, it does not guarantee that the pipe will not be damaged by floating logs or debris. This hazard must be evaluated from the conditions encountered.

Identify the schedule number of pipe by direct measurement

Add 3 in to the actual inside diameter; divide this by the wall thickness; the schedule number can then be identified by the following table:

Schedule	Range
30	40 to 50
40	29 to 39
60	25 to 29
80	20 to 23
100	16 to 18
120	13 to 15
140	11 to 13
160	9 to 11

Example. Pipe measures 12.5 in inside diameter, wall thickness 0.75 in; 15.5/0.75 = 20.2; pipe is Schedule 80, 14 in outside diameter.

The rule fails on small-diameter pipe (under 6 in), and also on schedules 10 and 20 because of the use of "standard" thicknesses in this range.

Determine buoyancy of bare steel pipe

Multiply the square of the outside diameter of the pipe, in inches, by 0.341, and subtract the pipe weight in lb/ft; the answer is the buoyancy per foot in fresh water.

Example. 12.750″ OD × 0.250 w.t. (33.#/ft)

$$\text{Bouyancy} = 12.750^2 \times 0.341 - 33.4$$
$$= 162.56 \times 0.341 - 33.4 = 55.4 - 33.4$$
$$= 22 \text{ lb/ft}$$

Because all of the coating materials in common use are heavier than water (but not as much so as steel), coated pipe will have less buoyancy than indicated by the rule. For sea water, use 0.350 for the constant.

Determine buoyancy of bare and concrete-coated steel pipe in water and mud

To find the buoyancy of bare steel and concrete coated pipe in water in lb/ft:
For bare pipe:

$$\text{Bouyancy} (B) = \frac{D}{3}(D - 32t) + 11t^2$$

For coated pipe:

$$\text{Bouyancy} (B) = \frac{D}{3}(D - 32t) + t_1 D \left(\frac{63 - W_c}{48}\right)$$

where

D = outside diameter of pipe, in,
t = wall thickness of pipe, in,
t_1 = thickness of concrete coating, in,
W_c = weight of concrete, lb/ft^3.

To find the buoyancy of bare steel and coated pipe in mud in lb/ft:
For bare pipe:

$$\text{Buoyancy} (B) = 10.7D \left[\left[\frac{DW_m}{2000} - t \right] + 11t^2 \right]$$

For coated pipe:

$$B = 10.7D \left[\left[\frac{DW_m}{2000} - t \right] + t_1 D \left[\frac{W_m - W_c}{48} \right] \right]$$

for coated pipe, where W_m is weight of mud, lb/ft^3.

Example 1. Find the buoyancy of steel pipe with 20″ OD × 0.500″ w.t. steel pipe in water. What thickness of concrete coating will be required to give this pipe a negative buoyancy of 100 lb./ft. ($W_c = 143$)?

$$\text{Buoyancy} (B) = \frac{D}{3}(D - 32t) + 11t_1^2$$

$$\text{Buoyancy} (B) = \frac{20}{3}\left(20 - 32^*0.5\right) + 11^*0.5^2 = 26.7 + 2.8$$

$$= 29.5 \text{ ft positive bouyancy}$$

To give this pipe a negative buoyancy of 100 lb./ft:

$$-100 = \frac{20}{3}\left(20 - 32^*0.5\right) + t_1 20 \left[\frac{63 - 143}{48}\right]$$

$$-100 = 26.7 - 33.3t_1$$

$$t_1 = 3.8 \text{ in of concrete coating.}$$

The error introduced by this method is about 15%, but would be less as the thickness of coating decreases.

Example 2. Bottom conditions at a certain crossing require the pipe to have a negative buoyancy of at least 50 lb./ft. Determine if 10.750″ pipe with 2″ of concrete ($W_c = 143$) has sufficient negative buoyancy

$$B = \frac{10}{3}\left(10 - 32^*0.5 + 2^*10\right)\left[\frac{63 - 143}{48}\right]$$

$$B = -20 - 33.3$$

$$= -53.3 \text{ lb/ft; so coated pipe meets design requirements.}$$

Weights of piping materials

The weight per foot of steel pipe is subject to the following tolerances:

Specification	Tolerance
A.S.T.M.A—53 A.S.T.M.A—120	STD WT +5%, −5%
	XS WT +5%, −5%
	XXS WT +10%, −10%
A.S.T.M.A — 106	SCH 10—120 + 6.5%, −3.5%
	SCH 140—160 + 10%, −3.5%
A.S.T.M.A—158 A.S.T.M.A—206 A.S.T.M.A—280	12″ and under +6.5%, −3.5%
	Over 12″ +10%, −5%
API 5L	All sizes +65%, −3.5%

$$\text{Weight of Tube} = F \times 10.6802 \times T \times (D - T) \text{ lb/ft}$$

where

T = wall thickness, in,
D = outside diameter, in,
F = relative weight factor.

The weight of tube furnished in these piping data is based on low carbon steel weighing 0.2833 lb/in^3

Relative weight factor *F* of various metals

Aluminum = 0.35
Brass = 1.12
Cast Iron = 0.91
Copper = 1.14
Lead = 1.44
Ferritic Stainless Steel = 0.95
Austenitic Stainless Steel = 1.02
Steel = 1.00
Tin = 0.93
Wrought Iron = 0.98

$$\text{Weight of contents of a tube} = G \times 0.3405$$
$$\times (D - 2T)^2 \text{ lb/ft}$$

where

G = Specific gravity of contents,
T = Tube wall thickness, in,
D = Tube outside diameter, in.

The weight of welding tees and laterals is for full size fittings. The weight of reducing fittings is approximately the same as for full size fittings.

The weight of welding reducers is for one size reduction and is approximately correct for other reductions.

Pipe covering temperature ranges are only intended as a guide and do not constitute a recommendation for specific thickness of material.

Pipe covering thicknesses and weights indicate average conditions and include all allowances for wire, cement, canvas, bands, and paint. The listed thicknesses of combination covering is the sum of the inner and the outer layer thickness. When specific inner and outer layer thicknesses are known, add them, and use the weight for the nearest tabulated thickness.

To find the weight of covering on fittings, valves, or flanges, multiply the weight factor (light faced subscript) by the weight per foot of covering used on straight pipe. All flange weights include the proportional weight of bolts or studs required to make up all joints.

Lap joint flange weights include the weight of the lap.

Welding neck flange weights are compensated to allow for the weight of pipe displaced by the flange. Pipe should be measured from the face of the flange.

All flanged fitting weights include the proportional weight of bolts or studs required to make up all joints.

To find the approximate weight of reducing-flanged fittings, subtract the weight of a full size slip-on flange and add the weight of reduced size slip-on flange.

Weights of valves of the same type may vary because of the individual manufacturer's design. Listed valve weights are approximate only. When it is possible to obtain specific weights from the manufacturer, such weights should be used.

To obtain the approximate weight of flanged end steel valves, add the weight of two slip-on flanges of the same size and series to the weight of the corresponding welding-end valves.

Allowable working pressure for carbon steel pipe

Based on formula No. 7 of the ASA Code for Pressure Piping, Section 3B, this table was computed for $6\frac{5}{8}$-in OD to 36-in. OD carbon steel seamless pipe, API Specifications 5L Grade A, 5L Grade B, and API 5LX Grade X-42.

The formula is:

$$t = \frac{PD}{2(S + YP)} + u + c$$

In the calculation of the tables, the values of Y, u, and c were:

$$Y = 0.4$$

$$u = 0.125t \left(u = 12\tfrac{1}{2}\% \text{ oft} \right)$$

$$c = 0.050 \text{ in.}$$

O.D.	6⅝	8⅝	10¾	12¾	14	16	18	20	22	24	26	28	30	32	34	36	O.D.
Wall Thickness																	**Wall Thickness**
0.188	895 1044 1253	685 799 959	548 640 768	.....	.....	.....	.....	.....	.....	.....	.....	.....	.....	.....	.....	.....	0.188
0.203	.....	1070	857	721	.....	.....	.....	.....	.....	.....	.....	.....	.....	.....	.....	.....	0.203
0.219	1110 1295 1554	849 991 1189	680 793 952	572 667 801	729	637	.....	.....	.....	.....	.....	.....	.....	.....	.....	.....	0.219
0.250	1327 1548 1858	1014 1184 1420	811 947 1136	683 797 956	621 725 870	543 633 760	482 562 675	434 506 607	551	505	466	433	404	378	356	336	0.250
0.281	.....	.....	.....	794 926 1111	722 842 1010	631 736 883	560 654 784	504 588 705	641	587	542	503	469	439	414	390	0.281
0.312	1765 2059 2471	1347 1572 1886	.....	905 1056 1267	823 961 1153	719 839 1007	638 745 894	574 670 804	521 608 730	478 557 669	441 514 617	573	534	501	471	445	0.312
0.344	1994 2326 2791	1520 1774 2128	1214 1416 1700	1021 1190 1429	928 1083 1299	810 946 1135	720 839 1007	647 755 906	587 685 823	538 628 753	695	645	602	564	530	501	0.344
0.375	2216 2568 3103	1689 1970 2364	.....	1133 1322 1586	1030 1202 1442	900 1049 1259	798 931 1118	718 837 1005	652 760 912	597 696 836	551 642 771	715	667	625	588	555	0.375
0.406	.....	.....	.....	1744	.....	.....	1228	1104	1002	918	847	786	733	687	646	610	0.406
0.438	.....	2034 2373 2848	1622 1892 2271	1362 1589 1907	1238 1444 1733	1081 1261 1513	959 1005 1342	862 1005 1206	782 913 1095	716 836 1003	661 771 925	858	801	750	706	666	0.438
0.500	.....	 3328	 2651	 2225	1444 1685 2022	1260 1470 1764	1118 1304 1565	1004 1171 1406	911 1063 1276	834 974 1168	770 898 1077	1000	932	873	822	775	0.500
Wall Thickness																	**Wall Thickness**
O.D.	6⅝	8⅝	10¾	12¾	14	16	18	20	22	24	26	28	30	32	34	36	O.D.

where

t = wall thickness, in,
P = allowable working pressure, psi,
D = outside diameter, in,
S = allowable stress, psi,
Y = a coefficient,
u = under-thickness tolerance, in,
c = allowance for threading, mechanical strength, and corrosion, in.

Substituting in formula and solving for P:

$$P = \frac{1.75S(t - 0.057)}{D + 0.04 - 0.7t}.$$

The S-values are:

Spec. API 5L Grade A—25,500 psi
Spec. API 5L Grade B—29,750 psi
Spec. API 5LX Grade X-42—35,700 psi

Example. Find pipe specification for a 12-in line operating at 1078 psi: In the $12\frac{3}{4}$-in column these conditions will be met using Grade X-42 pipe with 0.281-in wall thickness, Grade B pipe with 0.330-in wall thickness, or Grade A pipe with 0.375-in wall thickness.

Find the stress in pipe wall due to internal pressure

Multiply ½ by diameter by pressure and divide by wall thickness to get stress in psi.

Example. Find the wall stress due to internal pressure in a 24-in pipeline having a 1/4-in wall thickness if the pressure gauge reads 800 psi:

$$1/2 \times 24 \times 800 \div 1/4 = 38,400 \text{ psi}$$

What wall thickness should be specified for a 30-in discharge line where pressure runs 1000 psi and wall stress is to be limited to 30,000 psi?

$$1/2 \times 30 \times 1000 \div \text{thickness} = 30,000$$

Wall thickness = 1/2 in.

This method of calculating transverse tensile stress due to internal pressure on the pipe is called the Barlow formula. Actually, there are two accepted variations, and some designers use inside diameter of the pipe in calculating this stress. This gives a slightly lower figure for the transverse tensile stress and consequently gives less safety factor. The method used here is based on the assumption that maximum stress occurs at the outside diameter of the pipe. The results give a conservative value for safety factor calculations.

How to calculate stress in above ground/below ground transitions

Design of gas gathering/gas reinjection project in southern Iran provides solutions for deflection and anchor block forces in long underground lines

P. J. Schnackenberg, Design Engineer, Iran-Texas Engineering Co., Tehran, Iran.

Stresses and deflections occur in pipelines at the transition from the belowground (fully restrained) to the aboveground (unrestrained) condition.

Analysis of the stresses and deflections in transition areas, resulting from internal pressure/temperature change, is necessary in determining anchor block requirements and design. Longitudinal deflections are used to determine whether an anchor block is required. Anchor block forces required to maintain the pipe in a fully constrained condition are then determined.

Iran-Texas Engineering Co. recently completed an analysis for a gas gathering/compression/reinjection project in southern Iran under contract to Oil Service Co. of Iran. Each of the two fields, Bibi-Hakimeh and Rag-E-Safid have two gas/oil separation plants, and at present, the gas is being flared. The broad scope of the project is as follows:

- Provision of gas gathering lines from existing and new wellhead separation to a production unit.
- Design of four production units (each adjacent to the GOSP) comprising knockout facilities, compressor stations, and dehydration plant.
- Provision of transfer pipelines to send gas plus condensate as a two-phase flow to a natural gas liquid separation plant (under design by Foster Wheeler).

- Provision of reinjection system to inject lean gas back into a number of wells.
- Some 9.4 MMm³/d (330 MMscfd) of gas plus 15,000 bpd condensate will be going to the NGL plant. LPG will be used as feedstock and for export.

The following is a brief review of the analysis that resulted in more accurate solutions for deflection and anchor block forces. Sample calculations are for line sizes up to 41-cm (16-in) CD, pressure to 193 bars (2800 psig), and temperatures to 72 °C (162 °F).

Case I—no anchor. Considering the pipeline as shown in Figure 1, the section up to point A is fully restrained, the section A—B is a transition from fully restrained to unrestrained, and beyond point B is unrestrained. (Fully restrained is taken to be a condition of zero longitudinal strain.)

Analyzing a short section of fully restrained pipe, subject to internal pressure and a temperature change, there is a tensile stress due to a Poisson effect of

$$\sigma_{PO} = v\sigma \text{ where, } \sigma h = \frac{PDi}{2t}$$

Figure 1. Area A–B depicts transition from fully restrained (left) to fully unrestrained (right).

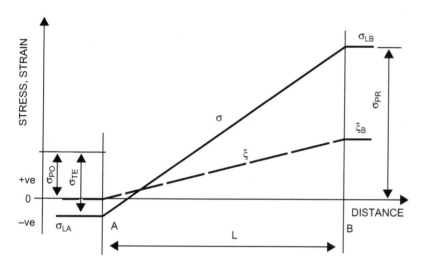

Figure 2. Transition of stress and strain between points *A* and *B* varies as linear function of length.

and a compressive stress due to a temperature change of $\sigma_{TE} = E\alpha\Delta T$.

The net longitudinal stress, therefore, at point A will be

$$\sigma_{LA} = v\sigma_H - E\alpha\Delta T \text{ psi} \tag{1}$$

and the strain at point A, being fully restrained, will be zero

$$\varepsilon_A = 0. \tag{2}$$

In a section of unrestrained pipe at B, the longitudinal stress caused by only internal pressure will be half the hoop stress.

$$\sigma_{LB} = \frac{\sigma_H}{2} \text{ psi} \tag{3}$$

The associated strain will be made up of a

(i) Temperature effect

$$\varepsilon_{TE} = \alpha\Delta T$$

(ii) Pressure effect

$$\varepsilon_{PR} = \frac{\sigma_{LB}}{E} = \frac{\sigma_H}{2E}$$

(iii) Poisson effect

$$\varepsilon_{PO} = \frac{v\sigma_H}{E}.$$

The net longitudinal strain at point *B* will therefore be

$$\varepsilon_B = \sigma\Delta T + \frac{\sigma_H}{E}(1/2 - v) \text{ in/in} \tag{4}$$

The transition of stress and strain between points *A* and *B* is assumed to vary as a linear function of length as shown in Figure 2.

In order to establish the length *L* over which the transition occurs, the longitudinal resistance of the soil needs to be known. For this part, it is assumed that any tendency to move will be counteracted by constant and opposite soil force. Wilbur[1] has recommended a design value for average soils of

$$F_s = 80\left(\frac{D_o}{12}\right) \text{ lbf/ft.} \tag{5}$$

Figure 3. Distribution of stress where anchor is required to contain longitudinal deflections.

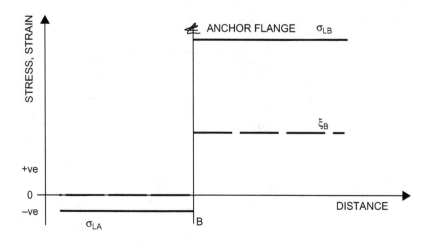

Between A and B, equilibrium of forces exists and therefore,

$$F_s L = A_m(\sigma_{LB} - \sigma_{LA})$$

from which

$$L = \frac{A_m(\sigma_{LB} - \sigma_{LA})}{F_s} \text{ ft.} \qquad (6)$$

Total movement at B will be the average strain between A and B over length L or

$$\delta = \frac{\varepsilon_B}{2}(12L) \text{ in.} \qquad (7)$$

Cell II—with anchor. Where an anchor is required to contain longitudinal deflections, the stress distribution will be as shown in Figure 3. The transition from being fully restrained to unrestrained will occur at the anchor. Resultant force on the anchor will simply be the difference in stress on each side times the pipe metal area (equilibrium of forces) or

$$F = (\sigma_{LB} - \sigma_{LA})A_m \text{ lb.} \qquad (8)$$

Note that for the case of an increase in wall thickness at and beyond the anchor block, the result will be essentially the same—the decrease in stress being compensated by an increase in pipe metal area.

It can be shown that this force is equal to that to produce a deflection of 2δ in Equation 7 over length L.

Table 1 is a summary of results for a number of high pressure gas transmission lines currently under design in Iran. Lines 1 through 4 are transfer lines from a number of production units to an NGL separation plant. Lines 5 through 7 are injection lines from the NGL separation plant to a number of wellheads. The following values were used for the various parameters:

Youngs modulus E 29×10^6, psi
Poisson's ratio υ, 0.3
Thermal expansion coefficient 6.5×10^{-6}, in/in °F
Installation temperature, Ti, 80 °F (minimum summer temperature when construction is in progress).

Negative stress values denote a compressive stress.

Discussion

As shown in Table 1, anchor block forces can be quite large; therefore, careful attention must be paid to the design. In order to minimize the size of the block, upper design limit

Table 1
Summary of Results for High Pressure Gas Transmission Lines Under Design in Iran

No.	Pipe (All API-5LX-60)	P psig	T_O °F	F_s lb./ft.	σ_H psi	σ_{LA} psi	σ_{LB} psi	L ft.	δ in.	F lb.
1	16″ OD × 0.312″ w.t.	1150	162	142.2	28,334	−6957	14,167	2285	9.81	324,890
2	20″ OD × 0.344″ w.t.	1000	162	222.2	28,070	−7036	14,035	2014	8.62	447,550
3	16″ OD × 0.281″ w.t.	1000	162	142.2	27,470	−7216	13,735	2045	8.71	290,800
4	16″ OD × 0.250″ w.t.	880	162	142.2	27,280	−7273	13,640	1819	7.73	258,690
5	16″ OD × 0.688″ w.t.	2800	150	142.2	29,758	−4268	14,879	4457	17.66	633,770
6	10″ NB × 0.500″ w.t.	2800	150	64.2	28,700	−4585	14,350	4728	18.60	304,850
7	6″ NB × 0.432″ w.t.	2800	150	24.4	18,670	−7594	9,335	5828	20.41	142,200

Figure 4. Sectional plan for anchor block design.

Figure 5. Sectional elevation for anchor block design.

values of lateral soil bearing pressure with due consideration to possible slip have been used.

For large blocks, friction between the block and the soil may also be figured in obtaining its resistance to imposed forces. The block should be reinforced concrete cast against undisturbed soil. Figures 4 and 5 indicate in broad detail the anchor block design.

Care should also be taken to ensure that connected surface piping has sufficient flexibility to absorb a degree of lateral anchor movement. Scraper traps should be installed such that they can move with the piping rather than being rigidly attached to blocks. Numerous examples of traps together with their support blocks being displaced a few inches are not uncommon.

References

1. Wilbur, W. E., *Pipe Line Industry*, February 1963.
2. Timoshenko, S., *Theory of Elasticity*.

Nomenclature

A_m	Pipe metal area $= \pi(D_o = t)t$, sq in
D_o	Pipe outside diameter, in
D_i	Pipe inside diameter, in
t	Pipe wall thickness, in
E	Young's modulus of elasticity, pi
F	Anchor force, lb
F_s	Soil resistance, lb/ft
L	Transition length, ft
P	Design pressure, psig
T_i	Installation temperature, °F
T_o	Operating temperature, °F
ΔT	Temperature difference $= (T_o - T_i)$, °F
α	Coefficient of thermal expansion, in/in °F
σ	Stress, psi
ϵ	Strain, in/in
υ	Poisson's ratio
δ	Deflection, in

Subscripts (for Greek letters only)

L	Longitudinal
H	Hoop
A	Point A, fully restrained

B	Point B, unrestrained
PR	Pressure
TE	Temperature
PO	Poisson

How to identify the series number of flanged fittings

Divide the flange thickness, expressed in sixteenths of an inch, by the nominal pipe size + 12. If the answer is less than 1.00, the flange is Series 15; if between 1.15 and 1.33, it is Series 30; if between 1.37 and 1.54, it is Series 40; and if it is between 1.60 and 1.85, it is Series 60.

Example. A 6-in flange has a thickness of $1^7/_{16}$ inch = $^{23}/_{16}$ in.

$^{23}/_{18} = 1.28$; the flange is series 30.

The rule is not applicable to fittings below 4 in in size; it also fails for a 4-in Series 60, which comes out 1.50.

Dimensions of three-diameter ells with tangents

Here are face-to-face and center-to-face dimensions of the three-diameter radius weld ells being used in natural gas compressor station piping.

Degree Ell	Outside Diameter	A (Center To Face)	A With 4" Tan	A With 6" Tan
30°	16″	1′-0⅞″	1′- 4⅞″	1′- 6⅞″
30°	20″	1′-4¹⁄₁₆″	1′- 8¹⁄₁₆″	1′-10¹⁄₁₆″
30°	24″	1′-7⁵⁄₁₆″	1′-11¹¹⁄₁₆″	2′- 1⁵⁄₁₆″
30°	26″	1′-8⅞″	2′- 0⅞″	2′- 2⅞″
30°	30″	2′-0⅛″	2′- 4⅛″	2′- 6⅛″
45°	16″	1′-7⅞″	1′-11⅞″	2′- 1⅞″
45°	20″	2′-0⅞″	2′- 4⅞″	2′- 6⅞″
45°	24″	2′-5¹³⁄₁₆″	2′- 9¹³⁄₁₆″	2′-11¹³⁄₁₆″
45°	26″	2′-8⁵⁄₁₆″	3′- 0⁵⁄₁₆″	3′- 2⁵⁄₁₆″
45°	30″	3′-1¼″	3′- 5¼″	3′- 7¼″
90°	16″	4′-0″	4′- 4″	4′- 6″
90°	20″	5′-0″	5′- 4″	5′- 6″
90°	24″	6′-0″	6′- 4″	6′- 6″
90°	26″	6′-6″	6′-10″	7′- 0″
90°	30″	7′-6″	7′-10″	8′- 0″

Spectacle blind thicknesses

The thickness of blinds is calculated using Formula 16, page 23, section 304.5.3 from Petroleum Refinery Piping ANSI B31.3-1973, sponsored by the American Society of Mechanical Engineers:

$$t = d\sqrt{\frac{3P}{16SE}} + \text{ca}$$

where

t = Thickness, inches,
D = Pipe inside diameter, inches,
P = Maximum pressure blind is to contain, psig,
SE = SMYS of blind material,
ca = Corrosion allowance.

Example:

Pipe OD	8.625 in
d	7.981 in
P	1440 psig
SE	42,000 psi
ca	0.125 in

$$t = 7.981\sqrt{\frac{3 \times 1440}{16 \times 42,000}} + 0.125$$

$t = 0.7649$ in.

Polypipe design data

Internal pressure

Polyethylene pipe for industrial, municipal, and mining applications is manufactured to specific dimensions as prescribed by applicable American Society for Testing and Materials (ASTM) standards. Piping outside diameters meet the IPS system. Wall thicknesses are based on the Standard Dimension Ratio (SDR) system, a specific ratio of the outside diameter to the minimum specified wall thickness. Selected SDR numbers are from the ANSI Preferred Number Series. Use of the SDR number in the ISO equation, recognized as an equation depicting the relationship of pipe dimensions'—both wall and o.d.—internal pressure carrying capabilities and tensile stress, in conjunction with a suitable safety factor (F), will give the design engineer confidence that the pipe will not prematurely fail due to internal pressurization.

A pressure differential is required to move material through a pipeline. For atmospheric systems (gravity flow), gravitational forces provide the impetus for movement of heavier-than-air mass. Internal pressure requirements must be recognized in the design stage in order to provide desired operational life. In some cases, a gravity flow system must be treated comparable to the design considerations of a pressure flow system.

Polyethylene pipe under pressure is time/temperature/pressure dependent. As the temperature increases and the pressure remains constant, the creep rupture strength decreases. Conversely, as the temperature decreases, the strength increases. Varying these parameters, in conjunction with the SDR, can provide the desired service within the capabilities of polyethylene pipe.

Calculations for determining the pressure rating of polyethylene pipe is based on the ISO equation as follows:

$$P = \frac{2S}{SDR - 1} \times F$$

where:

P = Internal pressure, psi,
S = Long-term hydrostatic strength, psi (1600 psi for polyethylene pipe),
SDR = Standard dimension ratio, (D/t),
D = Outside diameter, inches,
t = Minimum wall thickness, inches,
F = Design safety factor (0.5 for water @ 73 °F).

Use of other factors determined from Figures 1–3, will provide for a more accurate performance characteristic for systems operating at other temperatures for shorter periods of time.

The ISO equation now becomes:

$$P = \frac{2S}{SDR - 1} \times F \times F_1 \times F_2 \times F_3$$

F_1 = Operational life factor,
F_2 = Temperature correction,
F_3 = Environmental service factor (Figure 3).

Other design factors may be necessary in the design of systems where abrasion, corrosion, or chemical attack may be a factor.

SDR	7.3	9	11	13.5	15.5	17	21	26	32.5
Pressure Rating	250	200	160	125	110	100	80	64	50

Figure 1. Internal pressure ratings (psi) PE pipe at 73 °F for 50 years.

To use Figure 2 for determining operational life factor multiplier F_1, select expected operating life (less than 50 years) on the X axis and move upwards to intersect the diagonal line and then left to reading on Y axis.

Example. 5-year operational life will yield F_1 = 1.094

Figure 2. Operational life.

Figure 2
operational life
factor F$_1$

Substance	Service Factor F$_3$
Crude Oil	0.5
Wet Natural Gas	0.5
Dry Natural Gas	0.64

Figure 3. Environmental service factor F$_3$.

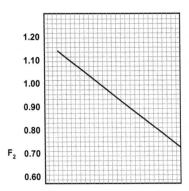

Figure 4. Temperature correction. Example: A polyethylene pipe operating at 125 °F would have an F$_2$ factor of 0.62.

Figure 5. Recommended liquid velocities.

	Pipe size, inches		
	3–10	10–28	28–48
Water and similar viscosity fluids			
	Velocity, fps		
Pump suction*	1 to 4	4 to 6	5 to 8
Pump discharge	3 to 5	4 to 8	6 to 12
Gravity drain system	3 to 5	4 to 8	6 to 10
Viscous liquids			
Pump suction*	0.5 to 2	1 to 4	2 to 5
Pump discharge	3 to 5	4 to 6	5 to 7

SDR	F	SDR	F
7	0.706	13.5	0.852
7.3	0.719	15.5	0.873
9	0.774	17	0.884
9.3	0.781	21	0.909
11	0.817	26	0.928
		32.5	0.944

Figure 6. ID to OD ratio.

Example. A horizontal process water line 2000 ft in length, 125 °F water, 60 psig pressure, with a desired life of five years is proposed. Determine the standard pipe SDR for the application

$$60 = \frac{2 \times 1600}{SDR-1} \times F \times F_1 \times F_2 \times F_3$$

where

F = design safety factor (use 0.5 for water @ 73 °F),
F_1 = operational life factor,
F_2 = temperature correction,
F_3 = environmental service factor (Figure 3).

Therefore, use SDR = 17.
Note: Using SDR 17 is conservative.

Pipe diameter

Volumetric flow, Q, can be determined from the continuity equation $Q = A \times V$ by selecting a suitable velocity from Figure 5 for a given pipe size. When modified for flow in gallons per minute, the equation becomes:

$$Q = 2.449 \times V \times d^2$$

where

Q = flow rate gpm,
V = velocity, in feet per second,
d = inside diameter, inches,
or $d = (Q \div 2.449 \times V)^{1/2}$ for pipe size determination.

Example. If the required water flow rate for a system is 2000 gpm and the flow velocity is to be maintained below 8 fps, the pipe diameter can be found as follows:

$$d = (Q \div 2.449 \times V)^{1/2}$$

$$d = (2000 \div 2.449 \times 8)^{1/2}$$

$$d = 101 \text{ in inside diameter}$$

$$OD = ID \div F(F \text{ from Figure 6})$$

Pressure drop

Refer to Section 13: Liquids—Hydraulics and use the Hazen-Williams equation. Use a C of 150.

Fitting equivalent length

The exact pressure loss through fittings cannot readily be calculated because of the geometry of the fitting. The values shown in Figure 7 are the result of general industry consensus.

Fitting type	Equivalent length of pipe
90 deg. elbow	30 D
60 deg. elbow	25 D
45 deg. elbow	16 D
45 deg. wye	60 D
Running tee	20 D
Branch tee	50 D
Gate valve, full open	8 D
Butterfly valve (3¢¢ to 14¢¢)	40 D
Butterfly valve [3] 14¢¢	30 D
Swing check valve	100 D
Ball valve full bore full open	3 D

Figure 7. Equivalent length of fittings.

Example. A running tee has an equivalent length of 20 D. For an 18-in SDR 11 line, this represents 20 × ID. From Figure 6, ID = OD × F = 18 × 0.817 = 14.7″ × 20 = 294 ÷ 12 and is equal to 24.5 ft. In calculating the total pressure drop in the system, 24.5 ft of 18-in pipe should be added to the total line length to account for the increased pressure loss due to the presence of the tee.

Published with permission—CSR™ PolyPipe

Reuse of polyethylene pipe in liquid hydrocarbon service

When polyethylene pipe has been permeated with liquid hydrocarbons, joining should be connected by use of mechanical connections. Joining by means of fusion bonding may produce a low-strength joint.

Gravity or nonpressure service above 180 °F is not recommended.
Pressure service above 140 °F is not recommended.

4: Electrical Design

Electrical design

Most pipeline facilities use electrical equipment of some sort or another, ranging from simple power circuits of a few amperes capacity to sophisticated supervisory control and data acquisition systems. It is beyond the scope of this manual to provide a complete section on electrical information for pipeline facilities. This section is intended to provide some basic data that will prove useful to field personnel responsible for electrical installations.

Electric motor selection

Based on known facts and calculations, the best selection is made after a close study of the installation, operation, and servicing of the motor. Basic steps in proper selection are numbered 1–8 and are briefly described.

1. **Power supply**

    a. Voltage—National Electrical Manufacturers Association (NEMA) has recommended the following standards.

Nominal Power System (V)	Motor Nameplate (V)
240	230
480	460
600	575

    b. Frequency—Motors rated 200 horsepower or less can vary not to exceed 5% above or below its rated frequency.
    c. Phases—Three-phase power supplies are found in most industrial locations; for most residential and rural areas, only single-phase power is available.

2. **HP and duty requirements**

    a. Continuous duty—means constant load for an indefinite period (about 90% of all motor applications).
    b. Intermittent duty—means alternate periods of load and no-load, or load and rest.
    c. Varying duty—means both the load and time operation vary to a wide degree.

3. **Speeds**—Single speed motors are the most common, but when a range of speeds is required multispeed motors will give two, three, or four fixed speeds.

4. **Service factors**—Open, general purpose motors have a service factor depending upon the particular rating of the motor (usually between 1.15 and 1.25).

5. **Selection of motor type**

There are three main types:

a. DC motors are designed for industrial drives requiring a controlled speed range and constant torque output on adjustable voltage control systems. Because the speed of rotation controls the flow of current in the armature, special devices must be used for starting DC motors.

b. Single-phase alternating-current motors also require some auxiliary arrangement to start rotation.

c. Polyphase motors are alternating-current types (squirrel cage or wound rotor).

   Applications—Some driven machines require a low-starting torque that gradually increases to full-load speed; others require higher-starting torque. NEMA code letters A, B, C, D, etc., on the motor nameplate designate the locked rotor kVA per horsepower of that particular motor design. The diagram shows representative speed-torque curves for poly-phase and single-phase NEMA design motors Types A–D.

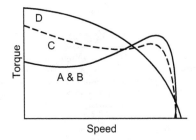

6. **Torque definitions**

    a. Locked-rotor torque, or "starting torque," is the minimum-torque that the motor will develop at rest for all angular positions of the rotor.

Code Letter	Locked Rotor (kVA/hp)
A	0–3.14
B	3.15–3.54
C	3.55–3.99
D	4.0–4.49
E	4.5–4.99
F	5–5.59

    b. Breakdown torque is the maximum-torque at rated voltage and frequency without abrupt speed drop.
    c. Full-load-torque is the torque necessary to produce rated horsepower at full-load speed.

d. Locked-rotor current is the steady-state current of the motor with the rotor locked at rated voltage and frequency.

7. **Selection of enclosure**—The two general classifications are open, which permits passage of air over and around the windings, and totally enclosed, which prevents exchange of air between inside and outside of frame (but is not strictly airtight).

a. Open Drip-Proof—means that liquid or solid particles falling on the motor at an angle not greater than 15° from vertical cannot enter the motor.

b. NEMA Type 1—a weather-protected machine with its ventilating passages constructed to minimize the entrance of rain, snow, and airborne particles to the electric parts and having its ventilating openings so constructed to prevent the passage of a cylindrical rod 3/4 in in diameter.

c. NEMA Type II—a weather-protected machine that has in addition to the enclosure defined for a Type I weather-protected machine, its ventilating passages at both intake and discharge so arranged that high velocity air and airborne particles blown into the machine by storms or high winds can be discharged without entering the internal ventilating passages directly leading to the electric parts of the machine itself. The normal path of the ventilating air that enters the electric parts of the machine shall be so arranged by baffling or separate housings as to provide at least three abrupt changes in direction, none of which shall be less than 90°. In addition, an area of low velocity, not exceeding 600 ft/min, shall be provided in the intake air path to minimize the possibility of moisture or dirt being carried into the electric parts of the machine.

d. Totally Enclosed Nonvent—means that a motor is not equipped for cooling by external means.

e. Totally Enclosed Fan-Cooled—means the motor has a fan integral with the motor but not external to the enclosed parts.

f. Explosion-Proof—means the enclosure is designed to withstand an explosion of a specified gas that may occur within the motor and to prevent the ignition of gas around the motor.

8. End shield mountings

Three types of end shields with rabbets and bolt holes for mounting are standard in the industry:

a. Type C Face provides a male rabbet and tapped holes.

b. Type D Flange has a male rabbet with holes for through bolts in the flange.

c. Type P base has a female rabbet and through bolts for mounting in the flange (used for mounting vertical motors).

Hazardous locations

Definition of class locations

Reprinted with permission from National Fire Protection Association (NFPA) 70-1990, the *National Electrical Code*®, Copyright © 1989, NFPA, Quincy, MA 02269. This reprinted material is not the complete and official position of the NFPA, on the referenced subject, which is represented only by the standard in its entirety.

The *National Electrical Code* and *NEC* are registered trademarks of the NFPA.

500—5. Class I Locations. Class I locations are those in which flammable gases or vapors are or may be present in the air in quantities sufficient to produce explosive or ignitible mixtures. Class I locations shall include those specified in (a) and (b) below.

a. **Class I, Division 1**. A Class I, Division 1 location is a location: (1) in which ignitible concentrations of flammable gases or vapors can exist under normal operating conditions; or (2) in which ignitible concentrations of such gases or vapors may frequently exist because of repair or maintenance operations or because of leakage; or (3) in which breakdown or faulty operation of equipment or processes might release ignitible concentrations of flammable gases or vapors and might also cause simultaneous failure of electric equipment.

(FPN): This classification usually includes locations where volatile flammable liquids or liquefied flammable gases are transferred from one container to another; interiors of spray booths and areas in the vicinity of spraying and painting operations where volatile flammable solvents are used; locations containing open tanks or vats of volatile flammable liquids; drying rooms or compartments for the evaporation of flammable solvents; locations containing fat and oil extraction equipment using volatile flammable solvents; portions of cleaning and dyeing plants where flammable liquids are used; gas generator rooms and other portions of gas manufacturing plants where flammable gas may escape; inadequately ventilated pump rooms for flammable gas or for volatile flammable liquids; the interiors of refrigerators and freezers in which volatile flammable materials are stored in open, lightly stoppered, or easily ruptured containers; and all other locations where ignitible concentrations of flammable vapors or gases are likely to occur in the course of normal operations.

b. **Class I, Division 2.** A Class I, Division 2 location is a location: (1) in which volatile flammable liquids or flammable gases are handled, processed, or used, but in which the liquids, vapors, or gases will normally be confined within closed containers or closed systems from which they can escape only in case of accidental rupture or breakdown of such containers or systems, or in the case of abnormal operation of equipment; or (2) in which ignitible concentrations of gases or vapors are normally prevented by positive mechanical ventilation and which might become hazardous through failure or abnormal operation of the ventilating equipment; or (3) that is adjacent to a Class I, Division 1 location, and to which ignitible concentrations of gases or vapors might occasionally be communicated unless such communication is prevented by adequate positive-pressure ventilation from a source of clean air, and effective safeguards against ventilation failure are provided.

(FPN No. 1): This classification usually includes locations where volatile flammable liquids or flammable gases or vapors are used, but which, in the judgment of the authority having jurisdiction, would become hazardous only in case of an accident or of some unusual operating condition. The quantity of flammable material that might escape in case of accident, the adequacy of ventilating equipment, the total area involved, and the record of the industry or business with respect to explosions or fires are all factors that merit consideration in determining the classification and extent of each location.

(FPN No. 2): Piping without valves, checks, meters, and similar devices would not ordinarily introduce a hazardous condition even if used for flammable liquids or gases. Locations used for the storage of flammable liquid or of liquefied or compressed gases in sealed containers would not normally be considered hazardous unless also subjected to other hazardous conditions.

Group A: Atmospheres containing acetylene.

Group B: Atmospheres containing hydrogen, fuel, and combustible process gases containing more than 30% hydrogen by volume or gases or vapors of equivalent hazard such as butadiene,[1] ethylene oxide,[2] propylene oxide,[2] and acrolein.[2]

Group C: Atmospheres such as cyclopropane, ethyl ether, ethylene, or gases or vapors of equivalent hazard.

Group D: Atmospheres such as acetone, ammonia,[3] benzene, butane, ethanol, gasoline, hexane, methanol, methane, natural gas, naptha, propane, or gases of vapors of equivalent hazard.

---

[1] Group D equipment may be used for this atmosphere if such equipment is isolated in accordance with Section 501-5(a) by sealing all conduits 1/2-in or larger.
[2] Group C equipment may be used for this atmosphere if such equipment is isolated in accordance with Section 501-5(a) by sealing all conduits 1/2-in or larger.
[3] For classification of areas involving ammonia atmosphere, see Safety Code for Mechanical Refrigeration, ANSI/ASHRAE 15-1978, and Safety Requirements for the Storage and Handling of Anhydrous Ammonia, ANSI/CGA G2.1-1981.

NEMA enclosure types[1]

Nonhazardous locations

Type	Intended Use and Description
1	Enclosures are intended for indoor use primarily to provide a degree of protection against limited amounts of falling dirt.
2	Enclosures are intended for indoor use primarily to provide a degree of protection against limited amounts of falling water and dirt.
3	Enclosures are intended for outdoor use primarily to provide a degree of protection against rain, sleet, windblown dust, and damage from external ice formation.
3R	Enclosures are intended for outdoor use primarily to provide a degree of protection against rain, sleet, and damage from external ice formation.
3S	Enclosures are intended for outdoor use primarily to provide a degree of protection against rain, sleet, windblown dust, and to provide for operation for external mechanisms when ice laden.
4	Enclosures are intended for indoor or outdoor use primarily to provide a degree of protection against windblown dust and rain, splashing water, hose-directed water, and damage from external ice formation.
4X	Enclosures are intended for indoor or outdoor use primarily to provide a degree of protection against corrosion, windblown dust and rain, splashing

water, hose-directed water, and damage from external ice formation.

5   Enclosures are intended for indoor use primarily to provide a degree of protection against settling airborne dust, falling dirt, and dripping noncorrosive liquids.

6   Enclosures are intended for indoor or outdoor use primarily to provide a degree of protection against hose-directed water, the entry of water during occasional temporary submersion at a limited depth, and damage from external ice formation.

6P   Enclosures are intended for indoor or outdoor use primarily to provide a degree of protection against hose-directed water, the entry of water during prolonged submersion at a limited depth, and damage from external ice formation.

12   Enclosures are intended for indoor use primarily to provide a degree of protection against circulating dust, falling dirt, and dripping noncorrosive liquids.

12K Enclosures with knockouts are intended for indoor use primarily to provide a degree of protection

against circulating dust, falling dirt, and dripping noncorrosive liquids.

13   Enclosures are intended for indoor use primarily to provide a degree of protection against dust, spraying of water, oil, and noncorrosive coolant.

Hazardous (classified locations)

7   Enclosures are intended for indoor use in locations classified as Class I, Groups A, B, C, or D, as defined in the National Electrical Code.

8   Enclosures are intended for indoor use in locations classified as Class I, groups A, B, C, or D, as defined in the National Electrical Code.

9   Enclosures are intended for indoor use in locations classified as Class II, groups E, F, or G, as defined in the National Electrical Code.

10   Enclosures are constructed to meet the applicable requirements of the Mine Safety and Health Administration.

Reference NEMA Standards Publications/No. 250-1991.
Used with permission—National Electrical Manufacturer's Association.

Size portable electric generators

A power plant must be properly sized for the required load. Electric motors are particularly difficult for portable plants, because they require two to three times motor nameplate amps or wattage.

A portable power plant's surge capacity is limited by engine horsepower and inertia of its rotating parts. Thus, a current surge of short duration can be supplied by a power plant, but a current demand of longer duration, such as a heavily loaded motor starting a high inertia load, can overload the power plant and possibly damage the power plant and the motor. A 3450 rpm air compressor is a prime example of this type of load.

Example:
Load requirements:

Electric heater	1000 W
11–100 W lamps	100 W
Motor	600 W
Total motor load	600 × 3 = 1800 W
Total other load	2100 W
Total load	3100 W
Add 25% for future	975 W
Total load for sizing power plant	4875 W

A generator capable of continuously supplying 5000 W will be adequate.

When more than one motor is connected to a power plant, always start the largest motor first. If the total load is connected to one receptacle on the power plant, be sure the ampere rating of the receptacle is not exceeded.

Typical wattages for tools and appliances

Equipment	Running Watts
1/2″ Drill	1000
1″ Drill	1100
6″ Circular saw	800
10″ Circular saw	2000
14″ Chain saw	1100
Radio	50–200
Television	200–500
Toaster, coffeemaker	1200
Water heater (storage type)	1100–5500
Hot plate or range (per burner)	1000
Range oven	10,000
Skillet	1200
Fan	50–200
Floodlight	1000
Water pump	500
Vacuum cleaner	200–300
Refrigerator (conventional)	200–300
Refrigerator (freezer-combination)	250–600
Furnace fan blower	500–700

Knockout dimensions

Conduit size, in	Knockout Diameter, in		
	Minimum	Nominal	Maximum
1/2	0.859	0.875	0.906
3/4	1.094	1.109	1.141
1	1.359	1.375	1.406
1¼	1.719	1.734	1.766
1½	1.958	1.984	2.016
2	2.433	2.469	2.500
2½	2.938	2.969	3.000
3	3.563	3.594	3.625
3½	4.063	4.125	4.156
4	4.563	4.641	4.672
5	5.625	5.719	5.750
6	6.700	6.813	6.844

Reference NEMA Standards Publications/No. 250-1991.
Used with permission—National Electrical Manufacturer's Association.

National electrical code tables

Table 430-150
Full-Load Current* Three-Phase Alternating-Current Motors

HP	Induction Type Squirrel-Cage and Wound-Rotor Amperes							Synchronous Type †Unity Power Factor Amperes			
	115V	200V	208V	230V	460V	575V	2300V	230V	460V	575V	2300V
½	4	2.3	2.2	2	1	.8					
¾	5.6	3.2	3.1	2.8	1.4	1.1					
1	7.2	4.1	4.0	3.6	1.8	1.4					
1½	10.4	6.0	5.7	5.2	2.6	2.1					
2	13.6	7.8	7.5	6.8	3.4	2.7					
3		11.0	10.6	9.6	4.8	3.9					
5		17.5	16.7	15.2	7.6	6.1					
7½		25.3	24.2	22	11	9					
10		32.2	30.8	28	14	11					
15		48.3	46.2	42	21	17					
20		62.1	59.4	54	27	22					
25		78.2	74.8	68	34	27		53	26	21	
30		92	88	80	40	32		63	32	26	
40		119.6	114.4	104	52	41		83	41	33	
50		149.5	143.0	130	65	52		104	52	42	
60		177.1	169.4	154	77	62	16	123	61	49	12
75		220.8	211.2	192	96	77	20	155	78	62	15
100		285.2	272.8	248	124	99	26	202	101	81	20
125		358.8	343.2	312	156	125	31	253	126	101	25
150		414	396.0	360	180	144	37	302	151	121	30
200		552	528.0	480	240	192	49	400	201	161	40

*These values of full-load current are for motors running at speeds usual for belted motors and motors with normal torque characteristics. Motors built for especially low speeds or high torques may require more running current, and multispeed motors will have full-load current varying with speed, in which case the nameplate current rating shall be used.

†For 90 and 80% power factor the above figures shall be multiplied by 1.1 and 1.25%, respectively.

The voltages listed are rated motor voltages. The currents listed shall be permitted for system voltage ranges of 110–120, 220–240, 440–480, and 550–600 V.

Adaptation reprinted with permission from NFPA 70-2002. The National Electrical Code® 1998. National Fire Protection Association, Quincy, MA 02269. This reprinted material is not the complete and official position of the National Fire Protection Association on the referenced subject, which is represented only by the standard in its entirety.

Table 310.16 Allowable Ampacities of Insulated Conductors Rated 0–2000 V, 60–90 °C (140–194 °F), Not More Than Three Current-Carrying Conductors in Raceway, Cable, or Earth (Direct Buried), Based on Ambient Temperature of 30 °C (86 °F)

Size AWG or kcmil	Temperature Rating of Conductor (See Table 310.13.)						Size AWG or kcmil
	60°C (140°F)	75°C (167°F)	90°C (194°F)	60°C (140°F)	75°C (167°F)	90°C (194°F)	
	Types TW, UF	Types RHW, THHW, THW, THWN, XHHW, USE, ZW	Types TBS, SA, SIS, FEP, FEPB, MI, RHH, RHW-2, THHN, THHW, THW-2, THWN-2, USE-2, XHH, XHHW, XHHW-2, ZW-2	Types TW, UF	Types RHW, THHW, THW, THWN, XHHW, USE	Types TBS, SA, SIS, THHN, THHW, THW-2, THWN-2, RHH, RHW-2, USE-2, XHH, XHHW, XHHW-2, ZW-2	
	Copper			Aluminium or Copper-Clad Aluminium			
18	—	—	14	—	—	—	—
16	—	—	18	—	—	—	—
14*	20	20	25	—	—	—	—
12*	25	25	30	20	20	25	12*
10*	30	35	40	25	30	35	10*
8	40	50	55	30	40	45	8
6	55	65	75	40	50	60	6
4	70	85	95	55	65	75	4
3	85	100	110	65	75	85	3
2	95	115	130	75	90	100	2
1	110	130	150	85	100	115	1
1/0	125	150	170	100	120	135	1/0
2/0	145	175	195	115	135	150	2/0
3/0	165	200	225	130	155	175	3/0
4/0	195	230	260	150	180	205	4/0
250	215	255	290	170	205	230	250
300	240	285	320	190	230	255	300
350	260	310	350	210	250	280	350
400	280	335	380	225	270	305	400
500	320	380	430	260	310	350	500
600	355	420	475	285	340	385	600
700	385	460	520	310	375	420	700
750	400	475	535	320	385	435	750
800	410	490	555	330	395	450	800
900	435	520	585	355	425	480	900
1000	455	545	615	375	445	500	1000
1250	495	590	665	405	485	545	1250
1500	520	625	705	435	520	585	1500
1750	545	650	735	455	545	615	1750
2000	560	665	750	470	560	630	2000

Correction Factors

Ambient Temp. (°C)	For ambient temperatures other than 30°C (86°F), multiply the allowable ampacities shown above by the appropriate factor shown below.						Ambient Temp. (°F)
21–25	1.08	1.05	1.04	1.08	1.05	1.04	70–77
26–30	1.00	1.00	1.00	1.00	1.00	1.00	78–86
31–35	0.91	0.94	0.96	0.91	0.94	0.96	87–95
36–40	0.82	0.88	0.91	0.82	0.88	0.91	96–104
41–45	0.71	0.82	0.87	0.71	0.82	0.87	105–113
46–50	0.58	0.75	0.82	0.58	0.75	0.82	114–122
51–55	0.41	0.67	0.76	0.41	0.67	0.76	123–131
56–60	—	0.58	0.71	—	0.58	0.71	132–140
61–70	—	0.33	0.58	—	0.33	0.58	141–158
71–80	—	—	0.41	—	—	0.41	159–176

*See 240.4(D).

Adaptation reprinted with permission from NFPA 70-2002. *National Electrical Code®* Copyright© 1998. National Fire Protection Association, Quincy, MA 02269. This reprinted material is not the complete and official position of the National Fire Protection Association on the referenced subject, which is represented only by the standard in its entirety.

Table C8(A) Maximum Number of Compact Conductors in Rigid Metal Conduit

		Metric Designator (Trade Size)											
		16	21	27	35	41	53	63	78	91	103	129	155
Type	Conductor Size (AWG/kcmil)	(½)	(¾)	(1)	(1¼)	(1½)	(2)	(2½)	(3)	(3½)	(4)	(5)	(6)
THW, THW-2, THHW	8	2	4	7	12	16	26	38	59	78	101	158	228
	6	1	3	5	9	12	20	29	45	60	78	122	176
	4	1	2	4	7	9	15	22	34	45	58	91	132
	2	1	1	3	5	7	11	16	25	33	43	67	97
	1	1	1	1	3	5	8	11	17	23	30	47	68
	1/0	1	1	1	3	4	7	10	15	20	26	41	59
	2/0	0	1	1	2	3	6	8	13	17	22	34	50
	3/0	0	1	1	1	3	5	7	11	14	19	29	42
	4/0	0	1	1	1	2	4	6	9	12	15	24	35
	250	0	0	1	1	1	3	4	7	9	12	19	28
	300	0	0	1	1	1	3	4	6	8	11	17	24
	350	0	0	1	1	1	2	3	5	7	9	15	22
	400	0	0	1	1	1	1	3	5	7	8	13	20
	500	0	0	0	1	1	1	3	4	5	7	11	17
	600	0	0	0	1	1	1	1	3	4	6	9	13
	700	0	0	0	1	1	1	1	3	4	5	8	12
	750	0	0	0	0	1	1	1	3	4	5	7	11
	1000	0	0	0	0	1	1	1	1	3	4	6	9
THHN, THWN, THWN-2	8	—	—	—	—	—	—	—	—	—	—	—	—
	6	2	5	8	13	18	30	43	66	88	114	179	258
	4	1	3	5	8	11	18	26	41	55	70	110	159
	2	1	1	3	6	8	13	19	29	39	50	79	114
	1	1	1	2	4	6	10	14	22	29	38	60	86
	1/0	1	1	1	4	5	8	12	19	25	32	51	73
	2/0	1	1	1	3	4	7	10	15	21	26	42	60
	3/0	0	1	1	2	3	6	8	13	17	22	35	51
	4/0	0	1	1	1	3	5	7	10	14	18	29	42
	250	0	1	1	1	2	4	5	8	11	14	23	33
	300	0	0	1	1	1	3	4	7	10	12	20	28
	350	0	0	1	1	1	3	4	6	8	11	17	25
	400	0	0	1	1	1	2	3	5	7	10	15	22
	500	0	0	0	1	1	1	3	5	6	8	13	19
	600	0	0	0	1	1	1	2	4	5	6	10	15
	700	0	0	0	1	1	1	1	3	4	6	9	13
	750	0	0	0	1	1	1	1	3	4	5	9	13
	1000	0	0	0	0	1	1	1	2	3	4	6	9
XHHW, XHHW-2	8	3	5	9	15	21	34	49	76	101	130	205	296
	6	2	4	6	11	15	25	36	56	75	97	152	220
	4	1	3	5	8	11	18	26	41	55	70	110	159
	2	1	1	3	6	8	13	19	29	39	50	79	114
	1	1	1	2	4	6	10	14	22	29	38	60	86
	1/0	1	1	1	4	5	8	12	19	25	32	51	73
	2/0	1	1	1	3	4	7	10	16	21	27	43	62
	3/0	0	1	1	2	3	6	8	13	17	22	35	51
	4/0	0	1	1	1	3	5	7	11	14	19	29	42
	250	0	1	1	1	2	4	5	8	11	15	23	34
	300	0	0	1	1	1	3	5	7	10	13	20	29
	350	0	0	1	1	1	3	4	6	9	11	18	25
	400	0	0	1	1	1	2	4	6	8	10	16	23
	500	0	0	0	1	1	1	3	5	6	8	13	19
	600	0	0	0	1	1	1	2	4	5	7	10	15
	700	0	0	0	1	1	1	1	3	4	6	9	13
	750	0	0	0	1	1	1	1	3	4	5	8	12
	1000	0	0	0	0	1	1	1	2	3	4	7	10

Definition: *Compact stranding* is the result of a manufacturing process where the standard conductor is compressed to the extent that the interstices (voids between strand wires) are virtually eliminated.

Table C9 Maximum number of conductors or fixture wires in rigid PVC conduit, schedule 80

		Metric Designator (Trade Size)											
		16	21	27	35	41	53	63	78	91	103	129	155
Type	Conductor Size (AWG/kcmil)	(½)	(¾)	(1)	(1¼)	(1½)	(2)	(2½)	(3)	(3½)	(4)	(5)	(6)
RHH, RHW, RHW-2	14	3	5	9	17	23	39	56	88	118	153	243	349
	12	2	4	7	14	19	32	46	73	98	127	202	290
	10	1	3	6	11	15	26	37	59	79	103	163	234
	8	1	1	3	6	8	13	19	31	41	54	85	122
	6	1	1	2	4	6	11	16	24	33	43	68	98
	4	1	1	1	3	5	8	12	19	26	33	53	77
	3	0	1	1	3	4	7	11	17	23	29	47	67
	2	0	1	1	3	4	6	9	14	20	25	41	58
	1	0	1	1	2	4	6	9	13	17	22	27	38
	1/0	0	0	1	1	1	3	5	8	11	15	23	33
	2/0	0	0	1	1	1	3	4	7	10	13	20	29
	3/0	0	0	1	1	1	3	4	6	8	11	17	25
	4/0	0	0	0	1	1	2	3	5	7	9	15	21
	250	0	0	0	1	1	1	2	4	5	7	11	16
	300	0	0	0	1	1	1	2	3	5	6	10	14
	350	0	0	0	1	1	1	1	3	4	5	9	13
	400	0	0	0	0	1	1	1	3	4	5	8	12
	500	0	0	0	0	1	1	1	2	3	4	7	10
	600	0	0	0	0	0	1	1	1	3	3	6	8
	700	0	0	0	0	0	1	1	1	2	3	5	7
	750	0	0	0	0	0	1	1	1	2	3	5	7
	800	0	0	0	0	0	1	1	1	2	3	4	7
	1000	0	0	0	0	0	1	1	1	1	2	4	5
	1250	0	0	0	0	0	0	1	1	1	1	3	4
	1500	0	0	0	0	0	0	1	1	1	1	2	4
	1750	0	0	0	0	0	0	0	1	1	1	2	3
	2000	0	0	0	0	0	0	0	1	1	1	1	3
TW, THHW, THW, THW-2	14	6	11	20	35	49	82	118	185	250	324	514	736
	12	5	9	15	27	38	63	91	142	192	248	394	565
	10	3	6	11	20	28	47	67	106	143	185	294	421
	8	1	3	6	11	15	26	37	59	79	103	163	234
RHH*, RHW*, RHW-2*	14	4	8	13	23	32	55	79	123	166	215	341	490
	12	3	6	10	19	26	44	63	99	133	173	274	394
	10	2	5	8	15	20	34	49	77	104	135	214	307
	8	1	3	5	9	12	20	29	46	62	81	128	184
RHH*, RHW*, RHW-2*, TW, THW, THHW, THW-2	6	1	1	3	7	9	16	22	35	48	62	98	141
	4	1	1	3	5	7	12	17	26	35	46	73	105
	3	1	1	2	4	6	10	14	22	30	39	63	90
	2	1	1	1	3	5	8	12	19	26	33	53	77
	1	0	1	1	2	3	6	8	13	18	23	37	54
	1/0	0	1	1	1	3	5	7	11	15	20	32	46
	2/0	0	1	1	1	2	4	6	10	13	17	27	39
	3/0	0	0	1	1	1	3	5	8	11	14	23	33
	4/0	0	0	1	1	1	3	4	7	9	12	19	27
	250	0	0	0	1	1	2	3	5	7	9	15	22
	300	0	0	0	1	1	1	3	5	6	8	13	19
	350	0	0	0	1	1	1	2	4	6	7	12	17
	400	0	0	0	1	1	1	2	4	5	7	10	15
	500	0	0	0	1	1	1	1	3	4	5	9	13
	600	0	0	0	0	1	1	1	2	3	4	7	10
	700	0	0	0	0	1	1	1	2	3	4	6	9
	750	0	0	0	0	0	1	1	1	3	4	6	8
	800	0	0	0	0	0	1	1	1	3	3	6	8
	900	0	0	0	0	0	1	1	1	2	3	5	7
	1000	0	0	0	0	0	1	1	1	2	3	5	7
	1250	0	0	0	0	0	1	1	1	1	2	4	5
	1500	0	0	0	0	0	0	1	1	1	1	3	4
	1750	0	0	0	0	0	0	0	1	1	1	3	4
	2000	0	0	0	0	0	0	0	1	1	1	2	3

(table continued on next page)

Table C9 Maximum number of conductors or fixture wires in rigid PVC conduit, schedule 80

		Conductors											
		Metric Designator (Trade Size)											
Type	Conductor Size (AWG/kcmil)	16 (½)	21 (¾)	27 (1)	35 (1¼)	41 (1½)	53 (2)	63 (2½)	78 (3)	91 (3½)	103 (4)	129 (5)	155 (6)
THHN, THWN, THWN-2	14	9	17	28	51	70	118	170	265	358	464	736	1055
	12	6	12	20	37	51	86	124	193	261	338	537	770
	10	4	7	13	23	32	54	78	122	164	213	338	485
	8	2	4	7	13	18	31	45	70	95	123	195	279
	6	1	3	5	9	13	22	32	51	68	89	141	202
	4	1	1	3	6	8	14	20	31	42	54	86	124
	3	1	1	3	5	7	12	17	26	35	46	73	105
	2	1	1	2	4	6	10	14	22	30	39	61	88
	1	0	1	1	3	4	7	10	16	22	29	45	65
	1/0	0	1	1	2	3	6	9	14	18	24	38	55
	2/0	0	1	1	1	3	5	7	11	15	20	32	46
	3/0	0	1	1	1	2	4	6	9	13	17	26	38
	4/0	0	0	1	1	1	3	5	8	10	14	22	31
	250	0	0	1	1	1	3	4	6	8	11	18	25
	300	0	0	0	1	1	2	3	5	7	9	15	22
	350	0	0	0	1	1	1	3	5	6	8	13	19
	400	0	0	0	1	1	1	3	4	6	7	12	17
	500	0	0	0	1	1	1	2	3	5	6	10	14
	600	0	0	0	0	1	1	1	3	4	5	8	12
	700	0	0	0	0	1	1	1	2	3	4	7	10
	750	0	0	0	0	1	1	1	2	3	4	7	9
	800	0	0	0	0	1	1	1	2	3	4	6	9
	900	0	0	0	0	0	1	1	1	3	3	6	8
	1000	0	0	0	0	0	1	1	1	2	3	5	7
FEP, FEPB, PFA, PFAH, TFE	14	8	16	27	49	68	115	164	257	347	450	714	1024
	12	6	12	20	36	50	84	120	188	253	328	521	747
	10	4	8	14	26	36	60	86	135	182	235	374	536
	8	2	5	8	15	20	34	49	77	104	135	214	307
	6	1	3	6	10	14	24	35	55	74	96	152	218
	4	1	2	4	7	10	17	24	38	52	67	106	153
	3	1	1	3	6	8	14	20	32	43	56	89	127
	2	1	1	3	5	7	12	17	26	35	46	73	105
PFA, PFAH, TFE	1	1	1	1	3	5	8	11	18	25	32	51	73
PFA, PFAH, TFE, Z	1/0	0	1	1	3	4	7	10	15	20	27	42	61
	2/0	0	1	1	2	3	5	8	12	17	22	35	50
	3/0	0	1	1	1	2	4	6	10	14	18	29	41
	4/0	0	0	1	1	1	4	5	8	11	15	24	34
Z	14	10	19	33	59	82	138	198	310	418	542	860	1233
	12	7	14	23	42	58	98	141	220	297	385	610	875
	10	4	8	14	26	36	60	86	135	182	235	374	536
	8	3	5	9	16	22	38	54	85	115	149	236	339
	6	2	4	6	11	16	26	38	60	81	104	166	238
	4	1	2	4	8	11	18	26	41	55	72	114	164
	3	1	2	3	5	8	13	19	30	40	52	83	119
	2	1	1	2	5	6	11	16	25	33	43	69	99
	1	0	1	2	4	5	9	13	20	27	35	56	80
XHH, XHHW, XHHW-2, ZW	14	6	11	20	35	49	82	118	185	250	324	514	736
	12	5	9	15	27	38	63	91	142	192	248	394	565
	10	3	6	11	20	28	47	67	106	143	185	294	421
	8	1	3	6	11	15	26	37	59	79	103	163	234
	6	1	2	4	8	11	19	28	43	59	76	121	173
	4	1	1	3	6	8	14	20	31	42	55	87	125
	3	1	1	3	5	7	12	17	26	36	47	74	106
	2	1	1	2	4	6	10	14	22	30	39	62	89

Table C9 Maximum number of conductors or fixture wires in rigid PVC conduit, schedule 80

		Conductors											
		Metric Designator (Trade Size)											
Type	Conductor Size (AWG/kcmil)	16 (½)	21 (¾)	27 (1)	35 (1¼)	41 (1½)	53 (2)	63 (2½)	78 (3)	91 (3½)	103 (4)	129 (5)	155 (6)
XHH, XHHW, XHHW-2	1	0	1	1	3	4	7	10	16	22	29	46	66
	1/0	0	1	1	2	3	6	9	14	19	24	39	56
	2/0	0	1	1	1	3	5	7	11	16	20	32	46
	3/0	0	1	1	1	2	4	6	9	13	17	27	38
	4/0	0	0	1	1	1	3	5	8	11	14	22	32
	250	0	0	1	1	1	3	4	6	9	11	18	26
	300	0	0	1	1	1	2	3	5	7	10	15	22
	350	0	0	0	1	1	1	3	5	6	8	14	20
	400	0	0	0	1	1	1	3	4	6	7	12	17
	500	0	0	0	1	1	1	2	3	5	6	10	14
	600	0	0	0	0	1	1	1	3	4	5	8	11
	700	0	0	0	0	1	1	1	2	3	4	7	10
	750	0	0	0	0	1	1	1	2	3	4	6	9
	800	0	0	0	0	1	1	1	1	3	4	6	9
	900	0	0	0	0	0	1	1	—	3	3	5	8
	1000	0	0	0	0	0	1	1	1	2	3	5	7
	1250	0	0	0	0	0	1	1	1	1	2	4	6
	1500	0	0	0	0	0	0	1	1	1	1	3	5
	1750	0	0	0	0	0	0	1	1	1	1	3	4
	2000	0	0	0	0	0	0	1	1	1	1	2	4

		Fixture Wires					
		Metric Designator (Trade Size)					
Type	Conductor Size (AWG/kcmil)	16 (½)	21 (¾)	27 (1)	35 (1¼)	41 (1½)	53 (2)
FFH-2, RFH-2, RFHH-3	18	6	11	19	34	47	79
	16	5	9	16	28	39	67
SF-2, SFF-2	18	7	14	24	43	59	100
	16	6	11	20	35	49	82
	14	5	9	16	28	39	67
SF-1, SFF-1	18	13	25	42	76	105	177
RFH-1, RFHH-2, TF, TFF, XF, XFF	18	10	18	31	56	77	130
RFHH-2, TF, TFF, XF, XFF	16	8	15	25	45	62	105
XF, XFF	14	6	11	20	35	49	82
TFN, TFFN	18	16	29	50	90	124	209
	16	12	22	38	68	95	159
PF, PFF, PGF, PGFF, PAF, PTF, PTFF, PAFF	18	15	28	47	85	118	198
	16	11	22	36	66	91	153
	14	8	16	27	49	68	115
HF, HFF, ZF, ZFF, ZHF	18	19	36	61	110	152	255
	16	14	27	45	81	112	188
	14	10	19	33	59	82	138
KF-2, KFF-2	18	28	53	88	159	220	371
	16	19	37	62	112	155	261
	14	13	25	43	77	107	179
	12	9	17	29	53	73	123
	10	6	11	20	35	49	82
KF-1, KFF-1	18	33	63	106	190	263	442
	16	23	44	74	133	185	310
	14	16	29	50	90	124	209
	12	10	19	33	59	82	138
	10	7	13	21	39	54	90
XF, XFF	12	3	6	10	19	26	44
	10	2	5	8	15	20	34

Note: This table is for concentric stranded conductors only. For compact stranded conductors, Table C9(A) should be used.

*Types RHH, RHW, and RHW-2 without outer covering.

National electrical code tables

Article 344—Rigid Metal Conduit

Metric Designator	Trade Size	Nominal Internal Diameter		Total Area 100%		2 Wires 31%		Over 2 Wires 40%		1 Wire 53%		60%	
		mm	in	mm²	in²	mm²	in²	mm²	in²	mm²	in²	mm²	in²
12	³/₈	—	—	—	—	—	—	—	—	—	—	—	—
16	½	16.1	0.632	204	0.314	63	0.097	81	0.125	108	0.166	122	0.188
21	³/₄	21.2	0.836	353	0.549	109	0.170	141	0.220	187	0.291	212	0.329
27	1	27.0	1.063	573	0.887	177	0.275	229	0.355	303	0.470	344	0.532
35	1¼	35.4	1.394	984	1.526	305	0.473	394	0.610	522	0.809	591	0.916
41	1½	41.2	1.624	1333	2.071	413	0.642	533	0.829	707	1.098	800	1.243
53	2	52.9	2.083	2198	3.408	681	1.056	879	1.363	1165	1.806	1319	2.045
63	2½	63.2	2.489	3137	4.866	972	1.508	1255	1.946	1663	2.579	1882	2.919
78	3	78.5	3.090	4840	7.499	1500	2.325	1936	3.000	2565	3.974	2904	4.499
91	3½	90.7	3.570	6461	10.010	2003	3.103	2584	4.004	3424	5.305	3877	6.006
103	4	102.9	4.050	8316	12.882	2578	3.994	3326	5.153	4408	6.828	4990	7.729
129	5	128.9	5.073	13050	20.212	4045	6.266	5220	8.085	6916	10.713	7830	12.127
155	6	154.8	6.093	18821	29.158	5834	9.039	7528	11.663	9975	15.454	11292	17.495

Reprinted with permission from NFPA 70-2002. *National Electrical Code*® Copyright © 1998. National Fire Protection Association, Quincy, MA 02269. This reprinted material is not the complete and official position of the National Fire Protection Association on the referenced subject, which is represented only by the standard in its entirety.

Electrical formulas

Single phase

kVA = VA/1000
hp = (VA × eff × PF)/746
kw = (VA × PF)/1000
PF = kW/kVA
Torque (ft-lb) = hp × 5250/rpm
Motor kVA = (hp × 0.746)/eff × PF
Motor kW = (hp × 0.746)/eff
V = line-to-line volts
A = amperes
eff = efficiency (decimal)
PF = power factor (decimal)

kVA = kilovolt amperes
kW = kilowatts
hp = horsepower
rpm = revolutions per minute

Three phase

$(3^{1/2}V \times A)/1000$
$(3^{1/2}VA \times PF \times eff)/746$
$(3^{1/2}VA \times PF)/1000$
$(kW \times 1000)/3^{1/2} \times VA$

Full-load currents—single-phase transformers

kVA Rating	Voltage—Volts					
	120	240	480	2400	4160	7200
1.0	8.3	4.2	2.1	0.4	0.2	0.1
3.0	25.0	12.5	6.3	1.3	0.7	0.4
5.0	41.7	20.8	10.4	2.1	1.2	0.7
7.5	62.5	31.3	15.6	3.1	1.8	1.0
10.0	83.3	41.7	20.8	4.2	2.4	1.4
15.0	125.0	62.5	31.3	6.3	3.6	2.1
25.0	208.3	104.2	52.1	10.4	6.0	3.5
37.5	312.5	156.3	78.1	15.6	9.0	5.2
50.0	416.7	208.3	104.2	20.8	12.0	6.9
75.0	625.0	312.5	156.3	31.3	18.0	10.4
100.0	833.3	416.7	208.3	41.7	24.0	13.9
125.0	1041.7	520.8	260.4	52.1	30.0	17.4
167.5	1395.8	697.9	349.0	69.8	40.3	23.3
200.0	1666.7	833.3	416.7	83.3	48.1	27.8
250.0	2083.3	1041.7	520.8	104.2	60.1	34.7
333.0	2775.0	1387.5	693.8	138.8	80.0	46.3
500.0	4166.7	2083.3	1041.7	208.3	120.2	69.4

Conduit size for combinations of cables with different outside diameters

The conduit size required for a group of cables with different outside diameters may be determined by calculating the equivalent outside diameter d'' and then finding the size required for this diameter in Table 1 below. For one to four cables:

$$d'' = \sqrt{\frac{n_1 d_1^2 + n_2 d_2^2 + n_m d_m^2 + \cdots}{n_1 + n_2 n_m + \cdots}}$$

where:

d'' = Equivalent diameter of same number of cables all of same outside diameter having total area equal to total area of group of cables of different sizes (a fictitious diameter appearing in the column corresponding to that number of cables ($n_1 + n_2 + n_m + \ldots$)
n_1 = number of cables of diameter d_1
n_2 = number of cables of diameter d_2
n_m = number of cables of diameter d_m, etc.

Example. Find the size conduit for two neoprene-sheathed cables having diameters of 1.20″ and one cable having a diameter of 0.63″.

$$d'' = \sqrt{\frac{2 \times (1.20)^2 + 1 \times (0.63)^2 \ldots}{2 + 1}}$$

$$d'' = 1.045$$

Refer to column 3 in Table 1 and note that the diameter of 1.045″ is between 0.901 and 1.120″. Therefore, a 3″ conduit will be required.

To determine the size conduit required for any number (n) of equal diameter cables in excess of four, multiply the diameter of one cable by $\sqrt{n/4}$. This will give the equivalent diameter of four such cables, and the conduit size required for (n) cables may then be found by using the column for four cables.

Table 1
Maximum Allowable Diameter (in Inches) of Individual Cables in Given Size of Conduit

Nonmetallic Jacketed Cable (All Cables of Same Outside Diameter)				
Nominal Size Conduit	**Number of Cables Having Same OD**			
	1	**2***	**3**	**4**
½	0.453	0.244	0.227	0.197
¾	0.600	0.324	0.301	0.260
1	0.763	0.412	0.383	0.332
1¼	1.010	0.542	0.504	0.436
1½	1.173	0.633	0.588	0.509
2	1.505	0.812	0.754	0.653
2½	1.797	0.970	0.901	0.780
3	2.234	1.206	1.120	0.970
3½	2.583	1.395	1.296	1.121
4	2.930	1.583	1.470	1.273
5	3.675	1.985	1.844	1.595
6	4.416	2.385	2.215	1.916

*Diameters are based on percent fill only. The jam ratio, conduit OD to cable ID should be checked to avoid possible jamming.

Minimum bending radius for insulated cables for permanent training during installation

These limits do not apply to conduit bends, sheaves, or other curved surfaces around which the cable may be pulled under tension while being installed. Larger radii bends may be required for such conditions to limit sidewall pressure. In all cases the minimum radii specified refer to the inner surface of the cable and not to the axis of the cable.

Power and control cables without metallic shielding or armor

The minimum bending radii for both single and multiple conductor cable with or without lead sheath and without metallic shielding or armor are given in Table 2.

Table 2

Thickness of Conductor Insulation, Inches	Overall Diameter of Cable, Inches		
	Minimum Bending Radius as a Multiple of Cable Diameter		
	1000 and Less	1.001–2.000	2.001 and Over
0.156 and less	4	5	6
0.157 to 0.315	5	6	7
0.316 and over		7	8

Twisted pair instrumentation cable

Table 3
Twisted Pair Instrumentation Cable

Type of Cable	Minimum Bending Radius as a Multiple of Cable Diameter
Armored, wire type, or corrugated sheath or interlocked type	7*
Nonarmored without shielded conductors	6
Nonarmored, metallized-polyester, or braid shielded	8
Nonarmored, flat, or corrugated tape shielded	12†

*With shielded conductors.
†For longitudinally applied corrugated shield with PVC jacket.

Power and control cables with metallic shielding or armor

Table 4
Power and Control Cables with Metallic Shielding or Armor

Type of Cable	Minimum Bending Radius as a Multiple of Cable Diameter	
	Power	Control
Armored, flat tape, or wire type	12	12
Armored smooth aluminum sheath, up to		
0.75 in cable diameter	10*	10*
0.76 in cable diameter	12	12
>1.5 in cable diameter	15	15
Armored corrugated sheath or interlocked type	7	7
With shielded single conductor	12	12
With shielded multiconductor	†	†
Nonarmored, flat, or corrugated		
Tape shielded single conductor	12	12
Tape shielded multiconductor	†	†
Multiconductor overall tape shield	12	12
LCS with PVC jacket	15	15
Nonarmored, flat strap shielded	8	—
Nonarmored, wire shielded	Refer to twisted pair instrumentation cable (Table 3)	

LCS = longitudinally applied corrugated shield.
*With shielded conductors, 12.
†12 times single conductor diameter or 7 times overall cable diameter, whichever is greater.

Rubber Jacketed Flexible Portable Power and Control Cables Used on Take-up Reels and Sheaves

Type of Cable	Minimum Bending Radius as a Multiple of Cable Diameter
0–5 kV	6
Over 5 kV	8

Full-load currents—three-phase transformers

kVA Rating	Voltage (Line-to-Line)				
	240	480	2400	4160	12,470
3	7.2	3.6	0.7	0.4	0.1
6	14.4	7.2	1.4	0.8	0.3
9	21.7	10.8	2.2	1.2	0.4
15	36.1	18.0	3.6	2.1	0.7
30	72.2	36.1	7.2	4.2	1.4
37.5	90.2	45.1	9.0	5.2	1.7
45	108.3	54.1	10.8	6.2	2.1
50	120.3	60.1	12.0	6.9	2.3
75	180.4	90.2	18.0	10.4	3.5
100	240.6	120.3	24.1	13.9	4.6
112.5	270.6	135.3	27.1	15.6	5.2
150	360.9	180.4	36.1	20.8	6.9
200	481.1	240.6	48.1	27.8	9.3
225	541.3	270.6	54.1	31.2	10.4
300	721.7	360.9	72.2	41.6	13.9
450	1082.6	541.3	108.3	62.5	20.8
500	1202.8	601.4	120.3	69.4	23.2
600	1443.4	721.7	144.3	83.3	27.8
750	1804.3	902.1	180.4	104.1	34.7
1000	2405.7	1202.8	240.6	138.8	46.3
1500	3608.5	1804.3	360.9	208.2	69.5
2000	4811.4	2405.7	481.1	277.6	92.6
2500	6014.2	3007.1	601.4	347.0	115.8
5000	12,028.5	6014.2	1202.8	694.0	231.5
10,000	24,057.0	12,028.5	2405.7	1387.9	463.0

Motor controller sizes

Polyphase motors

	Maximum Horsepower Full Voltage Starting	
NEMA Size	230 V	460–575 V
00	1.5	2
0	3	5
1	7.5	10
2	15	25
3	30	50
4	50	100
5	100	200
6	200	400
7	300	600

Single-phase motors

	Maximum Horsepower Full Voltage Starting (Two Pole Contactor)	
NEMA Size	115 V	230 V
00	1.3	1
0	1	2
1	2	3
2	3	7.5
3	7.5	15

Voltage drop on circuits using 600 V copper conductors in steel conduit

To determine the voltage drop, multiply the length in feet of one conductor by the current in amperes and by the number listed in the table for the type of system and power factor and divide the result by 1,000,000 to obtain the voltage loss. This table takes into consideration reactance on AC circuits and resistance of the conductor. Unless otherwise noted, the table is based on 60 Hz.

Wire Size	Direct Current	Three-Phase Lagging Power Factor					Single-Phase Lagging Power Factor				
		100%	90%	80%	70%	60%	100%	90%	80%	70%	60%
14	6100	5280	4800	4300	3780	3260	6100	5551	4964	4370	3772
12	3828	3320	3030	2720	2400	2080	3828	3502	3138	2773	2404
10	2404	2080	1921	1733	1540	1340	2404	2221	2003	1779	1547
8	1520	1316	1234	1120	1000	880	1520	1426	1295	1159	1017
6	970	840	802	735	665	590	970	926	850	769	682
4	614	531	530	487	445	400	614	613	562	514	462
3	484	420	425	398	368	334	484	491	460	425	385
2	382	331	339	322	300	274	382	392	372	346	317
1	306	265	280	270	254	236	306	323	312	294	273
0	241	208	229	224	214	202	241	265	259	247	233
00	192	166	190	188	181	173	192	219	217	209	199
000	152	132	157	158	155	150	152	181	183	179	173
0000	121	105	131	135	134	132	121	151	156	155	152
250M	102	89	118	123	125	123	103	136	142	144	142
300M	85	74	104	111	112	113	86	120	128	130	131
350M	73	63	94	101	105	106	73	108	117	121	122
400M	64	55	87	95	98	100	64	100	110	113	116
500M	51	45	76	85	90	92	52	88	98	104	106
600M	43	38	69	79	85	87	44	80	91	98	101
700M	36	33	64	74	80	84	38	74	86	92	97
750M	34	31	62	72	79	82	36	72	83	91	95
800M	32	29	61	71	76	81	33	70	82	88	93
900M	28	26	57	68	74	78	30	66	78	85	90
1000M	26	23	55	66	72	76	27	63	76	83	88

Determine the most economical size for electric power conductors

To quickly calculate the most economical size copper wire for carrying a specified current load, use the formula:

$$A = (59)(I)\sqrt{\frac{C_e \times t}{C_c \times F}}$$

where:

A = Conductor cross-sectional area (circular mils)
I = Current load (amperes)
C_e = Cost of electrical power (cents per kWh)
C_c = Cost of copper (cents per lb)
t = Hours of service per year
F = Factor for fixed charges (amortization, insurance, taxes, etc.)

Selection of the proper size electric conductor will depend on the load to be carried and the mechanical strength required, as well as economics. Once these are considered, the most economical size conductor is that for which the annual energy cost equals the copper cost. This method can be used for other metallic conductors.

Example. Determine the most economical size copper conductor for an installation operating 365 days a year, 8 h per day, with a 100 A load. Energy costs are $0.00875 per kWh. Fixed charges are 23%, considering a 5-year amortization with 3% for insurance and taxes. Copper costs $0.46 per lb.

$$A = (59)(100)\sqrt{\frac{0.00875 \times 365 \times 8}{0.46 \times 0.23}}$$

$$= 91,686 \text{ circular mils}$$

How to find the resistance and weight of copper wires

It is easy to mentally calculate the approximate resistance and weight of standard sizes of copper wires, by remembering the following rules:

Rule 1: Number 10 wire has a resistance of 1 Ω per 1000 ft. Number 2 wire weighs 200 lb. per 1000 ft.

Rule 2: An increase of 10 numbers in the size is an increase of 10 times in the resistance, and a decrease to one-tenth of the weight.

Rule 3: An increase of three numbers in the size doubles the resistance and halves the weight.

Rule 4: An increase of one number in the size multiplies the resistance by 5/4 and the weight by 4/5.

Always proceed from the known value (No. 10 for resistance, No. 2 for the weight) in the smallest number of jumps; that is, go by tens, then by threes, and finally by ones; the jumps do not all have to be in the same direction.

Example. What is the weight and resistance of 200 ft of No. 18? Apply the rules in order, as follows; resistance first:

Number 10 wire has a resistance of 1 Ω per 1000 ft (Rule 1)
Number 20 wire has a resistance of 10 Ω per 1000 ft (Rule 2)
Number 17 wire has a resistance of 5 Ω per 1000 ft (Rule 3)
Number 18 wire has a resistance of 6.25 Ω per 1000 ft (Rule 4)
Number 18 wire has a resistance of 1.25 Ω per 200 ft (Answer)
Number 2 wire weighs 200 lb per 1000 ft (Rule 1)
Number 12 wire weighs 200 lb per 1000 ft (Rule 2)
Number 22 wire weighs 2 lb per 1000 ft (Rule 2)
Number 19 wire weighs 4 lb per 1000 ft (Rule 3)
Number 18 wire weighs 5 lb per 1000 ft (Rule 4)
Number 18 wire weighs 1 lb per 200 ft (Answer)

The weight determined by the above rules is that of bare copper wire. Weight of insulated wire varies widely according to the type of insulation. Resistance values are correct for bare or insulated wire and for solid or stranded. Errors resulting from the use of the rules will rarely exceed 2%.

What you should remember about electrical formulas

Ohm's law

Easily **R**emember **I**t, or $E = RI$.

Power

DC **P**ower **I**s **E**asy, or $P = IE$.
It **R**eally **I**s, or $P = IRI = I^2R$

Other facts

The number of *watts in one hp* is equal to the year Columbus discovered America divided by two, or $1492/2 = 746$ W/hp.

The horsepower of a reciprocating engine necessary to drive a three-phase, 60-cycle electric generator can be determined by multiplying kW by 1.5.

How to calculate microwave hops on level ground

Rule of thumb. Excluding obstructions such as hills, the microwave (line-of-sight) distances between two towers will be twice the distance from the transmitter to the horizon. The line-of-sight distance in miles from the top of the tower to the horizon can be found by adding the square root of the tower height to the square root of the first square root (the latter, of course, is the fourth root).

$$x = \sqrt{h} + \sqrt{\sqrt{h}} \ \text{or} \ \sqrt{h} + \sqrt[4]{h}$$

Example. How far apart can 100-ft-high towers be spaced to provide line-of-sight transmission if there are no obstructions between them?

$X = \sqrt{100} + \sqrt{10}$

$X = 10 + 3 +$ or $+ 13$ miles

Distance between towers can be 2X or 26 miles.

This same formula can be used to estimate distances in an airplane.

Example. How far is the line-of-sight distance to the horizon in an airplane flying at an altitude of 4900 ft?

$$\left(\sqrt{4900} = 70\right)$$

$X = \sqrt{4900}\sqrt{70}$

$X \doteq 70 + 8$

Distance to horizon is 78 miles.

For quick determination of the horsepower per ampere for induction motors (3 phase) at different voltages

Voltage	hp/A
480	1
2400	5
4160	8

Chart of electric motor horsepower for pumping units

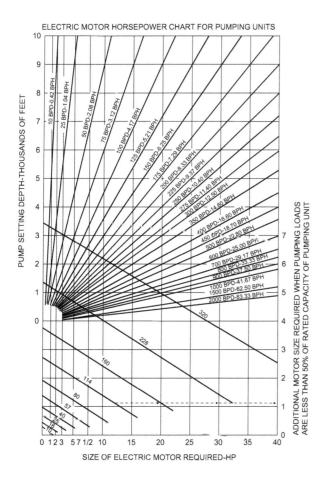

This chart provides a means of determining the power requirements for a beam pumping installation powered by an electric motor based on a fluid with a specific gravity of 1, with fluid at the pump. The power requirement determined by the chart includes a mechanical efficiency factor of 0.45 and a cyclic factor of 0.75, which are factors frequently applied to motors used in sucker rod pumping service.

An arrangement is available for correcting the power requirement in the case of an underloaded pumping unit.

The example shown in the chart is self-explanatory.

After the power requirements are determined from the chart, the next higher size of commercially available motor is used.

Pumping stations

Capacitor multipliers for kilowatt loads for different desired power factor improvements

Original Power Factor, Percent	DESIRED POWER FACTOR—Percent				
	100	95	90	85	80
50	1.732	1.403	1.248	1.112	0.982
51	1.687	1.358	1.203	1.067	0.937
52	1.643	1.314	1.159	1.023	0.893
53	1.600	1.271	1.116	0.980	0.850
54	1.559	1.230	1.075	0.939	0.809
55	1.518	1.190	1.035	0.898	0.769
56	1.480	1.151	0.996	0.860	0.730
57	1.442	1.113	0.958	0.822	0.692
58	1.405	1.076	0.921	0.785	0.655
59	1.369	1.040	0.885	0.749	0.619
60	1.333	1.004	0.849	0.713	0.583
61	1.299	0.970	0.815	0.679	0.549
62	1.266	0.937	0.782	0.646	0.516
63	1.233	0.904	0.749	0.613	0.483
64	1.201	0.872	0.717	0.581	0.451
65	1.169	0.840	0.685	0.549	0.419
66	1.138	0.809	0.654	0.518	0.388
67	1.108	0.779	0.624	0.488	0.358
68	1.078	0.749	0.594	0.458	0.328
69	1.049	0.720	0.565	0.429	0.299
70	1.020	0.691	0.536	0.400	0.270
71	0.992	0.663	0.508	0.372	0.242
72	0.964	0.635	0.480	0.344	0.214
73	0.936	0.607	0.452	0.316	0.186
74	0.909	0.580	0.425	0.289	0.159
75	0.882	0.553	0.398	0.262	0.132
76	0.855	0.526	0.371	0.235	0.105
77	0.829	0.500	0.345	0.209	0.079
78	0.802	0.473	0.318	0.182	0.052
79	0.776	0.447	0.292	0.156	0.026
80	0.750	0.421	0.266	0.130	....
81	0.724	0.395	0.240	0.104	....
82	0.698	0.369	0.214	0.078	....
83	0.672	0.343	0.188	0.052	....
84	0.646	0.317	0.162	0.026	....
85	0.620	0.291	0.136	....	....
86	0.593	0.264	0.109	....	....
87	0.567	0.238	0.083	....	....
88	0.540	0.211	0.056	....	....
89	0.512	0.183	0.028	....	....
90	0.484	0.155			
91	0.456	0.127			
92	0.426	0.097			
93	0.395	0.066			
94	0.363	0.034			
95	0.329	....			
96	0.292	....			
97	0.251	....			
98	0.203	....			
99	0.143	....			

Example

Assume total plant load is 100 kw at 60 percent power factor. Capacitor kvar rating necessary to improve power factor to 80 percent is found by multiplying kw (100) by multiplier in table (0.583), which gives kvar (58.3). Nearest standard rating (60 kvar) should be recommended.

Courtesy General Electric Company.

FLOODLIGHTING CONCEPTS

This material is intended as a helpful guide in the evaluation of lighting requirements and is intended to serve only as a guide.

Terms

Candela or Candlepower—Light sources do not project the same amount of light in every direction. The directional characteristic of a lamp is described by the candlepower in specific directions. This directional strength of light or luminous intensity is measured in candelas.

Lumen—Light quantity, irrespective of direction, is measured in lumens. Lamp lumens are the quantity of light produced by a lamp. Average light level calculations use total lamp lumens as a basis and then adjust for all factors that lower this quantity. The amount of useful light in a floodlight beam is measured in beam lumens.

Footcandle (fc)—Specifications are usually based on density of light or level of illumination, which is measured in footcandles. Footcandles are the ratio of quantity of light in lumens divided by the surface area in square feet on which the lumens are falling. A density of one lumen per square foot is one footcandle. One footcandle is equal to 10.76 lux.

Lux (lx)—The SI unit of illuminance. This metric measurement is based upon the density of lumens per unit surface area similar to footcandle, except one lux is one lumen per square meter. One lux is equal to 0.09 footcandle.

Light Loss Factor (LLF)—These factors are used to adjust lighting calculations from laboratory test conditions to a closer approximation of expected field results. The I.E.S. Lighting Handbook, 1984 Reference Volume, defines LLF as follows: "a factor used in calculating illuminance after a given period of time and under given conditions. It takes into account temperature and voltage variations, dirt accumulations on luminaire and room surfaces, lamp depreciation, maintenance procedures and atmosphere conditions."

Floodlighting calculations

Floodlighting encompasses many variations. Since the location of the floodlight relative to the object to be lighted can be in any plane and at any distance from the source, a floodlighting application is often considered the most complex and difficult of all lighting techniques.

The most commonly used systems for floodlight calculations are the point-by-point method and the beam-lumen method.

Point-by-point method

The point-by-point method permits the determination of footcandles at any point and orientation on a surface and the degree of lighting uniformity realized for any given set of conditions.

In such situations, the illumination is proportional to the candlepower of the source in a given direction and inversely proportional to the square of the distance from the source. See Fig. 1.

$$\text{Footcandles on plane (Normal to light ray)} = \frac{\text{Candlepower of light ray}}{\text{Distance in feet from Source to point} - \text{squared}}$$

$$E = 1/D^2 \tag{1}$$

When the surface on which the illumination to be determined is tilted, the light will be spread over a greater area, reducing the illumination in the ratio of the area of plane A to the area of plane B as shown in Fig. 2. This ratio is equal to the cosine of the angle of incidence; thus:

I = 1 Candle- E = 1 Foot- E = ¼ Foot- E = $\frac{1}{D^2}$
Power   Candle   Candle

Figure 1. Inverse square law.

$$\text{Footcandles on plane B} = \frac{\text{Candlepower of light ray}}{\begin{array}{c}\text{Distance in feet from}\\\text{Source to point} - \text{squared}\end{array}} \times \text{Cosine of angle } \beta$$

$$E = I/D^2 \times \cos\beta \tag{2}$$

Then, β equals the angle between the light ray and a perpendicular to the plane at that point.

Illumination E =
Flux in Lumens (F)
Area in Sq. Ft. (A)

Angle of Tilt
or Incident
Normal

Area of Plane = A/Cos β
E = F/A (Cos β)

Figure 2. Cosine of angle law.

Beam-lumen method

The beam-lumen method is quite similar to the method used for interior lighting except that the utilization factors must take into consideration the fact that floodlights are not usually perpendicular to the surface and therefore not all of the useful light strikes the surface.

Beam lumens are defined as the quantity of light that is contained within the beam limits as described as "beam spread." Beam lumens equal the lamp lumens multiplied by the beam efficiency of the floodlight.

Coverage. It is recommended that sufficient point-by-point calculations be made for each job to check uniformity and coverage.

Light Loss Factor. The maintenance or light loss factor is an allowance for depreciation of lamp output with age and floodlight efficiency due to the collection of dirt on the lamp, reflector, and cover glass. The total factor may vary from 0.65 to 0.85 depending on the type of lamp and luminaire used and may include losses due to lamp orientation, or "tilt."

Design procedure

Step 1—Determine the level of illumination. See Table 5 for some typical levels of illumination (fc). The basic formula is:

$$\text{fc} = \frac{N \times BL \times CBU \times LLF}{A} \tag{3}$$

where:

N = quantity of luminaries
A = area in square feet
BL = beam lumens
CBU = coefficient of beam utilization
LLF = light loss factor

Step 2—Determine type and location of floodlights. Regardless of light source, there are industry standards on beam spreads. See Table 6.

Step 3—Determine the coefficient of beam utilization. This factor, CBU, written as a decimal fraction, is expressed in the following ratio:

$$CBU = \frac{\text{Utilized lumens}}{BL} \tag{4}$$

The exact CBU can be graphically determined by projecting the outline of the area to be lighted upon the photometric data and totaling the utilized lumens. This procedure is detailed in the I.E.S. Lighting Handbook. See Table 7.

Table 5
Typical Illumination Levels

Area	fc
Building exteriors	
Entrances	
Active	5
Inactive	1
Vital locations	5
Catwalks	3
Stairs	5
Barge or truck unloading	5
Storage tanks	1
Pump areas	2
Parking	1
Substation	2
Storage yards	
Active	20
Inactive	1

Table 6
Outdoor Floodlight Luminaire Designations

Beam Spread, Degrees	NEMA Type
10–18	1
>18–29	2
>29–46	3
>46–70	4
>70–100	5
>100–130	6
>130 up	7

The following general principles apply in the choice of beam spread.
1. The greater the distance from the floodlight to the area to be lighted, the narrower the beam spread desired.
2. Since by definition the candlepower at the edge of a floodlight beam is 10% of the candlepower near the center of the beam, the illumination level at the edge of the beam is one-tenth or less of that at the center. To obtain reasonable uniformity of illumination, the beams of individual floodlights must overlap each other, as well as the edge of the surface to be lighted.
3. The percentage of beam lumens falling outside the area to be lighted is usually lower with narrow-beam units than with wide-beam units. Thus, narrow-beam floodlights are preferable where they will provide the necessary degree of uniformity of illumination and the proper footcandle level.

As an approximation, the average CBU of all the floodlights in an installation should fall within the range of

Table 7
Typical CBUs

Area	CBU	Area	fc
Baseball			
Infield	0.65	Tennis	0.75
Outfield	0.85	Large parking lots	0.80
Football	0.60	Building facades	0.65

0.6–0.9. If less than 60% of the beam lumens are utilized, a more economical lighting plan should be possible by using different locations or narrower beam floodlights. If the CBU is greater than 0.9, it is probable that the beam spread selected is too narrow and the resultant illumination will be spotty. An estimated CBU can be determined by experience or by making calculations for several potential aiming points and using the average figure thus obtained.

Step 4—Determine the quantity of floodlights (N) required. Rearrange the basic formula in Step 1 as follows:

$$N = \frac{A \times \text{fc}}{\text{BL} \times \text{CBU} \times \text{LLF}} \qquad (5)$$

Conductor size conversion chart—metric to AWG

Metric Size	AWG Size
2.5 mm^2	#14
4 mm^2	#12
6 mm^2	#10
10 mm^2	#8
16 mm^2	#6
25 mm^2	#31
35 mm^2	#2
35 mm^2	#1*
50 mm^2	#1/0
70 mm^2	#2/0
95 mm^2	#3/0
95 mm^2	#4/0*
120 mm^2	250 kcmil
150 mm^2	300 kcmil
185 mm^2	350 kcmil
185 mm^2	400 kcmil*
240 mm^2	500 kcmil
300 mm^2	600 kcmil
400 mm^2	700 kcmil
400 mm^2	750 kcmil
400 mm^2	800 kcmil*

*This AWG size is the closest equivalent to the metric size.

Commonly used switchgear device numbers

Number	Function
12	Overspeed—Usually a direct-connected speed switch that functions on machine overspeed.
15	Speed or frequency matching that functions to match and hold the speed or the frequency of a machine or system equal to or approximately equal to that of another machine, source, or system.
21	Distance relay that functions when the circuit admittance, impedance, or reactance changes beyond determined limits.
25	Synchronizing or synchronism check device that operates when two AC circuits are within the desired limits of frequency, phase angle, or voltage, to permit paralleling two circuits.
32	Directional power relay that functions on a desired value of power flow in a given direction or upon reverse power.
38	Bearing protective device that functions on excessive bearing temperature or other abnormal conditions associated with the bearing such as undue wear that may ultimately result in excessive bearing temperature or failure.
39	Mechanical condition monitor that functions when an abnormal mechanical condition (except conditions associated with bearings covered under device 38) such as excessive vibration, eccentricity, expansion, shock, tilting, or seal failure.
40	Field relay that functions on a given or abnormally low value or failure of machine field current, or due to excessive value of the reactive component of the armature current in an AC machine indicating abnormally low field excitation.
41	Field circuit breaker functions to apply or remove the field excitation of a machine
43	Manual Transfer or Selector device is a manually operated device that transfers the control circuits in order to modify the plan of operation of switching equipment or of some of the devices.
46	Reverse phase or phase balance current relay operates when the polyphase currents are of reverse phase sequence or when the currents are unbalanced or contain negative phase sequence components greater than a set value.
47	Phase sequence voltage relay operates upon a predetermined value of polyphase voltage in the desired sequence.
50	Instantaneous overcurrent or rate-of-rise relay instantaneously operates on an excessive value of current or on an excessive rate of current in an AC circuit exceeds a predetermined value.
51	AC time overcurrent relay that has a definite or inverse time characteristic that operates when the current in an AC circuit exceeds a predetermined value.
52	AC circuit breaker that is used to close and interrupt an AC power circuit under normal conditions or under fault or emergency conditions.
59	Overvoltage relay that operates on a given value of overvoltage.
60	Voltage or current balance relay that operates on a given difference in voltage or current input or output of two circuits.
62	Time delay stopping or opening relay that serves in conjunction with the device that initiates the shutdown, stopping, or opening operation in an automatic sequence or protective relay system.
64	Ground protective relay operates on failure of the insulation of a machine, transformer, or other apparatus to ground, or on flashover of a DC machine to ground. This device is assigned only to a relay that detects the flow of current from the frame of a machine or enclosing case or structure of a piece of apparatus to ground, or detects a ground on a normally ungrounded winding or circuit. It is not applied to a device connected in the secondary circuit of a current transformer, or in the secondary neutral of current transformers, connected in the power circuit of a normally grounded system.
65	Governor is the assembly of fluid, electrical, or mechanical control equipment used for regulating the flow of water, steam, or other medium to the prime mover for such purposes as starting holding speed, load, or stopping.
67	AC directional overcurrent relay that operates on a desired value of AC overcurrent flowing in a predetermined direction.
70	AC reclosing relay that controls the automatic reclosing and locking out of an AC circuit interrupter.
81	Frequency relay operates on a predetermined value of frequency (either under or over or on normal system frequency or rate of change of frequency.

Number	Function
86	Lock-out relay is an electrically operated hand or reset relay or device that operates to shutdown or hold an equipment out of service, or both, when an abnormal condition occurs.
87	Differential protective relay that operates on a percentage or phase angle or other quantitative difference of two currents or of some other electrical quantities.

Suffix letters permit a manifold multiplication of available function designations for the large number and variety of devices that may be used. They may also serve to denote individual or specific parts or auxiliary contacts of these devices or certain distinguishing features, characteristics, or conditions that describe the use of the device or its contacts in the equipment.

Suffix letters should be used only when they accomplish a specific purpose. For example, when all of the devices in an equipment are associated with only one kind of apparatus, such as a feeder or motor or generator, it is common practice in order to retain maximum simplicity in device function identification, not to add the respective suffix letter F or M or G to any of the device function numbers.

Some commonly used suffix designations are as follows:

A	Automatic
M	Manual
27R	Voltage restraint
27C	Voltage controlled
R	Reverse
RV	Reverse VARS
G	Generator or ground
T	Transformer

Bonding the grounding system to building and structure foundations

During the 2005 National Electrical Code (NEC) cycle, a subtle change was made to Section 250.50. The words "if available" were removed from the beginning of the paragraph to clarify the objective to include all concrete-encased electrodes in the foundations of new buildings and structures.

Looking at it from a strictly electrical standpoint, this seems like a good idea. Without a doubt, concrete-encased electrodes make excellent grounding electrodes. If the concrete stays moist and the reinforcing steel bars are conductive (that is, not coated with epoxy or anything else) the results, for grounding purposes, will be exceptional. Even without moist conditions, the concrete-encased system will provide a plane of equal potential when bonded to the building steel and other grounding electrodes indicated in Section 250.52. However, what appeared to be a step in the right direction as far as safety is concerned may have some potential drawbacks in some installations.

In large industrial refineries and petrochemical plants, the typical grounding electrode system usually consists of 2/0 AWG, 4/0 AWG, up to 500 kcmil bare copper cables laid out in a grid around structural steel, pumps, compressors, and other equipment. Multiple ground rods installed throughout the grid system are typically copper-clad steel 3/4 of an inch (1.9 cm in diameter in lengths of 10–20 ft (25–51 m). Nearly every piece of equipment is bonded to the grounding electrode system through taps of smaller-gauge cable, such as 2 AWG or 2/0 AWG. This allows the formation of an equipotential plane that prevents a lightning strike or an electrical fault from creating an unsafe condition resulting from the subsequent potential difference between neighboring vessels, motors, steel, and pipes.

In the past, the concrete-encased electrodes in the pipe rack piers and vessel footings were not usually bonded to the grid, even though the above-ground structural steel was. Blindly doing so now without considering the potential for corrosion of the structural rebar in the concrete can be costly.

Corrosion is an electrochemical process involving the flow of electrons and ions that occurs when a corrosion cell is formed. A corrosion cell consists of an anode, a cathode, a common electrolyte containing the anode and cathode, and a metallic path between the anode and cathode (see Fig. 1). Corrosion occurs at the anode. When the anode and cathode are of dissimilar metals, current is generated between the two due to the inherent difference in their refined energy levels. Electrons flow in the metallic path and ions flow in the electrolyte. This is the basis for the operation of the common dry cell or battery.

In a copper-based grounding system of a facility constructed of steel, the copper will behave as a cathode and the steel will behave as an anode. The resulting corrosion of the steel reduces the structural strength of the rebar and destroys the rebar/concrete bond. In addition, the corrosion products are physically larger than the original reactants and will thus exert tremendous pressure on the surrounding concrete, causing the concrete to fracture and fail. This is referred to as spalling.

For facilities constructed in compliance with the 2005 edition of the NEC, there are several corrosion-mitigation options available.

Option one

The first option is to eliminate the corrosion cell circuit by breaking the metallic path. This can be done by isolating the copper grounding system from underground steel objects such as piping and rebar. To eliminate the dissimilar-metal corrosion problem, the isolation must be guaranteed by proper construction practices and periodic testing. Isolation is normally maintained using insulating flange kits, dielectric unions, nonmetallic conduits and cable trays, dielectric pads and supports, and separate grounding systems. Care has to be taken in the instrumentation and electrical designs so as not to short the intended isolation.

This isolation option is also practiced when cathodically protected steel structures come into physical contact with nonprotected steel structures. However, isolating the copper grounding system in this manner does not comply with NEC Section 250.50, and the method should not be used where electrical systems are present and installation is within the scope of NEC Section 90.2(A). This option may be used in facilities, such as a utility power generation facility, that are not subject to NEC requirements.

As a practical matter, complete electrical isolation is very difficult to achieve and maintain.

Option two

The second option is to eliminate the voltage difference between the anode and cathode using components other than copper in the soil and concrete electrolyte. Copper grounding electrode cables can be insulated from the soil as they travel from rod to rod and all taps in between, or a material other than copper, such as zinc-clad or stainless steel, can be used in the rods. Of course, stainless steel or nonferrous ground rods less than 5/8 of an inch (1.6 cm) in diameter should be listed per NEC 250.52 (A)(5)(b). Listed zinc-galvanized ground rods are available. Stainless steel rods should only be used after a thorough soils investigation, as high-chloride content in the soil can badly corrode stainless steel.

One disadvantage of this option is the loss of contact between the copper cables and soil throughout the grid. Since insulated cables do not lower the overall resistance to ground, more rods will be needed to achieve the same ohmic levels. The cables in this case become nothing more than jumpers between the rods, allowing a convenient place to tap to the equipment.

This option may be the most economical and practical choice for many industrial facilities, provided a regimented maintenance plan is put into effect. Periodic testing of the grounding system should be conducted as the grounding components are no longer cathodically protected. This option also complies with NEC Section 250.50.

Option three

The third option is to provide cathodic protection for both the grounding system and structural steel (see Fig. 4). Cathodic protection reduces the corrosion of a metal surface by making the surface the cathode of an electrochemical cell. Before applying cathodic protection, corroding structures will have both cathodic and anodic areas where corrosion occurs. This is one reason grounding practitioners prefer to use copper: it appears to be very stable when, in fact, it is being cathodically protected by the steel anode in the corrosion cell circuit.

Because grounding system components are generally bare to maximize surface area and minimize the resistance to remote earth and because cathodic protection depends on an adequate current flow to all bare metal surfaces, protecting the facility in its entirety often requires a tenfold increase in the amount of cathodic protection current required.

In the case of a utility power generation facility, where the cost of such an installation can be justified and both grounding performance and integrity are of great importance, option three may be the best alternative.

In many industrial plants, however, combining options two and three may work best.

By using noncopper grounding rods interconnected with insulated conductors and including the grounding system in the cathodic protection system's zone of protection, the grounding system, buried steel piping, and concrete reinforcing steel are all protected. Eliminating exposed copper from the grounding system will reduce the required capacity of the cathodic protection system.

Will rebar corrosion be a problem in residential and commercial foundations? Probably not. These types of structures typically have limited amounts of copper grounding electrodes in the soil, and circuit resistance of the electrolytic path will be relatively high. Minor corrosion of rebar and limited foundation cracking and spalling will not seriously affect their design lives.

There is certainly more to be learned about corrosion and cathodic protection than can be covered here. The purpose of this article is to make the electrical design engineer aware of the possible adverse consequences and to suggest that a qualified corrosion expert be consulted when designing copper grounding systems that are bonded to structural steel rebar.

By Whitt Trimble and Edie Guidry

5: Hydrostatic Testing

THE BENEFITS AND LIMITATIONS OF HYDROSTATIC TESTING

John F. Kiefner and Willard A. Maxey, Adapted from Dr John Kiefner, Kiefner and Associates, Inc. Reprinted with permission.

The purpose of this discussion is to clarify the issues regarding the use of hydrostatic testing to verify pipeline integrity. There are those who say it damages a pipeline, especially if carried out to levels of 100% or more of the specified minimum yield strength (SMYS) of the pipe material. These people assert that if it is done at all, it should be limited to levels of around 90% of SMYS. There are those who insist that pipelines should be retested periodically to reassure their serviceability. The reality is that if and when it is appropriate to test a pipeline, the test should be carried out at the highest possible level that can feasibly be done without creating numerous test failures. The challenge is to determine if and when it should be done, the appropriate test level, and the test-section logistics that will maximize the effectiveness of the test.

The technology to meet these challenges has been known for 30 years. Nothing has arisen in the meantime to refute this technology. The problem is that people both within and outside the pipe industry either are not aware of the technology or have forgotten it, or for political reasons are choosing to ignore it.

In this discussion we show the following:

- It makes sense to test a new pipeline to a minimum of 100% of SMYS at the highest elevation in the test section.
- Pipe that meets the specified minimum yield strength is not likely to be appreciably expanded even if the maximum test pressure is 110% of SMYS.
- If hydrostatic retesting is to be conducted to revalidate the serviceability of a pipeline suspected to contain defects that are becoming larger with time in service, the highest feasible test pressure level should be used.
- If the time-dependent defects can be located reliably by means of an in-line inspection tool, using the tool is usually preferable to hydrostatic testing.

We also note the following as reminders:

- When a pipeline is tested to a level in excess of 100% of SMYS, a pressure—volume plot should be made to limit yielding.
- A test may be terminated short of the initial pressure target, if necessary, to limit the number of test breaks as long as the MOP guaranteed by the test is acceptable to the pipeline's operator.

And, we suggest that:

- Test-section length should be limited to prevent elevation differences within a test section from exceeding 300 ft.

The pressure level for verifying integrity can be higher than the level needed to validate the MOP of the pipeline, and the integrity test to a level above 1.25 times MOP, if used, needs to be no longer than ½ h.

Background

The concept and value of high-pressure hydrostatic testing of cross-country pipelines were first demonstrated by Texas Eastern Transmission Corporation. Texas Eastern sought the advice of Battelle in the early 1950s as they began to rehabilitate the War Emergency Pipelines and to convert them to natural gas service. Prior to testing, these pipelines exhibited numerous failures in service due to original manufacturing defects in the pipe. The Battelle staff recommended hydrostatic testing to eliminate as many of these types of defects as possible. After being tested to levels of 100—109% of SMYS, during which time "hundreds" of test breaks occurred, not one in-service failure caused by a manufacturing defect was observed. The news of this successful use of hydrostatic testing spread quickly to other pipeline operators, and by the late 1960s the ASA B31.8 Committee (forerunner of ASME B31.8) had established an enormous database of thousands of miles of pipelines that had exhibited no in-service ruptures from original manufacturing or construction defects after having been hydrostatically tested to levels at or above 90% of SMYS.[1] These data were used to establish the standard practice and ASA B31.8 Code requirement that, prior to service, each gas pipeline should be hydrostatically tested to 1.25 times its maximum allowable operating pressure. Later, a similar requirement for liquid pipelines was inserted into the ASME B31.4 Code. When federal regulations for pipelines came along, the precedent set by the industry of testing to 1.25 times the MOP was adopted as a legal requirement.

Both field experience and full-scale laboratory tests have revealed much about the benefits and limitations of hydrostatic testing. Among the things learned were the following:

- Longitudinally oriented defects in pipe materials have unique failure pressure levels that are predictable on the basis of the axial lengths and maximum depths of the defects and the geometry of the pipe and its material properties.[2]
- The higher the test pressure, the smaller will be the defects, if any, that survive the test.
- With increasing pressure, defects in a typical line-pipe material begin to grow by ductile tearing prior to failure. If the defect is close enough to failure, the ductile tearing that occurs prior to failure will continue even if pressurization is stopped and the pressure is held constant. The damage created by this tearing when the defect is about ready to fail can be severe enough that if pressurization is stopped and the pressure is released, the defect may fail upon a second or subsequent pressurization at a pressure level below the level reached on the first pressurization. This phenomenon is referred to as a pressure reversal.[3,4]
- Testing a pipeline to its actual yield strength can cause some pipe to expand plastically, but the number of pipes affected and the amount of expansion will be small if a pressure–volume plot is made during testing and the test is terminated with an acceptably small offset volume or reduction in the pressure-volume slope.[5]

Test-pressure-to-operating-pressure ratio

The hypothesis that "the higher the test-pressure-to-operating-pressure ratio, the more effective the test," is validated by Figure 1. Figure 1 presents a set of failure-pressure-versus-defect-size relationships for a specific diameter, wall thickness, and grade of pipe. A great deal of testing of line-pipe materials over the years has validated these curves.[2] Each curve represents a flaw with a uniform depth-to-wall-thickness ratio. Nine such curves are given (d/t ranging from 0.1 to 0.9).

Consider the maximum operating pressure (MOP) for the pipeline (the pressure level corresponding to 72% of SMYS). That pressure level is represented in Figure 1 by the horizontal line labeled MOP. At the MOP, no defect longer than 10 in. and deeper than 50% of the wall thickness can exist. Any such defect would have failed in service. Similarly, no defect longer than 4 in. and deeper than 70% of the wall thickness can exist, nor can one that is longer than 16 in. and deeper than 40% of the wall thickness.

By raising the pressure level above the MOP in a hydrostatic test, the pipeline's operator can assure the absence of defects smaller than those that would fail at the MOP. For example, at a test pressure level equivalent to 90% of SMYS, the largest surviving defects are determined in Figure 1 by the horizontal line labeled 90% of SMYS. At that level, the longest surviving defect that is 50% through the wall can be only about 4.5 in. Compare that length to the length of the longest possible 50%-through flaw at the MOP; it was 10 in. Alternatively, consider the minimum survivable depth at 90% SMYS for a 10-in.-long defect (the size that fails at the MOP if it is 50% through the wall). The survivable depth is only about 32% through the wall. By a similar process of reasoning, one can show that even smaller flaws are assured by tests to 100% or 110% of SMYS (the horizontal lines drawn at those pressure levels on Figure 1).

The point is that the higher the test pressure (above MOP), the smaller will be the possible surviving flaws. This fact means a larger size margin between flaw sizes left after the test and the sizes of flaws that would cause a failure at the MOP. If surviving flaws can be extended by operating-pressure cycles, the higher test pressure will assure that it takes a longer time for these smaller flaws to grow to a size that will fail at the MOP. Thus, Figure 1 provides proof of the validity of the hypothesis (i.e. the higher the test-pressure-to-operating-pressure ratio, the more effective the test).

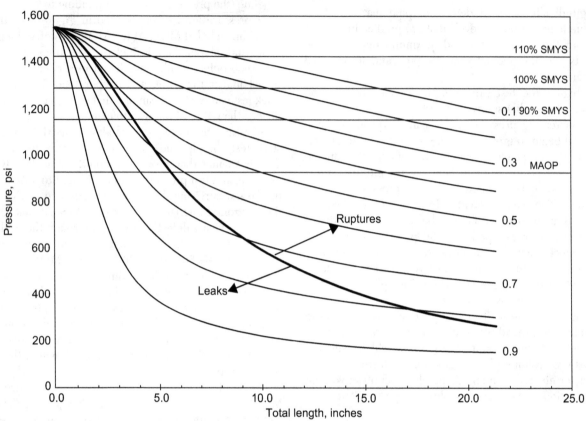

Rectangular surface flaw equation

Diameter, inc	30,000	Charpy, ft-lb	50.0
Wall thickness, inch	0.375	Charpy area, in^2	0.1240
SMYS, psi	52,000		

Figure 1. Impact of test pressure levels on margin of safety.

Testing to levels above 100% of SMYS

Given the previous argument for testing to the highest feasible level, one needs to consider practical upper limits. In the case of a new pipeline constructed of modern high-quality, high-toughness line pipe, the maximum test level can generally be in excess of 100% of SMYS. Reasons why this will not cause significant yielding of the pipe are as follows.

First, as shown in Figure 2, the average yield strength of an order of pipe is usually well above the minimum specified value. Very few pieces will have yield strengths low enough to cause yielding at 100% of SMYS.

Secondly, when a buried pipeline is pressurized, it is restrained by the soil from shortening in the axial direction. This causes an axial tensile stress equal to Poisson's ratio times the hoop stress (Poisson's ratio is 0.3 for steel). What this means to testing in excess of 100% of SMYS is shown in Figure 3.

The tensile test commonly used to assess the yield strength of line pipe is a transverse, flattened uniaxial specimen designed to test the circumferential (hoop) direction tensile properties. A test of such a specimen reveals a unique value of yield strength at a certain value of applied stress. In Figure 3, that value is represented as 1.0 on the circumferential tensile stress to uniaxial yield stress ratio (vertical) axis. If one were to test the same type of specimen using a longitudinal specimen, the unique yield strength measured in the test could be plotted on Figure 3 at 1.0 on the longitudinal tensile stress uniaxial yield strength (horizontal axis). Negative numbers on the horizontal axis represent axial compressive stress. The typical line-pipe material exhibits an elliptical yield-strength relationship for various combinations of biaxial stress.[4] As shown in Figure 3, this results in yielding at a higher value of circumferential tensile stress to uniaxial yield strength ratios than 1.0. In tests of pressurized pipes, the ratio for a buried pipeline (longitudinal tensile stress to circumferential tensile stress ratio of 0.3) was found to be about 1.09.[6] So, this effect also suppresses yielding in a hydrostatic test of a pipeline to a pressure level in excess of 100% of SMYS.

To resolve how much yielding actually takes place, Texas Eastern designed a gauging pig in the mid 1960s to measure diametric expansion.[5] In 300 miles of 30-in OD X52 pipe, tested to a maximum of 113% of SMYS, they found only 100 joints of pipe (out of 40,000) that had expanded as much as 1.0%. In 66 miles of 36-inch OD X60 pipe tested to a maximum of 113% of SMYS, they found 100 joints of pipe (out of 6600) that had expanded as much as 1.0%, still not a lot of expanded pipe.

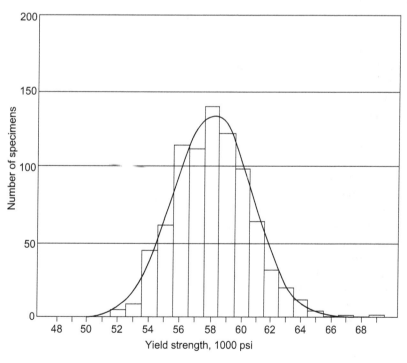

Normal curve fitted to frequency distribution of yield strength for 844 pipe specimens from 30-Inh by 0.375-inch X52 pipe

Figure 2.

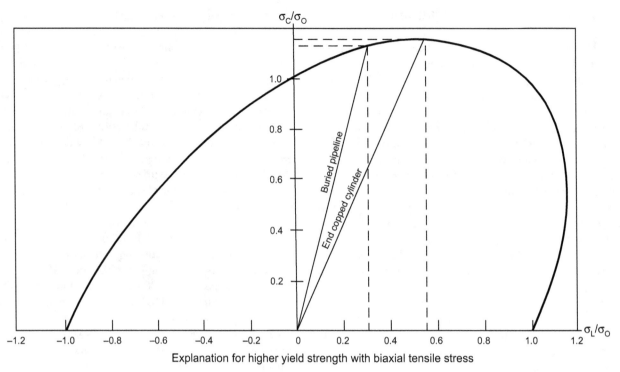

Explanation for higher yield strength with biaxial tensile stress

Figure 3.

Pressure–volume plots

Figure 4 shows a pressure-versus-pump-stroke plot of an actual hydrostatic test. The plot is created by recording the number of pump strokes of a positive displacement pump as each 10 psig increase of pressure is attained. Prior to beginning the plot, it is prudent to hold the test section at a constant pressure to assure that there are no leaks. After it is established that there is no leak, the plot should be started at a pressure level no higher than 90% of SMYS for the low elevation point in the test section in order to establish the "elastic" slope of the plot. By projecting the elastic slope lines across the plot as shown, one can then record pump strokes and compare the evolving plot to those slopes. If and when the actual plot begins to deviate from the elastic slope, either some pipe is beginning to yield or a leak has developed. The pressurization can be continued in any event until the "double-the-strokes" point is reached. This is the point at which it takes twice as many strokes to increase the pressure 10 psi as it did in the elastic range. Also, we suggest stopping at 110% of SMYS if that level of pressure is reached before the double-the-stroke point. Once the desired level has been reached, a hold period of 30 min should establish whether or not a leak has developed. Some yielding can be taking place while holding at the maximum pressure. Yielding will cease upon repeated repressurization to the maximum pressure, whereas a leak likely will not.

Testing an existing pipeline to a level at which yielding can occur may or may not be a good idea. It depends on the number and severity of defects in the pipe, the purpose of the test, and the level of maximum operating pressure that is desired. More will be said about this in the next section.

Finally, on the subject of testing to actual yield, the following statements apply:

- Yielding does not hurt or damage sound pipe. If it did, no one would be able to make cold-expanded pipe or to cold bend pipe.
- Yielding does not damage the coating. If it did, one could not field-bend coated pipe or lay coated pipe from a reel barge.
- Very little pipe actually undergoes yielding in a test to 110% of SMYS.
- Those joints that do yield do not affect pipeline integrity, and the amount of yielding is small.
- The only thing testing to a level in excess of 100% of SMYS may do is to void a manufacturer's warranty to replace test breaks if such a warranty exists.

Testing existing pipelines

Testing of an existing pipeline is a possible way to demonstrate or revalidate its serviceability. For a variety of reasons, retesting of an existing pipeline is not necessarily the best

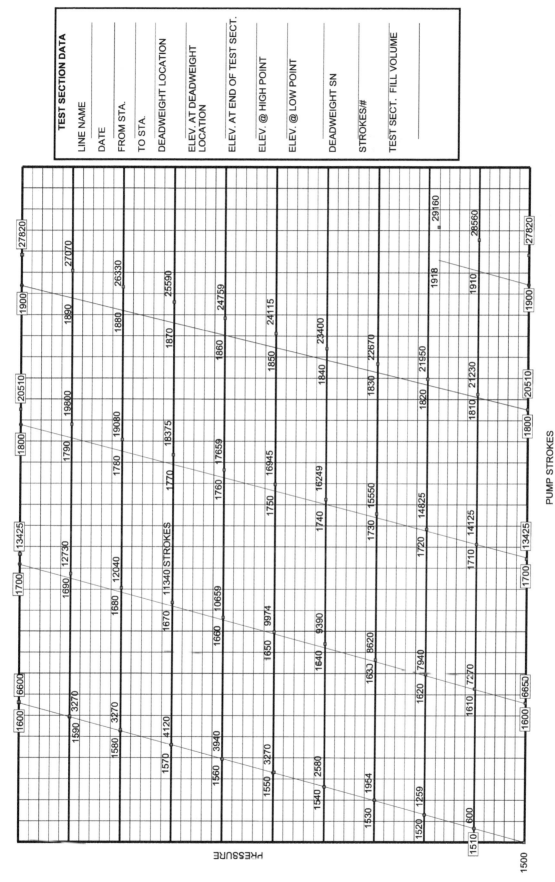

PUMP STROKES

Figure 4. Pressure volume plot.

means to achieve confidence in its serviceability, however. First, a pipeline operator who elects to retest a pipeline must take it out of service and purge it of product. The downtime represents a loss of revenue and a disruption to shippers. Second, the operator must obtain test water. To fill 30 miles of a 16-in. pipeline, an operator would need nearly 40,000 barrels of water. This is equivalent to a 100×100-ft pond, 22 ft in depth. For 30 miles of 36-in. pipe, the volume required would be five times as large. After the test, the water is considered a hazardous material because of being contaminated with product remaining in the pipeline. And, a test break, if one occurs, releases contaminated water into the environment. Aside from these issues, some problematic technical considerations exist.

The most important reason why a hydrostatic test may not be the best way to validate the integrity of an existing pipeline is that in-line inspection is often a better alternative. From the standpoint of corrosion-caused metal loss, this is almost certainly the case. Even with the standard resolution tools that first emerged in the late 1960s and 1970s this was true. Consider Figure 5. This figure shows the relative comparison

between using a standard resolution in-line tool and testing a pipeline to a level of 90% of SMYS. The assumption is made in this case that the operator excavates and examines all "severe" and "moderate" anomalies identified by the tool, leaving only the "lights" unexcavated. In terms of the 1970s technology, the terms "light," "moderate," and "severe" meant the following:

- Light indications: metal loss having a depth less than or equal to 30% of the wall thickness.
- Moderate indication: metal loss having a depth more than 30% of the wall thickness but less than 50% of the wall thickness.
- Severe indication: metal loss having a depth more than 50% of the wall thickness.

In Figure 5, the boundary between "light" and "moderate" is nearly at the same level of failure pressure as the 90% of SMYS test for long defects, and it is well above that level for short defects. Because even the standard resolution tools have some defect—length-indicating capability, an in-line inspection on the basis represented in Figure 5 gives a better

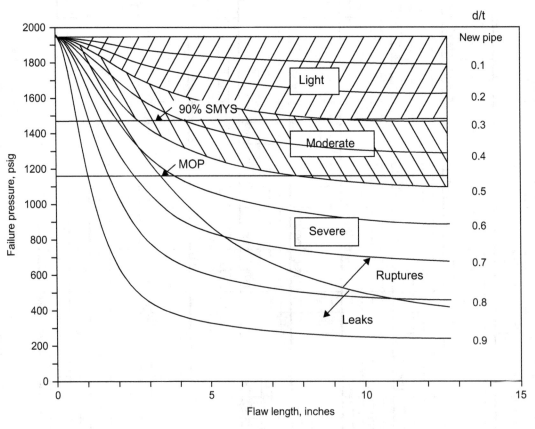

Sizes of flaws located by in-line inspection (correction)
(16 inch by 0.250 inch X52, blunt defect

Figure 5.

assurance of pipeline integrity than a hydrostatic test to 90% of SMYS. With the advent of high-resolution tools, the advantage shifts dramatically in favor of using in-line inspection instead of hydrostatic testing to validate the serviceability of a pipeline affected by corrosion-caused metal loss.

From the standpoint of other types of defects, the appropriate in-line inspection technology is evolving rapidly and, in some cases, it has proven to be more effective than hydrostatic testing. One example is the use of the elastic-wave tool for detecting seam-weld defects in submerged arc-welded pipe.[7] Another is the use of transverse-field magnetic-flux-leakage inspection to find seam anomalies alongside or in the seams of electric-resistance-welded (ERW) pipe.[8] In these cases, the particular tools revealed defects that were too small to have been found by a hydrostatic test to any reasonable level up to and including 110% of SMYS. When a tool has established this kind of track record, a pipeline operator can justify using the tool instead of hydrostatic testing.

The concept of using in-line tools to detect flaws invariably raises the question about defects possibly not being detected. The reasonable answer is that the probability of nondetection is small (acceptably small in the authors' opinion) but not zero. In the same context, one must also recognize that hydrostatic testing is not foolproof either. One issue with hydrostatic testing is the possibility of a pressure reversal (discussed later). The other issue is that because hydrostatic testing can leave behind defects that could be detected by in-line inspection, the use of hydrostatic testing often demonstrates serviceability for only a short period of time if a defect-growth mechanism exists. This possibility is discussed in our companion paper to this article.[9]

Pressure reversals

A pressure reversal is defined as the occurrence of a failure of a defect at a pressure level that is below the pressure level that the defect has previously survived due to defect growth produced by the previous higher pressurization and possible subsequent damage upon depressurization. Pressure reversals were observed long before their probable cause was identified.[3] The pipeline industry supported a considerable amount of research to determine the causes of pressure reversals. The most complete body of industry research on this subject is Kiefner, Maxey, and Eiber.[4] Figure 6, taken from that source, reveals the nature of experiments used to create and demonstrate pressure reversals. It shows photographs of highly magnified cross sections of the tips of six longitudinally oriented flaws that had been machined into a single piece of 36-in. OD by 0.390 in. w.t. X60 pipe. Each flaw had the same length, but each was of a different depth, giving a graduation in severities. When the single specimen containing all six flaws was pressurized to failure, the deepest flaw (Flaw 1) failed. By calculations based on their lengths

Table 1

Flaw Number	Test Pressure Level at Failure of Flaw 1 as a Percentage of the Calculated Failure Pressure of the Flaw
2	97
3	94
4	91
5	89
6	87

and depths, the surviving flaws were believed to have been pressurized to the percentages of their failure pressures shown in Table 1.

As one can see, the tips of Flaws 2, 3, and 4 exhibit some crack extension as a result of the pressurization to failure. The nearer the defect to failure, the more crack extension it exhibited. In fact, due to its extension during the test, Flaw 2 is now deeper than Flaw 1 was at the outset. Logic suggests that if we could have pressurized the specimen again, Flaw 2 would have failed at a level below that which it experienced during the testing of Flaw 1 to failure. Indeed, in similar specimens designed in a manner to allow subsequent pressurizations, that is exactly what often occurred. This type of testing led to an understanding of pressure reversals in terms of ductile crack extension occurring at near-failure pressure levels where the amount of crack extension is so great that crack closure upon depressurization does further damage, leading to the inability of the flaw to endure a second pressurization to the previous level. The pressure reversal is expressed as a percentage:

$$\text{Pressure reversal} = \frac{(\text{original pressure} - \text{failure pressure})}{(\text{original pressure} \times 100)}.$$

Once the cause was known, the next key question was: What is the implication of the potential for pressure reversals on confidence in the safety margin demonstrated by a hydrostatic test? This question has been answered in particular circumstances, and the answer comes from numerous examples of actual hydrostatic tests.[a] Figure 7 is an

---

[a] In actual hydrostatic tests, direct evidence that pressure reversals are the result of the type of flaw growth shown in Figure 6 has seldom been obtained. However, a few such cases have been documented and it is assumed that defect growth is responsible for all such cases.

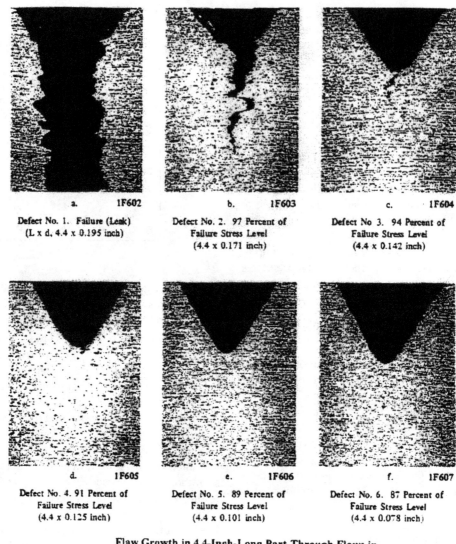

a. 1F602

Defect No. 1. Failure (Leak)
(L x d, 4.4 x 0.195 inch)

b. 1F603

Defect No. 2. 97 Percent of
Failure Stress Level
(4.4 x 0.171 inch)

c. 1F604

Defect No 3. 94 Percent of
Failure Stress Level
(4.4 x 0.142 inch)

d. 1F605

Defect No. 4. 91 Percent of
Failure Stress Level
(4.4 x 0.125 inch)

e. 1F606

Defect No. 5. 89 Percent of
Failure Stress Level
(4.4 x 0.101 inch)

f. 1F607

Defect No. 6. 87 Percent of
Failure Stress Level
(4.4 x 0.078 inch)

Flaw Growth in 4.4-Inch-Long Part-Through Flaws in
36 x 0.390-Inch X60 Pipe

Note: Loading consisted of —
1st cycle — 0 → 1330 psig with 30 sec hold
2nd cycle — 0 → 1300 psig with 30 sec hold
3rd cycle — 0 → 1230 psig with 30 sec hold

Figure 6.

example of an analysis of pressure reversals in a specific test. This figure is a plot of sizes of pressure increases or decreases (reversals) on subsequent pressurizations versus the frequency of occurrence of that size of increase or decrease. Among other things, these data show that upon pressurization to the target test-pressure level, a 1% pressure reversal (34 psi) can be expected about once in every 15 pressurizations, a 2% pressure reversal (68 psi) can be expected about once in every 100 pressurizations, and a 3% pressure reversal

(102 psi) can be expected about once in every 1000 pressurizations. For a target test pressure level of 1.25 times the MOP, the expectation of a 20% pressure reversal (enough to cause failure at the MOP) is off the chart, that is, it is an extremely low-probability event (but not an impossible event).

There have been a handful of pipeline service failures in which a pressure reversal is the suspected but unproven cause. There is also one case of a large pressure reversal

Figure 7.

(62%) that was unequivocally demonstrated because it occurred on the fifth cycle of a five-cycle hydrostatic test.[4] It should be noted that most of the experiences of numerous and large pressure reversals in actual hydrostatic tests have involved flaws associated with manufacturing defects in or near ERW seams, particularly in materials with low-frequency welded (generally pre-1970) ERW pipe. But in most cases where numerous reversals occurred, the sizes of the actual pressure reversals observed are small (less than 5%). One thing seems clear: if a hydrostatic test can be successfully accomplished without the failure of any defect, the likelihood of a pressure reversal will be extremely small. It is the tests in which numerous failures occur that have the highest probabilities of reversals. And, when the number of reversals becomes large, the probability of a reversal of a given size can be estimated as was done on the basis of Figure 7.

Practical considerations

For new pipeline materials made to adequate specifications with adequate inspection and pipe-mill testing, one does not expect test failures even at pressure levels corresponding to 100% or more of SMYS. Therefore, there is no reason not to test a pipeline constructed of such materials to levels in excess of 100% of SMYS. As has been shown, the higher the ratio of test pressure to operating pressure, the more confidence one can have in the serviceability of a pipeline. In the case of existing pipelines, especially the older ones, such test levels may be impossible to achieve, and if numerous test failures occur, the margin of confidence may become eroded by the potential for pressure reversals. Weighing against low test-pressure-to-operating-pressure ratios, on the other hand, is the fact that such tests, by definition, generate lower levels of confidence and buy less time between retests if the issue of concern is time-dependent defect growth. A discussion of some possible responses to this dilemma follows.

First and foremost, as has already been mentioned, the use of an appropriate in-line inspection tool is always to be preferred to hydrostatic testing if there is sufficient confidence in the ability of the tool to find the defects of significance. Most of the pipe in a pipeline is usually sound. Therefore, it makes sense to use a technique that will find the critical defects and allow their repair as opposed to testing the whole pipeline when it is not necessary. The industry now has access to highly reliable tools for dealing with corrosion-caused metal loss, and tools are evolving rapidly to detect and characterize cracks. As has been noted, some uses of these tools have already proven their value and, in those cases, their use in lieu of hydrostatic testing makes good sense.

There are still certain existing pipelines for which hydrostatic testing remains the best (in some cases the only) means to revalidate their serviceability. In those cases, the following advice may be useful. Determine the mill hydrostatic test level for the pipe. The mill-test certificates will show the level applied if such certificates can be found. Also, search the records for prior hydrostatic tests at or after the time of construction. Review the pressure levels and causes of mill-test or in-place test failures if they exist. If none of these records is available, look up the API 5L specification applicable to the time the pipe was manufactured. This will reveal the standard mill-test pressure for the pipe. Do not assume, if you do not know, that the pipe was tested in the mill to 90% of SMYS. This was not always the case especially for non-X grades and smaller diameter pipe materials. If you decide to test the pipe to a level in excess of its mill-test pressure for the first time ever, anticipate test failures. If you cannot tolerate test failures, consider testing to a level just below the mill-test pressure. This may mean, of course, that the MOP you validate is less than 72% of SMYS (the minimum test pressure must be at least 1.25 times the MOP for 4 h plus 1.10 times MOP for 4 h for a buried pipeline; see Federal Regulation Part 195). Alternatively, if you can tolerate at least one test failure, pressurize to a level as high as you wish or until the first failure, whichever comes first. If you then conduct your 1.25-times MOP test at a level at least 5% below the level of the first failure, a second failure will be highly improbable.

It is always a good idea to conduct an integrity test as a "spike" test. This concept has been known for many years,[4] but more recently it has been advocated for dealing with stress-corrosion cracking.[10] The idea is to test to as high a pressure level as possible, but to hold it for only a short time (5 min is good enough). Then, if you can live with the resulting MOP, conduct your 8-h test at a level of at least 5% below the spike-test level. The spike test establishes the effective test-pressure-to-operating-pressure ratio; the rest of the test is only for the purpose of checking for leaks and for meeting the requirements of Part 195.

Summary

To summarize, it is worth reporting the following:

- Test-pressure-to-operating-pressure ratio measures the effectiveness of the test.
- In-line inspection is usually preferable to hydrostatic testing.
- Testing to actual yield is acceptable for modern materials.
- Pressure reversals, if they occur, tend to erode confidence in the effectiveness of a test but usually not to a significant degree.
- Minimizing test-pressure cycles minimizes the chance for pressure reversals.

References

1. Bergman, S. A., "Why Not Higher Operating Pressures for Lines Tested to 90% SMYS?" *Pipeline and Gas Journal*, December 1974.
2. Kiefner, J. F., Maxey, W. A., Eiber, R. J., and Duffy, A. R., "Failure Stress Levels of Flaws in Pressurized Cylinders," *Progress in Flaw Growth and Toughness Testing*, ASTM STP 536, American Society for Testing and Materials, pp. 461–481, 1973.
3. Brooks, L. E., "High-Pressure Testing—Pipeline Defect Behavior and Pressure Reversals," ASME, 68-PET-24, 1968.
4. Kiefner, J. F., Maxey, W. A., and Eiber, R. J., *A Study of the Causes of Failure of Defects That Have Survived a Prior Hydrostatic Test*, Pipeline Research Committee, American Gas Association, NG-18 Report No. 111, November 3, 1980.
5. Duffy, A. R., McClure, G. M., Maxey, W. A., and Atterbury, T. J., *Study of Feasibility of Basing Natural Gas Pipeline Operating Pressure on Hydrostatic Test Pressure*, American Gas Association, Inc., Catalogue No. L30050, February 1968.
6. Maxey, W. A., Podlasak, R. J., Kiefner, J. F., and Duffy, A. R., "Measuring the Yield Strength of Pipe in the Mill Expander," Research report to A.G.A./PRC, A.G.A. Catalog No L22273, 61 pages text plus 2 appendices, 23 pages, December 1973.
7. Maxey, W. A., Mesloh, R. E., and Kiefner, J. F., "Use of the Elastic Wave Tool to Locate Cracks Along the DSAW Seam Welds in a 32-Inch (812.8 mm) OD Products Pipeline," *International Pipeline Conference*, Volume 1, ASME, 1998.

8. Grimes, K., "A Breakthrough in the Detection of Long Seam Weld Defects in Steel Pipelines," Pipeline Integrity International.

9. Kiefner, J. F., and Maxey, W. A., "Periodic Hydrostatic Testing or In-line Inspection to Prevent Failures from Pressure-Cycle-Induced Fatigue," Paper Presented at API's 51st Annual Pipeline Conference and Cybernetics Symposium, New Orleans, Louisiana, April 18–20, 2000.

10. Leis, B. N., and Brust, F. W., *Hydrotest Strategies for Gas Transmission Pipelines Based on Ductile-Flaw-Growth Considerations*, Pipeline Research Committee, American Gas Association, NG-18 Report No. 194, July 27, 1992.

Hydrostatic testing for pipelines

After construction of a pipeline is completed, or if pipe has been replaced or relocated, it is necessary to hydrostatically test the pipeline to demonstrate that the pipeline has the strength required to meet the design conditions, and to verify that the pipeline is leak free.

The U.S. Federal Safety Regulations for Pipelines require that pipelines used to transport hazardous or highly volatile liquids be tested at a pressure equal to 125% of the maximum allowable operating pressure (MAOP) for at least 4 continuous hours and for an additional 4 continuous hours at a pressure equal to 110% or more of the MAOP if the line is not visually inspected for leakage during the test. A design factor of 72% of the specified minimum yield strength (SMYS) of the pipe is used to determine the maximum allowable operating pressure. The requirement to test to 125% of the MAOP will therefore cause the pipe to be tested to a pressure equal to 90% of the SMYS of the pipe. See Section 3—Pipe Design for additional information on calculating the MAOP.

The regulations for gas lines specify design factors based on the class location of the pipeline. The class location is determined by the number of buildings in a specified area on either side of the pipeline. Refer to Part 192.111 of the Minimum Federal Safety Standards for Gas Lines for the details on how to determine the class location. See Section 3, Pipe Design for a listing of the design factors for the different class locations. The regulations specify that the test pressure must be maintained for at least 8 h and must be equal to at least 125% of the maximum allowable operating pressure. The regulations should be consulted for the specific testing requirements, as the regulations are subject to change.

Usually, operators will specify a test pressure range from 90% to 95% of the SMYS of the pipe. Some will allow test pressures as high as 100% of the SMYS of the pipe and some will test to slightly beyond the SMYS of the pipe. Specifying a test pressure at least equal to 90% of the SMYS of the pipe will qualify it for the maximum allowable operating pressure. In some cases, the test pressure will be based on the minimum yield strength determined from the mill test reports.

Hydrostatic testing a pipeline is certainly a major operation and should be carefully planned. Most companies have hydrostatic test manuals that detail the procedures to be followed to complete the test. Usually, this work is performed by a hydrostatic testing contractor hired by the pipeline owner, or hydrostatic testing may be included as a part of the main construction contract.

One of the first steps in planning the hydrostatic test operation is to examine the elevation gradient. The gradient, along with the location of the water source, and the pipe design data, will be used to determine the length and number of test sections. Figure 1 shows a typical pipeline elevation gradient.

Where the pipeline traverses hilly terrain, the elevation gradient must be carefully considered in selecting the pipeline test segments. Different companies have differing philosophies on how to do this. Some limit the amount of elevation difference while others may specify a range of allowable percentages of the SMYS of the pipe—i.e. 90–95% or 90–100% of the SMYS of the pipe. In any case, the test gradient should be plotted to be sure the test pressure falls within the specified pressure limits.

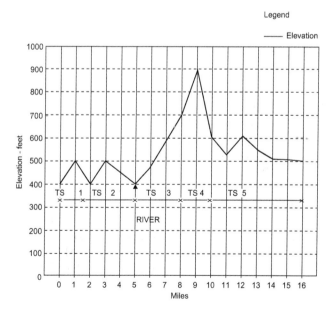

Figure 1. Pipeline elevation profile.

The test gradient must be based on water head in feet, if water is used as the test medium. Water pressure may be converted to head by dividing the pressure by 0.433. If sea water is being used as the test medium, a factor of 0.445 should be used. This assumes a specific gravity of 1.02 for sea water. In any case, the factor of 0.433 should be modified according to the specific gravity of the test medium.

Let's assume that we have a line that is to be tested and the elevation at the test site is 1000 feet and the elevation at the end of the line is 1200 ft. Fresh water will be used as the test medium, and it is desired to test the pipe to a minimum of 90% and a maximum of 96% of the SMYS of the pipe. The pipe is 30-in OD × 0.390″ wt API 5LX X60. There is a difference in elevation of 200 ft and this is equal to 86.6 psig (200 ft × 0.433 = 86.6 psig). A test pressure equal to 90% of SMYS is 1404 psig. Since the test site is lower than the high end of the line, the 86.6 psig is added to 1404 psig to obtain a test site pressure of 1491 psig. The pressure at the end of the line will be 1404 psig, which equates to 90% of SMYS. The pressure at the low point equates to 96% of SMYS.

Now, let's assume a line has a high point elevation of 1100 ft, a low elevation point of 1000 ft, and the elevation at the test site is 1050 ft. The test pressure at the high point will need to be 1404 psig in order to meet the 90% of SMYS requirement. The pressure at the low point will be 1447 psig, and the pressure at the test site will be 1426 psig.

When testing offshore lines, the pressure at or below the water surface will be the same as at the low elevation point due to the offsetting external subsea pressure head. If the line previously described was laid offshore, the test pressure at the water surface would be 1404 psig. The elevations are as follows:

Top of riser	+11 ft
Test site	+7 ft
Water surface	0 ft
Pipe depth	−168 ft

The test pressure at the test site would be 1404 − (7 × 0.445) or 1401 psig. At the top of the riser, the test pressure would be 1404 − (11 × 0.445) or 1399 psig.

The typical profile shown in Figure 1 represents a pipeline that requires testing. The pipeline crosses a river at approximately MP 5, and river water will be used to test the line. Test sections 1–5 have been chosen as indicated. The lengths of sections 3 and 4 are limited by elevation difference. This line was designed to operate at 936 psig or 72% of the SMYS of 30 in × 0.375 in wt API 5LX X52 pipe for Class I locations, and 60% of the SMYS of 30 in × 0.390 in wt API 5LX X60 pipe for Class II locations, and 50% of the SMYS of 30 in × 0.438 in wt API 5LX X65 for Class III locations.

The pipeline crosses one railroad, one highway, and the river. It also includes one main line block valve assembly. The

NOTE: Dimensions A thru D are based on the pipe dia. and pig length.

Figure 2. Typical pressure manifold. Courtesy Milbar Hydrotest.

valve assembly, river crossing, and road crossing were pretested before installation and are tested again after installation.

Four test manifolds were installed to facilitate filling the line and for isolating the test sections during the test operation. See Figure 2 for a typical pressure sectionalizing manifold. Two-way pigs were loaded at each of the intermediate manifolds. The pigs are moved by the fill water and are necessary to remove the air from the line.

It is generally a good idea to test the section most distant from the water source first. If it should rupture, then testing on intermediate sections can continue while repairs on the failed section are completed.

The pumps used to fill the line should have sufficient capacity to fill the line at a rate of about 1/2 mile per hour. Water filters should be used and are normally equipped with 100 mesh screens. See Figure 4 for a typical fill site arrangement. Some companies may specify a finer mesh screen. The filling unit should be equipped with a flow meter to measure the amount of water pumped into the pipeline. Temperature recorders will be used to record ambient temperature, the temperature of the pipe and water, and

Figure 3. Typical test manifold. Courtesy Milbar Hydrotest.

A - Suction Strainer
B - 6" Nominal Victaulic Pipe
C - Lowhead Pump
D - Water Filter
E - Flowmeter
F - Fill Pump
G - 6" Check Valve

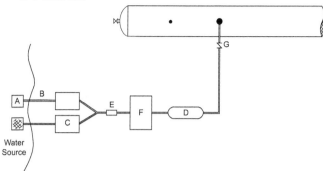

Figure 4. Typical fill site. Courtesy Milbar Hydrotest.

pressuring operation. Small high pressure units with a capacity of 6–30 gpm may be used for short sections of large diameter lines and smaller diameter lines.

Deadweight pressure and temperature readings are recorded after the prescribed test pressure has been reached and the pressures and temperatures have stabilized. A pressure vs. time plot may also be made. Readings are usually made at 15-min intervals for the first hour and at 30-min intervals thereafter. The procedure for accepting a leakage test will vary from company to company. Some will accept the leakage test if there is no pressure drop in a 3-h period.

A pressure–volume plot may be required, especially if the test pressures approach or exceed the SMYS of the pipe. The plot is made manually during the pressuring operation by recording pump strokes on the X-axis and pressure on the Y-axis. A straight line will be produced until the plastic range of the pipe is reached or until a leak occurs. This type of plot is also useful should a joint of lighter wall thickness or lower yield strength pipe be inadvertently placed in the pipeline.

Small leaks during the testing operation can be difficult to locate. A change in the water/pipe temperature may give the appearance of a leak. If the temperature of the pipe/water decreases, the test pressure will decrease. An increase in water/pipe temperature will cause the test pressure to increase. The effect of a temperature change may be estimated using the equations and data contained in Appendices A and B. To achieve any degree of accuracy in these

ground temperature. See Figure 5 for typical test thermometer placement. A pressure recorder will be used to provide a recording of the pipeline test pressure. A deadweight tester will be used to calibrate the pressure recorder. A high pressure pump capable of delivering 70–150 gpm at a pressure exceeding the required test pressure will be used in the final

T1-Thermometer in Thermo Well on Inlet Manifold Riser
T2-Thermometer in Backfill next to 48" Pipe line (Recording Thermometer)
T3-Thermometer in Protected Ambient Atmosphere (Recording Thermometer) - not shown

Figure 5. Test thermometer placement. Courtesy Milbar Hydrotest.

Figure 6. Typical pressure unit setup. Courtesy Milbar Hydrotest.

calculations, it will be necessary to have accurate temperature and pressure readings.

The pipeline test temperature may be affected by river/creek crossings along the pipeline route. Also, if cold river water is used, there will be a changing temperature gradient. Air, either trapped or entrained, will also affect the pressure-temperature calculations since the coefficient of expansion for air is not the same as for water. If the calculations are made and the decrease in pressure cannot be attributed to a temperature change, then it may be necessary to section-alize the test section by manifolding and repressuring the shorter segments to facilitate locating the leak. Once the leak has been located and repaired, testing operations can resume.

Air compressors will be required to remove the water once the testing is complete. The maximum head for this particular project was 450 ft (195 psig). Allowance will need to be included for friction loss in the cross-over manifolds. The compressors will need to have sufficient capacity to remove the water at about the same rate as the fill rate. Most portable compressors are limited to approximately 125 psig. In this case, booster compressors will be needed for the 195 + psig. See Figure 7 for a typical dewatering connection.

Upon completion of testing, the test pressure is bled off to leave approximately 10–20 psig at the high point in the test section. Dewatering may be accomplished using the air compressors discussed earlier. Some companies may elect to displace the test water with the material that will normally be

pumped through the line. This is usually the case with crude oil and refined product pipelines. Pipelines used to transport natural gas and certain chemicals are usually dewatered using compressed air. Care must be exercised in the dewatering operation to make sure that no air is introduced into the test section, thus minimizing the possibility of air locks. An air lock is probably the most severe problem involved in dewatering. Air locks are caused by air accumulating in the downhill leg and water accumulating in the uphill leg, which creates a manometer in the pipeline. In some instances, extremely high pressures may be required to overcome the manometer. Care should be exercised to be sure the maximum allowable pipeline operating pressure is not exceeded. It may also be necessary to tap the line and vent air at the high points.

If it is necessary to pipe the water away from the right-of-way, welded pipe should be used. The pipe must be securely anchored. The use of tractors and skids will probably provide sufficient anchoring.

Always use a valve to control the amount of water being bled from the pipeline.

If there will be a delay in placing the pipeline in operation, and it is decided to leave the test water in the line until operations begin, consideration should be given to the use of a corrosion inhibitor. Long-term storage of the line may necessitate the use of oxygen scavengers in addition to the inhibitor.

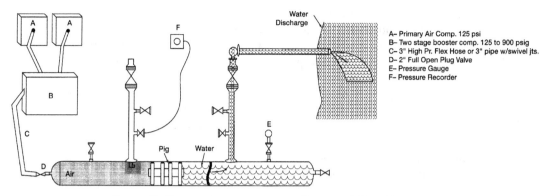

A– Primary Air Comp. 125 psi
B– Two stage booster comp. 125 to 900 psig
C– 3" High Pr. Flex Hose or 3" pipe w/swivel jts.
D– 2" Full Open Plug Valve
E– Pressure Gauge
F– Pressure Recorder

Figure 7. Dewatering arrangement. Courtesy Milbar Hydrotest.

FIELD PRESSURE & TEST REPORT

Page _____ of _____

Contractor or Company: _____

Pipe Line Description: _____

Section Tested: From: _____ To: _____

Test Section No.: _____ Length: _____

Type and Size of Pipe: _____ O. D. x _____ W.T. x Grade _____ Manufacturer: _____

Pressure Unit No.: _____ Gallons per Stroke: _____ Fill Unit No.: _____

Pressure Unit Location: _____ Water Source _____

Time and Date Test Started: _____ AM PM _____ Test Pressure (Maximum): _____ PSIG

Time and Date Test Ended: _____ AM PM _____ Pressure Volume Plot: Yes _____ No _____

Section Accepted _____ Section Leaking _____ Section Ruptured _____

DEADWEIGHT RECORDED READING (PSIG)

TIME	PRESSURE	TEMP DEG. F	TIME	PRESSURE	TEMP DEG. F	TIME	PRESSURE	TEMP DEG. F
Date: _____								
AM PM			AM PM			AM PM		

Strokes per 10 psi: _____ Remarks: _____

_____

_____

_____

Milbar Supt. _____ Inspector _____

Witness: 1. _____ 2. _____

Gauges and Recorders Last Tested _____ Deadweight Serial No. _____

Figure 8. Typical test report. Courtesy Milbar Hydrotest.

If after dewatering, residual moisture in the pipeline poses a problem, it will be necessary to clean and dry the pipeline. See Section 6 for drying details.

A permit to discharge the water may be required, especially if there is a possibility the water may enter a stream of water. This should be determined and permits obtained, if required, in the planning phase of the testing operation. A "splash" plate or bales of hay may be used when dewatering to prevent erosion.

The most important aspect of any test operation is complete and accurate documentation, since it will become a permanent record that must be retained for as long as the facility tested remains in operation. Federal Pipeline Safety Regulations for pipelines transporting liquids require that these records include the following information:

- the pressure recording charts;
- test instrument calibration data;
- the names of the operator, of the person responsible for making the test, and of the test company used, if any;
- the date and time of the test;
- the minimum test pressure;
- the test medium;
- a description of the facility tested and the test apparatus;
- an explanation of any pressure discontinuities, including test failures, that appear on the pressure recording charts; and,
- where the elevation differences in the section under test exceed 100 ft, a profile of the pipeline that shows the elevation and test sites over the entire length of the test section.

The records required by the Federal Pipeline Safety Regulations for pipelines transporting gas are similar to those required for pipelines used to transport liquids.

A typical test report is shown in Figure 8. If failures occur during the test, a report documenting the failure and the suspected reasons for the failure should be completed. A typical form is shown in Figure 9.

A daily operating log should be used to record activities associated with the test, operating status of the test equipment in use, engine rpm, and any unusual circumstances that occur during the test. A typical form is shown in Figure 10.

PIPE LINE FAILURE REPORT

1. Company _____

2. Section tested _____

3. Time of failure _____ AM, PM. Date _____

4. Location of failure: _____

5. Pressure, psig, at point of failure _____

6. Description of failure: Leak _____, Break _____, Length of failure _____

7. If leak, fill in blanks: _____ gallons lost per hour, _____ pounds lost per hour.

8. Describe any peculiarities or defects on failed part, such as corrosion or evidence of prior damage, etc.

9. Possible cause of failure _____

10. Pipe size: _____ OD x _____ Wt., Grade _____ Mfg. by _____

11. Repairs: Pipe installed _____ OD x _____ Wt., Grade _____

    Mfg. by _____, Length of joint or pup _____

12. Date repaired _____ 19 _____ By _____

13. Remarks (Damage to property, persons injured, etc.) _____

Figure 9. Pipeline failure report. Courtesy Milbar Hydrotest.

DAILY OPERATING REPORT

COMPANY & CONTRACTOR _____ DATE _____

LOCATION (City) _____ CONTRACT NO. _____

PIPE LINE DESCRIPTION: _____

_____

PERSONNEL: _____

_____

EQUIPMENT: _____

_____

_____

ACTIVITIES: _____

_____
_____
_____
_____
_____
_____
_____
_____
_____
_____
_____
_____
_____
_____
_____
_____
_____
_____
_____
_____
_____
_____
_____
_____
_____

_____   _____   _____
COMPANY REP.   CONTRACTOR REP.   MILBAR SUPT.

Figure 10. Daily Operation Report. Courtesy Milbar Hydrotest.

APPENDIX A

Volume of water required to fill test section

$$V = L \times (3.1416 \times d^2/4) \times 12/231 \quad (1)$$

$$V - 0.0408 \times d^2 \times L$$

where

$L =$ Length of test section, feet,
$d =$ Inside diameter of pipe, inches,
$V =$ Gallons required to fill @ O psig.

Volume required at test pressure

$$V_{tp} = V \times F_{wp} \times F_{pp} \times F_{pwt} \qquad (2)$$

V_{tp} = Gallons contained in test section at test pressure P, and temperature T—°F,

F_{wp} = Factor to correct for the compressibility of water due to increasing pressure from 0 psig to test pressure P, psig,

F_{pp} = Factor to correct for volume change in pipeline due to pressure increase from 0 psig to test pressure P, psig,

F_{pwt} = Factor to correct for the change in water volume and pipe volume due to change in pipe and water temperature from base of 60 °F to pipe and water test temperature, T—°F.

$$F_{wp} = 1 / \left[1 - \left(4.5 \times 10^{-5} \right) \times (P/14.73) \right] \qquad (3)$$

P = Test pressure, psig.

$$F_{pp} = 1 + \left[(D/t) \times \left(0.91P/30 \times 10^6 \right) \right]$$
$$+ \left[3.6 \times 10^{-6}(T - 60) \right] \qquad (4)$$

D = Outside diameter of pipe, inches,
t = Pipe wall thickness, inches,
T = Pipe temperature, °F.

$$F_{pwt} = F_{pt}/F_{wt} \qquad (5)$$

F_{pt} = Factor to correct for change in pipe volume due to thermal expansion of the pipe from base temperature of 60°F.

$$F_{pt} = 1 + \left[(T - 60) \times 18.2 \times 10^{-6} \right] \qquad (6)$$

F_{wt} = Factor to correct for thermal change in specific water volume from 60°F to test water temperature. Refer to Table 1.

Sample calculation:

Pipe Size: 10.750 in. OD × 0.279 in. w.t. X-52
Length: 5 miles
Test pressure: 2430 psig
Temperature: 50°F

Use equation 1 to determine the initial line fill volume, V.

$$V = 0.0408 \times (10.75 - (2 \times 0.279))^2 \times 5280 \times 5$$
$$= 111,888 \text{ gallons}$$

Volume required to achieve test pressure $= V_{tp}$

$$= V \times F_{wp} \times F_{pp} \times F_{pwt}.$$

$$F_{wp} = 1/1 - \left[\left(4.5 \times 10^{-5} \right) \times (2430/14.73) \right] = 1.007479.$$

$$F_{pp} = 1 + \left[(10.75/0.279) \times \left(0.91 \times 2430/(30 \times 10^6) \right) \right]$$
$$+ \left[3.6 \times 10^{-6} \times (50 - 60) \right]$$
$$= 1.002804$$

$$F_{pwt} = F_{pt}/F_{wt}$$

$$F_{pt} = 1 + \left[(50 - 60) \times 18.2 \times 10^{-6} \right] = 0.999818$$

$$F_{wt} = 0.9993061 \text{(from Table 1)}$$

$$F_{pwt} = 0.999818/0.9993061 = 1.000512$$

$$V_{tp} = 111,888 \times 1.007457 \times 1.002804 \times 1.000512$$
$$= 113,099 \text{ gals.}$$

Incremental volume required to reach test pressure, P.

$$= 113,099 - 111,888 = 1211 \text{ gals.}$$

After a period of time, the test pressure, P has decreased to 2418 psig and the temperature of the pipe and test water has decreased to 48°F. The calculation procedure previously described may be repeated to determine if the pressure decrease is attributable to the decrease in temperature.

V = 111,888 gals,
P_1 = 2422 psig,
T_1 = 48°F,
F_{wp1} = 1.007454,
F_{pp1} = 1.002796,
F_{pt1} = 0.999781,
F_{wt1} = 0.999217(from Table 1),

$F_{pwt1} = 0.999781/0.999217 = 1.000565,$

$V_{pt1} = 111,888 \times 1.007454 \times 1.002796$

$\qquad \times 1.00565 (@ \ 2422 \ \text{psig and } 48°\text{F}),$

$V_{tpl} = 113,101 (\text{Volume at reduced temperature—}$

$\qquad$ represents volume that would be required to raise

$\qquad$ test procedure to original value at 48°F),

$V_{tp} = 113,099 (\text{Volume at initial test conditions}).$

To restore the pressure to the original test pressure, 2 gals of water could be added to the test section. If more than two gallons are required to restore the test pressure, there is the possibility that a leak may exist.

When making these calculations, care must be exercised to be sure that accurate temperature and pressure data is available. Where test sections are long, it is very likely that the temperature will not be uniform throughout the test section and thus affect the accuracy of the calculated results.

Table 1

F_{wt}—Factor to correct for the thermal change in the specific volume of water from 60°F to the test water temperature

Temp. (°F)	F_{wt}	Temp. (°F)	F_{wt}
35	0.9990777	70	1.0010364
36	0.9990590	71	1.0011696
37	0.9990458	72	1.0012832
38	0.9990375	73	1.0014229
39	0.9990340	74	1.0015420
40	0.9990357	75	1.0016883
41	0.9990421	76	1.0018130
42	0.9990536	77	1.0019657
43	0.9990694	78	1.0021222
44	0.9990903	79	1.0022552
45	0.9991150	80	1.0024178
46	0.9991451	81	1.0025561
47	0.9991791	82	1.0027251
48	0.9992168	83	1.0028684
49	0.9992599	84	1.0030435
50	0.9993061	85	1.0031919
51	0.9993615	86	1.0033730
52	0.9994112	87	1.0035573
53	0.9994715	88	1.0037133
54	0.9995322	89	1.0039034
55	0.9996046	90	1.0040642
56	0.9996683	91	1.0042601
57	0.9997488	92	1.0044357
58	0.9998191	93	1.0046271
59	0.9999074	94	1.0047972
60	1.0000000	95	1.0050043
61	1.0000803	96	1.0052142
62	1.0001805	97	1.0053915
63	1.0002671	98	1.0056067
64	1.0003746	99	1.0057884
65	1.0004674	100	1.0060090
66	1.0005823	101	1.0061949
67	1.0006811	102	1.0064207
68	1.0008031	103	1.0066108
69	1.0009290	104	1.0068417

APPENDIX B

How to use charts for estimating the amount of pressure change for a change in test water temperature

Example.

Pipe data	18 OD × 0.375 in wt
Water temp. at beginning of test	70°F
Water temp. at time T	66°F
Test pressure	1800 psig

Calculate

$$D/t = 18/0.375 = 48$$

where

$D = $ Pipe OD, in,

$t = $ Pipe wall thickness, in.

Figure 1.

Figure 2.

Figure 3.

Figure 4.

Calculate

Average temperature $= (70 + 66)/2 = 68°F$

Use the chart in Figure 3. Enter at $68°F$ and at the intersection with D/t line representing 48, read 23 psig/°F.

Basis for chart development

$$\Delta P = \frac{\beta - 2\alpha}{\dfrac{D \times (1 - \nu^2)}{E \times t} + C}$$

where

$\Delta P =$ psig change per °C,
  $\beta =$ Coefficient of expansion for water,
  $\alpha =$ Coefficient of linear expansion for steel
       1.116×10^6 per °C,
  $D =$ Pipe OD inches,

$\nu =$ Poisson's ratio 0.3,
$E =$ Young's modulus for steel 30×10^6,
$t =$ Pipe wall thickness, inches,
$C =$ Compressibility factor for water, cu in/cu in/psig.

$$\beta \times 10^6 = -64.268 + (17.0105 \times T) - (0.20369 \times T^2)$$
$$+ (0.0016048 \times T^2)$$

$T =$ water temperature, °C.

Compressibility factor for water

Pressure range 1450 psig to 2925 psig

Temperature (°C)	Compressibility factor cu in/cu in/psig
0	3.35×10^{-6}
10	3.14×10^{-6}
20	3.01×10^{-6}
50	2.89×10^{-6}

From *PIPECALC 2.0: PRACTICAL Pipeline Hydraulics*, Gulf Publishing Co., Houston, Texas.

Hydrostatic test records

At a minimum, the following records should be made for a hydrostatic test.

- Calculations supporting basis for hydrostatic test. This should take into account the high and low point elevations in the test section.
- Pressure recording charts.
- Test instrument calibration. This should include the pressure recorder and temperature recorder. Record the serial numbers of the instruments used for the test.
- Date and time of the test.
- Temperature of the test medium.
- Test medium used.

- Explanation of any pressure or temperature discontinuities.
- Explanation of any failures that occurred during the test.
- Weather conditions during test.
- Description of pipe in the test section.
- Any permits required for the test.
- Permits for discharge of test water.
- Any other records required by agencies having jurisdiction over the testing/operation of pipelines.
- All records should be signed and dated by the engineer in charge of the test.

6: Pipeline Drying

PIPELINE DEWATERING, CLEANING, AND DRYING*

Newly constructed pipelines are typically hydrostatically tested, using water as the test medium, upon completion of construction. Older lines already in service may be re-tested either to confirm an earlier test or to qualify the line for a higher operating pressure. Once the hydrostatic testing has been completed, it is necessary to remove the water from the line and place the line in service. Dewatering can be a simple process or, if the procedure is not properly planned, a difficult one. Pipelines used to transport crude oil and/or refined products will probably only require removal of the test water before the line is placed in service. If the pipeline will be used to transport materials that must meet a specified dryness requirement, the pipeline will need to be dewatered, cleaned, and dried. Pipelines used to transport natural gas will need some drying, depending on the operating pressure and the location of the line, to prevent the formation of hydrates. Other pipelines may require drying to protect the pipe from internal corrosion caused by the formation of corrosive acids, such as carbonic acid in the case of carbon dioxide pipelines.

* Based on a paper entitled "The State of the Art of Drying Large Diameter Gas Pipelines after Hydrotest," by Marvin D. Powers, Pipeline Dehydrators, Inc., Houston, Texas.

Dewatering

Dewatering is considered to commence with the running of the first pig after hydrostatic testing is completed and begins with the insertion of a displacer, commonly referred to as a pig, in the pipeline. The dewatering pig may be pushed through the pipeline with crude oil or other petroleum product if no drying is required. If the pipeline is to be cleaned and/or dried, the pig will be pushed by either compressed air or gas. In either case, proper precautions must be taken to be sure the test water is properly disposed of and that any required water discharge permits are obtained ahead of the dewatering operation.

Several types of pigs may be used for the dewatering phase, and pig selection should depend upon the design characteristics of the pipeline and the degree of cleaning, if any, that is desired. Ideally, the pig will form a perfect seal with the inner periphery of the pipe and will not allow any of the material behind the pig to leak past the pig and co-mingle with the test water ahead of the pig.

Normally, the pig will move through the pipeline without difficulty. However, there are many opportunities for problems to develop, causing the pig to stick or even disintegrate.

Sticking may be caused by any or a combination of the following things:

- Pig is incorrect length and cannot negotiate check valves, tees, and bends;
- Pig is too large for heavy wall sections of the line;
- Pipeline may be equipped with reduced opening valves;
- Excess debris or construction material may be left in the pipeline;

- Air/gas may bypass the pig and create an air lock condition; and,
- Other unknown reasons.

Air locks are more likely to occur in hilly country than in flat land. Air locks occur when the accumulated static heads are greater than the available displacing pressure. In some cases, a pressure greater than the maximum allowable pipeline pressure would be required to overcome the air lock. Usually, the air lock occurs where air/gas has bypassed the pig and the downhill legs of the pipeline are filled with air and the uphill legs of the pipeline are filled with water. The air/gas can get in front of the dewatering pig in several ways, such as:

- Poor filling techniques;
- Poor dewatering procedure such as draining water from the line at low points;
- Air/gas bypassing the pig because the pig is too small for the pipeline;
- Air/gas bypassing the pig before it is launched;
- Air/gas bypassing the pig in a fitting such as in a tee, or a steel shaft pig in a short radius bend; and,
- Other unknown reasons.

When an air lock condition occurs, it is necessary to either increase the displacing pressure or remove air/gas through existing vents or other connections at high points in front of the pig.

For these reasons, it is important that the dewatering process be carefully planned, especially for pipelines located in hilly country.

Cleaning pipelines

It is believed there are no written specifications to define the degree of cleaning for a pipeline since there is a question of the definition of a clean pipeline, and furthermore, a method for measuring the cleanliness has not been developed. It is known, however, that cleaning a pipeline does at least four good things. Cleaning the pipeline will:

- Improve flow efficiency because of a smoother pipe wall
- Reduce product contamination and formation of hydrates

- Reduce abrasive damage to pipeline appurtenances such as valves and instruments
- Facilitate pipeline drying

Internal cleaning of the pipeline may be accomplished by any of a combination of the following methods:

- Running a brush pig with air, gas, or liquid
- Internal sand blasting
- Chemical cleaning
- Purging with air or gas followed by a liquid flush

Brush pig run with gas

One brush pig pushed by gas or air will not usually remove a significant amount of small debris from the pipeline, particularly if the debris is wet. However, the scratching action of a good brush pig will more evenly distribute the loose debris and leave a smoother pipe wall, which will improve pipeline flow efficiency.

Brush pig run with liquid

A brush pig displaced by water is a very efficient means for removing debris from the pipeline provided certain precautions are taken. Experience has shown that a pig velocity greater than 3 ft per second is desired. A pig, such as the Power Brush Pig, that allows some of the liquid to bypass the pig to keep the bristles clean and move the loose debris into suspension in the liquid in front of the pig is also desirable. The debris must be free of any sticky material such as oil, grease, or paraffin.

It is not necessary that the pipeline be completely filled with liquid. Short slugs of liquid batched between two pigs are usually preferable as long as there is enough water to keep the pigs apart and maintain a velocity greater than 3 ft per second. Short slugs will minimize the water/debris disposal

problems and will allow for higher velocities with lower pressures.

Disposal of the dirty liquid may present a problem, depending upon the location. As indicated earlier, environmental aspects should be investigated thoroughly during the pre-planning phase. Usually, rust will settle out of still water in about 1 hour, leaving an almost clear liquid. If the dumping of semi-clear water is acceptable, then the problem may be solved by the use of holding tanks.

The debris-holding ability of the propelling fluid is proportional to the viscosity/density of the propelling fluid. The greater the viscosity/density, the higher the holding ability. It has been reported that gels have been developed and used that are supposedly more effective than water.

Internal sand blasting

Internal sand blasting is an extremely effective method for cleaning pipelines. This method was extensively used when natural gas was inexpensive and there were no environmental controls. Some short, small-diameter pipelines are still being internally sand blasted by using air or nitrogen to propel the sand. This procedure, if properly executed, will leave the pipeline dry upon completion.

After the pipeline has been dewatered, clay is blown into short sections of the pipeline with a dry gas to absorb the remaining water. The correct type of sand is then blown through the dust dry section at a high velocity to remove the clay, rust, and mill scale. Caution must be taken to ensure that short sections are cleaned each time so that high velocities may be maintained; that large quantities of dry air or gas are available to sustain the high velocities; that the pipeline is dust dry; that all sand has been removed from the pipeline after the sand blasting is complete; and, that the pipeline facilities will not be damaged by the high-velocity sand.

Chemical cleaning

Cleaning with chemicals is typically used for smaller diameter lines. A properly planned chemical cleaning operation can produce a very clean pipeline. Chemical cleaning will remove rust from any internal pits. Chemical cleaning is accomplished by pushing a batch of hydrochloric acid and water for rust removal, a neutralizer, and a passivator, all separated by pigs, through the pipeline. This method will leave a bright metal finish on the interior of the pipeline. If a pipeline is being converted for a different service, such as a crude oil line being converted to chemical service, a batch of detergent can be added behind the cleaning process to remove any traces of the crude oil or other product.

Typically, chemical cleaning is more involved than other cleaning processes in that some special equipment will be required to safely handle the chemicals. Disposing of the spent material will also present a problem. Care must be exercised in the selection of pigs to be sure they will not be consumed by the cleaning solution. Personnel safety must be carefully considered when choosing this method of pipeline cleaning.

Purging with gas. This method is seldom used for pipeline cleaning, because the amount of debris that will be removed is small. A high gas velocity is required.

Pipeline drying

Pipelines used to transport petrochemicals such as propylene and ethylene must be dried in order for the delivered product to meet moisture specifications. Natural gas pipelines are usually dried to a lesser extent to prevent the formation of hydrates. It is not unusual for a petrochemical line to be dried to a dew point of $-80°F$. A typical dew point for a propylene pipeline will be $-70°F$. A carbon dioxide pipeline might be typically dried to a $-40°F$ dew point. The natural gas industry specifies dryness in pounds of water per million standard cubic feet of gas. The table at the end of this section will enable you to convert from one method to the other.

Dew point, by definition, is the temperature at which water vapor begins to condense out of a gas at atmospheric pressure. For example, at atmospheric pressure, water vapor begins to condense out of a gas that has a moisture content of 7 lb. of water per million standard cubic feet at $-39°F$. Therefore, a gas that has a moisture content of 7 lb. of water per million standard cubic feet has a dew point of $-39°F$.

Air at $+70°F$ with a humidity of 100% has a dew point of $+70°F$ and holds approximately 1,253 pounds of water (150 gallons) per million standard cubic feet.

The most common methods for drying pipelines are as follows:

- Drying with super dry air
- Drying with methanol

- Drying with inert gas such as nitrogen
- Internal sand blasting
- Drying with the medium to be transported
- Vacuum drying

All of these methods may be applied to pipeline drying depending on the particular line and amount of dryness required. No single method can be considered ideal for all situations. Many times, a combination of two or more methods will be used to achieve a dry pipeline at the least cost. The first three methods are probably the most economical and technically feasible for most pipeline drying applications.

Before any type of drying operation commences, it will be necessary to clean the line using one of the previously described processes. If the rust and mill scale are not removed from the pipe wall, moisture will remain trapped and will bleed out over a long period of time. A pipeline can be dried without cleaning; however, the cost and time required will be great.

In *drying with super dry air*, soft foam pigs pushed by dry air are used to absorb any free water remaining in the pipeline after dewatering. After the line is dust dry, wire brush pigs are run to remove any water bearing debris from the pipe wall. The wire brush pigs are then followed by soft foam pigs to absorb the loosened debris. Near the end of this phase, the pigs may be weighed prior to insertion and after removal to monitor the amount of debris that is being removed. The first pigs run through the line will naturally weigh much more than their clean weight. As the cleaning and drying progress, the pig weights will approach their original weight. Pig color will also give some indication of how the debris removal is proceeding.

Dew-point readings will need to be made to determine when the line has been dried to the specified dew point.

It will be necessary to give special consideration to laterals, by-passes, and valve body cavities as any free water trapped here could affect the final dew point readings.

Drying with super dry air provides internal corrosion protection if the line is to remain out of service for some period of time before it is placed in service.

Methanol drying relies on the hygroscopic effect of the methanol. Any remaining moisture in the line will be absorbed by batches of methanol pushed through the line with either gas or dry air. Pigs are used to separate the methanol batches from the displacing medium.[1] Methanol drying usually requires fewer pig runs and, consequently, less line cleaning is accomplished. Pure methanol is expensive, and sometimes a 96% methanol/water mix is used. Since the methanol mix contains water, some water will be left in the pipeline. Some of the methanol will vaporize in the pipeline and will be absorbed by the displacing medium. Toward the end of the line, the moisture content of the methanol will increase, which in turn reduces the amount of water that it can absorb. If the pressure used in the drying operation is too high, hydrate formation can occur, usually at the far end of the line. If natural gas is being used to push the methanol batches, it will probably be necessary to flare some of the line fill volume to be sure that no methanol impurities are contained in the gas.[1]

Methanol run with a dry gas will absorb most of the water and facilitate the vaporization of the remaining water. Soft swabs run through a line with a dry purge gas will accelerate the evaporation of remaining methanol/water solution.

Also of great concern with the methanol method of drying is the fact that explosive mixtures can easily be formed, whether gas or air is used to displace the methanol batches. It may be desirable to use an inert gas such as nitrogen to buffer the methanol batch from the air or gas used to displace the methanol batch. However, many pipelines have been dried without the use of nitrogen buffers with no adverse results. The air/methanol mixture is also highly poisonous and corrosive.[1]

Plans will need to be made for proper disposal of the spent methanol.

If internal corrosion protection is desired, then another drying method should be considered. If the pipeline is to be used to transport sour gas, the methanol drying method should be carefully evaluated before it is used.

Drying with nitrogen is accomplished in much the same manner as when using super dry air. Nitrogen drying will cost more than super dry air by a factor of approximately 50%. Also, super dry air is plentiful and non-polluting.

Internal sand blasting will leave the pipeline clean and dry. The procedure for internal sand blasting is described earlier in this section.

Drying with natural gas requires large volumes of gas. This method is slow and not very effective unless the line is thoroughly cleaned by one of the cleaning processes previously described. If, however, the gas being used to dry with can be blended with another dry gas stream and sold or used, then this is an economical method for pipeline drying. The cost of the gas that will be used to purge the line during the drying process should be weighed against the cost of using super dry air. The Btu value of a volume of natural gas is approximately 15 times greater than that of the fuel required to produce an equal volume of super dry air.

Vacuum drying is a slow process. All free water should be removed from the pipeline before drying begins. This method appears to be used infrequently, and perhaps only offshore.

If the pipeline has been properly cleaned by the water slug method using brush pigs run with liquid, drying can be accomplished by running soft foam pigs with dry air or gas to remove any free water left in the pipeline. This will usually produce a pipeline dry enough for natural gas operations. If additional drying is desired, it can be accomplished by using methanol or super dry air.

Reference

1. Kopp, *Pipe Line Industry*, October 1981, January 1986.

Moisture content of air

DEW POINT °F @ 14.7 PSIA	DEW POINT °C @ 14.7 PSIA	PPM BY VOLUME	PPM BY WEIGHT	LBS. WATER MMSCF DRY AIR	DEW POINT °F @ 14.7 PSIA	DEW POINT °C @ 14.7 PSIA	PPM BY VOLUME	PPM BY WEIGHT	LBS. WATER MMSCF DRY AIR	DEW POINT °F @ 14.7 PSIA	DEW POINT °C @ 14.7 PSIA	PPM BY VOLUME	PPM BY WEIGHT	LBS. WATER MMSCF DRY AIR
-110	-78.9	0.63	0.39	.032	-70	-56.7	16.6	10.3	.83	-30	-34.4	235	146	11.8
-109	-78.3	0.69	0.43	.034	-69	-56.1	17.9	11.1	.96	-29	-33.9	250	155	12.5
-108	-77.8	0.75	0.47	.038	-68	-55.6	19.2	11.9	1.0	-28	-33.3	265	165	13.3
-107	-77.2	0.82	0.51	.041	-67	-55.0	20.6	12.8	1.03	-27	-32.8	283	176	14.2
-106	-76.7	0.88	0.55	.044	-66	-54.4	22.1	13.7	1.1	-26	-32.2	300	186	15.0
-105	-76.1	1.00	0.62	.050	-65	-53.9	23.6	14.7	1.2	-25	-31.7	317	197	15.9
-104	-75.5	1.08	0.67	.054	-64	-53.3	25.6	15.9	1.3	-24	-31.1	338	210	16.9
-103	-75.0	1.18	0.73	.059	-63	-52.8	27.5	17.1	1.4	-23	-30.6	358	223	17.9
-102	-74.4	1.29	0.80	.065	-62	-52.2	29.4	18.3	1.5	-22	-30.0	378	235	18.9
-101	-73.9	1.40	0.87	.070	-61	-51.7	31.7	19.7	1.6	-21	-28.4	400	249	20.0
-100	-73.3	1.53	0.95	.077	-60	-51.1	34.0	21.1	1.7	-20	-28.9	422	212	21.1
-99	-72.8	1.66	1.03	.083	-59	-50.6	36.5	22.7	1.8	-19	-28.3	448	278	22.4
-98	-72.2	1.81	1.12	.091	-58	-50.0	39.0	24.2	2.0	-18	-27.8	475	295	23.8
-97	-71.7	1.96	1.22	.098	-57	-49.4	41.8	26.0	2.1	-17	-27.2	500	311	25.0
-96	-71.1	2.15	1.34	.108	-56	-48.9	44.6	27.7	2.2	-16	-26.7	530	329	26.5
-95	-70.6	2.35	1.46	.118	-55	-48.3	48.0	30	2.4	-15	-26.1	560	348	28.0
-94	-70	2.54	1.58	.127	-54	-47.8	.51	32	2.6	-14	-25.6	590	367	29.5
-93	-69.4	2.76	1.72	.138	-53	-47.2	55	34	2.8	-13	-25.0	630	391	31.5
-92	-68.9	3.00	1.86	.150	-52	-46.7	59	37	2.9	-12	-24.4	660	410	33.0
-91	-68.3	3.28	2.04	.164	-51	-46.1	62	39	3.1	-11	-23.9	700	435	35.0
-90	-67.8	3.53	2.19	.177	-50	-45.6	67	42	3.4	-10	-22.3	740	460	37.0
-89	-67.2	3.84	2.39	.192	-49	-45.0	72	45	3.6	-9	-22.8	780	485	39.0
-88	-66.7	4.15	2.58	.221	-48	-44.4	76	47	3.8	-8	-22.2	820	509	41.0
-87	-66.1	4.50	2.80	.225	-47	-43.9	82	51	4.1	-7	-21.7	870	541	43.5
-86	-65.6	4.78	2.97	.239	-46	-43.3	87	54	4.4	-6	-21.1	920	572	46.0
-85	-65.0	5.30	3.30	.265	-45	-42.8	92	57	4.6	-5	-20.6	970	603	48.5
-84	-64.4	5.70	3.50	.285	-44	-42.2	98	61	4.9	-4	-20.0	1020	634	51.0
-83	-63.9	6.2	3.9	.31	-43	-41.7	105	65	5.3	-3	-19.4	1087	675	54.4
-82	-63.3	6.6	4.1	.33	-42	-41.1	113	70	5.7	-2	-18.9	1153	716	57.7
-81	-62.8	7.2	4.5	.36	-41	-40.6	119	74	6.0	-1	-18.3	1220	758	61.0
-80	-62.2	7.8	4.8	.39	-40	-40.0	125	80	6.4					
-79	-61.7	8.4	5.2	.42	-39	-39.4	136	85	6.8					
-78	-61.1	9.1	5.7	.46	-38	-38.9	144	89	7.2					
-77	-60.6	9.8	6.1	.49	-37	-38.3	153	95	7.7					
-76	-60.0	10.5	6.5	.53	-36	-37.9	164	102	8.2					
-75	-59.4	11.4	7.1	.57	-35	-37.2	174	108	8.7					
-74	-58.9	12.3	7.6	.62	-34	-36.7	185	115	9.3					
-73	-58.3	13.3	8.3	.67	-33	-36.1	196	122	9.8					
-72	-57.9	14.3	8.9	.72	-32	-35.6	210	131	10.5					
-71	-57.2	15.4	9.6	.77	-31	-35.0	222	140	11.1					

(table continued on next page)

Dew Point °F @ 14.7 PSIA	Dew Point °C @ 14.7 PSIA	PPM by Volume	PPM by Weight	Lbs. Water MMSCF Dry Air	Dew Point °F @ 14.7 PSIA	Dew Point °C @ 14.7 PSIA	PPM by Volume	PPM by Weight	Lbs. Water MMSCF Dry Air	Dew Point °F @ 14.7 PSIA	Dew Point °C @ 14.7 PSIA	PPM by Volume	PPM by Weight	Lbs. Water MMSCF Dry Air
0	-17.8	1299	807	64.9	40	4.4	8874	5514	444	80	26.7	34844	21650	1742
1	-17.2	1378	856	68.9	41	5.0	9272	5761	464	81	27.2	36138	22454	1807
2	-16.7	1458	906	74.3	42	5.6	9670	6008	484	82	27.8	37432	23258	1872
3	-16.1	1537	955	76.9	43	6.1	10068	6256	503	83	28.3	38727	24062	1936
4	-15.6	1616	1004	80.8	44	6.7	10466	6503	522	84	28.9	40021	24866	2001
5	-15.0	1695	1053	84.8	45	7.2	10824	6725	541	85	28.4	41315	25670	2066
6	-14.4	1775	1103	88.8	46	7.8	11182	6948	559	86	30.0	42609	26474	2130
7	-13.9	1854	1152	92.7	47	8.3	11540	7170	577	87	30.6	43903	27278	2195
8	-13.3	1933	1201	96.7	48	8.9	11898	7393	595	88	31.1	45198	28083	2260
9	-12.8	2053	1276	103	49	9.4	12255	7614	613	89	31.7	46492	28887	2325
10	-12.2	2174	1351	109	50	10.0	12613	7837	631	90	32.2	48146	29914	2407
11	-11.7	2294	1425	115	51	10.6	12971	8059	649	91	32.8	49801	30943	2490
12	-11.1	2415	1501	121	52	11.1	13329	8282	666	92	33.3	51455	31971	2573
13	-10.6	2535	1575	127	53	11.7	13687	8504	684	93	33.9	53109	32998	2655
14	-10.0	2656	1650	133	54	12.2	14268	8865	713	94	34.4	54764	34027	2738
15	-9.4	2776	1725	139	55	12.8	14849	9226	742	95	35.0	56419	35055	2821
16	-8.9	2897	1800	145	56	13.3	15430	9587	772	96	35.6	58023	36083	2904
17	-8.3	3017	1875	151	57	13.9	16011	9948	801	97	36.1	59728	37111	2986
18	-7.8	3197	1986	160	58	14.4	16593	10310	830	98	36.7	61382	38139	3069
19	-7.2	3376	2098	169	59	15.0	17174	10671	859	99	37.2	63477	39440	3174
20	-6.7	3556	2209	178	60	15.6	17754	11031	889	100	37.9	65571	40741	3279
21	-6.1	3735	2321	187	61	16.1	18336	11393	917	101	38.3	67666	42043	3383
22	-5.6	3915	2432	196	62	16.7	18917	11754	946	102	38.9	69760	43344	3488
23	-5.0	4094	2544	205	63	17.2	19685	12231	984	103	39.4	71855	44646	3593
24	-4.4	4274	2656	214	64	17.8	20452	12707	1023	104	40.0	73950	45948	3698
25	-3.9	4453	2767	223	65	18.3	21220	13185	1061	105	40.6	76045	47249	3802
26	-3.3	4633	2879	232	66	18.9	21987	13661	1099	106	41.1	78139	48550	3907
27	-2.8	4883	3034	244	67	19.4	22755	14138	1138	107	41.7	80234	49852	4012
28	-2.2	5133	3189	357	68	20.0	23522	14615	1176	108	42.2	83861	51484	4193
29	-1.7	5383	3345	269	69	20.6	24290	15092	1215	109	42.8	85489	53117	4274
30	-1.1	5633	3400	282	70	21.1	25057	15569	1253	110	43.3	88116	54749	4406
31	-0.56	5883	3655	294	71	21.7	25825	16046	1291					
32	0	6133	3811	307	72	22.2	26827	16668	1341					
33	0.56	6383	3966	319	73	22.8	27829	17291	1391					
34	1.1	6633	4121	332	74	23.3	28831	17914	1442					
35	1.7	6883	4277	344	75	23.9	29833	18536	1492					
36	2.2	7281	4524	364	76	24.4	30836	19159	1542					
37	2.8	7679	4771	384	77	25.0	31838	19782	1592					
38	3.3	8077	5019	404	78	25.6	32840	20405	1642					
39	3.9	8475	5266	424	79	26.1	33842	21027	1692					

Commissioning petrochemical pipelines

Pipelines carrying a variety of gaseous and volatile liquid products are extensively employed in today's industry as a safe and efficient means of transportation. New pipelines must be placed into service initially (i.e., commissioned), and older lines occasionally are taken out of service in order to perform some desired maintenance and are then recommissioned. Among the reasons for taking a pipeline out of service (i.e., decommissioning) are: hydrostatic testing to recertify or upgrade the pipeline's ability to be used at higher operating pressures; performance of construction work on the pipeline; and a change in the product transported by the pipeline. Pipelines, or sections of a pipeline, may need to be relocated because of highway work, the necessity to deepen a canal, or because of increases in the population surrounding the pipeline. It may also be necessary to replace valves, fittings or a damaged section of the pipeline or add a new connection to service a customer or supplier.

Typically, decommissioning and recommissioning a petrochemical pipeline will include the steps of decommissioning by removing the product from the pipeline, and flaring any remaining residual product. Any necessary construction, upgrading, or cleaning of the pipeline can then be performed. Usually, the pipeline is then filled with water for hydrostatic testing. After pressure testing, the water is removed, the pipeline is cleaned and dried to a specific low dewpoint (to avoid the problems of water contamination of the product), and the pipeline is inerted with nitrogen for recommissioning. Refer to the section on Pipeline Drying for drying techniques.

In order to recommission the pipeline, the nitrogen must be displaced by the desired petrochemical. Before returning to service, product purity must be established and the line safely filled to operating pressure. The terms "commissioning" and "recommissioning" are used interchangeably to refer to a process whereby a first inerting gas, normally nitrogen, in a pipeline is replaced with the desired product at the desired purity and pressure.

Safety and economics are two primary concerns for any proposed pipeline operation. Thus, a commissioning process that brings the purity and pressure of potentially explosive products such as ethylene or propylene up to specification quickly, but at the risk of damage to the pipeline or reduced safety to operating personnel, is not acceptable. Similarly, a process that uses large quantities of product to push nitrogen from a line and results in wasted product/nitrogen mixtures that must be flared or otherwise disposed of would also be unacceptable. Improvement over known methods requires consideration of its safety and economic benefits (considering both the cost of wasted product and the cost attributed to the time the line must

remain out of service) as well as its ability to bring the line back into service with product at desired pressures and purities.[1]

The main problems posed by recommissioning are (1) the fast and economical purging of nitrogen so as to obtain uncontaminated products in the line, and (2) possible damage to the pipeline due to cold temperatures to which it may be subjected during the process. The latter can be a problem when the pipeline is to be recommissioned with a product at a pressure significantly higher than that of the nitrogen inerted line. The pressure drop of a petrochemical product entering a pipeline can cause large drops in temperature and, consequently, potential damage to the carbon steel pipeline.

Due to the demands of the expanding petrochemical industry, and the aging of the present pipeline system, the need for decommissioning, repairing, cleaning, drying, and recommissioning pipelines is increasing. Given the value of today's petrochemical products, a recommissioning procedure, that brings the product purity to acceptable levels as quickly as possible, is desirable. Furthermore, processes and equipment which reduce the possibility for damage to the pipeline due to cold temperature conditions which can occur during the recommissioning process are also highly desirable.[2]

Due to the unusual properties of ethylene, decommissioning and recommissioning ethylene pipelines require special consideration. Ethylene is a colorless, flammable, gaseous, unsaturated hydrocarbon obtained by the pyrolysis of petroleum hydrocarbons. The relative density of ethylene is 0.9686 (air = 1), and its molecular weight is 28.054. The molecular weight of air is 28.964, and the molecular weight of nitrogen is 28.013.

Ethylene decomposition

Ethylene is subject to thermal decomposition (self propagating reaction zone, explosion, or flame) under certain circumstances. Thermal decomposition occurs when the temperature of a substance is raised above the value needed to cause it to begin to self-heat because of a change in molecular structure at a rate high enough to result in combustion. An external source of ignition is not required to precipitate this process, nor is a source of oxygen needed. Thermal decomposition is usually initiated by a source of heat from events such as sudden compression, external source of heat, or the thermal runaway of a heater.[4] Heaters are sometimes used in ethylene measurement stations to maintain the ethylene at an optimum temperature for measurement.

Ethylene is routinely handled under conditions where decomposition may be initiated. The point at which decomposition begins is dependent upon system parameters such as line or vessel size, operating pressure, and operating temperatures. Generally, ethylene decomposition is initiated in pipeline facilities by a rapid rise in temperature associated with sudden compression in the presence of a diatomic gas such as nitrogen or oxygen, exposure of the pipe to flame, possibly hot tapping operations.

The autoignition temperature (AIT) for ethylene is 914°F at atmospheric pressure.[3] Experience of others suggests that at 1,000 to 1,500 psig, the AIT could drop to as low as 400° to 700°F.[3,4] As the concentration of oxygen increases, the decomposition initiation temperature decreases.[4] From this, it may be concluded that a mixture of low oxygen concentration and ethylene could be very hazardous. For this reason, ethylene pipelines, which have been evacuated for maintenance or repairs, are usually filled with nitrogen to displace oxygen from the pipe. Compressing nitrogen with ethylene may result in high nitrogen temperatures—especially if the compression is rapid. Also, when the pressure of the ethylene source is high, refrigeration across the throttling valve used to control the ethylene may result in extremely low temperatures and could cause the valve to fail due to excessively low temperature.

Commissioning an ethylene pipeline with warm ethylene (±70°F) at a low pressure (±50 psig) is quick, easy, inexpensive, and very safe when the ethylene is heated before it is admitted to the pipeline being commissioned.

A patented process used by Pipeline Dehydrators can be used to safely commission ethylene pipelines. The process uses a shell and tube heat exchanger to heat the high pressure ethylene before the pressure is reduced to the low pressure desired for line commissioning.

The methods currently used to commission ethylene pipelines are potentially hazardous, expensive, and time consuming. The two most severe problems are potential damage to the carbon steel in the pipeline due to the cold temperature of the expanded ethylene and contamination of the ethylene from the nitrogen used to inert the pipeline during commissioning.

An ethylene pipeline that is being commissioned has usually been dried to a −70°F dew point or more and inerted with nitrogen. The nitrogen pressure left on the line for commissioning varies from 10 to 1,000 psig depending on the pressure of the ethylene source and the commissioning procedure used and the pipeline owner's preference.

The source of ethylene used for commissioning an ethylene pipeline is usually more than 900 psig and may be up to 2,200 psig.

If the pressure of the ethylene source used for commissioning is high and the nitrogen pressure in the pipeline being commissioned is low, the resulting temperature of the ethylene, due to the pressure drop, may be well below the −20°F design temperature of most carbon steel pipelines.

The carbon steel in some pipelines is subject to becoming brittle at temperatures below −20°F and may fail at a very low internal pressure because of the low temperature.

At low pressures, the very cold ethylene will be much more dense than nitrogen, thereby producing an extensive product interface during commissioning or trapping pockets of the less dense nitrogen at each high place in the pipeline.

In this case, the amount of ethylene burned to produce product purity may be expensive and time consuming and may create environmental problems in addition to subjecting the steel in the pipeline to sub-design temperatures.

If the pressure of the ethylene source used for commissioning is high, and the nitrogen pressure in the pipeline being commissioned is also high, cold temperatures will not be encountered. However, even at warm temperatures, high-pressure ethylene is still much more dense than nitrogen at the same pressure. Therefore, the extensive product interface and the tapped nitrogen at the high points will still occur, resulting in an expensive and time consuming purification process plus the cost of the large volume of nitrogen. Also, there is a large inventory of ethylene in a pipeline at high pressure. If a pipeline blow-down is required because of a leak, lack of purity, or for whatever reason, it will be a major product loss and expense.

The Pipeline Dehydrators system utilizes a tube and shell heat exchanger and a water–glycol mix that is used as the heat transfer fluid. A pump forces the heat transfer fluid

through the fired heater and then through the shell side of the exchanger. The ethylene passes through the tubes of the heat exchanger and is safely and properly warmed by the heat transfer fluid without entering the direct fired heater, thereby eliminating the possibility of an ethylene decomposition due to heater thermal runaway.

The heater will produce ethylene at ground temperature (±70°F) at ±50 psig corresponding to the conditions of the nitrogen in the pipeline. Under these conditions, ethylene and nitrogen have basically the same density; therefore, the product interface during commissioning will be insignificant; no nitrogen will be trapped at the high points in the pipeline; the steel in the pipeline will not be subjected to cold temperatures; purification will be simple, short, and inexpensive; and the ethylene inventory in the pipeline will be minimal in case a pipeline blow-down is required. The savings on nitrogen will also be significant.

Two pressure drops will be taken on the ethylene, one on each side of the tube-shell heat exchanger. The pressure will be dropped to approximately 600 psig prior to the ethylene entering the heat exchanger. The 600 psig ethylene will be warmed to approximately 140°F. Then a second pressure drop will occur, allowing the pressure and temperature of the ethylene to equalize with the pressure and temperature of the nitrogen in the pipeline (±50 psig @ ±70°F). Under these conditions, PLD's 6 million BTU per hour heater can handle up to 45,000 lb. of ethylene per hour. This high flow rate will normally be used only during the pressuring up phase of the process and not used during the purification process. The warm ethylene at low pressure, traveling about 10 mph in the small pipelines that are typically used to transport ethylene, will produce excellent results during the purification phase of the process.

The minimum ethylene temperature encountered in this commissioning process is well above 0°F, and the maximum temperature is 160°F. This is a very safe temperature range to handle ethylene in carbon steel piping systems.

Purity can be established at 50 psig or less, with very little loss of product, and the system can be checked for leaks with ethylene in the pipeline while the pressure is low and the dollar value of ethylene in the pipeline is still small.

After purity has been established at ±50 psig and potential leak points have been checked, the major remaining effort will be to warm enough ethylene to bring the pipeline up to approximately 600 psig. Beyond this point, the ethylene can be throttled into the pipeline without heating, because the cooling due to pressure drop will no longer be great enough to cause damage to the pipeline and equipment. During this pressure-up phase of the commissioning operation, the ethylene does not have to be heated to +70°F temperature and the pipeline pressure will be increasing, both resulting in a reduction in the amount of heat required per pound of ethylene. Therefore, for this phase of the commissioning operation, PLD's new ethylene heater will handle considerably more than 45,000 lb. of ethylene per hour.

This system provides excellent conditions for commissioning an ethylene pipeline quickly, inexpensively, and safely.

This heater can also be utilized to commission other product pipelines where cold temperatures are encountered due to pressure drops such as with propylene, ethane, propane, and carbon dioxide.[2]

Acknowledgement

The editor wishes to thank Marvin D. Powers of Pipeline Dehydrators for permission to use material from a paper on *Ethylene Heater For Commissioning Pipelines*.

Literature cited

1. US Patent No. 5,027,842 July. 2, 1991 Powers.

2. Powers, Marvin D. *"Ethylene Heater For Commissioning Pipelines"* 5–89 Pipeline Dehydrators.

3. Hilado, Carlos J. and Clark, Stanley W. *"Autoignition Temperatures of Organic Chemicals"* Chemical Engineering September 4, 1972, p 76.

4. McKay, F. F., Worrell, G. R., Thornton, B. C., and Lewis, H. L. *"How To Avoid Decomposition In Ethylene Pipelines,"* Pipe Line Industry August 1978, p 80.

Vacuum drying

The vacuum drying process is shown graphically in Figure 1 and consists of three separate phases. Corrosion is generally inhibited at relative humidity (R. H.) levels below 30%, but in the presence of hygroscopic dirt (present in millscale) corrosion can occur at R.H. levels of 20%. Therefore, systems should be thoroughly drained and then vacuum dried to lower than 20% R.H.

Target Dewpoint (D.P.)

$$R.H = \frac{\text{Water vapor pressure}}{\text{Saturated Vapor Pressure (SVP)}}$$

for a constant temperature. Assume the lowest average temperature of the system is 0°C (during the winter months), which gives

$$SVP = 0.6 \text{ KPaA}$$

The vacuum level required for water vapor at 20% R.H. is:

$$0.2 \times SVP = 0.2 \times 0.6 = 0.12 \text{KPaA}$$

This corresponds to a dewpoint of −18°C. A safety allowance should be provided for some desorption from the pipe walls; and therefore a dewpoint of −20°C would be used.

Phase 1–Evacuation

During this phase, the pressure in the pipeline is reduced to a level where the ambient temperature of the pipeline will cause the free water to boil and change to water vapor. This pressure level corresponds to the saturated vapor pressure of the free water in the pipeline, which is dependent upon the ambient temperature of the pipeline.

The approximate pressure value is calculated in advance but is easily recognized on site by a fall in the rate of pressure reduction, which is noted from the plot of pressure against time.

At some convenient point in time a "leak test" is carried out by stopping the vacuum equipment and observing the pressure, usually for a period of 4 hours. Any "air-in" leaks on flanges, fittings, or hoses are rectified at this time, although leaks are not a common occurrence.

Phase 2–Evaporation

Once the saturated vapor pressure has been reached, then evaporation of the free water into water vapor will commence. During this phase, the vacuum equipment is carefully controlled to maintain the pressure at a constant

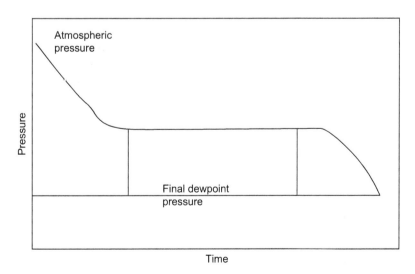

Figure 1. Vacuum drying pressure vs time curve.

level until all the free water has been converted into water vapor. This phase may take several days to complete depending on: (1) the amount of water to be evaporated; (2) the size of the vacuum equipment; and (3) the ambient temperature of the pipeline. The end of the evaporation phase will be observed on site by a noticeable decrease in pressure.

At this time it is prudent to carry out a "soak test" to ensure that all the free water has in fact evaporated. The vacuum equipment is temporarily isolated from the pipeline, usually for a period of 12 hours, and a careful note made of the pressure. If all free water has evaporated then the pressure will remain constant and the final drying phase can be commenced.

Phase 3—Final drying

Once the free water has been converted into water vapor, the majority of it must be removed from the pipeline in order to reach the required dryness level. This is achieved by reducing the pressure in the pipeline still further which has the effect of drawing the water vapor out of the pipeline through the vacuum equipment. Obviously, the more water vapor removed, the drier the pipeline will become.

During this phase a careful watch is kept on the slope of the final drying line to ensure that it follows the calculated value, since a shallower slope would indicate the continuing presence of some free water still remaining in the pipeline.

The necessary calculations to determine the time required for each phase of vacuum drying are detailed in Figure 2.

What is dryness?

The dryness of a pipeline is measured in terms of dewpoint, which is the temperature at which mist or dew will begin to form. A convenient method of measuring dewpoint is to use an instrument called a mirror hygrometer where the water vapor is passed across a polished surface that is slowly cooled until dew forms. The temperature at which the dew forms is the dewpoint of the water vapor and is normally expressed in degrees centigrade. The drier the air, the lower the temperature at which dew will form.

In terms of a pipeline being vacuum dried, the lower the pressure in the pipeline, the lower the dewpoint will be. For example, at a pressure level of 0.26 kPaA, the equivalent dewpoint of the pipeline would be $-10°C$. If the pressure were further reduced to 0.104 kPaA, then the dewpoint would be $-20°C$.

For gas pipelines a dewpoint level of $-20°C$ is generally considered to be adequate and the 0.10 kPaA bar pressure level required to achieve this dewpoint is readily attainable using the portable vacuum equipment described previously.

Evacuation:

$$t = \frac{V}{Se} \log e \frac{Pi - Pv}{Pf - Pv}$$

$$SE = \frac{1}{C} + \frac{1}{Sp}$$

Evaporation:

$$t = \frac{M}{SeDv}$$

Final drying:

$$t = \frac{V}{Se} \log e \frac{Pv}{Preq}$$

Pi = Initial pressure
Pf = Final pressure
Pv = Vapor pressure
Se = Effective pump speed
V = System volume
Sp = Pump speed
C = Conductance
M = Mass of water
Dv = Vapor density
Preq = Required dewpoint pressure

Figure 2. Vacuum drying calculations.

For example, consider a 100-mile-long 36-inch-diameter pipeline that, prior to drying, contained 10,000 gallons of water as a film on the inside surface of the pipe. At a dewpoint of $-10°C$, the quantity remaining would be reduced to only 46.5 gallons and at $-20°C$ to 19.7 gallons. Also, this water would not be free water but rather water vapor, and it would only revert to free water if the ambient temperature of the pipeline were further reduced. This water vapor can subsequently be removed from the pipeline during the purging operation.

The relations between pressure and dewpoint are shown in Figure 3 and Table 1.

Proving the dryness

Immediately following the final drying phase, a dry gas purge using atmospheric air or nitrogen is carried out to prove the dryness of the pipeline. It is possible, under certain circumstances, for a small amount of free water to still remain in the pipeline. Usually this water will have turned to ice due to the

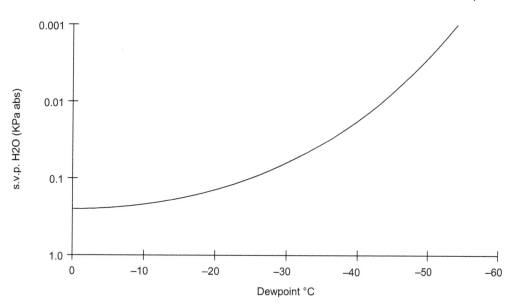

Figure 3. Dewpoint temperature versus saturated vapor pressure.

Table 1
Water Vapor Pressure Table

T°C	SVP (kPaA) (mb × K 10⁻¹)	Vapor Density (gm m³)
−50	0.0039 (0.039 mb)	0.038
−45	0.0128	0.119
−35	0.0223	0.203
−30	0.0308	0.339
−25	0.0632	0.552
−20	0.1043	0.884
−15	0.1652	1.387
−10	0.2597	2.139
−5	0.4015	3.246
0	0.6108	4.847
5	0.8719	6.797
10	1.2270	9.399
15	1.7040	12.830
20	2.3370	17.300
25	3.1670	23.050
30	4.2430	30,380

Once this pressure level has been reached, the vacuum equipment is started and that pressure level maintained. This has the effect of drawing gas through the pipeline under vacuum at a relatively constant dewpoint equal to the final dewpoint required.

At some point in time the purge gas, now under vacuum at a dewpoint of say, −20°C, will reach the vacuum equipment and be pulled through it. The dewpoint at both ends of the pipeline is carefully monitored and compared. If there is no free water remaining in the pipeline then the dewpoint at the vacuum equipment end will be the same as the dewpoint at the remote end. However, if there is any free water present then the dry air passing through the pipeline under vacuum will absorb the water hygroscopically. The dry gas purge operation must then continue to remove the remaining free water until the dewpoints at both ends are equal, at which time purging is discounted. The pipeline has now been vacuum dried to the required dewpoint level, and the dryness proved.

Purging and commissioning

Once the dryness has been attained and proved, the pipeline is ready for commissioning. It is possible to introduce the gas directly into the vacuum or to relieve the vacuum using dry nitrogen.

chilling effect of the vacuum drying process and may not be apparent during the final drying phase or soak test.

Nitrogen or atmospheric air is allowed to enter the pipeline through a valve at the end remote from the vacuum equipment until the pressure has risen to the SVP equivalent of the target dewpoint.

7: Control Valves

Control valve sizing formulas

Liquid formulas (noncompressible fluids)

Volume basis:

$$C_v = q\sqrt{G/\Delta P} \qquad (1)$$

Weight basis:

$$C_v = W/(500\sqrt{\Delta P \times G}) \qquad (2)$$

Equations 1 and 2 exclude control valve calculations where flashing, high-viscosity, two-phase flow, or high pressure drops are involved.

Gas and vapors (other than steam). Volume basis:

$$C_v = (Q/1360)\sqrt{2GT/[\Delta P(P_1 + P_2)]} \qquad (3)$$

Equation 3 does not take into consideration gas compressibility, mixed phase flow or high pressure drop. Compressibility is considered in Equations 12 and 13. Usable pressure drop for valve sizing calculations for gas, vapor, and steam flows is always limited to half of the absolute inlet pressure of the control valve. At that approximate value, critical (sonic) velocity is reached and any further reduction of the downstream pressure will not increase the velocity through a valve, thus, there will be no increase in flow.

Steam formula.

$$C_v = \frac{W(1.0 + 0.0007\,\text{Tsh})}{2.1\sqrt{\Delta P(P_1 + P_2)}} \qquad (4)$$

Special conditions

Equations are given below to correct for flashing, cavitation, high-viscosity fluids, and mixed phase flow, where the simplified equations may lead to inaccurate sizing. These additional expressions, together with the FCI equations, provide satisfactory solutions for most control valve sizing applications.

Liquids

Flashing service. When flashing occurs, the simplified liquid sizing equation gives erroneous results. Flashing occurs when liquids enter a valve at or near their boiling points, and they begin vaporizing as their pressures decrease, due to conversions of static pressure to velocity heads. With

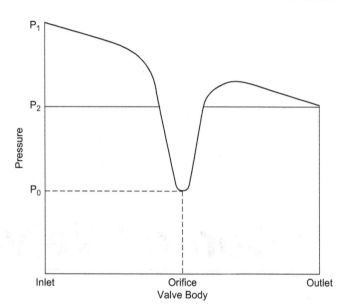

Figure 1. Typical control valve pressure gradient. The ΔP between P_1 and P_2 is much less for high recovery valves.

the advent of high recovery valves, this condition has been aggravated. High recovery valves take a much higher pressure drop from inlet to valve orifice than standard control valves (see Figure 1). When the pressure at the orifice drops to the vapor pressure of the fluid at flowing temperature, the maximum valve capacity is reached. This is true, because flashing occurs across the valve orifice and a flow condition similar to the critical flow of vapors exists.

Two approaches take into account the flashing condition:

1. Hans D. Baumann, in a paper presented in *ISA Transaction*, Volume 2, Number 2 (April, 1963), introduced an equation using a C_f factor to help determine the valve capacity when flashing occurs. The equation, based also on the use of liquid vapor pressure (P_v), is

$$C_v = (Q/C_f)\sqrt{G/(P_1 - P_v)} \qquad (5)$$

2. C. B. Schuder, Fisher Controls, in a paper, "How to Size High Recovery Valves Correctly," proposed an equation to solve for the maximum pressure drop to be used for liquid sizing:

$$\Delta P_m = K_m(P_1 - r_c P_v) \qquad (6)$$

(See Figure 2 for values for r_c.)

Figure 2. Critical pressure ratios (r_c) for use with Equation 6.

$$R = (10,000 \times q)/\sqrt{C_v} \times cs \tag{7}$$

This equation is used for viscosity values between 100 and 200 SSU; the viscosity must be converted to centistokes. When viscosities exceed 200 SSU, Equation 8 applies:

$$R = (46,500 \times q)/\sqrt{C_v} \times \text{SSU} \tag{8}$$

In either equation, the following steps should be followed:

1. Solve for C_v, assuming no viscosity effects.
2. Solve for factor R using Equation 7 or 8.
3. From the viscosity correction curve, Figure 3, read the correction factor at the intercept of factor R on the curve.
4. Multiply the C_v obtained in Step 1 by the correction factor obtained in Step 3 for the corrected C_v.

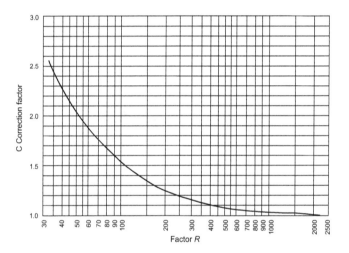

Figure 3. Viscosity correction curve for use with Equation 8.

When this limiting pressure drop from Equation 6 is used in the liquid flow formula, Equation 1 gives a valve size in which flashing will not occur unless the pressure drop increases above the calculated ΔP_m. When the ΔP exceeds ΔP_m, flashing and possible cavitation can occur. Cavitation is the effect caused by the sudden condensing of vapors when the static pressure increases to a point above the boiling point of the liquid.

When the possibility of cavitation exists, the process should be reviewed and the condition eliminated if at all possible. A high recovery valve should not be used in this application. This second approach is suggested as a good solution to the flashing problem.

Viscosity Correction. The liquid sizing formulas in general use have no factor applied to correct for viscosity. This correction is a complex function involving the Reynolds number, which varies with the valve size, port area, velocity, and viscosity. No correction is required for a viscosity value below 100 SSU. Above viscosity values of 100 SSU, two equations are given to obtain correction factors:

Gases and vapors

The simplified equation for gases and vapors (Equation 3) favorably agrees with experimental data when used for conventional globe valves, but agreement with experimental data is poor when used with high recovery valves.

A universal equation proposed by C. B. Schuder, Fisher Controls, in *Instrumentation Technology*, February, 1968, reportedly showed excellent agreement with experimental results covering a wide range of body plug configurations and pressure drop ratios.

Volume basis:

$$Q = \sqrt{\frac{520}{CT}} C_1 C_2 C_v P_1 \sin \left[\frac{3417}{C_1 C_2} \sqrt{\frac{\Delta P}{P_1}} \right] \text{Deg.} \tag{9}$$

Weight basis:

$$W = 1.06\sqrt{dP_1 Z}\, C_1 C_2 C_v \sin\left[\frac{3417}{C_1 C_2}\sqrt{\frac{\Delta P}{P_1}}\right] \text{Deg.} \qquad (10)$$

The C_1 factor in Equations 9 and 10 is currently available from Fisher Controls for their valves. Other companies presumably do not publish this factor. (The C_2 factor is shown in Figure 4.)

Figure 4. C_2 factor applicable only to Equations 9 and 10, the Fisher universal equations for gas and vapor flows.

An equation for gases and vapors proposed by the Foxboro Company (Foxboro TI Bulletin 31-4K) is identical with the FCI equation, except that P_2 is used instead of $(P_1 + P_2)/2$:

$$C_v(Q/1360)\sqrt{GT/\Delta P(P_2)} \qquad (11)$$

A comparison of the Fisher Universal, FCI, and Foxboro equations is made in the TI Bulletin 31-4K. A plot is given showing the results of these three equations using the same data (Figure 5).

The Foxboro equation gives C_v values essentially the same as the universal equation at pressure drops up to 25% of P_1 and gives 7% larger values at a pressure drop of 50% of P_1. At the higher end of $\Delta P/P_1$ ratios, a slightly larger valve should be helpful to allow for the expanding vapors. The same agreement is claimed for steam flows, although no plot is given.

To correct for deviation from the universal gas law, the compressibility factor is inserted in Equation 11, becoming:

$$C_v = (Q/1360)\sqrt{G_b TZ/(\Delta P) \times P_2} \qquad (12)$$

On a weight basis, this equation becomes:

$$C_v = (W/104)\sqrt{TZ/(\Delta P \times P_2 \times G_b)} \qquad (13)$$

Mixed phase flow. Mixed phase flow is generally experienced when:

1. Liquid entrainment is carried by a gaseous flow.
2. Liquid at or near its boiling point vaporizes as it flows through a line and its restrictions.

In the discussion under "Flashing Service," it is explained that flashing can be reduced or eliminated by limiting the ΔP across the valve. When two-phase flow already exists or when it occurs in the control valve, two methods are given to size the valve.

1. A rule of thumb that has been used for years is to calculate the C_v based on liquid only and increase the valve body to the next size. This is satisfactory in many cases, but it obviously does not take into account varying ratios of liquid to vapor.
2. A preferred method is to determine the amount of vapor and liquid, then calculate and add the C_v of each phase to get the overall C_v requirement. The percent vaporization can be calculated by the responsible engineer if it is not furnished by the process people.

Example.

Fluid—Propane

100 gpm @ 300 °psia and 130 °F
s.g. @ 130 °F = 0.444
s.g. @ 60 °F = 0.5077
s.g. @ P_2(200 psia) = 0.466 @ 104 °F
ΔP = 100 psi
% vaporization = 15.05%
M.W. = 44
Z = 0.8
G_b = 44/29 = 1.518
Flow = 100 gpm × 0.444 = 22,200 lb/h
Vapor = 22,200 × 0.1505 = 3340 lb/h
Liquid = 22,200 − 3340 = 18,860 lb/h

COMPARATIVE Cv VALUES
FROM VARIOUS GAS EQUATIONS

Figure 5. Curves showing the close agreement (particularly for low pressure drops) of the Foxboro, FCI, and Fisher universal equations for gas flows. Courtesy of the Foxboro Co.

Using Equation 13:

$$C_v(\text{vapor}) = (W/104)\sqrt{(T \times Z)/(\Delta P \times G_b \times P_2)}$$
$$= (3340/104)\sqrt{(564)(.8)/100(1.518)200}$$
$$= 3.92$$

Using Equation 2:

$$C_v(\text{liquid}) = (W/500)/(18,860/\sqrt{\Delta P \times G})$$
$$= 18,860/(500\sqrt{100 \times 0.466}) = 5.53$$

$$C_v(\text{total}) = 3.92 + 5.53 = 9.45$$

The value of P must never exceed $P_2/2$ for sizing calculation.

Nomenclature

C_v = Valve flow coefficient
G = Specific gravity of fluid at flowing temperature (water = 1.0)
P = Pressure drop $(P_1 - P_2)$, psi
W = Flow of fluid, lb/hr
q = Flow rate, gpm

T = Flowing temperature, °R
T_{sh} = Degrees of superheat, °F
C_f = Critical flow factor, from valve manufacturer
P_m = Maximum useable P (choked flow occurs), psi
K_m = Valve recovery coefficient
r_c = Theoretical critical pressure ratio at P_v
P_v = Fluid vapor pressure at flowing temperature, psia
R = Factor for use with Figure 3
cs = Viscosity at flowing temperature, centistokes
SSU = Viscosity at flowing temperature, Saybolt Seconds Universal
C_1 = Valve sizing factor
C_2 = Specific heat correction factor (see Figure 4)
P_1 = Valve upstream pressure, psia
P_2 = Valve downstream pressure, psia
d = Density of vapors at valve inlet lb/ft^3
d_1 = Density of vapors at valve outlet lb/ft^3
Z = Compressibility factor
G_b = Specific gravity of gas or vapor at base conditions

Source

Applied Instrumentation in the Process Industries, Gulf Publishing Company, Houston, Texas.

Sizing control valves for throughput

These useful charts give data essential to calculation of pressure drops through regulating stations.

W. G. Birkhead, Gas Measurement Engineer, United Gas Pipeline Co., Shreveport, Louisiana.

Pressure losses through piping fittings, once given little consideration in the sizing of control valves, have become important as more efficient control valves have been developed. With the development of high capacity valves and the desire for a more exacting flow formula and efficiency, designers and users look to consideration of minor losses through elbows, tees, swages, or reduced ported valves. The capacity of control valves and regulators was once calculated using the basic hydraulic head equation with an assumed efficiency. This is no longer the case.

The head loss in any closed filled conduit has been accepted as proportional to its length, diameter, and fluid velocity head, and can be calculated as

$$H_L = f \frac{L}{D} \frac{V^2}{2g} \tag{1}$$

where,

H_L = Head loss of flowing fluid (ft of fluid)
f = Friction factor
L = Length of pipe, ft
D = Diameter of pipe, ft
V = Average fluid velocity, ft/s
g = Acceleration of gravity, ft/s/s.

Sometimes L/D is referred to as the "equivalent length in diameters." Valves, fittings, etc., have often been expressed in equivalent diameters. If the friction factor f is combined with L/D, a constant K is derived for any pipe, valve, or fitting. If $K = f(L/D)$, then

$$H_L = K \left(\frac{V^2}{2g} \right) \tag{2}$$

The term K is called the "head loss factor" or "pressure drop factor."

Most control valves are rated with a capacity term called C_v. C_v is the number of gallons of water per minute that will flow through the valve with a 1 lb/in^2 pressure drop across the valve.

$$C_v = Q / \sqrt{\Delta P} \tag{3}$$

Q = Quantity of water in gpm
ΔP = Pressure drop in psi

Rearranging Equation 3:

$$\Delta P = Q^2 / (C_v)^2 \tag{3a}$$

This equation may be rearranged so that C_v in terms of our K factor may be expressed as:

$$C_v = \frac{29.9(d)^2}{\sqrt{K}} \tag{4}$$

where d = inside diameter of the pipe or fitting, inches.

The head loss or pressure drop of a fluid flowing through various pipes, valves, or fittings in series can be stated:

$$(\Delta P)_{\text{Total}} = (\Delta P)_1 + (\Delta P)_2 + (\Delta P)_3 + \cdots \tag{5}$$

Substituting Equation 3a in Equation 5:

$$\frac{1}{(C_v)^2_{\text{Total}}} = \frac{1}{(C_v)^2_1} + \frac{1}{(C_v)^2_2} + \frac{1}{(C_v)^2_3} + \cdots \tag{6}$$

This equation is unwieldy if several fittings are used. By equating Equation 4, substituting in Equation 6 and dividing through by $[29.9(d)^2]^2$:

$$K_{\text{Total}} = K_1 + K_2 + K_3 + \cdots = \sum K \tag{7}$$

Substituting in Equation s4 for K_T an overall C_v can be derived:

$$C_v = \frac{29.9(d^2)}{\sqrt{\sum K}}$$

K factors for various valves and fittings

For full open valves, pipes or fittings, the K factor is approximately equal to $f(L/D)$ or Equation 1, where

L = Equivalent length of valve or pipe or fitting, inches (or feet)
D = Diameter of the valve, inches (or feet)
f = Friction factor.

Some of the recognized K factors for various welding fittings are shown in Table 1.

Other fitting K factors can be obtained from any fluid mechanics, hydraulics, or piping handbook.

Low, average, and high values of K given in Table 1 point out caution necessary in selecting specific types of fittings. Pressure losses in valves and fittings can be more accurately calculated by consulting the references.

Standard Welding Reducer C_V

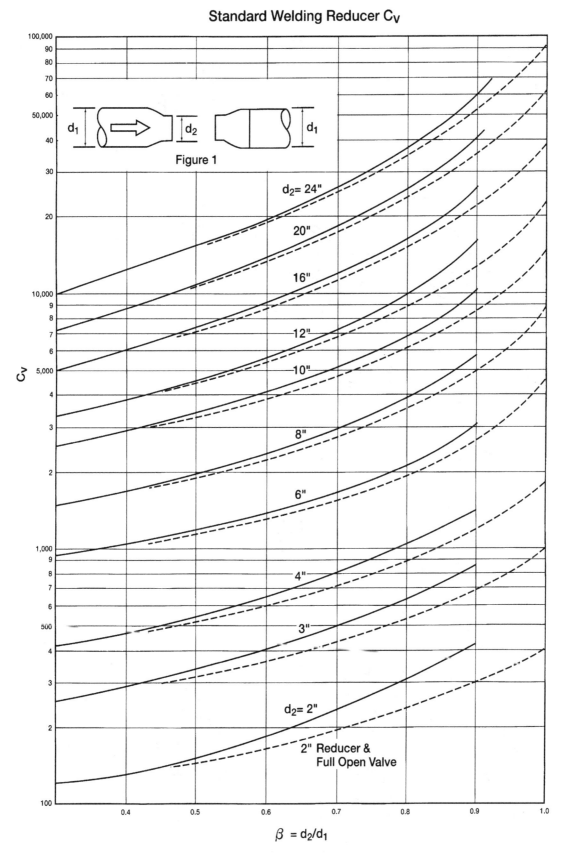

Figure 1. C_v for standard welding reducers.

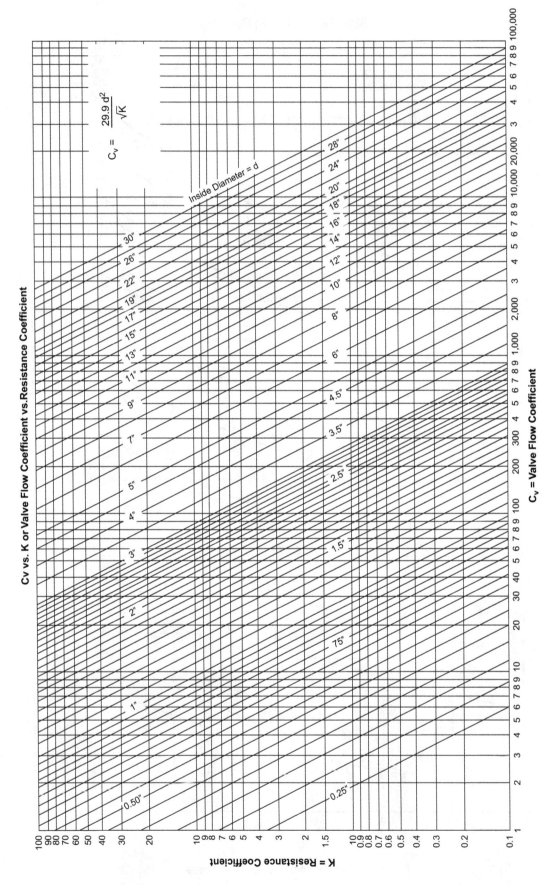

Figure 2. C_v versus K values for valves.

Table 1

K Factor			
Low	Average	High	Type of Welding Fittings and Valves
0.10	0.20	0.40	90° Elbow
0.05	0.12	0.25	45° Elbow
0.25	0.50	1.00	Tee (Enter run, out branch)
0.75	1.00	1.20	Tee (Enter branch, out run)
0.05	0.10	0.15	Tee (Enter run, out run)
See Figs. 1 and 2			Standard Reducer-Increaser
0.25	0.50	1.50	Plug Valves (See manufacturer's bulletin)

All K factors, L/D or C_v, must be converted to the same basis using the same diameter d when using an overall C_v for a piping run. This can be accomplished with the following equations:

$$K_a = K_b (f_a/f_b)(d_a/d_b)^4 \tag{8}$$

$$(L/D)_a = (L/D)_b (d_a/d_b)^4 \tag{9}$$

$$(C_v)_a = (C_v)_b (d_b/d_a)^2 \tag{10}$$

Examples

Refer to Figure 3 for a schematic of a typical control valve station using an old-style control valve. Flow is from a header through a 12 in by 6 in reducer, a 6 in isolating valve, 6 in globe body control valve, 6 in pipe spool, a second isolating valve, a 6 in by 12 in increaser, a 12 in welding elbow, and into a header. Note should be made of Figure 1 in that the C_v from the solid curve is for the complete reducer–increaser combination with regard to the ratio of diameters. In this case, the Beta ratio is 6/12 or 0.5 and C_v is 1200 for the small diameter of 6 in Since C_v cannot be simply added, they must be converted to K factors To do this, refer to Figure 2, which converts C_v to K.

Also, the isolating valves and control valves are considered. Resistance coefficients, C_v or pressure drop factors are published by most valve manufacturers and may be converted to a K factor by using Figure 2. In this case, 6 in K_v is 0.05.

Pipe resistance must also be considered (Figure 5). The L/D ratio must be calculated with reference to the same units of L and D. In the example, 12 ft of 6 in pipe has a K_p equal to 0.36.

All resistances must be converted to the same base diameter (in this case 6 in). The entrance factor K_e is 0.05 for 12 in. Referring to Figure 6, the 6 in K_e is about 0.04; the 12 in elbow coefficient converted to 6 in, 6 in K_{el} is 0.02, and the 12 in exit coefficient, 6 in K_x, is 0.8. A tabulation and sum of these coefficients as determined from Figure 3 would be:

12 in Entrance	6 in K_e = 0.04
6 in reducer–increaser	6 in K_n = 0.80
2–6 in valves	6 in K_v = 0.10
6 in control valve	6 in K_v = 10.00
12 ft spool of 6 in pipe	6 in K_p = 0.36
12 in elbow	6 in K_{el} = 0.02
12 in exit	6 in K_x = 0.08
	$\sum K$ = 11.40

Refer to Figure 2 for a conversion of K to C_v. With the above values noted, the overall 6 in C_v is 320, approximately a 6% reduction of the original C_v of 340.

Now refer to Figure 4. This is an identical control valve run with Figure 3, except that a 6 in × 4 in × 6 in high-capacity ball valve has been substituted for the 6 in singleported globe body control valve. Note that the ΣK is now 1.93, which converted to an overall C_v is 775. This is a reduction of about 48% from the original 6 in × 4 in × 6 in control valve C_v of 1495.

If, in Figure 4, the 6 in × 4 in × 6 in valve were substituted with a full open 6 in ball valve, the overall C_v is 895. This represents a reduction in C_v of about 81% from the original 6 in ball valve C_v of 4700.

The methods and graphs presented in this article are proposed for sizing and calculating pressure drops in a control station. The assumption of assigning values to certain unusual piping configuration must be recognized. The author has some field data confirmation on similar control station designs.

Figure 3. Piping configuration.

Figure 4. Piping configuration.

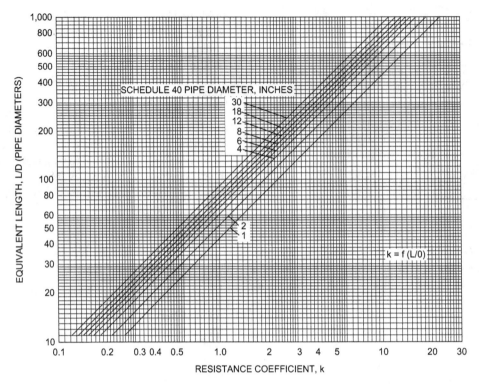

Figure 5. Resistance coefficients versus equivalent length.

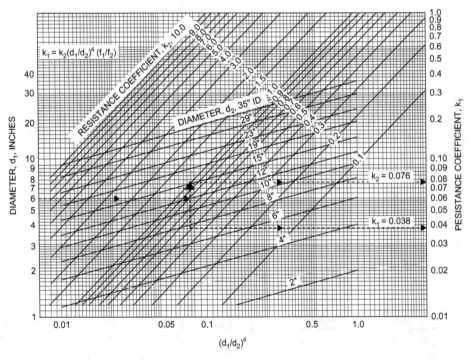

Figure 6. Equivalent resistance coefficients versus diameter ratios.

Control valve selection

A control valve may be thought of as a section of pipe with a variable resistance. Control valves are typically used to match pumps to the pipeline system characteristics. They are not normally used with reciprocating compressors but may be used with centrifugal compressors for preventing surge.

Liquid pipeline applications usually involve controlling the flow rate, suction and discharge pressure, and other variables for centrifugal pumps. Control valves may also be used with reciprocating pumps to regulate the pump discharge pressure by recycling fluid from the pump discharge to the pump suction when the discharge pressure approaches the control point. The control valve may also be used to unload the pump during the startup by opening just prior to startup and slowly closing after the pump driver is up to operating speed.

Other applications for control valves include:

- Pressure reduction
- Pressure relief
- Back pressure control
- Control delivery flow rates

Ideally, the pump(s) for a system will be sized such that a minimum amount of pressure throttling will be required during normal operations. System startup, system upsets, and unit startup and shutdown require that a means be provided to regulate pressures, flow rates, and in some cases motor load. This may be accomplished by using variable speed motors, variable speed couplings, or a control valve.

The control valve, when used with centrifugal pumping units, is installed in the pump discharge line. If multiple pumping units are used, only one control valve will be required.

Table 1 lists typical values of C_v for conventional and ball control valves.

When conventional globe-type control valves are used in pipeline applications for pump control, valves with equal percentage trim are the best choice. Figure 1 shows the typical control valve static characteristic curves for quick opening, linear, and equal-percentage valve trims.

Curves such as those shown in Figure 1 are based on maintaining a constant differential pressure of 1 psig across the valve for varying flow rates. In actual practice, the pressure differential across the control valve will vary according to the needs of the system. Figure 2 shows a comparison of the static and dynamic characteristics for an equal percentage control valve. The dynamic characteristic curve was developed by taking the pressure drop across a 12 in double ported valve with equal percentage trim in series with 10 miles of 12 in pipeline. A maximum flow rate of 6000 bbl/h of 0.86 sp. gr. crude oil at a constant input pressure of 1000 psig was used to calculate the pressure loss in the pipeline. The pressure drop

Table 1
Typical Values of C_v for Control Valves* (Equal Percentage Trim)

Valve Size	Double Ported Globe	Single Ported Globe	Ball Valve Full Bore[†]
2	48	46	406
4	195	195	1890
6	450	400	4580
8	750	640	8900
10	1160	1000	14,600
12	1620		22,800
14	2000		28,700
16	2560		38,900
20			66,000
24			98,600
26			119,800

*These values are suitable for estimating only. The valve manufacturer should be consulted for the actual C_v for the valve selected.
[†]Courtesy: Thunderco–Houston.

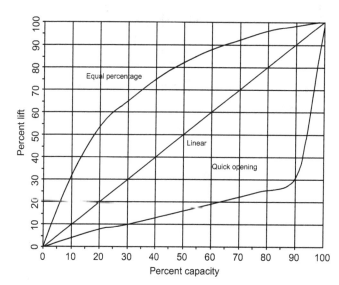

Figure 1. Control valve characteristics.

across the control valve was calculated for various flow rates while maintaining a constant system input pressure. The C_v was then calculated for each of the flow conditions and converted to percent of valve lift.

Figure 2. Control valve characteristics.

Figure 3. Control valve characteristics.

Figure 3 shows a comparison of the static and dynamic characteristics for an 8 in × 8 in × 8 in ball valve under the same conditions of flow and pressure. A 6 in × 6 in × 6 in ball valve with conical reducers has a C_v of 2100, which would have satisfied the pressure drop requirements; however, the velocity through the 6 in valve would have been approximately 45 ft/s. Experience has shown that the valve velocity should be limited to about 25 ft/s. The velocity through the 8 in valve will be 27 ft/s. The C_v for the 8 in valve with standard

welding reducers is 2600, and with conical reducers, the C_v is 4100. The choice of using standard concentric welding reducers was made because that combination came closest to satisfying the C_v needs. When sizing and selecting valves, it is important that the valve not be oversized, as this will always result in poor system control. Reduced-size trim is available for conventional control valves, and ported ball valves are available.

In this particular application, Figures 2 and 3 show that for a 10% change in system rate, the conventional globe valve must change its position by 50% (assuming that it was operating in the full open position). The ball valve would be required to change its position by 30% under the same conditions.

In pipeline applications, it is desirable to have virtually no pressure drop across a control valve when it is wide open. Any built-in residual pressure drop across a control valve will result in added fuel costs; however, in certain situations, some residual pressure drop is unavoidable.

In process control, some pressure drop across the valve is desired to improve the control of the process. Ball valves have gained wide acceptance in pipeline operations because of their high C_v factor, which in turn results in very low pressure drop across the valve when wide open. As can be seen from Table 1, a 16 in double ported valve (equal percentage trim) has a maximum C_v of 2560. Based on a maximum pressure drop of 5 psig with the valve fully open, the maximum flow rate for 0.86 sp. gr. crude oil would be 212,000 bpd for a single valve. For flow rates greater than this, multiple valves may be used or one ball valve may be used. Very large ball valves of up to at least 36 in in size have been successfully used for control valve applications.

The ideal control valve and control system will be quick to act when needed, and when controlling, will be stable at all control points. As shown by the dynamic characteristic curves, the valve must be capable of moving quickly from the wide open position to the control point without overshoot. Certain types of control systems may prove to be unstable on control when required to move the valve at high speeds.

High flow rates and high pressure drops across a control valve will require a valve operator with sufficient power to move the valve quickly and positively. Hydraulic systems have gained wide acceptance for this type of application. Air motors may also be used; however, the cost of the air compressor and air dryer will usually equal or exceed the cost of a hydraulic system. Other factors favoring a hydraulic system include the use of variable vane pumps, which reduces the power consumption during normal control operations. With a hydraulic system, a small hand pump may be used during electrical power outages. Control valves with air motors may be equipped with a handwheel for manual operation.

RELIEF VALVE SIZING, SELECTION, INSTALLATION, AND TESTING

Effectiveness of a Safety Relief Valve Depends on Proper Sizing and Type of Equipment Selected

Gary B. Emerson, Manager SVR Products, Anderson-Greenwood USA, Bellaire, Texas

A safety relief valve is an essential and important piece of equipment on virtually any pressured system. Required by the ASME-UPV Code, among others, it must be carefully sized to pass the maximum flow produced by emergency conditions.

After it is installed, the user hopes that it will never have to operate, which makes the safety relief valve rather unique as compared to other types of process equipment.

Sizing

The sizing formulas for vapors and gases fall into two general categories, based on the flowing pressure with respect to the discharge pressure. When the ratio of P_1 (set pressure plus allowable accumulation) to P_2 (outlet pressure) is greater than two, the flow through the relief valve is sonic; the flow reaches the speed of sound for the particular flowing medium. Once the flow becomes sonic, the velocity remains constant; it cannot go supersonic. No decrease of P_2 will increase the flow rate.

Sonic flow

In accordance with API RP 520, Part 1, Section 3.3, the formulas used for calculating orifice areas for sonic flow are:

$$A = \frac{W\sqrt{TZ}}{CKP_1\sqrt{M}}$$

$$A = \frac{Q\sqrt{Z}}{18.3KP_1(C/356)\sqrt{520/T}\sqrt{29/M}}$$

where
$A =$ Calculated orifice area (in^2)
$W =$ Flow capacity (lb/h)
$Q =$ Flow capacity (scfin)
$M =$ Molecular weight of flowing media
$T =$ Inlet temperature, absolute (°F $+$ 460)
$Z =$ Compressibility factor
$C =$ Gas constant based on ratio of specific heats (Table 1)
$K =$ Valve coefficient of discharge
$P_1 =$ Inlet pressure, psia (set pressure $+$ accumulation $+$ atmospheric pressure).

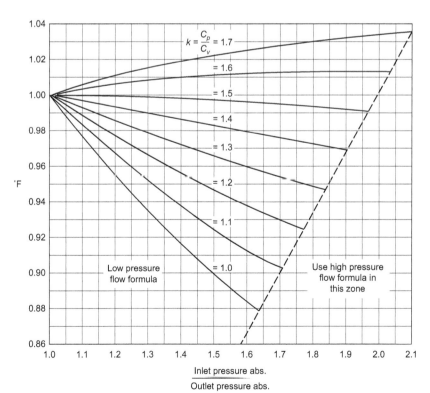

Figure 1. Low pressure and high pressure flow formulas.

Subsonic flow

The second general category for vapor or gas sizing is generally when P_2 is greater than half of P_1 (backpressure greater than half of inlet pressure).

Using k (ratio of specific heats) and P_1/P_2 (absolute), confirm from Figure 1 that the subsonic ("low pressure") flow formula is required. If so, then determine "F" factor. If not, use above sonic flow formula.

$$A = \frac{Q\sqrt{GI}}{863KF\sqrt{(P_1 - P_2) \times P_2}}$$

where

G = Specific gravity
F = Factor from Table 1
I = Temperature of relieving gas
P_2 = Outlet pressure, psia (backpressure + atmospheric pressure).

After determining the calculated orifice area, select the next largest standard orifice size from the relief valve manufacturer's catalog.

Selection

The fundamental selection involves the consideration of the two basic types of relief valves more commonly used: conventional spring-loaded type and pilot operated relief valves.

The advantages of the spring-loaded relief valves are as follows:

- Competitive price at lower pressures and in smaller sizes.
- Wide range of chemical compatibility.
- Wide range of temperature compatibility, particularly at higher temperatures.

The disadvantages are as follows:

- Metal-to-metal seat not tight near set pressure and usually after valve relieves.
- Sensitive to conditions that can cause chatter and/or rapid cycling.
- Protection against effects of backpressure is expensive, pressure limited, and creates possible additional maintenance problems.
- Testing of set pressure not easily accomplished.

Advantages of pilot operated type relief valves are as follows:

- Seat tight to set pressure (Figure 2).
- Ease of setting and changing set and blowdown pressures.
- Can achieve short blowdown without chatter.
- Pop or modulating action available.
- Easy maintenance.

Table 1
Gas Constant Based on Ratio of Heats

Gas	Mol. Wt.	C_p/C_v	°F
Acetylene	26	1.26	343
Air	29	1.40	356
Ammonia	17	1.31	348
Argon	40	1.67	378
Benzene	78	1.12	329
Butadiene	54	1.12	324
Carbon dioxide	44	1.28	345
Carbon monoxide	28	1.40	356
Ethane	30	1.19	336
Ethylene	28	1.24	341
Freon 22	86.47	1.18	335
Helium	4	1.66	377
Hexane	86	1.06	322
Hydrogen	2	1.41	357
Hydrogen sulphide	34	1.32	349
Methane	16	1.31	348
Methyl mercapton	48.11	1.20	337
N-Butane	58	1.09	326
Natural gas (0.60)	17.4	1.27	344
Nitrogen	28	1.40	356
Oxygen	32	1.40	356
Pentane	72	1.07	323
Propane	44	1.13	330
Propylene	42	1.15	332
Sulfur dioxide	64	1.29	346
VCM	62.5	1.18	335

Figure 2. Setting seat pressure to set flow pressure.

- Easy verification of set pressure without removing relief valve from service.
- Flexibility: offers options for remote operation, backpressure protection, valve position indication.

The disadvantages are as follows:

- Maximum temperature limitations.
- Should not be used in extremely dirty or polymerizing type service.

Installation procedures

Safety relief valve installation requires that careful consideration be given to inlet piping, discharge piping, pressure sense lines (where used), and bracing if required. Any marginal installation practice can render the safety relief valve inoperable, at worst, or severely restrict its design capacity at best. Either condition compromises the safety of the installation.

Inlet piping

The proper design of inlet piping is extremely important.

Very often, safety relief valves are added to an installation at the most physically convenient location, with little regard to flow considerations. Pressure loss during flow in a pipe always occurs. Depending upon the size, geometry, and inside surface condition of the pipe, the pressure loss may be large (20, 30, or 40%) or small (less than 5%).

API RP 520, Part 2, recommends a maximum inlet pipe pressure loss to a safety relief valve of 3%. This pressure loss shall be the sum total of the inlet loss, line loss, and when a block valve is used, the loss through it. The loss should be calculated using the maximum rated flow through the pressure relief valve.

The 3% maximum inlet loss is a commendable recommendation but often very difficult to achieve. If it cannot be achieved, then the effects of excessive inlet pressure should be known. These effects are rapid or short cycling with direct spring operated valves or resonant chatter with pilot operated relief valves. In addition, on pilot operated relief valves, rapid or short cycling may occur when the pilot pressure sensing line is connected to the main valve inlet. Each of these conditions results in a loss of capacity.

Pilot operated valves can tolerate higher inlet losses when the pilot senses the system pressure at a point not affected by inlet pipe pressure drop. However, even though the valve satisfactorily operates, reduced capacity will still occur because of inlet pipe pressure losses. The sizing procedure should consider the reduced flowing inlet pressure when the required orifice area, "A," is calculated.

A conservative guideline to follow is to keep the equivalent L/D ratio (length/diameter) of the inlet piping to the relief valve inlet to 5 or less.

Rapid cycling

Rapid valve cycling with direct spring operated valves occurs when the pressure at the valve inlet excessively decreases at the start of relief valve flow. Figure 3 is a schematic of system pressures that is shown before and during flow. Before flow begins, the pressure is the same in the tank and at the valve inlet. During flow, the pressure at the valve inlet is reduced

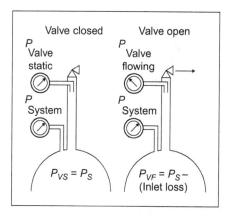

Figure 3. System pressures before and after flow.

due to the pressure loss in the inlet pipe to the valve. Under these conditions, the valve will cycle at a rapid rate.

The valve responds to the pressure at its inlet. When that pressure decreases during flow to below the valve's reseat point, the valve will close. However, as soon as the flow stops, the inlet pipe pressure loss becomes zero and the pressure at the valve inlet once again rises to tank pressure. If the tank pressure is still equal to or greater than the relief valve set pressure, the valve will open and close again. Rapid cycling reduces capacity and is sometimes destructive to the valve seat in addition to subjecting all of the moving parts in the valve to excessive wear.

On pilot operated relief valves with the pilot valves' sense line connected at the main valve inlet, rapid cycling can also occur for the same reasons described earlier; namely the pressure at the relief valve inlet is substantially reduced during flow (Figure 4). With the pilot sense line located at the main valve inlet, the pilot valve responds to the pressure at the location and therefore closes the main valve. If the pressure loss during flow is small (less than the reseat pressure setting of the pilot), the main valve may or may not partially close depending on the type of pilot design used.

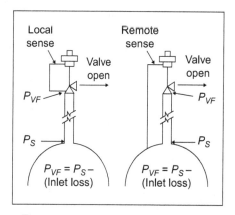

Figure 4. Schematic of rapid cycling.

In actual application, because of the long response time of a pilot operated valve compared with a direct spring operated valve, the main valve may only go into partial life since the inlet pipe pressure loss occurs before the pilot has sufficiently vented the dome area for full piston travel.

As with direct spring operated valves, capacity is reduced but under severe inlet pressure loss conditions, resonant chatter may be induced in pilot operated relief valves. This chatter is extremely destructive and may result in complete valve failure.

On existing installations, the corrective action is limited. For direct spring operated valves, increasing the blowdown will minimize or eliminate rapid cycling if the blowdown can be adjusted to below the initial flowing pressure.

Resonant chatter

Resonant chatter occurs with pilot operated relief valves when the inlet piping produces excessive pressure losses at the valve inlet. The higher the set pressure or the greater the inlet pipe pressure loss, the more likely resonant chatter will occur.

Unlike the rapid cycling noted earlier, resonant chatter is uncontrolled; that is, once started, it cannot be stopped unless the pressure is removed from the valve inlet. In actual application, however, the valve will self-destruct before a shutdown can take place because of the magnitude of the forces involved in a resonant mode.

Discharge piping

Discharge piping for direct spring operated valves is more critical than for pilot operated valves. As with inlet piping, pressure losses occur in discharge headers with large equivalent L/D ratios. Excessive backpressure will reduce the lift of a direct spring operated valve, and enough backpressure (15–20% of set + overpressure) will cause the valve to reclose. As soon as the valve closes, the backpressure in the discharge header decreases and the valve opens again with the result that rapid cycling can occur.

Pilot operated relief valves with the pilot vented to the atmosphere or with a pilot balanced for backpressure are not affected by backpressure. However, if the discharge pressure can even exceed the inlet pressure, a back flow preventer must be used.

The valve relieving capacity for either direct or pilot operated relief valves can be affected by backpressure if the flowing pressure with respect to the discharge pressure is below critical (subsonic flow).

Balanced bellows valves (direct spring operated) have limitations on maximum permissible backpressure due to the collapse pressure rating of the bellows element. This limitation will, in some cases, be less than the backpressure limit of a conventional valve. Manufacturer's literature should be consulted in every case. If the bellows valve is used for systems with superimposed backpressure, the additional built up backpressure under relieving conditions must be added to arrive at maximum backpressure.

The performance of pressure relief valves, both as to operation and flow capacity, can be achieved through following good discharge piping practices:

- Discharge piping must be at least the same size as the valve outlet connection and may be increased when necessary to larger sizes.
- Flow direction changes should be minimized, but when necessary use long radius elbows and gradual transitions.
- If a valve has a drain port on the outlet side, it should be vented to a safe area. Avoid low spots in discharge piping, preferably pitch piping away from the valve outlet to avoid liquid trap at the valve outlet.
- Proper pipe supports to overcome thermal effects, static loads due to pipe weight, and stresses that may be imposed due to reactive thrusts forces must be considered.

Reactive force

On large orifice, high-pressure valves, the reactive forces during valve relief are substantial and external bracing may be required (Figure 5).

API RP 520, Part 2 gives the following formula for calculating this force.

$$F = \frac{Q_h}{366}\sqrt{\frac{kT}{(k+1)M}} \tag{1}$$

Where

F = Reactive force at valve outlet centerline (lb)
Q_h = Flow capacity (lb/h)
k = Ratio of specific heats (C_p/C_v)
T = Inlet temperature, absolute (°F + 460)
M = Molecular weight of flowing media.

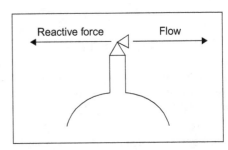

Figure 5. Reactive forces during valve relief.

If bracing is not feasible, a dual outlet valve (available in some pilot operated safety relief valves) can be used. The reactive forces are equal but opposite, resulting in zero force on the valve outlet, but if redirected can still impose loads that must be reacted in some manner.

An approximate method for calculating the reactive force is to multiply the following pressure (psig) by the orifice area (in^2); the reactive force obtained will be on the high side.

Example	A-G Type 22312R68 "R" Orifice (16.00 in^2)
Set pressure—	1000 psi, 10% overpressure
Flowing media—	Natural gas, 60 °F
	$M = 17.4(G = 0.60)$
	$C = 344\ (k = 1.27)$
Reactive force =	10,143 lb., using Equation 1
Reactive force =	17,600 lb., using approximate method

Testing

Using DOT Title 49, Part 192.739 as a guideline, each valve should be inspected at least once a year to determine that it:

- Is in good mechanical condition
- Will operate at the correct set pressure
- Has adequate capacity and is operationally reliable
- Is properly installed and protected from dirt, liquids, or other conditions that might prevent proper operation

The effectiveness of a safety relief valve installation greatly depends on proper sizing and selection of the valve type, suitable installation conditions, and proper and timely testing of the valve.

Rupture disk sizing

A review of the ASME Section VIII requirements

Changes to Section VIII, Division 1 of the ASME Code have impacted the methods used to size pressure relief systems that include rupture disks. The objective of this article is to clarify the sizing methodologies.

Key terminology

- K_D (Coefficient of Discharge)—This is a unitless factor used to derate the theoretical flowing capacity of a device or system. For rupture disks, this value is defined by the Code as 0.62.
- K_R (Resistance to Flow Factor)—This is a unitless factor used to characterize the resistance to flow of a rupture disk device. The larger the K_R value, the more restrictive the device. The K_R value has no direct correlation to the K_D, coefficient of discharge.
- Combination Capacity Factor—This is a unitless factor used to derate the capacity of a pressure relief valve, when a rupture disk is installed upstream. The combination capacity factor may be the default value of 0.9, or a higher value, if the disk/valve combination has been tested.
- MNFA (Minimum Net Flow Area)—This is a derived area based on the area of the device, and/or the piping area,

and is validated during ASME certification testing. The MNFA is used only in the coefficient of the discharge method.

Sizing methodologies

The code describes three methods to provide adequate relieving capacity for a pressure vessel. These methodologies are applied based on the characteristics of the relieving system, not individual components.

When a rupture disk is installed upstream of a pressure relief valve, the code requires the capacity of the valve to be derated by a factor of 0.9, or by a *combination capacity factor* determined by test for that disk/valve combination. This method assumes that the disk has a capacity equal to or greater than the valve and does not interfere with the normal functioning of the valve. From the rupture disk standpoint, this means the rupture disk should be the same nominal size as the relief valve inlet (or larger) and that the disk be of a full opening design. This method is unchanged from the previous versions of Section VIII.

The *coefficient of discharge method* (K_D) is not a rupture disk sizing method but a method for sizing a relatively simple relief system. The calculations for this method remain unchanged from previous versions of the code, but guidelines for use have been added. This method calculates the theoretical capacity of system and derates it by the coefficient of discharge, $K_D = 0.62$. This method is used only when each of the following conditions is true:

(1) the rupture disk discharges to atmosphere
(2) the rupture disk is installed within eight pipe diameters of the vessel
(3) there are no more than five pipe diameters of discharge piping on the outlet of the device.

The reason this is considered a system analysis method is because it takes into account the entrance conditions from the vessel to the discharge piping, eight pipe diameters of piping to the rupture disk, and five pipe diameters of piping, from the rupture disk to the atmosphere.

The *resistance to flow method* (K_R) is used if neither of the first two methods apply, such as systems with long runs of piping, discharging into manifolds, or flare systems. The resistance to flow method is not a rupture disk sizing method, but a system sizing method. This analysis takes into account the resistance to flow of all of the elements within the relieving system, including the resistance of the rupture disk. The code then requires that the system capacity then be derated by a factor of 0.9 to account for uncertainties in the method.

The code does not describe the detailed analysis method, other than to say "accepted engineering practices."

Commonly used references for this type of analysis are: API RP521 Guide for Pressure Relieving and Depressuring Systems of Crane Paper 410.

As a point of reference, the following table is an excerpt from API RP521 listing typical K values for various piping components. The rupture disk K_R values available from the manufacturer are used in the same way as the K values for these common piping components. In many cases the rupture disk is not a limiting or even significant factor in the capacity of a relieving system. Note the API generic rupture disk K value listed here is not necessarily conservative and actual tested values may be significantly higher or lower than 1.5.

Fitting	K
Globe valve, open	9.7
Angle valve, open	4.6
Swing check valve, open	2.3
90° mitered elbow	1.72
Rupture disk	1.5
Welded tee flowing through branch	1.37
15 feet of 3 in sch 40 pipe	1.04
90° std threaded elbow	0.93
90° long sweep elbow	0.59

Reprinted with permission—The Fike Corp. Blue Springs, Missouri.

Rupture disk sizing using the resistance to flow method (K_R)

An excerpt from Fike Technical Bulletin TB8102:

This is a sizing method for relief systems that have a rupture disk as the relief device. Rupture disk devices can be characterized as to their respective resistance to fluid flow. The K_R value represents the velocity head loss due to the rupture disk device. This head loss is included in the overall system loss calculations to determine the capacity of the relief system.

ASME PTC25 provides standardized test methods to measure the K_R of rupture disk devices. By quantifying this performance characteristic, rupture disk devices may be

accounted for in piping system sizing calculations in the same way as piping and piping components (exit nozzles on vessels, elbows, tees, reducers, valves).

Sample calculation using K_R method

The example shown is based on Crane TP-410 methods. It assumes a steady-state relieving condition where the vessel volume is large relative to the relieving capacity.

Piping Component or Feature	Flow Resistance Value (K)	Reference
Entrance—sharp edged	$K_1 = 0.50$	Crane 410 Pg A-29
1 ft. of 3 in sch 40 pipe	$K_2 = 0.07$	$K = fL/D$; $f = 0.018$ (Crane 410 Pg A-26) $L = 1$ ft, ID = 3.068 in/12 ft.
Fike 3 in SRX-GI Rupture Disk	$K_R = 0.99$	National Board Cert. No. FIK-M80042
20 ft of 3 in sch 40 Pipe	$K_3 = 1.41$	$K = fL/D$; $f = 0.018$ (Crane 410 Pg A-26) $L = 20$ ft., ID = 3.068 in/12 ft.
3 in sch 40 standard 90° elbow	$K_4 = 0.54$	Crane 410 Pg A-29
40 ft. of 3 in sch 40 Pipe	$K_5 = 2.82$	$K = fL/D$; $f = 0.018$ (Crane 410 Pg A-26) $L = 40$ ft., ID = 3.068 in/12 ft.
Pipe exit—sharp edged	$K_6 = 1.00$	Crane 410 Pg A-29
Total system flow resistance	$K_T = 7.33$	$K_T = K_1 + K_2 + K_R + K_3 + K_4 + K_5 + K_6$

Given information:

1. Pressure vessel MAWP = 1000 psig.
2. Relieving pressure as allowed by ASME Section VII Div. 1 + 110% × MAWP = 1114.7 psia = $P'1$
3. Back pressure (outlet pressure) = 14.7 psia = $P2$
4. Working fluid—air ($k = Cp/Cv = 1.4$)
5. Air temperature at disk rupture = 500 °F = 960 °R = $T1$
6. Maximum flow rate into the vessel = 40,000 SCFM-Air
7. Rupture disk—Fike 3 in SRX-GI $K_R = 0.99$

The Darcy equation defines the discharge of compressible fluids through valves, fittings, and pipes. Since the flow rate into the example vessel is defined in SCFM, the following form of the Darcy equation is used:

Crane Equation 3–20

$$q'_m = 678 \times Y \times d^2 \sqrt{\frac{\Delta P \times P'_1}{K T_1 S_g}} \, \text{SCFM}$$

$q'_m =$ Rate of flow in cubic feet per minute at standard conditions (14.7 psia and 60 °F)

$Y =$ Net expansion factor for compressible flow through orifices, nozzles, and pipes (Crane 410 Pg A-22)

$\Delta d =$ Internal diameter of pipe (inches)

$\Delta P =$ Change in pressure from state 1 to 2 (entrance to exit).

$P'_1 =$ Pressure at state 1 (entrance)—psia

$K =$ Loss coefficient

$T_1 =$ Absolute temperature at state 1 (entrance) in degrees Rankine (°R)

$Sg =$ Specific gravity relative to air = 1.0 for this example

To determine Y, first it must be determined if the flow will be sonic or subsonic. This is determined by comparing the *actual* Δ/P'_1 to the *limiting* Δ/P'_1 for sonic flow. Crane Table A-22 shows limiting factors for $k = 1.4$ for sonic flow at a known value of K_T. If $(\Delta P/P'_1)_{sonic}$ $(\Delta/P'_1)_{actual}$, then the flow will be sonic.

Limiting Factors for Sonic Velocity ($k = 1.4$). Excerpted from Crane 410 P4 A-22.

K	$\Delta P/P_1'$	Y
1.2	0.552	0.588
1.5	0.576	0.606
2.0	0.612	0.622
3	0.622	0.639
4	0.697	0.649
6	0.737	0.671
8	0.762	0.685
10	0.784	0.695
15	0.818	0.702
20	0.839	0.710
40	0.883	0.710
100	0.926	0.710

For this example: $(\Delta P/P_1')_{actual} = \dfrac{114.7 - 14.7}{1114.7} = 0.9868$

From Table A-22 at $K_T = 7.33$

$(\Delta P/P_1')_{sonic} = 0.754$

Since $(\Delta P/P_1')_{sonic}(\Delta P/P_1')_{actual}$, the flow is sonic in the piping. Thus, we will use $Y_{sonic} = 0.680$ in the flow calculations (Crane 410 Eq. 3–20).

Since $(\Delta P/P_1')_{sonic} = 0.754$, then $\Delta P = 0.754*P_1'$
$$= 0.754*1114.7 = 840.5 \text{ psi}.$$

Calculating the system capacity is completed by substituting the known values into Crane 410 Equation 3–20.

$$q_m' = 678 \times Y \times d^2 \sqrt{\frac{\Delta P \times P_1'}{KT_1S_g}}$$

$$q_m' = 678 \times 0.680 \times (3.068)^2 \sqrt{\frac{840.5 \times 1114.7}{7.33 \times 960 \times 1}}$$

$$q_m' = 50,074 \text{ SCFM—Air}$$

The ASME Pressure Vessel Code, Section VIII, Division 1, paragraph UG-127(a)(2)(b), also requires that the calculated system capacity using the resistance to flow method must also be multiplied by a factor of 0.90 to account for uncertainties inherent with this method.

$$q_{m-ASME}' = 50,074 \times 0.90 = 45,066 \text{ SCFM}_{Air}$$

Thus, the system capacity is greater than the required process capacity (40,000 SCFM_{Air}).

Reprinted with permission—The Fike Corp. Blue Springs, Missouri and *Flow Control Magazine*.

Variable orifice rotary control valves

Flow coefficient (C_v). The flow coefficient, which describes the flow capacity of a valve, is an empirically determined constant equal to the valve's flow capacity in U.S. gallons per minute of water at 60 °F when a pressure drop of 1 psi exists across the valve in its most wide-open position. Table 1 lists C_v for all sizes of DANTROL-270 valves, with and without trim.

The flow capacity of a valve in a partially open position may be calculated by selecting the appropriate C_v from Table 1 and multiplying it times the percent of maximum C_v at the desired position on the selected flow curve in Figure 2.

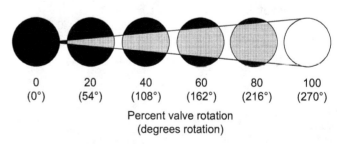

0	20	40	60	80	100
(0°)	(54°)	(108°)	(162°)	(216°)	(270°)

Percent valve rotation
(degrees rotation)

Figure 1. Orifice opening.

Table 1
Maximum Flow Coefficients (C_v)

Valve Size	Ball Dia.	Slot Length	3No. Plates Each Slot	C_v w/Trim	C_v w/o Trim
1	4.25	7.40	8	45	53
1 ½	5.50	9.58	16	98	120
2	6.50	11.32	16	195	220
3	8.00	13.93	18	410	480
4	9.75	16.58	20	765	860
6	13.25	23.08	25	1900	1990

Figure 2. Percent of maximum C_v versus valve rotation.

Sizing valves for liquid flow

These equations for single-phase Newtonian liquid flow assume that the liquid is nonviscous and neither cavitating nor flashing. (Factors and constants are subsequently outlined.)

$$C_v = \frac{Q}{N_1 F_P E_R} \sqrt{\frac{G_F}{(P_1 - P_2)}}$$

or

$$C_v = \frac{W}{N_6 F_P F_R \sqrt{(P_1 - P_2)\gamma}}$$

The piping geometry factor (F_P)

The piping geometry factor (F_P) is equal to 1.0 for line-size valves. With reducers, use the value of F_P from the "Valve Reference Data"—Table 2 or 4.

The Reynold's number factor (F_R)

The Reynold's number factor (F_R) has a value of 1.0 for most flows in the process industry. However, sizing errors due to low Reynold's numbers (RE_v) can occur when the fluid viscosity is high or flow velocities low. In either case, RE_v can be calculated from these equations:

$$RE_v = \frac{N_3 W}{v\sqrt{F_{LP}C_v}} [\frac{(F_{LP}C_v)^2}{N_2 D^4}]^{0.25}$$

or

$$RE_v = \frac{N_3 Q}{v\sqrt{F_{LP}C_v}} [\frac{(F_{LP}C_v)^2}{N_2 D^4}]^{0.25}$$

Using the calculated RE_v, the value of F_R can be found from the graph in Figure 3. Substitute F_L for F_{LP} with line-size valves. (Values for F_L and F_{LP} are in the "Valve Reference Data"—Table 2 or 4.)

Choked flow

Choked flow is a limiting flow that occurs as a result of liquid vaporization when the internal valve pressure falls below the liquid's vapor pressure. Cavitation has then fully developed within the valve, and an increase in pressure drop across the valve will not produce an increase through the valve if

$$\Delta P \geq F_{LP}^2 (P_1 - F_F P_v).$$

(Substitute F_L for F_{LP} with line-size valves.) If the valve must be operated under choked conditions, use the following choked-flow equations in place of the basic equation for valve flow capacity:

$$C_v = \frac{Q}{N_1 F_{LP}} \sqrt{\frac{G_F}{(P_1 - F_F P_v)}}$$

or

$$C_v = \frac{Q}{N_6 F_{LP}} \sqrt{\frac{W}{(P_1 - F_F P_v)\gamma}}$$

where,

$$F_F = 0.96 - 0.28 \frac{P_v}{P_c}$$

(use F_L for F_{LP} for line-size valves).

The cavitation index (K_c)

The cavitation index (K_c) relates to a condition just prior to flow choking where cavitation has only partially developed and flow rate is no longer proportional to the square root of pressure drop. This condition occurs when:

$$\Delta P_c \geq K_c(P_1 - P_v)$$

Values for the cavitation index (K_c) can be found in the Valve Reference Data—Table 2 or 4.

Anticavitation trim

Generally, operating a valve at conditions corresponding to K_c or choked flow will result in damaging cavitation, noise, and vibration.

If cavitation exists, a valve with anticavitation trim will prevent or lessen the effects of cavitation. When anticavitation trim is used, recalculate the flow-capacity coefficient (C_v) using factors for "With Trim" in the Valve Reference Data—Table. If noise is a concern, see the procedure under "Hydrodynamic Noise Prediction" in the manufacturer's literature.

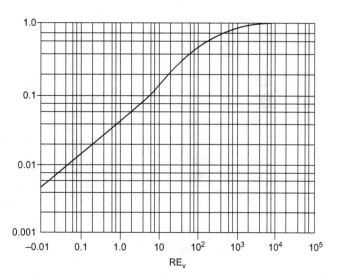

Figure 3. Reynolds number factor.

SIZING VALVES FOR GAS AND VAPOR

Basic valve flow-capacity coefficient (C_v)

The basic valve flow coefficient (C_v) for gas or vapor flow through a variable orifice valve may be calculated using any of the following equivalent formulas for nonchoked flow:

$$C_v = \frac{Q}{N_7 F_P Y} \sqrt{\frac{G_G T_1 Z}{X}}$$

or

$$C_v = \frac{W}{N_6 F_P Y \sqrt{X P_1 \gamma_1}}$$

or

$$C_v = \frac{Q}{N_9 F_P P_1 Y} \sqrt{\frac{M T_1 Z}{X}}$$

$$C_v = \frac{W}{N_8 F_P P_1 Y} \sqrt{\frac{T_1 Z}{X M}}$$

where

$$F_k = \frac{K}{140} \quad \text{and} \quad X = \frac{P_1 - P_2}{P_1}$$

and

$$Y = \frac{1 - X}{3 F_K X_{TP}} \leq 0.667^{a}$$

[a]See "Choked Flow" for explanation.

Substitute X_T for X_{TP} with line-size valves. Values for F_P, X_T, and X_{TP} are found in the "Valve Reference Data"—Table 2 or 4.

The compressibility factor (Z)

The compressibility factor (Z) can be found from Figure 4 after the reduced pressure (P_R) and reduced temperature.

(T_R) have been calculated from these equations:

$$P_R = \frac{\text{Inlet Pressure(Absolute)}}{\text{Critical Pressure(Absolute)}}$$

$$T_R = \frac{\text{Inlet Temperature(Absolute)}}{\text{Critical Temperature(Absolute)}}$$

$(°R = °F + 460, °K = °C + 273)$

Choked flow

Gas flow through valves can choke when the gas velocity reaches the velocity of sound. The pressure drop ratio at this point has the value:

$$X = F_K X_{TP}$$

Substitute X_T for X_{TP} with line-size valves. If this valve is to be used under choked conditions, Y is fixed at 0.667 regardless of the pressure drop. Substitute 0.667 for Y and $F_K X_T$ or $F_K X_{TP}$ for X in a basic value flow-capacity coefficient equation.

Valve outlet velocity

Sonic velocity at the valve's outlet will cause shock waves that produce vibration and noise problems. Therefore, you should make sure the outlet velocity does not reach sonic velocity for gas and steam service. The actual valve outlet diameter must be larger than the calculated diameter from the following equations:

$$d = N_{10}\sqrt{\frac{Q\sqrt{G_G T}}{P_2}} \text{ for gas}$$

$$d = N_{11}\sqrt{\frac{W(1 + N_{12}T_{SH})}{P_2}} \text{ for steam}$$

Low-noise trim

When the pressure-drop ratio (X) is near or exceeds the minimum value for choked flow, check the valve for excessive noise using the procedure under "Aerodynamic Noise Prediction." If noise is excessive, use antinoise trim; recalculate the flow-capacity coefficient (C_V) using factors from "With Trim" in the "Valve Reference Data"—Table 4. For procedures on noise-level prediction, refer to the manufacturer's literature.

Figure 4. Compressibility versus reduced pressure.

Table 2
Valve reference data (no trim)

Value	Percent Valve Rotation												D/d
	10	20	30	40	50	60	70	80	85	90	95	100	
F_L	0.95	0.94	0.93	0.91	0.89	0.85	0.78	0.69	0.65	0.60	0.56	0.51	
X_T	0.86	0.85	0.83	0.80	0.75	0.65	0.49	0.32	0.24	0.17	0.10	0.08	1.00
K_c	0.78	0.77	0.73	0.69	0.61	0.55	0.48	0.41	0.36	0.31	0.25	0.18	
F_P	1.0	1.0	1.0	1.0	1.0	0.99	0.98	0.95	0.91	0.86	0.78	0.62	
F_{LP}	0.95	0.94	0.93	0.91	0.88	0.83	0.76	0.64	0.59	0.52	0.45	0.35	1.25
X_{TP}	0.86	0.85	0.83	0.80	0.75	0.65	0.50	0.33	0.26	0.20	0.14	0.10	
F_P	1.0	1.0	1.0	1.0	1.0	0.99	0.97	0.92	0.89	0.83	0.74	0.56	
F_{LP}	0.95	0.94	0.93	0.91	0.88	0.82	0.75	0.64	0.57	0.50	0.42	0.34	1.33
X_{TP}	0.86	0.85	0.83	0.80	0.75	0.65	0.50	0.35	0.27	0.21	0.16	0.12	
F_P	1.0	1.0	1.0	1.0	1.0	0.99	0.96	0.91	0.86	0.76	0.64	0.50	
F_{LP}	0.95	0.94	0.93	0.91	0.88	0.82	0.74	0.63	0.56	0.49	0.40	0.32	1.50
X_{TP}	0.86	0.85	0.83	0.80	0.75	0.65	0.51	0.36	0.29	0.25	0.19	0.16	
F_P	1.0	1.0	1.0	1.0	1.0	0.99	0.94	0.85	0.78	0.68	0.56	0.44	
F_{LP}	0.95	0.94	0.93	0.91	0.88	0.81	0.73	0.62	0.55	0.46	0.38	0.29	1.67
X_{TP}	0.86	0.85	0.83	0.80	0.75	0.65	0.49	0.32	0.24	0.17	0.10	0.08	
F_P	1.0	1.0	1.0	1.0	1.0	0.97	0.93	0.83	0.76	0.65	0.50	0.40	
F_{LP}	0.95	0.94	0.93	0.91	0.88	0.80	0.72	0.60	0.53	0.45	0.36	0.27	2.00
X_{TP}	0.86	0.85	0.83	0.80	0.75	0.66	0.53	0.40	0.35	0.32	0.29	0.28	
F_P	1.0	1.0	1.0	1.0	1.0	0.95	0.89	0.75	0.64	0.46	0.38	0.33	
F_{LP}	0.95	0.94	0.93	0.91	0.88	0.80	0.72	0.60	0.53	0.45	0.36	0.24	2.67
X_{TP}	0.86	0.85	0.83	0.80	0.75	0.66	0.54	0.46	0.41	0.38	0.36	0.34	

Table 3
Constants in liquid sizing equations

N		W	Q	ΔP, P	D, d	γ	v
N_1	0.865	—	m³/h	bar	—	—	—
	1.000	—	gpm	psia	—	—	—
N_2	0.00214	—	—	—	mm	—	—
	890	—	—	—	in	—	—
N_3	76.0	kg/h	—	—	mm	—	cSt
	34.5	lb/h	—	—	in	—	cSt
N_4	76,000	—	m³/h	—	mm	—	cSt
	17,300	—	gpm	—	in	—	cSt
N_6	27.3	kg/h	—	bar	—	kg/m³	—
	63.3	lb/h	—	psia	—	lb/ft³	—

Table 4
Valve reference data (with trim)

Value	Percent Valve Rotation								D/d
	10–50	60	70	80	85	90	95	100	
F_L	1.0	1.0	1.0	0.92	0.85	0.77	0.67	0.54	
X_T	1.0	1.0	1.0	0.87	0.76	0.61	0.37	0.20	1.00
K_c	1.0	1.0	1.0	0.90	0.83	0.69	0.46	0.27	
F_P	1.0	1.0	1.0	0.98	0.97	0.95	0.92	0.87	
F_{LP}	1.0	1.0	1.0	0.90	0.84	0.75	0.63	0.48	1.25
X_{TP}	1.0	1.0	1.0	0.89	0.80	0.65	0.41	0.22	
F_P	1.0	1.0	1.0	0.98	0.97	0.95	0.92	0.87	
F_{LP}	1.0	1.0	1.0	0.90	0.84	0.75	0.62	0.47	1.33
X_{TP}	1.0	1.0	1.0	0.89	0.80	0.66	0.44	0.23	
F_P	1.0	0.99	0.98	0.95	0.93	0.89	0.85	0.78	
F_{LP}	1.0	1.0	1.0	0.89	0.83	0.74	0.61	0.46	1.50
X_{TP}	1.0	1.0	1.0	0.89	0.81	0.68	0.47	0.27	
F_P	1.0	0.99	0.98	0.94	0.91	0.86	0.80	0.71	
F_{LP}	1.0	1.0	1.0	0.89	0.83	0.74	0.61	0.45	1.67
X_{TP}	1.0	1.0	1.0	0.89	0.81	0.69	0.50	0.30	
F_P	1.0	0.99	0.98	0.94	0.90	0.84	0.77	0.66	
F_{LP}	1.0	1.0	1.0	0.89	0.83	0.73	0.60	0.44	2.00
X_{TP}	1.0	1.0	1.0	0.90	0.82	0.71	0.55	0.36	
F_P	1.0	0.99	0.98	0.94	0.89	0.83	0.75	0.64	
F_{LP}	1.0	1.0	1.0	0.89	0.83	0.73	0.59	0.43	2.33
X_{TP}	1.0	1.0	1.0	0.90	0.82	0.72	0.57	0.39	
F_P	1.0	0.99	0.98	0.93	0.88	0.81	0.73	0.61	
F_{LP}	1.0	1.0	1.0	0.88	0.81	0.71	0.57	0.42	2.50
X_{TP}	1.0	1.0	1.0	0.92	0.87	0.80	0.71	0.60	
F_P	1.0	0.99	0.98	0.92	0.86	0.79	0.70	0.59	
F_{LP}	1.0	1.0	1.0	0.88	0.81	0.70	0.55	0.41	2.67
F_{TP}	1.0	1.0	1.0	0.93	0.88	0.82	0.75	0.65	

Table 5
Constants in gas/vapor sizing equations

N		W	Q	ΔP, P	D, d	γ	T	T_{SH}
N_6	27.3	kg/h	—	bar	—	kg/m^3	—	—
	63.3	lb/h	—	psia	—	lb/ft^3	—	—
N_7	417	—	m^3/h	bar	—	—	°K	—
	1360	—	scfh	psia	—	—	°R	—
N_8	94.8	kg/h	—	bar	—	—	°K	—
	19.3	lb/h	—	psia	—	—	°R	—
N_9	2250	—	m^3/h	bar	—	—	°K	—
	7320	—	scfh		—	—	°R	—
N_{10}	0.25	—	m^3/h	bar	mm	—	°K	—
	63.3	—	lb/h	psia	in	—	°R	—
N_{11}	37.1	kg/h	—	bar	mm	—	—	°C
	0.119	lb/h	— —	psia	in	—	—	°F
N_{12}	0.00126	kg/h	—	bar	mm	—	—	°C
	0.0007	lb/h	—	psia	in	—	—	°F

Easily visualize pump and control valve interaction

By **G. Unnikrishnan,** Kuwait Oil Company, Kuwait.

Control valves are the final control elements of a process control system and play the key role of controlling the process variables within the safe operational envelope. It is the dynamic element in the system that translates the control system logic to mechanical action. The fact is that little attention is usually paid to this aspect while designing the process plant. During the typical project design stage for a pump and control system, the process design engineers usually assign a certain pressure drop for the control valve and develop the pump discharge pressures in a pumping circuit. Instrument design engineers select a suitable control valve based on the process data furnished by the process design engineers, the most important being the parameters at minimum, normal, and maximum flows. The required C_v value is calculated from the process parameters and the control valve is selected such that the ratio of calculated C_v to selected control valve C_v is in the 0.6 to 0.75 range. Final selection and confirmation is usually from the manufacturer.

One of the most important factors that govern dynamic behavior of the control valve is the control valve gain. Gain is defined as:

Gain = Change in the flow rate/Change in valve position.

Ideally this should not change much; it should be constant at least in the normal control operating region. In other words, the valve travel when the flow changes from, say, 30–40% should be same as when the flow changes from 50% to 60%. The dynamic nature of the valve gain can be seen from the valve equations and formula for valve position shown in Table 1. Further certain key points to understand control valve dynamics are given in Table 2. Once the valve gain is computed and presented in a suitable form, as a graph, control valve behavior can be easily visualized. This will save the time required for full-scale control system modeling.

This article shows a way to solve the problem using well-established graphical methods. The capabilities of popular spreadsheets make the effort easy. Following are the main steps in easily visualizing and understanding the behavior of typical pump and control valve interaction.

The graphs and spreadsheet are set up in the following sequence:

Table 1
Valve Types and Equations

Valve type	Valve equation	Valve position, %
Linear $f(L) = L$	$Q = Cvsel\ L\sqrt{\frac{\Delta P}{sg}}$	$L = \frac{Cvinst}{Cvsel}$
Equal-percentage $f(L) = Rn^{(L-1)}$	$Q = Cv\ Rn^{(L-1)}\sqrt{\frac{\Delta P}{sg}}$	$L = 100\left[1 + \frac{\ln(B)}{\ln(Rn)}\right]^2$

Table 2
Recap of key points to understand control valve dynamics

1. The Cv_1 values found in the control valve manufacturer's catalog refer to the coefficients when there is a constant pressure drop of 1 psi across the control valve and nothing else. This is the inherent Cv of the control valve.

2. The inherent Cv will be modified when the control valve interacts with the system in which it operates in a dynamic way depending on the type of control valve selected—linear, equal-percentage, quick-opening, etc. The installed characteristics will be different from the inherent characteristics. For example, a valve with an equal-percentage inherent characteristic will not show the same behavior when installed. In fact, equal-percentage valves exhibit more linear characteristics when interacting with the system.

3. Control valve gain ideally should be the same, that is, it should be constant for the flows around the operating region. Literature says that the gain at any flow should not be less than 0.50.

4. Try to match the valve type, its characteristic and gain with the process system gain under consideration. There is extensive literature available on the subject. The results are worth the effort.

5. Control valves are mechanical devices and need time to react to the commands from the controllers that act quite fast. The controllers do not wait to see the results of the control valve action and issue further commands to the valve. Effectively, the control valve struggles to catch up with the controller commands resulting in continuing cycles of opposite movements by the control valve, resulting in the so called "induced limit cycle." An oversized control valve has high process gain and small changes to inputs tend to make large changes in the process parameter. Interested readers can see Ref. [3].

6. The whole control loop including the control valve behaves nonlinearly. For critical control loops, review of the nonlinear behavior will definitely be beneficial.

A The pump and system curves

A.1 Pump curve. A1.1 The characteristic for a preliminary selected pump is plotted in a spreadsheet graph. The plot will show pump head on the y axis and flow rate on the x axis.

Basic pump parameters are calculated while designing the system. The rated flow rate and pump differential pressure are known (this includes the pressure loss assigned to the control valve at rated flow). If the pump curve is not available, a preliminary curve can be plotted by fixing two more points. The shutoff head at zero flow is usually taken as 1.25 times the

head at the rated flow. One more point for pump head can be fixed at the other end of the curve as a suitable fraction of the head at maximum flow expected over and above that of the rated flow. We now have three points for the head versus flow relationship for the pump to plot the graph.

A.2 System curve. A.2.1 Fix the static head. This will have to be added as a constant to the variable pressure loss through the system comprising piping, pipe fittings, and other equipment. The variation in pressure drop with flow can be ascertained by a detailed calculation or can be simplified by using the relationship

$$Q_1/Q_2 = K(\Delta P_1/\Delta P_2)^2$$

A reasonable approximation is enough at this stage. The system curve does not include the pressure drop for the control valve. Get at least three points for the pressure loss through the system.

Plot these points on the same graph (Figure 1).

A.3 Fitting equations to the curves

Use the "Add Trend Line" feature available in the spreadsheet program to fit polynomial equations to both these curves (two coefficient equations will be enough to give a good fit). Use the Options tab to display the equations and the R^2 values. The R^2 values give a measure of the goodness of fit. That is, the closer the values of R^2 to 1, the better the fit. The fitted equations are displayed on the graph (Figure 1). Now set up a spreadsheet table as shown in Table 3. The spreadsheet layout shows the type of entry in each column. Note these fitted equations and enter the same in columns 2 and 3 of Table 3.

Figure 1. Pump and system curves.

Table 3
Spreadsheet Layout

1 Q	2 P_p	3 P_s	4 ΔP	5 Cvinst	6 L%	7 Q%	8 Gain	9 L%	10 Q%	11 Gain
Manual entry	Formula from equation for trend line for pump curve or manual entry	Formula from equation for trend line for system curve or manual entry	Difference of columns 2 and 3 (Step B.1)	Formula for installed Cv, Cvinst (Equation 2)	Valve opening for equal-percentage control valve, % (Equation 6)	Flow rate as percentage of maximum flow (Step B.3)	Formula for gain for equal-percentage control valve obtained by differentiating equation for trend line (Figure 2)	Valve opening for linear control valve, %	Flow rate as percentage of maximum flow	Formula for gain for linear control valve obtained by differentiating equation for trend line

B Control valve positions for various flows

This is where the graphing capabilities of the spreadsheet can be used. It consists of two parts.

B.1 Computing the absorbed pressure drop.
The control valve on the pump discharge will absorb the difference in pressures between that developed by the pump and that required by the system. This difference can be calculated with the equations obtained from Step A.3 by entering the difference in column 4. Fill the blank cells for different flow rates by copying the formulas down. (Note: If sufficient data points are available, Step A.3 can be avoided. Just plot the graphs and write the formula for the absorbed pressure directly in the table.)

While computing the differences, omit the intersection point of the pump and system curves. At this point, the valve is fully open theoretically with zero pressure drop but in actuality it will still drop some pressure across it depending on the internals. Therefore, it is better to understand this point as a system with no valve.

B.2 Preliminary selection of control valve.
Preliminary selection of a suitable control valve has to be done at this stage. Using the process parameters available from process design, calculate the required Cv (denoted as Cvr) of the valve at the *rated* flow—at this point a reasonable pressure drop has to be assigned to the control valve based on the pressure differences available from the table. Select a suitable size control valve based on calculated Cvr from the manufacturer's catalogs such that the ratio of the Cvr to the $Cvsel$ of the selected valve is in the 0.60 to 0.75 range. Remember that this $Cvsel$ value is based on a 1 psi pressure drop across the valve and represents the inherent characteristic of the valve. Add one more column (column 5) in Table 3 for Cv at installed conditions ($Cvinstal$) at different flow rates. The values can be calculated by manipulating the formula for flow through the control valve to get $Cvinstal$:

$$Q = Cinstal\sqrt{\Delta P/sg} \qquad (1)$$

$$Cvinstal = \frac{Q}{\sqrt{\Delta P/sg}} \qquad (2)$$

The formula can be copied down the column for various flow rates. The calculations are implemented in the spreadsheet table to show the valve opening for various flows.

B.2.1 Calculating the control valve opening.
Add one more column (column 6) for the percentage valve opening based on the installed characteristic equation. The characteristic generic equation for a control valve is typically:

$$Q = Cusel\,f(L)\sqrt{\Delta P/sg} \qquad (3)$$

$f(L)$ is a function of the valve opening, L (from 0 fully closed to $L = 1$ full open). The valve behavior is altered as a result of varying flow and pressure drop across the valve.

As an example, let us highlight the steps for a control valve with equal-percentage trim. Equation 3 will have to be modified for the equal-percentage valve. For an equal-percentage valve, $f(L) = Rn^{(L-1)}$. Rn = rangeability (ratio of maximum to minimum flow—usually in the 20 to 50 range for globe valves).

The equations satisfying the flow, pressure drop, and valve opening for a particular flow Q will be:

$$Q = Cvsel\,Rn^{(L-1)}\sqrt{\Delta P/sg} = Cvinst\sqrt{\Delta P/sg} \qquad (4)$$

Thus,

$$R^{(L-1)} = Cvinst/Cvsel \qquad (5)$$

Manipulate the equation to get valve opening in percentage as a function of the pressure drop through the valve as:

$$L = 100[1 + \ln(B)/\ln(Rn)] \qquad (6)$$

$$B = Cvinst/Cvsel \qquad (7)$$

Enter this formula in column 6 of Table 3 and copy it down to get the opening for various flow rates.

This is the equation that will be implemented in the graph from which the valve gain will be obtained.

Similar equations and steps for a linear type valve are shown in Table 3.

B.3 Flow rates as percentage of maximum. Convert the flow rates as a percentage of maximum ($100Q/Q_o$) and add a column (column 7) for the same. Now we have the percentage valve opening values for different percentage flow rates.

B.4 Plotting percentage valve opening. Plot the percentage flow rates and percentage valve opening values in a graph with flow rate on the y axis and valve opening on the x axis using the values in columns 6 and 7 of Table 3 (Figure 2). Again, use the "Add Trend Line" feature to fit the equation to the graph. Use a three-coefficient equation for a good fit. Note the equation.

C. Calculating the control valve gain

C.1 Remember that control valve gain is the ratio of flow rate change to valve position change. We can derive this by *differentiating the equation obtained* in Step B.4 with respect to L. Implement the differentiated equation in Table 3 by adding another column (column 8) and copying down the formula. We now have a set of values for valve gain with respect to percentage flow rates.

C.2 Plot these values from C.1 in a spreadsheet graph (Figure 3). [For comparison, Step B.4 can be repeated for a linear valve by using the equations in Table 1. Readers can compare the valve behavior for the two types (Figures 4 and 5)].

Figure 3. Gain versus percent flow for an equal-percentage valve.

Figure 4. Percent flow versus percent valve opening for a linear valve.

Figure 2. Percent flow vx percent valve opening for an equal percentage valve.

Figure 5. Gain versus percent flow for a linear valve.

Table 4
Spreadsheet

Fluid sg				1	Selected valve Cvsel @ rated flow equal percentage			Selected valve Cvsel @ rated flow linear		
Values used for plotting				Cvr			136			148
Q	Pp	Ps	ΔP	85						
0.00	196.00	112.00	84.00							
400.00	170.00	150.00	20.00							
460.00	162.00	152.00	10.00							

Spreadsheet (Refer to Table 3 for layout)					Selected valve equal %R=		50	Selected valve linear		
1	2	3	4	5	6	7	(8) gain	9	10	(11) gain
Q	Pp	Ps	ΔP	Cvinst	L%	Q%	dQ/dL	L%	Q%	dQ/dL
	Values from graph									
0.00	196.00	112.00	84.00	0.00	0.00	0.00	−0.95	0.00	0.00	2.85
50.00	195.47	119.18	76.29	5.72	19.02	10.87	0.13	3.87	10.87	2.62
100.00	194.44	125.84	68.58	12.08	38.10	21.74	0.99	8.16	21.74	2.37
150.00	192.91	132.04	60.87	19.23	49.99	32.61	1.42	12.99	32.61	2.10
200.00	190.88	137.72	53.16	27.43	59.08	43.48	1.69	18.53	43.48	1.81
250.00	188.35	142.90	45.45	37.08	66.78	54.35	1.88	25.06	54.35	1.49
300.00	185.32	147.58	37.74	48.83	73.82	65.22	2.02	33.00	65.22	1.14
350.00	181.79	151.76	30.03	63.87	80.68	76.09	2.13	43.15	76.09	0.74
400.00	177.76	155.44	22.32	84.67	87.89	86.96	2.22	57.21	86.96	0.30
450.00	173.23	158.62	14.61	117.73	96.31	97.83	2.28	79.55	97.83	−0.17
460.00	172.26	159.20	13.07	127.25	98.30	100.00	2.29	85.98	100.00	−0.24

These graphs will show valve behavior for different flow rates. Ideally, the valve gain should be constant near the operating flow rates. Also, one has to particularly look for the behavior at minimum and maximum flows. Check if this behavior is reasonable with respect to the process requirements. Otherwise, the valve selection, pumping pressure, and line size need to be reviewed at this stage.

With some effort, the steps can be implemented for a gas compressor and its control valve. Table 4 shows a spreadsheet with a numerical example and Table 5 describes steps to be followed that are described in the article and the corresponding equations.

Nomenclature

Cvr = Control valve capacity coefficient required at rated flow
$Cvsel$ = Capacity coefficient of selected control valve
$Cvinst$ = Control valve capacity coefficient at installed condition
ΔP_1 and ΔP_2 = System pressure drop at flow rates $Q1$ and $Q2$
K = Constant multiplier
L = Valve opening, fraction
DP = Pressure drop across control valve, psi
P_p = Pump head, psi
P_s = System head, psi
Q = Flow rate, gpm
Q_o = Flow rate when control valve is full open, gpm
Q_1 and Q_2 = Flow rate at particular operating conditions 1 and 2, gpm
$Q\%$ = Flow rate as percentage of maximum
Rn = Rangeability
sg = Specific gravity

Bibliography

1. Connell, J. R., "Realistic Control Valve Pressure Drops," *Chemical Engineering*, September 1987, pp. 123–127.
2. Driskell, L., *Control Valve Selection and Sizing*, Instrument Society of America, 1983.
3. EnTech Control Engineering Inc., *Control Valve Dynamic Specification*, November 1998.
4. Fehr, M., "Improve Valve and Pump Sizing," *Hydrocarbon Processing*, July 1998, pp. 53–57.

Reprinted with permission *Hydrocarbon Processing*, August 2007 issue, Houston, Texas.

Table 5
Numerical Example (Step Numbers are the Same as in the Text)

A Pump and system curve

A.1 Pump curve

Assuming that the actual pump curve is not available, rated flow and differential head are calculated as $Q = 400$ gpm at $H = 170.0$ psi

Head at shut-off head at $Q = 0 = 1.25 \times 170.00 = 196.00$ psi

Head at maximum flow at $Q = 460 = 0.95 \times 170.00 = 162.00$ psi

The above factors can be varied based on the actual context

Plot the points in a spreadsheet (Figure 1).

A.2 System curve

Calculate values of flow versus pressure drop through the system (excluding the control valve) for:

Pressure at $Q = 0$ static head only $= 112$ psi

Pressure required at $Q = 400$ gpm $= 150$ psi

Pressure required at $Q = 460$ gpm $= 152$ psi

Plot the points in Figure 1.

A.3 Fitting equations to the curves

Right click on the curve to get the "Add Trend Line" feature. Select a two-polynomial equation. Also, check the boxes for show equation and R^2 values. OK all of the above to get the equation in terms of $y = f(x)$ displayed on the graph. Repeat the same for the system curve. The equations are:

$Y_p = -0.0001\ X^2 - 0.0056X + 196$ pump curve

$Y_s = -0.0001\ X^2 + 0.1486X + 112$ system curve

Divide the flow rate in suitable increments to get sufficient number of points and enter them manually at Column 1 of Table 1.

For column enter the system head values manually. Implement the difference of equations obtained in A.2 ($Y_p - Y_s$) in Column 3 to get the pressure drop.

B Control valve positions for various flows

B.1 Pressure drop absorbed by the control valve is obtained as noted above

B.2 Select control valve:

Calculate Cvr of the control valve at $Q = 400$ gpm and ΔP of 22 psi as 85 using formula:

$Cvr = \dfrac{Q}{\sqrt{\Delta P/sg}}$, which will be $= \dfrac{400}{\sqrt{22/2}} = 85$ (taking sg as 1,00)

B.2.1 From a manufacturer's catalog obtain the $Cvsel$ for a 4 in equal-percentage valve for about 72% valve opening as 136. Check on the ratio:

$Cvr/Cvsel = 85/136 = 0.63$

Implement Equation 6 for opening of equal-percentage control valve L in %.

Copy the formula down to fill all rows (Column 6).

B.3 Convert the flow rates as percentages of maximum flow and Implement in Column 7.

B.4 Plot another graph (Figure 2) with flow Q% on the y axis and L% opening on the x axis.

Repeat the "Add Trend Line" procedure and get a three-coefficient equation as:

$Y = -0.0001\ X^3 + 0.0312\ X^2 - 0.9472\ X + 18.826$ which is

$Q = -0.0001\ L^3 + 0.0312\ L^2 - 0.9472\ L + 18.826$ in terms of Q and L.

This equation gives flow rate, $Q\%$ (Y) as a function of valve opening, L% (X).

C. Calculating the control valve gain

C.1 Differentiate the above equation with respect to L to get dQ/dL (rate of change of flow with respect to rate of change of valve opening) to get gain as:

$dQ/dL = -0.0003\ L^2 + 0.0624\ L - 0.9472$

The right-hand side of the above equation is written on top of Figure 2.

Implement the above equation in Column 8 and fill all rows with gain values at different flow rates.

C.2 Plot Figure 3 using the values of flow percentage, $Q\%$, on the x axis and gain on the y axis to visualize the behavior of the equal-percentage-type control valve.

The above steps are repeated for a linear-type valve with the equations shown in Table 1 in Columns 9, 10, and 11 to get the gain versus flow rates (Figures 4 and 5).

Avoid cavitation in butterfly valves

E. Casado Flores, EEAA, Madrid, Spain.

Butterfly valves may be used to control flow in pipes if they are throttled in angles where the valve works stable without noise or vibration, i.e. cavitation free. The initial cavitation coefficients, C_i, presented in this article are compared with those predicted by Refs [1] and [2], suggesting the C_i values for different valve sizes and the maximum allowable pressure drop and flow rates through the valves to run without cavitation.

They are based on tests performed in a Spanish nuclear power plant on 24 in butterfly valves installed in the service water system. The goal was to know the throttling positions free of cavitation, to have the system running safely and to obtain the head-flow curves of the new pumps.

Tests performed on 24 in valves. To determine that the throttling positions were free of cavitation and the butterfly valves were completely quiet, we performed different tests for valve opening angles from 25° to 90° (valve 100% open). Figure 1 shows one of the tested valves. In some throttling positions the valves were quiet. In other positions, significant noise was generated. Finally, in the rest of the throttling positions, the typical cavitation noise (like stones

Figure 1. More than 50 tests were performed for different valve openings on 24 in butterfly valves. The data collected during the tests were: Pressures upstream and downstream of the valves with 0.5% accuracy manometers and 0–10 kg/cm² range; flow through the valves with a 1.12% accuracy Annubar and 0–5000 m³/h range.

dragged inside the pipe) began to be heard. In these positions, a slight opening of the valve (approximately 1°) eliminated the noise. More than 50 tests were performed for different valve openings. The cases where the cavitation noise begins correspond to the initial cavitation.

Results obtained for C_i. The coefficient, C_i when cavitation noise begins is defined as

$$C_i = \frac{P_1 - P_v}{P_1 - P_2}$$

where

P_2 = downstream valve pressure, kg/cm² absolute
P_1 = upstream valve pressure, kg/cm² absolute
P_v = fluid vapor pressure, kg/cm² absolute.

It is possible to obtain the different C_i values from the pressure measurements. The C_i values are represented in Figure 1 as a function of C_d with a 10% conservatism.

C_d is the discharge valve coefficient defined as

$$C_d = \frac{1}{[(17183.71/C_p)^2 + 1]^{\frac{1}{2}}}$$

C_v is the valve flow coefficient defined as

$$C_v = \frac{1.17Q}{(P_1 - P_2)^{\frac{1}{2}}}$$

where Q = water flow, m³/h.

Water temperature during the tests was 20 °C, so P_v is approximately 0.

NOTE: There is no cavitation in the valve if the values of P_1, P_2, and P_v during its running fulfill the following expression: $(P_1 - P_v)/(P_1 - P_2)$ C_i.

Comparison with the data. In Ref. [1], a 6 in butterfly valve was tested at $P_1 = 5.8$ kg/cm² and presented the corresponding C_i values as a function of C_d similar to Figure 2. For other pressures and valve sizes, Ref. [1] proposes a group of correction equations to account for the scale effects.

Applying this to the cavitation level curve of Ref. [1], named as critical cavitation, it is possible to obtain the critical cavitation values for 24 in valves and the test pressure predicted by Ref. [1] as shown in Figure 3. The critical cavitation of Ref. [1] is very similar to the C_i values in Figure 2.

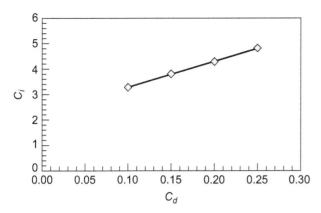

Figure 2. C_i as a function of C_d.

assumed to be in the curve parallel to that of Ref. [1]. The result is that the tested C_i values are practically the same as that of Ref. [1] for the critical level and higher than that predicted by Ref. [2], but taking into account, in this case, that they correspond to an incipient damage level.

Another conclusion is that the incipient cavitation level proposed in Ref. [1] (higher values than critical cavitation level) may be very conservative to determine the industrial butterfly valves' throttling limits to perform without cavitation. For this reason, based on my experience with these types of valves, I proposed the C_i curves shown in Figure 4 as adequate to establish the initial cavitation level.

In Ref. [2], the author predicts the cavitation coefficient for 24 in butterfly valves starting from the data measured in 6 in valves for the incipient damage level and calculated sound level using 85 dBA as a limiting condition. The proposed cavitation values for 24 in butterfly valves according to Ref. [2] are also shown in Figure 3.

So, Figure 3 shows the comparison of the C_i tested values with those theoretically predicted by Refs [1] and [2] for 24 in butterfly valves. The C_i values for C_d greater than 0.3 are

C_i proposed values. Based on the C_i values obtained in the tests for 24 in butterfly valves—since they are slightly higher than the critical cavitation values of Ref. [1]—I assume that, for 6 in valves, the corresponding C_i values will also be very similar to the critical cavitation values of Ref. [1] for this size. So, it is possible by interpolation for other sizes to obtain the C_i curves shown in Figures 4 and 5.

The curves of Figures 4 and 5 are base reference curves for water and liquids with P_1 (pressure before the valve) equal to or below 10 kg/cm^2 abs and vapor pressures, P_v, equal to or greater than 0.

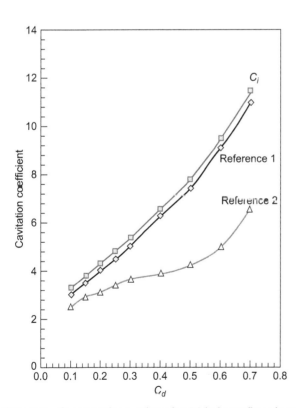

Figure 3. C_i comparison values for 24 in butterfly valves.

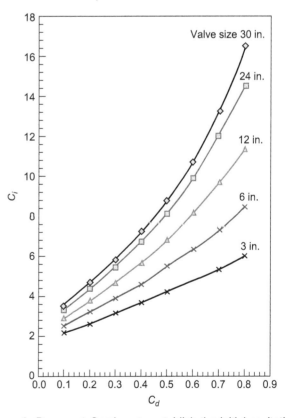

Figure 4. Proposed C_i values to establish the initial cavitation.

Figure 5. C_i values as a function of valve opening angle.

From the C_i curves in Figure 5, it is possible to obtain the maximum admissible pressure drop and flow for the different opening angles of the valve as follows:

Pressure drop, $DP = P_1 - P_2 = (P_1 - P_v)/C_i$
For the base reference values, $P_1 = 10 \text{ kg/cm}^2$ abs and $P_v = 0$,
$DP = 10/C_i$
Maximum admissible flow, $Q = 0.85 \, C_v \, DP^{1/2} = 0.85 \, C_v \, (10/C_v)^{1/2}$
The curves for DP and Q correspond to Figures 6 and 7.

Figure 6 shows the maximum allowable pressure drop through the valves without cavitation as a function of the opening angle.

Figure 7 shows the maximum allowable flow through the valves corresponding to the pressure drops of Figure 6.

Corrections for pressures. The corrections factors deduced from Ref. [1] are:

PV for $P_v > 0$ is $1 - P_v/P_1$

PSE for $P_1 > 10 \text{ kg/cm}^2$ abs is $(P_1/10)^{0.3} = 0.5 \, P_1^{0.3}$

These factors are applied to the C_i reference values, maximum allowable pressure drop, DP, and maximum

Figure 6. Maximum allowable pressure valves through the valves without cavitation.

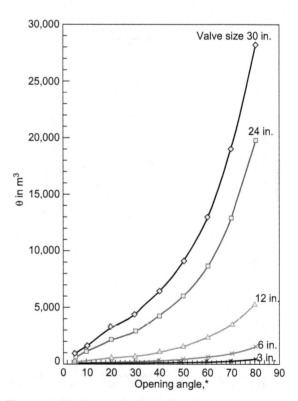

Figure 7. Maximum allowable flow through the valves.

allowable flow, Q, respectively obtained from Figures 5–7 to get the corrected ones: C_{ip}, DP', and Q':

$$C_{ip} = 1 + PSE(C_i - 1) = 1 + 0.5\,P_1^{0.3}\,(C_i - 1)$$

$$DP' = \frac{PVP_1}{1 + PSE(10/DP - 1)}$$

$$Q' = Q(PV)^{0.5}[\frac{P_1}{DP(1 - PSE) + 10PSE}]^{0.5}$$

where DP' is kg/cm^2 and Q' is m^3/h.

The C_i proposed values for butterfly valves have sufficient conservatism to avoid cavitation in their throttling positions.

Figures 4 and 5 directly allow obtaining the C_i values for different valve sizes, avoiding the use of the size corrections proposed in Ref. [1].

Figures 6 and 7 help plant operators know how much they may throttle the butterfly valves to work without cavitation.

Literature Cited

1 Tullis, J. P., "Cavitation Guide for Control Valves," *NUREG/CR-6301*, April 1993.
2 Baumann, H. D., "Preventing Cavitation in Butterfly Valves," *Chemical Engineering*, March 1985.

Reprinted with permission *HYDROCARBON PROCESSING* AUGUST 2006 issue, Houston, Texas.

How to read a regulator flow curve

By **Bill Menz**, Swagelok Company.

The best way to select a regulator for your application is to examine its flow curve, which is often provided by the manufacturer. "Flow curve" is a misleading name. You could easily call it a "pressure curve" instead, since a regulator controls pressure, not flow. The curve represents the range of pressures that a regulator will maintain given certain flow rates in a system.

When selecting a regulator, you are not just looking for the right size. You are looking for a set of capabilities, which is a function of the regulator's design. A flow curve illustrates the regulator's range of capabilities at one glance. Once you understand a flow curve—and we will explain how—it is easy and very quick to read.

Unfortunately, a more common way to select a regulator is to consult its flow coefficient (C_v). If the system flow is within range of the C_v, some people may believe that the regulator is the right "size." But that is not necessarily true. The C_v represents the regulator's maximum flow capacity. At maximum flow, a regulator can no longer control pressure. If you are expecting flow rates to reach the regulator's C_v, it is probably not the right regulator for your system.

Let us discuss how to read a flow curve. We will cover the basics first and then some of the complexities, including droop, choked flow, seat load drop or lock-up, hysteresis, and accumulation.

The Basics

A regulator's main purpose is to maintain a constant pressure on one side of the regulator even though there is a different pressure or fluctuating pressure on the other side. In the case of a pressure-reducing regulator, you control the downstream pressure. In the case of a backpressure regulator, you control the upstream pressure.

For now, let us talk in terms of pressure-reducing regulators since they are more common. Later, we will provide some direction for selecting backpressure regulators.

A flow curve illustrates a regulator's performance in terms of outlet pressure (Y axis) and flow rate (X axis). Flow is not controlled by the regulator. It is controlled downstream by a valve or flow meter. The curve shows you how a regulator will respond as flow in the system changes.

Let us look at the top curve in Figure 1. The curve starts at 400 pounds per square inch gauge (psig) (27.5 bar). This is the original set pressure for the regulator. No adjustments are made to the regulator, yet the curve shows a change in pressure.

The regulator is making adjustments, trying to maintain the original set pressure as flow changes. But, it is not perfect. No regulator is. As system flow increases, pressure downstream of the regulator drops. The important question is: How much does it drop?

When reading a flow curve, first identify the range of flows that you can expect to see in your system. Mark these on the graph. Then, look to see what the corresponding changes in outlet pressure will be. Is that range of pressures acceptable to you? If not, you need to look for a different regulator.

Ideally, you want to operate the regulator on the flattest part of the curve. There, the regulator will maintain relatively constant pressures even with significant changes in flow. At the far ends of the curve, there are steep drops, where

Figure 1. Manufacturers often provide multiple flow curves for the same regulator at different inlet pressures to provide a range of the regulator's operating capabilities.

pressure will dramatically change with even the slightest change in flow. You do not want to operate the regulator at these locations.

Any given regulator can produce a nearly infinite number of curves, so you need to make sure you are looking at the right one. For every set pressure, there will be a different curve. In Figure 1, there are two main sets of curves: one based on a set pressure of 400 psig (27.54 bar) and one for a set pressure of 200 psig (13.7 bar). It is helpful when a manufacturer provides more than one set of curves representing the range of set pressures possible with a particular regulator. If your set pressure lies between the curves, you can interpolate. Note that the two curves are close to the same shape but in different locations on the graph.

There is one additional variable that affects the shape of a curve—inlet pressure (i.e. pressure going into a pressure-reducing regulator on the upstream side). Note that for each of the two sets of curves in Figure 1, there are three curves representing a range of inlet pressures.

In sum, to locate the right curve for your system, look for (1) the right set pressure, (2) the right inlet pressure, and (3) the right range of flows.

Finally, make sure you are looking at the right units. Pressure readings are provided most commonly as psig or bar. Flow rate units vary depending on the system media, so be sure to note whether the regulator is rated for liquid or gas service. Liquid flow is typically expressed as gallons per minute (gal/min) or liters per minute (L/min), while gas flow is conveyed as standard cubic feet per minute (std ft³/min) or standard liters per minute (std L/min).

Curves are usually created using air or nitrogen (for gas service) or water (for liquid service). If your system media is

a gas, you may need to make an adjustment in the manufacturer's curve. Gases compress at different rates, so you may need to multiply the flow curve's volume units by a gas correction factor. For example, hydrogen's correction factor is 3.8; meaning 3.8 molecules of hydrogen have the same volume as one molecule of air. Therefore, the point on a flow curve showing an air flow volume of 100 std ft³/min (2831 std L/min) indicates a comparable hydrogen flow of 380 std ft³/min (10 760 std L/min). The curve stays the same, but the flow scale will change.

For liquids, the difference in flow between water and a different medium is not as dramatic due to the incompressibility of liquids.

Droop, Choked Flow, and Other Complexities

As mentioned above, it is best to operate along the flattest—or most horizontal—part of a flow curve. Indeed, the ideal flow curve would be a flat line. However, no regulator can produce a perfectly flat line over the full range of pressures because the internal components of the regulator have limitations. In a springloaded regulator, longer springs produce flow curves with broader horizontal sections. Domeloaded regulators, which use a trapped volume of gas instead of a spring, produce even broader horizontal sections. Electrically controlled pneumatic-loaded regulators and domeloaded regulators with external feedback produce the broadest horizontal sections.

Typically, a flow curve consists of three parts: (1) a relatively flat part in the middle, (2) a steep drop on the far left, and (3) a steep drop on the far right. The flat part in the middle is not perfectly flat. Usually, it slopes downward, and this is called droop. As flow increases, pressure will drop some—or a lot depending on the regulator design. While droop is relatively modest along the flat part of the curve, it is quite steep at the far ends of the curve.

Choked flow occurs on the far right of a curve. See the choked-flow area in Figure 2, where pressure begins to droop sharply at 140 std ft³/min (3964 std L/min). Eventually, near 150 std ft³/min (4247 std L/min), pressure drops to zero. At this point, the flow demand has exceeded the pressure-controlling capabilities of the regulator. Here, the regulator is wide open and is no longer regulating pressure. It has essentially changed from a pressure-controlling device to an open orifice. Increasing downstream flow to this point or beyond renders the regulator ineffective. It is not advisable to operate a regulator in the choked-flow area due to the sharp pressure drop.

Note that C_v is measured at the regulator's fully choked point, and that is why it is not an especially useful measurement. It is a poor indicator of a regulator's overall performance.

Seat load drop occurs on the far left of the regulator curve (Figure 2), where there is initially a steep drop in pressure. If

Figure 2. Typical flow curve for a pressure reducing regulator demonstrates several phenomena, including the ideal operating are, droop, choked flow, seat load drop or lock-up, and flow coefficient (*Cv*).

you are reading the curve from left to right, imagine that the system is in a no-flow state. The regulator is set to a certain pressure, but there is no flow. Then, imagine that an operator slowly opens a downstream valve to initiate flow. Right away, there is a sharp drop in pressure because it is difficult for a regulator to maintain pressure at this location. If your regulator is operating along this steep drop in the curve, you may hear chattering or pulsating as the regulator fluctuates between flow and no-flow conditions.

Now let us read the curve from right to left. Imagine that the system is operating along the flat part of the curve. Then, imagine that an operator is slowly closing a downstream valve, reducing flow to near zero. We are moving up the curve. As we near the no-flow state, the regulator has difficulty maintaining the set pressure. Again, you may hear chattering. Eventually, the regulator snaps shut, stopping flow. This is called **lock-up**.

The terms seat load drop and lock-up are essentially interchangeable. Sometimes, people use lock-up to describe both conditions. It is not advisable to operate a regulator in this area.

Hysteresis

It matters whether you are reading a flow curve from left to right or right to left. When reading left to right, flow is increasing. And the reverse is true when reading right to left. Depending on whether flow is increasing or decreasing, the curve will be slightly different. Outlet pressure will not follow the same "droop line," nor end at the original set pressure. This phenomenon, which is illustrated in Figure 3, is called

hysteresis. The upper line is for decreasing flow, while the bottom line is for increasing flow.

Hysteresis, which results from dynamic friction forces within the regulator, is usually not an issue when evaluating the performance of a regulator. However, it can be a point of confusion during system operation. Suppose an operator sets up a system to deliver an outlet pressure of 50 psig (3.4 bar) at 110 ft^3/min (3115 L/min). The next day, he notices that the pressure is now 50.5 psig (3.48 bar), but the flow is still 110 ft^3/min (3115 L/min). Hysteresis is to blame. It is likely that something in the system temporarily created more flow demand downstream. Moving from left to right on the curve, the temporary flow increase slightly reduced the outlet pressure. Then, as the flow demand returned to 110 ft^3/min (3115 L/min), hysteresis caused the outlet pressure to return to a point slightly higher than the initial set point.

This scenario illustrates the need for operators to periodically check systems to ensure regulators are set to the proper outlet pressures. It is recommended to approach your set pressure from a lower pressure. Another best practice is to employ pressure gauges in a system to help fine tune regulator settings to achieve desired operating pressures.

Backpressure Regulators and Accumulation

Flow curves for backpressure regulators reveal a phenomenon known as accumulation (see Figure 4), which is the opposite of droop. Because a backpressure regulator controls inlet pressure, pressure will go up as flow increases instead of down as in a pressure-reducing regulator. This phenomenon occurs because the backpressure regulator acts as

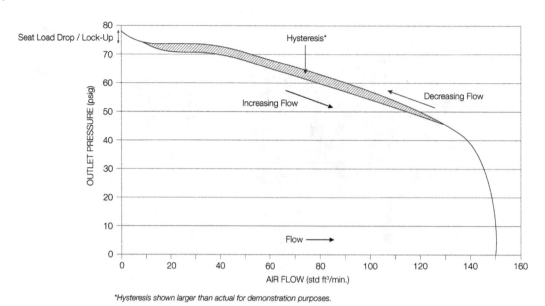

Figure 3. Phenomenon of hysteresis reveals which will be higher at the same flow volume when decreasing flow compared to when increasing flow. (NOTE: Hysteresis shown larger than actual for demonstration purposes.)

Figure 4. Flow curve for a backpressure regulator indicates accumulation, which is the opposite of droop.

a restriction and essentially limits flow at the inlet. Your objective is the same as it is with pressure-reducing regulators: to operate on the flat part of the curve.

Conclusion

When selecting a regulator for your application, consult the flow curve, not the C_v. Make sure you are looking at the right curve. Does it reflect the pressure you will set your regulator to (your "set pressure")? Does it reflect the right inlet pressure range? Are you working with the right units? Do you need to calculate any adjustments if your system media is a gas?

Once these issues have been addressed, you can start reading your curve. On the curve, identify the range of flows you can expect in your system. Given that range, the curve will tell you what pressures you can expect the regulator to maintain. Ideally, you want the regulator to operate along the relatively flat part of the curve. That is where the regulator will perform most consistently. You want to avoid operating the regulator on the far ends of the curve where undesirable conditions like lock-up and choked flow occur.

If a flow curve is not available or if you need help selecting a regulator, consult with a knowledgeable sales and service representative who can provide guidance on properly sizing a regulator for your application.

Sidebar: regulator flow curve terminology

- **Accumulation**: The increase in inlet pressure required to obtain a specified flow rate for a backpressure regulator.
- **Choked-Flow Area**: The area of a flow curve in which flow demand is greater than the pressure-controlling capabilities of the regulator.

- **Droop**: The outlet pressure change from the set pressure that occurs as flow rate increases for a pressure-reducing regulator.
- **Hysteresis**: The difference between flow curves for increasing and decreasing flow.

- **Lock-Up**: The outlet pressure increase that occurs as flow is decreased to zero.

- **Seat Load Drop**: The initial pressure drop experienced when a regulator starts from a no-flow state.

About the author

As manager, field engineering, Bill Menz leads a team of engineers dedicated to supporting application, service, and training of Swagelok products. He has worked closely with distributors and end users to define market trends, growth opportunities, new product needs, and applications for Swagelok's analytical and process instrumentation products. Bill has more than 13 years of experience gained through a variety of engineering and marketing roles with Rosemount Analytical. Bill graduated from the University of Cincinnati with a bachelor of science degree in chemical engineering. He can be reached at Bill.Menz@Swagelok.com.

Used with permission of Swagelok Company—This article originally appeared in the August 2011 Issue of Flow Control Magazine.

8: Corrosion/Coatings

National Association of Pipe Coating Applications (NAPCA) specifications[a]

RECOMMENDED SPECIFICATION DESIGNATIONS FOR ENAMEL COATINGS

These are recommended enamel coating system designations for coal tar or asphalt coatings, except for specialty requirements.

*PART #1	**PART #2	SHOT BLAST	PRIMER	93.75 MILS MIN. ENAMEL	FIBERGLASS WRAP	#15 FELT WRAP	SEAL COAT	GLASS MAT	POLYETHYLENE WITH KRAFT	HEAVY KRAFT PAPER	ELECTRICAL INSPECTION
TF–		X	X	X		X				X	X
AF–		X	X	X		X				X	X
TG–		X	X	X	X					X	X
AG–		X	X	X	X					X	X
TGF–		X	X	X	X	X				X	X
AGF–		X	X	X	X	X				X	X
TGSF–		X	X	X	X	X	X			X	X
AGSF–		X	X	X	X	X	X			X	X
TGM–		X	X	X				X		X	X
AGM–		X	X	X				X		X	X
TGMF–		X	X	X		X		X		X	X
AGMF–		X	X	X		X		X		X	X
TGMP–		X	X	X				X	X		X
AGMP–		X	X	X				X	X		X

*PART #1 SYMBOLS—
ENAMEL & WRAPPERS

T—Coal Tar   G—Fiberglass
A—Asphalt   S—Seal Coat
F—#15 Felt   GM—Glass Mat
P—Polyethylene wrapper
   w/kraft paper
O—Outer wrap
   To specify Outer Wrap, use
   letter O in place of F.

**PART #2 SYMBOLS
GRADES OF ENAMEL

3— Fully-Plasticized Coal Tar
   (ordered by penetration
   dictated by ambient temperature)
4— High Temperature Surface Coal Tar
7— Asphalt

EXAMPLE: 93.75 mils (minimum) fully-plasticized coal tar enamel wrapped with fiberglass, felt and kraft — the symbol would be TGF-3.

Reprinted courtesy of National Association of Pipe Coating Applicators.

---

[a] *NOTE:* These are not all the specifications approved by NAPCA. For other specifications, consult an NAPCA member.

STANDARD APPLIED PIPE COATING WEIGHTS FOR NAPCA COATING SPECIFICATIONS

Refer to NAPCA 1-65-83

Spec. Symbol		TF 3-4	AF-7	TG 3-4	AG-7	TGF -3-4	AGF-7
PIPE SIZE		Weight in Pounds—Per 100 Lineal Feet					
1/2"	I.D.	21	18	17	14	21	18
3/4"	"	26	22	21	17	26	22
1"	"	32	27	26	22	33	28
1 1/4"	"	40	34	33	27	41	35
1 1/2"	"	46	39	38	31	46	40
2 3/8"	O.D.	57	48	47	39	57	49
2 7/8"	"	68	58	57	47	69	59
3 1/2"	"	83	71	70	58	84	72
4"	"	95	81	80	66	96	82
4 1/2"	"	106	91	90	74	108	92
5 9/16"	"	130	111	111	91	132	113
6 5/8"	"	155	132	132	109	157	134
8 5/8"	"	201	171	171	141	204	174
10 3/4"	"	250	213	214	176	254	216
12 3/4"	"	296	252	253	209	300	256
14"	"	325	276	278	229	329	281
16"	"	372	317	318	262	377	322
18"	"	419	356	357	295	424	362
20"	"	463	393	397	327	469	400
22"	"	510	433	436	360	516	440
24"	"	555	472	476	392	563	480
26"	"	601	511	516	425	610	519
28"	"	648	550	555	458	656	559
30"	"	692	588	595	490	702	598
32"	"	739	627	634	523	749	638
34"	"	785	667	674	556	796	678
36"	"	831	706	714	589	842	717
42"	"	970	824	832	687	983	837

Spec. Symbol		TGM 3-4	AGMP 7	TGMP 3-4	AGM 7	TGMF 3-4	AGMF 7	TGSF 3-4	AGFS 7
PIPE SIZE		Weight in Pounds — Per 100 Lineal Feet							
1/2"	I.D.	17	14	17	14	21	18	26	22
3/4"	"	21	17	21	17	26	22	33	28
1"	"	26	22	26	22	33	28	41	35
1 1/4"	"	33	27	33	27	41	35	51	43
1 1/2"	"	38	31	38	31	46	40	58	50
2 3/8"	O.D.	47	39	47	39	57	49	72	61
2 7/8"	"	57	47	57	47	69	59	87	74
3 1/2"	"	70	58	70	58	84	72	106	90
4"	"	80	66	80	66	96	82	121	102
4 1/2"	"	90	74	90	74	108	92	136	115
5 9/16"	"	111	91	111	91	132	113	167	141
6 5/8"	"	132	109	132	109	157	134	198	168
8 5/8"	"	171	141	171	141	204	174	258	218
10 3/4"	"	214	176	214	176	254	216	321	271
12 3/4"	"	253	209	253	209	300	256	380	321
14"	"	278	229	278	229	329	281	417	352
16"	"	318	262	318	262	377	322	477	403
18"	"	357	295	357	295	424	362	537	453
20"	"	397	327	397	327	469	400	594	501
22"	"	436	360	436	360	516	440	654	552
24"	"	476	392	476	392	563	480	713	602
26"	"	516	425	516	425	610	519	772	651
28"	"	555	458	555	458	656	559	832	702
30"	"	595	490	595	490	702	598	889	750
32"	"	634	523	634	523	749	638	949	800
34"	"	674	556	674	556	796	678	1008	850
36"	"	714	589	714	589	842	717	1067	900
42"	"	832	687	832	687	983	837	1245	1051

All weights provide for standard cutbacks.

Weights per 100 lineal feet are figured on the following material weights per square foot of pipe surface:

	Weight Lbs.—Per Sq. Ft.		Weight Lbs.—Per Sq. Ft.
Coal Tar Primer & Enamel—1/32" Min.	.2461	#15 Pipeline Felt (Tar or asphalt saturated)	.1271
Coal Tar Primer & Enamel—2/32" Min.	.4850	Outerwrap	.1225
Coal Tar Primer & Enamel—3/32" Min.	.7238	.010 . . . . . . Fiberglass Wrap	.0110
Asphalt Primer & Enamel—1/32" Min.	.2000	Kraft Paper	.0200
Asphalt Primer & Enamel—2/32" Min.	.3950	Glass Mat	.0114
Asphalt Primer & Enamel—3/32" Min.	.5912	Polyethylene	.0450

Reprinted courtesy of National Association of Pipe Coating Applicators.

NAPCA SPECIFICATIONS PIPELINE FELTS

PROPERTY	#15 TAR N.R.	#15 TAR R.F.	#15 ASPH. N.R.	#15 ASPH. R.F.
Saturated Felt — Not Perforated				
Weight (Min.) lbs. per 100 s.f. (ASTM-D-146-59 Sec. 1-9)	12	12	12	12
Saturation (Percent) MIN. (ASTM-D-146-59 Sec. 18 MAX. by extraction)	18 28	18 28	22 40	22 40
Ash Content (MIN.) (Percent) (ASTM-D-146-59 Sec. 21)	70	70	70	70
Caliper (Min.) (Inches) (ASTM-D-645-64T Method C)	.021	.018	.021	.018
Tensile Strength (Min.) (lbs. per inch of width) (ASTM-D-146-59 Sec. 12a) With fiber grain Across fiber grain	35 12	37 No. spec.	35 12	37 No. spec.
Tear Strength (Min.) (Grams) (ASTM-D-689-62) Across fiber grain With fiber grain	400 240	Not applicable	400 240	Not applicable
Loss on heating (Max.) (Percent) (AWWA C203-66 Sec. 2.6.6)	10	10	10	10
Pliability (AWWA C203-66 Sec. 2.6.4)	No cracking	No cracking	No cracking	No cracking
Saturated and Perforated Felt				
(1/16" perforations on 1" Staggered Centers)				
Tensile Strength (Min.) (lbs. per inch of width) (ASTM-D-146-59 Sec. 12a) With fiber grain Across fiber grain	30 10	35 No. spec.	30 10	35 No. spec.
Tear Strength (MIN.) (Grams) (ASTM-D-689-62) Across fiber grain With fiber grain	348 190	Not applicable	348 190	Not applicable
Unsaturated Felt				
Asbestos Content (%) (MIN.) (ASTM-D-1918-67)	85	85	85	85

N.R. — NON REINFORCED
R.F. — REINFORCED LENGTHWISE WITH PARALLEL FIBER GLASS YARNS ON NOMINAL ¼" CENTERS.

Reprinted courtesy of National Association of Pipe Coating Applicators.

Minimum test voltages for various

Coating Thicknesses		
32nd Inch	Coating Thicknesses Mils	Test Volts
—	16	5000
1	31	7000
2	62	9800
3	94	12,100
4	125	14,000
5	156	15,000
6	188	17,100
16	500	28,000
20	625	31,000
24	750	35,000

Reprinted courtesy of National Association of Pipe Coating Applicators.

How much primer for a mile of pipe?

Multiply pipe O.D. by $2\frac{3}{4}$ to get gallons of primer per mile.

Examples. Pipe 24 in in diameter will require $24 \times 2\frac{3}{4}$ or 66 Gal of primer per mile; Pipe 16 in in diameter will require $16 \times 2\frac{3}{4}$ or 44 Gal of primer per mile; and

Pipe 4½ in in diameter will take $4\frac{1}{2} \times 2\frac{3}{4}$ or 123/8 Gal of primer per mile (12 Gal in round figures).

The given multiplier is applicable only where the pipe is new and in good condition. For rough pipe the multiplying factor should be upped from $2\frac{3}{4}$ to 3.45. In other words, rough 20-in pipe will require 20×3.45 or 69 Gal of primer per mile.

It is hard to estimate the amount of spillage and waste. Temperature and relative humidity affect the way the primer goes on the pipe.

How much coal-tar enamel for a mile of pipe?

For an average cover of $3/32$ in, multiply outside diameter of pipe by 0.6. The answer is in tons of enamel per mile. (Add 15% to take care of gate valves, drips, bends, etc.)

Examples. How many tons of coal-tar enamel will be required to apply a $3/32$-in coat to a 26-in pipeline?

$$26 \times 0.6 = 15.6 \text{ tons per mile}$$

Add 15% to take care of gate valves, bends etc.

$$15.6 + 2.3 = 17.9 \text{ or } 18 \text{ tons per mile}$$

How many tons of coal-tar enamel will be required to apply a $3/32$-in coat to an 8.625-in pipeline?

$$8.625 \times 0.6 = 5.2 \text{ tons per mile}$$
$$5.2 \times 15\% = 0.8$$
$$= 6 \text{ tons per mile}$$

The following factors strongly affect the amount of coating used per mile on pipelines: 1. Temperature of the pipe and ambient temperatures. (It is assumed that the temperature of the enamel can be controlled). 2. The type of dope machine. 3. Humidity. 4. The number of bends per mile. 5. The number of valves, crossovers, etc. 6. The condition of the pipe.

This latter point has a great effect on the amount of enamel required for a mile of pipeline. In fact, where reconditioned pipe is being coated, it is recommended that the amount of enamel be increased by about 20%.

How much wrapping for a mile of pipe?

For a 14-in to 48-in pipe, multiply the diameter in inches by 15 to get the number of squares (100 ft^2). For pipe sizes of $12\frac{3}{4}$ in and smaller, add 1 to the nominal size of pipe and multiply by 15.

Examples. How many squares of asbestos felt would be required to wrap 1 mile of 30-in pipeline?

$$30 \times 15 = 450 \text{ squares of felt}$$

How many squares of felt would be required to wrap 1 mile of 4-in pipe?

$$(4 + 1) \times 15 = 75 \text{ squares of felt}$$

These figures are conservative and do not allow for wrapping drips, patching holidays, and for short ends that are usually discarded. They allow for machine wrapping using $\frac{3}{4}$-in lap on diameters up to $10\frac{3}{4}$ in and using 1-in lap on larger diameters.

For smaller diameters of pipe, this figure should be increased about 5% to take care of additional wastage and patching.

Estimating coating and wrapping materials required per mile of pipe

Primer and coal-tar enamel

Gallons of primer and tons of coal-tar enamel are based on covering new pipe in good condition. For rough pipe, increase primer quantities by 25% and the coal-tar quantities by 20%. No allowance is included for spillage or waste.

Wrappings

Squares are based on machine wrapping, allowing $\frac{3}{4}$-in laps on pipe sizes through 10 in and 1-in laps on pipe sizes greater than 10 in No allowance is included for wrapping drips, patching holidays, or short ends, which are usually discarded.

Nominal Pipe Size	Primer Gallons	3/32″ Coal-tar Enamel Tons	Glass Wrap Squares	Kraft Paper Squares	15-Pound Felt Squares
4	11	3	75	75	75
6	17	4	105	105	105
8	22	6	135	135	135
10	28	7	165	165	165
12	33	8	195	195	195
14	39	10	210	210	210
16	44	11	240	240	240
18	50	13	270	270	270
20	55	14	300	300	300
24	66	17	360	360	360
30	83	21	450	450	450
36	99	25	540	540	540
42	116	29	630	630	630

Coefficient of friction for pipe coating materials

Tests indicate that previous values appear to be valid for thin-film epoxies and conservative for coal tars

J. B. Ligon, Assistant Professor, Mechanical Engineering-Engineering Mechanics Dept., Michigan Technological University, Houghton, Mich., and **G. R. Mayer,** Project Engineering Manager, Bechtel, Inc., San Francisco

A major factor in the stress analysis of buried pipelines is the movement that pipe undergoes in the presence of temperature and pressure differentials during its life. This movement is highly dependent upon friction resistance of the soil.

Although ample information is available on the static coefficient of friction for many materials, there is a lack of data on friction between soils and various coatings used in the pipeline industry. In the past, friction coefficient information was extrapolated from data in the literature that were believed to have a similarity to the external pipe coating to soil interface.

However, with the development of thin-film epoxy resin coating systems and the increasing use of these systems in the pipeline industry, a change from a conventional coal tar felt coating to a thin-film epoxy coating would indicate a significant change in the friction coefficient design criteria due to the extreme contrast in the surface texture of the two materials.

To evaluate the effect of the difference in surface texture on a pipeline system, test procedures were developed to determine the coefficient of friction for both coal tar felt and thin-film epoxy to various soils and to obtain more reliable information for future pipeline designs using these coating materials.

Static friction tests were conducted to find the coefficient of friction between coal tar felt and thin-film epoxy pipe coating and eight representative backfill soil samples from typical locations along a pipeline right of way. The results indicate that the friction coefficients are significantly larger than those previously extrapolated from literature and that coal tar has a higher friction resistance in respect to anchorage of a pipeline.

For the coal tar felt coating, the coefficient varies from 0.59 to 0.91 depending on the soil and moisture content. The thin-film epoxy coating varies from 0.51 to 0.71 under the same conditions.

Definitions

Static coefficient of friction. The theoretical longitudinal soil force acting on the pipe surface can be calculated from the relation:

$$F = \mu \int_A p \, dA$$

where:

F = Longitudinal soil friction force (lb)
μ = Coefficient of friction (dimensionless)
p = Normal soil pressure acting on the pipe surface (psi)
dA = Soil to pipe differential contact area (in^2)
$\int_A p \, dA$ = Total normal soil force on pipe surface (lb).

The above relation is independent of the pressure per unit area of the mating surfaces as long as very high contact pressures are not encountered, which is the case for a buried pipeline.

In terms of a buried pipeline supporting a soil burden up to three pipe diameters of depth, the soil force relation becomes:

$$F = \mu \left[2D\gamma \left(H - \frac{D}{2} \right) + Wp \right]$$

where:

Wp = Weight of pipe and contents (lb/in)
D = Pipe diameter (in)
H = Depth of the pipe centerline (in)
γ = Specific weight of the soil (lb/in^3)

The sensitivity of the soil friction forces to the coefficient of friction is readily apparent from the above relation. Since this force is inversely proportional to the active length that a pipeline moves as a result of temperature and pressure expansion, the coefficient of friction becomes a major factor in pipeline stress design.

Test system. The theoretical soil friction force acting on the surface of a coated plate can be calculated from the relation:

$$F = \mu N$$

where:

N = Normal force acting on the plate surfaces (lb)
W = Total weight of plate and external load (lb).

Since the coefficient of the friction parameter is the same in all of the previous expressions, the plate test system can be used to simulate the soil-to-pipeline interface.

Coal tar felt. The coal tar coating, enamel, and felt combination, was prepared to simulate an actual coated pipeline.

The surface of a steel plate was coal tar coated and then wrapped in a 15-lb felt. Therefore, the soil surface contact was predominantly with the felt material.

Thinfilm epoxy. The epoxy coating simulated a thin-film epoxy resin pipeline coating applied by the fusion method.

Plates used in the tests were coated by the fluidized bed method. However, the surface texture was the same as that expected by coating plant production equipment.

Results

Coal tar to soils. Figure 1 illustrates that the friction coefficient varies from 0.59 to 0.91 for the various soils tested at field moisture content.

Figure 1. Coefficient of friction for coal tar coating to soil.

Figure 2. Coefficient of friction for thin film epoxy to soil.

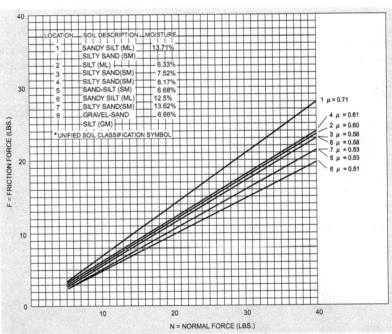

The mechanism by which the friction coefficient changes among the various soils is not apparent from the results of these tests, since the soil samples used did not vary considerably in description. However, the range over which the factor changes is considerably greater than that previously extrapolated from the literature, as follows:

Soil Description	Coefficient of Friction Commonly Used
Silt	0.3
Sand	0.4
Gravel	0.5

The tests also indicate that the moisture content alters the friction factor to some extent, as would be expected.

In order to investigate the influence of temperature, tests were conducted with coal tar felt wrapping heated to 120 °F. Only a slight softening beneath the coating surface was observed, and it is believed that temperatures up to this range will not significantly affect the magnitude of the friction coefficient.

Thinfilm epoxy to soils. As expected, results shown in Figure 2 indicate that the friction factor range of 0.51 to 0.71 for the epoxy is somewhat lower than that of the coal tar.

Again, the mechanism is not clear, but the results are fairly consistent with the friction coefficient increasing in most instances in the same order of magnitude as for coal tar. Since such information on this coating was not available in the literature, no comparison to similar tests could be made.

Temperatures in the 120 °F range should have little or no effect on the coefficient of friction of epoxy to soil.

Conclusions

Although it is virtually impossible to precisely simulate the surface contact situation of a pipeline in a back-filled trench, the test procedure and apparatus reported here are a means of approximating it. The results indicate that coal tar coatings have a higher friction resistance than epoxy coatings as far as anchoring capabilities of the soil are concerned.

The selection of a coating, based on its soil friction resistance, could be of economic value in reducing those situations where extreme expansions call for reinforced fittings or elaborate culverts to overcome excessive stress levels in a pipeline system.

The test results show that previous values commonly used for the coefficient of friction were conservative for similar soils, and it is suggested that some conservatism still be incorporated in future analyses.

Since the tests indicated that the friction coefficient of the epoxies was similar to those previously used for coal tar coatings, their continued use for epoxy-coated pipelines should be valid. A more conservative approach to these values is recommended in soils where the presence of excessive moisture could substantially change the friction factor when interfaced with a smooth epoxy coating.

Troubleshooting cathodic protection systems: Magnesium anode system

The basic technique is the same as that outlined earlier; more measurements are required because of the multiplicity of drain points. First, the current output of stations nearest the point of low potential should be checked; if these are satisfactory, a similar check should be extended in both directions until it is clear that the trouble must be on the line. When a given anode group shows a marked drop in current output, the cause may be drying out, shrinkage of backfill, or severed or broken lead wires. If the current is zero, the pipe-to-soil potential of the lead wire will show whether it is still connected to the pipe or the anode, and thus indicate the direction to the failure. If the current is low, there may be a loss of one or more anodes by a severed wire; a pipe-to-soil potential survey over the anodes will show which are active, just as in the case of rectifier anodes.

Trouble indicated on the line, rather than at the anode stations, is tracked down in the same manner as that used for

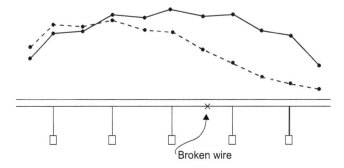

Broken wire

Figure 1. Locating idle anodes by surface potentials. The solid line shows the potentials found along a line of anodes when all are delivering current; the dotted line exhibits the change when there is a break in the anode lead at the point indicated. Single disconnected anodes may also be located by this method. A driven ground rod, a pipe lead, or even a rectifier terminal may be used for the reference ground; all readings should be referred to the same reference.

line protected by rectifiers—first, investigating locations where something that has been done might be responsible; next, checking potentials; and finally, making a line current survey.

Cathodic protection for pipelines

Estimate the rectifier size required for an infinite line. Refer to Figure 1. If coating conductance tests have not been performed, pick a value from Table 1.

Table 1
Typical Values of Coating Conductance

Micromhos/ft²	Coating Condition
1–10	Excellent coating—high resistivity soil
10–50	Good coating—high resistivity soil
50–100	Excellent coating—low resistivity soil
100–250	Good coating—low resistivity soil
250–500	Average coating—low resistivity soil
500–1000	Poor coating—low resistivity soil

Use a value of -0.3 volts for ΔE_x. This is usually enough to raise the potential of coated steel to about -0.85 volts.

Use a value of 1.5 volts for AE at the drain point. Higher values may be used in some circumstances; however, there may be a risk of some coating disbondment at higher voltages.

Calculate I at the drain point.

$$\Delta I_A = amp/in \times \Delta E_x/0.3 \times D,$$

where D = pipe OD, inches.

Example. 30-in OD line with coating conductivity $= 100$ micromhos/ft². What is the current change at the drain point,

Based on 0.25-inch wall pipe and steel resistivity of 18.8 microhm-cm

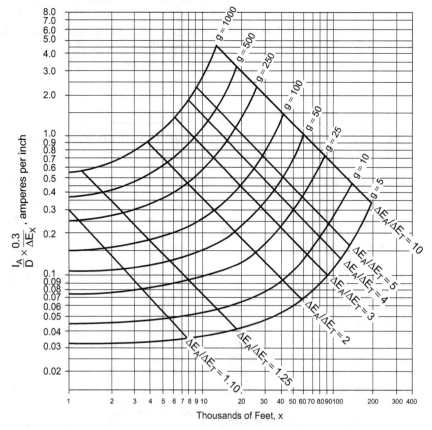

Figure 1. Drainage current versus distance for a coated infinite line.

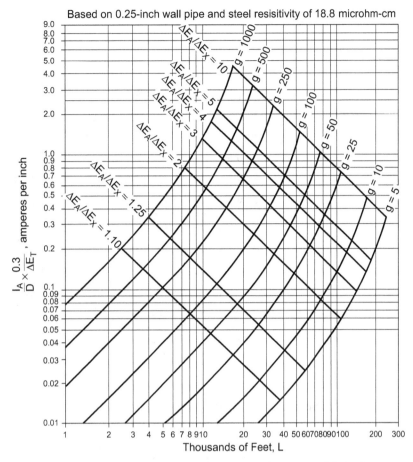

Based on 0.25-inch wall pipe and steel resisitivity of 18.8 microhm-cm

Figure 2. Drainage current versus distance for a coated finite line.

and how far will this current protect an infinitely long pipeline?

$\Delta E_A = 1.5$

$\Delta E_X = 0.3$

$\Delta E_A/\Delta E_X = 5.0$

Refer to Figure 1 and read

$L = 30,000$ ft

$\Delta I_A/D \times 0.3/\Delta E_x = 0.7$ (from Figure 1)

$\Delta I_A = 0.7 \times 30 = 21$ amp.

21 amps will protect the line for 30,000 ft in either direction from the drain point.

Estimate rectifier size for a finite line. Refer to Figure 2.

A 3½-in OD line 20,000 ft long is protected by insulating flanges at both ends and has a coating conductivity of 500 micromhos/ft². What is $\Delta E_A/E_T$ and what ΔI_A is required? Assume that

$E_T = 0.3$.

From Figure 2, read

$\Delta E_A/E_T = 5.8$

$\Delta E_A = 1.74$

$\Delta I_A/D \times 0.3/E_T = 1.8$ amp/in

$\Delta I_A A = 1.8 \times 3.5 = 6.3$ amp.

Estimate ground bed resistance for a rectifier installation.

Example. A ground bed for a rectifier is to be installed in 1000 ohm-cm soil. Seven 3-in × 60-in vertical graphite anodes with backfill and a spacing of 1oft will be used. What is the resistance of the ground bed?

Refer to Figure 3. The resistance is 0.56 ohms.

Figure 3. Anode bed resistance versus number of anodes. 3-in × 60-in vertical graphite anodes in backfill.

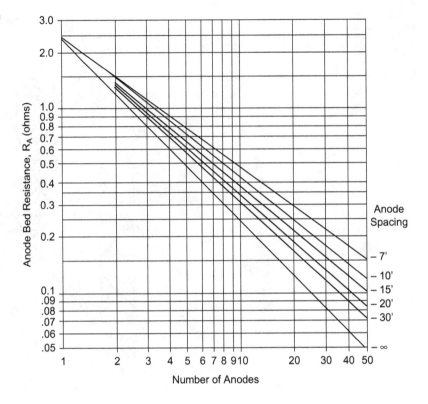

Figure 4. Anode bed resistance versus number of anodes. 2-in × 60-in vertical bare anodes.

Figure 3 is based on a soil resistivity of 1000 ohms-cm. If the soil resistivity is different, use a ratio of the actual soil resistivity divided by 1000 and multiply this by the reading obtained from Figure 3.

A rectifier ground bed is to be composed of $10 - 2 \times 60$-in bare "Duriron" silicon anodes spaced 20 ft apart. The soil resistivity is 3000 ohms-cm. What is the ground bed resistance? Refer to Figure 4. The resistivity from the chart is 0.55 ohms. Since the chart is based on soil resistivity of 1000 ohms-cm, the ground bed resistivity is 3000/1000 × 0.55 or 1.65 ohms.

The resistance of multiple anodes installed vertically and connected in parallel may be calculated with the following equation:

$$R = 0.00521 \, P/NL \times (2.3 \log 8L/d - 1$$
$$+ \, 2L/S \log 0.656N) \qquad (1)$$

where:

R = Ground bed resistance, ohms
P = Soil resistivity, ohm-cm
N = Number of anodes
d = Diameter of anode, feet
L = Length of anode, feet
S = Anode spacing, feet.

If the anode is installed with backfill such as coke breeze, use the diameter and length of the hole in which the anode is installed. If the anode is installed bare, use the actual dimensions of the anode.

Figure 5 is based on Eq. 1 and does not include the internal resistivity of the anode. The resistivity of a single vertical anode may be calculated with Eq. 2.

$$R = 0.00521 \, P/L \times (2.3 \log 8L/d - 1) \qquad (2)$$

If the anode is installed with backfill, calculate the resistivity using the length and diameter of the hole in which the anode is installed. Calculate the resistivity using the actual anode dimensions. The difference between these two values is the internal resistance of the anode. Use the value of *P*, typically about 50 ohm-cm, for the backfill medium.

Figure 5 is based on 1000 ohm-cm soil and a 7-ft × 8-in hole with a 2-in × 60-in anode.

Example. Determine the resistivity of 20 anodes installed vertically in 1500 ohm-cm soil with a spacing of 20 ft.

Read the ground bed resistivity from Figure 5.

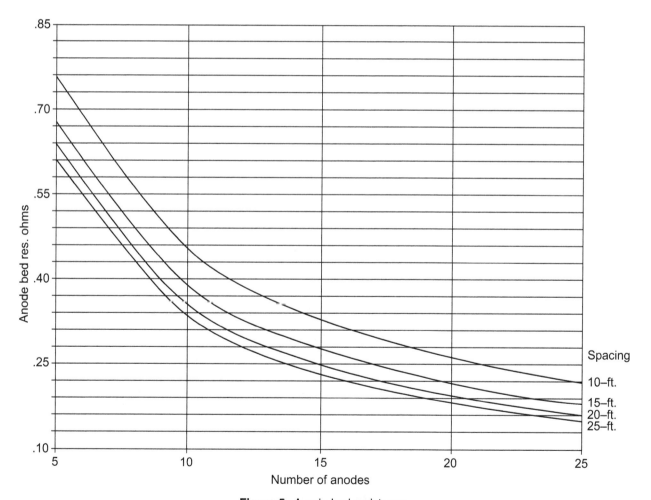

Figure 5. Anode bed resistance.

R = 0.202 ohm.

Since the anodes are to be installed in 1500 ohm-cm soil and Figure 5 is based on 1000 ohm-cm soil, multiply R by the ratio of the actual soil resistivity to 1000 ohm-cm.

R = 0.202 × 1500/1000

R = 0.303 ohm.

The internal resistivity for a single 2-in × 60-in vertical anode installed in 50 ohm-cm backfill (7 ft × 8-in hole) is 0.106 ohm.

Since 20 anodes will be installed in parallel, divide the resistivity for one anode by the number of anodes to obtain the internal resistivity of the anode bank.

0.106/20 = 0.005 ohm

The total resistivity of the 20 anodes installed vertically will therefore be 0.308 ohm (0.303 + 0.005).

Galvanic anodes

Zinc and magnesium are the most commonly used materials for galvanic anodes. Magnesium is available either in standard alloy or high purity alloy. Galvanic anodes are usually prepackaged with backfill to facilitate their installation. They may also be ordered bare if desired. Galvanic anodes offer the advantage of more uniformly distributing the cathodic protection current along the pipeline, and it may be possible to protect the pipeline with a smaller amount of current than would be required with an impressed current system, but not necessarily at a lower cost. Another advantage is that interference with other structures is minimized when galvanic anodes are used.

Galvanic anodes are not an economical source of cathodic protection current in areas of high soil resistivity. Their use is generally limited to soils of 3000 ohm-cm except where small amounts of current are needed.

Magnesium is the most-used material for galvanic anodes for pipeline protection. Magnesium offers a higher solution potential than zinc and may therefore be used in areas of higher soil resistivity. A smaller amount of magnesium will generally be required for a comparable amount of current. Refer to Figure 6 for typical magnesium anode performance data. These curves are based on driving potentials of −0.70 volts for H-1 alloy and −0.90 volts for Galvomag working against a structure potential of −0.85 volts referenced to copper sulfate.

The driving potential with respect to steel for zinc is less than for magnesium. The efficiency of zinc at low current levels does not decrease as rapidly as the efficiency for magnesium. The solution potential for zinc referenced to

Figure 6. Magnesium anode current.

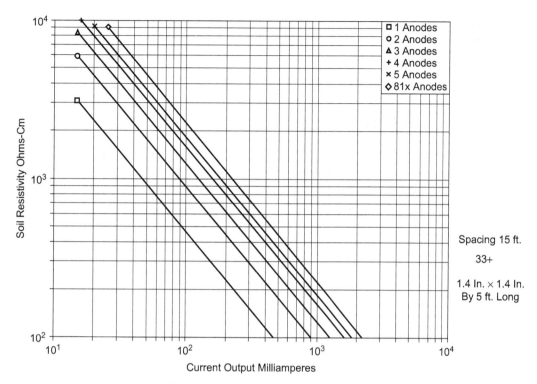

Figure 7a. Current output zinc anodes.

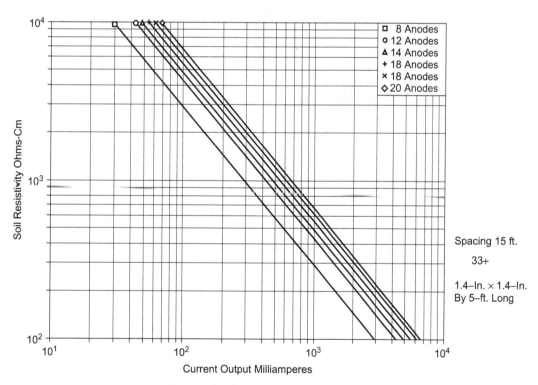

Figure 7b. Current output zinc anodes

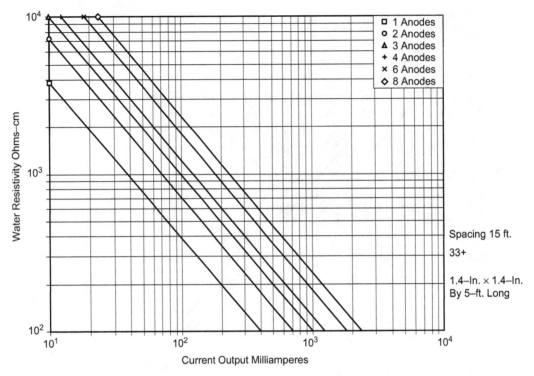

Figure 8a. Current output zinc anodes.

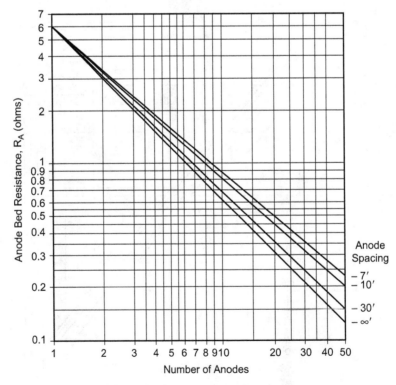

Figure 8b. Current output zinc anodes.

a copper sulfate cell is −1.1 volts; standard magnesium has a solution potential of −1.55 volts; and high purity magnesium has a solution potential of −1.8 volts.

If, for example, a pipeline is protected with zinc anodes at a polarization potential of −0.9 volts, the driving potential will be −1.1 − (−0.9) or −0.2 volts. If standard magnesium is used, the driving potential will be −1.55 − (−0.9) or −0.65 volts. The circuit resistance for magnesium will be approximately three times as great as for zinc. This would be handled by using fewer magnesium anodes, smaller anodes, or using series resistors.

If the current demands for the system are increased due to coating deterioration, contact with foreign structures, or by oxygen reaching the pipe and causing depolarization, the potential drop will be less for zinc than for magnesium anodes. With zinc anodes, the current needs could increase by as much as 50% and the pipe polarization potential would still be about 0.8 volts. The polarization potential would drop to about 0.8 volts with only a 15% increase in current needs if magnesium were used.

The current efficiency for zinc is 90%, and this value holds over a wide range of current densities. Magnesium anodes have an efficiency of 50% at normal current densities.

Magnesium anodes may be consumed by self-corrosion if operated at very low current densities. Refer to Figures 7(a) and 7(b) and Figures 8(a) and 8(b) for zinc anode performance data. The data in Figures 7(a) and 7(b) are based on the anodes being installed in a gypsum-clay backfill and having a driving potential of −0.2 volts. Figures 8(a) and 8(b) are based on the anodes being installed in water and having a driving potential of −0.2 volts.[1]

Example. Estimate the number of packaged anodes required to protect a pipeline.

What is the anode resistance of a packaged magnesium anode installation consisting of nine 32-lb anodes spaced 7 ft apart in 2000 ohm-cm soil?

Refer to Figure 9. This chart is based on 17# packaged anodes in 1000 ohm-cm soil. For nine 32-lb anodes, the resistivity will be:

$$1 \times 2000/1000 \times 0.9 = 1.8 \text{ ohm.}$$

See Figure 10 for a table of multiplying factors for other size anodes.

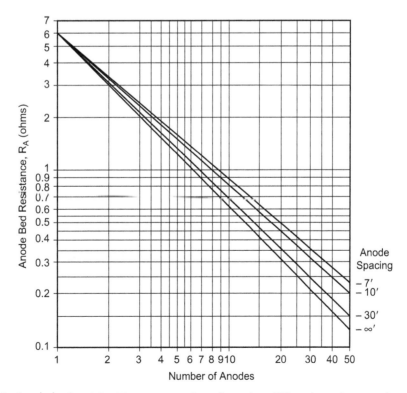

Figure 9. Anode bed resistance versus number of anodes; 17# packaged magnesium anodes.

Chart based on 17-lb. magnesium anodes installed in 1,000 ohm-cm soil in groups of 10 spaced on 10-ft centers.

For other conditions multiply number of anodes by the following multiplying factors:

For soil resistivity: MF $= \dfrac{p}{1,000}$ For 9-lb. anodes: MF = 1.25

For conventional magnesium: MF = 1.3   For 32-lb. anodes: MF = 0.9

Figure 10. Number of anodes required for coated line protection.

Example. A coated pipeline has a coating conductivity of 100 micromhos/sq ft and is 10,000 ft long, and the diameter is $10\frac{3}{4}$ in How many 17-lb magnesium anodes will be required to protect 1000 ft? Refer to Figure 7 and read 2 anodes per 1000 ft a total of twenty 17 lb anodes will be required for the entire line.

Source

Pipeline Corrosion and Cathodic Protection, Gulf Publishing Company, Houston, Texas.

Reference

1. From data prepared for the American Zinc Institute.

Estimate the pounds of sacrificial anode material required for offshore pipelines

This rule of thumb is based on the following assumptions:

- 5% of pipeline considered bare
- 3 mA/ft^2 required for protection
- 30-year design life

Basis for rule: sq ft of surface area of pipeline/linear ft × % of bare pipe × 3.0 mA/ft^2 × lb/amp-year × design life in years = the lb of galvanic anode material that will be required.

Multiplier $=$ OD $\times$ lb/amp-yr $\times$ (12 PI/144) $\times$ 30 $\times$ 0.05 $\times$ 0.003 $\times$ 1000 or:

Pipe OD $\times$ lb/amp-yr $\times$ 1.1781 = lb of anode material/1000 linear ft of pipeline.

Example. How many pounds of zinc will be required for a 30-in OD line installed in seawater and operating at 80 °F? Pounds of zinc per ampere-year = 24.8 (seawater).

30 $\times$ 24.8 $\times$ 1.1781 = 877 lb/1000 linear ft of pipeline.

How many pounds of Galvalum® III will be required for a 30-in OD line installed in seawater and operating at 80 °F? Pounds of Galvalum® III per ampere-year in seawater = 7.62 (32–95 °F).

30 $\times$ 7.62 $\times$ 1.1781 = 269 lb/1000 linear ft of pipeline.

How many pounds of Galvalum® III will be required for a 30-in OD line installed in saline mud and operating at 80 °F? Pounds of Galvalum® III per ampere-year in saline mud = 9.22 (32–95 °F).

30 $\times$ 9.22 $\times$ 1.1781 = 326 lb/1000 linear ft of pipeline.

How many pounds of Galvalum® III will be required for a 30-in OD line installed in saline mud and operated at a temperature of 160 °F? Pounds of Galvalum® III per ampere-year in saline mud = 21.9 (140–170 °F).

30 $\times$ 21.9 $\times$ 1.1781 = 774 lb/1000 linear ft of pipeline.

It is always a good idea to check with the bracelet vendor to determine the nearest size mold available to avoid the need for a special size mold. Also, the mold size will be a factor in determining anode spacing.

These calculations assume a utilization factor of 100%, and the designer should select the appropriate utilization factor for the specific application. For estimating, you may want to use 85%. Temperatures used in the examples are assumed to be the anode operating temperature.

Generally, zinc anodes should not be used if the pipeline will be operated at temperatures above 90 °F. See ASTM B418-80 and NACE Paper No. 311 by Elwood G. Haney presented at CORROSION/85.

See Table 3 for typical anode performance data. All performance data shown here are suitable for estimating. For final design purposes, the anode manufacturer or distributor should be consulted for actual anode performance data required for a specific application.

Galvalum® III is a registered trademark of Oronzio De Nora S. A., Lugano (Switzerland). Galvomag® is a registered trademark of Dow Chemical Company.

Typical anode specifications

Tables 1–4 are guideline-type information suitable for general use. The characteristics of a specific application must be considered, and it may be necessary to contact the anode manufacturer or distributor to obtain detailed anode performance data before finalizing a design.

Table 1
Zinc Anodes

Material	Mil. Spec. MIL-A-1800 1H	ASTM Type 1	ASTM Type II
Lead	0.006 max	—	—
Iron	0.005 max	0.005 max	0.0014 max
Cadmium	0.025–0.15 Range	0.03–0.1	0.003 max
Copper	0.005 max	—	—
Aluminum	0.10–0.50 max	0.1–0.4	0.005 max
Silicon	0.125 max	—	—
Zinc	Remainder	Remainder	Remainder

Table 2*
Cast Magnesium Anodes

Material	H-1 Grade A Type III	H-1 Grade B Type II	H-1 Grade C Type I	Galvomag®
Al	5.3–6.7	6.3–6.7	5.0–7.0	0.010 max
Zn	2.5–3.5	2.5–3.5	2.0–4.0	0.05 max
Mn	0.15 min	0.15 min	0.10 min	0.5–1.3
Fe	0.003 max	0.003 max	0.003 max	0.03 max
Ni	0.002 max	0.003 max	0.003 max	0.001 max
Cu	0.02 max	0.05 max	0.10 max	0.02 max
Si	0.10 max	0.30 max	0.30 max	0.05 max
Others	0.30 max	0.30 max	0.30 max	0.05 max, each 0.30 max, total

*Reprinted courtesy of Dow Chemical U.S.A.

Table 3
Typical Anode Performance Data

Anode Type	Amp-h/lb-in Seawater	Potential Negative Volts Ag/AgCl	Approx. Current Density mA/ft^2
Zinc	354	1.03	36–600
Galvalum® 111	1150–1204 (32–95 °F) in saline mud	1.10	100–1000
	950 (32–95 °F)	1.10	100
	930 (104 °F)	1.08	100
	600 (140 °F)	1.05	100
	400 (176 °F)	1.04	100
	400 (221 °F) in earth	1.02	100
Magnesium	400–500	1.53	36–600
Galvomag®	500–540	1.75	36–600

Table 4
Typical Environments for Sacrificial Anodes

Material	Driving Potential Volts	Environment
Magnesium	0.9	Underground Freshwater 200–10,000 ohm-cm
Aluminum	0.25	Seawater Brine Saline mud <200 ohm-cm
Zinc	0.25	Seawater Brine Underground <1500 ohm-cm

Comparison of other reference electrode potentials with that of copper-copper sulfate reference electrode at 25 °C

Type of Comparative Reference Electrode	Structure-to-Comparative Reference Electrode Reading Equivalent to −0.85 volt with Respect to Copper Sulfate Reference Electrode	To Correct Readings between Structure And Comparative Reference Electrode to Equivalent Readings with Respect to Copper Sulfate Reference Electrode
Calomel (saturated)	−0.776 volt	Add −0.074 volt
Silver-silver chloride (0.1 N KCl solution)	−0.822	Add −0.028
Silver-silver chloride (silver screen with deposited silver chloride)	−0.78	Add −0.07
Pure zinc (special high grade)	+0.25*	Add −1.10

*Based on zinc having an open circuit potential of −1.10 volt with respect to copper sulfate reference electrode. (From A. W. Peabody, Chapter 5, p. 59, *NACE Basic Corrosion Course*, Houston, TX, 1973.)

Chart aids in calculating ground bed resistance and rectifier power cost

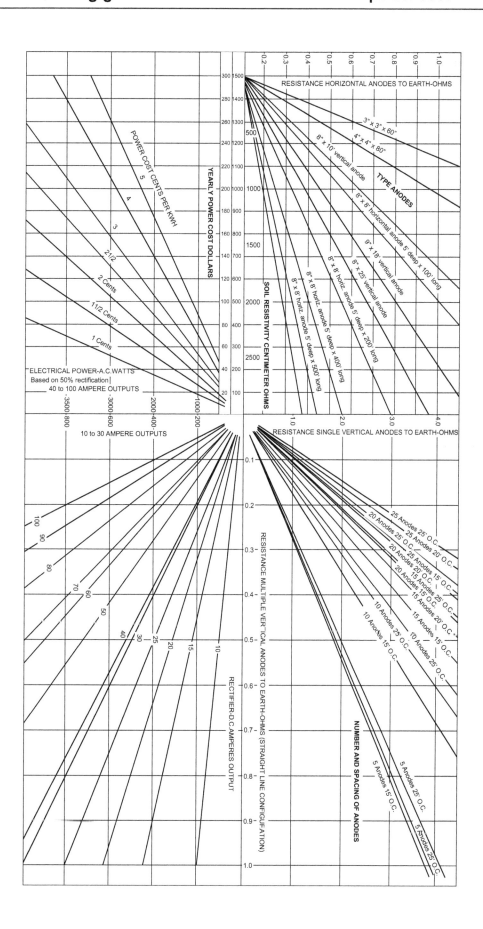

For conventional-type ground beds, this chart permits a rapid determination of the resistance to earth of various ground beds and the yearly power costs based on 50% efficient rectification. These curves are specifically designed for standard 3-in by 60-in and 4-in by 80-in graphite anodes centered in a vertical or horizontal excavation and backfilled to the dimensions indicated with well-tamped coal coke breeze.

Example.

> Soil resistivity = 1000 ohm-cm
> Power rate = 3 cents per kwh
> Rectifier rating = 20-amp

To determine the optimum size ground bed, enter the chart at the "Soil Resistivity-ohm-cm" line and extend a vertical line from the 1000-ohm-cm point to intersect all the anode types. Proceed horizontally to the left for the resistance to earth of horizontal type anodes and horizontally to the right for vertical type anodes. Various groups of vertical anodes with 15-, 20-, and 25-ft spacings are represented by curves.

Resistance values for horizontal anodes obtained from the ordinate of the upper left-hand section should be transferred to the abscissa of the upper right-hand section to determine power consumption and annual costs from the lower right-hand and lower left-hand section. In obtaining power costs for the two lower sections, the ordinate shows power consumption in AC watts based on 50% efficient rectification, and the abscissa of points on the lower left-hand section indicates annual power costs in dollars.

How can output of magnesium anodes be predicted?

For a single 17-lb. magnesium anode, working against a fully protected pipe, divide the soil resistivity in ohm-cm units into 180,000; the answer is the current output in mA.

Example. Soil resistivity 3200 ohm-cm
180,000/3200 = 56.25

Current output will be about 56 mA.

The rule works for *single* anodes against fully protected pipe; anode groups will have lower current output per anode, and if the pipe is not fully protected, the current output will be greater. For 32-lb. anodes use 195,000; for 50-lb anodes use 204,000.

How to determine the efficiency of a cathodic protection rectifier?

Multiply DC voltage output times DC amperes times 100 and divide by the AC watts input. The answer is % efficiency.

Example. A rectifier has an output of 50 amps DC at 15 volts. The wattmeter gives a direct reading of an AC input of 1155 watts.

$$\text{Efficiency} = \frac{(\text{DC volts})(\text{DC amperes})(100)}{(\text{AC watts input})}$$

$$= \frac{(15)(50)(100)}{(1,155)} = 65\%$$

On those rectifier units that do not have a wattmeter, the AC input may be obtained from the watt hour meter mounted on the pole. To do this, count the number of revolutions made by the watt hour meter in so many seconds, multiply by the meter constant stamped on the meter times 3600, and divide by the number of seconds timed. This gives AC input in watts. Then, go ahead and use the formula as above.

Example. A rectifier unit does not have a wattmeter. The wheel of the watt hour meter revolves 11 times in one minute. Meter constant is shown as 1.8. Output of the rectifier is the same as in the first case, 50 amps at 15 volts DC.

$$\text{AC watts} = \frac{K \times N \times 3600}{T}$$

where K is the meter constant and N is the number of meter revolutions in T seconds.

$$\text{AC watts} = \frac{(1.8)(11)(3600)}{60} = 1188 \text{ watts AC}$$

$$\text{Efficiency} = \frac{(15)(50)(100)}{1188} = 63\%$$

How to calculate the voltage drop in ground bed cable quickly

Use the following formula for calculating the voltage drop in a copper ground bed cable:

$$V_T = \frac{R_{1000}}{1000}[I_T D_{GR} + D_A(N-1)(I_T)$$
$$- I_A(1+2+3+4+ \ldots +N-1)]$$

where:

V_T = Total voltage drop in volts
R_{1000} = Cable resistance per 1000 ft
I_T = Total current in amps
$I_A = I_T/N$ current per anode
D_{GB} = Distance from rectifier to first anode in ft
D_A = Distance between anodes in ft
N = Number of anodes in ground bed.

Example. What is the voltage drop in a copper wire ground bed cable in a cathodic protection system having a rectifier output of 225 amps? The distance from rectifier to ground bed is 100 ft, and 16 anodes are spaced on 20-ft centers. The cable used is No. 4 copper having a resistance of 0.2487 ohm per 1000 ft.
Then:

$R_{1000} = 0.2487$ ohm
$I_T = 225$ amps
$I_A = 225/16 = 14$ amps
$D_{GB} = 100$ ft
$D_A = 20$ ft
$N = 16$

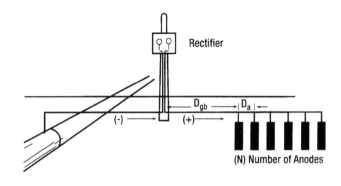

$$V_T = \frac{0.2487}{1000}[(225)\ (100) + 20\ [(15)\ (225) - (14)\ (120)]]$$

$$V_T = \frac{0.2487}{1000}[22500 + 20\ (1695)]$$

$$V_T = \left(\frac{0.2487}{1000}\right)(56,400)$$

$$V_T = (0.2487)\ (56.4)$$

$$V_T = 14.00 \text{ volts drop.}$$

What is the most economical size for a rectifier cable?

Choosing the economical size for a rectifier cable involves balancing the first cost of the cable against the power losses. A rule that does not include all of the factors, but which usually gives a very close approximation, is as follows:

$$a = 2280\ I\sqrt{P}$$

where a is the area of the cable in circular mils (refer to a wire table for size), I is the current to be carried, in amperes, and P is the cost of power in cents per kilowatt-hour. This formula is based on the present prices of copper cables and assumes the use of a good grade of direct burial cable.

Example. Suppose
$I = 60$ amps
$P = 4$ cents
$a = (2,280)\ (60)\ (\sqrt{4})$
$a = 273,600$ circular mils

Cable is sized 250,000 and 300,000 circular mils. Use the latter.

How to estimate the number of magnesium anodes required and their spacing for a bare line or for a corrosion "hot spot"?

Soil and pipeline conditions will vary widely, but where these given assumptions hold, an estimate can be made for test installations using this method. First, find the area of pipe to be protected, multiply by the milliamperes per square feet necessary to protect the line, and divide by design output of an anode in this type soil. This will give the number of anodes required; to find spacing, divide number of anodes into the length of line.

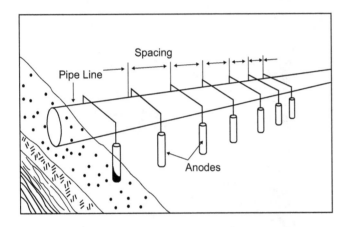

Example. Find the number of 17-lb. magnesium anodes needed and spacing used to protect 10,000 ft of bare 4-in pipeline in a corrosive soil having a resistance of 1000 ohm per cubic centimeter. Assume a current requirement of 1 mA/ft^2, and that the anode output curve shows an output of 100 mA per anode in this type soil.

$$\text{Area of pipe} = \frac{4 \times 3.14}{12} \times 10,000 = 10,500 \text{ ft}^2$$

Total current requirement $= 10,500 \times 1.0 = 10,500$ mA

$$\text{Anodes required} = \frac{10,500 \text{ mA}}{100 \text{ mA/anode}}$$

$$= 105 \text{ anodes distributed}$$

$$\text{Spacing} = \frac{10,000}{105} = 95 \text{ ft}$$

Similarly, find the number of anodes required to protect ten 100-ft long "hot-spots" on a 4-in line assuming a 100-mA output from a 17-lb magnesium anode. Estimate as follows:

$$\text{Area of each hot} - \text{spot} = \frac{4 \times 3.14}{12} = 105 \text{ ft}^2$$

$$\text{Anodes required} = \frac{105 \text{ ft}^2 \times 1.0 \text{ mA/ft}^2}{100 \text{ mA per anode}} = 1.05$$

One 17-lb magnesium anode will produce approximately enough current in 1000-ohm soil to protect one of these "hot-spots."

Ten anodes are required for "hot-spots" on this section of line.

Note: For these rules, the protective criterion is -0.85 volts pipe-to-soil potential using a copper-copper sulfate reference electrode; or a negative change of potential of as much as -0.25 volts. Soil and pipeline conditions vary widely, so consult an experienced corrosion engineer in the design of a complete cathodic protection system.

How can resistivity of freshwater be determined from chemical analysis?

Divide the total solids expressed in parts per million into 500,000; the result is the resistivity in ohm-centimeter units.

Example. Total solid content is 400 ppm.
500,000 ÷ 400 = 1250 ohm centimeters resistivity.

This rule of thumb will give very accurate results for any naturally occurring freshwaters; severely contaminated waters may differ markedly, however, depending on the kind of contamination.

What will be the resistance to earth of a single graphite anode?

For the most common size, 3-in by 60-in, installed in 10 ft of coke breeze backfill in an 8-in hole, multiply the soil resistivity in ohm-cm by 0.002; this will give the resistance to earth in ohm.

Example. 1800 ohm-cm soil

$1800 \times 0.002 = 3.6$ ohm

This assumes that the resistivity is uniform to considerable depth; if it increases with depth, the resistance will be slightly higher.

How to estimate the monthly power bill for a cathodic protection rectifier?

Multiply the rectifier output voltage and current together, and multiply this by 11/2; the result is the power consumption for one month, in kilowatt-hours. If the unit is on a meter of its own, figure the bill by the utilities rate schedule; if it represents an additional load, use the rate per kilowatt-hour for the highest bracket reached.

Example. The terminal voltage of a rectifier is 10.2 volts, and the current output is 21 amps.

$10.2 \times 21 \times 1.5 = 321.2$

The power consumption will be about 320 kwh per month; at 3 cents this would amount to $9.60, at 1.1 cents it would be $3.52, etc.

Note that the actual current and voltage values are used; if the rating of the unit is used, the result will be the maximum power consumption within the range of the equipment. The formula is approximate; it is based on an efficiency of about 48%.

What will be the resistance to earth of a group of graphite anodes, in terms of the resistance of a single anode?

When the spacing is 20 ft (the most common value), the resistance of the group will be equal to the resistance of a single anode, divided by the square root of the number of anodes.

Example. Resistance of 4 anodes = ½ resistance of single; resistance of 16 anodes = ¼ resistance of single, etc.

This rule is reasonably accurate for soils that are uniform to a considerable depth; if the resistivity increases with depth, the "crowding" effect will be greater, whereas if it decreases, the individual anodes will not affect each other so much.

How can the current output of magnesium rod used for the cathodic protection of heat exchanger shells be predicted?

Divide the total dissolved solids content of the water by 13; the answer is the current output per ft of 1.315 in (1-in pipe size) magnesium rod to be expected.

This is an approximation to the current flow that will be found after stabilized conditions have been reached, if

protection is obtained. If insufficient magnesium is installed, the current flow will probably be greater, and if too much is installed, the current will be less.

What spacing for test leads to measure current on a pipeline?

Multiply the pipe weight, in lb/ft, by 4; the result is a spacing that will give a resistance of 0.001 ohm, or 1 amp per millivolt IR drop.

Example. Pipe weight is 40.48 lb/ft
$40.5 \times 4 = 162$
Install test leads at 162 ft spacing.

The rule works for welded lines of grade B seamless; variations in steel may introduce errors as great as 15%. It is always better to calibrate a line section if accurate results are to be had, but if the above rule is used, the spacing will not be too small for satisfactory measurement, or too great for convenience in installation.

How many magnesium anodes are needed for supplementary protection to a short-circuited bare casing?

For 17-lb anodes, use A ρ/50,000 to give the number needed for the casing alone; A is the area of the casing, in square feet, and ρ is the soil resistivity in ohm-centimeters. The life of the anodes may be estimated from ρ/180. If the life in years, as given by this expression, is too short, refigure for 32-lb anodes, using A ρ/55,000 for the number and ρ/100 for the life. If this is still too short, figure for 50-lb anodes; number is A ρ/60,000 and life is ρ/65.

Such an installation, grouped on both sides of the line at each end of the casing, will enable the main line protection unit to carry beyond the casing; the values obtained above are approximate only, and should be checked by potential measurements after installation.

Example. Suppose the pipeline has shorted out to 45 ft of 24-in bare casing. Soil resistivity is 1100 ohm-cm.

How many 17-lb anodes would be required for supplementary protection?

$$\text{Number of anodes} = \frac{A\rho}{50,000}$$

$$A = 2 \times 3.14 \times 45 = 283$$

$$\text{Number of anodes} = \frac{(284)(1100)}{50,000}$$

$$= 6.21$$

Place three anodes at each end of casing. Their life expectancy will be:

$$\frac{1100}{180} = 6 \text{ years}$$

Group installation of sacrificial anodes

When anodes are installed in groups, a spacing correction factor must be taken into account to determine the current output. The chart below shows the correction factor for multiple anode installations.

Example. Single anode 8-in by 60-in in 1,500 ohm soil.
Resistance = 4.849 ohm
Potential = 0.85 volts

$$\text{Current output} = \frac{E}{R} = \frac{17.5 - 0.85}{4.849} = 0.185 \text{ amps}$$

Current output of 5 anodes on 15 ft spacing under above conditions.

$$\text{Combined resistance of group} = \frac{4.849}{5} \times 1.22$$

Correction factor = 1.17 ohm

$$\text{Total current output} = \frac{1.75 - 0.85}{1.17} = 0.769 \text{ amps}$$

How can the life of magnesium anodes be predicted?

Multiply the anode weight in pounds by 57, and divide by the current output in mA.

Example. A 17-lb anode has a current output of 83 mA. What is its projected life?
17 × 57/83 = 11.7.

The projected life is 11.7 years.

The rule is based on 50% current efficiency, and on stabilized conditions. If more anodes are to be added, or if time has not been allowed for polarization (1–4 weeks), then the current output will probably fall off, and the actual life will be 10 to 40% longer than indicated by the rule.

How to find the voltage rating of a rectifier if it is to deliver a given amount of current through a given ground bed (graphite or carbon)?

Multiply the desired current, in amperes, by the total resistance of the ground bed, in ohm (either calculated or measured); add 2 volts to overcome the galvanic difference between graphite and protected steel.

Example. A current of 20 amps is desired from a ground bed whose resistance is 0.9 ohm.

(20 × 0.9) + 2 = 20 volts

Specify the next larger size, which will probably be (depending upon the manufacturer) 24 volts.

Note: Where the resistance is measured rather than calculated, and is the total loop resistance between pipe and ground bed, this method is quite precise; for calculated values, caution should be observed.

Determining current requirements for coated lines

Survey along the line after a polarization run of three or four h, and locate the two points whose pipe-to-soil potentials are 1.0 volt and 0.8 volt (see Figure 1). The section of line lying between these two points will then be at an average potential of about 0.9 volt, which is approximately the average of a line protected with distributed magnesium anodes. Then, if, the line current is measured at each of these two points, the difference between these two line currents will be the total amount of current picked up on the line in the section. This quantity, divided by the length of the section, will then give the current requirements of the line, in amperes per mile. Naturally this figure can only be used for an entire line if

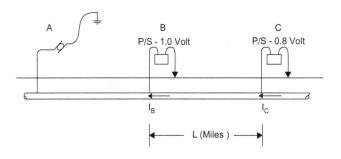

the section chosen is truly representative; judgment must be used at this point.

Determine current requirements for coated lines when pipe-to-soil potential values are estimated

If it is not possible or convenient to find the points that have the exact values of 1.0 and 0.8, then readings can be taken that approximate these values, and both potential and line

current can then be adjusted to find the desired quantity; this is illustrated in Figure 1; readings were taken at points "A" and "B," and, by assuming a static potential of 0.6 volt and

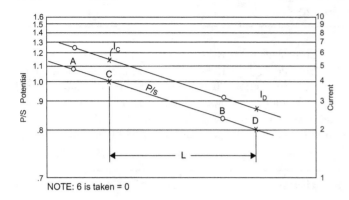

NOTE: 6 is taken = 0

linear attenuation, it is determined that points "*C*" and "*D*" have the values of 1.0 and 0.8; since line current varies according to the same laws as potential, it is possible to determine the current values at the two latter points, and so to arrive at the desired quantity.

It is not to be expected that the estimate thus made will be highly accurate; it is intended to be merely preliminary.

HVDC effects on pipelines

Latest methods to calculate quantitative characteristics of the earth's gradient and high voltage direct current effects on pipelines.

Thomas B. Hays, George W. Silkworth, Robert D. Schilling, Cornell, Howell, Hayes & Merryfield, Inc.

A pipeline collects, conducts, and discharges electric current of High Voltage Direct Current (HVDC) origin as a result of the electric field established by the DC current transmitted through the earth and by the electrical properties of the pipeline itself.

The electric field

It is assumed that the electrical properties of the earth and pipeline are uniform and that the HVDC effects are isolated from other stray or induced current effects. This assumption makes possible the analysis of this portion of the total protection problem and assists in visualizing the unique effects of HVDC earth currents.

Ultimately, the total electrical current conducted by a pipeline will be the sum of all the various components. This includes protective current, telluric current, and other stray or interference currents.

Conduction of a direct current through the earth is an electrolytic phenomena, characterized by an electric field that surrounds each of the two earth electrodes, and is directly proportional to the HVDC earth current (Figure 1). When the electrodes are spaced several hundred miles apart, the field appears to be concentric around each ground electrode.

The fields surrounding the opposite electrodes are of the opposite polarity. If the polarity of the ground current reverses, the surrounding electric fields will also reverse. The potential at any point in the electric field with respect to remote earth (*V*), due to an electrode current (*I*) at a distance

Figure 1. Conduction of a direct current through the earth is an electrolytic phenomena, characterized by an electric field that surrounds each of the two earth electrodes, and is directly proportional to the HVDC earth current.

(r) from an electrode in a homogeneous earth of the resistivity (ρ) is found by the relationship:

$$V = \frac{\rho I}{2\pi r}$$

This relationship shows a simplified representation of what actually occurs. The earth's resistivity is not uniform at all depths or horizontal distances, which causes irregularities to the ideal voltage distribution existing around an electrode in a homogeneous earth.

To refine the earth current analysis and to account for different resistivities, computer techniques have been developed to estimate the earth's gradients existing for conditions of multilayers of earth of different resistivities. Pipeline effects have also been predicted by this method. Reasonable agreement between these models and field data has been reported. Techniques are available to calculate the quantitative characteristics of both the earth's gradient and HVDC effects on pipeline.

HVDC effects on a pipeline

Whether the HVDC component of the total current existing on the line causes a corrosion problem depends on the level of protection along the line and the effect of the added component to the level of protection. Figure 2(a) illustrates the HVDC component of the total conducted current along the pipeline for a given separation distance under the condition where the HVDC electrode is collecting current from the earth.

In determining protection requirements for the pipeline, the characteristic of the current transfer relationship is important (Figure 2(b)). In those regions of the pipeline where the current transfer is from the pipeline to the soil—along the center in this case—corrosion could occur if the HVDC component of the total transfer current to the soil is great enough to exceed any local level of protection against corrosion.

The pipeline HVDC arrangement shown in Figure 2(a) will serve as a basis to develop and illustrate the performance requirements for protecting a pipeline from the detrimental effects of HVDC earth currents. The hypothetical problem, as shown, is one that may be encountered in the field. In this example, the following assumptions are made:

- The pipeline is long, straight, electrically continuous, and uniform in wall thickness, diameter, material, and coating.
- It is approximately 30 miles away from the nearest HVDC earth electrode.
- The earth's resistivity is uniform over the general area.
- The maximum value of the HVDC earth current in either polarity causes the pipeline effects.

While these conditions may not exist in any actual pipeline HVDC system situation, they will serve to identify and evaluate the requirements for protection against HVDC effects in an actual situation.

Figure 3 shows the range or envelope of effects as the HVDC varies from its maximum value in one polarity to the maximum value of the other polarity. The coordinates on this graph are DISTANCE along the X axis and VOLTAGE on the Y axis.

The voltage corresponds to the effect that would usually be measured on an actual pipeline: pipe-to-soil voltage measured with a copper-copper sulfate half-cell. The value of the voltage effect at any point on the line is the product of the current transfer, and the electrical resistance between the pipe material and earth over the same area.

Figure 3 also shows reasonable hypothetical scale values illustrating the various combinations of protection requirements that may exist in an actual installation. The voltage values give the change in pipe-to-soil reading that would be caused by the HVDC transfer current. If this HVDC is

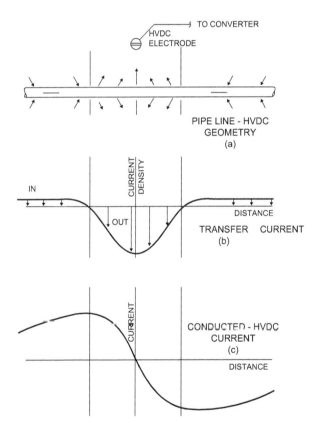

Figure 2. (a) Geometric plan between the HVDC electrode and the pipeline. Arrows indicate the direction of the HVDC component current. These would be reversed if the polarity of the electrode were reversed. (b) Curve shows characteristics of the transfer of HVDC into and out of the pipeline along its length. The transfer current is equivalent to current density. (c) Conducted current in the pipeline, which results in the summation of the transfer current along the pipe.

considered as the only source of corrosion, then a pipe-to-soil voltage reading with a copper–copper sulfate half-cell would be approximately 0.85 volts at no transfer current. This scale is shown on the left side of the figure. Corrosion occurs for any reading below this axis, since the current is leaving the pipeline and entering the soil to cause this change in the pipe-to-soil reading.

At locations on the pipeline where HVDC causes conditions that increase the volt-meter reading, the current is collecting on the pipeline. The horizontal axis, therefore, represents a corrosion limit. Above this axis, a coating disbonding limit can exist where the density of the current transferred from the earth to the pipeline causes gas evolution, which may blister the coating.

In this example, a pipe-to-soil reading with a copper–copper sulfate half-cell of 1.5 volts is shown as the coating disbonding limit. The difference between these limits gives an assumed safe operating band width of 0.65 volts. Values other than these may be used for specific applications. These limits, however, will serve to illustrate the principles.

Protective system limits

The requirements for protection along the line from either corrosion or disbonding are developed from examination of Figure 3, and the results are plotted in the subsequent figures. The upper limit of protection is the maximum

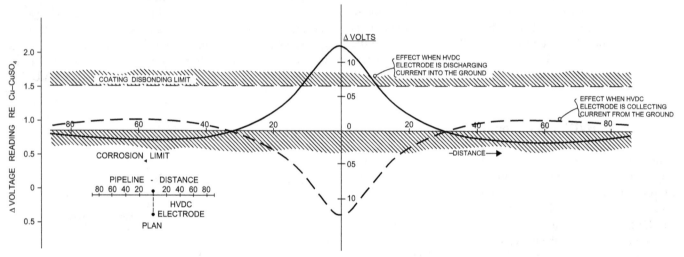

Figure 3. Range or envelope of effects which an HVDC component can have along the length of a pipeline as the HVDC varies from its maximum value in one polarity to the maximum value of the other polarity.

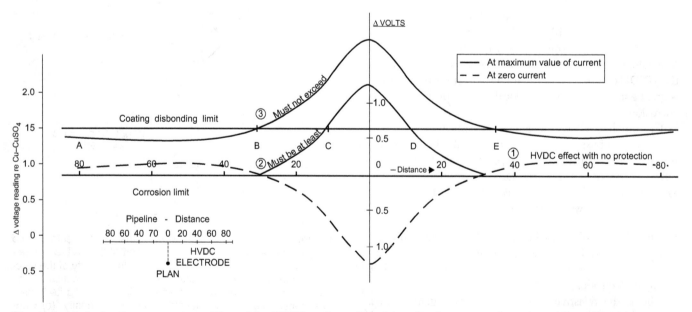

Figure 4. Limits for the assumed condition of the HVDC system when it is collecting current from the earth at the highest value possible.

amount of current that can be collected by the pipeline without exceeding the disbonding limit. The lower limit is the minimum amount of current that can be collected by the pipeline to keep it above the corrosion limit.

At this value of HVDC (Figure 4), the uncorrected pipeline effect is shown by Curve 1. No protection is needed along either end of the pipeline in Regions *A-B* or *E-F*. Corrosion protection is needed between *B* and *E*. The minimum protection effect along the line is shown by Curve 2. The resultant of the HVDC and this protection effect lies on the corrosion limit. The maximum protective effect that can be applied along the line, without exceeding the coating disbonding limit, is shown by Curve 3. Under this maximum value and polarity of HVDC current, the region between these two protective limits, therefore, defines the range of acceptable protective system performance at any location on the pipeline.

When the HVDC effect decreases, due to a reduction in the ground current, the required level of protection on the line also decreases. Along the pipeline length from *C* to *D*, the protection *must* be reduced to avoid exceeding the disbonding limit. At zero HVDC effect, the protective system effect must fall between the corrosion and disbonding limits all along the line.

Examination of the limits for these two values of the HVDC system operation shows that the protective effect must be varied in response to the HVDC effect over the length of the pipeline from *C* to *D*: automatic regulation of the protective current is needed over this length of the

pipeline. With only this polarity of HVDC current, a fixed level of protection can be applied along the pipeline from *A* to *C* and from *D* to *F*.

When the earth's current can reverse polarity, as with the Pacific Northwest (PNW)-Pacific Southwest (PSW) intertie, then the same process that was used to develop the limits, shown in Figure 4, can be repeated for the opposite HVDC polarity. The limits on protection along the pipeline when the HVDC electrode is discharging current, at its maximum value, into the earth are shown in Figure 5.

The combination of these two sets of limiting characteristics is shown in Figure 6. The graph shows the length along the pipeline where a fixed level of cathodic protection can be used and those regions where controlled protection is required for any value of HVDC earth current. In this particular example, variable protection is required along the pipeline for approximately 20–30 miles on either side of the point closest to the HVDC electrode. Cathodic protection is required when the HVDC electrode is collecting current from the earth. That is, the HVDC component of current, conducted by the pipeline, is discharging from the line near the earth electrode. The required protective current along the pipeline to counteract the HVDC effect under this condition is show as Region 1 in Figure 6.

When the HVDC polarity is reversed and the magnitude of the ground current is increased to the maximum value, a different problem is encountered. The electric field which is established around the electrode and pipeline causes current to enter the pipeline through the coating. In this example, the

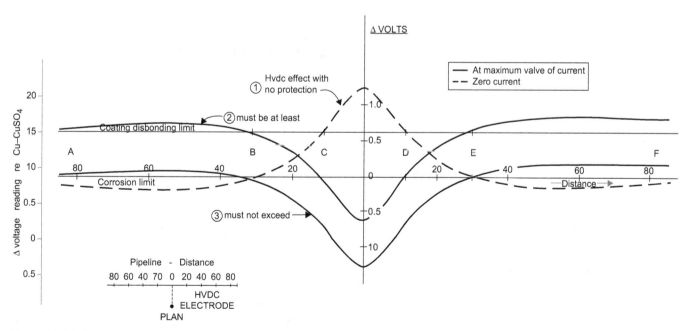

Figure 5. Limits on protection along the pipeline when the HVDC electrode is discharging current, at its maximum value, into the earth.

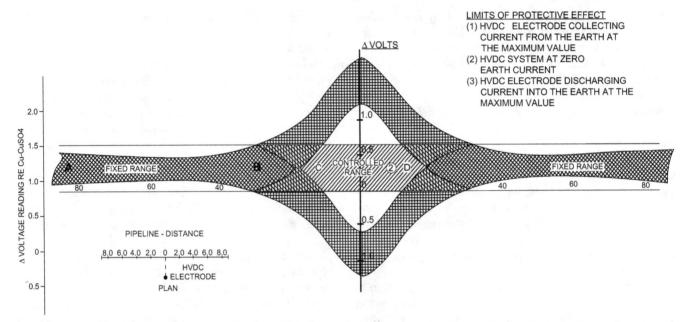

Figure 6. Combination of Figures 4 and 5. The length along the pipeline where a fixed level of cathodic protection can be used and those regions where controlled protection is required for any value of HVDC earth current. In this example, variable protection is required along the pipeline for approximately 20–30 miles on either side of the point closest to the HVDC electrodes.

disbonding limit is exceeded over the length of the pipeline from *C* to *D* at the maximum value of HVDC earth current. The required correction for this detrimental effect is similar to the effect for the previous HVDC polarity: namely, the uncorrected electric field must be modified by an added field

to give an acceptable total. Theoretically, in this situation, a portion of the HVDC earth current can be collected in a protective ground bed and conducted to the pipeline to satisfactorily modify the original field into an acceptable one.

Troubleshooting cathodic protection systems: Rectifier-ground bed

When the monitor, or a pipe-to-soil potential survey, indicates inadequacy of protection, the first place to look is at the protective unit. The current output of the rectifier should be checked; if it is normal, the trouble is on the line itself; if it is high, and accompanied by low voltage, the trouble is certainly on the line, and is caused either by increased current demand or by a short-circuit to parasitic metal. If the current output is low, with normal or high voltage, the trouble is in the ground bed or connecting cables. A pipe-to-soil potential over the ground bed will show a peak over every anode that is working; a disconnected anode will not show at all. This test is particularly useful if comparison can be made to a similar test made at the time of the original installation. When inactive anodes are found, only digging will uncover the cause.

If the rectifier and its anode bed appear to be performing satisfactorily, the source of the low potentials must be

sought on the line itself. Any area in which work has been recently done should be investigated; for example, if a new lateral has been connected, the insulation should certainly be checked. If investigation of such suspected sources discloses nothing, then a more detailed search must be made. First, the pipe-to-soil potentials should be studied, to see if the failure seems to be localized. A more thorough, but slower, approach is that of making a detailed line current survey; find out where the drained current is coming onto the line. This will be much more easily interpreted if a similar survey, made when the line was in satisfactory condition, is on record; a comparison will often quickly locate the offending parasite. On the other hand, if the current collected in each section is greater than on the earlier survey, with no pronounced differences, then the trouble is simply that of increased overall current requirement, probably due to coating deterioration.

How to control corrosion at compressor stations

Eight factors design engineers should consider

R. E. Hodge, Superintendent of Corrosion Control, Natural Gas Pipeline Co. of America

Corrosion failures at compressor stations result from carelessness on the part of the user, or poor choice of material/configuration by the designer.

Design engineers should consider the following:

- Coat everything underground, except ground rods, with a coating properly selected for the conditions.
- Use proper coating application and inspection.
- Use ground rods anodic to steel and insulated (coated) ground wires.
- Avoid shielding as much as possible and allow space for corrosion control additions.
- Use a minimum of insulating fittings.
- Protect against water system corrosion.
- Select materials carefully, with corrosion in mind, as well as the more obvious considerations.
- Do not retire facilities in place—take them out.

Coatings

Coatings are one of the most important considerations for control of underground corrosion. Generally, all underground metallic structures, except ground rods, should be coated. This includes gas piping, control lines, tubing, water lines, conduit, air lines, braces, etc.

There is enough buried conduit around a compressor station to soak up large amounts of cathodic protection if not coated. Nonmetallic carriers can be used in some cases instead of coating.

Proper selection and use of coatings are important. The best of coatings are no good if used in the wrong place or improperly applied. The emphasis here is on the selection of coating for the temperature and environment expected.

For gas discharge lines, temperature becomes a dominant consideration. Be sure to get a coating that will withstand gas discharge temperature. Be careful, because sag temperatures listed in coating literature are not maximum operating temperatures. Very few coatings on the market can withstand operating temperatures above 120° without damage. Any bare metal underground becomes a drain on the cathodic protection system and a potential shielding problem.

There are other areas for coating consideration besides underground (e.g., tank interiors and atmospheric coatings). These must also be matched to the situation.

Grounding

Grounds should be of materials anodic to steel. These might include magnesium or zinc anodes, or galvanized ground rods.

Copper ground rods should not be used since it is cathodic to steel. Some compressor stations have been installed with massive bare copper grounding systems. These are large galvanic cells with the steel gas piping as the anode (corroding element). Bare ground wires and rods also take cathodic protection current. This bad situation can be avoided by using galvanized rods, for example, and using copper ground wire with insulation. The junction of copper ground wire and galvanized ground rod must be coated. Grounding systems should be checked periodically for proper functioning.

Shielding

Shielding poses a major problem at compressor stations. It could be described as the blocking of cathodic protection due to geometric configuration.

Congested areas, where buried metal concentration is high, are difficult to cathodically protect. Piping, or other metal structures, will tend to shield cathodic protection currents from reaching all of the buried structures to be protected.

Cathodic protection ground beds of various styles have been used at compressor stations. Remote conventional beds and deep beds are popular for many applications. However, we have found that distributed anodes seem to be the best way to get adequate cathodic protection to all the shielded areas in a station. This type of bed can be described as a ground bed with the anodes distributed in a nongeometrical pattern, interspersed through with piping.

Designers should allow space between lines where possible; avoid congestion. This would cause less shielding and allow space for distributed anode installations where required. It would also allow space for access to work equipment for anode installations.

Insulation

Use insulating fittings (flanges, couplings, unions, etc.) only where absolutely necessary.

Hold insulating fittings to a minimum. These can often cause more problems than they solve. You might want to insulate water wells from the adjacent piping for control of cathodic protection current to the wells. This may not even be necessary if distributed ground beds are used.

Some companies insulate compressor stations from the lines entering and leaving the station and other companies do not. This is a matter of individual company preference.

Water systems

Each water system (jacket water, piston water, boiler water, etc.) should have corrosion monitoring provisions designed into the system, such as coupons or corrosion rate probes. If there is a gas treating plant in conjunction with the compressor station, corrosion monitoring provisions should be designed into that system also.

Water treatment for corrosion control would have to be considered, depending on individual circumstances. Some treatment facilities may be designed into the systems.

Provisions should be considered for cathodic protection of the internal surfaces of storage tanks and water softeners. One problem often encountered on smaller tanks is adequate openings. Enough openings of sufficient size should be allowed on any tank for possible future work.

Water should be kept out of some systems. For instance, hair dryers may be used on control lines.

Materials selection

Proper selection of materials is important throughout design.

Two key considerations from a corrosion standpoint are: (1) avoid the use of dissimilar metals; (2) match the material to its environment. Examples of these would include the use of a brass valve in a buried steel water line or use of mechanically strong bolts in the wrong electrolyte, resulting in subsequent failure from stress corrosion cracking.

Select materials carefully, with corrosion in mind, as well as the more obvious considerations. If design engineers do this, many subsequent failures can be eliminated.

Retirements

Many compressor station designs involve retiring something that is already there.

If something must be retired, especially underground, take it out. Many times a company unnecessarily pays to continue protection of facilities retired in place.

Even if the retired facilities can be completely disconnected from the protective systems, shielding is still a problem. Shielding is a problem by foundations also, so if they are retired, they should be removed.

Thus, attacking the corrosion problems requires a truly interdisciplinary effort. In fact, corrosion offers one of the richest fields for interdisciplinary activity.

(Based on a paper, "*Construction Design to Facilitate Corrosion Control—Compressor Stations,*" presented by the author at AGA Distribution Conference, May 8–10, 1972, Atlanta, Ga.)

Project leak growth

The numbers of leaks in older pipeline systems tend to increase exponentially. Exponential growth of leaks will fit the following equation:

$$N = N_1 e^{nA}$$

where:

N = Number of leaks for the study area in any year considered
N_1 = Number of leaks in the first year for which leak data are available
e = 2.718
n = nth year for which the number of leaks is to be estimated
A = Leak growth-rate coefficient

The coefficient A can be calculated from the graph drawn of the leak data for the study period. Preferably, the study period should be in the range of 5–10 years. Any data less

than this will not give enough points to accurately determine the coefficient A. Before a graph is drawn, leak report quantities must be adjusted for any changes to the system during the study period. In other words, if a township was the study area, then pipe replaced, added, abandoned, or installed should be tallied. If this is not done, errors will result. For example, if leak data were collected on a township basis, and during the study period considered, 10% of the township's pipe was renewed, leak frequency in the old area will continue to rise while in the renewed area there should be none, possibly giving an overall number of leaks less than that of the year prior to the replacement. For an accurate picture and a determination of the leak growth-rate coefficient, the number of leaks in the years following the replacement should be accounted for as follows:

Let x = Percentage of replacement completed.
Let y = Actual number of leaks for the remainder of the township for the year considered

Adjusted number N = y/(1 − x).

The leak growth rate for some pipelines may not follow the exponential curve.

For example, in a particular case, a linear increase with time may be predicted. In that case, use a linear equation instead of the exponential form.

Example.

N_1 = Number of leaks in first year (i.e., 1979) = 140
N = Number of leaks in year considered (1986) = 310
n = 86−79 = 7
$310 = 140e^{7A}$

Hence, coefficient A = 0.113. The estimated number of leaks in 1990 will be:

$$N = 140e^{11 \times 0.113} = 485 \text{ leaks}$$

Source

Ahmad, Hayat, *Gas Pipeline Renewal Insertion Technology*, 3 and 4 (1990).

ADVANCES IN PIPELINE PROTECTION[b]

Specialized corrosion surveys for buried pipelines: methods experience

M. D. Allen, CEng, MIM, MICorrST; **N. R. Barnes**, BTech, CEng, MIM, MiCorrST Spencer and Partners, U.K.

Synopsis

The external corrosion protection of a pipeline is commonly achieved by a system comprising an insulating coating and cathodic protection. The performance of this system is normally assessed by regular monitoring of pipe to soil potentials at selected intervals along the pipeline. Three more detailed pipeline survey procedures are available to provide additional monitoring information and the case histories described indicate the experience and value gained from their selective use.

Introduction

The external corrosion protection of a buried steel pipeline is generally achieved by a system comprising an insulating coating and cathodic protection. On modern pipelines, the coating is regarded as the primary means of protection with the cathodic protection systems providing control of corrosion where the coating has failed or has been damaged.

The performance of the cathodic protection system is usually assessed by regular monitoring of pipe-to-soil potentials at selected intervals along the pipeline.

From these results, conclusions may be drawn concerning the level of cathodic protection being achieved and by inference the performance of the insulating coating. Under most circumstances, such measurements at selected locations will provide an acceptable indication of the overall level of corrosion protection being achieved on the pipeline.

These measurement locations, however, commonly at least 1 km apart, are generally selected for ease of access and thus only provide a valid assessment where pipeline coating is consistent, ensuring a constant attenuation of cathodic protection levels between each measurement point. Where the coating is variable in quality, it cannot be assumed that satisfactory levels of protection are maintained between measurement locations and a deficiency in the protection system may remain undetected allowing corrosion to occur.

Coating defect surveys have also been used to identify areas where pipeline coating quality is questionable. Historically, these surveys have almost universally taken the form of Pearson-type surveys. These surveys have entailed a two-man survey team walking the length of a pipeline identifying coating defects and subsequently marking the points with pegs for further investigation.

Surveys of this type have two inherent disadvantages, namely that the survey procedures are time consuming and no indication of cathodic protection levels at the defect location is obtained.

With these shortcomings in mind, two other survey procedures have been developed, the signal attenuation coating survey and the close interval potential survey. The former provides a rapid assessment and record of pipeline coating condition and more readily identifies areas where a defect or defects are present. Identified areas may then be surveyed in detail by close interval potential techniques that cannot only confirm coating defect locations, but also establish the status of the cathodic protection system at close intervals throughout the chosen areas.

---

[b] Paper first presented at the 7th International Conference on the Internal and External Protection of Pipes. Held in London, England. Organized and sponsored by BHRA, The Fluid Engineering Centre © BHRA, The Fluid Engineering Centre, Cranfield, Bedford, MK43 OAJ, England 1987.

Methods of locating coating defects

Pearson survey

The Pearson survey is an aboveground technique used to locate coating defects on buried pipelines. An AC signal is injected onto the pipeline from a power source connected between a cathodic protection test post or valve and an earth pin. The AC frequency selected is dependent upon the type of pipeline coating in order to minimize current loss from the pipeline through sound coating. At coating defects, the AC signal leaks to earth and flows via the soil to the earth spike to complete the circuit. The localized current flow from the pipeline at the defect creates a potential gradient through the soil, and this is detected by measurement at a ground surface.

In practice, this is achieved by two surveyors walking in tandem over the entire pipeline, or section of pipeline. Earth contacts, either ski poles or foot cleats, are connected to a receiver via a cable harness. As the first surveyor approaches a coating fault, an increased voltage gradient is observed, either audibly through headphones, or visually on the receiver signal level meter. As the lead surveyor passes the defect, the signal fades, and then increases again as the rear surveyor's contacts pass over the defect.

The defect may be recorded on a record sheet, prepared beforehand, complete with a measured distance from a fixed reference point or indicated by a marker peg or nontoxic paint. A recent development in Pearson equipment has been the introduction of a system that continuously records, through data logging equipment, the potential difference between each earth contact. In this way, data may be graphically presented.

Signal attenuation coating survey

This method is similar in many respects to the Pearson survey, the essential difference being the method of defect detection. The AC current injected onto the pipeline creates a magnetic field around the pipe, the strength of this field being proportional to the amplitude of the AC signal. Measurements of the field strength are made by recording the voltage generated in a sensing coil and indicated on a dB scale, when the coil is held over the pipeline. Loss of current from a coating defect creates a change in the field strength due to a change in the pipe current and this is detected as a change in output from the sensing coil.

In practice, this is achieved by one surveyor measuring field strengths at discrete, preset intervals. At each location the field strength and pipeline chainage are recorded. A graph of distance versus field strength may then be produced. Sections of pipeline having poor coating quality are indicated by the steepest gradient on the line plot. The results may also be expressed as histograms to aid interpretation. Typical results are indicated in Figure 1. Any area indicating high signal losses may then be resurveyed using the same procedures to more closely locate the coating defect(s).

Close interval potential survey

The close interval potential survey is an aboveground technique used to provide a detailed profile of the pipe-to-soil potential with distance. This profile can be used to determine detailed information concerning the performance of the cathodic protection system, the coating system, and to a lesser degree, interaction effects. The advent of portable microprocessor-based measuring equipment and computer-assisted data handling, together with accurate synchronized high-speed current interruption devices, have allowed these surveys to provide accurate, practical data and collection methods.

To eliminate errors in the pipe-to-soil potentials recorded, arising from the inclusion of volt drops caused by the flow of cathodic protection current in the soil, synchronized current interrupters are employed in the cathodic protection units. Pipe-to-soil potentials are then instantly measured after the current is interrupted, before significant chemical depolarization has occurred, thus providing an "error free" polarized potential.

In practice, the precision current interrupters are best fitted in the DC output circuit of the cathodic protection power supplies, prior to commencing the survey to interrupt the cathodic protection current for short cyclic periods. The cycle time is chosen with regard to pipeline characteristics, although a range of "ON" and "OFF" periods and "ON/OFF" ratios is available.

For this survey, a variety of purpose built-voltmeter/microprocessors are available that record and store the pipe to soil potential data with an additional facility to store coded entries of topographical or pipeline features. This information is transferred to floppy disc storage from which graphs of ON and OFF pipe-to-soil potentials against pipeline chainage are produced. A typical close interval potential graph is illustrated in Figure 2.

It will be apparent from the description of the pipeline surveys that the form of results is different for each survey.

Pearson

Results are not generally presented in graphical form. Coating defects are marked in the field; in addition, the detection of defects is dependent upon operator skill. Data can only be presented in graphical form if the recording Pearson technique is used. At present, experience of this technique is limited.

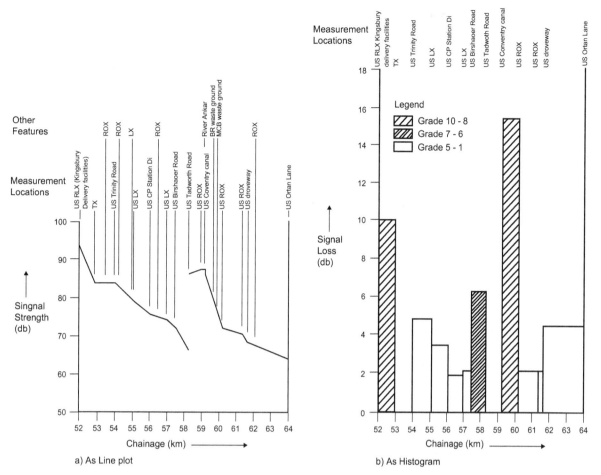

Figure 1. Typical signal attenuation coating graphs.

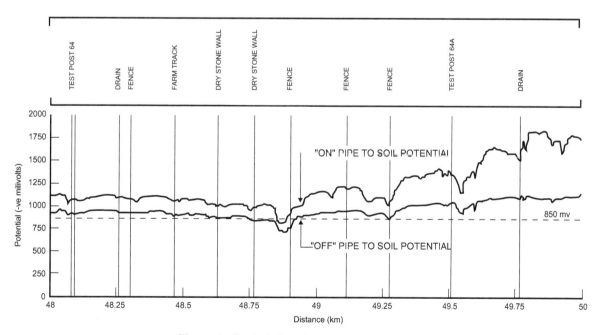

Figure 2. Typical close interval potential graph.

Signal attenuation coating survey

Results are presented in graphical form, although quick analysis may be performed in the field from raw data. Areas of poor coating quality may be readily identified from final graphs.

Following the production of graphs and histograms, an evaluation system employed in pipeline coating quality assessment may assign values to the level of signal loss and the measurement distance interval. The sum of the two values referred to as the "dB factor" and the "km factor" may be used to indicate the future action required and the level of priority for each section surveyed.

The evaluation of the coating quality is illustrated in Figure 1. It should be noted, however, that gradings assigned to one particular pipeline cannot be compared to those assigned to another.

Close interval potential survey

Results are shown in graphical form. Defect location is undertaken after the survey and requires experienced personnel to interpret the results.

The achievement of cathodic protection on the pipeline is indicated by the "ON" and "OFF" potential, both being more negative than the accepted criteria.

Pipeline locations with inadequate cathodic protection will be indicated by the graph "ON" and "OFF" potentials shifting to a less negative level and become cause for concern if the "OFF" potentials are more positive than, say, −0.85 V.

It is considered that a localized coating defect would normally be indicated by a sharp positive shift of the "ON" potential plot coupled with a plot shift of smaller amplitude of the "OFF" potential, i.e., that the potential difference (IR drop) between the "ON" and "OFF" reading decreases, indicating a reduced coating resistance at that point.

The "OFF" potential plot change would not be as large as that of the "ON" potential, as the "OFF" potential would be the polarized (or partially polarized) potential of the bare steel. This differs from the case where insufficient current for the achievement of polarization is available.

In all these cases, the IR drop measured between the "ON" potentials and the "OFF" potentials includes, along with the resistance of the coating, the resistance of the soil. The resistance of the coating is assumed to be the largest component of this resistance value.

Current is also a factor of the IR value. The current density impressed on the pipeline will increase as the cathodic protection stations are approached, giving the largest "I" value of the IR drop near the location of the negative connection.

In order to interpret the data fully obtained, the anomalies observed in the plotted results may be classified into a number of categories relating to "point" anomalies or "general area" anomalies. For example, the following categories may be used.

Classification	Definition
Type I	Both the "ON" and "OFF" potentials are more positive than preset criteria. There is little or no "IR" voltage. The potentials are unsatisfactory, and the locations are not cathodically protected.
Type II	The "OFF" potential is more positive than a preset limit, the "ON" potential more negative. There is a reduced level of protection indicated at "point" locations, which may be improved by increased output from the nearest cathodic protection station, although where this occurs over long pipeline lengths, it indicates high-resistance coating or soils.
Type III	Positive change in the "ON" and "OFF" potential plot where both remain more negative than the selected criteria. This indicates increased current demand at these locations, but protection is maintained at a satisfactory level.
Type IV	Unusual potential variations due to outside influences, e.g., stray currents, station earthing, AC induced voltages.
Type V	A change in level of the "ON" potential at a test post due to the change in the length of that part of the measuring circuit passing current, i.e., increasing "IR" volt drop due to increasing nine length in circuit.

Of the five categories listed above, the first three are indicative of coating defects. Types IV and V are commonly observed on close interval potential survey results and require careful interpretation.

It is important that these considerations are born in mind when selecting the survey or combination of surveys to be employed and, in addition, the choice of which survey to employ and when to employ it is dependent on a number of factors, namely:

- the purpose of the survey
- the condition of the pipeline
- the access over the pipeline
- the time allowed.

As the protection system relies on both the external coating and a cathodic protection system, these may be considered as first and second lines of defense against corrosion. It is possible to take the position that concern over damaged coating is not warranted as backup from the cathodic protection system is available, and therefore the integrity of the pipeline is not at risk. If this attitude is taken, then obviously close interval potential surveys would seem the most appropriate and perhaps the only survey system to consider.

Experience has shown, however, that the repair of significant coating defects will benefit the overall level of cathodic protection achieved, possibly permitting reduction in output levels and/or increasing the distance protected from anyone protection station. In addition, the maintenance of a "two-level" protection system also ensures the highest integrity is maintained for the pipeline. With this purpose in mind, the advantages of the coating surveys become apparent.

If we consider the requirements of a newly laid pipeline, it has been considered important to ensure as near perfect a coating as is possible at the start of the pipeline's life. It has therefore been common practice to "Pearson" survey the pipeline after installation. As described earlier, this survey requires a two-man team and commonly progresses at some 4–6 km per day. This procedure may be accelerated by the use of the signal attenuation coating survey. Bearing in mind the high-quality coating of a new pipeline, and the comparison basis employed by the signal attenuation coating survey, even small coating defects will become readily noticeable.

At the other extreme, with poorly coated old pipeline, a Pearson survey may have an almost continuous response, making it impossible to discern individual defects or the most significant coating breakdown. The signal attenuation coating survey would, however, be able to give a rapid assessment of coating condition and indicate where limited rewrapping would be of greatest benefit.

In general, therefore, experience has shown that it is cost effective for the more detailed surveys of coating condition and cathodic protection to be undertaken in conjunction with a signal attenuation coating survey. A number of case histories are described below.

Case histories

1. | | |
   |---|---|
   | Pipeline | Multiproducts UK |
   | Length | 115 kilometers |
   | Built | 1982 |
   | Coating | Coal tar reinforced enamel with self-adhesive tape at field joints |
   | Survey | Signal attenuation coating survey |
   | Survey period | Nine days (four days for disconnection of all "temp" magnesium anodes, five days for survey) |

Survey results:
Survey indicated that over some 80 km of the pipeline the coating was satisfactory and further survey work was not recommended for those areas. Signal losses over these sections ranged between 1.5–3 dB/km. Some nine sections, amounting to a total of some 10 km were given first priority for further investigation. Signal losses were as high as 17.6 dB over distances of 1.5 km. Two further sections, amounting to some 21 km, were classified as second-priority items.

It was decided that this information gathered be correlated with the results obtained from a proposed intelligent pig survey before further action.

2. | | |
   |---|---|
   | Pipeline | White Oil products, UK |
   | Length | 115 kilometers |
   | Built | Approx 1963 |
   | Coating | Coal tar enamel, tape, or bitumen flood field joints |
   | Survey | Signal attenuation coating survey Pearson survey |
   | | Close interval potential survey |

Survey period	10 days—Signal attenuation coating survey
	1 day—Pearson survey
	4 days—Close interval potential survey

Survey results:
Initial signal attenuation coating survey identified 30 locations of significantly high signal losses; of these, 12 were attributed to interpipeline bonds and sacrificial anodes, and the remaining 18 were subject to a more detailed investigation.

Close interval potential surveys showed that a satisfactory level of cathodic protection was being achieved at each of the locations where defects had been pinpointed using Pearson equipment.

Three particular locations were excavated; in the first, a coating defect of some 800 mm associated with a welding arc burn was found within an area previously showing a signal loss of 6 dB over less than 100 m section. In the second area, showing similar signal attenuation, a number of small coating defects were found together with a box of welding rods touching the pipe. At the third location, a badly poured bitumen field joint had been indicated by a signal attenuation loss of 10 dB.

3. | | |
   |---|---|
   | Pipeline | Oil, UK |
   | Length | 210 kilometers |
   | Built | Approx. 1970 |
   | Coating | Coal tar enamel with tape-wrapped field joints. |
   | Survey | Close interval potential survey Recording both "ON" and "OFF" potentials at 5 meter intervals |
   | Survey period | 40 days for three man team |

Survey results:

Review of all the close interval potential results highlighted some 28 locations where both the "ON" and "OFF" potentials were less negative than the protection criteria. Of these 28 locations, 80% would not have been apparent from potential measurements recorded at adjacent test posts, thereby allowing the integrity of the pipeline to be put at risk where these unsatisfactory levels of cathodic protection coincided with coating defects.

4. Pipeline | Aviation Fuel 6″, UK
   Length | 5 kilometers
   Built | 1960
   Coating | Asphalt
   Survey | Signal attenuation survey
   Survey period | Two days

Survey results:

The survey showed consistently high signal losses, indicating a generally deteriorated coating. Signal loss values were of the order of 40 dB per km.

More than 50% of the total signal loss was concentrated over four areas, amounting to some 1720 meters of pipe. Correlating close interval potential survey results over the same areas allowed 315 meters to be selected for rewrapping.

5. Pipeline | Crude Oil, Middle East
   Length | 132 kilometers
   Built | 1986
   Coating | Extruded polyethylene
   Survey | Signal attenuation coating survey
   Survey period | Four days

Survey results:

Generally, signal losses were found to average less than 1 dB per kilometer. One location was found to indicate a signal loss of 14 dB over 40 m. More detailed signal attenuation survey work located two possible coating defects with signal losses of 10 dB over 20 and 48 dB over 20 m. A subsequent Pearson survey and two excavations revealed five small defects in a shrink sleeve weld joint coating for the 10 dB loss and one defect less than 100 mm^2 in the factory-applied coating for the 4.8 dB.

Conclusion

It is considered that the discussion and case histories illustrated have shown that each of the aboveground survey procedures offers certain advantages and disadvantages, and that many of the technical and economic penalties of choosing any one specific survey may be overcome if used in conjunction with another.

References

J. Pearson, *Petroleum Engineer* 12(10) 1941.
CCEJV "Cathodic Protection Monitoring of Buried Pipelines"—State of the Art Report by Task Group E4-2.
N. Frost, "Monitoring Pipeline Corrosion." *Pipes and Pipelines International*, May 1986.

Source

Jones, Glyn, and Thron, Joanna, *Advances in Pipeline Protection*, Gulf Publishing Co., 1990.
C. J. Argent et al. "Coating defect location by above ground surveys"—proceedings of UK corrosion, 1986.

Estimate the number of squares of tape for pipe coating (machine applied)

$$\text{Squares of coating} = \frac{W \times A}{(W - O) \times 100 \text{ ft}^2/\text{square}}$$

where:

W = Width of coating, inches
A = Pipe surface area, square feet
O = Overlap, inches

Example.

Pipe OD = 6 ⅝″
Length = 1 mile
Tape width = 6″
Overlap = 1″
Pipe surface area = 1.73442 ft^2/ft (Table 7 Pipe Data Section 3)
Total surface area = 1.73442 × 5280 = 9157 ft^2

$$\text{Squares of coating} = \frac{6 \times 9157}{(6 - 1) \times 100}$$

$$= 110 \text{ squares}$$

Estimate the amount of primer required for tape

Multiply pipe surface area by 0.001047 to obtain the number of gallons of primer required per mile of pipe.

Example. Pipe OD = 6 ⅝ ″
Length = 1 mile

Pipe surface area = $1.73442 \text{ ft}^2/\text{ft}$
Total surface area = $1.73442 \times 5280 = 9157 \text{ ft}^2$
Gals of primer = $9157 \times 0.001074 = 9.8$ gals (Use 10 gals).

Tape requirements for fittings

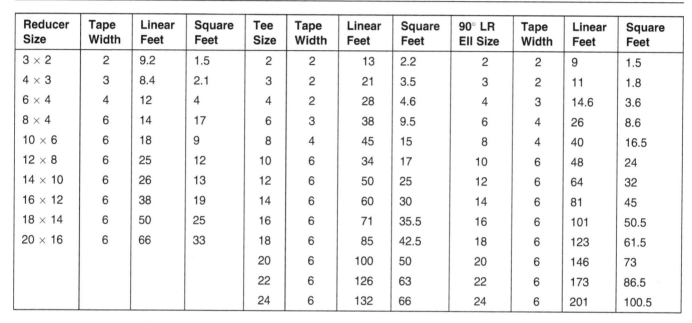

Reducer Size	Tape Width	Linear Feet	Square Feet	Tee Size	Tape Width	Linear Feet	Square Feet	90° LR Ell Size	Tape Width	Linear Feet	Square Feet
3 × 2	2	9.2	1.5	2	2	13	2.2	2	2	9	1.5
4 × 3	3	8.4	2.1	3	2	21	3.5	3	2	11	1.8
6 × 4	4	12	4	4	2	28	4.6	4	3	14.6	3.6
8 × 4	6	14	17	6	3	38	9.5	6	4	26	8.6
10 × 6	6	18	9	8	4	45	15	8	4	40	16.5
12 × 8	6	25	12	10	6	34	17	10	6	48	24
14 × 10	6	26	13	12	6	50	25	12	6	64	32
16 × 12	6	38	19	14	6	60	30	14	6	81	45
18 × 14	6	50	25	16	6	71	35.5	16	6	101	50.5
20 × 16	6	66	33	18	6	85	42.5	18	6	123	61.5
				20	6	100	50	20	6	146	73
				22	6	126	63	22	6	173	86.5
				24	6	132	66	24	6	201	100.5

Tape requirements for fittings, continued

90° SR Ell Size	Tape Width	Linear Feet	Square Feet	45° Ell Size	Tape Width	Linear Feet	Square Feet
2	2	7	1.5	2	2	5.6	0.93
3	2	10	1.8	3	2	9	1.5
4	3	13	3.25	4	3	13	3.2
6	4	21	7	6	4	17.3	5.75
8	4	38	13	8	4	36	12
10	6	46	23	10	6	34	17
12	6	60	30	12	6	48	22
14	6	76	38	14	6	52	26
16	6	96	48	16	6	64	32
18	6	118	59	18	6	76	38
20	6	136	68	20	6	96	48
22	6	166	83	22	6	110	55
24	6	190	95	24	6	126	63

INDUCED AC VOLTAGES ON PIPELINES MAY PRESENT A SERIOUS HAZARD

by Earl L. Kirkpatrick, PE., ELK Engineering Associates, Inc., Fort Worth, TX, and Kirk Engineering Co., Oneonta, AL

The problem of induced AC voltages on pipelines has always been with us. Early pipeline construction consisted of bare steel or cast iron pipe, which was very well grounded. Bell and spigot, mechanical, or dresser-style joint couplings were often used, creating electrically discontinuous pipelines that are less susceptible to AC induction. Although induced, AC affects any pipeline parallel to a high-voltage alternating current (HVAC) power line, the effects were not noticeable on bare pipelines. With the advent of welded steel pipelines, modern Cathodic Protection (CP) methods and materials, and the vastly improved quality of protective coatings, induced AC effects on pipelines have become a significant consideration on many pipeline rights-of-way.

In the past 2–3 decades, we have been seeing much more joint occupancy of the same right-of-way by one or more pipelines and power lines. As the cost of right-of-way and the difficulty in acquisition, particularly in urban areas, have risen, the concept of joint occupancy rights-of-way has become more attractive to many utility companies. Federal and state regulations usually insist on joint-use right-of-way when a utility proposes crossing regulated or publicly owned lands, wherever there is an existing easement. Such joint use allows the induced AC phenomena to occur and may create electrical hazards and interference to pipeline facilities. Underground pipelines are especially susceptible if they are well-coated and electrically isolated for CP.

Induced AC mechanisms

There are three coupling modes of primary concern by which voltages are induced in pipelines.

Electrostatic coupling, commonly known as "capacitive" coupling, is caused by the electrostatic field surrounding the energized conductor. This electrostatic field is proportionally produced by the voltage in the overhead conductor. The closer the pipeline is to the energized overhead conductor, the stronger the electrostatic voltage induced in the pipeline.

Electrostatic coupling is of primary concern when a pipeline is under construction near an overhead HVAC system. The length of the exposed conductor is also a factor in the electrostatic charge induced in the structure. A relatively short conductor isolated from ground, such as a single joint of line pipe supported by a nonmetallic sling or on a rubber-tired vehicle on the right-of-way, will become charged like a capacitor. (A rubber-tired vehicle on the right-of-way should have drag chains to ground the vehicle).

Once the pipeline is backfilled, electrostatic coupling is no longer of any real concern because the pipeline coating will allow sufficient leakage to earth through pinholes or holidays, which will effectively ground out the electrostatic effects.

Electromagnetic coupling

Electromagnetic coupling is also known as "transformer action." When current flows in an energized conductor, it produces an electromagnetic field at right angles to the conductor. In AC power systems, electric current flowing in the conductor changes direction 120 times per second for 60-Hz and 100 times per second to 50-Hz systems. Thus, the electromagnetic field surrounding the energized conductor is constantly expanding and contracting.

Whenever electromagnetic lines of force cut through another conductor, a voltage is induced in that conductor. This is the principle upon which power transformers, alternators, and generators function. A well-insulated pipeline parallel to an overhead HVAC line then becomes the secondary of an air core transformer. With sufficient parallel length, significant voltages and currents can be inducted into the pipeline. These voltages can be hazardous to anyone who comes in contact with the exposed pipeline or appurtenances. The parallel power lines may induce sufficient power into the pipeline to damage the pipeline or related facilities.

Pipeline hazards

A products pipeline failed because of lightning and fault current actually penetrating the pipeline wall. This was a clear case of arc burn penetration of the pipe wall. There are other case histories where actual pipe wall penetration has occurred with or without a fire being started. People have been killed by contacting an energized pipeline under construction. We must be aware of these electromagnetic and electrostatic effects on influenced pipelines and know how to avoid injury and damage. An excellent source of information on this subject is NACE Standard RP0177-95. Additional information is contained in IEEE Standard 80.

Resistive coupling

Resistive coupling is another mode of electrical coupling between a pipeline and a parallel overhead power system. This is a concern during "fault conditions" on a power system. If lightning strikes one of the energized conductors on the overhead power line, the resulting voltage rise on the wire will exceed the Breakdown Insulation Level (BIL) of the insulator at the nearest tower. When the BIL is exceeded, a flashover will occur from the energized conductor to the tower and then fault current will flow through the tower to the tower ground. Current will flow from the energized conductor to the tower structure via the ionized gases (plasma) generated by the lightning. Fault current will flow through the tower and the tower ground into the earth for a fraction of a second, until the circuit protection device has a chance to operate.

As a result of these fault currents, the current will radiate from the tower foundation and grounds in all directions from the faulted tower. A severe potential gradient will occur across the earth, radial to the faulted tower. If there is a nearby pipeline in the earth, the gradient field will be distorted and accentuated. This effect is greatest with bare pipelines and pipelines that are near the tower foundation. Consequently, a significant portion of the fault current may flow in the pipeline as a result of resistive coupling.

Whenever a coated pipeline and an HVAC transmission circuit are near each other, the magnetic field associated with the currents in the power transmission line will induce a voltage in the pipeline.

Physical separation between the pipeline and the power line tower is an important factor in determining the extent of damage that will occur on the pipeline as the result of a fault.

Predicting AC voltages on a pipeline

Induced AC voltage on a pipeline can be measured by methods similar to those used to conduct a DC pipe-to-soil cathodic protection potential survey. A multimeter is placed on the appropriate AC voltage range and a steel pin is used in place of the copper/copper-sulfate reference electrode. A detailed AC pipe-to-soil potential survey should be conducted over any area where induced AC voltages are suspected. It is important to note the time of each reading because many power companies maintain a record of currents in the circuit. The voltage induced in the pipeline is directly proportional to the line currents in the overhead conductors. This can provide valuable information to assist in calculating the peak voltages that can be anticipated at each location under maximum line loading conditions.

Whenever a coated pipeline and an HVAC transmission circuit are near each other, the magnetic field associated with the currents in the power transmission line will induce a voltage in the pipeline. The actual magnitude of the induced AC voltage depends on many factors, including the overall configuration of all the structures involved, soil resistivity, pipe coating effectiveness or resistance to remote earth, pipeline propagation constant, magnitude of the line currents in the power circuit(s), and any current imbalance between the phases.

The magnitude of steady-state AC potentials induced on an underground pipeline by parallel high-voltage transmission lines can be accurately estimated using the appropriate mathematical formulas. The formulas characterize the circuit in terms of "steady-state" line currents, phase relationships, pipeline-to-conductor distances, pipeline propagation constants, characteristic impedances, soil resistivity, and other factors. The technique can predict, with reasonable accuracy, the areas where the maximum AC potentials will occur and approximate the actual induced voltage at that point. These calculations require the use of advanced mathematics, which can best be done on a computerized workstation by an experienced professional.

Mitigation techniques

Mitigation techniques that can control induced voltages on an influenced pipeline include the following:

- Supplemental grounding of the pipeline with sacrificial anodes or other grounding means.
- Bonding the pipeline to individual power line pole grounds or towers through the use of polarization cells.
- Installation of parallel mitigation wires bonded to the pipeline at regular intervals.
- Bonding the pipeline to purposefully designed "made" grounds.
- Changing phase relationships between multiple power line circuit conductors.
- Use of Faraday cages with sacrificial anodes.
- Relocation of the pipeline or power line to provide greater separation from the influencing power system.
- Installation of a nonmetallic pipeline such as high-density polyethylene pipe, if design pressures permit.
- Installation of gradient control electrodes or mats at all aboveground appurtenances.
- Security fencing around aboveground appurtenances does not mitigate induced voltages, but they do limit access to the structure.

Whenever there is a pipeline closely parallel to a power line for any significant distance (e.g., closer than 500 ft or more than 2000 ft), further investigation is warranted. If five volts AC or more is measured at any point on the parallel section, this voltage should be related to line loading to estimate the pipeline-induced AC voltage at peak line-loading conditions. The induced voltage should be mitigated if the calculated values approach or exceed 15 volts AC. Fault conditions and step-and-touch voltages also must be considered.

Personnel safety

On any construction or maintenance project, safety is an attitude. This attitude is developed by proper training. A valuable source for such information is Section 4 of NACE Standard RPO177. NACE also offers an audiovisual presentation that deals with the effects of electric shock in such situations. RP0177 should be included in the construction specifications whenever a pipeline is built or exposed for maintenance on an energized HVAC right-of-way. One of the inspectors should be designated in charge of electrical safety. This individual must be familiar with and properly equipped to test for safe levels of induced voltage in the pipeline. A safety meeting should be held prior to construction with all construction employees to discuss electrical safety requirements.

The right-of-way must be vacated if thunderstorms come within 10 miles of the pipeline right-of-way or the power line(s) that are influencing the pipeline. Never work on an influenced pipeline during a thunderstorm because of the potential for a direct lightning strike. Power system fault currents will flow in the earth if lightning causes flashover of a power line insulator. A person on the right-of-way is in danger, even if they are not touching the pipeline.

Corrosion and maintenance personnel should be very cautious about stringing test lead wire on the right-of-way where the lead wire parallels the influencing HVAC line. Extremely hazardous voltages can be induced on a significant length of test lead wire laid on top of the ground.

The practice of using insulating rubber gloves should also be discouraged unless the work is being performed by a trained power company employee. "Rubber Goods" must be specially cared for and tested to ensure their reliability.

The only safe alternative in the pipeline environment is to test the level of induced AC on the pipeline before contacting anything that may be a conductor. If voltages are safe, normal measurement or repair and maintenance techniques may be employed. If unacceptably high voltages are encountered, one must work on a ground mat. A ground mat can be as simple as a piece of 6-ft road mesh laid on top of the ground and bonded to the pipeline or appurtenance with an automotive jumper cable. Since the individual, the ground mat, and the pipeline appurtenance are all at equal potential, it matters little as to how high the actual measured voltage is on the pipeline.

Reprinted with permission—*Pipeline & Gas Journal.*

MEASURING UNWANTED ALTERNATING CURRENT IN PIPE

William H. Swain, William H. Swain Co.

AC flowing on a pipeline can be a portent of trouble, such as corrosion or harm to persons. Clamp-on ammeters represent an accurate way to measure AC. AC voltage drop measurements on a pipeline can be seriously erroneous. This article describes the use of clamp-on AC ammeters for interference and fault location, corrosion control, and personnel safety purposes.

AC is best measured with a clamp-on AC ammeter. Accuracy is important in terms of pipe current and stray pickup, especially in a common corridor shared with an overhead electrical transmission line.

AC flowing in a pipe can also be detected by using the mV drop method wherein the pipe itself is used as a shunt—the voltage drop indicating current flowing in the pipe. An advantage of this method is simplicity. A disadvantage is that steel pipe is both resistive and inductive. The inductive component can cause critical errors.[1] Moreover, accurate calibration is difficult, if not impossible.

Structure of a clamp-on AC ammeter

Figure 1 is an outline of a clamp-on AC ammeter. The aperture of the clamp may range from $\frac{3}{4}$ to 60 in (2–152 cm) in diameter.

Reference 1 gives the equations relating input current (i_i), winding turns (N_s), and output voltage (v_o), along with further information about clamp-on ammeters.

An undesired current will likely have an identifying characteristic (i.e., a "signature"). An oscilloscope connected across v_o in Figure 1 will show the waveform of i_i. The frequency of i_i can be measured on a portable meter, such as a Fluke 87[c], or estimated with earphones. Another output is a precision rectifier. It is used for driving a data logger to record the change of i_i over time.

CP is pulsed at a fairly high frequency. This should have a characteristic sound when earphones are used. If the current is pulsed, this will show on an oscilloscope. Frequency and duty factor can be measured on the meter.

A foreign water pipe contact or Underground Residential Distribution (URD) interference will likely have a strong signal—60 Hz in the U.S. or 50 Hz in some other nations. Either headphones or the meter will measure frequency.

---

[c]Trade name.

One owner reported sharp increases and decreases in 60 Hz i_i in a large-diameter gasoline pipeline. An intermittent short to a water pipe carrying 60 Hz ground current was finally located. Every time a heavy truck went over the pipe, contact to a water line was made, then broken. A fairly deep pit was found in the gasoline pipe at the point of contact.

Accuracy of a clamp-on ammeter

A typical measurement is accurate at 50–60 Hz to within $\pm 1\%$ of reading, ± 3 least significant digits, and $\pm$ stray pickup. Frequency response varies with range sensitivity and clamp size. Figure 2 shows results on a 12-in (30-cm) clamp, and Figure 3 shows results on a 49-in (125-cm) clamp.

In the top curve of Figure 2, the sensitivity or transfer resistance (R_T) at 25 mA full-scale input is flat within $\pm 2\%$ (0.2 dB) from 17 to 5500 Hz. This was measured on a 12-in clamp with an analog indicator.

The lower curve in Figure 2 is similar, but the sensitivity is reduced so that the full-scale i_i is now 250 mA. The frequency response is flat within $+1\%$ and -3% from 17 to 5500 Hz.

Figure 3 is similar to Figure 2 except that the clamp was much larger. When tested with an analog indictor,

Figure 1. Clamp-on AC ammeter structure.

Figure 2. Transfer resistance (R_T) versus frequency, 12-in clamp.

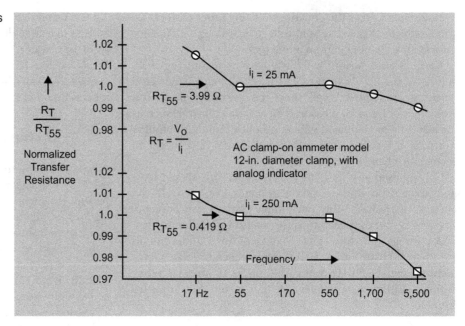

Figure 3. Transfer resistance (R_T) versus frequency, 49-in clamp.

the frequency response was flat within ±1% from 17 to 5500 Hz.

The top curve in Figure 3 is for an i_i of 80 mA passing through the aperture of the 49-in clamp. Flat frequency responses to within ±1% enable the user to get a more accurate picture of the current waveform with an oscilloscope, etc. The lower curve shows the response at 800 mA.

Stray pickup

The clamp-on AC ammeter shown in Figure 1 accurately measures i_i, but it is not perfect. Especially when i_i is small, the clamp may have an output current as a result of stray alternating magnetic fields. Generally, this stray pickup is small and represents an error <1% of measured current.

Figure 4. Structure of the mV drop method for AC.

Reference 1 provides specific data and a discussion of the likely effect of stray pickup. It is probably less than a fraction of an ampere of equivalent i_i in clamps on a pipe in a common corridor with a three-phase transmission line having up to 1000 A in the lines.

Uses of a clamp-on AC ammeter

Interference and fault location, corrosion control, and personnel safety are important considerations when using clamp-on ammeters.

Interference and fault location. A corrosion control professional may find that there is DC interference flowing in a line, but does not know where it is coming from. The waveform or audio tone of the interfering current is a clue to the probable source.

A 24-h record of i_i will likely show a constant interference current if it is from a foreign CP system. Recurring peaks and valleys can point to interference from an electrified railroad. A 25-Hz current, for example, has been observed on a pipeline near an AC-operated railroad.

Corrosion control. Gummow has written about corrosion resulting from alternating current.[2] He notes that studies in the mid 1980s found that, above a certain minimum level of AC current density, normal levels of CP will not provide acceptable levels of AC corrosion control.

There are also concerns over corrosion problems in copper and steel freshwater piping systems associated with unwanted AC caused by grounding practices. It was reported that in a large hospital, pipe replacements were required after only two to three years because AC current was found on the piping; a pipe current >75 mA rms was deemed a hazard.

The DC offset associated with nonlinear loads may have caused the problem.

Personnel safety. Gas and oil transmission pipelines are frequently laid in a corridor adjacent to high-voltage transmission lines. There can be considerable current induced in pipe in such a corridor. This can be a hazard to both property and personnel.

mV drop method

Figure 4 illustrates the mV drop method. Here, the pipe itself is used as a shunt for measuring AC flowing in the pipe. This is similar to the method that is used for measuring DC flowing in the pipe, except that the mV potential is measured using an AC millivoltmeter.

If the resistance of the pipe over the span is 0.5 mΩ and the mV meter reads 1.1 mV, at first one may think that the pipe current is 2.2 A. This would be correct for DC, but there is inductance—skin effect and stray pickup need to be considered.

Inductance

If the reactance of the $\sim$100-ft (30.5-m) measurement span is 0.3 mΩ at 60 Hz, this adds in quadrature to the 0.5 mΩ resistance (R), so the impedance (Z) becomes 0.58 mΩ—a 16% increase. At 60 Hz, skin effect may add even more Z.

At 120 Hz, the major ripple frequency for most full-wave rectifiers, the reactance is likely doubled to 0.6 mΩ, so Z increases to 0.78 mΩ, a 56% increase. Skin effect is also greater.

Figure 5. Impedance (Z) of cast-iron lab sample pipe versus frequency of current in pipe.

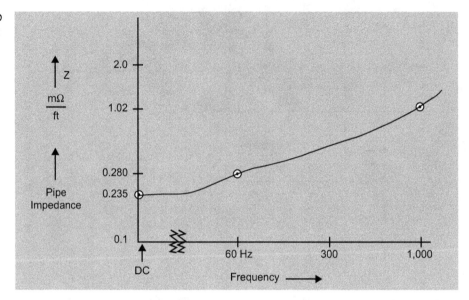

Figure 6. Normalized impedance (Z) of cast-iron lab sample versus frequency of current in pipe.

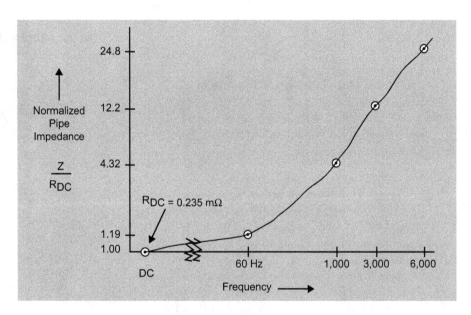

A 3-in (76-mm) diameter cast iron pipe with 3/16-in (4.8-mm) wall thickness was tested in the laboratory. The effect of high frequency on pipe impedance was evident. Figure 5 is a log-log graph of Z against the frequency of the current flowing in the pipe.

Figure 6 is a log-log plot of normalized Z against frequency. In the cast-iron pipe sample, the 60-Hz Z was 19% greater than the DC resistance.

Sheehan[3] found that the 60-Hz normalized impedance of very large-diameter steel pipe ranged from 160 to 310 times

that of the DC resistance. These numbers are within the spread calculated using several values of steel permeability.

Stray pickup

The 60-Hz magnetic field set up by the current flowing in a transmission line over a pipe will induce voltage in a properly oriented loop of wire beside the pipe. In Figure 4, the AC mV induced in the meter wire will depend on how it is laid with respect to the pipe.

General conclusion

Clamp-on AC ammeters, as described in this article, are accurate for measuring from 0.1 A to more than 20 A. The typical overall accuracy is ±3% from 17 to 5000 Hz. This is for clips and clamps from $\frac{3}{4}$ - to 60-in diameter aperture.

Clamp-on AC ammeters are practical tools for measuring and locating current that can cause corrosion or endanger humans. The resolution is 1 mA or better, even up to 60-in diameter pipe. Stray pickup is generally unimportant, even in a common corridor under an AC power transmission line.

The mV drop method, wherein the pipe itself is used as a current shunt, is much less sensitive and accurate, even at 60 Hz.

References

1. W. H. Swain, "Measuring Unwanted Alternating Current in Pipe," W. H. Swain Co. Web site: www.SwainMeter.com.

2. R. A. Gummow, et al., "AC Corrosion—A Challenge to Pipeline Integrity," MP 38, 2 (1999).

3. Private communication from B. Sheehan at San Diego County Water Authority, August 28, 2000.

WILLIAM H. SWAIN is President of William H. Swain Co., 239 Field End St., Sarasota, FL 34240. He founded the company in 1964 and has invented and marketed DC and AC amp clamps, meters, and other products for corrosion control on lines up to 5-ft (1.5-m) diameter. He has an M.S. degree in electrical engineering and is a professional engineer in the state of Florida. He has been a NACE member for more than 16 years. **MP**

Minimizing shock hazards on pipelines near HVAC lines

As right-of-way for pipelines becomes more difficult to obtain, especially in congested areas, rights-of-way are being shared by pipelines and high voltage transmission lines. Sharing a common right-of-way can pose problems during operation, and especially during pipeline construction unless special precautions are taken.

Considerable voltage can be induced in pipelines aboveground on skids unless the section of pipeline is grounded. Without grounding, a lethal shock hazard to construction personnel can occur. To minimize the voltage on the pipe, drive ground rods at each end of the section of aboveground pipe. The ground rods should be driven at least 4 ft in the ground and connected to the aboveground section of pipe with insulated copper wire and a C-clamp. If the aboveground section is longer than ½ mile, additional ground rods should be driven and connected to the pipe.

Work should be discontinued if thunderstorms are occurring in the vicinity (approximately 10 miles) of the pipeline construction. If lightning should strike the power line, even at a distant location, while personnel are in contact with the pipe, they could suffer a severe shock or even electrocution. To eliminate this danger, personnel should always stand on a 3-ft × 3-ft (minimum) steel mat made of 1″ square (maximum) hardware cloth that is electrically well grounded to the nearest electrical tower ground via a copper cable. The mat should be fabricated so that personnel can easily move it from tower to tower. This can be done by bolting a suitable length of flexible welding cable firmly to the metallic mat. The other end of the cable should be equipped with a suitable copper or brass clamp that can easily be connected to the tower ground. Personnel should always place the mat on the ground and stand on it while connecting the cable clamp to the tower ground. When work is complete, personnel should remain standing on the mat while disconnecting the cable clamp from the tower ground wire. This will make it safe to carry the mat to the next tower ground connection site.

When tie-ins are made, additional personnel protection is required. Use a metallic mat or driven ground rod connected to the pipe at each end of the tie-in, and connect an insulated copper cable (bond) across the tie-in point. The pipe and ground mat will be essentially at the same potential if a high-voltage surge occurs during the tie-in operation. The bond will prevent buildup of voltage between the unconnected ends at the tie-in location.

After the pipeline has been buried and placed in service, personnel should always connect a bonding cable across any location where the pipe is to be separated, i.e., a valve to be removed or a section of pipe is to be removed and replaced. The bonding cable should remain in place until all connections have been completed.

Equipment with steel tracks will not require any special grounding procedures for traveling along the power line right-of-way, but machinery with rubber tires should have a heavy-duty chain attached to the vehicle frame with a good electrical connection so that approximately 3 ft of chain will drag along the ground when moving or standing along the power line right-of-way. Any time there is a possibility of

a lightning strike, personnel should not grasp any rubber-tired machinery while standing on the ground. Personnel should hop on or off when mounting or dismounting. Care should be exercised to prevent injury.

Cathodic protection test point installations

Test points provide the best means of electrically examining a buried pipeline to determine whether it is cathodically protected. They also help to ensure that other tests associated with corrosion control are working. To serve their purposes over the years, test points must be convenient to use and constructed in a manner that minimizes future damage to the test station.

Types of test points

Figure 1 illustrates test point types by function. The types shown do not necessarily represent any particular standard, but they are intended to represent the variety that may be encountered. A color code is indicated to illustrate a system whereby leads may be identified. Whatever color code is adopted should be made standard throughout a pipeline system.

The two-wire potential test point is the one used most frequently. Two wires make it possible to check pipe-to-earth potential with one while test current is being applied to the line (if desired) using the other. The four-wire insulated joint test point permits measuring pipe-to-earth potentials on each side of an insulated joint. The second pair of heavier gauge wires is available for inserting a resistance or solid bond across the insulated joint if necessary.

The four-wire calibrated line current test point permits accurate measurement of pipeline current flow. The six-wire combination insulated joint and line current test point is useful, particularly at terminal insulated flanges, because it permits positive measurement of current flow through an insulated flange should the flange become totally or partially shorted. Likewise, it will measure the current flowing through a solid or resistance bond should such measurement be necessary. One heavier-gauge wire is

provided on each side of the insulated joint for bonding purposes (if required).

An indicating voltmeter test point is installed at key points on some systems. These meters may be read by operating personnel on a routine basis and the indicated values recorded and reported to the corrosion engineer. A voltmeter may be connected between the pipe and a reference electrode suitable for underground service.

The foreign line crossing test point provides two wires to each line. One wire is used for potential measurements or other tests as required and the other (the heavier-gauge wire) is available for installing a bond when needed. Test wires should never be attached to another company's pipeline unless the pipeline owner authorizes it. Further, many companies will allow such attachments only if they are made by their own personnel or if their own representatives are present while the attachments are being made.

The galvanic anode test point is typically used in connection with anodes at one location.

Adapted by Joram Lichtenstein from Peabody's Control of Pipeline Corrosion, 2 ed., *R. Bianchetti, ed. (Houston, Texas; NACE, 2001), pp. 239–241.* Reproduced with permission–NACE International, Houston, Texas. All rights reserved. Published in the November 2003 issue of *Materials Performance.* Copyright © NACE International 2003.

CORROSION OF LOW-VELOCITY, HIGH WATER CUT OIL EMULSION PIPELINES

Michael F. McDonald, CORR_X Engineering Ltd., Calgary, AB, Canada

In addition to water cut, the effect of fluid velocity has long been recognized as a key contributing factor to corrosion in oil and gas pipelines. Low velocities encourage water hold-up and the accumulation of solid debris. This condition can cause accelerated attack from concentration cells such as under-deposit corrosion. Likewise, high velocities can lead to Flow-Induced Localized Corrosion (FILC).[1] In this case, corrosion is caused by the formation of corrosion products, in combination with critical sheer stresses when the "erosional velocity"[2] is reached and exceeded. This mechanism is another problem and it is the author's experience that for many years, corrosion personnel have pondered how to determine a criterion for selecting pipelines in low-velocity, high water cut oil emulsion systems in which to apply corrosion mitigation measures. Ideally, utilizing a well-designed, field-wide corrosion monitoring program will greatly assist in determining the corrosion potential of the lines and assist in prioritizing lines in terms of failure risk. It is often difficult to justify the cost of such a monitoring program, however, because of monitoring

costs, the number of lines in a field, and the average oil production for each line.

For fields in which pipeline corrosion is a serious issue and requires remedial action, most operators select a simple solution, usually based on failure history and water cut; they apply a threshold water cut value in which all pipelines at or over this value receive corrosion mitigation measures. Investigation of internal corrosion failures in several oil systems indicates that applying a water cut criterion to prioritize corrosion-prone pipelines is an overly simplistic and conservative approach to cost-effective corrosion control.

Significance of fluid velocity

Low velocities encourage water hold-up and the accumulation of solid debris. This condition can cause accelerated attack from concentration cells such as under-deposit corrosion. Likewise, high velocities can lead to FILC.[1] In this case, corrosion is caused by the formation of corrosion products, in combination with critical sheer stresses when the "erosional velocity"[2] is reached and exceeded. This mechanism is another problem and it is the author's experience that corrosion caused by very low velocities in oil pipelines is a more common problem in onshore production facilities than corrosion in high-velocity pipelines. The challenge, therefore, is to combine both parameters—water cut and velocity—into a practical corrosion mitigation model for prioritizing pipelines in a field with similar fluid properties.

Development of the strategy

Several years ago, the author was involved in developing a corrosion mitigation strategy for a large oil-gathering system in southeast Alberta. The field had experienced more than 10 internal corrosion pipeline failures during a three to four year period. The field consists of 10 batteries and ~150 producing wells, with the attendant flow lines and group lines, the majority of which are internally bare steel pipelines. The fluid properties are fairly common throughout the field and are characterized by fairly high levels of the HCO_3^- anion (~5000 mg/L) and low levels of the Cl^- anion (~2500 mg/L). This observation supports the premise that the corrosion being experienced in the oil flow lines and group lines in the field was caused by HCO_3^- anions and carbon dioxide (CO_2) partial pressure (~30 psig).

Although the Cl^- anion concentration level is relatively low, thereby reducing the overall electrolyte strength, the HCO_3^- anion concentration is roughly five to 10 times the concentration typically found in oilfield produced waters. Work by Schmidt, et al.[3] reported that bicarbonate ions will actually initiate and accelerate the CO_2 corrosion mechanism believed to be the primary cause of corrosion in the field. It is

believed that the high concentration of bicarbonate anions cause localized precipitation of scales at the bottom of the pipe and facilitates an under deposit mode of attack.

The previous corrosion mitigation strategy employed by the operator involved installing High-Density Polyethylene (HDPE) liners in pipelines that experienced internal corrosion failures, but were still considered economically viable. The operator wished to upgrade corrosion control in this field by utilizing a proactive, yet cost-effective approach and requested the pipeline integrity consultant to create a "liner priority matrix" for the ~150 internally bare steel pipelines in the field. This "liner priority matrix" consisted of a schedule in which 10–12 pipelines would be budgeted for lining every quarter. This staged lining approach permitted acceptable budgetary planning within operating expense limits; if engineered and implemented correctly, it was intended to provide adequate corrosion mitigation combined with tolerable cost control in a large, mature gathering system with a serious internal corrosion control problem.

The challenge in developing a sound, reliable method for prioritizing the lines was to establish a relationship between the parameters that are variable from line to line. These variable parameters were water cut, fluid velocity, and whether or not a particular line had failed from internal corrosion. All other corrosion risk factors such as fluid properties and the conditions and age of the lines were fairly similar and as such would be considered irrelevant. A series of graphs were prepared showing the water cuts and fluid velocities of the lines that were known to have failed from internal corrosion over the last four to five years. The graphs were plotted using various scales and normal, semi-log, and log formats. Figure 1 shows the graph scale and format deemed to be most revealing. This graph shows a failure

Figure 1. Failure risk curve.

curve that is similar to a simple exponential curve. The area in which the curve changes direction represents the region where the dominant parameter in the corrosion mechanism changes from fluid velocity to water cut.

Next, the water cut and velocities of the other internally bare steel lines in the field were plotted on this graph. The lines that fell outside the "failure risk curve" were judged to be at high risk of internal corrosion failure. These lines were designated "high risk priority 1" (Figure 2). The lines that fell directly on the curve were designated "high risk priority 2" (Figure 3). The lines that fell just inside the curve were designated "moderate risk priority 3" (Figure 4). From Figures 2–4, a lining priority matrix was prepared and presented to the operator. Two pipelines on Figure 3 (high risk priority 2) lie outside the curve. According to the rules set out by this strategy, these lines should have been designated "high risk priority 1" rather than "high risk priority 2." These two lines were designated "high risk priority 2" to account for the fact that they were installed and put into service about five years later than all other "high risk" and "moderate risk" lines and were consequently age adjusted.

Figure 2. Failure risk curve—high-risk lines, priority 1.

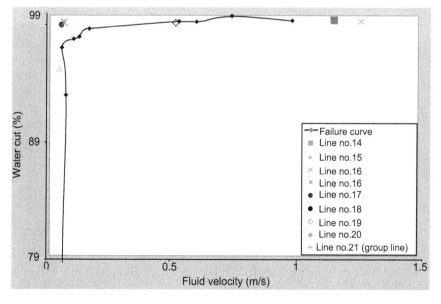

Figure 3. Failure risk curve–high-risk lines, priority 2.

Figure 4. Failure risk curve—high-risk lines, priority 3.

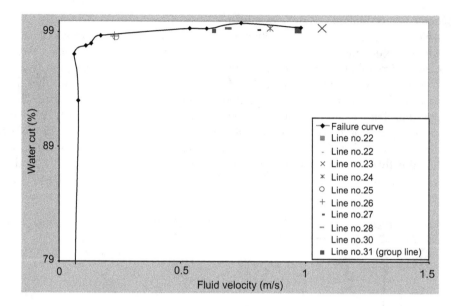

The long-term plan was to review the production of all wells in the field every one to two years to determine if the updated water cuts (which generally increase) and fluid velocities (which usually decrease provided no changes occur, such as installing new pumps) of the lines alter their position on the "failure risk curve."

Effect of hydrogen sulfide and water chemistry

The operator had hydrogen sulfide (H_2S) measurements taken on the solution gas from most wellheads in the field, including the fluid from all failed lines. The range of H_2S concentrations varied from 20 to 200 mg/L and no correlation between failures and H_2S levels was in evidence. It is strongly suspected that the higher than normal bicarbonate levels (typically 5000 mg/L) in the produced water is the dominant species in contributing to the corrosion mechanism. The effect of the bicarbonate ion in the corrosion process is well documented.[3]

Validation of the corrosion mitigation strategy

The development of the "failure risk curve" and "liner schedule matrix" were accepted by the operator. When the internal corrosion potential for this field was revisited by the author several years later, it was discovered that of the 13 lines originally deemed "high risk priority 1," seven lines were later shut-in because of declining production, three lines were lined with HDPE, and three lines failed. These results validated the strategy (virtually 100% correlation) and the relationship among water cut, fluid velocity, and failures. This approach has since been applied to several smaller fields, and

the results have indicated that, to the author's knowledge, one of these fields has since experienced a single failure; this occurred on a pipeline that fell into the "high risk priority 1" category.

Limitations of the strategy

It is not proposed to utilize this strategy for all oil fields with a documented internal corrosion failure history. This strategy can be applied only to fields that have similar fluid properties and conditions throughout the field, and the ages of the pipelines should ideally be about the same. If the ages of the lines vary considerably, a certain amount of judgment is required in prioritizing the lines. Furthermore, if a pigging program is in place, then all lines in question should have the same pigging frequency. In addition, this strategy should be employed with great care if any type of corrosion inhibition program is being applied. If, however, a failure risk curve can be generated that looks somewhat predictable (such as an exponential curve) regardless of any previous or current corrosion mitigation measures in place, then the operator can predict with some degree of confidence the corrosion behavior of all pipelines in the field that have similar fluid properties.

The above strategy has been applied only to systems under primary production. It is not known if the corrosion behavior of lines under secondary or water flood production would be as predictable as described in this article. If, however, a predictable "failure risk curve" can be generated for the lines (including group lines), this strategy may be employed, with some caution initially, to prioritize the lines as to failure risk.

Other potential applications of this strategy

The application of a similar strategy on other types of systems, such as low-velocity sweet gas lines, where it is not economical to employ a comprehensive, field-wide monitoring program, should be investigated. Although fluid velocity will undoubtedly be one of the parameters, the parameter representing water production would need to be thought out and a certain amount of trial and error experimentation would be required.

References

1. G. Schitt, C. Werner, M.J. Schoning, CORROSION/2002, paper no. 280 (Houston, TX: NACE International, 2002).
2. API 14E, "Recommended Practice for Design and Installation of Offshore Production Systems" (Washington, DC: API).
3. G. Schmitt, B. Rothmann, Werkstoffe, Korrosion 28, 826 (1977).

MICHAEL F. MCDONALD is president and owner of CORR-X Engineering, Ltd., 127 Discovery Ridge Bay, S.W., Calgary, AB T3H 5T7, Canada. He also is a contract consultant to Broadsword Corrosion Engineering, Ltd., headquartered in Calgary, AB, Canada. He has 39 years of experience in corrosion, materials, and inspection in the oil and gas industry. McDonald is a Registered Professional Technologist (Engineering) with APEGGA. His defined scope of practice is: "The design and management of corrosion control programs and preparation of related material specifications for oil and gas wells, gathering facilities, and associated facilities." He has been a NACE member since 2000.

INTERNAL STRAY CURRENT INTERFERENCE FORM AN EXTERNAL CURRENT SOURCE

W. Brian Houtsbaum, CC Technologies Canada, Calgary, AB, Canada

Stray current interference occurs from current traveling in an unintended path (Figure 1). Current will follow all paths inversely proportional to the resistance of each path; should a structure be available as a parallel path, some of the current will choose this path. The area of current pickup is not of concern, but should that current leave the metal via some electrolyte to return to its source, then corrosion will be accelerated in this section. Stray current interference testing and monitoring is, therefore, an integral part of a CP commissioning or annual survey.

Stray current interference related to pipelines often occurs at pipeline crossings (Figure 1(a)) or across isolating features such as between a well casing and pipeline (Figure 1(b)).

In Figure 1(a), current has collected through the soil on Pipe B from the Pipe A CP system to the left of the figure. To return to its source, the current has selected a path through the soil electrolyte at the crossing to return to Pipe A, and then back to the rectifier. Current from the pipeline CP system in Figure 1(b) has collected on the outside of the casing and has elected to return to the pipeline through the soil electrolyte at the wellhead. In both cases, no corrosion occurs as a result of the current pickup but corrosion is accelerated where the current is discharged into the soil electrolyte. The above concepts on stray current interference are generally well understood.[1-3] Note that in the above examples, the interference effects occur in the same electrolyte as that in which the current is being impressed. Is this always the case? No, there is an exception.

Internal stray current

Internal corrosion has occurred at an isolating flange. The only evidence is that corrosion exists on the inside near one flange face. Without further investigation, the conclusion may be that it is caused by a natural corrosion mechanism. If three conditions exist, internal interference should be added to the list of causes for consideration. These three conditions include:

- The corrosion is at an isolating fitting or high-resistance coupling.
- An electrolyte (especially of low resistivity) flows through the isolator.
- A voltage exists across the flange or fitting.

If the stray current returning along the foreign metallic path encounters an electrical isolating feature or high-resistance joint, it may bypass the electrical isolation by going through the soil (Figure 1) or it may bypass the isolation by going through an electrolyte that bridges the isolation inside (Figure 2). Note that corrosion only occurs on the side of current discharge.[4-5]

Figure 1.

Figure 2.

Detection of internal interference

Stray current on the external side of a pipeline can be detected by structure-to-electrolyte potentials if the reference electrode is placed at the applicable locations.[1-3,6] With the reference electrode located on the soil side of the pipeline, internal stray current will not be detected. If current is bypassing an effective above-grade electrical isolation, it can be measured with a clamp-on ammeter (Figure 3). Interference current through the soil will not be included in this measurement, however. The following tests can then be completed to predict internal interference:

- Confirm that the electrical isolation is effective.

- Measure the voltage across the electrical isolation.
- With an aboveground isolation, measure the current with a clamp-on ammeter.

Internal stray current interference

Stray current will continue through the metallic contact in a shorted isolation without causing interference. It is therefore important to confirm the status of the isolation. Without a voltage across an effective isolation, internal stray current will not occur because a driving voltage is required.

Since an effective electrical isolation is intended to interrupt the current in the pipe wall, a current measured by

Figure 3.

a clamp-on ammeter with the coil at the isolation fitting indicates that internal interference exists. The current measured in this manner does not include the current exchange through the soil path (Figure 3). With an effective isolation, a current measured at the isolating fitting will represent the internal interference current.

Prevention of internal stray current interference

Depending on the circumstances, a possible method to avoid internal stray current interference may be one or more of the following:

- Eliminate the isolation.
- Increase the resistance of the internal path by an internally lined or coated pipe spool.
- Eliminate the collection of low-resistivity water across the electrical isolation.
- Minimize the potential difference across the isolation.

Eliminate the isolation

Where the CP system has the capacity to provide protection with the two structures electrically, eliminating the isolation is the most assured solution. Otherwise, one of the other cures must be considered.

Increase the resistance of the internal path

This can be accomplished by either installing a nonmetallic spool section or an internally coated spool piece. The spool piece must be long enough to significantly increase the resistance across the section of isolation. In the first case, the nonmetallic section must meet the internal pressure, and in both cases the nonmetallic liner material or coating must be compatible with the product. If a nonmetallic section is used, then no isolating feature is required. When an internally coated section is installed, then an isolating feature must be installed at one end, such that the coating will be where stray current pickup occurred (Figure 4), unless one can be assured of a "holiday-free" coating. If not, a rapid penetration of the pipe wall will occur where current discharges from a coating holiday.

Eliminate the collection of low-resistivity electrolyte

This solution is applicable to products such as an oil line where brine tends to settle out in inadvertent liquid traps such as across a horizontal flange. Installing the isolating feature on a vertical riser or a section of pipe that slopes sufficiently to prevent the collection of brine will

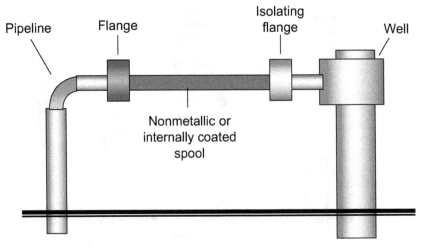

Figure 4.

eliminate the path of internal stray current (Figure 5). This assumes that the product itself is not a low-resistivity electrolyte.

Minimizing potentials across the isolation

If practical, equalizing the potential difference across the isolating feature will serve to reduce the stray current inside the pipeline.

Summary

Consider the fact that an internal stray current condition may exist inside a pipeline at an isolating fitting if a low-resistivity electrolyte exists across the fitting. Prevention may include eliminating the isolation, installing a nonmetallic or internally coated section on the noncorroding side of the isolating fitting, installing the isolating fitting on a vertical or sloping section of pipeline, or minimizing potentials across the isolation.

Figure 5.

References

1. A.W. Peabody, R.L. Bianchetti, ed., Control of Pipeline Corrosion, 2nd Ed. (Houston, TX: NACE International, 2001), p. 211–236.
2. M.E. Parker, E. Peattie, Pipeline Corrosion and Cathodic Protection (Houston, TX: Gulf Publishing), p. 100–124.
3. W.B. Holtsbaum, Stray Current Interference and Underground Corrosion, ASM Handbook, Vol. 3C (Materials Park, OH: ASM International).
4. J.F. Chmilar, "Cathodic Protection for a 2000 Kilometer Pipeline Network".

IMPROVEMENTS TO THE EXTERNAL CORROSION DIRECT ASSESSMENT PROCESS

David H. Kroon, Olagoke Olasisi, Larry G. Rankin, James T. Carroll, Dale D. Lindemuth, and Marlane L. Miller, Corrpro Companies, Inc., Houston, Texas

For cased pipe, improvements to the External Corrosion Direct Assessment (ECDA) process include identification of applicable new technologies, development of techniques for the application of existing technologies, and creation of a threat matrix for cased pipes. The research led to a recommended ECDA methodology for cased pipes.

The purpose of the cased pipe study was to determine the applicability of existing and emerging technologies to assess cased pipes for external corrosion using ECDA. This included examination of existing ECDA processes, best practices of pipeline operators, and emerging technologies. The project findings are significant for gas transmission pipeline operators in the United States because the integrity of all pipe in High-Consequence Areas (HCA) must be assessed by December 17, 2012, including those segments of pipe in casings. There is an industry need for a methodology to assess cased pipe where in-line inspection and pressure testing are neither possible nor practical. There are currently a number of emerging technologies that could be used on cased pipe such as:

- Electromagnetic wave
- Pulsed eddy current
- Conformable array
- Magnetometer electrochemical impedance spectroscopy
- Robotic cameras and sensors

Although these technologies are promising, they will likely take years to develop. Both past and current research and testing have demonstrated that a CP current can reach cased pipes when an electrolyte is present in the annulus between the pipe and casing, even if the casing is electrically shorted to the pipe.[1] Consequently, standard CP and protective coating survey techniques should provide valuable information regarding the condition of the protective coating and the level of CP on pipe within a casing, recognizing the limitations that may be created by shorted conditions, coated casings, and no filling, intermittent filling, or partial filling of the annulus with groundwater.

A matrix was developed to identify external corrosion threats for pipes in casings. The threat matrix can be used to prioritize cased pipe for assessment. It incorporates construction, operations, and maintenance parameters such as:

- Electrical isolation
- CP
- Groundwater level and fluctuations
- Annulus atmosphere
- Age
- Pipe coating
- Casing coating
- High temperature

The developed cased-pipe ECDA methodology follows the general steps in the process delineated in NACE SP0502-2008.[2]

Preassessment

The preassessment step requires compiling and analyzing construction, operation, and maintenance information for the cased pipe under study. Annual CP survey, including pipe-to-soil (P/S) and casing-to-soil potential measurements and Close Interval Survey (CIS) should be reviewed. Possible external corrosion mechanisms should be identified and ECDA regions defined. A feasibility analysis needs to be performed to include a critical determination of conditions that could prevent or alter electrical measurements on the pipe within the casing, such as casings that are electrically shorted to the pipe, coated casings, and lack of electrolyte within the casing. The ECDA Indirect Inspection (IDI) tools must then be selected to accommodate pipe and casing conditions.

Indirect inspection

The IDI tools for cased pipe ECDA are the same as those used for direct buried pipe. A minimum of two tools are required. At least one tool should address the protective coating condition and defects, and the other should evaluate CP levels and detect CP indications with CIS. IDI tools used for cased pipe are primarily CIS, direct current voltage gradient (DCVG), alternating current (AC) attenuation (ACCA), and AC voltage gradient (ACVG). The methodology requires that IDI be performed over the entire length of cased pipe plus 300 ft (91 m) beyond both ends.

For this project, data were collected and analyzed for more than 200 cased pipes and adjacent uncased pipe for six transmission and distribution system operators. Figures 1–3 illustrate the data collected from three of these surveys.

Figure 1 contains data from a pipe that is electrically isolated from the casing as demonstrated by a 700-mV difference pipe and casing potentials, and by the identical on and instant-off current potentials on the casing. There were no DCVG indications within the casing, but the ACCA survey results indicated a change in signal strength within the casing. Both on and instant-off CIS data suggest CP levels that satisfy the criteria in NACE SP0169-2008.[3] There were no changes in potential at the ends of the casing and valid measurements were collected above the casing at grade. (Data were not collected over the paved areas.) The ACCA indication corresponds to an area where CIS data could not be collected.

Figure 2 shows data from an electrically isolated cased crossing. Note the ACCA indication dips in both on and instant-off potential measurements as the casing is approached, and continues as data are collected over the casing. There was a DCVG indication at the upstream end of the casing. All IDI inspection data are normal with no apparent abnormalities associated with data collection over the cased pipe.

Figure 3 illustrates data from a cased pipe where the electrical isolation of the casing is unknown (no vents or test leads connected to the casing). CIS and DCVG data correlate

Figure 1. Survey data from a pipe electrically isolated from the casing.

Figure 2. Survey data of an electrically isolated casing.

Figure 3. Survey data from a cased pipe where the electrical isolation of the casing is unknown.

in demonstrating coating defects as the cased pipe is approached. Another coating defect was also identified inside the casing which correlated with the location of a dent found by ILI.

The aforementioned are examples only. The data from the 200 cased pipes that were tested clearly indicate that IDI surveys can be used for ECDA of pipelines within casings,

subject to certain limitations. Further, the research data were used to develop the IDI severity ranking.

Direct examination

Referencing Table 2, all cased pipes with an action priority of "immediate" are to be examined within the period to time

Table 1
IDI Severity Ranking

IDI Survey Tools	Severity Classifications			
	None	Minor	Moderate	Severe
AC attenuation survey	Uniform attenuation profile with no significant changes	Small change in attenuation profile over short length of pipe inside or near casing	Moderate change in attenuation profile over short length of pipe inside or near casing	Large change in attenuation profile over short length of pipe inside or near casing
DC or AC voltage gradient survey	No indications on adjacent buried pipe and no indications on cased pipe	Few indications on adjacent buried pipe but no indications on cased pipe	Several indications on adjacent buried pipe but no indications on cased pipe	Numerous indications on adjacent buried pipe or indications on cased pipe
Close interval potential survey	Uniform potential profile with no significant depressions and all potentials more negative than −850 mV	Minor potential depressions but all potentials more negative than −850 mV	Moderate potential depressions but all potentials more negative than −850 mV	Large potential depressions or any potentials less negative than −850 mV

Table 2
Action Priorities for IDI Indications

Action Prioritization for Cased Pipe Segments Based on IDI Survey Severity Classifications	CP Severity Classifications Based on Close Interval Potential Survey Results		No Indications	Minor Indications	Moderate Indications	Severe Indications	
Coating condition severity classifications	Based on AC current attenuation survey results	No indications	No Action	Monitor	Schedule	Immediate	
		Minor indications	Monitor	Monitor	Schedule	Immediate	
		Moderate indications	Monitor	Schedule	Schedule	Immediate	
		Severe Indications	Schedule	Schedule	Immediate	Immediate	
	Based on DC or AC voltage gradient survey results	No indications	No Action	Monitor	Schedule	Immediate	
		Minor Indications	Monitor	Monitor	Schedule	Immediate	
		Moderate indications	Monitor	Schedule	Schedule	Immediate	
		Severe Indications	Schedule	Schedule	Immediate	Immediate	

commensurate with the indicated conditions. Cased-pipe locations prioritized as "schedule" are to be examined before the next integrity assessment. Cased-pipe locations with action priority "monitor" are to be monitored at appropriate time intervals until the next integrity assessment is performed. No action is necessary for those locations indicated as such. Direct examination techniques include visual inspection, ILI, pressure test, Guided Wave Ultrasonic Testing (GWUT), electromagnetic wave, and other acceptable techniques either individually or in combinations.

Post assessment

Evaluations of the direct examination results of cased pipes with action priorities of "immediate" and "schedule" determine the need to perform additional direct examinations on those cased-pipe segments with a priority of "monitor." Reassessment intervals are then set independently of adjacent, uncased, buried pipe. A very rigorous evaluation of the effectiveness of the ECDA methodology for the cased-pipe segments studied should then be performed.

Conclusions

The cased-pipe portion of the project concluded:

- ECDA can be used to evaluate the integrity of cased pipe in accordance with a recommended methodology.
- The methodology makes use of ECDA IDI survey techniques currently being used for uncased, buried pipe, as part of the process for identifying and ranking direct examination priorities and selecting the most effective assessment tools.
- Direct examination techniques include visual inspection, ILI, pressure test, GWUT, and electromagnetic wave either individually or in combinations.
- It is possible for cased pipe with a water-filled annulus to be categorized as "no action," required, or "monitor."
- There are limitations to the use of ECDA as a tool for integrity management of cased pipes that need to be recognized and understood.

Acknowledgments

We would like to acknowledge the PHMSA funding and support for this work and the contribution of time and effort from our team members—El Paso, ExxonMobil, Southern Union, and the Texas Gas Association.

References

1. L. R. Rankin, H.M. Al Manrous, "External Corrosion Probability Assessment for Carrier Pipes inside Casings," Gas Technology Institute, GRI 05/0020, March 2005.
2. NACE Standard SP0200-2008 "Steel-Cased Pipeline Practices" (Houston, TX: NACE International, 2008).
3. NACE Standard SP0169-2007 "Control of External Corrosion on Underground or Submerged Metallic Piping Systems" (Houston, TX: NACE , 2007).

Part 2

A study sponsored by the U. S. Department of Transportation Pipeline & Hazardous Materials Safety Administration (PHMSA) was conducted to determine the applicability of existing and emerging technologies to assess buried pipelines for external corrosion using external corrosion using ECDA. This included examination of existing ECDA processes, best practices of pipeline operators, and emerging technologies.

The project findings are significant for gas transmission pipeline operators in the United States because the integrity of all pipe in High-Consequence Areas (HCAs) must be assessed by December 17, 2012, including those segments of pipe in casings. There is an industry need for a methodology to assess cased pipe where ILI and pressure testing are either not possible or not practical.

Table 1
Enhanced Severity Classification Criteria of Indirect Inspections (IDI) (Modification of Table 3 in NACE SP0502-2010)

IDI Classification			
Measure	**Minor**	**Moderate**	**Severe**
IDI Tool = Close Interval Potential Survey			
A = off (polarized) potential (mV)	−950 mV < A < −850 mV OR	−950 mV < A < −650 mV OR	−650 mV < A AND
B = on potential (mV)	−1.000 mV < B < −950 mV	−950 mV < B < −850 mV	−850 mV < B
C = on/off convergence (mV)	50 mV < C < 70 mV	30 mV < C 10 mV	10 mV < C
D = on and/or off profile depression within 100 ft (30.5 m) (mV/span)	50 mV/span < C < 100 mV/span	100 mV/pan < D < 200 mV/span	200 mV/span < D
IDI Tool = AC Current Attenuation			
E = current 98 Hz frequency signal loss (−) (mdB(mA)/ft)	7 mdb(mA/ft < E < 3 mdb/ft AND/OR	12 mdb(mA)/ft < E < 7 mdb(mA)/ft AND/OR	12 mdb(mA)/ft < E AND/OR
F = current 4 Hz frequency signal loss (−) (mdb(mA)/ft	20 mdb(mA)/ft < F < 40 mdb(mA)/ft	40 mdb(mA)/ft < F < 60 mdb(mA)/ft	60 mdb(mA)/ft < F
CP level modifier	Adequate CP level	Adequate to marginal CP level	All indications with inadequate CP level
IDI Tool = AC Voltage Gradient			
G = voltage signal loss (−) 9 dB(mV))	44 dB(mV) < G < 60 dB(mV) AND	60 dB(mV) < G < 78 dB(mV) AND	78 dB(mV) < G OR
CP level modifier	Adequate CP	Adequate to marginal CP level	All indications with inadequate CP
IDI Tool = DC Voltage Gradient			
H = coating defect size (% IR)	5% IR < H < 20% IR AND	20% IR < h < 50% IR OR	50% IR < H OR
I = corrosion state assessment (normal operating conditions)	I = Cathodic/cathodic or cathodic/neutral AND	All indications 5% IR < H < 50% IR where I = cathodic/anodic AND	All indications where I = anodic/anodic OR
CP level modifier	Adequate CP level	Adequate to marginal CP level	All indications with inadequate CP level

Severity ranking

The purpose of the severity ranking portion of the study was to enhance the existing Tables 3 and 4 in NACE SP0502-2010.[1] The existing tables in the standard are very general, which result in varying interpretations and inconsistencies in application under the current practice. The project goals were to identify improvements that could be made and develop an enhanced severity ranking methodology.

Data from five transmission and distribution system operators were compiled, sorted, and analyzed. This included the results of 400 direct examinations with complete, applicable data sets, including soil analysis. Fifty percent (200) of the data sets used in the study demonstrated measurable external corrosion. ILI data were also analyzed, which covered 14,000 joints of pipe where CIPS and ACCA surveys had been performed. These data included 4000 joints of pipe with measureable corrosion and 100 excavations. To capture best current industry practices, this phase of the project was discussed with 10 qualified and experienced operator and service provider professionals with a total of over 300 years of experience.

In developing improvements to Tables 3 and 4 in NACE SP0502-2010, specific numerical criteria were developed, which covered a wide range of definable conditions. The work included analysis of Rupture Pressure Ratio (RPR) and percent wall loss relative to above grade measurements and individual IDI indications. The enhancement of existing Tables 3 and 4 appear as Tables 1 and 2 herein. During the course of evaluating soil data at IDI indications, it was noticed that soil conditions appeared to correlate with the severity ranking as defined by the numerical ranges. This, of course, is consistent with what we have all observed, but have only loosely been able to tie to specific soil properties such as soil resistivity, pH, active ion concentration, and moisture content. A broader characterization of soil "texture" was postulated to provide a better indicator of corrosive conditions. This correlation was investigated using the U.S. Department of Agriculture (USDA) Web Soil Survey,[2] which provides significant detail regarding soil texture and physical/chemical properties across the United States. It has the additional advantage of being easily accessible on the Internet, free for anyone to use. The investigation concluded that percent clay content (expressed as a percentage of total composition) was a parameter that correlated with the presence of external pipeline corrosion. Data from the 14,000 joints of pipe (4000 with measurable external corrosion) representing 188 soil types were plotted (Figure 1).

The data were then further analyzed by ranking the leak and rupture hazards as a percent of clay for the data set. The threat of leaks was indicated by wall loss while the threat of rupture was expressed as rupture pressure ratio. The data clearly illustrated that as the percent clay in the soil increases, so does the threat of both rupture and leak. A soil modifier was then applied to the Severity Classification Criteria of Indirect Inspections and the Prioritization Criteria for Indirect Indications as shown on the bottom of Table 1 and Table 2, respectively.

Conclusions concerning severity ranking

Improved tables for Severity Classification and Prioritization Criteria for Indirect Inspection Indications were developed, which provide a more consistent assessment of the external corrosion threat.

Table 2
Enhanced Prioritization Criteria for Indirect Inspection Indications (Modification of Table 4 in NACE SP0502-2010)

USDA Soil Texture Modifier	IDI Tool 2 Classification	IDI Tool 1 Classification		
		Severe	**Moderate**	**Minor**
Severe	Severe	Immediate	Immediate	Scheduled
Severe	Moderate	Immediate	Scheduled	Scheduled
Severe	Minor	Scheduled	Scheduled	Monitored
Moderate	Severe	Immediate	Scheduled	Scheduled
Moderate	Moderate	Scheduled	Scheduled	Monitored
Moderate	Minor	Monitored	Monitored	Monitored
Minor	Severe	Immediate	Scheduled	Monitored
Minor	Moderate	Scheduled	Monitored	Monitored
Minor	Minor	Monitored	Monitored	Monitored

Methodologies developed represent an enhancement to NACE SP O5O2-2010 for quantification and qualification of IDI indications, effective use of available soils data, and introduction of a soil texture modifier.

The new methodologies quantified and verified both the project research data and industry knowledge and experience.

Potentials in paved areas

Current industry practices for collecting potential measurements in paved areas are to drill through the pavement, collect potentials offset from the location of the pipeline, surface wetting, or simply skipping data collection in paved areas. The purpose of the potential in the paved areas portion of the project was to develop a methodology to collect more reliable data in a more user friendly, efficient, and safe manner. The goal was for the methodology to be applicable to both transmission and distribution systems, and ultimately to provide for more data collection in paved areas, thereby enhancing pipeline integrity assessment.

We considered gravel, asphalt, and concrete surfaces. Variability in thickness, aggregate, sub-base, and construction yields a nearly infinite number of conditions. If basic electrical measurements could characterize a pavement, then decisions and guidelines could be developed regarding the validity of potential measurements with reference electrode placement on the pavement. As the research progressed, this postulation was tested and refined. The result was a simple test procedure that can be used at the onset of a potential survey to determine if on-paving measurements can be made accurately.

The research approach consisted of reviewing prior work, running large scale laboratory tests as illustrated in Figure 2, and collecting field data on operating distribution and transmission pipelines.

Resistance measurements to characterize the pavement were made using a digital meg-ohmmeter having a maximum 1000-V DC source specifically manufactured for high-resistance circuits. Most measurements were made with one terminal of the meg-ohmmeter connected to an 8-in × 8-in (203-mm × 203-mm) metal plate electrode on the paved surface and the other terminal collected to a nearby electrical ground used as an earth electrode. No surface wetting was done for these measurements, other than to use a damp towel directly under the metal plate electrode. Figure 3 shows surface resistance values for asphalt pavement with and without visible cracking.

Many CIS surveys were performed in the field that compared current on and instant-off pipe-to-soil (P/S) potential measurements with the cell placed on dry pavement, wet pavement, or drilled holes through the pavement. Figure 4 contains results from a survey on weathered asphalt where the pavement contact resistance was $2 \times 10^8 \ \Omega \ \text{ft}^2$ measured as described above. The data on drilled holes are consistently accurate, whereas there are great inaccuracies in the data on the asphalt surface as evidenced by the evidence of the extreme data scatter in both the positive and negative directions.

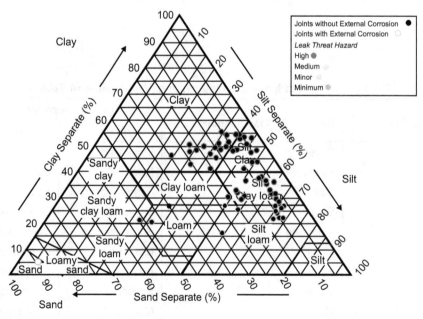

Figure 1. Observed corrosion by soil texture.

(a)   (b)

Figure 2. Large-scale lab testing of potentials in paved areas: (a) steel plate electrode and (b) copper/copper sulfate (Cu/CuSO4).

Figure 5 shows data collected on concrete pavement, and potentials on concrete pavement. Using the data from the drilled holes as the basis, potentials on pavement were first more positive and then became more negative as the survey progressed down the pipeline. This was the case even though the pavement contact resistance was a very low 100 Ω ft^2.

For gravel and asphalt, a procedure has been developed for measurement of the resistance through the pavement using a metallic electrode on the paved surface and a MΩ resistance meter. Figure 6 shows the correlation between accurate P/S potential data and the surface resistance measurement for 61 surveys on asphalt pavement. Based on analysis of the data collected, a threshold normalized resistance of 2×10^5 Ω-ft^2 has been established. That is, when gravel or asphalt paving exhibits a resistance of 200,000 Ω-ft^2 or less, a reliable potential measurement

can be made with the reference electrode on the pavement.

For concrete pavement, the research concludes there is no clear, consistent method for making reliable P/S potential measurements without placing the reference electrode in direct contact with the underlying soil (e.g., by drilling holes through the pavement), P/S potentials with the reference electrode on a concrete surface are either more negative or more positive than when in contact with the underlying soil. While P/S potential measurements are not valid with a reference electrode on the concrete pavement, DCVG measurements may be.

Conclusions concerning potentials in paved areas

For gravel and asphalt pavement:

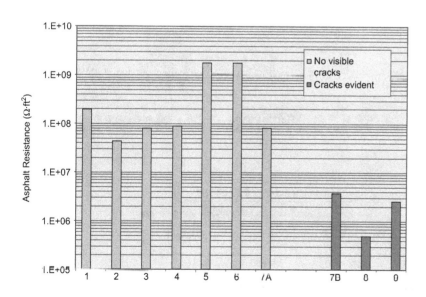

Figure 3. Asphalt resistance measurements.

A simple, straightforward, pre-survey surface resistance measurement can be used to determine if potentials recorded with the reference electrode placed on the pavement will provide accurate data.

A threshold of 200,000 Ω-ft^2 has been identified, below which potentials on pavement demonstrated accuracy.

A standard 3-in (76-mm) diameter reference electrode with a wetted towel or sponge is adequate to minimize contact resistance.

For concrete pavement:

No clear, consistent method for recording accurate potential measurements on concrete pavement was identified except by drilling through the pavement to facilitate reference electrode contact with the underlying soil.

Acknowledgments

We would like to acknowledge the PHMSA funding and support for this work and the contribution of time and effort from our team members—El Paso, ExxonMobil, Southern Union, and the Texas Gas Association.

References—Part 1

1. L. R. Rankin, H.M. Al Manrous "External Corrosion Probability Assessment for Carrier Pipes Inside Casings," Gas Technology Institute , GRI 05/0020, March 2005.
2. NACE Standard SP0200-208, "Steel Cased Pipeline Practices" (Houston, Texas, NACE International 2008).
3. NACE Standard SP0169-2007 "Control of External Corrosion on Underground or Submerged Metallic Piping Systems" (Houston, Texas NACE, 2007).

References—Part 2

1. NACE SP0502-2010, "Pipeline External Corrosion Direct Assessment Methodology, (Houston, Texas NACE International, 2010).

Figure 4. P/S potential measurements on weathered asphalt.

Figure 5. P/S potential measurements on weathered concrete.

2. U.S. Department of Agriculture Web Soil Survey, http://websoilsurvey.nrcs.usda.gov/app/HomePage.htm.

Bibliography

Carroll, J., D. Lindemuth, M. Miller, O. Olabisi, L. Rankin. "PHMSA-Sponsored Research; Improvements to ECDA Process Severity Ranking." Corrosion 2010, paper no. 10054, Houston, Texas NACE 2010.

"External Corrosion Probability Assessment for Carrier Pipes Inside Casings (Casing Corrosion Direct Assessment—CCDA)" gri-05/0020, Des Plaines, Illinois, Gas Technology Institute.

Carroll, J., D. Lindemuth, M. Miller, O. Olabisi, L. Rankin. "PHMSA-Sponsored Research; Improvements to ECDA Process Severity Ranking." Corrosion 2010, paper no. 10055 Houston, Texas NACE 2010.

NACE SP0200-2008. "Steel-Cased Pipeline Practices." Houston, Texas NACE 2008. Carroll, J., D. Lindemuth, M. Miller, O. Olabisi, L. Rankin. "PHMSA-Sponsored Research; Improvements to ECDA Process Severity Ranking." Corrosion 2010, paper no. 10056, Houston, Texas NACE 2010.

USDA Natural Resources Conservation Service. Http://websoilsurvey.nrcs.usda.gov/app/Homepage.htm., 2009

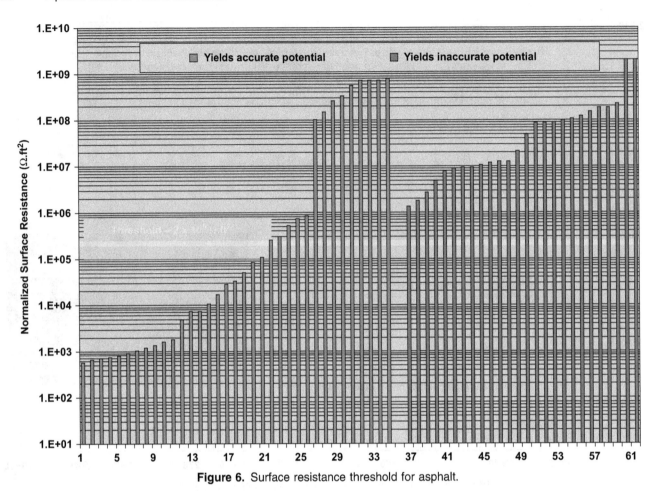

Figure 6. Surface resistance threshold for asphalt.

Causes of Flange Isolation Gasket, Sleeve, and Washer Failure

Problems with gasket and flange isolation material failures appear to be more far reaching than many users realize. A number of companies have experienced gasket failures at rates not seen since the change from flat flange isolation gaskets to pressure energized gaskets. The reasons for these failures are still being studied. This article presents the findings from evaluations of four recent failures of self-energizing gaskets. In each case, alternative "equivalents" were substituted for the specified PSI Line Backer[1] gaskets. Sleeve and washer failures are discussed also.

Gasket Failures—The gasket in Figure 1 appears to have a seamless molded sealing element with inconsistent quality control of the volume-to-void relationship. The groove depth for the sealing element also exhibits its unacceptable variances.

The retainer material for the gasket in Figure 2, an alternative to the specified gasket, was not made from a sheet size large enough for an 84-in (2-m) pipe, so the manufacturer spliced it together. The high visibility of the splice indicates that it was not well made; this should have been detected during field inspection prior to installation. Personnel responsible for installing flange isolation kits are not always adequately trained in component inspection criteria, however. The fabrication shop producing the alternative gaskets could not get them to pass hydrostatic pressure testing, but the originally specified gaskets installed on the same flanges worked without exception.

Figure 3 illustrates a gasket that failed upon hydrotest. The quality of the laminate, sealing element, and groove dimensions are all questionable. The user successfully replaced all such gaskets with the specified seals.

The gasket in Figure 4 appears to have split or cracked at the sealing element location from improper dimensioning or

---

[1] Trade Name

durometer testing of the elastomer. Although the original specification is absent, the gasket obviously was not properly made for the intended application.

Factors to consider when specifying flange isolation kits

Failures have occurred with flange isolation kits on oil and gas, product, natural gas, and potable water lines. The following factors should be considered when specifying such kits.

Gaskets—Gasket quality has a significant impact on performance, as does the design of the sealing element itself. Overlapping seams on the sealing elements have been identified as the culprit. Literally hundreds of thousands of properly designed and manufactured gaskets have been in service for more than 30 years without failures. Extruded O-rings are frequently offered as a solution to concerns about overlapping seams, but they must have seams. The issue with this type of gasket is not the seams, but rather control of tolerances and holding to exacting design, quality, and raw material standards.

With the globalization of sourcing, inexpensive raw materials that supposedly meet NEMA standards for phenolic or G-10 classification are being offered to manufacturers to reduce costs. These materials may or may not be effective as gaskets. For example, there are numerous options for phenolic that technically meet NEMA standards, but have not been properly evaluated for suitability as a gasket material. Some phenolics are more porous than others.

Sleeves and Washers—Specifications for flange isolation kit sleeve material frequently call for the sleeve to be "full length," an undefined measure for which there is no known industry standard. A better specification would require the minimum standard length of the sleeve to extend at least halfway into the back-up steel washers.

Figure 2. The gasket for a water project failed the initial hydrotest.

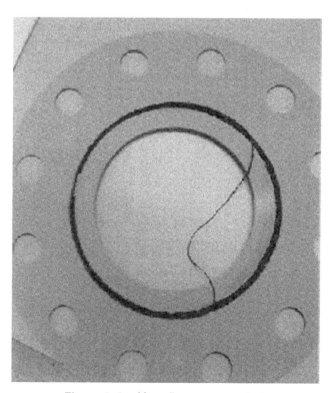

Figure 1. Leaking oil company gasket.

Figure 3. A gasket for a natural gas project.

Figure 4. A gasket from a treated water dump.

Figure 5. Bolt sleeve failure from shear stresses.

In Figure 5, the 45-degree failure angle with respect to the bolt's axis indicates that the sleeve failed in shear. The contractor ordered G-10 isolation washers and sleeves from one supplier and the steel washers from another. Normally, the isolation and steel washers are supplied as part of a kit. In this case, the steel washers had a slightly smaller diameter (ID) than the G-10 washers, so there was no clearance between the steel washers and the sleeve. As the nuts were tightened, the steel washers grabbed the sleeves and twisted them. To avoid this problem, both the steel and isolation washers must be manufactured to precise dimensions with matching ID and outside diameters for uniform loading to allow the sleeve material to fit within the ID of both washers.

The white isolation washer in Figure 6 was squashed from the compressive load of the blue steel washer. In this case, the problem was caused by using the wrong washer material for the loads being applied and the dimensional variance between the steel and isolation washers. It can be caused by overloading the compressive strength of the washer.

Gasket life and documentation

Companies often have little knowledge regarding the type of gaskets in their pipelines or the materials they are made of. Many have been operating the same lines for more than 30 years, carrying fluids with different additives, chemical inhibitors, odorants, and methyl tri-butyl ethylene. The gaskets in Figures 1–3 have no markings that indicate their

material, date of manufacture, pipe size, and pressure rating. This information should appear on all gaskets and flange isolation kits. Existing pipelines should be surveyed for the type of gaskets presently in service, their expected longevity, and length of time in service to determine the need for replacement.

Conclusions

Gasket, sleeve, and washer failures can be dramatically reduced by specifying higher grade products in new installations, and ensuring that the product being used is the same as the one specified. In addition, the specifics of gasket material should be documented for each flange, so new gaskets introduced in the system can be evaluated for

Figure 6. Over-compressed white isolation washer.

their compatibility with different media or additives in the line.

The importance of initial quality cannot be overstated. Unfortunately, "garage shop or equivalents" are often being used without the operators' or engineers' knowledge. Isolation is performance-critical even though its cost is insignificantly relative to the systems in which it is installed. As such, low-cost, high-consequence products deserve the attention they warrant.

Author—John T. Keldsen. Reproduced with permission from NACE International, Houston, TX. All rights reserved. Published in the August 2011 issue of JOURNAL© NACE International 2011.

9: Gas—General

Know the gas laws

The symbols that will be used in the mathematical expression of the various gas laws are:

V_1 = Volume of the gas in cubic ft with original conditions
V_2 = Volume of gas in cubic ft under the second set of conditions
T_1 = Initial temperature of the gas in °R (°F + 460)
T_2 = Temperature of gas under second conditions °R
P_1 = Pressure of gas in psia under original conditions
P_2 = Pressure of gas in psia under second conditions

Any other symbols used in mathematically expressing the gas laws will be explained at the time of their introduction.

Boyle's law. If the temperature remains the same, the volume of a given quantity of gas will vary inversely as the absolute pressure. This may be expressed mathematically as:

$$\frac{V_1}{V_2} = \frac{P_2}{P_1}$$

Inasmuch as in the application of Boyle's Law we are generally interested in the volume at a second set of pressure conditions, a rearrangement of the formula more readily used is:

$$V_2 = V_1 = \frac{P_1}{P_2} \tag{1}$$

Example. A quantity of gas at a pressure of 42 psig has a volume of 1,000 cubic ft. If the gas is compressed to 100 psig, what volume would it occupy? Assume the barometric pressure to be 14.2 psia and the temperature to remain the same. Substituting in the second arrangement of the formula for Boyle's Law above would give:

$$V_2 = 1,000 \times \frac{42 + 14.2}{100 + 14.2}$$

$$= 492.1 \text{ cubic ft}$$

Charles' law. (sometimes called Gay-Lussac's Law). If the pressure remains the same, the volume of a given quantity of gas will vary directly as the absolute temperature.
This may be expressed mathematically as:

$$\frac{V_1}{V_2} = \frac{T_1}{T_2}$$

Again, since we are usually more interested in the volume at a second set of temperature conditions than any other

information, an arrangement of the formula that would be handy is:

$$V_2 = V_1 \times \frac{T_2}{T_1} \tag{2}$$

A second part of Charles' Law is: If the volume of a quantity of gas does not change, the absolute pressure will vary directly as the absolute temperature.
Expressed mathematically this is:

$$\frac{P_1}{P_2} = \frac{T_1}{T_2}$$

In this instance, we would probably be more interested in the pressure at a second temperature condition, which could be expressed as:

$$P_2 = P_1 \times \frac{T_2}{T_1} \tag{3}$$

Example. A given weight of gas has a volume of 450 cubic ft when the temperature is 45°F and the pressure is 10 psig. If the pressure remains the same, but the temperature is changed to 90°F, what will be the volume of the gas?

**Table 1
Molecular Weights of Hydrocarbons and Other Compounds Associated with Natural Gas**

Compound	Atomic Formula	Molecular Weight
Methane	CH_4	16.043
Ethane	C_2H_6	30.070
Propane	C_3H_8	44.097
Butane	C_4H_{10}	58.124
Pentane	C_5H_{12}	72.151
Hexane	C_6H_{14}	86.178
Heptane	C_7H_{16}	100.205
Carbon Dioxide	CO_2	44.011
Nitrogen	N_2	28.016
Oxygen	O_2	32.00
Water	H_2O	18.016
Air		28.967

Note: Molecular weights based on following values of atomic weights: hydrogen 1.008, carbon 12.011, nitrogen 14.008, oxygen 16.000, and argon 39.944. Air was assumed to consist of 78.09% nitrogen, 20.95% oxygen, 0.93% argon, and 0.03% carbon dioxide.

Substitution in the formula for the first part of the Charles' Law gives:

$$V_2 = 450 \times \frac{(90 + 460)}{(45 + 460)}$$

$$= 490 \text{ cubic ft}$$

It is desired to determine what the new pressure would be for the gas in the above example if the volume remains the same and the temperature changes from 45°F to 90°F as indicated. (Atmospheric pressure is 14.4 psia.)

Substitution in the formula gives:

$$P_2 = \left(10 + 14.4\right) \times \frac{(90 + 460)}{(45 + 460)}$$

$$= 26.6 \text{ psia or } 12.2 \text{ psig}$$

A convenient arrangement of a combination of Boyle's and Charles' laws that is easy to remember and use can be expressed mathematically as:

$$\frac{P_1 V_1}{T_1} = \frac{P_2 V_2}{T_2} \qquad (4)$$

One can substitute known values in the combination formula and solve for any one unknown value. In cases where one of the parameters, such as temperature, is not to be considered, it may be treated as having the same value on both sides of the formula, and consequently it can be cancelled out.

Avogadro's law. This law states that equal volumes of all gases at the same pressure and temperature conditions contain the same number of molecules.

From this it may be seen that the weight of a given volume of gas is a function of the weights of the molecules and that there is some volume at which the gas would weigh, in pounds, the numerical value of its molecular weight.

The volume at which the weight of the gas in pounds is equal to the numerical value of its molecular weight (known as the "mol-volume") is 378.9 cubic ft for gases at a temperature of 60°F and a pressure of 14.73 psia. Table 1 gives the atomic formula and molecular weights for hydrocarbons and other compounds frequently associated with natural gas. Reference to the table reveals that the molecular weight for methane is 16.043. Going back to the mol-volume explanation shows that 378.9 cubic ft of methane at 60°F and a pressure of 14.73 psia would weigh 16.043 lb.

Avogadro's law ties in closely with what is usually known as the ideal gas law.

The ideal gas law. Although expressed in many slightly different arrangements, this law is most frequently expressed as:

$$pV = nRT$$

where: p = Pressure of the gas
V = Volume of the gas
n = Number of lb-mols of gas
R = The universal gas constant, which varies depending upon the units of pressure, volume, and temperature employed

Since the number of lb-mols of a gas would be equal to the weight of the gas divided by the molecular weight of the gas, we can express the ideal gas law as:

$$pV = 10.722 \times \frac{W}{M} \times T \qquad (5)$$

where: p = Pressure of the gas, psia
V = Volume of the gas, cubic ft
W = Weight of the gas, lb
M = Molecular weight of the gas
T = Temperature of the gas, °R

The constant 10.722 is based upon the generally used value for the universal gas constant of 1,544 when the pressure is expressed in lb/sq. ft absolute.

This formula can be used in many arrangements. An arrangement that may be used to determine the weight of a quantity of gas is:

$$W = 0.0933 \times \frac{MVp}{T} \qquad (6)$$

when the symbols and units are as above.

Example. It is desired to find the weight of a gas in a 1,000-cubic-ft container if the gas is at a pressure of 150 psig and a temperature of 90°F. The molecular weight of the gas is known to be 16.816, and the barometric pressure is 14.3 psia.

Substitution in the formula gives:

$$W = \frac{0.933 \times 16.816 \times 1,000 \times (150 + 143)}{(90 + 460)}$$

$$= 468.7 \text{ lb}$$

The formula above may be used when the molecular weight of a gas is known; however, at times it is desirable to determine the weight of a given volume of gas when the molecular weight is not known or cannot be readily determined. This may be accomplished if the specific gravity of the gas is known by using the formula:

$$W = 2.698 \times \frac{GV_p}{T}$$

where G = the specific gravity of the gas (air equals 1.000) and the other symbols and units are as previously given.

Example. If the molecular weight of the gas in the preceding example was unknown, but the specific gravity was known to be 0.581, substitution in the formula would give:

$$W = 2.698 \times \frac{0.581 \times 1,000 \times (150 + 143)}{(90 + 460)}$$
$$= 468.3 \text{ lb}$$

Calculate gas properties from a gas analysis

In most practical applications involving gas calculations, the gas will consist of a mixture of components. Properties of the components are known; however, the properties of the mix must be determined for use in other calculations, such as compressor performance calculations. The gas molecular weight, K-value (isentropic exponent), and compressibility of the gas mix will be determined in the calculation to follow. Other data that may be needed for compressor performance determination will likely come from the process design and from the manufacturer's data.

Sample calculation 1

Given: A gas mixture at 30 psia and 60°F consisting of the following components:

Table 1

Gas	Mol %	Mol wt	P_c	T_c	cp lbm-mol-°R
Ethane	5	30.070	706.5	550	12.27
Propane	80	44.097	616.0	666	17.14
n-Butane	15	58.123	550.6	766	22.96
Total	100				

Table 2
Individual Component Contributions

Gas	Mol %	Mol wt	P_c	T_c	cp
Ethane	5	1.50	35	28	0.61
Propane	80	35.28	493	533	13.71
n-Butane	15	8.72	83	115	3.44
Total	100	45.5	611	676	17.76

Therefore, the properties of the mix will be:
Molecular weight = MW = 45.5
Critical pressure = P_c = 611
Critical temperature = T_c = 676
Specific heat at constant pressure = cp = 17.76

The ratio of specific heats for the mixture are calculated from Equation 1.

$$k_{mix} = cp_{mix}/(cp_{mix} - 1.986) \tag{1}$$

$$k_{mix} = 17.76/(17.76 - 1.986)$$
$$= 1.126$$

Determine the compressibility of the mix by calculating the reduced pressure and reduced temperature by using Equations 2 and 3.

$$P_r = P/P_c \tag{2}$$

$$T_r = T/T_c \tag{3}$$

$$P_r = 30/611$$
$$= 0.049$$

$$T_r = (60 + 460)/676$$
$$= 0.769$$

Refer to Figure 1 to determine the compressibility using the values of P_r and T_r. Z = 0.955.

If the mixture of components is given in mols of component or mass flow of components, these must be converted to mol % before calculating the properties of the mix.

Sample calculation 2

Mols of components are given:

Ethane	400 mols/hr
Propane	4,800 mols/hr
n-Butane	900 mols/hr
Total Flow	6,100 mols/hr

Mol % ethane = 400/6,100 = 6.56%

Table 3

Component	Mol/hr	Mol %
Ethane	400	6.56
Propane	4,800	78.69
n-Butane	900	14.75
Totals	6,100	100.00

The mol % values listed in Table 4 can be used to calculate the properties of the mixture following the procedure detailed in Sample Calculation 1.

Sample calculation 3

Mass flow of components is given.

Table 4

Component	Mass flow (kg/h)	Mol wt	Mol flow (kmol/h)	Mol %
Ethane	15,000	30.070	499	11.08
Propane	150,000	44.097	3,402	75.55
n-Butane	35,000	58.123	602	13.37
Totals			4,503	100.00

Mol flow of ethane $= 15,000/30.07$
$$= 499$$

Mol % of ethane $= 499/4,503$
$$= 11.08\%$$

When a dry gas is saturated with water, it will be necessary to consider the effects of the water on the molecular weight of the mixture. The water will also affect the total mass flow of the mixture.

Sample calculation 4

Given: Compress 1,000 lbm/min dry CO_2, which is initially water saturated. MW = 44.01, P_{inlet} = 16 psia, T_{inlet} = 100°F.

Calculate the molecular weight of the mixture and the total mass flow for the required 1,000 lbm/min of dry CO_2.

$$\text{Mol fraction of water vapor} = y = P_w/P \tag{4}$$

P_w = partial pressure of water vapor. It is exactly equal to the saturation pressure if the gas is saturated.

$$MW_{mix} = yMW_{mix} + (1 - y) \times MW_{dg} \tag{5}$$

where w = water vapor

   dg = dry gas

$$m_{total} = m_{dg}\left[1 + (MW_w - P_w)/\left(MW_{dg} \times P_{dg}\right)\right] \tag{6}$$

Saturated pressure of water vapor at 100°F is equal to 0.95 psia, and since the mixture is saturated, the partial pressure is also equal to 0.95 Asia.
Calculate mol % of water vapor using Equation 4.
y = 0.96/16 = 5.94 mol % water vapor
Calculate MW_{mix} using Equation 5.

$$MW_{mix} = 0.0594(18.02) + (1 - 0.059) \times 44.01 = 42.48$$

Calculate the mass flow of the mix using Equation 6.

$$M_{total} = 1,000[1 + (18.02 \times 0.95)/(44.01 \times (16 - 0.95))]$$
$$= 1,026 \text{ lbm/min}$$

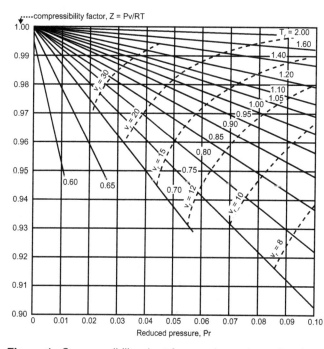

Figure 1. Compressibility chart for very low values of reduced pressure. Reproduced by permission of *Chemical Engineering*, McGraw Hill Publications Company, July 1954.

Figure 2. Compressibility chart for low values of reduced pressure. Reproduced by permission of *Chemical Engineering*, McGraw Hill Publications Company, July 1954.

Figure 3. Compressibility chart for low to high values of reduced pressure. Reproduced by permission of *Chemical Engineering*, McGraw Hill Publications Company, July 1954.

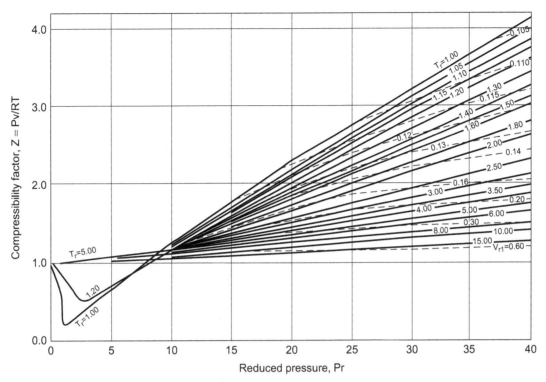

Figure 4. Compressibility chat for low to very high values of reduced pressure. Reproduced by permission of *Chemical Engineering*, McGraw Hill Publications Company, July 1954.

Figure 5. Compressibility chart for natural gas. Reprinted by permission and courtesy of Ingersol Rand.

Physical properties of selected hydrocarbons and other chemicals and gases

Physical Constants *See the Table of Notes and References.

Number	Compound	Formula	A. Molar mass (molecular weight)	B. Boiling point, °F 14.696 psia	Vapor pressure, psia 100 °F	C. Freezing point, °F 14.696 psia	D. Refractive index, n_D 60°F	Critical constants Pressure, psia	Critical constants Temperature, °F	Critical constants Volume, ft³/lbm	Number
1	Methane	CH₄	16.043	−258.73	(5000)*	−296.44*	1.00042*	666.4	−116.67	0.0988	1
2	Ethane	C₂H₆	30.070	−127.49	(800)*	−297.04*	1.20971*	706.5	89.92	0.0783	2
3	Propane	C₃H₈	44.097	−43.75	188.64	−305.73*	1.29480*	616.0	206.06	0.0727	3
4	Isobutane	C₄H₁₀	58.123	10.78	72.581	−255.28	1.3245*	527.9	274.46	0.0714	4
5	n-Butane	C₄H₁₀	58.123	31.08	51.706	−217.05	1.33588*	550.6	305.62	0.0703	5
6	Isopentane	C₅H₁₂	72.150	82.12	20.445	−255.82	1.35631	490.4	369.10	0.0679	6
7	n-Pentane	C₅H₁₂	72.150	96.92	15.574	−201.51	1.35992	488.6	385.8	0.0675	7
8	Neopentane	C₅H₁₂	72.150	49.10	36.69	2.17	1.342*	464.0	321.13	0.0673	8
9	n-Hexane	C₆H₁₄	86.177	155.72	4.9597	−139.58	1.37708	436.9	453.6	0.0688	9
10	2-Methylpentane	C₆H₁₄	86.177	140.47	6.769	−244.62	1.37387	436.6	435.83	0.0682	10
11	3-Methylpentane	C₆H₁₄	86.177	145.89	6.103	———	1.37888	453.1	448.4	0.0682	11
12	Neohexane	C₆H₁₄	86.177	121.52	9.859	−147.72	1.37126	446.8	420.13	0.0667	12
13	2,3-Dimethylbutane	C₆H₁₄	86.177	136.36	7.406	−199.38	1.37730	453.5	440.29	0.0665	13
14	n-Heptane	C₇H₁₆	100.204	209.16	1.620	−131.05	1.38989	396.8	512.7	0.0691	14
15	2-Methylhexane	C₇H₁₆	100.204	194.09	2.272	−180.89	1.38714	396.5	495.00	0.0673	15
16	3-Methylhexane	C₇H₁₆	100.204	197.33	2.131	———	1.39091	408.1	503.80	0.0646	16
17	3-Ethylpentane	C₇H₁₆	100.204	200.25	2.013	−181.48	1.39566	419.3	513.39	0.0665	17
18	2,2-Dimethylpentane	C₇H₁₆	100.204	174.54	3.494	−190.86	1.38446	402.2	477.23	0.0665	18
19	2,4-Dimethylpentane	C₇H₁₆	100.204	176.89	3.293	−182.63	1.38379	396.9	475.95	0.0668	19
20	3,3-Dimethylpentane	C₇H₁₆	100.204	186.91	2.774	−210.01	1.38564	427.2	505.87	0.0662	20
21	Triptane	C₇H₁₆	100.204	177.58	3.375	−12.81	1.39168	428.4	496.44	0.0636	21
22	n-Octane	C₈H₁₈	114.231	258.21	0.53694	−70.18	1.39956	360.7	564.22	0.0690	22
23	Diisobutyl	C₈H₁₈	114.231	228.39	1.102	−132.11	1.39461	360.6	530.44	0.0676	23
24	Isooctane	C₈H₁₈	114.231	210.63	1.709	−161.27	1.38624	372.4	519.46	0.0656	24
25	n-Nonane	C₉H₂₀	128.258	303.47	0.17953	−64.28	1.40746	331.8	610.68	0.0684	25
26	n-Decane	C₁₀H₂₂	142.285	345.48	0.06088	−21.36	1.41385	305.2	652.0	0.0679	26
27	Cyclopentane	C₅H₁₀	70.134	120.65	9.915	−136.91	1.40896	653.8	461.2	0.0594	27
28	Methylcyclopentane	C₆H₁₂	84.161	161.25	4.503	−224.40	1.41210	548.9	499.35	0.0607	28
29	Cyclohexane	C₆H₁₂	84.161	177.29	3.266	43.77	1.42862	590.8	536.6	0.0586	29
30	Methylcyclohexane	C₇H₁₄	98.188	213.68	1.609	−195.87	1.42538	503.5	570.27	0.0600	30
31	Ethene(Ethylene)	C₂H₄	28.054	−154.73	(1400)*	−272.47*	(1.228)*	731.0	48.54	0.0746	31
32	Propene(Propylene)	C₃H₆	42.081	−53.84	227.7	−301.45*	1.3130*	668.6	197.17	0.0689	32
33	1-Butene(Butylene)	C₄H₈	56.108	20.79	62.10	−301.63*	1.3494*	583.5	295.48	0.0685	33
34	cis-2-Butene	C₄H₈	56.108	38.69	45.95	−218.06	1.3665*	612.1	324.37	0.0668	34
35	trans-2-Butene	C₄H₈	56.108	33.58	49.87	−157.96	1.3563*	587.4	311.86	0.0679	35
36	Isobutene	C₄H₈	56.108	19.59	63.02	−220.65	1.3512*	580.2	292.55	0.0682	36
37	1-Pentene	C₅H₁₀	70.134	85.93	19.12	−265.39	1.37426	511.8	376.93	0.0676	37
38	1,2-Butadiene	C₄H₆	54.092	51.53	36.53	−213.16	———	(653.)*	(340.)*	(0.065)*	38
39	1,3-Butadiene	C₄H₆	54.092	24.06	59.46	−164.02	1.3975*	627.5	305.	0.0654	39
40	Isoprene	C₅H₈	68.119	93.31	16.68	−230.73	1.42498	(558.)*	(412.)*	(0.065)*	40
41	Acetylene	C₂H₂	26.038	−120.49*		−114.5*		890.4	95.34	0.0695	41
42	Benzene	C₆H₆	78.114	176.18	3.225	41.95	1.50396	710.4	552.22	0.0531	42
43	Toluene	C₇H₈	92.141	231.13	1.033	−139.00	1.49942	595.5	605.57	0.0550	43
44	Ethylbenzene	C₈H₁₀	106.167	277.16	0.3716	−138.966	1.49826	523.0	651.29	0.0565	44
45	o-Xylene	C₈H₁₀	106.167	291.97	0.2643	−13.59	1.50767	541.6	674.92	0.0557	45
46	m-Xylene	C₈H₁₀	106.167	282.41	0.3265	−54.18	1.49951	512.9	651.02	0.0567	46
47	p-Xylene	C₈H₁₀	106.167	281.07	0.3424	55.83	1.49810	509.2	649.54	0.0570	47
48	Styrene	C₈H₈	104.152	293.25	0.2582	−23.10	1.54937	587.8	(703.)*	0.0534	48
49	Isopropylbenzene	C₉H₁₂	120.194	306.34	0.1884	−140.814	1.49372	465.4	676.3	0.0572	49
50	Methyl alcohol	CH₄O	32.042	148.44	4.629	−143.79	1.33034	1174.	463.08	0.0590	50
51	Ethyl alcohol	C₂H₆O	46.069	172.90	2.312	−173.4	1.36346	890.1	465.39	0.0581	51
52	Carbon monoxide	CO	28.010	−312.68	———	−337.00*	1.00036*	507.5	−220.43	0.0532	52
53	Carbon dioxide	CO₂	44.010	−109.257*	———	−69.83*	1.00048*	1071.	87.91	0.0344	53
54	Hydrogen sulfide	H₂S	34.08	−76.497	394.59	−121.88*	1.00060*	1300.	212.45	0.0461	54
55	Sulfur dioxide	SO₂	64.06	14.11	85.46	−103.86*	1.00062*	1143.	315.8	0.0305	55
56	Ammonia	NH₃	17.0305	−27.99	211.9	−107.88*	1.00036*	1646.	270.2	0.0681	56
57	Air	N₂+O₂	28.9625	−317.8	———		1.00028*	546.9	−221.31	0.0517	57
58	Hydrogen	H₂	2.0159	−422.955*	———	−435.26*	1.00013*	188.1	−399.9	0.5165	58
59	Oxygen	O₂	31.9988	−297.332*	———	−361.820*	1.00027*	731.4	−181.43	0.0367	59
60	Nitrogen	N₂	28.0134	−320.451	———	−346.00*	1.00028*	493.1	−232.51	0.0510	60
61	Chlorine	Cl₂	70.906	−29.13	157.3	−149.73*	1.3878*	1157.	290.75	0.0280	61
62	Water	H₂O	18.0153	212.000*	0.9501	32.00	1.33335	3198.8	705.16	0.0497s	62
63	Helium	He	4.0026	−452.09	———		1.00003*	32.99	−450.31	0.2300	63
64	Hydrogen chloride	HCl	36.461	−121.27	906.71	−173.52*	1.00042*	1205.	124.77	0.0356	64

12/12/86

NOTE: Numbers in this table do not have accuracies greater than 1 part in 1,000; in some cases extra digits have been added to calculated values to achieve consistency or to permit recalculation of experimental values.

Physical Constants

*See the Table of Notes and References.

Number	E. Density of liquid 14.696 psia, 60°F			F. Temperature coefficient of density, 1/°F	G. Acentric factor, ω	H. Compressibility factor of real gas, Z 14.696 psia, 60°F	I. Ideal gas 14.696 psia, 60°F			J. Specific Heat 60°F 14.696 psia Btu/(lbm·°F)		Number
	Relative density (specific gravity) 60°F/60°F	lbm/gal	gal./lb mole				Relative density (specific gravity) Air = 1	ft³ gas/lbm	ft³ gas/gal liquid	c_p, Ideal gas	c_p, Liquid	
1	(0.3)*	(2.5)*	(6.4172)*	——	0.0104	0.9980	0.5539	23.654	(59.135)*	0.52669	——	1
2	0.35619*	2.9696*	10.126*	——	0.0979	0.9919	1.0382	12.620	37.476*	0.40782	0.97225	2
3	0.50699*	4.2268*	10.433*	-0.00162*	0.1522	0.9825	1.5226	8.6059	36.375*	0.38852	0.61996	3
4	0.56287*	4.6927*	12.386*	-0.00119*	0.1852	0.9711	2.0068	6.5291	30.639*	0.38669	0.57066	4
5	0.58401*	4.8690*	11.937*	-0.00106*	0.1995	0.9667	2.0068	6.5291	31.790*	0.39499	0.57272	5
6	0.62470	5.2082	13.853	-0.00090	0.2280	——	2.4912	5.2596	27.393	0.38440	0.53331	6
7	0.63112	5.2617	13.712	-0.00086	0.2514	——	2.4912	5.2596	27.674	0.38825	0.54363	7
8	0.59666*	4.9744*	14.504*	-0.00106*	0.1963	0.9582	2.4912	5.2596	26.163*	0.39038	0.55021	8
9	0.66383	5.5344	15.571	-0.00075	0.2994	——	2.9755	4.4035	24.371	0.38628	0.53327	9
10	0.65785	5.4846	15.713	-0.00076	0.2780	——	2.9755	4.4035	24.152	0.38526	0.52732	10
11	0.66901	5.5776	15.451	-0.00076	0.2732	——	2.9755	4.4035	24.561	0.37902	0.51876	11
12	0.65385	5.4512	15.809	-0.00076	0.2326	——	2.9755	4.4035	24.005	0.38231	0.51367	12
13	0.66631	5.5551	15.513	-0.00076	0.2469	——	2.9755	4.4035	24.462	0.37762	0.51308	13
14	0.68820	5.7376	17.464	-0.00068	0.3494	——	3.4598	3.7872	21.729	0.38447	0.52802	14
15	0.68310	5.6951	17.595	-0.00070	0.3298	——	3.4598	3.7872	21.568	0.38041	0.52199	15
16	0.69165	5.7664	17.377	-0.00070	0.3232	——	3.4598	3.7872	21.838	0.37882	0.51019	16
17	0.70276	5.8590	17.103	-0.00069	0.3105	——	3.4598	3.7872	22.189	0.38646	0.51410	17
18	0.67829	5.6550	17.720	-0.00070	0.2871	——	3.4598	3.7872	21.416	0.38594	0.51678	18
19	0.67733	5.6470	17.745	-0.00073	0.3026	——	3.4598	3.7872	21.386	0.39414	0.52440	19
20	0.69772	5.8170	17.226	-0.00067	0.2674	——	3.4598	3.7872	22.030	0.38306	0.50138	20
21	0.69457	5.7907	17.304	-0.00068	0.2503	——	3.4598	3.7872	21.930	0.37724	0.49920	21
22	0.70696	5.8940	19.381	-0.00064	0.3977	——	3.9441	3.3220	19.580	0.38331	0.52406	22
23	0.69793	5.8187	19.632	-0.00067	0.3564	——	3.9441	3.3220	19.330	0.37571	0.51130	23
24	0.69624	5.8046	19.679	-0.00065	0.3035	——	3.9441	3.3220	19.283	0.38222	0.48951	24
25	0.72187	6.0183	21.311	-0.00061	0.4445	——	4.4284	2.9588	17.807	0.38246	0.52244	25
26	0.73421	6.1212	23.245	-0.00057	0.4898	——	4.9127	2.6671	16.326	0.38179	0.52103	26
27	0.75050	6.2570	11.209	-0.00073	0.1950	——	2.4215	5.4110	33.856	0.27199	0.42182	27
28	0.75349	6.2819	13.397	-0.00069	0.2302	——	2.9059	4.5090	28.325	0.30100	0.44126	28
29	0.78347	6.5319	12.885	-0.00065	0.2096	——	2.9059	4.5090	29.452	0.28817	0.43584	29
30	0.77400	6.4529	15.216	-0.00062	0.2358	——	3.3902	3.8649	24.940	0.31700	0.44012	30
31					0.0865	0.9936	0.9686	13.527	——	0.35697	——	31
32	0.52095*	4.3432*	9.6889*	-0.00173*	0.1356	0.9844	1.4529	9.0179	39.167*	0.35714	0.57116	32
33	0.60107*	5.0112*	11.197*	-0.00112*	0.1941	0.9699	1.9373	6.7636	33.894*	0.35446	0.54533	33
34	0.62717*	5.2288*	10.731*	-0.00105*	0.2029	0.9665	1.9373	6.7636	35.366*	0.33754	0.52980	34
35	0.60996*	5.0853*	11.033*	-0.00106*	0.2128	0.9667	1.9373	6.7636	34.395*	0.35574	0.54215	35
36	0.60040*	5.0056*	11.209*	-0.00117*	0.1999	0.9700	1.9373	6.7636	33.856*	0.37690	0.54839	36
37	0.64571	5.3834	13.028	-0.00089	0.2333	——	2.4215	5.4110	29.129	0.36351	0.51782	37
38	0.65799*	5.4857*	9.8605*	-0.00101*	0.2540	(0.969)	1.8677	7.0156	38.485*	0.34347	0.54029	38
39	0.62723*	5.2293*	10.344*	-0.00110*	0.2007	(0.965)	1.8677	7.0156	36.687*	0.34120	0.53447	39
40	0.68615	5.7205	11.908	-0.00082	0.1568	——	2.3520	5.5710	31.869	0.35072	0.51933	40
41	(0.41796)	(3.4842)	(7.473)	——	0.1949	0.9930	0.8990	14.574	——	0.39754	——	41
42	0.88448	7.3740	10.593	-0.00067	0.2093	——	2.6971	4.8581	35.824	0.24296	0.40989	42
43	0.87190	7.2691	12.676	-0.00059	0.2633	——	3.1814	4.1184	29.937	0.26370	0.40095	43
44	0.87168	7.2673	14.609	-0.00056	0.3027	——	3.6657	3.5744	25.976	0.27792	0.41139	44
45	0.88467	7.3756	14.394	-0.00052	0.3942	——	3.6657	3.5744	26.363	0.28964	0.41620	45
46	0.86875	7.2429	14.658	-0.00053	0.3257	——	3.6657	3.5744	25.889	0.27427	0.40545	46
47	0.86578	7.2181	14.708	-0.00056	0.3216	——	3.6657	3.5744	25.800	0.27471	0.40255	47
48	0.91108	7.5958	13.712	-0.00053	(0.2412)	——	3.5961	3.6435	27.675	0.27110	0.41220	48
49	0.86634	7.2228	16.641	-0.00055	0.3260	——	4.1500	3.1573	22.804	0.29170	0.42053	49
50	0.79626	6.6385	4.8267	-0.00066	0.5649	——	1.1063	11.843	78.622	0.32316	0.59187	50
51	0.79399	6.6196	6.9595	-0.00058	0.6438	——	1.5906	8.2372	54.527	0.33222	0.56610	51
52	0.78939*	6.5812*	4.2561*	——	0.0484	0.9959	0.9671	13.548	89.163*	0.24847	——	52
53	0.81802*	6.8199*	6.4532*	-0.00583*	0.2667	0.9943	1.5196	8.6229	58.807*	0.19911	——	53
54	0.80144*	6.6817*	5.1005*	-0.00157*	0.0948	0.9846	1.1767	11.135	74.401*	0.23827	0.50418	54
55	1.3974*	11.650*	5.4987*	——	0.2548	0.9802	2.2118	5.9238	69.012*	0.14804	0.32460	55
56	0.61832*	5.1550*	3.3037*	——	0.2557	0.9877	0.5880	22.283	114.87*	0.49677	1.1209	56
57	0.87476*	7.2930*	3.9713*	——	——	1.0000	1.0000	13.103	95.557*	0.23988	——	57
58	0.071070*	0.59252*	3.4022*	——	-0.2202	1.0006	0.06960	188.25	111.54*	3.4038	——	58
59	1.1421*	9.5221*	3.3605*	——	0.0216	0.9992	1.1048	11.859	112.93*	0.21892	——	59
60	0.80940*	6.7481*	4.1513*	——	0.0372	0.9997	0.9672	13.546	91.413*	0.24828	——	60
61	1.4244*	11.875*	5.9710*	——	0.0878	(0.9875)	2.4482	5.3519	63.554*	0.11377	——	61
62	1.00000	8.33712	2.1609	-0.00009	0.3443	——	0.62202	21.065	175.62	0.44457	0.99974	62
63	0.12510*	1.0430*	3.8376*	——	0.	1.0006	0.1382	94.814	98.891*	1.2404	——	63
64	0.85129*	7.0973*	5.1373*	-0.00300*	0.1259	0.9923	1.2589	10.408	73.869*	0.19086	——	64

12/12/86

NOTE: Numbers in this table do not have accuracies greater than 1 part in 1,000; in some cases extra digits have been added to calculated values to achieve consistency or to permit recalculation of experimental values.

Physical Constants

*See the Table of Notes and References.

Number	Compound	K. Heating value, 60°F Net — Btu/ft³ Ideal gas, 14.696 psia	Btu/lbm Liquid	K. Heating value, 60°F Gross — Btu/ft³ ideal gas 14.696 psia	Btu/lbm Liquid	Btu/gal Liquid	L. Heat of vaporization 14.696 psia at boiling point, Btu/lbm	M. Air required for combustion ideal gas ft³(air)/ft³(gas)	Flammability limits, vol % in air mixture — Lower	Higher	ASTM octane number — Motor method D-357	Research method D-908	Number
1	Methane	909.4	———	1010.0	———	———	219.45	9.548	5.0	15.0	———	———	1
2	Ethane	1618.7	20277.*	1769.6	22181.*	65869.*	211.14	16.710	2.9	13.0	+0.05	+1.6*	2
3	Propane	2314.9	19757.*	2516.1	21489.*	90830.*	183.01	23.871	2.0	9.5	97.1	+1.8*	3
4	Isobutane	3000.4	19437.*	3251.9	21079.*	98917.*	157.23	31.032	1.8	8.5	97.6	+0.1*	4
5	n–Butane	3010.8	19494.*	3262.3	21136.*	102911.*	165.93	31.032	1.5	9.0	89.6*	93.8*	5
6	Isopentane	3699.0	19303.	4000.9	20891.	108805.	147.12	38.193	1.3	8.0	90.3	92.3	6
7	n–Pentane	3706.9	19335.	4008.9	20923.	110091.	153.57	38.193	1.4	8.3	62.6*	61.7*	7
8	Neopentane	3682.9	19235.*	3984.7	20822.*	103577.*	135.58	38.193	1.3	7.5	80.2	85.5	8
9	n–Hexane	4403.8	19232.	4755.9	20783.	115021.	143.94	45.355	1.1	7.7	26.0	24.8	9
10	2–Methylpentane	4395.2	19202.	4747.3	20753.	113822.	138.45	45.355	1.18	7.0	73.5	73.4	10
11	3–Methylpentane	4398.2	19213.	4750.3	20764.	115813.	140.05	45.355	1.2	7.7	74.3	74.5	11
12	Neohexane	4384.0	19163.	4736.2	20714.	112916.	131.23	45.355	1.2	7.0	93.4	91.8	12
13	2,3–Dimethylbutane	4392.9	19195.	4745.0	20746.	115246.	136.07	45.355	1.2	7.0	94.3	+0.3	13
14	n–Heptane	5100.0	19155.	5502.5	20679.	118648.	136.00	52.516	1.0	7.0	0.0	0.0	14
15	2–Methylhexane	5092.2	19133.	5494.6	20657.	117644.	131.58	52.516	1.0	7.0	46.4	42.4	15
16	3–Methylhexane	5096.0	19146.	5498.6	20671.	119197.	132.10	52.516	(1.01)	(6.6)	55.8	52.0	16
17	3–Ethylpentane	5098.3	19154.	5500.7	20679.	121158.	132.82	52.516	(1.00)	(6.5)	69.3	65.0	17
18	2,2–Dimethylpentane	5079.6	19095.	5481.9	20620.	116606.	125.12	52.516	(1.09)	(6.8)	95.6	92.8	18
19	2,4–Dimethylpentane	5084.2	19111.	5486.7	20635.	116526.	126.57	52.516	(1.08)	(6.8)	83.8	83.1	19
20	3,3–Dimethylpentane	5086.4	19119.	5488.8	20643.	120080.	127.20	52.516	(1.04)	(7.0)	86.6	80.8	20
21	Triptane	5081.2	19103.	5483.5	20628.	119451.	124.21	52.516	(1.08)	(6.8)	+0.1	+1.8	21
22	n–Octane	5796.1	19096.	6248.9	20601.	121422.	129.52	59.677	0.8	6.5	———	———	22
23	Diisobutyl	5780.5	19047.	6233.5	20552.	119586.	122.83	59.677	(0.92)	(6.3)	55.7	55.2	23
24	Isooctane	5778.8	19063.	6231.7	20568.	119389.	112.94	59.677	0.95	6.0	100.0	100.0	24
25	n–Nonane	6493.2	19054.	6996.5	20543.	123634.	124.36	66.839	0.7	5.6	———	———	25
26	n–Decane	7189.6	19018.	7742.9	20494.	125448.	119.65	74.000	0.7	5.4	———	———	26
27	Cyclopentane	3512.1	18825.	3763.7	20186.	126304.	167.33	35.806	(1.48)	(8.3)	84.9*	+0.1	27
28	Methylcyclopentane	4199.4	18771.	4501.2	20132.	126467.	148.54	42.968	1.0	8.35	80.0	91.3	28
29	Cyclohexane	4179.7	18675.	4481.7	20036.	130873.	153.03	42.968	1.2	8.35	77.2	83.0	29
30	Methylcyclohexane	4863.6	18640.	5215.9	20002.	129071.	136.30	50.129	1.1	6.7	71.1	74.8	30
31	Ethene(Ethylene)	1499.1	———	1599.8	———	———	207.41	14.323	2.7	36.0	75.6	+0.03	31
32	Propene(Propylene)	2181.8	(19858.)	2332.7	(21208.)	(92113.)	188.19	21.484	2.0	11.7	84.9	+0.2	32
33	1–Butene(Butylene)	2878.7	19309.*	3079.9	20670.*	103582.*	167.96	28.645	1.6	10.	80.8*	97.4	33
34	cis–2–Butene	2871.0	19241.*	3072.2	20602.*	107724.*	178.89	28.645	1.6	10.	83.5	100.0	34
35	trans–2–Butene	2866.8	19221.*	3068.0	20582.*	104666.*	174.37	28.645	1.6	10.	———	———	35
36	Isobutene	2859.9	19182.*	3061.1	20543.*	102830.*	169.47	28.645	1.6	10.	———	———	36
37	1–Pentene	3575.0	19184.	3826.5	20545.	110602.	154.48	35.806	1.3	10.	77.1	90.9	37
38	1,2–Butadiene	2789.0	19378.*	2939.9	20437.*	112111.*	191.88	26.258	(1.62)	(10.3)	———	———	38
39	1,3–Butadiene	2729.0	18967.*	2879.9	20025.*	104717.*	185.29	26.258	2.0	12.5	———	———	39
40	Isoprene	3410.8	18832.	3612.1	19953.	114141.	163.48	33.419	(1.12)	(8.5)	81.0	99.1	40
41	Acetylene	1423.2	(20887.)	1473.5	(21613.)	(75204.)	151.90	11.935	1.5	100.	———	———	41
42	Benzene	3590.9	17256.	3741.8	17989.	132651.	169.24	35.806	1.2	8.0	+2.8	———	42
43	Toluene	4273.6	17421.	4475.0	18250.	132661.	154.83	42.968	1.2	7.1	+0.3	+5.8	43
44	Ethylbenzene	4970.5	17593.	5222.2	18492.	134387.	144.02	50.129	1.0	8.0	97.9	+0.8	44
45	o–Xylene	4958.2	17544.	5209.9	18444.	136036.	149.10	50.129	1.0	7.6	100.0	———	45
46	m–Xylene	4956.3	17541.	5207.9	18440.	133559.	147.24	50.129	1.0	7.0	+2.8	+4.0	46
47	p–Xylene	4957.1	17545.	5208.8	18444.	133131.	145.71	50.129	1.0	7.0	+1.2	+3.4	47
48	Styrene	4829.8	17414.	5031.1	18147.	137841.	152.85	47.742	1.1	8.0	+0.2	>+3.*	48
49	Isopropylbenzene	5660.9	17709.	5962.8	18662.	134792.	134.24	57.290	0.8	6.5	99.3	+2.1	49
50	Methyl alcohol	766.1	8559.	866.7	9751.	64731.	462.58	7.161	5.5	44.0	———	———	50
51	Ethyl alcohol	1448.1	11530.	1599.1	12770.	84539.	359.07	14.323	3.28	19.0	———	———	51
52	Carbon monoxide	320.5	———	320.5	———	———	92.77	2.387	12.50	74.20	———	———	52
53	Carbon dioxide	0.0	———	0.0	———	———	246.47*	———	———	———	———	———	53
54	Hydrogen sulfide	586.8	6337.*	637.1	6897.*	46086.*	235.63	7.161	4.30	45.50	———	———	54
55	Sulfur dioxide	0.0	———	0.0	———	———	167.22	———	———	———	———	———	55
56	Ammonia	359.0	———	434.4	———	———	589.48	3.581	15.50	27.00	———	———	56
57	Air	0.0	———	0.0	———	———	88.20	———	———	———	———	———	57
58	Hydrogen	273.8	———	324.2	———	———	192.74	2.387	4.00	74.20	———	———	58
59	Oxygen	0.0	———	0.0	———	———	91.59	———	———	———	———	———	59
60	Nitrogen	0.0	———	0.0	———	———	85.59	———	———	———	———	———	60
61	Chlorine	———	———	———	———	———	123.75	———	———	———	———	———	61
62	Water	0.0	*	*	0.0	0.0	970.18	———	———	———	———	———	62
63	Helium	0.0	———	0.0	———	———	———	———	———	———	———	———	63
64	Hydrogen chloride	———	———	———	———	———	190.43	———	———	———	———	———	64

12/12/86

NOTE: Numbers in this table do not have accuracies greater than 1 part in 1,000; in some cases extra digits have been added to calculated values to achieve consistency or to permit recalculation of experimental values.

(Cont'd)
NOTES AND REFERENCES FOR THE TABLE OF PHYSICAL CONSTANTS

Number	Compound	Formula	A. Molar mass (molecular weight)	B. Boiling point, °F 14.696 psia	Vapor pressure, psia 100 °F	C. Freezing point, °F 14.696 psia	D. Refractive index, n_D 60 °F	Critical constants Pressure, psia	Temperature, °F	Volume, ft^3/lbm	Number
1	Methane	CH_4	19	4	a,b	c,57	i,57	4	4	4	1
2	Ethane	C_2H_6	19	35	a,b	c,57	h,55	35	35	35	2
3	Propane	C_3H_8	19	34	34	c,57	h,55	34	34	34	3
4	Isobutane	C_4H_{10}	19	30	32	57	h,55	32	32	32	4
5	n—Butane	C_4H_{10}	19	33	33	57	h,55	33	33	33	5
6	Isopentane	C_5H_{12}	19	57	57	57	57	43	43	43	6
7	n—Pentane	C_5H_{12}	19	57	57	57	57	43	43	43	7
8	Neopentane	C_5H_{12}	19	57	57	57	h,57	43	43	43	8
9	n—Hexane	C_6H_{14}	19	57	57	57	57	43	43	43	9
10	2—Methylpentane	C_6H_{14}	19	57	57	57	57	43	43	43	10
11	3—Methylpentane	C_6H_{14}	19	57	57	f,22,29	57	43	43	43	11
12	Neohexane	C_6H_{14}	19	57	57	57	57	43	43	43	12
13	2,3—Dimethylbutane	C_6H_{14}	19	57	57	57	57	43	43	43	13
14	n—Heptane	C_7H_{16}	19	57	57	57	57	43	43	43	14
15	2—Methylhexane	C_7H_{16}	19	57	57	57	57	43	43	43	15
16	3—Methylhexane	C_7H_{16}	19	57	57	f,38	57	43	43	43	16
17	3—Ethylpentane	C_7H_{16}	19	57	57	57	57	43	43	43	17
18	2,2—Dimethylpentane	C_7H_{16}	19	57	57	57	57	43	43	43	18
19	2,4—Dimethylpentane	C_7H_{16}	19	57	57	57	57	43	43	43	19
20	3,3—Dimethylpentane	C_7H_{16}	19	57	57	57	57	43	43	43	20
21	Triptane	C_7H_{16}	19	57	57	57	57	43	43	43	21
22	n—Octane	C_8H_{18}	19	57	57	57	57	43	43	43	22
23	Diisobutyl	C_8H_{18}	19	57	57	57	57	43	43	43	23
24	Isooctane	C_8H_{18}	19	57	57	57	57	43	43	43	24
25	n—Nonane	C_9H_{20}	19	57	57	57	57	43	43	43	25
26	n—Decane	$C_{10}H_{22}$	19	57	57	57	57	43	43	43	26
27	Cyclopentane	C_5H_{10}	19	57	57	57	57	57	57	57	27
28	Methylcyclopentane	C_6H_{12}	19	57	57	57	57	57	57	57	28
29	Cyclohexane	C_6H_{12}	19	57	57	57	57	57	57	57	29
30	Methylcyclohexane	C_7H_{14}	19	57	57	57	57	57	57	57	30
31	Ethene(Ethylene)	C_2H_4	19	63	b	c,63	h,26	63	63	63	31
32	Propene(Propylene)	C_3H_6	19	5	5	5	h,26	q,5	44	57	32
33	1—Butene(Butylene)	C_4H_8	19	57	57	57	h,26	57	57	57	33
34	cis—2—Butene	C_4H_8	19	57	25	57	h,26	57	57	57	34
35	trans—2—Butene	C_4H_8	19	57	25	57	h,26	57	57	57	35
36	Isobutene	C_4H_8	19	57	57	57	h,26	57	57	57	36
37	1—Pentene	C_5H_{10}	19	57	57	57	26	57	57	57	37
38	1,2—Butadiene	C_4H_6	19	57	57	56		a	a	a	38
39	1,3—Butadiene	C_4H_6	19	57	25	56	h,26	57	57	57	39
40	Isoprene	C_5H_8	19	57	57	56		a	a	a	40
41	Acetylene	C_2H_2	19	d,1	b	c,56		57	57	57	41
42	Benzene	C_6H_6	19	14	57	14	57	14	14	14	42
43	Toluene	C_7H_8	19	57	57	15	57	57	57	57	43
44	Ethylbenzene	C_8H_{10}	19	57	57	15	57	57	57	57	44
45	o—Xylene	C_8H_{10}	19	57	57	15	57	57	57	57	45
46	m—Xylene	C_8H_{10}	19	57	57	15	57	57	57	57	46
47	p—Xylene	C_8H_{10}	19	57	57	15	57	57	57	57	47
48	Styrene	C_8H_8	19	56	13	56	57	56	a,11	56	48
49	Isopropylbenzene	C_9H_{12}	19	57	r,57	57	57	57	57	57	49
50	Methyl alcohol	CH_4O	19	58	58	58	58	58	58	58	50
51	Ethyl alcohol	C_2H_6O	19	58	58	58	58	1	58	58	51
52	Carbon monoxide	CO	19	58	b	c,58	i,58	1	1	1	52
53	Carbon dioxide	CO_2	19	d,3	b	c,3	i,58	3	3	3	53
54	Hydrogen sulfide	H_2S	19	31	31	c,31	i,58	31	31	31	54
55	Sulfur dioxide	SO_2	19	1	23	c,58	i,58	58	58	58	55
56	Ammonia	NH_3	19	1	23	c,58	i,58	1	1	1	56
57	Air	N_2+O_2	19,40	45	b		i,60	45	45	45	57
58	Hydrogen	H_2	19	p,58	b	p,c,58	i,58	p,58	p,58	p,58	58
59	Oxygen	O_2	19	u,10	b	c,u,10	i,58	46	46	46	59
60	Nitrogen	N_2	19	6	b	c,6	i,58	6	6	6	60
61	Chlorine	Cl_2	19	1	2	c,58	h,58	2	2	2	61
62	Water	H_2O	19	u,10	37	58	58	37	37	37	62
63	Helium	He	19	7	b		i,58	7	s,7	7	63
64	Hydrogen chloride	HCl	19	1	23	c,58	i,58	58	58	58	64

12/12/86

(Cont'd)
NOTES AND REFERENCES FOR THE TABLE OF PHYSICAL CONSTANTS

Number	E. Density of liquid 14.696 psia, 60°F — Relative density (specific gravity) 60°F/60°F	E. — lbm/gal.	E. — gal./lb mole	F. Temperature coefficient of density, 1/°F	G. Acentric factor, 3	H. Compressibility factor of real gas, Z 14.696 psia, 60°F	I. Ideal gas 14.696 psia, 60°F — Relative density (specific gravity) Air = 1	I. — ft³ gas/lbm	I. — ft³ gas/gal. liquid	J. Specific Heat 60°F 14.696 psia Btu/(lbm·°F) — C_p, Ideal gas	J. — C_p, Liquid	Number
1	b,a	b,a,v	b,a	b,a	4	57			b,a	8	b	1
2	h,35	h,v,35	h,35	h,35	57	57				8	53	2
3	h,34	h,34	h,34	h,34	34	57			h	8	34	3
4	h,n,32	h,n,32	h,n,32	h,n,32	32	57			h	30	32	4
5	h,33	h,33	h,33	h,33	33	57			h	8	33	5
6	57	57	57	57	57					8	36	6
7	57	57	57	57	57					8	47	7
8	h,57	h,57	h,57	h,57	57	57			h	8	24	8
9	57	57	57	57	57					8	47	9
10	57	57	57	57	57					8	57	10
11	57	57	57	57	57					8	57	11
12	57	57	57	57	57					8	57	12
13	57	57	57	57	57					8	57	13
14	57	57	57	57	57					57	47	14
15	57	57	57	57	57					57	57	15
16	57	57	57	57	57					57	38	16
17	57	57	57	57	57					57	57	17
18	57	57	57	57	57					57	57	18
19	57	57	57	57	57					57	57	19
20	57	57	57	57	57					57	57	20
21	57	57	57	57	57					57	57	21
22	57	57	57	57	57					57	47	22
23	57	57	57	57	57					57	57	23
24	57	57	57	57	57					57	57	24
25	57	57	57	57	57					57	47	25
26	57	57	57	57	57					57	47	26
27	57	57	57	57	57					8	22,59	27
28	57	57	57	57	r,57					57	22,59	28
29	57	57	57	57	57					8	9	29
30	57	57	57	57	57					57	22,59	30
31	b	b	b	b	57	57			b	8	b	31
32	h,5	h,5	h,5	h,5	5	57			h	8	17	32
33	h	h	h	h	57	57			h	57	24	33
34	h,r	h,r	h,r	h,r	25	57			h	57	17	34
35	h,r	h,r	h,r	h,r	25	57			h	57	r,17	35
36	h,r	h,r	h,r	h,r	57	57			h	57		36
37	57	57	57	57	57					57	57	37
38	h,r	h,r	h,r	h,r	r,57	a			h	57	57	38
39	h,r	h,r	h,r	h,r	25	a			h	57	54,59	39
40	r	r	r	r	57					57	48	40
41	b	b	b	b	r,57	57			b	8		41
42	57	57	57	57	57					57	15	42
43	57	57	57	57	57					16	15	43
44	57	57	57	57	57					16	15	44
45	57	57	57	57	57					16	15	45
46	57	57	57	57	57					16	15	46
47	57	57	57	57	57					16	15	47
48	j,50	j,50	j,50	j,52	a,51					57		48
49	57	57	57	57	57					61	42	49
50	58	58	58	58	58					62	62	50
51	62	62	62	62	58					62	62	51
52	k,39	k,39	k,39		39				t	8	b	52
53	h,3	h,3	h,3	h,3					h	8		53
54	h,31	h,v,31	h,31	h,31	31	58			h	8	31	54
55	h	h	h		23	58			h	8		55
56	h	h	h		23	58			h	58		56
57	k,45	k,45	k,45		57				t	45		57
58	k,p,21	k,p,21	k,p,21		p,23	58			t	8	b	58
59	k,46	k,46	k,46		q,58	58			t	8	b	59
60	k,6	k,6	k,6		6	58			t	8	b	60
61	h	h	h		2	a			h	58		61
62	37	37	37	41	r,58					27	28	62
63	k	k	k		58	58			t	8	b	63
64	h	h	h	h	58	58			h	58		64

12/12/86

(Cont'd)
NOTES AND REFERENCES FOR THE TABLE OF PHYSICAL CONSTANTS

Number	Compound	K. Heating value, 60°F — Net Btu/ft³ Ideal gas, 14.696 psia	Net Btu/lbm Liquid	Gross Btu/ft³ ideal gas, 14.696 psia	Gross Btu/lbm Liquid	Gross Btu/gal. Liquid	L. Heat of vaporization 14.696 psia at boiling point, Btu/lbm	M. Air required for combustion ideal gas ft³(air)/ft³(gas)	Flammability limits Lower	Flammability limits Higher	ASTM octane Motor method D-357	ASTM octane Research method D-908	Number
1	Methane	57	b	57	b	b	4		x	x			1
2	Ethane	57	h,57	57	h,57	h,57	35		x	x		e,g	2
3	Propane	57	h,57	57	h,57	h,57	34		x	x	e	e,g	3
4	Isobutane	57	h,57	57	h,57	h,57	32		x	x		e,g	4
5	n-Butane	57	h,57	57	h,57	h,57	33		x	x	g	g	5
6	Isopentane	57	57	57	57	57	57		x	x			6
7	n-Pentane	57	57	57	57	57	57		x	x	g	g	7
8	Neopentane	57	h,57	57	h,57	h,57	57		x	x			8
9	n-Hexane	57	57	57	57	57	57		x	x			9
10	2-Methylpentane	57	57	57	57	57	57		x	x			10
11	3-Methylpentane	57	57	57	57	57	57		x	x			11
12	Neohexane	57	57	57	57	57	57		x	x			12
13	2,3-Dimethylbutane	57	57	57	57	57	57		x	x		e	13
14	n-Heptane	57	57	57	57	57	57		x	x			14
15	2-Methylhexane	57	57	57	57	57	57		x	x			15
16	3-Methylhexane	57	57	57	57	57	57		a,x	a,x			16
17	3-Ethylpentane	57	57	57	57	57	57		a,x	a,x			17
18	2,2-Dimethylpentane	57	57	57	57	57	57		a,x	a,x			18
19	2,4-Dimethylpentane	57	57	57	57	57	57		a,x	a,x			19
20	3,3-Dimethylpentane	57	57	57	57	57	57		a,x	a,x			20
21	Triptane	57	57	57	57	57	57		a,x	a,x	e	e	21
22	n-Octane	57	57	57	57	57	57		x	x			22
23	Diisobutyl	57	57	57	57	57	57		a,x	a,x			23
24	Isooctane	57	57	57	57	57	57		x	x			24
25	n-Nonane	57	57	57	57	57	57		x	x			25
26	n-Decane	57	57	57	57	57	57		a,x	a,x			26
27	Cyclopentane	57	57	57	57	57	57		a,x	a,x	g	e	27
28	Methylcyclopentane	20	20	20	20	20	57		x	x			28
29	Cyclohexane	57	57	57	57	57	57		x	x			29
30	Methylcyclohexane	20	20	20	20	20	57		x	x			30
31	Ethene(Ethylene)	57	b	57	b	b	63		x	x		e	31
32	Propene(Propylene)	57	a	57	a	a	57		x	x		e	32
33	1-Butene(Butylene)	57	h,57	57	h,57	h,57	57		x	x	g		33
34	cis-2-Butene	57	h,57	57	h,57	h,57	57		a,x	a,x			34
35	trans-2-Butene	57	h,57	57	h,57	h,57	57		a,x	a,x			35
36	Isobutene	57	h,57	57	h,57	h,57	57		a,x	a,x			36
37	1-Pentene	57	57	57	57	57	57		x	x			37
38	1,2-Butadiene	57	h,57	57	h,57	h,57	56		a,x	a,x			38
39	1,3-Butadiene	57	h,57	57	h,57	h,57	56		x	x			39
40	Isoprene	57	57	57	57	57	56		a,x	a,x			40
41	Acetylene	57	a	57	a	a	56		x	x			41
42	Benzene	57	57	57	57	57	14		x	x	e		42
43	Toluene	57	57	57	57	57	57		x	x	e	e	43
44	Ethylbenzene	57	57	57	57	57	57		x	x		e	44
45	o-Xylene	57	57	57	57	57	57		x	x			45
46	m-Xylene	57	57	57	57	57	57		x	x	e	e	46
47	p-Xylene	57	57	57	57	57	57		x	x	e	e	47
48	Styrene	20	20	20	20	20	56		x	x	e	e	48
49	Isopropylbenzene	57	57	57	57	57	57		x	x		e	49
50	Methyl alcohol	62	62	62	62	62	58		x	x			50
51	Ethyl alcohol	62	62	62	62	62	58		x	x			51
52	Carbon monoxide	8	b	8	b	b	58		x	x			52
53	Carbon dioxide						58						53
54	Hydrogen sulfide	18	h,58	18	h,58	h,58	31		x	x			54
55	Sulfur dioxide						56						55
56	Ammonia	58		58			58		x	x			56
57	Air						45						57
58	Hydrogen	58		58			p,58		x	x			58
59	Oxygen						46						59
60	Nitrogen						58						60
61	Chlorine						56						61
62	Water	8	w,8	w,8			37						62
63	Helium												63
64	Hydrogen chloride						56						64

12/12/86

(Cont'd)

NOTES FOR THE TABLE OF PHYSICAL CONSTANTS

a. Values in parentheses are estimated values.

b. The temperature is above the critical point.

c. At saturation pressure (triple point).

d. Sublimation point.

e. The + sign and number following specify the number of cm^3 of TEL added per gallon to achieve the ASTM octane number of 100, which corresponds to that of Isooctane (2,2,4-Trimethylpentane.)

f. These compounds form a glass. An earlier investigator [12] reported a value of -244°F for the freezing point of 3-Methylpentane.

g. Average value from octane numbers of more than one sample.

h. Saturation pressure and 60°F.

i. Index of refraction of the gas.

j. The temperature difference correction from [51].

k. Densities of the liquid at the normal boiling point.

m. Heat of sublimation.

n. Equation 2 of the reference was refitted to give: $a = 0.7872957; b = 0.1294083; c = 0.03439519.$

p. Normal hydrogen (25% para, 75% ortho).

q. The vapor pressure [5] at the critical temperature [43].

r. An extrapolated value.

s. The value for the critical point of helium depends upon the temperature scale chosen. See reference 7.

t. Gas at 60°F and the liquid at the normal boiling point.

u. Fixed points on the 1968 International Practical Temperature Scale (IPTS-68).

v. Densities at the normal boiling point are: Ethane, 4.540 [34]; Propane, 4.848 [33]; Propene, 5.083 [5]; Hydrogen Chloride, 9.948 [56]; Hydrogen Sulfide, 7.919 [30]; Ammonia, 5.688 [56]; Sulfur Dioxide, 12.20 [56].

w. Technically, water has a heating value in two cases: net (-1060. Btu/lbm) when water is liquid in the reactants and gross (+50.313 Btu/ft^3) when water is gas in the reactants. The value is the ideal heat of vaporization (enthalpy of the ideal gas less the enthalpy of the saturated liquid at the vapor pressure. This is a matter of definition; water does not burn.

x. Extreme values of those reported by ref. 24.

A. Molar mass (molecular weight) is based upon the following atomic weights: C = 12.011; H = 1.00794; O = 15.9994; N = 14.0067; S = 32.06, Cl = 35.453. The values were rounded off after calculating the molar mass using all significant figures in the atomic weights.

B. Boiling point: the temperature at equilibrium between the liquid and vapor phases at 14.696 psia.

C. Freezing point: the temperature at equilibrium between the crystalline phase and the air saturated liquid at 14.696 psia.

D. The refactive index reported refers to the liquid or gas and is measured for light of wavelength corresponding to the sodium D-line (589.26 nm).

E. The relative density (specific gravity): $\rho(\text{liquid, } 60°F)/\rho(\text{water, } 60°F)$. The density of water at 60°F is 8.33718 lbm/gal.

F. The temperature coefficient of density is related to the expansion coefficient by:

$$\frac{1}{\rho}\left(\frac{\partial \rho}{\partial T}\right)_P = -\frac{1}{V}\left(\frac{\partial V}{\partial T}\right)_P , \text{ in units of } 1/°F$$

G. Pitzer acentric factor:

$$\omega = -\log_{10}(P/P_c) - 1, \quad P \text{ at } T = 0.7\, T_c$$

H. Compressibility factor of the real gas, $Z = (PV)/(RT)$, is calculated using the second virial coefficient equation.

I. The density of an ideal gas relative to air is calculated by dividing the molar mass of the of the gas by 28.9625, the calculated average molar mass of air. See ref. 39 for the average composition of dry air. The specific volume of an ideal gas is calulated from the ideal gas equation. The volume ratio is: $V(\text{ideal gas})/V(\text{liquid in vacuum})$.

J. The liquid value is not rigorously C_P, but rather it is C_S, the heat capacity at saturated conditions, defined by:

$$C_S = C_P - T\left(\frac{\partial V}{\partial T}\right)_P \frac{dP_s}{dT}$$

When dealing with liquids far from the critical point, $C_S \approx C_P$.

K. The heating value is the negative of the enthalpy of combustion at 60°F and 14.696 psia in an ideal reaction (one where all gasses are ideal gasses). For an arbitrary organic compound, the combustion reaction is:

$$C_n H_m O_h S_j N_k (s, l, or\ g) + (n + \frac{m}{4} - \frac{h}{2} + j)O_2(g) \longrightarrow nCO_2(g) + \frac{m}{2}H_2O(g\ or\ l) + \frac{k}{2}N_2(g) + jSO_2(g)$$

where s, l, and g denote respectively solid, liquid and ideal gas. For gross heating values, the water formed is liquid; for net heating values, the water formed is ideal gas. Values reported are on a dry basis. To account for water in the heating value, see GPA 2172. The Btu/lbm or gal. liquid column assumes a reaction with the fuel in the liquid state, while the Btu/ft^3 ideal gas column assumes the gas in the ideal gas state. Therefore, the values are not consistent if used in the same calculation, e.g., a gas plant balance.

L. The heat of vaporization is the enthalpy of the saturated vapor at the boiling point at 14.696 psia minus the enthalpy of the saturated liquid at the same pressure.

M. Air required for the combustion of ideal gas for compounds of formula $C_n H_m O_h S_j N_k$ is:

$$V(air)/V(gas) = (n + \frac{m}{4} - \frac{h}{2} + j)/0.20946$$

COMMENTS

Units - all dimensional values are reported in terms of the following three units, which are defined in terms of the corresponding SI units:

mass - pound (avdp), lbm = 0.45359337 kg

length - foot, ft = 0.3048 m

temperature - degree fahrenheit, (t/°F) = 32 + [1.8(t/°C)]

Temperature of the Celsius scale is defined by the International Practical Temperature of 1968 (IPTS-68).

Other derived units are:

volume - cubic foot, ft^3 = 0.02831685 m^3

gallon = 231 in^3 = 0.003785412 m^3

pressure - pound per square inch absolute = 6894.757 Pa

energy - British thermal unit (I.T.), Btu = 251.9958 cal (I.T.) = 1055.056 J

Physical constants:

gas constant, R = 1.98598 Btu(I.T.)/(°R·lb mol)

10.7316 ft^3 psia/(°R·lb mol)

8.31448 J/(K · mol)

Conversion factors:

1 ft^3 = 7.480520 gal.

1 lbm/ft^3 = 0.1336806 lbm/gal = 16.018462 kg/m^3

1 psia = 0.06804596 atm

1 atm = 14.69595 psia = 760 Torr

1 Btu (I.T.) = 252.1644 cal$_{th}$

REFERENCES FOR THE TABLE OF PHYSICAL CONSTANTS

1. Ambrose, D., National Physical Laboratory, Teddington, Middlesex, England: Feb. 1980, NPL Report Chem 107.
2. Ambrose, D.; Hall, D. J.; Lee, D. A.; Lewis, G. B.; Mash, C. J., J. Chem. Thermo., 11, 1089 (1979).
3. Angus, S.; Armstrong, B.; de Reuck, K. M., Eds. "Carbon Dioxide. International Thermodynamic Tables of the Fluid State-3"; Pergamon Press: Oxford, 1976.
4. Angus, S.; Armstrong, B.; de Reuck, K. M., Eds. "Methane. International Thermodynamic Tables of the Fluid State-5"; Pergamon Press: Oxford, 1978.
5. Angus, S.; Armstrong, B.; de Reuck, K. M., "Propylene (Propene). International Thermodynamic Tables of the Fluid State-7"; Pergamon Press: Oxford, 1980.
6. Angus, S.; de Reuck, K. M.; Armstrong, B., Eds. "Nitrogen. International Thermodynamic Tables of the Fluid State-6"; Pergamon Press: Oxford, 1979.
7. Angus, S.; de Reuck, K. M.; McCarthy, R. D., Eds. "Helium. International Thermodynamic Tables of the Fluid State-4"; Pergamon Press: Oxford, 1977.
8. Armstrong, G. T.; Jobe, T. L., "Heating Values of Natural Gas and its Components," NBSIR 82-2401, May 1982.
9. Aston, J. G.; Szasz, G. J.; Fink, H. L., J. Am. Chem. Soc., 65, 1135 (1943).
10. Barber, C. R., Metrologia 5, 35 (1969).
11. Boundy, R. H.; Boyer, R. F., (Eds.), "Styrene, Its Polymers, Copolymers and Derivatives," A.C.S. monograph No. 115, Reinholt, N.Y., 1952.
12. Bruun, J. H.; Hicks-Bruun, M. M., J. Res. NBS, 5, 933 (1930).
13. Chaiyavech, P.; Van Winkle, M., J. Chem. Eng. Data, 4, 53 (1959).
14. Chao, J., "Bensene," Key chemical data books, Thermodynamics Research Center, Texas A & M University, College Station, Texas (1978).
15. Chao, J., Hydrocarbon Processing, 59(11), 295 (1979).
16. Chao, J.; Hall, K. R., Thermochimica acta, 72, 323 (1984).
17. Chao, J.; Hall, K. R.; Yao, J., Thermochimica acta, 64, 285 (1983).
18. CODATA Task Group on Key Values for Thermodynamics, CODATA Special Report No. 7, 1978.
19. Commission on Atomic Weights and Isotopic Abundances, Pure and Appl. Chem. 49, 1102 (1983).
20. Cox, J. D.; Pilcher, G., "Thermochemistry of Organic and Organometallic Compounds," Academic Press, London, 1970.
21. Dean, J. W., "A Tabulation of the Properties of Normal Hydrogen from Low Temperature to 300 K and from 1 to 100 Atmospheres," NBS Tech. Note 120, November 1961.
22. Douslin, D. R.; Huffman, H. M., J. Am. Chem. Soc., 68, 1704 (1946).
23. Edwards, D. G., "The Vapor Pressure of 30 Inorganic Liquids Between One Atmosphere and the Critical Point," Univ. of Calif., Lawrence Radiation Laboratory, UCRL-7167. June 13, 1963.
24. Engineering Sciences Data Unit, "EDSU, Engineering Sciences Data," EDSU International Ltd., London.
25. Flebbe, J. L.; Barclay, D. A.; Manley, D. B., J. Chem. Eng. Data, 27, 405 (1982).
26. Francis, A. W., J. Chem. Eng. Data, 5, 534 (1960).
27. Friedman, A. S.; Haar, L., J. Chem. Phys. 22, 2051 (1954).
28. Ginnings, D. C.; Furukawa, G. T., J. Am. Chem. Soc. 75, 522 (1953).
29. Glasgow, A. R.; Murphy, E. T.; Willingham, C. B.; Rossini, F. D., J. Res. NBS, 37, 141 (1946).
30. Goodwin, R. D., "Isobutane: Provisional Thermodynamic Functions from 114 to 700K at Pressures to 700 Bar," NBSIR 79-1612, July 1979.
31. Goodwin, R. D., "Hydrogen Sulfide Provisional Thermochemical Properties from 188 to 700 K at Pressures to 75 MPa," NBSIR 83-1694, October 1983.
32. Goodwin, R. D.; Haynes, W. M., "Thermophysical Properties of Isobutane from 114 to 700 K at Pressures to 70 MPa," NBS Tech. Note 1051, January 1982.
33. Goodwin, R. D.; Haynes, W. M., "Thermophysical Properties of Normal Butane from 135 to 700 K at Pressures to 70 MPa," NBS Monograph 169, April 1982.
34. Goodwin, R. D.; Haynes, W. M., "Thermophysical Properties of Propane from 85 to 700 K at Pressures to 70 MPa," NBS Monograph 170, April 1982.
35. Goodwin, R. D.; Roder, H. M.; Straty, G. C., "Thermophysical Properties of Ethane, 90 to 600 K at Pressures to 700 bar," NBS Tech. Note 684, August 1976.
36. Guthrie, G. B.; Huffman, H. M., J. Am. Chem. Soc., 65, 1139 (1943).
37. Haar, L.; Gallagher, J. S.; Kell, G. S., "NBS/NRC Steam Tables," Hemisphere Publishing Corporation, Washington, 1984.
38. Huffman, H. M.; Park, G. S.; Thomas, S. B., J. Am. Chem. Soc., 52, 3241 (1930).
39. Hust, J. G.; Stewart, R. B., "Thermodynamic Property Values for Gaseous and Liquid Carbon Monoxide from 70 to 300 Atmospheres," NBS Technical Note 202, Nov. 1963.
40. Jones, F. E., J. Res. NBS, 83, 419 (1978).
41. Kell, G. S., J. Chem. Eng. Data, 20, 97 (1975).
42. Kishimoto, K.; Suga, H.; Syuzo, S., Bull. Chem. Soc. Japan, 46, 3020 (1973).
43. Kudchadker, A. P.; Alani, G. H.; Zwolinski, B. J., Chem. Rev. 68, 659 (1968).
44. Marchman, H.; Prengle, H. W.; Motard, R. L., Ind. Eng. Chem., 41, 2658 (1949).
45. "The Matheson Unabridged Gas Data Book"; Matheson Gas Products; New York, 1974.
46. McCarty, R. D.; Weber, L. A., "Thermophysical Properties of Oxygen from the Freezing Liquid Line to 600 R for Pressures to 5000 Psia", NBS Technical Note 384, July 1971.
47. Messerly, J. F.; Guthrie, G. B.; Todd, S. S.; Finke, H. L., J. Chem. Eng. Data, 12, 338 (1967).
48. Messerly, J. F.; Todd, S. S.; Guthrie, G. B., J. Chem. Eng. Data, 15, 227 (1970).
49. Miller, A.; Scott, D. W., J. Chem. Phys., 68(3), 1317 (1978).
50. Miller, L.P.; Wachter, N. W.; Fried, V., J. Chem. Eng. Data, 20, 417 (1975).
51. Ohe, S., "Computer Aided Data Book of Vapor Pressure," Data Book Publishing Co., Tokyo, Japan, 1976.
52. Riddick, J. A.; Bunger, W. B., "Organic Solvents, Physical Properties and Methods of Purification," Wiley-Interscience: New York, 1970.
53. Roder, H. M., "Measurements of the Specific Heats, C_s, and C_v, of Dense Gaseous and Liquid Ethane," J. Res. Nat. Bur. Stand. (U.S.) 80A, 739 (1976).
54. Scott, R. B.; Meyers, C. H.; Rands, R. D.; Brickwedde, F. G.; Bekkedahl, N., J. Res. NBS, 35, 39 (1945).
55. Sliwinski, P., Z. Physik. Chem. (Frankfurt) 63, 263 (1969).
56. Stull, D. R.; Westrum, E. F.; Sinke, G. C., "The Chemical Thermodynamics of Organic Compounds," John Wiley & Sons, Inc., New York, 1969.
57. "TRC Thermodynamic Tables – Hydrocarbons," Thermodynamics Research Center, Texas A&M University System: College Station, Texas.
58. "TRC Thermodynamic Tables–Non-Hydrocarbons," Thermodynamics Research Center, Texas A&M University System: College Station, Texas.
59. Vargaftik, N. B.; "Tables on the Thermophysical Properties of Liquids and Gases," Hemisphere Publishing Corporation, Washington D.C., 1975.
60. Weast, R. C., "Handbook of Chemistry and Physics," The Chemical Rubber Co: Cleveland, Ohio, 1969.
61. Wilhoit, R. C.; Aljoe, K. L.; Adler, S. B.; Kudchadker, S. A., "Isopropylbenzene and 1-Methyl-2-, -3-, and -4-isopropylbenzene," API Monograph Series No. 722, American Petroleum Institute, 1984.
62. Wilhoit, R. C.; Zwolinski, B. J., J. Phys. Chem. Ref. Data, 2, Suppl. no. 1 (1973).
63. Younglove, B. A., J. Phys. Chem. Ref. Data, 11, Suppl. no. 1 (1982).

Nomograph for calculating density and specific volume of gases and vapors

The density of a gas or vapor can be calculated from the equation for an ideal gas:

$$\rho = \frac{144P}{RT} = \frac{MP}{10.72T} = \frac{2.70\,PSg}{T}$$

where: ρ = Density of gas, lb/cu ft

P = Absolute pressure, psi (psig + 14.7)
R = Individual gas constant (10.72/M)
M = Molecular weight of gas
T = Temperature, °R
Sg = Specific gravity of individual gas relative to air
(= to ratio of gas molecular weight to that of air)

The above equation and corresponding nomograph (Figure 1) can be used to calculate the density (or specific volume) of gases at low pressures and high temperatures, where the ideal gas law holds. Note that the pressure scale is calibrated in psi gauge so that the correction of 14.7 psi is not needed when using the nomograph.

Example. What is the density of dry methane if its temperature is 100°F and its pressure is 15 psig?

Connect	With	Mark or Read
M = 16	t = 100°F	Index
Index Mark	P = 15 psig	ρ = 0.08 lb/cu ft

Source

"Fluid Flow Through Pipe, Fittings and Valves," *Technical Paper No. 410*, A-11, Crane Company, Chicago, IL (1957).

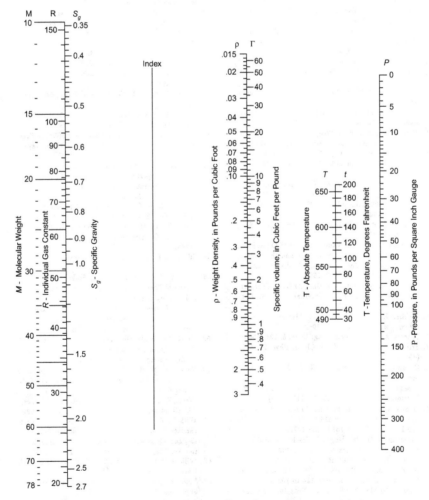

Figure 1. Weight density and specific volume of gases and vapors.

CONSIDERATIONS FOR SELECTING ENERGY MEASUREMENT EQUIPMENT

Keeping Track of Gas Heating Value is an Important Phase of Today's Market

Paul E. Kizer, Applied Automation, Inc., Houston, Texas

With natural gas energy determination becoming an integral part of the transmission and distribution business in today's market, companies are placing increased emphasis on choosing the correct measuring device. Most current contracts have a Btu specification, and many others use MMBtu instead of gas volume.

Here is what happens at the burner tip. The following is the exothermic reaction of methane and oxygen (80% to 95% of natural gas):

$$CH_4 + 2O_2 \rightarrow CO_2 + 2H_2O + \text{Heat } (1010 \text{ Btu/cf})[1]$$

for the case of ethane (up to 3% of natural gas):

$$2C_2H_6 + 7O_2 \rightarrow 4CO_2 + 6H_2O + \text{Heat } (1769 \text{ Btu/cf})[1]$$

and for the case of propane (usually up to 0.5%, except where propane/air peak shaving is used):

$$C_3H_8 + 5O_2 \rightarrow 3CO_2 + 4H_2O + \text{Heat } (2516 \text{ Btu/cf})[1]$$

Since natural gas is a mixture of these three hydrocarbons plus small amounts of other compounds, these three chemical reactions account for 90% to 99% of the heat generated when the gas is burned. However, one mole (a mole is a standard number of molecules [6.023×10^{23}, Avogadro's number])[17] of propane will generate more than twice the heat of one mole of methane ($CH_4 = 1010$ Btu/cf vrs $C_3H_8 = 2516$ Btu/cf at 60°F and 414.696 psia). A small error in the heavier or longer chain hydrocarbon components of natural gas leads to a larger error in the Btu calculation. Consequently, a small shift in the composition of any component can affect the heating value of a cubic foot of carefully measured gas.

The term MMBtu first came into wide use after the 1978 Natural Gas Policy Act changed the method of custody transfer from Mcf to MMBtu. This method was upheld in a court case involving FERC Order 356 (Jan. 1984).[2]

How much is it worth to keep track of the Btu? Postulate a ±5% error between doing a lab determination of the heating value on a spot sample of gas and on-line (nearly continuous) monitor of heating value, which results in a ±50 Btu difference at a 1,000 Btu/cf level. A thousand Btu/cf is fairly average for natural gas.

On a station that has 50 MMcf/d at $2.50 per Mcf or MMBtu, this is $125,000 worth of gas per day. Five percent of this is $6,250 per day. If a process chromatograph, $50,000 installed cost, is used to determine the energy content, a payout of less than 10 days is obtained on a 50 MMcf/d station. This explains why most major interconnects have Btu measurement on-line today.

General methods

First, consider the direct methods of energy determination, since this is the oldest method in use today. All direct methods involve the complete oxidation (burning) of gas to determine its energy content. There are two basic classes of direct measurement devices, calorimeter and thermtitrator.

Early on, the term representing 100,000 Btus was one therm of energy. One dekatherm (10 therms) is equal to 1,000,000 Btu or 1 MMBtu.[21] Perhaps this is where the name thermtitrator came from. However, as discussed earlier, MMBtu[2] has supplanted the therm unit of measurement for now.

The calorimeter was adapted from a laboratory physical chemistry method called "bomb calorimetry." These were laboratory instruments that yield precise results when used under controlled conditions. However, each determination required 4 to 8 hours to complete. This is fine for a research project, but in a production setting it is easy to see how gas sample cylinders would start piling up. The American calorimeter[3] provided laboratory type results in less time, but the lab conditions had to be tightly controlled with continually manned equipment. For on-line energy content measurement, the Cutler-Hammer Recording Calorimeter stood alone for decades.[3,4]

While this device still requires rigid environmental conditions, it runs unattended for long periods of time (weeks). It also provides a recorder output, which allows the correlation of the heating value with the volume of gas. The recording calorimeter's stated precision is ±5 Btu out of 1,000 Btus, although many operators can do better.

With the introduction of solid-state electronics, micro calorimetry devices have been tried, though not with much commercial success. Hart Scientific, Provo, Utah, has used a catalyst bed to perform a controlled isothermal oxidation on a very precise volume of gas. Initial results seem to show ±0.5 Btu at 1,000 Btu precision.[5]

The thermtitrator devices utilize a more elegant chemical method, stoichiometry—that is, air to gas ratios. One instrument, the PMI thermtitrator, uses the temperature difference between two flames to maintain the gas flow ratios at stoichiometric conditions.[6,7]

Accuracy is around ±1 Btu at 1,000 Btus and compares favorably with the recording calorimeter.[8] The other stoichiometric device is the Honeywell HVT. It uses a zirconium oxide oxygen sensor to determine stoichiometric conditions and automatically adjusts air/gas ratios to minimize excess O.

Both devices yield rapid reaction to step changes in Btu content and are useful in process control applications such as boiler feed forward controls.

There are two classes of indirect Btu or energy measurement, both chromatographic devices, relying on the separation, identification, and quantification of all the components that make up the natural gas mixture.[9] The two classes are: laboratory chromatographs and on-line or process chromatographs. Through the years a number of different people have compared calorimetry to chromatography with varied results.[10,11]

The laboratory chromatograph (Figure 1) is just that. It is located in a laboratory. Somehow a gas sample must be obtained and transported in a cylinder to the lab to be analyzed.

How the sample is obtained is the main difference between lab and on-line gas chromatographs (GCs). The natural gas samples analyzed in a lab today usually are called spot or grab samples. A cleaned vessel (cylinder), sometimes N_2 filled, is purged and filled at least three times for reliable sampling. The pressured-up bottle is carried to a lab to be analyzed. More and more samples today are being collected using a continuous sampler. This device takes a small aliquot of gas at a predetermined time interval so that the bottle is filled at the end of the sampling cycle required by the gas contract.

With recent developments, these devices have been tied to electronic flow computers to give flow-weighted composite samples rather than time-weighted ones. An example of such an installation is in Figure 2.

Flow weighting generally gives even better results than time-weighted samples if there are variations in the flowrate at the sampled station. Great care should be exercised when analyzing a sample cylinder of natural gas obtained by any of the previous methods. For instance, the sample bottle should be heated to 120°F for at least 2 hours before the analysis is made to ensure that all the heavier components are analyzed. A complete discussion of the problems associated with the recommendations for bottle sample analysis have been published previously.[9,12]

The process or on-line gas chromatograph (GC) (Figure 3) is designed to continually purge a small amount of the gas flowing through the station or pipeline to provide a representative sample at the time the GC is ready to make a new analysis. Typically, this is once every 6 minutes.

The main advantage of continuous sampling and analysis is the documentation of the duration and the amount of deviation of the energy content that has occurred during the measurement period. It is possible to assign a heating value to a specific volume of gas if the on-line GC is tied directly to an electronic flow computer. This, of course, allows actual billing on an MMBtu custody transfer basis.

Figure 1. Laboratory chromatograph used to measure gas samples.

Figure 2. Example of on-line sampler connected to electronic flow computers.

Figure 3. Process sampler continually purges gas moving through a station or pipeline.

GCs—lab or on-line—using the computing power of today's modern microprocessor are capable of repeatability values of ±0.5 Btu at 1,000 Btu/cf. While the precision of a GC makes it well suited for custody transfer applications, the 6- to 10-minute analysis time makes it difficult to use for burner feed forward control. Generally, the on-line analyzers do not require much environmental protection at all. Most require protection from rain.

How chromatographs work

The way a modern microprocessor controlled on-line chromatograph works has been thoroughly discussed in previous papers.[13–15] However, it would be helpful to review briefly just how a GC works. Again, except for the sample conditioning system, this discussion generally applies to lab GCs as well as on-line GCs. For purposes of this discussion, the on-line GC has been divided into the following four sections:

- Sample conditioning system
- Gas chromatograph oven
- Chromatograph controller
- Input/output chassis

The sequence also is the route the sample takes from the pipeline to the printer. This makes it a logical flow for discussion as well.

Figure 4. Bypass loop reduces sampling lag time.

Sample conditioning system

To obtain a representative sample of gas, a probe should be inserted into the middle third of the pipeline. This procedure is as essential for good sampling for on-line GCs as it is for spot or composite samples. A major difference, however, is that the gas pressure is reduced to 15 psig at the probe for transport to the analyzer to ensure no heavy component (C^6+) is lost. With spot or composite samples, care is taken to maintain pipeline conditions. Sometimes it is useful to use a sample bypass loop to reduce sampling lag times, as shown in Figure 4.

Note that filtering and liquid coalescing is accomplished, if needed, to protect the integrity of the analyzer columns. Sometimes double block and bleed sample switching is used to prevent cross-contamination by the standard or other sample streams.

Gas chromatograph oven

Once the gas enters the oven, a repeatable gas sample is measured by a sample inject valve. The chromatograph

Figure 5. Block diagram illustrates gas chromatograph oven.

Figure 6. Separated components move through cell detectors for quantifying sample.

controller then blocks the sample flow and allows it to equilibrate to atmospheric pressure. The sample is then swept onto a set of carefully selected chromatographic columns by a stream of helium gas for sequential separation of each component in the natural gas mixture. A block diagram is shown in Figure 5.

As each column does its specified separation job, it is switched out of the helium stream. The separated components make their way through the thermal conductivity cell detector where they can be quantified. An example of a detector, output is shown in Figure 6.

These peaks must be gated and the area under each peak integrated to be quantified, which is the job of the chromatograph controller.

Chromatograph controller

All the timing functions such as sample inject, column reverse, auto-calibration, and log results must be performed by the microprocessor controller board. In addition, this device must detect each peak, open and then close integration gates, identify the peak, and assign the correct response factor. The controller must then calculate the mole fraction of each component. Since these peaks are quite arbitrary in size due to equipment variability setup and environment, high-quality standards should be used to calibrate these instruments. This calibration assures accurate determination of mole fraction possible. All Btu determination methods require high quality standard samples.

On a Btu GC, the peaks also must be normalized to compensate for barometric pressure changes (which cause sample inject size variations). This allows the calculation of Btu to repeatability of ± 0.5 Btu at 1,000 Btu/cf. This processor must also calculate the specific gravity, compressibility, condensable liquids, wobbie index, as well as the Btu. Calculation results must be made available to the outside world. This is the job of the input/output chassis.

Input/output chassis

The device houses communications option cards such as printer interface cards, analog output cards, and computer interface cards. These usually can be configured based on the customer's needs. However, caution is advised as specific computer protocols may or may not be supported by one vendor or another. One function for this input/output chassis, for example, is to hold an RS232 output card. This card would log results to an analyzer interface unit (AIU), which would then be connected to up to 30 flow computers (Fig. 7).

Figure 7. Output card logs results in an analyzer interface unit connected to at least 30 computers.

How Btu is calculated

The methods for calculation of the natural gas Btu from analysis data are well documented. Calculations have been published previously.[2,3,9,16,18] Nevertheless, it is good to review these methods. Although they look complicated, they really are straightforward. Below is the general formula for the Btu calculation:

$$Btu/cf = X_1 * C_1 + X_2 * C_2 + X_3 * C_3 \cdots X_{11} * C_{11}$$

where:

$X_1 =$ Mole fraction of C_1 or methane
$X_2 =$ Mole fraction of C_2 or ethane
$X_3 =$ Mole fraction of C_3 or propane

            "   "
            "   "
            "   "

$X_{11} =$ Mole fraction of C_6+ hexanes and heavier

and:

$C_1 =$ Molar heating value of C_1, or methane (1010 Btu/cf)
$C_2 =$ Molar heating value of C_2, or ethane (1769 Btu/cf)

            "   "
            "   "
            "   "

$C_{11} =$ Molar heating value of C_6+ or hexanes and heavier (4943 Btu/cf)

The calculation of specific gravity and condensable liquids is handled in the same manner. For example:

$$\text{Specific gravity} = X_1 * G_1 + X_2 * G_2 + \cdots X_{11} * G_{11}$$

where: X_1, X_2, is the mole fraction of each natural gas component and: G_1, G_2, is the molar gravity contribution of each respective component.

This calculation yields gross or total heating value and is not compensated for compressibility Z. A short method for calculation of Z is:

$$Z = 1.00369 - (0.010 * S.G.) + (0.007 * X_{inert})$$

Total heating value would then be expressed by the formula:

$$\frac{Btu/cf}{Z} = Btu/cf\ Dry$$

in order to convert Dry Btu to saturated:

$$Btu\ saurated = \frac{Btu\ Dry}{1120\ saturation\ factor\ (1.0177)}$$

Some process gas chromatographs calculate all the above values at a pressure base of 14.696 psia only. Others allow user selection of a different contract pressure base, and all the constants are then factored by this new pressure base.

Care should be exercised when comparing Btu, SG, and others that all these values are calculated at the same temperature base and pressure base.

Conclusions

There are valid reasons for choosing one energy measurement device over the other. Thermtitrators are fast if control capability is needed. They are not suited for hazardous locations and require more rigorous attention to their operational environment. The modern microprocessors used on most chromatographs make them extremely capable in terms of computer interfaces.

Chromatographs are suitable for hazardous area installations. They are somewhat slower reacting, having 6-minute cycle times. However, they generally require less maintenance than calorimeters even though they seem more technically complicated. With the AGA-8 supercompressibility calculations, there will most likely be a requirement for a total analysis of the gas to do this part of the gas volume calculations. Thermtitrators cannot do this. GCs that only present the flow computer with four analog signals, proportional to Btu, SG, CO_2, and N_2, may not be much more help.

One company has an interface unit that will pick up and store hourly averages of all 12 components from a gas chromatograph. These hourly averages can be stored for up to three streams for 35 days each so that each hour's flow in the flow computer can be correlated with each hour's gas composition. This allows composition data for not only MMBtu computation but also AGA-8 volume calculations. From an audit standpoint, this allows downstream corrections and verification that has heretofore not been available.

References

1. GPA Standard 2145–84, p. 2.
2. GPA Reference Bulletin, Heating Value As a Basis for Custody Transfer of Natural Gas (1984 Draft).
3. AGA Gas Measurement Manual, Section 11A.2 Determination of Heating Value of Gas, p. IIA2.1 ff.
4. Installation and Operation of Recording Calorinators, A. E. Kersey ISHM #64, 1989, p. 93 ff.
5. Field Deployable Natural Gas Energy Meter, Lewis and Greefield, Sales Information 1989.
6. "Determination of Calorific Values of Natural Gas by Combustion Instruments," R. L. Howard, ISHM #64, 1989, p. 509.
7. "Fundamentals of Energy Management," R. L. Howard, GCMSC #24, 1989, p. 43.
8. "The Therm-titrator, A Comparison with You-Know-What," John Light, GPA #60, 1981.
9. "Analysis of Natural Gas by Gas Chromatography," ASTM D1945, p. 133 ff.
10. "Comparison of Calculated and Measured Heating Values of Natural Gas," Donald C. Melrose, Appalachian Gas Measurement Short Course, 1975.
11. "Measuring Gas; Chromatography Bests Calorimetry," Leisey, Potter, and McKoy, *Oil and Gas Journal*, July 1977.
12. "Sampling and Analysis of Natural Gas," Schepens, Kilmer and Bernos, GPA proceedings #61.
13. "Energy Measurement Utilizing On-Line Chromatographs," L. N. Cox, Prod. of Inter. School of Hydrocarbon Meas. #64, 1989, p. 469 ff.
14. "Btu Measurement in Natural Gas Using Process Gas Chromatography," Kizer and Sund, Proc. of Gulf Coast Meas. Soc. #24, 1989, p. 72 ff.
15. "How to Measure Btu Effectively," Staff, *Gas Industries*, August 1989.
16. "Calculations of Gross Heating Value, Relative Density and Compressibility Factor for Natural Gas Mixtures from Compositional Analysis," GPA Std. 2171–86.
17. *College Chemistry*, Bruce H. Mahan, Addison Wesley, 1966, p. 23.
18. "Gas Heating Value: What It Is and How To Measure, Calculate," D. N. McClanahan, *Oil and Gas Journal*, Feb. 20, 1967, p. 84 ff.
19. Campbell, "436 Pushes Measurement into the Electronics Age," *American Oil and Gas Report*, May 1987, p. 15 ff.
20. Hagar, "Pipelines, Producers Cope with Problems under FERC Order 436," *Oil and Gas Journal*, May 1987, p. 17 ff.
21. Paul, "A Multi Btu Approach to Fuels Procurement," Nymex Energy in the News, Spring, 1990, p. 4.

Facts about methane and its behavior

A study of the physical, transport, and thermodynamic properties of methane

Dr. Robert O. Parker and Dr. William H. Kapfer, New York University, New York, NY

Interest in liquefied natural gas (LNG) requires that the physical, transport, and thermodynamic properties of methane, the principal constituent of natural gas, be readily available. This article will present values most frequently used in process calculations and in other studies associated with the storage and processing of natural gas. In addition, sources of more detailed information are listed.

Methane, CH_4, is the hydrocarbon of lowest molecular weight (16.042). The molecule is best visualized as a regular tetrahedron with the carbon atom at the geometric center and the hydrogen atoms at each of the four apices. Methane is an end product of the anaerobic (without air) decay of plants, that is, of the breakdown of certain very complicated molecules. As such, it is the major constituent (up to 97%) of natural gas. It is the firedamp of the coal mines and can be seen as marsh gas bubbling to the surface of swamps.

The physical, transport, and thermodynamic properties required are:

Physical

1. Critical constants
2. Equation of state (PVT relationships)
3. Vapor pressure, including by and fp
4. Vapor and liquid density

Transport

1. Viscosity of liquid and gas
2. Thermal conductivity of liquid and gas
3. Diffusivity

Thermodynamic

1. Heat capacity
2. Latent heat
3. Enthalpy
4. Entropy

They are required for process calculations: material and energy balances; heat transfer, mass-transfer and pressure loss calculations.

Critical Constants

$T_c = -115.8°F\ (-82.1°C)[1]$

$P_c = 673\ psia\ (45.8\ atm)[2]$

$d_c = 10.17\ lb/ft^3\ (0.1630\ g/cc)[2]$

The main utility of critical constants to the engineer is related to the fact that correlations of many of the physical, transport, and thermodynamic properties of hydrocarbons are best expressed in terms of reduced temperature, T_R, and pressure, P_R:

$$\left(T_R = \frac{T}{T_c},\ P_R = \frac{P}{P_c} \right)$$

This is simply an application of the "principle of corresponding states" and is recommended only in the absence of sufficient reliable experimental data.

Equations of state
(pressure–volume–temperature behavior)

For ordinary conditions, low to moderate pressures and temperatures not too close to the normal boiling point, i.e., when inspection of a compressibility factor chart shows it to be nearly unity, the ideal gas law is:

$$PV = \frac{W}{M}RT \tag{1}$$

where: P = Pressure, psia
      V = Volume, ft^3
      W = Weight of gas, lb
      M = Molecular weight (here 16.042)
      R = Gas constant = 10.73 psia (ft^3)/lb mole (°R)
      T = Temperature, °R

is sufficient. For most other conditions, Equation 1 is modified by inclusion of the compressibility factor, Z, in the right-hand member, viz:

$$PV = \frac{W}{M}ZRT \tag{1a}$$

The writers recommend compressibility factors of Nelson and Obert.[3] If an analytical equation of state is preferred to Equation 1a the writers recommend the 8 constant equation of Benedict, Webb, and Rubin.[4]

$$P = RTd + (B_0RT - A_0 - C_0/T^2)d^2$$
$$+ (bRT - a)d^3 + a\alpha d^6 + \frac{cd^3(1 + \gamma d^2)e^{-\gamma d^2}}{T^2}$$

where d = density, g-moles/l

in which the constants[4] for methane are:

$B_o = 0.042600$
$A_o = 1.85500$
$C_o = 22,570.0$
$b = 0.00338004$
$a = 0.0494000$
$c = 2,545.00$
$\gamma = 0.0060000$
$\alpha = 0.000124359$

With these constants, $R = 0.0820$ liter-atm/(°K) (gram mole), P is in atmospheres, T is in °K, and e is the base of natural logarithms. Constants in this equation are available for ethane, ethylene, propane, propylene, n-butane, i-butane, i-butene, n-pentane, i-pentane, n-hexane, and n-heptane.[4] Using Equation 2 to predict the pressure corresponding to 89 data points,[5,6] it was found that the average absolute and maximum deviations for methane were 0.376% and +2.25% respectively. Reference 6 discusses a generalization of the Benedict–Webb–Rubin equation.

Vapor pressure[2]

Freezing point in air at 1 atm = −296.46°F (182.48°C)
Boiling point at 29.921 in. Hg = −258.68°F (−161.49°C)
Vapor pressures are tabulated on p. 121 of Reference 2 from 10 mm to 1,500 mm Hg and are represented by the Antoine equation.

$$\log_{10}P = A - B/(C + t) \tag{3}$$

or:

$$t = B/(A - \log_{10}P) - C \tag{3a}$$

with P in mm Hg and t in °C.

$A = 6.61184$
$B = 389.93$
$C = 266.00$

Transport properties

Viscosity (liquid)[2]

t°c	μ, centipoises
−185	0.226 (subcooled)
−180	0.188
−175	0.161
−170	0.142
−165	0.127
−160	0.115 (at sat. pressure)

Viscosity (vapor)[7] at 1 atm pressure:

t°c	μ, centipoises
−181.6	0.00348
−78.5	0.00760
0	0.01026
20	0.01087
100	0.01331
200.5	0.01605
284	0.01813
380	0.02026
499	0.02264

Additional data[8] are available.

Thermal conductivity (liquid)

	t°F	K, Btu/hr ft°F
	−280	0.0841
n.b.p	−258.68	0.0790
	−200	0.0568
crit. temp	−115.8	0.029 (reference[10])

Values for liquid thermal conductivity represent estimates by the authors using the method of Viswanath,[9] which seems to be quite reliable for organic liquids (halogenated compounds excepted).

Thermal conductivity (vapor)[7]

t°F	K, Btu/hr ft°F
−300	0.00541
−200	0.00895
−100	0.01265
−40	0.0149
−20	0.0157
0	0.01645
20	0.01725
40	0.01805
60	0.0190

t°F	K, Btu/hr ft°F
80	0.01985
100	0.0208
120	0.0216
200	0.0258

Diffusivity

The transport property, mass diffusivity (D), is of importance in calculations usually entering computations in combination with viscosity and density as the Schmidt No., μ/ρ D. Reliable estimates of D_{12}, the diffusivity of species 1 through species 2, are best made by the method of Hirschfelder, Bird, and Spotz.[11]

$$D_{12} = \frac{0.001858 T^{3/2}[M_1 + M_2]/M_1 M_2]^{1/2}}{P\sigma_{12}^2 \Omega_D} \qquad (4)$$

where: D = Diffusivity, cm^2/sec

T = Temperature, °K

M = Molecular weight, M_1 of species 1, M_2 of species 2

P = Pressure, atm

$\sigma_{12} = (\sigma_1 + \sigma_2)/2$ = collision diameter, A°

Ω_D = Collision integral = function of KT/ε_{12}

K = Boltzmann's constant

ε = A force constant

$\varepsilon_{12} = \sqrt{\varepsilon_1 \varepsilon_2}$

Force constant and collision diameters for air and methane are:

	ε/K, °K	σ, A°
Air	97	3.617
Methane	136.5	3.822

These are known as the Lennard—Jones force constants, and an extensive table is given.[11] Ω_D, which depends upon the temperature through KT/ε, is tabulated in the reference.[12] A few values are:

KT/ε	Ω_D
0.30	2.662
0.50	2.066
1.00	1.439
1.50	1.198
2.00	1.075
4.00	0.8836
10	0.7424
50	0.5756
100	0.5130
400	0.4170

If the calculation requiring diffusivities is a crucial one, we recommend that the engineer consult the principal literature reference.[11]

Thermodynamic properties

The most extensive, reliable compilation of the thermodynamic properties is the one by Matthews and Hurd.[13] In addition to a pressure—enthalpy diagram, tabulations are provided for saturated methane from −280°F to −115.8°F, the critical temperature, and for superheated methane to 500°F and 1,500 psia. This tabulated information is reproduced in the Chemical Engineers' Handbook 3rd Ed.[14] In the 4th Ed. (1963) superheat tables are omitted.

Corresponding to the pressure—temperature pair, values are given for:

a. Ratio fugacity (where fugacity[13] refers to "escaping tendency") to pressure
b. Specific volume of liquid and vapor
c. Enthalpy of liquid, vaporization, and vapor
d. Entropy of liquid and vapor

A few numbers of immediate interest to the process engineer are:

a. Latent heat of vaporization at normal boiling point = 219.22 Btu/lb
b. Liquid density at normal boiling point = 26.6 lb/ft^3
c. Vapor density at normal boiling point = 0.113 lb/ft^3

The liquid density is required to compute the volume of a diked area to contain a spill. Latent heat is useful in estimating the time required for a spill to be vaporized. The vapor density points up the fact that, although methane has a specific gravity referred to air of about 0.55, vapors generated at a spill are about 1.5 times as heavy as ambient air; a matter of importance in various studies.

Heat capacities

If heat capacities are required for process calculations, they can be extracted from the thermodynamic tables mentioned above by the following procedure:

$$Cp = \frac{H_1 - H_2}{T_1 - T_2}, Btu/lb°F \qquad (5)$$

Example. Find the heat capacity for superheated methane at 15 psia and 100°F.

Figure 14. Cold box where liquefaction of natural gas is accomplished.

From the Matthews and Hurd superheat table:[13,14]

at 15 psia

| $T_1 = 110$ | $H_1 = 426.6$ |
| $T_2 = 90$ | $H_2 = 415.8$ |

Then

$$C_P = \frac{426.6 - 415.8}{110 - 90} = 0.54 \ \text{Btu/lb}°\text{F}$$

The procedure given above is adequate for liquid heat capacities when enthalpy values are taken from the table for saturated methane. This works because the effect of pressure on liquid heat capacity is essentially negligible.

Alternatively, instantaneous values of heat capacity may be determined.[15]

$$C_P^o = 0.213 \ + \ 1.11 \\ \times \ 10^{-3} \ T - 2.59 \ \times \ 10^{-7} T^2 \ \text{Btu/lb} \ °\text{F} \tag{6}$$

Note carefully that Equation 6 yields the heat capacity at low pressure, 1 atm; at elevated pressures a correction may be required. If Equation 5 above is used, no pressure correction is necessary.

Heat capacity ratio

The ratio $C_p/C_v = \gamma$, useful in process calculations (adapted from reference 14), is for atmospheric pressure:

t°F	$\gamma = C_p/C_v$
−150	1.39
−100	1.36
−50	1.33
0	1.31
50	1.28
100	1.27
150	1.25

Flammability characteristics

Since methane is a combustible gas, consideration of any process involving its use or storage must include studies of the associated fire hazard. Basic data are flammable limits and the autoignition temperature. For methane in air these are[16,17] at atmospheric pressure:

a. Lower Flammable, Limit @ 25°C about 5% by vol
b. Upper Flammable, Limit @ 25°C about 15% by vol
c. Autoignition Temperature 1,004°F

These figures mean that mixtures of methane and air having less than 5% methane by volume are too lean to burn, those with more than 15% are too rich. The autoignition temperature is the minimum temperature at which ignition can occur. Figures on flammable limits refer to "homogeneous" mixtures. Industrially, few mixtures are completely homogeneous. In safety computations, therefore, the engineer must establish that the "highest local concentration" is below the lower flammable limit by a suitable margin before saying there is no fire hazard. A method dealing with LNG spills has been proposed by Parker and Spata.[18]

References

1. Kobe, K. A. and R. E. Lynn, *Chem. Revs.* 52, 117 (1953).
2. American Petroleum Institute, "Selected Values of Physical and Thermodynamic Properties of Hydrocarbons and Related Compounds," Project 44, Carnegie Press, Pittsburgh, Pa. 1953 (NBS—C—461).
3. Nelson, L. C. and E. F. Obert, *Trans.* ASME 76, 1057 (1954).
4. Benedict, M., G. B. Webb, L. C. Rubin and L. Friend, *Chem. Eng. Prog.* 47, 419, 449, 571, 609 (1951).
5. Kvalnes, H. M. and V. L. Gaddy, *J. Ann. Chem. Soc.* 53, 394 (1931).
6. Edmister, W. C., J. Vairogs and A. J. Klekers, *AIChE Journal* 14, No. 3, p. 479 (1968).
7. *Handbook of Chemistry and Physics, 47th Edition*, 1966—67.
8. Huang, E. T. S., G. W. Swift and F. Kurata, *AIChE Journal* 12, p. 932 (1966).
9. Viswanath, D. S., *AIChE Journal* 13, 850 (1967).
10. Gamson, B. W., *Chem. Eng. Prog.* 45, 154 (1949).
11. Hirschfelder, J. O., R. B. Bird and E. L. Spotz, *Trans. ASME* 71, 921 (1949).
12. Hirschfelder, J. O., C. F. Curtiss and R. B. Bird, *Molecular Theory of Gases and Liquids*, John Wiley and Sons, Inc., New York, 1954.
13. Matthews, C. S. and C. O. Hurd, *Trans. AIChE 42*, p. 55 (1946); ibid., pp. 78 (1009).
14. Perry, J. H. (ed.) *Chemical Engineers Handbook, 3rd ed.*, McGraw-Hill, New York, 1950.
15. Reid, R. C. and T. K. Sherwood, *The Properties of Gases and Liquids*, McGraw-Hill, New York (1958).
16. Coward, H. F. and G. W. Jones, U.S. Bureau of Mines, Bulletin 503 (1952).
17. Zabetakis, M. G., U.S. Bureau of Mines, Bulletin 627 (1965).
18. Parker, R. O. and J. K. Spata, Proc. 1st Intern. Conf. on LNG, Chicago, April 7—12, 1968, Paper No. 24.

Conversion table for pure methane

		Long tons Liquid	ft³ Liquid	ft³ gas	m³ liquid	m³ gas	U.S. bbl liquid	Therms	Thermies
long ton liquid	=	1	84.56	52,886	2.394	1,419.7	15.06	535.2	13,491
ft₃ liquid	=	0.011826	1	625.43	0.028311	16.789	0.1781	6.329	159.54
ft₃ gas × 10⁰	=	18.91	1,599	1 × 10₆	45.27	26,847	284.785	10,121	255,115
m₃ liquid	=	0.4177	35.315	22,090	1	503	6.29	223.55	5,635.2
m₃ gas × 10⁹	=	704,374	59.562 × 10⁶	37.252 × 10⁹	1.686 × 10⁶	1 × 10⁹	10.608 × 10⁶	0.377 × 10⁹	9.503 × 10⁹
U.S. bbl liquid	=	0.0664	5.615	3,512	0.15896	94.268	1	35.54	895.8
Therm × 10⁶	=	1,868	157,958	98.791 × 10⁶ 4,472	2.652 × 10⁶ 28,132	1 × 10⁶	25.201 × 10⁶		
Thermie × 10⁶	=	74.12	6,268	3.92 × 10₆	177.44	105,228	1,116.2	39,669	1 × 10⁶

Notes

1. ft³ measured at 60°F, 14.696 lbf/in², dry
2. m³ measured at 0°C, 760 mmHg, dry
3. Gross heat content of methane is 1012 Btu/ft³ in gaseous forms: Specific gravity 0.554 (Air = 1) Boiling point −258.7°F (14.696 lbf/in²)
4. 1 × 10⁶ ft³ gas = 24.42 long tons (0.4417 b/d) fuel oil equivalent
5. The above factors are based on the properties of pure methane; LNG as shipped in practice will vary in composition according to the presence of heavier hydrocarbons (C_2, C_3, C_4,) in the gas stream.

Categories of natural gas and reserves terminology

Associated and non-associated natural gas

Natural gas is found in underground structures similar to those containing crude oil. There are three types of natural gas reservoirs:

1. Structures from which only gas can be produced economically—called *non-associated gas* (or unassociated gas).
2. Condensate reservoirs which yield relatively large amounts of gas per barrel of light liquid hydrocarbons. Although many condensate reservoirs, are produced primarily for gas, there are cases where gas is re-injected or "re-cycled" to improve liquid recovery, particularly if no gas market is yet available. This gas also is termed **non-associated.**
3. Reservoirs where gas is found dissolved in crude oil (solution gas) and in some cases also in contact with underlying gas-saturated crude (**gas-cap gas**). Both are called **associated gas.** (Gas-cap gas is almost never produced until most of the economically recoverable oil has been yielded.)

In such fields, gas production rates will depend on oil output, with the oil usually representing the major part in terms of energy equivalents.

Natural gas reserves

There are no generally accepted definitions of natural gas reserves, and wide variations both in terms and principles may be met. The following definitions are the most commonly used:

Expected Ultimate Recovery is the total volume of hydrocarbons, gas and/or oil, which may be expected to be recoverable commercially from a given area at current or anticipated prices and costs, under existing or anticipated regulatory practices and with known methods and equipment. This total can be expressed as the sum of the "Ultimate Recovery from Existing Fields" and the "Expectation from Future Discoveries."

Ultimate Recovery from Existing Fields is the sum of the "Cumulative Production" from such fields and the "Reserves from Existing Fields."

Reserves from Existing Fields are the sum of "Proven Reserves," "Discounted (i.e., 50% of) Probable Reserves" and "Discounted (i.e., 25% of) Possible Reserves."

Proven Reserves represent the quantities of crude oil and/or natural gas and natural gas liquids that geological and engineering data demonstrate with reasonable certainty to be recoverable in the future from known oil and/or gas reservoirs. They represent strictly technical judgments and are not knowingly influenced by attitudes of conservatism or optimism.

Discounted (i.e., 50% of) **Probable Reserves** are those quantities of crude oil and/or natural gas and natural gas liquids for which there exists a probability of 50% that they will materialize. Such reserves are usually allocated to some conjectural part of a field or reservoir as yet undrilled or incompletely evaluated where, while characteristics of the productive zone and fluid content are reasonably favorable, other uncertain features may have equal chances of being favorable or unfavorable for commercial production.

Discounted (i.e., 25% of) **Possible Reserves** are those quantities of crude oil and/or natural gas and natural gas liquids thought to be potentially producible but where the chance of their being realized is thought to be of the order of 25%. Reserves in this category are usually allocated to possible extensions of existing fields where, although geological data may be favorable, uncertainties as to the characteristics of the productive zone and/or fluid content are such as to preclude a probability greater than 25%.

Terminology of Reserves

It should be noted that the terms "gas-in-place" and "oil-in-place" represent estimates of the total hydrocarbons present which, in conjunction with various engineering and economic criteria, provide the basis for the "reserve" estimates. They should never be confused with "reserves," which implies that physical recovery is possible.

Glossary of common gas industry terms

(italics denote cross reference)

absolute pressure *gauge pressure plus atmospheric pressure.*

associated gas see preceding page.

atmospheric pressure the pressure of the weight of air and water vapor on the surface of the earth. Approximately 14.7 lbf/in.^{2} at sea level.

blue water gas also called "water gas." Made in a cyclic process in which an incandescent bed of coke or coal is alternately subjected to blasts of air and steam. The gas consists mainly of equal proportions of carbon monoxide and hydrogen and has a gross heat content of about 300 Btu/ft^3 (see also *carbureted water gas*).

British thermal unit the heat required to raise the temperature of 1 lb of water 1°F.

burner capacity the maximum Btu/hr that can be released by a burner while burning with a stable flame.

calorie the quantity of heat required to raise the temperature of 1 gram of water 1°C (centigrade).

calorific value the quantity of heat released by the combustion, in a calora meter at constant pressure of 1 atmosphere, of a unit quantity of fuel measured under given conditions. In the case of gas, the conditions of temperature, pressure, and water vapor content under which the unit volume was measured affect the calorific value.

calorimeter an instrument for determining the heating value of a fuel.

carbureted water gas (carbureted blue gas) the gas resulting from the enrichment of *blue water gas* during its manufacture by a simultaneous process of light distillate, gas oil, or fuel oil gasification. The gas has a gross heat content of about 500 to 550 Btu/ft^3.

city gate a measuring station, which may also include pressure regulation, at which a distributing gas utility receives gas from a natural gas pipeline company or the transmission system.

coal gas or coke oven gas a *manufactured gas* made by destructive distillation ("carbonization") of bituminous coal in a gas retort or by-product coke oven. Its chief components are methane (20 to 30%) and hydrogen (about 50%). This gas generally has a gross heating value of 500 to 550 Btu/ft^3. When the process takes place in a closed oven (with gas as a by-product of coke production) it is generally designated as "Coke Oven Gas," and when produced in retorts it is called "Coal Gas."

commodity charge that portion of the charge or rate based upon the total volume of gas consumed or billed.

condensate a liquid/hydrocarbon mixture of 45 to 65° API gravity, which may be recovered at the surface from some *non-associated gas* reservoirs.

connection charge an amount to be paid by the customer in a lump sum, or in installments, for connecting the customer's facilities to the supplier's facilities.

contract demand the volume that the supplier of gas service agrees to deliver and, in general, the amount that the customer agrees to take or pay for.

cubic foot and cubic meter (gas) common units of measurement of gas volume. They are the amounts of gas required to fill a volume of 1 cubic ft or 1 cubic meter under stated conditions of temperature, pressure, and water vapor content.

customer charge a fixed amount to be paid periodically by the customer without regard to *demand or commodity* consumption.

cycling or recycling the process by which a *non-associated gas* reservoir is produced only for the recovery of condensate. The gas, after removal of the condensate, is reinjected into the formation.

daily average send-out the total volume of gas delivered during a period of time divided by number of days in the period.

daily peak the maximum volume of gas delivered in any one day during a given period, usually a calendar year.

degree day a measure of the extent to which the mean daily temperature falls below an assumed base, say 65°F. Thus, each degree by which the mean temperature for any day is less than 65°F would represent one degree day. (In Continental Europe, °C are used instead of °F, and the assumed base temperature is generally taken as 16°C equivalent to 60.8°F.)

demand (load) rate of flow of gas in a specified time interval, usually expressed in cubic ft (or cubic meters) per hour, per day, or per year.

demand charge that portion of the charge or rate for gas service based upon the customer's *demand* characteristics.

distribution company or gas utility a company that obtains the major portion of its gas operating revenues from the operation of a retail gas *distribution system* and which operates no transportation system other than incidental connections to a transportation system of another company. For purposes of American Gas Association (AGA) statistics, a distribution company obtains at least 95% of its gas operating revenues from the operation of its retail gas distribution system (see also *transmission company*).

distribution system feeders, mains, services, and equipment which carry or control the supply of gas from the point or points of local supply (usually the *city gate* station) to and including the consumer *meters.*

diversity factor the ratio of the sum of the non-coincident maximum demands of two or more loads to their coincident maximum demand for the same period.

dyne the standard centimeter-gram-second unit of force, equal to the force that produces an acceleration on one centimeter per second on a mass of one gram.

erg the centimeter-gram-second unit of work or energy, equal to the work done by a force of 1 dyne when its point of application moves through a distance of 1 centimeter in the direction of the force.

flat rate schedule a rate schedule that provides for a specified charge irrespective of the quantity of gas used or the *demand.*

gauge pressure the pressure generally shown by measuring devices. This is the pressure in excess of that exerted by the atmosphere (*see absolute pressure*).

gross calorific (heating) value the gross calorific value at constant pressure of a gaseous fuel is the number of heat units produced when unit volume of the fuel, measured under standard conditions, is burned in excess air in such a way that the materials after combustion consist of the gases carbon dioxide, sulfur dioxide, oxygen, and nitrogen, water vapor equal in quantity to that in the gaseous fuel and the air before combustion, and liquid water equal in quantity to that produced during combustion, and that the pressure and temperature of the gaseous fuel, the air, and the materials after combustion are 1 standard atmosphere and 25°C.

inch of mercury a unit of pressure (equivalent to 0.491154 lbf/in.^{2}) by which 1 lb/in.^{2} equals 2.036009 in. of mercury column at 0°C.

inch of water a unit of pressure (equivalent to 0.03606 lbf/in.^{2}) by which 1 lb/in.^{2} equals 27.68049 in. of water column at 4°C.

interruptible gas gas made available under agreements which permit curtailment or cessation of delivery by the supplier.

joule the work done when the point of application of a force of 1 newton is displaced through a distance of one meter in the direction of the force. It is equal to 107 *ergs*, and in electrical units is the energy dissipated by one watt in a second.

kilocalorie 1,000 *calories.*

kilowatt-hour a unit of energy equivalent to the energy transferred or expanded in 1 hour by 1 kilowatt of power.

line pack a method of *peak-shaving* by withdrawing gas from a section of a pipeline system in excess of the input into that section, i.e., normally the difference between the actual volume of gas in the pipeline at low flow (increased pressure) and that at normal flow.

liquefied natural gas (LNG) natural gas that has been liquefied by cooling to minus 258°F (−161°C) at atmospheric pressure.

liquefied petroleum gas (LPG) any hydrocarbon mixture, in either the liquid or gaseous state, the chief components of which consist of propane, propylene, butane, iso-butane, butylene, or mixtures thereof in any ratio.

load curve a graph in which the *send-out* of a gas system, or segment of a system, is plotted against intervals of time.

load density the concentration of gas load for a given area, expressed as gas volume per unit of time and per unit of area.

load duration curve a curve of loads, plotted in descending order of magnitude, against time intervals for a specified period. The coordinates may be absolute quantities or percentages.

load factor the ratio of the average load over a designated period to the *peak load* occurring in that period. Usually expressed as a percentage.

low pressure gas distribution system a gas *distribution system,* or the mains of a segment of a distribution system, operated at low pressures of less than 15 *inches water column.*

LPG-air mixtures mixtures of *liquefied petroleum gas* and air to obtain a desired heating value and capable of being distributed through a *distribution system* also used for stand-by and *peak-shaving* purposes by gas utilities.

manufactured gas combustible gases derived from primary energy sources by processes involving chemical reaction. For instance, gas produced from coal, coke or liquid hydrocarbons (see also *town gas*).

meter (gas) a mechanical device for automatically measuring quantities of gas. Some of the more important types are:

Diaphragm Meter a gas meter in which the passage of gas through two or more chambers moves diaphragms that are geared to a volume-indicating dial. Sizes up to 20,000 cubic ft per hour and 1,000 lbf/in.^{2} working pressure are available.

Mass Flowmeter a type of meter in which the gas flow is measured in terms of mass by utilizing the velocity and density of the gas stream to calculate the mass of a gas flowing in unit time.

Orifice Meter meter for measuring the flow of fluid through a pipe or duct by measurement of the pressure differential across a plate having a precision machined hole in its center. Ratings up to very high throughputs and pressures are available.

Rotary Displacement Meter the positive-pressure blower principle in reverse is used in this meter in which gas pressure turns twin matching impellers and the quantity of gas passing through is proportional to the number of revolutions. These meters are made in the range of 3,000 to 1,000,000 cubic ft per hour and for pressures up to 1,200 lbf/in.^{2}

Turbine Flowmeter a type of meter in which there is a turbine wheel or rotor made to rotate by the flowing gas. The rate or speed at which the turbine wheel revolves is a measure of the velocity of the gas.

Venturi Meter a meter in which the flow is determined by measuring the pressure drop caused by the flow through a Venturi throat. The flow rate is proportional to the square root of the pressure drop across the throat.

minimum charge a clause providing that the charge for a prescribed period shall not be less than a specified amount.

natural gasoline those liquid hydrocarbon mixtures containing essentially pentanes and heavier hydrocarbons that have been extracted from natural gas.

natural gas liquids a term used for those mixtures of hydrocarbon fractions that can be extracted in liquid form from natural gas. In the widest sense, they can comprise any combination of the following: ethane, propylene, propane, butylene, isobutane, n-butane, pentanes, hexanes, heptanes, and octanes.

net calorific (heating) value the net calorific value at constant pressure of a gaseous fuel is the number of heat units produced when unit volume of the fuel, measured under standard conditions, is burned in excess air in such a way that the materials after combustion consist of the gases carbon dioxide, sulfur dioxide, oxygen, nitrogen, and water vapor, and that the pressure and temperature of the gaseous fuel, the air and the materials after combustion are 1 standard atmosphere and 25°C. This value is derived from the measured gross figure.

non-associated gas see preceding section on "reserves terminology."

off-peak the period during a day, week, month, or year when the load being delivered by a gas system is not at or near the maximum volume delivered by that system.

peak or peak load the maximum load consumed or produced by a unit or group of units in a stated period of time.

peak shaving the practice of augmenting the normal supply of gas during peak or emergency periods from another source where gas may have either been stored during periods of low demand or manufactured specifically to meet the peak demand.

pressure maintenance (repressing) a process in which natural gas is injected into a formation capable of producing crude petroleum to aid in maintaining pressure in an underground reservoir for the purpose of assisting in the recovery of crude.

producer gas a gas manufactured by burning coal or coke with a regulated deficiency of air, normally saturated with steam. The principal combustible component is carbon monoxide (about 30%), and the gross heat content is between 120 and 160 Btu/ft^3.

recycling see *cycling*.

reforming the process of thermal or catalytic cracking of natural gas, *liquefied petroleum gas, refinery gas* or gas from oil, resulting in the production of a gas having a different chemical composition.

refinery gas a gas resulting from oil refinery operations consisting mainly of hydrogen, methane, ethylene, propylene, and the butylenes. Other gases such as nitrogen and carbon dioxide may also be present. The composition can be highly variable, and the heat content can range from 1,000 to 2,000 Btu/ft^3.

reserves see preceding section.

send-out the quantity of gas delivered by a plant or system during a specified period of time.

specific gravity of gas the ratio of the density of gas to the density of dry air at the same temperature and pressure.

standard metering base standard conditions, plus agreed corrections, to which all gas volumes are corrected for purposes of comparison and payment.

summer valley the decrease that occurs in the summer months in the volume of the daily load of a gas *distribution system*.

therm 100,000 Btus.

thermie the heat required to raise the temperature of 1 metric ton of water 1°C (centigrade).

town gas gas piped to consumers from a gas plant. The gas can comprise both *manufactured gas* (secondary energy) and natural gas (primary energy) used for enrichment.

transmission company a company that operates a natural gas transmission system and which either operates no retail distribution system, or, as defined by the American Gas Association (AGA), receives less than 5% of its gas operating revenues from such retail distribution system (see also *distribution company* or *gas utility*).

unaccounted for gas the difference between the total gas available from all sources and the total gas accounted for as sales, net interchange and company use. The difference includes leakage or other actual losses, discrepancies due to *meter* inaccuracies, variations of temperature, and pressure and other variants.

unassociated gas see preceding section.

utilization factor the ratio of the maximum *demand* on a system or part of a system to the rated capacity of the system or part of the system under consideration.

water gas see *blue water gas*.

watt the meter-kilogram-second unit of power, equivalent to one joule per second and equal to the power in a circuit in which a current on one ampere flows across a potential difference on one volt.

wobbe index the gross calorific value of the gas divided by the square root of the density of the gas as compared with air.

10: Gas—Compression

Compressors

The following data are for use in the approximation of horsepower needed for the compression of gas.

Definitions

The "**N value**" of gas is the ratio of the specific heat at constant pressure (C_p) to the specific heat at constant volume (C_v).

If the composition of gas is known, the value of N for a gas mixture may be determined from the molal heat capacities of the components. If only the specific gravity of the gas is known, an approximate N value may be obtained by using the chart in Figure 1.

Piston displacement of a compressor cylinder is the volume swept by the piston with the proper deduction for the piston rod. The displacement is usually expressed in cubic feet per minute.

Clearance is the volume remaining in one end of a cylinder with the piston positioned at the end of the delivery stroke for this end. The clearance volume is expressed as a percentage of the volume swept by the piston in making its full delivery stroke for the end of the cylinder being considered.

Ratio of compression is the ratio of the absolute discharge pressure to the absolute inlet pressure.

Actual capacity is the quantity of gas compressed and delivered, expressed in cubic feet per minute, at the intake pressure and temperature.

Volumetric efficiency is the ratio of actual capacity, in cubic feet per minute, to the piston displacement, in cubic feet per minute, expressed in percent.

Adiabatic horsepower is the theoretical horsepower required to compress gas in a cycle in which there is no transfer of sensible heat to or from the gas during compression or expansion.

Isothermal horsepower is the theoretical horsepower required to compress gas in a cycle in which there is no change in gas temperature during compression or expansion.

Indicated horsepower is the actual horsepower required to compress gas, taking into account losses within the compressor cylinder, but not taking into account any loss in frame, gear, or power transmission equipment.

Figure 1. Ratio of specific heat (*N*-value).

Compression efficiency is the ratio of the theoretical horsepower to the actual indicated horsepower required to compress a definite amount of gas. The efficiency, expressed in percent, should be defined in regard to the base at which the theoretical power was calculated, whether adiabatic or isothermal.

Mechanical efficiency is the ratio of the indicated horsepower of the compressor cylinder to the brake horsepower delivered to the shaft in the case of a power-driven machine. It is expressed in percent.

Overall efficiency is the product, expressed in percent, of the compression efficiency and the mechanical efficiency. It must be defined according to the base, adiabatic isothermal, which was used in establishing the compression efficiency.

Piston rod gas load is the varying, and usually reversing, load imposed on the piston rod and crosshead during the operation, by different gas pressures existing on the faces of the compressor piston.

The maximum piston rod gas load is determined for each compressor by the manufacturer, to limit the stresses in the frame members and the bearing loads in accordance with mechanical design. The maximum allowed piston rod gas load is affected by the ratio of compression and also by the cylinder design, i.e. whether it is single- or double-acting.

Performance calculations for reciprocating compressors

Piston displacement

Single-acting compressor:

$$P_d = \left[S_t \times N \times 3.1416 \times D^2\right]/[4 \times 1728] \qquad (1)$$

Double-acting compressor without a tail rod:

$$P_d = \left[S_t \times N \times 3.1416 \times \left(2D^2 - d^2\right)\right]/[4 \times 1728] \qquad (2)$$

Double-acting compressor with a tail rod:

$$P_d = \left[S_t \times N \times 3.1416 \times 2 \times \left(D^2 - d^2\right)\right]/[4 \times 1728] \qquad (3)$$

Single-acting compressor compressing on the frame end only:

$$P_d = \left[S_t \times N \times 3.1416 \times \left(D^2 - d^2\right)\right]/[4 \times 1728] \qquad (4)$$

where: P_d = Cylinder displacement, cu ft/min,
S_t = Stroke length, in,
N = Compressor speed, number of compression strokes/min,
D = Cylinder diameter, in,
d = Piston rod diameter, in.

Volumetric efficiency

$$E_v = 0.97 - \left[(1/f)r_p^{1/k} - 1\right]C - L \qquad (5)$$

where: E_v = Volumetric efficiency,
f = Ratio of discharge compressibility to suction compressibility Z_2/Z_I,
r_p = Pressure ratio,
k = Isentropic exponent,
C = Percent clearance,
L = Gas slippage factor.

Let $L = 0.3$ for lubricated compressors
Let $L = 0.07$ for nonlubricated compressors $\qquad (6)$

These values are approximations, and the exact value may vary by as much as an additional 0.02−0.03.

Note: A value of 0.97 is used in the volumetric efficiency equation rather than 1.0 since even with 0 clearance, the cylinder will not fill perfectly.

Cylinder inlet capacity

$$Q_1 = E_v \times P_d \qquad (7)$$

Piston speed

$$P_S = [2 \times S_t \times N]/12 \qquad (8)$$

Discharge temperature

$$T_2 = T_1\left(r_p^{(k-1)/k}\right) \qquad (9)$$

where: T_2 = Absolute discharge temperature °R,
T_1 = Absolute suction temperature °R.

Note: Even though this is an adiabatic relationship, cylinder cooling will generally offset the effect of efficiency.

Power

$$W_{cyl} = \left[144P_1Q_1/33,000n_{cyl}\right] \times [k/(k-1)] \times \left[r_p^{k-1/k} - 1\right] \qquad (10)$$

where: n_{cyl} = Efficiency,
W_{cyl} = Cylinder horsepower.

See Figure 1 for the curve of the efficiency versus pressure ratio. This curve includes a 95% mechanical efficiency and a valve velocity of 3000 ft per minute. Tables 1 and 2 permit a correction to be made to the compressor horsepower for specific gravity and low inlet pressure. While it is recognized that the efficiency is not necessarily the element affected, the desire is to modify the power required per the criteria in these figures. The efficiency correction accomplishes this. These corrections become more significant at the lower pressure ratios.

Reciprocating compressor efficiencies plotted against pressure ratio with a valve velocity of 3,000 fpm and a mechanical efficency of 95 percent.

Figure 1. Reciprocating compressor efficiencies.

Figure 2. Volume bottle sizing.

Inlet valve velocity

$$V = 288 \times P_d/A \qquad (11)$$

where

V = Inlet valve velocity,

A = Product of actual lift and the valve opening periphery; it is the total for inlet valves in a cylinder expressed in square in. (This is a compressor vendor furnished number.)

Example. Calculate the following:

Suction capacity Horsepower
Discharge temperature Piston speed

Given:

Bore	= 6 in,
Stroke	= 12 in,
Speed	= 300 rpm,
Rod diameter	= 2.5 in,
Clearance	= 12%,
Gas	= CO_2,
Inlet pressure	= 1720 psia,
Discharge pressure	= 3440 psia,
Inlet temperature	= 115°.

Calculate piston displacement using Equation 2.

$$P_d = 12 \times 300 \times 3.1416$$
$$\times \left[2(6)^2 - (2.5)^2\right] / 1728 \times 4 = 107.6 \text{ cfm}$$

Calculate volumetric efficiency using Equation 5. It will first be necessary to calculate f, which is the ratio of discharge compressibility to suction compressibility.

$$T_1 = 115 + 460 = 575 \,°\text{R}$$

$$T_r = T/T_c$$
$$= 575/548$$
$$= 1.05$$

where

T_r = Reduced temperature,
T_c = Critical temperature = 548° for CO_2,
T = Inlet temperature.

$$P_r = P/P_c$$
$$= 1720/1071$$
$$= 1.61$$

where

P_r = Reduced pressure,
P_c = Critical pressure = 1071 psia for CO_2,
P = Suction pressure.

From the generalized compressibility chart

$$Z_1 = 0.312$$

Determine discharge compressibility. Calculate discharge temperature by using Equation 9.

$$r_p = 3440/1720 = 2$$

$$k = c_p/c_v = 1.3 \text{ for } CO_2$$

$$\begin{aligned} T_2 &= 575 \times \left[2.0^{13-1/1.3}\right] \\ &= 674.7 \,^\circ R = 674.7 - 460 \\ &= 214 \,^\circ F \end{aligned}$$

$$T_r = 674.7/548 = 1.23$$

$$P_r = 3440/1071 = 3.21$$

From the generalized compressibility chart

$$Z_2 = 0.575$$

$$f = 0.575/1.61 = 1.843$$

$$\begin{aligned} E_v &= 0.97 - \left[(1/1.843) \times 2.0^{1/1.3} - 1\right] \times 0.12 - 0.05 \\ &= 0.929 \\ &= 93\% \end{aligned}$$

Note: A value of 0.05 was used for L because of the high differential pressure.

Calculate suction capacity.

$$\begin{aligned} Q_1 &= E_v \times P_d \\ &= 0.93 \times 107.6 \\ &= 100.1 \text{ cfm} \end{aligned}$$

Calculate piston speed.

$$\begin{aligned} P_S &= [2 \times S_t \times N]/12 \\ &= [2 \times 12 \times 300]/12 \\ &= 600 \text{ ft/min} \end{aligned}$$

Table 1
Efficiency Multiplier for Low Pressure

	Pressure psia							
r_p	10	14.7	20	40	60	80	100	150
3.0	0.990	1.00	1.00	1.00	1.00	1.00	1.00	1.00
2.5	0.980	0.985	0.990	0.995	1.00	1.00	1.00	1.00
2.0	0.960	0.965	0.970	0.980	0.990	1.00	1.00	1.00
1.5	0.890	0.900	0.920	0.940	0.960	0.980	0.990	1.00

Source: Modified courtesy of the Gas Processors Suppliers Association and Ingersoll-Rand.

Table 2
Efficiency Multiplier for Specific Gravity

	SG				
r_p	1.5	1.3	1.0	0.8	0.6
2.0	0.99	1.0	1.0	1.0	1.01
1.75	0.97	0.99	1.0	1.01	1.02
1.5	0.94	0.97	1.0	1.02	1.04

Source: Modified courtesy of the Gas Processors Suppliers Association.

Estimate suction and discharge volume bottle sizes for pulsation control for reciprocating compressors

Pressure surges are created as a result of the cessation of flow at the end of the compressor's discharge and suction stroke. As long as the compressor speed is constant, the pressure pulses will also be constant. A low pressure compressor will likely require little, if any, treatment for pulsation control; however, the same machine with increased gas density, pressure, or other operational changes may develop a problem with pressure pulses. Dealing with pulsation becomes more complex when multiple cylinders are connected to one header or when multiple stages are used.

API Standard 618 should be reviewed in detail when planning a compressor installation. The pulsation level at the outlet side of any pulsation control device, regardless of type, should be no more than 2% peak-to-peak of the line pressure of the value given by the following equation, whichever is less.

$$P\% = 10/P_{line}^{1/3} \qquad (12)$$

When a detailed pulsation analysis is required, several approaches may be followed. An analog analysis may be performed on the Southern Gas Association dynamic compressor simulator, or the analysis may be part of the compressor purchase contract. Regardless of who performs the analysis, a detailed drawing of the piping in the compressor area will be needed.

The following equations are intended as an aid in estimating bottle sizes or for checking sizes proposed by a vendor for simple installations—i.e., a single cylinder connected to a header without the interaction of multiple cylinders. The bottle type is the simple unbaffled type.

Example. Determine the approximate size of suction and discharge volume bottles for a single-stage, single-acting, lubricated compressor in natural gas service.

Cylinder bore	= 9 in,
Cylinder stroke	= 5 in,
Rod diameter	= 2.25 in,
Suction temp.	= 80 °F,
Discharge temp.	= 141 °F,
Suction pressure	= 514 psia,
Discharge pressure	= 831 psia,
Isentropic exponent, k	= 1.28,
Specific gravity	= 0.6,
Percent clearance	= 25.7%.

Step 1. Determine suction and discharge volumetric efficiencies using Equations 5 and 13.

$$E_{vd} = [E_v \times f]/r_p^{1/k} \qquad (13)$$

$$r_p = 831/514 = 1.617$$

$$Z_1 = 0.93$$

$$Z_2 = 0.93$$

$$f = 0.93/0.93 = 1.0$$

Calculate suction volumetric efficiency using Equation 5:

$$E_v = 0.97 - \left[(1/1) \times (1.617)^{1/1.28} - 1\right] \times 0.257 - 0.03$$
$$= 0.823$$

Calculate discharge volumetric efficiency using Equation 13:

$$E_{vd} = 1 \times [0.823]/\left[1.617^{1/1.28}\right]$$
$$= 0.565$$

Calculate volume displaced per revolution using Equation 1:

$$P_d/N = \left[S_t \times 3.1416 \times D^2\right]/[1728 \times 4]$$
$$= \left[5 \times 3.1416 \times 9^2\right]/[1728 \times 4]$$
$$= 0.184 \text{ cu ft or } 3.18 \text{ cu in.}$$

Refer to Figure 2, volume bottle sizing, using volumetric efficiencies previously calculated, and determine the multipliers.

Suction multiplier = 13.5,
Discharge multiplier = 10.4,
Discharge volume = 318 × 13.5 = 3308 cu in,
Suction volume = 318 × 10.4 = 4294 cu in.

Calculate bottle dimensions. For elliptical heads, use Equation 14.

$$\text{Bottle diameter } d_b = 0.86 \times \text{volume}^{1/3} \qquad (14)$$

Volume = suction or discharge volume

$$\text{Suction bottle diameter} = 0.86 \times 4294^{1/3}$$
$$= 13.98 \text{ in}$$

$$\text{Discharge bottle diameter} = 0.86 \times 3308^{1/3}$$
$$= 12.81 \text{ in}$$

$$\text{Bottle length} = L_b = 2 \times d_b \qquad (15)$$

$$\text{Suction bottle length} = 2 \times 13.98$$
$$= 27.96 \text{ in}$$

$$\text{Discharge bottle length} = 2 \times 12.81$$
$$= 25.62 \text{ in}$$

Source

Compressors—Selection & Sizing, Gulf Publishing Company, Houston, Texas.

Compression horsepower determination

The method outlined below permits the determination of approximate horsepower requirements for the compression of gas.

1. From Figure 1, determine the atmospheric pressure in psia for the altitude above sea level at which the compressor is to operate.
2. Determine intake pressure (P_s) and discharge pressure (P_d) by adding the atmospheric pressure to the corresponding gauge pressure for the conditions of compression.
3. Determine the total compression ratio $R = P_d/P_s$. If the ratio R is more than 5 to 1, two or more compressor stages will be required. Allow for a pressure loss of approximately 5 psi between stages. Use the same ratio for each stage. The ratio per stage, so that each stage has the same ratio, can be approximated by finding the nth root of the total ratio, when n = number of stages. The exact ratio can be found by trial and error, accounting for the 5 psi interstage pressure losses.
4. Determine the N value of gas.
5. Figure 3 gives horsepower requirements for compression of 1 million cu ft per day for the compression ratios and N values commonly encountered in oil-producing operations.
6. If the suction temperature is not 60°F, correct the curve horsepower figure in proportion to the absolute temperature. This is done as follows:

$$\text{HP} \times \frac{460° + T_s}{460° + 60°\text{F}}$$
$$= \text{hp (corrected for suction temperature)}$$

where T_s is suction temperature in °F.

7. Add together the horsepower loads determined for each stage to secure the total compression horsepower load. For altitudes greater than 1500 ft above sea level, apply a multiplier derived from the following table to determine the nominal sea level horsepower rating of the internal combustion engine driver.

Figure 1. Atmospheres at various atmospheric pressures. *From Modern Gas Lift Practices and Principles*, Merla Tool Corp.

Figure 2. Ratio of specific heat (*n*-value).

N = RATIO OF SPECIFIC HEATS C_P/C_V
PS = SUCTION PRESSURE IN P.S.I.A.
PD = DISCHARGE PRESSURE IN P.S.I.A.
R = COMPRESSION RATIO P_d/P_s

NOTE:
FOR "R" GREATER THAN 4.5 TO 5 USE
MULTI-STAGE COMPRESSION, ALLOWING
FOR PRESSURE DROP BETWEEN STAGES,
USUALLY 5 P.S.I.. MAKE CORRECTION
FOR INTERSTAGE SUCTION TEMPERATURE.

Figure 3. Brake horsepower required for compressing natural gas.

Altitude—Multiplier		Altitude—Multiplier	
1500 ft	1.000	4000 ft	1.12
2000 ft	1.03	4500 ft	1.14
2500 ft	1.05	5000 ft	1.17
3000 ft	1.07	5500 ft	1.20
3500 ft	1.10	6000 ft	1.22

8. For a portable unit with a fan cooler and pump driven from the compressor unit, increase the horsepower figure by 7½%.

The resulting figure is sufficiently accurate for all purposes. The nearest commercially available size of compressor is then selected.

The method does not take into consideration the super-compressibility of gas, and it is applicable for pressures up to 1000 psi. In the region of high pressure, neglecting the deviation of gas behavior from that of the perfect gas may lead to substantial errors in calculating the compression horse-power requirements. The enthalpy–entropy charts may be used conveniently in such cases. The procedures are given in References 1 and 2.

Example. What is the nominal size of a portable compressor unit required for compressing 1,600,000 standard cu ft of gas per 24 h at a temperature of 85°F from 40 psig pressure to 600 psig pressure? The altitude above sea level is 2500 ft. The *N* value of gas is 1.28. The suction temperature of stages, other than the first stage, is 130°F.

Solution

$P_s = 53.41$ psia

$P_d = 613.41$ psia

$R_d = P_d/P_s = 11.5 = (3.4)^2 = (2.26)^3$

Try a solution using a 3.44 ratio and two stages.

1st stage : 53.41 psia $\times 3.44 = 183.5$ psia discharge

2nd stage : 178.5 psia $\times 3.44 = 614$ psia discharge

Horsepower from curve, Figure 3 $= 77$ hp for 3.44 ratio

$$\frac{77\,\text{hp} \times 1,600,000}{1,000,000} = 123.1 \;(\text{for } 60°\text{F suction temp.})$$

1st stage : 123.1 hp $\times \dfrac{460 + 85°}{460 + 60°} = 129.1$ hp

2nd stage : 123.1 hp $\times \dfrac{460 + 130°}{460 + 60°} = 139.7$ hp

$1.05(129.1\,\text{hp} + 139.7\,\text{hp}) = 282$ hp

1.075×282 hp

Nearest nominal size compressor is 300 hp.

Centrifugal compressors

The centrifugal compressors are inherently high-volume machines. They have extensive application in gas transmission systems. Their use in producing operations is very limited.

References

1. *Engineering Data Book*, Natural Gasoline Supply Men's Association, 1957.
2. Dr. George Granger Brown: "A Series of Enthalpy–Entropy Charts for Natural Gas," *Petroleum Development and Technology*, Petroleum Division AIME, 1945.

Generalized compressibility factor

The nomogram (Figure 1) is based on a generalized compressibility chart.[1] It is based on data for 26 gases, excluding helium, hydrogen, water, and ammonia. The accuracy is about 1% for gases other than those mentioned.

To use the nomogram, the values of the reduced temperature (T/T) and reduced pressure (P/P_c) must be calculated first.

Where:

$T = $ Temperature in consistent units
$T_c = $ Critical temperature in consistent units
$P = $ Pressure in consistent units
$P_c = $ Critical pressure in consistent units

Example. $P_r = 0.078$, $T_r = 0.84$. What is the compressibility factor, z? Connect P_r with T_r and read $z = 0.948$.

Source

Davis, D. S., *Petroleum Refiner*, 37, No. 11, (1961).

Reference

1. Nelson, L. C., and Obert, E. F., *Chem. Engr.*, 203 (1954).

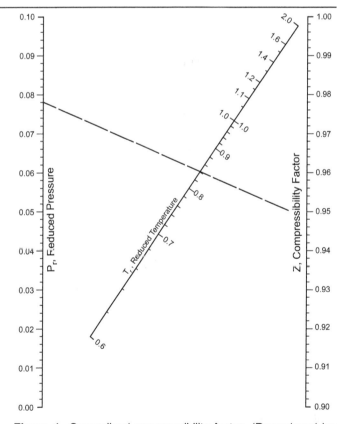

Figure 1. Generalized compressibility factor. (Reproduced by permission, *Petroleum Refiner*, Vol. 37, No. 11, copyright 1961, Gulf Publishing Co., Houston.)

Nomograph aids in diagnosing compressor cylinder ills

B. W. Robertson, Tennessee Gas Pipeline Co., Houston.

Compressor cylinders, like people, develop a fever when they are sick. Unlike humans, however, the normal temperature (suction) is not constant. The trick of using temperatures to determine malfunctions lies in determining the correct starting position.

To determine the starting position, it is necessary to know the true k value of the product being pumped. The k value may be obtained from measured pressures and temperatures according to the formula:

$$T_2/T_1 = (P_2/P_1)\frac{k-1}{k}$$

where the subscript 2 refers to discharge conditions and the 1 refers to suction, and all values of P and T are absolute.[1]

The preferred time to obtain this k would be with new equipment or from data obtained when the equipment was new. The next best time would be with newly overhauled equipment. Lacking either of these data, any values of k must be considered to contain some malfunctions, and any unnecessarily high numbers should be discarded.

Nomograph 1 has been successfully used for a number of years, although it is slowly giving way to electronic equipment. Nomograph 2 may be finished by completing the suction and discharge temperature scales according to the following formulas:

Suction : $t_i = 8.557\rho \, \text{Log}(t_s + 460) - 23.135\rho$

$\qquad\qquad + 2.51$ in above base line

Discharge : $t = 23.067\rho \, \text{Log}(t_d + 460)$

$\qquad\qquad - 62.65\rho$ in above base line

where

$\rho = \dfrac{k}{k-1}$ and logs are to the base 10.

References

1. Virgil Moring Faires, *Applied Thermodynamics*, 11th Printing, 1946, p. 43.

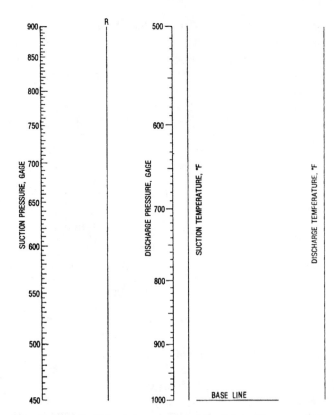

Nomograph 1. To determine starting position, it is necessary to know the true "k" value of the product being pumped. The "k" value may be obtained from measured pressure and temperatures according to the formula given.

Nomograph 2. This nomograph may be finished by completing the suction and discharge temperature scales according to the formulas given.

CENTRIFUGAL COMPRESSOR DATA

Centrifugal compressor performance calculations

Centrifugal compressors are versatile, compact, and generally used in the range of 1000–100,000 inlet cu ft per minute (ICFM) for process and pipeline compression applications.

Centrifugal compressors can use either a horizontal or a vertical split case. The type of case used will depend on the pressure rating with vertical split casings generally being used for the higher pressure applications. Flow arrangements include straight through, double flow, and side flow configurations.

Centrifugal compressors may be evaluated using either the adiabatic or polytropic process method. An adiabatic process is one in which no heat transfer occurs. This does not imply a constant temperature, only that no heat is transferred into or out of the process system. Adiabatic is normally intended to mean adiabatic isentropic. A polytropic process is a variable-entropy process in which heat transfer can take place.

When the compressor is installed in the field, the power required from the driver will be the same whether the process is called adiabatic or polytropic during design. Therefore, the work input will be the same value for either process. It will be necessary to use corresponding values when making the calculations. When using adiabatic head, use adiabatic efficiency, and when using polytropic head, use polytropic efficiency. Polytropic calculations are easier to make even though the adiabatic approach appears to be simpler and quicker.

The polytropic approach offers two advantages over the adiabatic approach. The polytropic approach is independent of the thermodynamic state of the gas being compressed, whereas the adiabatic efficiency is a function of the pressure ratio and therefore is dependent upon the thermodynamic state of the gas.

If the design considers all processes to be polytropic, an impeller may be designed, its efficiency curve determined, and it can be applied without correction regardless of pressure, temperature, or molecular weight of the gas being compressed. Another advantage of the polytropic approach is that the sum of the polytropic heads for each stage of compression equals the total polytropic head required to get from state point 1 to state point 2. This is not true for adiabatic heads.

Sample performance calculations

Determine the compressor frame size, number of stages, rotational speed, power requirement, and discharge temperature required to compress 5000 lb-in/min of gas from 30 psia at 60 °F to 100 psia. The gas mixture molar composition is as follows:

Ethane	5%
Propane	80%
n-Butane	15%

This is the same gas mixture that was considered in Sample Calculation 1, where the gas properties were calculated. The properties of this mixture were determined to be as follows:

$MW = 45.5$
$P_c = 611$ psia
$T_c = 676°R$
$c_p = 17.76$
$k_1 = 1.126$
$Z_1 = 0.955$

Before proceeding with the compressor calculations, let us review the merits of using average values of Z and k in calculating the polytropic head.

The inlet compressibility must be used to determine the actual volume entering the compressor to approximate the size of the compressor and to communicate with the vendor via the data sheets. The maximum value of θ is of interest and will be at its maximum at the inlet to the compressor where the inlet compressibility occurs (although using the average compressibility will result in a conservative estimate of θ).

Compressibility will decrease as the gas is compressed. This would imply that using the inlet compressibility would be conservative since as the compressibility decreases, the head requirement also decreases. If the variation in compressibility is drastic, the polytropic head requirement calculated by using the inlet compressibility would be practically useless. Compressor manufacturers calculate the performance for each stage and use the inlet compressibility for each stage. An accurate approximation may be substituted for the stage-by-stage calculation by calculating the polytropic head for the overall section using the average compressibility. This technique results in overestimating the first half of the impellers and underestimating the last half of the impellers, thereby calculating a polytropic head very near that calculated by the stage-by-stage technique.

Determine the inlet flow volume, Q_1:

$$Q_1 = m_1[(Z_1 R T_1)/(144 P_1)]$$

where

> m = Mass flow,
> Z_1 = Inlet compressibility factor,
> R = Gas constant = 1545/MW,
> T_1 = Inlet temperature °R,
> P_1 = Inlet pressure.

$$Q_1 = 5000[(0.955)(1545)(60+460)/(45.5)(144)(30)]$$
$$= 19,517 \text{ ICFM}$$

Refer to Table 1 and select a compressor frame that will handle a flow rate of 19,517 ICFM. A Frame C compressor will handle a range of 13,000–31,000 ICFM and would have the following nominal data:

$$H_{p \text{ nom}} = 10,000 \text{ ft—lb/lbm (nominal polytropic head)}$$

$$n_p = 77\% \text{ (polytropic effectively)}$$

$$N_{\text{nom}} = 5900 \text{ rpm}$$

Determine the pressure ratio, r_p.

$$r_p = P_2/P_1 = 100/30 = 3.33$$

Determine the approximate discharge temperature, T_2.

$$n/n - 1 = [k/k - 1]n_p = [(1.126/1.000)](0.77) = 6.88$$

$$T_2 = T_1 (r_p)^{(n-1)/n} = (60+460)(3.33)^{1/6.88} = 619\,°R$$
$$= 159\,°F$$

where

> T_1 = Inlet temp.

Determine the average compressibility, Z_a.

$$Z_1 = 0.955 \text{ (from gas properties calculation)}$$

where

> Z_1 = Inlet compressibility.

$$(P_r)_2 = P_2/P_c = 100/611 = 0.614$$

$$(T_r)_2 = T_2/T_c = 619/676 = 0.916$$

Table 1
Typical Centrifugal Compressor Frame Data[*]

Frame	Nominal Inlet Volume Flow		Nominal Polytropic Head		Nominal Polytropic Efficiency (%)	Nominal Rotational Speed (RPM)	Nominal Impeller Diameter	
	English (ICFM)	Metric (m³/h)	English(ft-lbf/lbm)	Metric(k-Nm/kg)			English (in)	Metric (mm)
A	1000–7000	1700–12,000	10,000	30	76	11,000	16	406
B	6000–18,000	10,000–31,000	10,000	30	76	7700	23	584
C	13,000–31,000	22,000–53,000	10,000	30	77	5900	30	762
D	23,000–44,000	39,000–75,000	10,000	30	77	4900	36	914
E	33,000–65,000	56,000–110,000	10,000	30	78	4000	44	1120
F	48,000–100,000	82,000–170,000	10,000	30	78	3300	54	1370

[*]While this table is based on a survey of currently available equipment, the instance of any machinery duplicating this table would be purely coincidental.

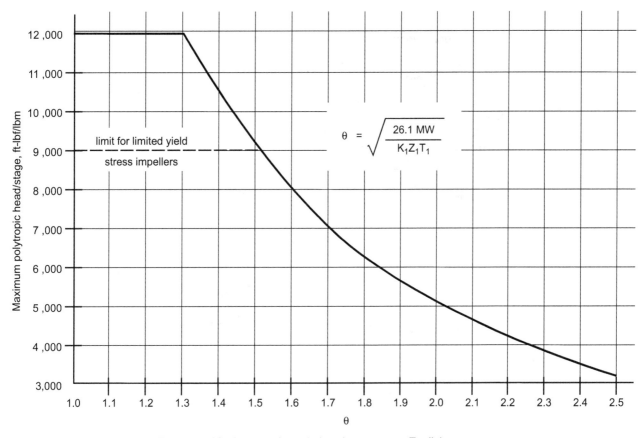

Figure 1. Maximum polytropic head per stage—English system.

As described earlier, find discharge compressibility Z_2.

$$Z_2 = 0.925$$

$$Z_a = (Z_1 + Z_2)/2 = 0.94$$

Determine average k-value. For simplicity, the inlet value of k will be used for this calculation. The polytropic head equation is insensitive to the k-value (and therefore the n-value) within the limits that k normally varies during compression. This is because any errors in the $n/(n-1)$ multiplier in the polytropic head equation tend to balance corresponding errors in the $(n-1)/n$ exponent. Discharge temperature is very sensitive to k-value. Since the k-value normally decreases during compression, a discharge temperature calculated by using the inlet k-value will be conservative, and the actual temperature may be several degrees higher—possibly as much as 25–50 °F Calculating the average k-value can be time-consuming, especially for mixtures containing several gases, since not only must the mol-weighted c_p of the mixture be determined at the inlet temperature but also at the estimated discharge temperature.

The suggested approach is as follows:

1. If the k-value is felt to be highly variable, one pass should be made at estimating discharge temperature based on the inlet k-value; the average k-value should then be calculated using the estimated discharge temperature.
2. If the k-value is felt to be fairly constant, the inlet k-value can be used in the calculations.
3. If the k-value is felt to be highly variable, but sufficient time to calculate the average value is not available, the inlet k-value can be used (but be aware of the potential discrepancy in the calculated discharge temperature).

$$k_1 = k_a = 1126$$

Determine average $n/(n-1)$ value from the average k-value. For the same reasons discussed earlier, use $n/(n-1) = 6.88$.

Table 2
Approximate Mechanical Losses as a Percentage
of Gas Power Requirement[*]

Gas Power Requirement		Mechanical Losses, L_m (%)
English (hp)	**Metric (kW)**	
0–3000	0–2500	3
3000–6000	2500–5000	2.5
6000–10,000	5000–7500	2
10,000+	7500+	1.5

[*]There is no way to estimate mechanical losses from gas power requirements. This table will, however, ensure that mechanical losses are considered and yield useful values for estimating purposes.

Determine polytropic head, H_p

$$H_p = \left[Z_a RT_1 (n/n - 1) r_p^{(n-1)/n} - 1 \right]$$
$$= (0.94)(1545/45.5)(520)(6.88)\left[(3.33)^{1/6.88} - 1\right]$$
$$= 21,800 \text{ ft—lbf/lbm}$$

Determine the required number of compressor stages, θ:

$$\theta = \left[(26.1 \text{ Mw})/k_1 Z_1 T_1 \right]^{0.5}$$
$$= \left[(26.1)(45.5)/(1.126)(0.955)(520) \right]^{0.5} = 1.46.$$

max H_p/stage from Figure 1 using $\theta = 1.46$.

Number of stages $= H_p/\text{max}.H_p/\text{stage} = 21,800/9700$
$$= 2.25 = 3 \text{ stages.}$$

Determine the required rotational speed:

$$N = N_{\text{nom}}\left[H_p/H_{p \text{ nom}} \times \text{no. stages} \right]^{0.5}$$
$$= 5900[21,800/(10,000)(3)]^{0.5} = 5030 \text{ rpm.}$$

Determine the required shaft power:

$$PWR_g = mH_p/33,000\, n_p$$
$$= (5000)(21,800)/(33,000)(0.77) = 4290 \text{ hp}$$

Mechanical losses $(L_m) = 2.5\%$ (from Table 2)

$$L_m = (0.025)(4290) = 107 \text{ hp.}$$

$$PWR_s = PWR_g + L_m = 4290 + 107 = 4397 \text{ hp.}$$

Determine the actual discharge temperature:

$$T_2 = T_1 (r_p)^{(n-1)/n} = 520(3.33)^{1/6.88} = 619°R$$
$$= 159°F.$$

The discharge temperature calculated in the last step is the same as that calculated earlier only because of the decision to use the inlet k-value instead of the average k-value. Had the average k-value been used, the actual discharge temperature would have been lower.

Source

Estimating Centrifugal Compressor Performance, Gulf Publishing Company, Houston.

Nomographs for estimating compressor performance[a]

Convert SCFM to inlet CFM

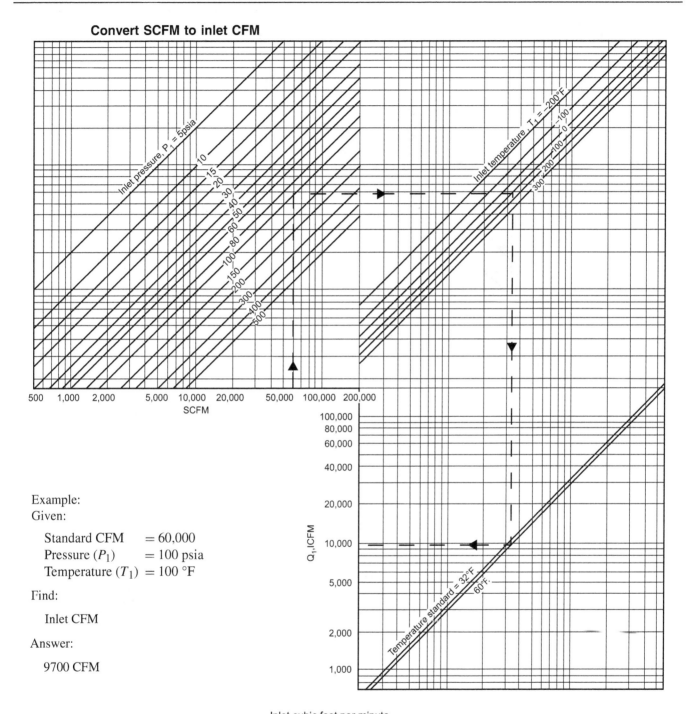

Example:

Given:

Standard CFM $= 60,000$
Pressure (P_1) $= 100$ psia
Temperature (T_1) $= 100\ °F$

Find:

Inlet CFM

Answer:

9700 CFM

Inlet cubic feet per minute

[a] Reprinted by permission of Gas Processors Suppliers Association.

Convert weight flow to inlet CFM

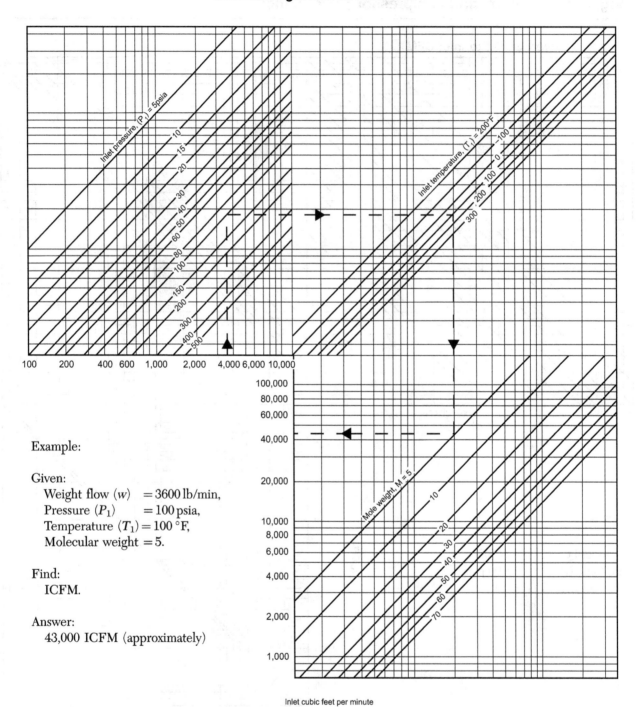

Inlet cubic feet per minute

Example:

Given:
   Weight flow (w) $= 3600\,\text{lb/min}$,
   Pressure (P_1) $= 100\,\text{psia}$,
   Temperature $(T_1) = 100\,°\text{F}$,
   Molecular weight $= 5$.

Find:
   ICFM.

Answer:
   43,000 ICFM (approximately)

Convert weight compression ratio to head

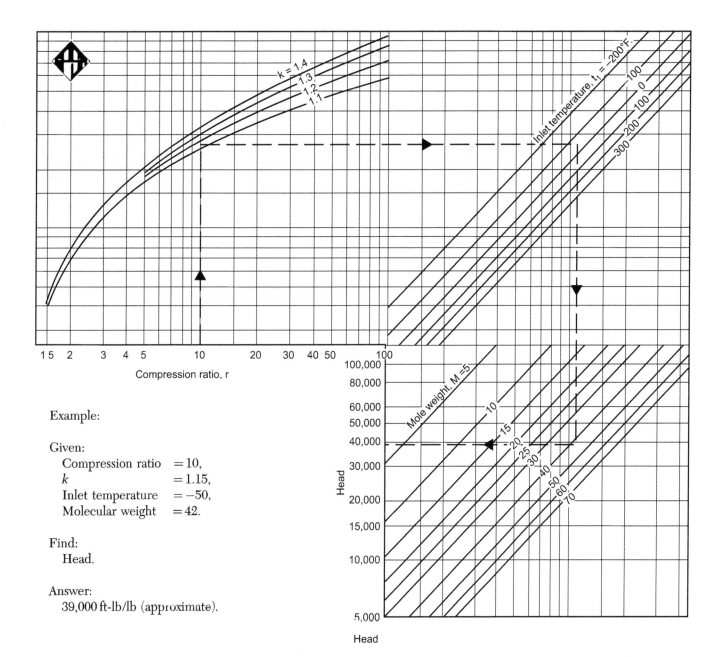

Compression ratio, r

Head

Head

Example:

Given:
    Compression ratio   = 10,
    k = 1.15,
    Inlet temperature   = −50,
    Molecular weight   = 42.

Find:
    Head.

Answer:
    39,000 ft-lb/lb (approximate).

Convert compression ratio to discharge temperature

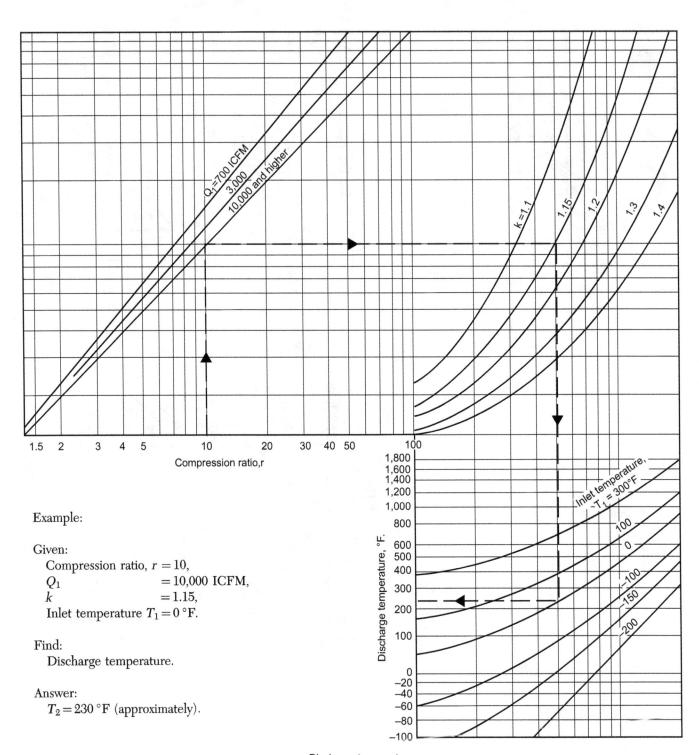

Discharge temperature

Example:

Given:
Compression ratio, $r = 10$,
Q_1 $\qquad = 10,000$ ICFM,
k $\qquad = 1.15$,
Inlet temperature $T_1 = 0\,°F$.

Find:
Discharge temperature.

Answer:
$T_2 = 230\,°F$ (approximately).

Horsepower determination

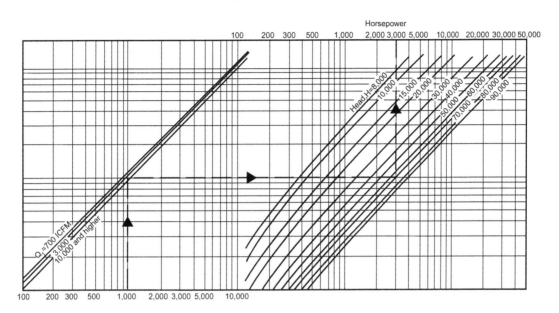

Weight flow, w, lb/min.
  Example:
  Given:

Weight flow, $w = 1000$ lb/min,
Head $= 70,000$ ft-lb/lb.

Find:
Horsepower.
Answer:
$H_p = 3000.$

Efficiency conversion

Estimate hp required to compress natural gas

To estimate the horsepower to compress a million cubic ft of gas per day (MMcfd), use the following formula:

$$BHP/MMcfd = \frac{R}{R + RJ}\left(\frac{5.16 + 124 \, \text{Log} \, R}{0.97 - 0.03R}\right)$$

where

R = Compression ratio. Absolute discharge pressure divided by absolute suction pressure,
J = Supercompressibility factor—assumed 0.022 per 100 psia suction pressure.

Example. How much horsepower should be installed to raise the pressure of 10 million cu ft of gas per day from 185.3 psi to 985.3 psi?

This gives absolute pressures of 200 and 1000.

$$R = \frac{1000}{200} = 5.0$$

Substituting in the formula:

$$BHP/MMcfd = \frac{5.0}{5.0 + 5 \times 0.044} \times \frac{5.15 + 124 \times .699}{.97 - .03 \times 5}$$

$$= 106.5 \, \text{hp} = BHP \text{ for 10 MMcfd}$$

$$= 1065 \, \text{hp}$$

where the suction pressure is about 400 psia, and the brake horsepower per MMcfd can be read from the chart.

The previous formula may be used to calculate horsepower requirements for various suction pressures and gas physical properties to plot a family of curves.

Estimate compressor hp where discharge pressure is 1000 psi

Rule. Use the following formula for reciprocating compressors where the compressor ratio is between 1.20 and 1.40.

Brake horsepower per MMcfd

$$= 11 + (42.5)(R - 1.20) \quad R \text{ is compression ratio.}$$

Example. How much horsepower would have to be installed in a field compressor station where the discharge pressure would be 1055 psia and suction would be 800 psia? The station must handle 10,000,000 cu ft per day.

Solution. The compression ratio is

$$\frac{1055}{800} = 1.32$$

Substituting in the formula:

Brake horsepower per MMcfd $= 11 + (42.5)(1.32 - 1.20)$

$$= 11 + 5.1 = 16.1 \, \text{hp}.$$

The station would require 161 hp.

For a centrifugal compressor the formula would be: Brake horsepower per MMcfd $= 11 + (42.5) \, (R - 1.20)$

Example. How much horsepower should be installed in a centrifugal compressor station handling 100,000,000 cu ft per day if the suction and discharge are the same as above?

Solution. Brake horsepower per MMcfd $= 10 + (49)(0.12) = 15.9$

For 100 MMcfd the horsepower would be 1590.

Calculate brake horsepower required to compress gas

The approximate brake horsepower per million cubic feet of gas to be compressed may be obtained from this table. The total brake horsepower required is calculated by multiplying the total million cubic feet of gas compressed. In this estimation, gas volumes are expressed at 14.7 psia and 60 °F, specific gravity as 0.6, and K (the ratio of specific heats) as 1.27. The suction temperature is assumed as 80 °F and interstage temperature as 130 °F. The relative humidity of the air equals 100%.

Example. What is the brake horsepower required to compress 5.0 million cu ft of gas from 200 psig suction pressure to 1000 psig discharge pressure?

On the chart, read opposite a suction pressure of 200 and under a discharge pressure of 1000 psig, that 100 brake horsepower is required per million cubic feet of gas compressed. Multiply 5.0 times 100 = 500 bhp.

Horsepower Selection Chart											
Natural Gas Compression											
		Discharge Pressure									
		250 psig	375 psig	500 psig	750 psig	1000 psig	1250 psig	1500 psig	2000 psig	3500 psig	6000 psig
Suction Pressure	0 psig	205	235	260	280	300	320	330	360	390	—
	10 psig	165	195	215	250	260	275	290	310	345	390
	25 psig	130	155	175	205	230	250	255	275	320	370
	50 psig	95	125	140	170	190	205	215	245	280	325
	75 psig	70	95	120	145	165	180	195	215	250	300
	100 psig	50	75	95	120	145	160	175	195	230	275
	200 psig	—	—	55	80	100	120	130	150	190	240
	300 psig	—	—	30	55	75	90	105	120	160	205
	500 psig	—	—	—	25	40	55	65	85	125	165
	750 psig	—	—	—	—	15	30	40	60	100	140
	1000 psig	—	—	—	—	—	—	25	40	75	125
	1500 psig	—	—	—	—	—	—	—	—	50	90
	2000 psig	—	—	—	—	—	—	—	—	—	70
	2500 psig	—	—	—	—	—	—	—	—	—	55

How can we find the size of a fuel gas line for a compressor station?

Suppose that an 8000 hp compressor building is added to an existing compressor plant. What size fuel line should be run from the meter building 1600 ft distant? Take-off pressure at the meter building will be 50 lb/in², and it is desirable to maintain 25 psi pressure on the large engine fuel header in the building. Assume that the gas is heated prior to expansion so that outlet temperature is 60 °F.

The first step is to estimate the ultimate compressor-building horsepower. Assume in the foreseeable future the building will be expanded to 12,000 hp.

Next, estimate the fuel needs.

$Q = 12,000 \times 240$ cu ft per day

$= 2,880,000$ cu ft per day

To estimate the size of the line that must be run for fuel, use this formula:

$$Q = \frac{(36,000)d^3\sqrt{P_1^2 - P_2^2}}{\sqrt{1}}$$

where

Q = Cubic feet of gas per 24 h
d = Pipe ID in inches
l = Length of pipe in feet
P_1 = Psi (abs) at starting point
P_2 = Psi (abs) at ending point

Using this formula, try several pipe diameters to determine the approximate size. For instance, if 6-in ID pipe were used, the approximate flow could be found as follows:

$$Q = \frac{36,300 \times 6^3 \sqrt{50 + 15^2 - (25 + 15)^2}}{\sqrt{1600}}$$

$$Q = \frac{36,300 \times 216\sqrt{65^2 - 40^2}}{40}$$

$Q =$ approximately $10,036,000$ cu ft per 24 h.

From this it is obvious that 6-in pipe would be considerably oversized, especially if the fuel gas header in the compressor building is of large diameter.

Try substituting 4-in diameter pipe in this formula to find an approximate size pipe.

$$Q = \frac{36,300 \times 4^3 \sqrt{50 + 15^2 - (25 + 15)^2}}{\sqrt{1600}}$$

$$Q = \frac{36,300 \times 64\sqrt{65^2 - 40^2}}{40}$$

$Q = 2,973,700$ cu ft per 24 h.

Apparently 4-in pipe will be the correct size. To check it, use this more accurate formula, which is based on the Weymouth formula:

$$Q = \frac{63,320 \times d^{8/3}\sqrt{P_1^2 - P_2^2}}{\sqrt{1}}$$

$$Q = \frac{63,320 \times 41\sqrt{65^2 - 40^2}}{\sqrt{1600}}$$

Four-in pipe is adequate.

$Q = 3,322,000$ cu ft per day

Note. This latter formula is based on the Weymouth formula, for gas with a specific gravity of 0.60 and a temperature of 60 °F. For every 10° variation in temperature, the answer will be approximately 1% in error. For every 0.01 variation in specific gravity, the answer will be three-fourths percent in error.

Estimate engine cooling water requirements

This equation can be used for calculating engine jacket water requirements, as well as lube oil cooling water requirements:

$$GPM = \frac{H \times BHP}{500\Delta t}$$

where $H =$ Heat dissipation in Btu per BHP/h. This will vary for different engines; where they are available, the manufacturers' values should be used. Otherwise, you will be safe in substituting the following values in the formula: For engines with water-cooled exhaust manifolds—engine jacket water $= 2200$ Btu per BHP/h. Lube oil cooling water $= 600$ Btu per BHP/h.

For engines with dry-type manifolds (so far as cooling water is concerned), use 1500 Btu/BHP/h for the engine jackets and 650 Btu/BHP/h for lube oil cooling water requirements.

BHP = Brake Horsepower Hour

$\Delta t =$ Temperature differential across engine. Usually manufacturers recommend this not exceed 15 °F; 10 °F is preferable.

Example. Find the jacket water requirements for a 2000-hp gas engine that has no water jacket around the exhaust manifold.

Solution.

$$GPM = \frac{1500 \times 2000}{500 \times 10}$$

$$GPM = \frac{3,000,000}{5000} = 600 \text{ gallons per min}$$

The lube oil cooling water requirements could be calculated in a like manner.

Estimate fuel requirements for internal combustion engines

When installing an internal combustion engine at a gathering station, a quick approximation of fuel consumptions could aid in selecting the type of fuel used.

Using Natural Gas: Multiply the brake hp at drive by 11.5 to get cubic feet of gas per hour.

Using Butane: Multiply the brake hp at drive by 0.107 to get gallons of butane per hour.

Using Gasoline: Multiply the brake hp at drive by 0.112 to get gallons of gasoline per hour.

These approximations will give reasonably accurate figures under full load conditions.

Example. Internal combustion engine rated at 50 bhp—three types of fuel available.

Natural Gas: $50 \times 11.5 = 575$ cu ft of gas per hour.
Butane: $50 \times 0.107 = 5.35$ gallons of butane per hour.
Gasoline: $50 \times 0.112 = 5.60$ gallons of gasoline per hour.

Estimate fuel requirements for compressor installation

Rule. Multiply installation horsepower by 240 to obtain cubic ft of fuel gas required per day.

Example. Estimate the amount of fuel required to supply a 1000-hp field compressor installation.

$$1000 \times 240 = 240 \text{ Mcfd}$$

Performance testing guidelines for centrifugal compressors

By **E. Wilcox,** Chevron Energy Technology Company, Houston, Texas.

Detailed performance analysis of centrifugal compressors in the field is essential to evaluate their existing condition. Current performance of a compressor can also be a valuable tool in evaluating its reliability. A decrease in compressor performance can be an excellent indication of internal wear or fouling, which, if allowed to continue, may result in unscheduled outages or reduced throughput. In contrast, perceived performance problems may be a result of a compressor operating far from its original design. However, obtaining accurate performance data in the field can be very challenging.

This article will explain the relative importance that different process variables have in performance calculations, as well as specify the necessary instrumentation to obtain process data with an acceptable uncertainty. Since original design conditions almost never match actual operating conditions, how to compare actual field data to design data using nondimensional head and efficiency will be demonstrated. Likewise, the limits on these comparisons will be outlined for users. Several example field performance evaluations are discussed with some common pitfalls that can be avoided. Examples of the effects of inaccurate process data are also included in the discussion.

Obtaining accurate field data. Accurate centrifugal compressor performance measurement depends on the quality of the field data. In general, testing should follow the conditions set forth in ASME PTC 10–1997. However, most petrochemical plants do not have the instrumentation specified in this test code. Compressor piping should be designed to accommodate flowmeter runs and to meet location requirements for pressure and temperature as well. Small inaccuracies (in certain areas) can make a large difference between the measured versus actual conditions.

The field data required for an accurate performance evaluation are:

• Suction and discharge pressures,
• Suction and discharge temperatures,
• Mass flowrate,
• Gas composition,
• Rotational speed,
• Driver load.

The most important factors in obtaining accurate field data are:

• Pressure and temperature transmitters should be located within a few feet of the compressor case at least 10 pipe diameters from any valve, tee, elbow, or other obstruction.
• Transmitters (both pressure and temperature) should be calibrated prior to the test.
• Data should only be taken during steady-state conditions. This is usually specified by requiring the discharge temperature to remain constant over a 15-min interval.
• RTDs should be used instead of thermocouples, if possible, due to improved accuracy.
• Test points at different flowrates or speeds are incredibly valuable. Inaccuracies in flow, pressure, or temperature may be difficult to determine with only one test point. But a map of points will usually reveal the problem.

Flowrate. The flowrate reported by the plant DCS is usually not absolutely correct. The meter factor, K, which is used to convert the measured differential pressure into a flowrate, is always a function of the gas pressure, temperature, and molecular weight (Equation 1). The flowrate reported by the meter must be corrected for the actual process conditions. However, the correction depends on the type of flow units. The correction equations for standard volume, mass, and actual volume flows are given in Equations 2–4, respectively. It is common for flowmeters to be "compensated" in the DCS for the actual pressure and temperature. However, it is rare for the compensation to include the gas molecular weight, since this requires an online gas analyzer. The flowmeter design and location should meet the requirements of the guidelines established by ASME PTC-10.

$$Q = K\sqrt{\Delta P}$$

where

$$K = f(P, T, Mw) \tag{1}$$

$$Q_A = Q\sqrt{\frac{P_A T_D Z_D Mw_D}{P_D T_A Z_A Mw_A}} \text{ for std. volume flow} \tag{2}$$

$$\dot{m}_A = \dot{m}\sqrt{\frac{P_A T_D Z_D Mw_A}{P_D T_A Z_A Mw_D}} \text{ for mass flow} \tag{3}$$

$$Q_A = Q\sqrt{\frac{P_A T_A Z_A Mw_D}{P_A T_D Z_D Mw_A}} \text{ for actual volume flow} \tag{4}$$

Example—incorrect flowmeter. An incorrect flow measurement will cause the compressor to appear either low or high in head because the operating point is marked incorrectly on the performance map. The best method to determine if a flow measurement is incorrect is to obtain several data points to compare against the entire curve. As can be seen in Figure 1, the maximum head should remain the same. The curve is just shifted to the right or left.

Gas composition. The gas composition is the most important data required to evaluate a compressor's performance. Likewise, in a wet gas application, an accurate gas composition is also the most difficult piece of test data to obtain. At least two gas samples should be taken during the performance test. Two samples are required to validate the gas analysis (for comparison) as well as when one of the samples is lost or invalid (i.e. without the gas composition, the other process data are worthless). The sample should be obtained using a free-flowing arrangement (Figure 2 shows two examples). An insertion probe should be used, if possible, to obtain the gas sample instead of a wall tap. Samples obtained from wall taps will generally be leaner in the higher molecular weight components due to the boundary layer

Figure 1. Effects of incorrect flow measurement.

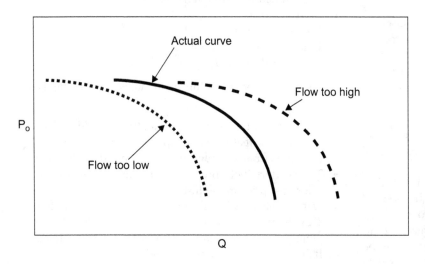

Figure 2. The sample should be obtained using a free-flowing arrangement.

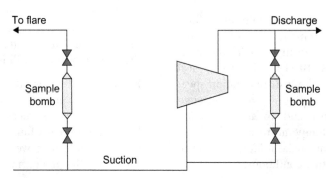

effect. The sample bomb should be heated to the process temperature before it is analyzed. This is to prevent any condensation of liquids that could alter the gas composition. Obviously, this is much more important for higher molecular weight services such as FCC wet gas compared to reformer or hydrocracker recycle hydrogen.

As an example, a sample taken from the discharge of a coker wet gas compressor was analyzed at the lab ambient temperature (approximately 75 °F) and at 275 °F (sample temperature, Table 1). As can be seen, the incorrect gas composition has a pronounced effect on the calculated gas horsepower. This effect is magnified because the molecular weight is used to calculate the head as well as correct the flow.

The gas composition should be reviewed to determine if it is feasible. One of the most common methods used to determine the gas sample composition is a gas chromatograph. A gas chromatograph determines the components by burning them in the presence of a carrier gas. For this reason, a gas chromatograph will not show any water vapor (i.e. water will not burn). However, it is very common for process gases to be saturated with water. For this reason, the measured gas composition must be adjusted if it is indeed saturated with water. The mole fraction of any constituent in a mixed gas stream is defined by Equation 5. Since the gas stream temperature is known, the water vapor partial pressure is equal to its saturation pressure at the given temperature. After the water vapor mole fraction is found, the normalized gas composition must be adjusted so that the total mole fraction equals 1.0.

$$\text{mole fraction} = \frac{\text{constituent partial pressure}}{\text{mixture total pressure}} \quad (5)$$

Likewise, for a multisection compressor, the gas composition will typically become leaner as liquids are condensed out in the intersection knockouts. If the gas samples do not reflect this, there is usually a problem. Also, certain components (i.e. H_2 or H_2S) are normally noncondensable. The mass fractions of these components should actually increase as the heavier liquids are removed from the gas stream.

Example—incorrect gas composition. If the measured gas composition is lower than the actual gas composition in the compressor, it will have the most pronounced effect on both the corrected flowrate and calculated polytropic head. An incorrect low molecular weight will cause the corrected flowrate (if it is measured in standard cubic feet) to be higher (Equation 2). Likewise, the low molecular weight will cause the calculated polytropic head to be higher than it actually is for the given compression ratio. A molecular weight that is too low will also cause the calculated polytropic efficiency to be higher, though not as pronounced. Additionally, the lower molecular weight will lower the calculated mass flow, which will in turn lower the calculated horsepower. Note, these are the effects produced by an incorrect gas composition being used as the input for a field test. These are not the results of a compressor that is operating in a gas that is actually lower in molecular weight than its design. If the measured gas composition is too high, the results are just the inverse of the above (i.e. the flow, head and efficiency are all lower). A summary is shown in Table 2.

Calculating and evaluating performance parameters. Once the field performance data are obtained, the correct performance parameters must be accurately calculated and evaluated to ensure that the field data are realistic. The most critical step in calculating performance parameters is determining the inlet and outlet density, enthalpy, and entropy. For hydrocarbon gas mixtures, performance programs that use equations of state, such as Lee—Kesler, Redlich—Kwong, or Peng—Robinson, will provide much better results than approximations from Mollier diagrams or ideal gas relationships. Process simulators can be used as well. Once the gas properties are calculated, the correct parameters must be selected to adequately evaluate the compressor performance. Evaluating the results of these calculations should be made to determine the validity of both the calculations and the field data (see Schultz for a listing of the equations for these parameters).

- Shaft horsepower—If a torque meter is available, the shaft horsepower may be calculated directly. However, since this is not commonly the case, the horsepower can be calculated by the heat balance method, as given in ASME PTC-10 and compared to the calculated driver horsepower. This comparison between driver and driven power is the most important indicator of accuracy of a performance test (i.e. if the two values are not close, something is wrong with the test data).

Table 1
Effects of Incorrect Gas Composition on Corrected Flow and Calculated Horsepower

	Correct Gas Sample, 275 °F	Incorrect Gas Sample, 75 °F
Molecular weight	48	34
Flow, MMscfd	27.4	32.5
Shaft horsepower, hp	5775	7036

Table 2
Effects of Incorrect Gas Composition on Calculated Performance Parameters

Mw	H_P	Effy.	hp	mf
↑	↓	↓	↑	↑
↓	↑	↑	↓	↓

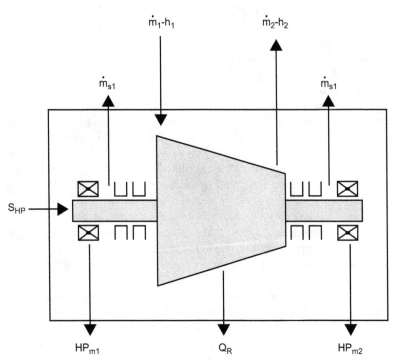

Figure 3. The first law of thermodynamics is applied to the control volume around the compressor.

Applying the first law of thermodynamics to the control volume around the compressor is shown in Figure 3.

$$Shp = \frac{(\dot{m}_1 - \dot{m}_{S1})(h_2 - h_1) + Q_R}{2545} + hp_{MECH} \qquad (6)$$

$$Shp = hp_{GAS} + hp_{MECH}; \; hp_{MECH} = hp_{M1} + hp_{M2} \qquad (7)$$

The radiant heat loss, Q_R, is normally negligible, but it can be approximated by dividing the compressor case into axial sections and approximating the heat transfer from each section. Likewise, the seal leakage on the inlet is normally less than 1%, but it can be easily calculated at the orifice in the vent off the seal pots. Note, the internal seal losses (i.e. balance piston and impeller labyrinth seals) do not affect the calculated shaft horsepower. However, they do affect the calculated head and efficiency, which in turn affects the discharge pressure and temperature. The effects of seal losses will be discussed later.

- Polytropic head—This value is limited to approximately 10,000–12,000 ft-lbf/lbm per impeller for most closed impellers. The maximum is set by the impeller tip speed which is limited by the gas sonic velocity as well as the impeller yield stress. Impellers designed to operate in highly corrosive processes (H_2S, CO_2) often require a maximum yield stress of 90 kpsi and Rockwell C of less than 22, which limits them to approximately 9000 ft-lbf/lbm. For example, if a performance test is done on a compressor and the calculated head per impeller is 20,000 ft-lbf/lbm, then either the measured compression ratio is too high or the measured molecular weight is too low.

- Polytropic efficiency—The maximum value depends on the impeller specific speed, but is limited to approximately 75–78% for older 2D impellers with vaneless diffusers and 80–85% for newer 3D impellers with vaned diffusers.

Losses. Losses are generally grouped into three distinct areas: mechanical, seal, and aerodynamic. Mechanical losses include power dissipated through bearings, oil or gas seals, shaft-driven lube oil pumps, and gearboxes. Seal losses are the decrease in the amount of energy available to convert into pressure head due to internal recirculation inside the compressor. Aerodynamic losses include effects such as friction and pressure losses in the impellers and diffusers. For the purpose of this article, aerodynamic losses are included in the impeller efficiencies and not discussed.

Mechanical losses. Mechanical losses are mostly a function of size and speed. Larger bearings and seals at higher speeds dissipate more power. Mechanical losses are simply added to the calculated gas horsepower. These losses can be approximated from:

- Measuring the flowrate and lube oil temperature increase and using Equation 8. This may be difficult due to many different lube oil return lines and pressure controllers that spill back to the reservoir.
- OEM-supplied curves for different bearings and seals.
- Bearing rotordynamic computer models that calculate horsepower losses as well.

Table 3
Mechanical Losses

Gas Power Requirement	Mechanical Losses, %
0–3000	3.0
3000–6000	2.5
6000–10,000	2.0
10,000+	1.5

Table 4
Gearbox Efficiencies

Gear Type	Efficiency, %
Helical	97–99
Herringbone	6–99
Straight bevel	95–98
Spiral bevel	96–98

- Tables based on compressor gas horsepower (Table 3). Approximate gearbox efficiencies are given in Table 4.

$$\text{hp}_{\text{MECH}} = \frac{\dot{m}c_p \Delta T_{\text{oil}}}{33,000} \tag{8}$$

Seal losses. Balance piston or division wall leakage is the only seal loss evaluated in this discussion since they are usually much larger than impeller labyrinth seal leakage. Seal losses are much more difficult to estimate than mechanical losses because they are not just added to the calculated gas horsepower. Leakage through the balance piston seal to the compressor suction increases the volume flow through the impellers as well as increases the inlet temperature, both of which decrease the compressor discharge pressure (Figure 4). Balance piston leakage causes the measured head

and efficiency to decrease, but does not increase the calculated gas horsepower. This does not mean that increased balance piston seal leakage does not affect compressor power. In the operating world, compressors have to put up a required discharge pressure to pump forward. As the balance piston seal leakage increases, either the compressor speed or suction pressure must be increased to maintain the desired discharge pressure and flow. This causes the power consumed by the compressor to increase.

A feel for the amount of balance piston leakage can be established by measuring the differential pressure across the balance piston line. Most OEMs design for this to be less than 2 or 3 psid. Anything above this usually means that a balance piston seal is leaking excessively. A differential pressure gauge is almost always required to make this measurement due to the small differential.

Seal leakage rate can be estimated by using a leakage equation. Equation 9 applies for adjacent teeth in the labyrinth. This is used to iteratively solve for the labyrinth cavity pressures and seal leakage rate. If the flow is choked, Equation 10 is used for the last labyrinth.

$$\dot{m} = \mu_0 \mu_1 H \sqrt{\frac{P_{i-1}^2 - p_i^2}{ZRT}} \tag{9}$$

$$\dot{m}_{NT} = \frac{0.51 \mu_0 P_{NC} H}{\sqrt{ZRT}} \tag{10}$$

Once the leakage rate is determined, the measured field data must be adjusted to find the actual conditions. In the case of a straight-through compressor with a balance piston seal, the hot balance piston seal leakage is mixed with the gas at the inlet using a mass and enthalpy balance to determine the actual inlet temperature to the first impeller (Figure 5). This will change the polytropic head and efficiency calculations. Adding in the model for the balance piston seal leakage allows us to see why the compressor efficiency is not as designed.

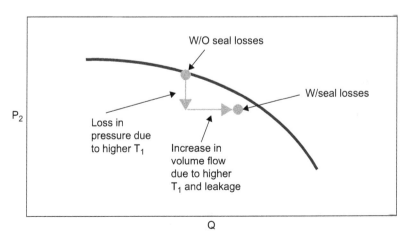

Figure 4. Effect of balance piston leakage on discharge pressure.

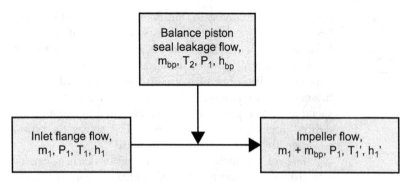

Figure 5. Mass and heat balances for balance piston seal leakage.

Comparing measured field performance to shop test or predicted performance. Since field conditions never exactly match the original design, certain nondimensional parameters must be calculated so that the field performance can be compared to the OEM shop test or predicted performance data. While these nondimensional parameters will enable "apple-to-apple" comparisons for different conditions, they have very real limitations based on the aerodynamic characteristics of the impellers.

These nondimensional parameters include: Polytropic head coefficient, μ_P

$$\mu_P = \frac{H_P}{(\pi DN)^2} \quad (12)$$

Polytropic efficiency, η_P

$$\eta_P = \frac{H_P}{h_s - h_1} \quad (13)$$

Flow coefficient, Φ

$$\Phi = \frac{Q}{ND^3} \text{ or } \frac{Q}{N} \quad (14)$$

Impeller Mach number, M

$$M = \frac{\pi DN}{\sqrt{k_1 Z_1 R T_1}} \quad (15)$$

The first step in any comparison is to obtain a set of nondimensional curves for the shop test or predicted data. If they were not provided by the OEM, they can be obtained by iteratively calculating the μ_P and η_P values from the given values of discharge pressure and shaft horsepower. This can be accomplished by guessing a discharge temperature for the given discharge pressure until the correct shaft horsepower has been reached. Once the correct discharge temperature is known, the polytropic head coefficient and efficiency can be calculated to give a set of nondimensional curves (Figure 6). These curves will predict the compressor performance for the given Mach number. They can be used to compare against the existing operating conditions if the field Mach number is close enough to the Mach number for the curves. The ASME

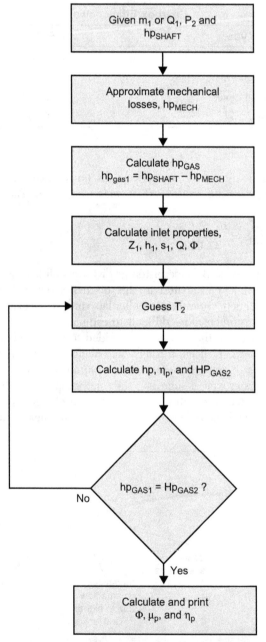

Figure 6. The polytropic head coefficient and efficiency can be calculated to give a set of nondimensional curves.

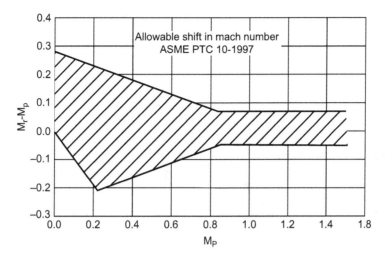

Figure 7. Allowable shift in test Mach number from design Mach number.

PTC-10 test code defines the maximum shift in Mach number ratio for a certified shop performance test (Figure 7). These limitations are good to apply in the field as well. If the Mach number shift is too large, the comparison may be inaccurate. If this is the case, a new set of performance curves, that match the actual inlet conditions, should be obtained from the OEM.

The polytropic head coefficient and efficiency at the existing operating conditions can be plotted on the nondimensional graphs to determine if the compressor is operating on its curve. Also, the nondimensional curves can be used to calculate the field discharge conditions (pressure, temperature, horsepower, etc.) based on the field inlet conditions and the performance curves (Figure 8). Even though seal losses do not affect the calculated horsepower (i.e. they are included in the lower measured efficiency), they do greatly affect the predicted horsepower. The seal losses increase the calculated value of F, which moves the predicted operating point farther to the right on the performance curves, which always increases the horsepower.

Example performance evaluation. Reformer hydrogen recycle compressor configuration:

- Six impellers, straight-through, barrel-type,
- Vaneless diffusers,
- Interlocking aluminum labyrinth balance piston seal,
- Motor-driven through a speed-increasing gearbox,
- Contact-type oil face seals.

Original design:

- 6370 hp,
- 7940 rpm,
- Inlet flow = 14,600 cfm,
- Molecular weight = 5.18,
- $T_1 = 100\ ^\circ F$,
- $P_1 = 187$ psia,
- $P_2 = 275$ psia.

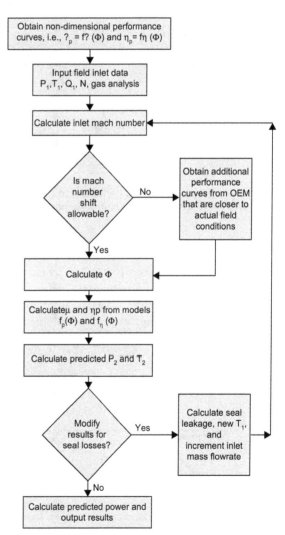

Figure 8. The nondimensional curves can be used to calculate the field discharge conditions.

Figure 9. The OEM supplied this performance curve showing predicted discharge pressure and shaft horsepower.

Balance piston labyrinth dimensional data:

Stationary labyrinth: nine teeth, 0.375-in equal spacing, 0.1875 in tall, 14.595 in internal diameter.
Rotating labyrinth: eight teeth, 0.375 in equal spacing, 0.1875 in tall, 14.937 in outside diameter.
Calculated seal leakage rate = 120 lbm/min.

Measured performance data:

$P_1 = 150$ psia,
$T_1 = 80\ °F$,
$P_2 = 248$ psia,
$T_2 = 191\ °F$,
Balance chamber pressure = 156 psia,
Flow = 190 MMscfd,
Speed = 7949 rpm,
Molecular weight = 7.1.

Driver data:

Amperage = 978,
Voltage = 4000,
Speed = 1784,
Pf = 0.92,
E = 0.957.

1) Calculate the horsepower supplied to the compressor

$$hp_{motor} = \frac{VI_e P_f}{431} = \frac{4000 \times 978 \times 0.957 \times 0.92}{431} = 7988\ hp$$

If the gearbox efficiency is assumed to be 97%, the calculated shaft horsepower delivered to the compressor is:

$$hp_{comp} = 7988 \times 0.97 = 7750\ hp$$

The OEM supplied a performance curve showing predicted discharge pressure and shaft horsepower (Figure 9). These curves are valid for only the design conditions listed previously (i.e. you cannot plot the measured discharge pressure and shaft horsepower on these curves). Likewise, they are only predicted curves (i.e. the compressor was not shop performance tested). These curves must be converted into a nondimensional form so that they can be used to predict the existing field performance (Figure 10).

These curves must be converted into a nondimensional form so that they can be used to predict the existing field performance (Figure 10).

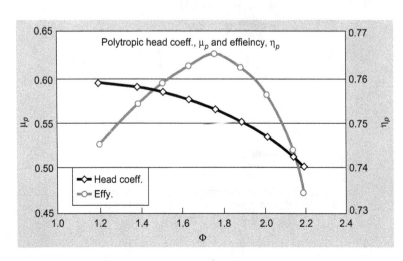

Figure 10. Nondimensional polytropic head coefficient and efficiency.

Table 5
Reformer Hydrogen Recycle Compressor Performance before Cleaning

Parameter	Measured Field Performance	Predicted Performance (w/o seal losses)	Predicted Performance (w/seal losses)
P_2	248	258	252
T_2	191	176	182
Head, lbf-ft/lbm	64,504	68,898	66,417
Efficiency, %	65	79	76
Shp	7647	6729	7030

Table 6
Example Test and Repredicted Field Performance Data

Title Hydrogen Recycle Compressor						
Test Data			**Corrected Test Data**		**Predicted Data**	
Suction Conditions:		**Discharge Conditions:**	Inlet	Outlet	Inlet	Discharge
Press, psia	150	248.0	150.0	248.0	150.0	252.1
Temp, °F	80	191.0	84.9	191.0	84.6	182.1
Flow	190	MMscfd (only one flow is required)				
	0	lcfm				
	0	lbm/min				
Flowmeter design data	1 Location (1—suction, 2—discharge)					
	0 Mole weight					
	0 Temp., °F					
	0 Press., psia					
Speed	7940 rpm				7940	
Impeller	Number 6	Diameter, 23 in.				
Calculated data						
Mole weight	7.2					
Properties	**Inlet**	**Outlet**	**Inlet**	**Outlet**	**Predicted Inlet**	**Predicted Outlet**
Compressibility	1.0026	1.0059	1.0026	1.0059	1.0026	1.0059
Enthalpy, Btu/lbm	556.1	683.4	561.6	683.4	561.3	673.0
Specific volume, ft³/lbm	5.39	3.94	5.44	3.94	5.44	3.83
Specific volume, ft³/lbm	5.39	3.94	5.44	3.94	5.44	3.83
C_p/C_v	1.33	1.32	1.33	1.32	1.33	1.32
Entropy, Btu/lbm, °F	3.76	3.84	3.77	3.84	3.77	3.82
Flow (corrected)			**Flow (corrected)**		**Flow (predicted)**	
MMscfd	190.0					
lcfm	13,456.9	Seal losses, lbm/min	114.0		120.1	
lbm/min	2496.0	Impeller flow, lbm/min	2610.0		2616.1	
Q/N	1.695	Impeller q/n	1.789		1.792	
	Performance		**Performance (corrected)**		**Predicted Performance**	
Polytropic exponent	1.61		1.56		1.48	
Polytropic head	64,509		64,795		66,417	
Polytropic head coeff.	0.545		0.547		0.561	
Polytropic wf.	1.000		1.000		1.000	
Polytropic effy.	0.65		0.68		0.76	
Gas horsepower	7497		7497		6892	
Shaft horsepower	7647		7647		7030	
Mach no.	0.357		0.359		0.355	
Volume ratio	1.367		1.379		1.421	

The discharge conditions are repredicted with and without seal losses to compare against the measured field results (Tables 5 and 6). The horsepower for the existing field performance data must be close to the horsepower supplied by the driver for the results to be considered valid. As you can see, the compressor efficiency is considerably lower than what is predicted by the OEM performance curve. Adding the seal losses (approximately 4—5% of the total flow) to the predicted curves brings the predicted and actual conditions closer together. However, the measured efficiency is just too low to be a balance piston problem alone. Likewise, the thrust bearing temperature and axial position were relatively low. Based on the past fouling history in this compressor as well as the high balance piston differential pressure, the loss in efficiency was thought to be a result of fouling. The compressor was still meeting the desired discharge pressure, but the low

efficiency was causing excessive horsepower consumption, which was limiting unit charge rate. Because the motor was oversized and the loss in efficiency had been gradual, operations was unaware that the problem was in the compressor and not the motor (i.e. they thought the motor was dirty).

The compressor was pulled because it could not be washed in place due to a lack of adequate case drains. A large amount of ammonia chloride buildup was found in the compressor stationary components (the diffuser channels had approximately 40% blockage, Figures 11 and 12). Note: The synchronous vibration amplitudes were relatively low (<1 mil) because the fouling was mostly on the stationary components; the only fouling on the rotor was on the inside diameter of the impeller eyes.

After the compressor was reinstalled, the measured field performance was within 3% of the predicted.

Figure 11. Fouled inlet guide vanes on hydrogen recycle compressor.

Figure 12. Diaphragm from hydrogen recycle compressor.

Nomenclature

C_p specific heat, constant pressure, Btu/lbm, °F
C_v specific heat, constant volume, Btu/lbm, °F
D impeller diameter, in
h enthalpy, Btu/lbm
H head, lbf-ft/lbm
hp horsepower
k specific heat ratio, c_p/c_v
K meter factor
m mass flow, lbm/s
M Mach number, non-dim.
$MMscfd$ million standard cubic feet per day
MW mole weight, lbm/lb-mole
N speed, rpm
NS specific speed, non-dim.
P pressure, psia
Q volume flow, cfm
Q_R radiative heat transfer, Btu
R gas constant
T temperature, °F
v specific volume, ft3/lbm
Z compressibility

Subscripts

1 inlet conditions
2 outlet conditions
bp balance piston
D design conditions
$MECH$ mechanical
M mechanical
P polytropic state
s isentropic state
S corresponds to seal conditions

Symbols

D differential
h efficiency, non-dim.
F sonic velocity ratio, non-dim.
m head coefficient, non-dim.
$m_{0,1}$ labyrinth seal entrance coefficient, non-dim.

Bibliography

1. Aicher, W., "Test of Process Turbocompressors Without CFC Gases," *Proceedings of the Twenty-Second Turbomachinery Symposium, Turbomachinery Laboratory*, Texas A&M University, College Station, Texas, 1993, pp. 179–184.
2. Bartos, J. C., "Predicting Centrifugal Compressor Performance from Impeller Dimensional Data," Power Machinery and Compression Conference, 1990, Houston, Texas.
3. Boyce, M. P., "Principals of Operation and Performance Estimation of Centrifugal Compressors," *Proceedings of the Twenty-Second Turbomachinery Symposium,* Turbomachinery Laboratory, Texas A&M University, College Station, Texas, 1993, pp. 161–175.
4. Clark, W. A., "Measure Power by the Simplest Method and Use for Onstream Computer Monitoring," *Proceedings of the Tenth Turbomachinery Symposium*, Turbomachinery Laboratory, Texas A&M University, College Station, Texas, 1981, pp. 151–153.
5. Davis, H. M., "Compressor Performance, Field Data Acquisition and Evaluation," *Proceedings of the Second Turbomachinery Symposium*, Turbomachinery Laboratory, Texas A&M University, College Station, Texas, 1973, pp. 15–22.
6. Gresh, M. T., Compressor Performance—Selection, Operation, and Testing of Axial and Centrifugal Compressors, Butterworth-Heinemann, Boston, 1991.
7. Heftmann, E., ed., Chromatography—A Laboratory Handbook of Chromatographic and Electrophoretic Methods, Van Nostrand Reinhold Co., New York, 1975.
8. Hill, D. J., "Centrifugal Compressor Rerates and Redesigns," Power Machinery and Compression Conference, Houston, Texas, 1995.
9. Hunt, J. W., "Field Testing of Centrifugal Compressors," Proceedings of the Tenth Turbomachinery Symposium, Turbomachinery Laboratory, Texas A&M University, College Station, Texas, 1981.
10. Japikse, D., "Efficiency: The Silent Partner of Machine Performance," Proceedings of the Eleventh Turbomachinery Symposium, Turbomachinery Laboratory, Texas A&M University, College Station, Texas, 1982, pp. 105–111.
11. Kerlin, T. W. and E. M. Katz, "Temperature Measurement in the 1990s," INTECH, April 1990, pp. 40–43.
12. Lapina, R. P., Estimating Centrifugal Compressor Performance, Gulf Publishing, Houston, Texas, 1982.
13. Lee, B. I. and M. G. Kelser, "A Generalized Thermodynamic Correlation Based on Three-Parameter Corresponding States," AIChE Journal, Vol. 21, No. 3, May 1975, pp. 510–527.
14. Matthews, T., "Field Performance Testing to Improve Compressor Reliability," *Proceedings of the Tenth Turbomachinery Symposium*, Turbomachinery Laboratory, Texas A&M University, College Station, Texas, 1981, pp. 165–167.
15. Nathoo, N. S. and W. G. Gottenberg, "Measuring the Thermodynamic Performance of Multistage Compressors Operating on Mixed Hydrocarbon Gases," *Proceedings of the Tenth Turbomachinery Symposium*, Turbomachinery Laboratory, Texas A&M University, College Station, Texas, 1981, pp. 15–22.
16. Pais, R. B. and G. J. Jacobik, "Field Performance Testing of Centrifugal Compressors," *Proceedings of the Third Turbomachinery Symposium,* Turbomachinery

Laboratory, Texas A&M University, College Station, Texas, 1974, pp. 84–90.

17. Schultz, J. M. "The Polytropic Analysis of Centrifugal Compressors," Journal of Engineering for Power, Vol. 1, pp. 69–83.

18. ASME PTC 10–1997 Performance Test Code on Compressors and Exhausters, American Society of Mechanical Engineers, New York.

19. Van Lanigham, F. L., "Guidelines for Performance Testing Centrifugal Compressors," *Proceedings of the Tenth Turbomachinery Symposium*, College Station, Texas, 1981, pp. 169–171.

20. "Bulletin TS-102," Minco Product Catalog.

Reprinted with permission *HYDROCARBON PROCESSING*, August 2007 issue, Houston, Texas.

11: Gas—Hydraulics

Gas pipeline hydraulics calculations

Equations commonly used for calculating hydraulic data for gas pipelines are as follows

Panhandle A.

$$Q_b = 435.87 \times (T_b/P_b)^{1.0788} \times D^{2.6182} \times E$$

$$\times \left[\frac{P_1^2 - P_2^2 - \frac{\left(0.0375 \times G \times (h_2 - h_1) \times P_{avg}^2\right)}{T_{avg} \times Z_{avg}}}{G^{0.8539} \times L \times T_{avg} \times Z_{avg}}\right]^{0.5394}$$

Panhandle B.

$$Q_b = 737 \times (T_b/P_b)^{1.020} \times D^{2.53} \times E$$

$$\times \left[\frac{P_1^2 - P_2^2 - \frac{\left(0.0375 \times G \times (h_2 - h_1) \times P_{avg}^2\right)}{T_{avg} \times Z_{avg}}}{G^{0.961} \times L \times T_{avg} \times Z_{avg}}\right]^{0.51}$$

Weymouth.

$$Q_b = 433.5 \times (T_b/P_b) \times \left[\frac{P_1^2 - P_2^2}{GLTZ}\right]^{0.5} \times D^{2.667} \times E$$

$$P_{avg} = \frac{2}{3}\left[P_1 + P_2 - \frac{P_1 \times P_2}{P_1 + P_2}\right]$$

Nomenclature for Panhandle equations

Q_b = Flow rate, SCFD
P_b = −Base pressure, psia
T_b = Base temperature, °R
T_{avg} = Average gas termperature, °R
P_1 = Inlet pressure, psia
P_2 = Outlet pressure, psia
G = Gas specific gravity (air = 1.0)
L = Line length, miles
Z = Average gas compressibility
D = Pipe inside diameter, in
h_2 = Elevation at terminus of line, ft
h_1 = Elcvation at origin uf line, ft
P_{avg} = Average line pressure, psia
E = Efficiency factor
E = 1 for new pipe with no bends, fittings, or pipe diameter changes

$E = 0.95$ for very good operating conditions, typically through first 12−28 months
$E = 0.92$ for average operating conditions
$E = 0.85$ for unfavorable operating conditions

Nomenclature for Weymouth equation

Q = Flow rate, SCFD
T_b = Base temperature, °R
P_b = Base pressure, psia
G = Gas specific gravity (air = 1)
L = Line length, miles
T = Gas temperature, °R
Z = Gas compressibility factor
D = Pipe inside diameter, in
E = Efficiency factor. (See Panhandle nomenclature for suggested efficiency factors)

Sample calculations

Q = ?
G = 0.6
T = 100 °F
L = 20 miles
P_1 = 2000 psia
P_2 = 1500 psia
Elev diff = 100 ft
D = 4.026 in
T_b = 60 °F
P_b = 14.7 psia
E = 1.0

$$P_{avg} = 2/3\left[2000 + 1500 - \frac{2000 \times 1500}{2000 + 1500}\right]$$

$$P_{avg} = 1762 \text{ psia}$$

As described earlier, in *Pipecalc 4.02*, Z at 1762 psia and 100 °F = 0.835.

Panhandle A.

$$Q_b = 435.87 \times (520/14.7)^{1.0788} \times 4.026^{2.6182} \times 1$$

$$\times \left[\frac{(2000)^2 - (1500)^2 - \frac{\left(0.0375 \times 0.6 \times 100 \times (1762)^2\right)}{560 \times 0.835}}{(0.6)^{0.8539} \times 20 \times 560 \times 0.835}\right]^{0.5394}$$

$$Q_b = 16,577 \text{ MCFD}$$

Panhandle B.

$$Q_b = 737 \times (520/14.7)^{1.020} \times (4.026)^{2.53} \times 1$$

$$\times \left[\frac{(2000)^2 - (1500)^2 - \frac{(0.0375 \times 0.6 \times 100 \times (1762)^2}{560 \times 0.835}}{(0.6)^{0.961} \times 20 \times 560 \times 0.835} \right]^{0.51}$$

$$Q_b = 17,498 \text{ MCFD}$$

Weymouth

- $Q = 0.433 \times (520/14.7) \times [(2000)^2 - (1500)^2/(0.6 \times 20 \times 560 \times 0.835)]^{1/2} \times (4.026)^{2.667}$

  $Q = 11,101 \text{ MCFD}$

Source

Pipecalc 4.02, Gulf Publishing Company, Houston, Texas. *Note:* Pipecalc 4.02 will calculate the compressibility factor, the minimum pipe inner diameter (ID), the upstream pressure, the downstream pressure, and the flow rate for Panhandle A, Panhandle B, Weymouth, AGA, and Colebrook–White equations. The flow rates calculated in the previous sample calculations will differ slightly from those calculated with Pipecalc 4.02, since the viscosity used in the examples was extracted as described earlier. Pipecalc uses the method from Dranchuk et al. for calculating gas compressibility.

Equivalent lengths for multiple lines based on Panhandle A

Condition I

Condition I is a single pipeline that consists of two or more different diameter lines.

Let

L_E = Equivalent length

$L_1, L_2, \ldots, L_n$ = Length of each diameter

$D_1, D_2, \ldots, D_n$ = Internal diameter of each separate line corresponding to $L_1, L_2, \ldots, L_n$

D_E = Equivalent internal diameter

$$L_E = L_1 \left[\frac{D_E}{D_1}\right]^{4.8539} + L_2 \left[\frac{D_E}{D_1}\right]^{4.8539} + \ldots + L_n \left[\frac{D_E}{D_n}\right]^{4.8539}$$

Example. A single pipeline 100 miles in length consists of 10 miles $10\frac{3}{4}$-in outer diameter (OD); 40 miles $12\frac{3}{4}$-in OD, and 50 miles of 22-in OD lines.

Find equivalent length (L_E) in terms of 22-in OD pipe.

$$L_E = 50 + 40 \left[\frac{21.5}{12.25}\right]^{4.8539} + 10 \left[\frac{21.5}{10.25}\right]^{4.8539}$$

$$= 50 + 614 + 364$$

$$= 1028 \text{ miles equivalent length of 22-in OD}$$

Condition II

Condition II is a multiple pipeline system consisting of two or more parallel lines of different diameters and different length.

Let

L_E = Equivalent length

$L_1, L_2, L_3, \ldots, L_n$ = Length of various looped sections

$d_1, d_2, d_3, \ldots, d_n$ = ID of the individual line corresponding to length $L_1, L_2, L_3,$ and L_n

$$L_E = L_1 \left[\frac{d_E^{2.6182}}{d_1^{2.6182} + d_2^{2.6182} + d_3^{2.6182} + \ldots + d_n^{2.6182}}\right]^{1.8539} + \ldots$$
$$+ L_n \left[\frac{d_E^{2.6182}}{d_1^{2.6182} + d_2^{2.6182} + d_3^{2.6182} + \ldots + d_n^{2.6182}}\right]^{1.8539}$$

Let

L_E = Equivalent length

$L_1, L_2, L_3,$ and L_n = Length of various looped sections

$d_1, d_2, d_3,$ and d_n = Internal diameter of individual line corresponding to lengths $L_1, L_2, L_3,$ and L_n

$$L_E = L_1 \left[\frac{d_E^{2.6182}}{d_1^{2.6182} + d_2^{2.6182} + d_3^{2.6182} + \ldots + d_n^{2.6182}}\right]^{1.8539}$$
$$+ \ldots$$
$$+ L_n \left[\frac{d_E^{2.6182}}{d_1^{2.6182} + d_2^{2.6182} + d_3^{2.6182} + \ldots + d_n^{2.6182}}\right]^{1.8539}$$

when

L_1 = Length of unlooped section

L_2 = Length of single looped section

L_3 = Length of double looped section, and

$d_E = d_1 = d_2$.

Then:

$$L_E = L_1 + 0.27664L_2 + L_3 \left[\frac{d_E^{2.6182}}{2d_1^{2.6182} + d_3^{2.6182}} \right]^{1.8539}$$

when $d_E = d_1 = d_2 = d_3$
then $L_E = L_1 + 0.27{,}664L_2 + 0.1305L_3$

Example. A multiple system consisting of a 15 miles section of $3-8\frac{5}{8}$ -in OD lines and $1-10\frac{3}{4}$ -in OD line, and a 30 mile section of $2-8\frac{5}{8}$ -in lines and $1-10\frac{3}{4}$ -in OD line. Find the equivalent length in terms of single 12-in ID line.

$$L_E = 15 \left[\frac{12^{2.6182}}{3(7.981)^{2.6182} + 10.02^{2.6182}} \right]^{1.8539}$$

$$+ 30 \left[\frac{12^{2.6182}}{2(7.981)^{2.6182} + 10.02^{2.6182}} \right]^{1.8539}$$

$$= 5.9 + 18.1 = 24.0 \text{ miles equivalent of 12-in ID pipe}$$

Example. A multiple system consisting of a single 12-in ID line 5 miles in length and a 30 mile section of 3×12-in ID lines.
Find equivalent length in terms of a single 12-in ID line.

$$L_E = 5 + 0.1305 \times 30$$

$$= 8.92 \text{ miles equivalent of single 12-in ID line}$$

Determine pressure loss for a low-pressure gas system

Use the Spitzglass equation for systems operating at less than 1 psig.

$$Q_h = 3550 \sqrt{\frac{\Delta h_w d^5}{S_g L(1 + (3.6/d) + 0.03d)}}$$

where:

Q_h = Rate of flow, in cubic feet per hour at standard conditions (14.7 psia and 60 °F), scfh
h_w = Static pressure head, in inches of water
S_g = Specific gravity of gas relative to air = the ratio of the molecular weight of the gas to that of air
d = Internal diameter of the pipe, inches, and
L = Length of the pipe, feet.

Example. Given the following conditions, find the flow in the system:

h_w = 50 in of water
S_g = 0.65
L = 500 ft, and
d = 6.187 in.

$$Q_h = 3550 \sqrt{\frac{50 \times 6.187^5}{0.65 \times 500(1 + (3.6/6.187) + 0.03 \times 6.187)}}$$

$$Q_h = 99.723 \text{ scfh}$$

Nomograph for determining pipe-equivalent factors

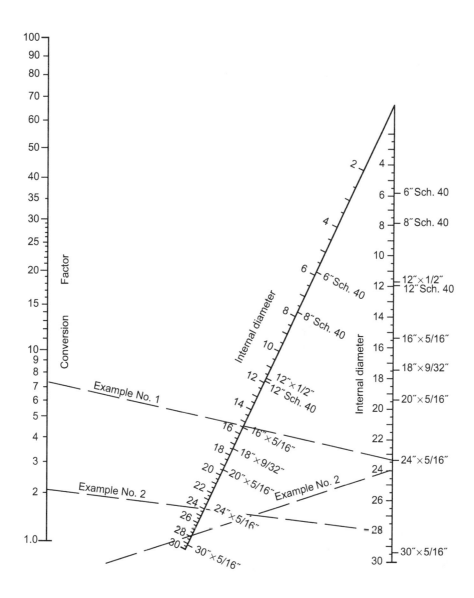

For turbulent flow, this handy nomograph will save a great deal of time in flow calculations where different sizes of pipe are connected. The advantage over a table of values is in having all sizes of pipe expressed in relation to any other size.

Example 1. 16 in × $\frac{5}{16}$ in terms of 24 in $\frac{5}{16}$ in.

$$\left(\frac{23.375}{15.375}\right)^{4.735} = 1.5203^{4.735} = 7.26$$

Example 2. Reciprocal values can be found when chart limits are exceeded. 28-in ID in terms of 24-in ID falls beyond the lower limits of the chart; however, 24-in ID in terms of 28-in ID gives a conversion factor of 2.1.

How much gas is contained in a given line section?

Multiply the square of the inside diameter, in inches, by the gauge pressure, in lb/in^2; multiply this by 0.372; the answer is the approximate number of cubic feet of gas (standard conditions) in 1000 ft of line.

Example. How much gas in 1000 ft of 16-in schedule 30 pipe at 350 lb pressure?

$$15.25 \times 15.25 \times 350 \times 0.372 = 30,280 \text{ cu. ft of gas.}$$

Example. Approximately how much gas in 8 miles of 0.250 wall 24-in pipe if the pressure is 400 psi?

$$\text{Cu. ft} = (23.5)(23.5)(400)(8)(5.28)(0.372)$$
$$= 3,470,000 \text{ cu. ft}$$

When a section of line is blown down from one pressure to another, the total gas lost may be computed by the difference in the contents at the two pressures, using the above rule.

How to estimate equivalent length factors for gas lines

Internal Diameter (ID)	Nominal Diameter	36″ 35.250	34″ 33.250	30″ 29.375	26″ 25.375	24″ 23.375	24″ 23.000	22″ 21.500	20″ 19.500	18″ 17.500	16″ 15.500	12¾″ 12.250	10¾″ 10.250	8⅝″ 8.125
35.250″	36″	1	0.758	0.416	0.204	0.137	0.126	0.092	0.056	0.034	0.019	0.006	0.002	0.0008
33.250	34	1.32	1	0.55	0.270	0.181	0.167	0.121	0.075	0.044	0.025	0.008	0.003	0.001
29.375	30	2.41	1.82	1	0.492	0.330	0.305	0.22	0.136	0.081	0.045	0.014	0.006	0.002
25.375	26	4.90	3.71	2.03	1	0.671	0.622	0.45	0.278	0.164	0.092	0.029	0.012	0.004
23.375	24	7.25	5.54	3.04	1.49	1	0.922	0.67	0.415	0.246	0.137	0.044	0.018	0.006
23.000	24	7.90	6.00	3.29	1.62	1.08	1	0.72	0.450	0.267	0.148	0.048	0.020	0.0064
21.500	22	10.93	8.31	4.55	2.24	1.50	1.37	1	0.624	0.369	0.205	0.065	0.027	0.009
19.500	20	17.60	13.33	7.31	3.59	2.41	2.22	1.60	1	0.590	0.328	0.105	0.044	0.014
17.500	18	29.8	22.5	12.35	6.08	4.075	3.78	2.72	1.69	1	0.555	0.177	0.074	0.024
15.500	16	52.8	40.5	22.2	10.9	7.345	6.75	4.88	3.04	1.80	1	0.319	0.134	0.043
12.250	12¾	168	127	69.8	34.3	23.01	21.2	15.35	9.55	5.65	3.13	1	0.420	0.136
10.250	10¾	400	303	167	81.7	54.67	50.6	36.5	22.7	13.4	7.47	2.38	1	0.324
8.125	8⅝	1235	935	513	252	168.8	156.2	112.5	70.5	41.5	23.0	7.35	3.08	1
6.125	6⅝	4860	3680	2020	992	666	615	443	276	163	90.7	29.0	12.2	3.94

This table shows the equivalent length factors for pipelines of different diameters. It is based on the Panhandle formula, and the factors were obtained by varying the lengths of the pipe and keeping the other variables constant.

Example. Find the equivalent length of a 24-in pipeline as compared with a 12¾-in line.

Enter the chart at the top—24 in—(23.00 ID) and proceed downward to the 12.250 parameter. The equivalent length is 21.2.

This means that under the same conditions of temperatures, pressures, specific gravities, etc., 1 mile of $12\frac{3}{4}$-in pipe will flow the same amount of gas that 21.2 miles of 24-in pipe will flow. Another way of putting it, the pressure drop in 1 mile of 24-in pipe will be the same as that in 0.048 miles of $12\frac{3}{4}$-in pipe if the pressures, volumes, etc., are the same.

Where a line is composed of several sizes of pipe diameters, or where the lines are looped, this chart should prove useful. In solving problems of this sort, the line is reduced to an equivalent length of some arbitrarily selected size pipe and treated as a single line of uniform size.

Estimating comparative capacities of gas pipelines

To estimate capacity of one diameter line in terms of another, read downward in the appropriate column to the desired diameter line.

Example. A 24-in (23.375-in ID) pipeline has a capacity 160% that of a 20-in (19.500-in ID) line.

This chart is based on diameter factors using the Panhandle equation. Equivalent inlet and outlet pressures, lengths of line section, and uniform flow are assumed for both lines.

(Figures give capacity of one diameter pipe as a percent of another.)

ID	Nominal Diameter	36″ 35.250″	34″ 33.250″	30″ 29.3125″	26″ 25.375″	24″ 2.375″	24″ 23.000″	20″ 19.500″	18″ 17.500″	16″ 15.500″	$12\frac{3}{4}$″ 12.250″
35.250″	36″	100	85.7	62.7	43.6	35.2	33.7	22.0	16.5	12.0	6.5
33.250″	34″	116.8	100	73.1	50.8	41.0	39.3	25.7	19.2	14.0	7.6
29.3125″	30″	159.7	137	100	69.6	56.1	53.8	35.7	26.3	19.2	10.4
25.375″	26″	229	197	149	100	80.7	77.5	50.2	37.8	27.6	14.8
23.375″	24″	285	244	178	124	100	96.0	62.3	46.9	34.2	18.5
23.000″	24″	298	255	186	129	104	100	65.1	49.0	35.7	19.2
19.500″	20″	458	392	287	199	160	154	100	75.3	35.0	29.6
17.500″	18″	608	519	379	264	213	204	132	100	73.2	39.3
15.500″	16″	835	715	522	363	293	281	182	127	100	53.8
12.250″	$12\frac{3}{4}$″	1545	1322	968	673	543	521	238	254	185	100

Determination of leakage from gas line using pressure drop method

To determine the leakage in Mmcf per year at a base pressure of 14.4 psia and a temperature of 60 °F, take as a basis a 1 mile section of line under 1 h test. Use the formula:

$$L_y = 9 \times D^2 \left[\frac{P_1}{460 + t_1} - \frac{P_2}{460 + t_2} \right]$$

where:

D = Inside diameter of the pipe, inches
P_1 = Pressure at the beginning of the test in psia
t_1 = Temperature at the beginning of the test in °F
P_2 = Pressure at end of the test, and
t_2 = Temperature at end of the test.

The error in using this formula can be about 4%.

Example. A section of a 10-in pipeline 1 mile in length is tested using the pressure drop method. ID is equal to 10.136 in. Pressure and temperature at the start of the test were 190.5 psig and 52 °F. After 1 h, the readings were 190.0 psig and 52 °F. Calculate the leakage in Mmcf/year.

The atmospheric pressure for the area is 14.4 psia. Thus:

$$P_1 = 190.5 + 14.4 = 204.9 \, \text{psia}$$

$$P_2 = 190.0 + 14.4 = 204.4 \, \text{psia}$$

$$L_y = 9(10.136)^2 \left[\frac{204.9}{406 + 52} - \frac{204.4}{460 + 52} \right]$$

$$= \frac{(9)(102.7)(0.5)}{512}$$

$$= 0.903 \, \text{Mmcf per mile per year.}$$

A quick way to determine the size of gas gathering lines

Here is a shortcut to estimate gas flow in gathering lines. For small gathering lines, the answer will be within 10% of that obtained by more difficult and more accurate formulas.

$$Q = \frac{(500)(d^3)\sqrt{P_1^2 - P_2^2}}{\sqrt{L}}$$

where:

> Q = Cubic feet of gas per 24 h
> d = Pipe ID in inches
> P_1 = Psi (abs) at starting point
> P_2 = Psi (abs) at ending point, and
> L = Length of line in miles.

Example. How much gas would flow through four miles of 6-in ID pipe if the pressure at the starting point is 485 psi and if the pressure at the downstream terminus is 285 psi?

Answer.

$$Q = \frac{(500)(6^3)\sqrt{(485 + 15)^2 - (285 + 15)^2}}{\sqrt{4}}$$

$$Q = \frac{(500)(216)\sqrt{500^2 - 300^2}}{2}$$

$$Q = \frac{(500)(216)(400)}{2}$$

$$Q = 21,600,000 \text{ cu. ft per 24 h}$$

Using a more accurate formula (a version of Weymouth's):

$$Q = \frac{(871)(d^{8/3})\sqrt{P_1^2 - P_2^2}}{\sqrt{L}}$$

$$Q = \frac{(871)(118.94)(400)}{2}$$

$$Q = 20,719,417 \text{ cu. ft per 24 h}$$

The error in this case is 880,582.8 cu. ft per 24 h, or about 4.25%.

Energy conversion data for estimating

The accompanying table gives the high heating values of practically all commonly used fuels along with their relation one to the other.

Example. If light fuel oil will cost a potential industrial customer 18 cents per gallon and natural gas will cost him 75 cents per Mcf, which fuel will be more economical?

Solution. Assume the operator would use 10,000 gallons of light fuel oil per year at 18 cents per gallon. The cost would be $1800. From the table, 10,000 gallons of fuel oil are equivalent to $10,000 \times 134 = 1,340,000$ cu. ft of natural gas. At 75 cents per Mcf this would cost $1340 \times 0.75 = \$1005$.

The burning efficiency of both fuels is the same. The savings by using natural gas would come to $795 per year.

Fuel	Heating Value	PERCENT BURNING EFFICIENCY			Unit	NUMBER OF CUBIC FEET OF NATURAL GAS EQUIVALENT TO 1 UNIT OF OTHER FUEL		
		Domestic	Commercial	Industrial		Domestic	Commercial	Industrial
Natural Gas	1000 B.T.U. per Cubic Foot	75	75	78	Cubic Foot	1	1	1
Propane	91,600 B.T.U. per U.S. gallon	75	75	78	Gallon	91.5	91.5	91.5
Butane	102,000 B.T.U. per U.S. gallon	.....	75	78	Gallon	.....	102	102
Light Fuel Oil	134,000 B.T.U. per U.S. gallon	65–75	75	78	Gallon	117–134	134	134
Heavy Fuel Oil	146,000 B.T.U. per U.S. gallon	.....	75	78	Gallon	.....	146	146
Hydro-electricity	3,412 B.T.U. per Kilowatt Hour	100	100	100	Kilowatt Hour	4.549	4.549	4.549
*Coal	8,700–11,000 B.T.U. per pound	55–60	65	73	Pound	6.37–8.07	7.55–9.57	8.45–10.7

*Coal qualities vary considerably. The figures given are average for Western Coals; slack, run-of-the-mine; potentially available at lowest cost.

How to estimate time required to get a shut-in test on gas transmission lines and approximate a maximum acceptable pressure loss for new lines

These two rules may prove helpful for air or gas testing of gas transmission lines. They are not applicable for hydrostatic tests. Values of the constants used by different transmission companies may vary with economic considerations, line conditions, and throughputs.

To determine the minimum time required to achieve a good shut-in test after the line has been charged and becomes stabilized, use the following formula:

$$H_m = \frac{3 \times D^2 \times L}{P_1}$$

where:

H_m = Minimum time necessary to achieve an accurate test in hours
D = Internal diameter of pipe in inches
L = Length of section under test in miles, and
P_1 = Initial test pressure in lb/in^2 gauge.

Example. How long should a 14 mile section of 26-in (25.375-in ID) pipeline be shut in under a 1050 psi test pressure to get a good test?

$$H_m = \frac{3 \times (25.375)^2 \times 14}{1,050}$$
$$= 25.75 \text{ h or } 25 \text{ h and } 45 \text{ min}$$

When a new line is being shut in for a test of this duration or longer, the following formula may be used to evaluate whether or not the line is "tight":

$$P_{ad} = \frac{H_t \times P_1}{D \times 949}$$

where:

P_{ad} = Acceptable pressure drop in lb/in^2 gauge
H_t = Shut-in test time in hours
D = Internal diameter of pipe in in, and
P_1 = Initial test pressure in lb/in^2 gauge.

Example. What would be the maximum acceptable pressure drop for a new gas transmission line under air or gas test when a 44 mile section of 30-in (29.250-in ID) pipeline has been shut in for 100 h at an initial pressure of 1100 psig?

$$P_{ad} = \frac{100 \times 1100}{29.250 \times 949} = 4 \text{ lb/in}^2$$

If the observed pressure drop after stabilization is less than 4 psig, the section of line would be considered "tight."

Note: Corrections must he made for the effect of temperature variations upon pressure during the test.

How to determine the relationship of capacity increase to investment increase

Example. Determine the relationship of capacity increase to investment increase when increasing pipe diameter, while keeping the factors P (in), P (out), and L constant.

Solution. The top number in each square represents the percent increase in capacity when increasing the pipe size from that size shown in the column at the left to the size shown by the column at the bottom of the graph. The lower numbers represent the percent increase in investment of the bottom size over the original pipe size. All numbers are based on pipe capable of 1000 psi working pressure.

Pipe size	4	6	8	10	12	14	16	18	20	22	24	26	28	30	32	34	36
34																	17/12
32																17/13	37/27
30															19/13	39/28	62/43
28														20/11	43/26	67/41	
26													23/11	47/22	75/39		
24												24/11	52/23	82/37			
22											26/14	56/26	91/40				
20										29/17	62/33	101/47					
18									34/14	72/33	117/51						
16								38/23	85/39	138/63							
14							44/14	98/39	165/58								
12						29/16	85/32	156/62									
10					61/22	108/42	198/61										
8				82/26	193/53	278/78											
6			105/22	274/54	502/87												
4	0/0	190/38	497/69	988/112													

Top number represents % increase in capacity when increasing pipe size form size on left to size at bottom

Lower number represents % increase in investment for pipe size at bottom over pipe size at left.

Estimate pipe size requirements for increasing throughput volumes of natural gas

Reflects the capacity-to-pipe size relationship when L (length), P_1 (inlet pressure), and P_2 (outlet pressure) are constant.

EXAMPLE

An 8" line will handle "X" quantity Find pipe size required for 4.5"X". Begin at 4.5 on left scale, move horizontally to 8" curve, then down to nominal pipe size. Use next largest pipe size.

CAPACITY MULTIPLIER

NOMINAL PIPE SIZE

Calculate line loss using cross-sectional areas table when testing mains with air or gas

Initial volume $= \dfrac{P_i}{P_B}(AL)$

Final volume $= \dfrac{P_f}{P_B}(AL)$

where:

P_i = Initial pressure, or pressure at start of test, psia
P_f = Final pressure, or pressure at end of test, psia
P_B = Base pressure, psia
A = Cross-sectional area of inside of pipe in ft^2, and
L = Length of line in ft.

The loss then would be the difference of the volumes, found by the above formula, which could be written:

Loss during test $= AL\dfrac{\left(P_i - P_f\right)}{P_B}$

To facilitate calculations, Table 1 contains the cross-sectional areas in ft^2 for some of the more popular wall thickness pipe from 1–24 in nominal size.

Example. Assume 3562 ft of 8.125 in inside diameter pipe is tested with air for a period of 26 h. The initial pressure was 116 psig, the final pressure 112 psig, and the barometer was 14.4 psia in both instances. It is desired to calculate the volume of loss during the test using a pressure base of 14.65 psia.

If the area "A" is obtained from Table 1, and information as given above is substituted in the formula, we would have:

Loss during test $= 0.360 \times 3562$

$$\times \frac{[(116 + 14.4) - (112 + 14.4)]}{14.65}$$

$$= 350.1 \text{ cu. ft}$$

Table 1
Characteristics of Pipe

Nominal Pipe Size D_n	Outside Diameter	Wall Thickness	Inside Diameter	"A" Inside Area in Sq. Feet
1	1.315	.133 / .179	1.049 / .957	.006 / .005
1¼	1.660	.140 / .191	1.380 / 1.278	.010 / .009
1½	1.900	.145 / .200	1.610 / 1.500	.014 / .012
2	2.375	.154 / .218	2.067 / 1.939	.023 / .021
2½	2.875	.203 / .276	2.469 / 2.323	.033 / .029
3	3.500	.125 / .156 / .188 / .210 / .250	3.250 / 3.188 / 3.124 / 3.008 / 3.000	.058 / .055 / .053 / .051 / .049
3½	4.000	.125 / .156 / .188 / .226 / .250	3.750 / 3.688 / 3.624 / 3.548 / 3.500	.077 / .074 / .072 / .069 / .067
4	4.500	.125 / .156 / .188 / .219 / .237 / .250	4.250 / 4.188 / 4.124 / 4.062 / 4.026 / 4.000	.099 / .096 / .093 / .090 / .088 / .087
5	5.563	.156 / .188 / .219 / .258	5.251 / 5.187 / 5.125 / 5.047	.150 / .147 / .143 / .139
6	6.625	.188 / .219 / .250 / .280	6.249 / 6.187 / 6.125 / 6.065	.213 / .209 / .205 / .201
8	8.625	.188 / .219 / .250 / .277	8.249 / 8.187 / 8.125 / 8.071	.371 / .366 / .360 / .355

Nominal Pipe Size D_n	Outside Diameter	Wall Thickness	Inside Diameter	"A" Inside Area in Sq. Feet
10	10.750	.188 / .219 / .250 / .279 / .307	10.374 / 10.312 / 10.250 / 10.192 / 10.136	.587 / .580 / .573 / .567 / .560
12	12.750	.219 / .250 / .281 / .312 / .330	12.312 / 12.250 / 12.188 / 12.126 / 12.090	.827 / .818 / .810 / .802 / .797
14	14.000	.250 / .281 / .312 / .344 / .375	13.500 / 13.438 / 13.376 / 13.312 / 13.250	.994 / .985 / .976 / .966 / .958
16	16.000	.250 / .281 / .312 / .344 / .375	15.500 / 15.438 / 15.376 / 15.312 / 15.250	1.310 / 1.300 / 1.289 / 1.279 / 1.268
18	18.000	.250 / .281 / .312 / .344 / .375	17.500 / 17.438 / 17.376 / 17.312 / 17.250	1.670 / 1.658 / 1.647 / 1.635 / 1.623
20	20.000	.250 / .281 / .312 / .344 / .375	19.500 / 19.438 / 19.376 / 19.312 / 19.250	2.074 / 2.061 / 2.048 / 2.034 / 2.021
22	22.000	.312 / .344 / .375	21.376 / 21.312 / 21.250	2.492 / 2.477 / 2.463
24	24.000	.312 / .344 / .375	23.376 / 23.312 / 23.250	2.980 / 2.964 / 2.948

Flow of fuel gases in pipelines

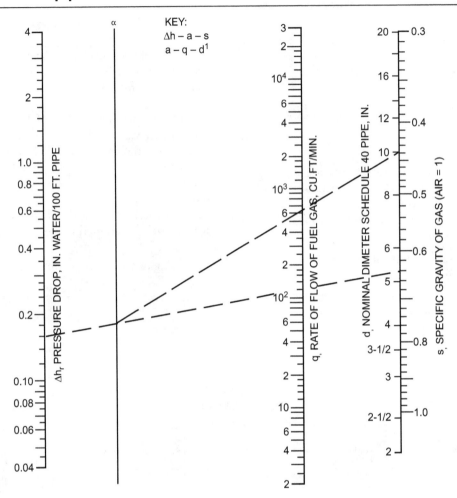

D. S. Davis, Special Projects, Inc., Bailey Island, Maine

The rate of flow of fuel gases at low pressures, ordinary temperatures, and under turbulent conditions in pipelines is reflected by an equation[1] based on sound theoretical considerations:

$$q = \frac{3.185 \Delta h^{0.543} d^{2.63}}{s^{0.468}}$$

where:

q = Rate of flow of gas, cu. ft/min
Δh = Pressure drop, inches of water per 100 ft of pipe
d = Actual ID of Schedule 40 steel pipe, inches, and
s = Specific gravity of gas at the prevailing temperature and pressure relative to air at 68 °F and 30 in of mercury.

The equation can be solved readily and accurately by means of the accompanying monograph constructed in accordance with methods described previously.[2]

The use of the chart is illustrated as follows. At what rate will fuel gas with a specific gravity of 0.640 flow in a 10-in Schedule 40 pipeline if the pressure is 0.160 in of water per 100 ft of pipe? Following the key, connect 0.160 on the Δh scale and 0.640 on the s scale with a straight line and mark the intersection with the α axis. Connect this point with 10 on the d' scale for nominal diameters and read the rate of flow at 624 cu. ft/min on the q scale.

Some authorities give the exponent of s as 0.408 when, for the data given, the rate of flow as read from the chart would be decreased by 2.7% (17 cu. ft/min) to attain 607 cu. ft/min.

References

1. Perry, J. H., ed., *Chemical Engineers'* Handbook, 3rd ed., p. 1589, McGraw-Hill Book Co., Inc., New York (1950).
2. Davis, D. S., *Nomography and Empirical Equations*, 2nd ed., Chapter 6, Reinhold Publishing Corp., New York, 1962.

Calculate the velocity of gas in a pipeline

$$V = \frac{144ZQT}{d^2P}$$

where:

V = Velocity of gas, ft/s
Z = Compressibility
Q = Volume, million cu. ft/h at standard conditions
T = Operating temperature, °R
d = ID of pipe, in
P = Operating pressure, psia

Example.

$Q = 5$ Mmcfh
$T = 60\,°F$
$d = 15.5$ in

$P = 500$ psig (514.7 psia)
$Z = 0.94$ (from compressibility chart)
$T = 520\,°R$

$$V = \frac{1440 \times 0.94 \times 5 \times 520}{240.25 \times 514.7}$$

$$V = 28\,\text{ft/s}$$

Consideration should be given to acceptable noise levels (60—80 ft/s) and erosion, especially where the gas may contain solid particles. Refer to API RP 14E for erosion and corrosion considerations. If the gas contains liquids, the minimum velocity should be limited to about 10 ft/s to help sweep liquids from the pipe and prevent slugging.

Determining throat pressure in a blow-down system

In designing a gas blow-down system, it is necessary to find the throat pressure in a blow-off stack of a given size with a given flow at critical velocity. The accompanying nomograph is handy to use for preliminary calculations. It was derived from the equation:

$$P_t = \frac{W}{d^2} \times \frac{\sqrt{\frac{RT_1}{K(K-1)}}}{11,400}$$

where:

P_t = Pressure in the throat, psia
W = Weight of flowing gas, lb/h
d = Actual ID of stack, in
R = Gas constant, ft lb/°F
K = C_P/C_V
C_P = Specific heat at constant pressure, Btu/lb/°F
C_V = Specific heat at constant volume, Btu/lb/°F
T_1 = Absolute temperature of gas flowing,
°R = (°F + 460)

Using the following values for natural gas, $R = 88.888$, and $K = 1.31$, the above equation reduces to:

$$P_t = \frac{W}{d^2} \times 0.0012978 \sqrt{T_1}$$

Example. To use the accompanying nomograph, assume that the rate of flow is known to be 2.00×10^6 lb/h, the ID of the stack is 12 in, and the flowing temperature is 480 °R. Start from the abscissa at the rate of flow, W, and proceed upward to the ID of the pipe. From this intersection proceed to the right to the flowing temperature of 480 °R. Turn upward from this intersection to the P_t scale and read the answer 395 psia.

Estimate the amount of gas blown off through a line puncture

To calculate the volume of gas lost from a puncture or blowoff, use the equation:

$$Q = D^2 P_1$$

where:

Q = Volume of gas in Mcf/h at a pressure of 14.7 psi, 60 °F, and with a specific gravity of 0.60
D = Diameter of the nipple or orifice in inches
P = Absolute pressure in lb/in^2 at some nearby point upstream from the opening

Example. How much gas will be lost during a 5 min blowoff through a 2-in nipple if the upstream pressure is 1000 psi absolute?

$$Q = D^2 P_1$$
$$Q = 2^2 \times 1000$$
$$Q = 4000 \text{ Mcf/h, and}$$
$$Q_{5 \text{ min}} = 4000 \times 5/60 = 333 \text{ Mcf}$$

A practical way to calculate gas flow for pipelines

Here is a shortcut to calculate gas flow in pipelines. It is based on the Weymouth formula. At 60 °F and specific gravity of 0.60, the answer will be accurate. For every 10° variation in temperature, the answer will be 1% error. For every 0.01 variation in specific gravity, the answer will be three-fourths percent in error.

Formula:

$$Q = \frac{(871)(d^{8/3})\sqrt{P_1^2 - P_2^2}}{\sqrt{L}}$$

where:

Q = Cu ft of gas per 24 h; 871 is a constant
d = Pipe ID in
P_1 = Psi (abs) at starting point
P_2 = Psi (abs) at ending point, and
L = Length of line in miles.

Example. How much gas would flow through 1 mile of 6-in ID pipe if the pressure at the starting point is 485 psi and if the pressure at the downstream terminus is 285 psi?

Solution.

$$Q = \frac{(871)(6^{8/3})\sqrt{(485+15)^2 - (285+15)^2}}{\sqrt{L}}$$

$$= \frac{(871)(120)\sqrt{500^2 - 300^2}}{\sqrt{L}}$$

$$= (871)(120)(400)$$

$$= 41,800,000 \text{ cu. ft per 24 h}$$

Example. What would be the flow through 4 miles of $10\frac{3}{4}$ in quarter-inch wall pipe using the same pressures?

$$Q = \frac{(871)(10.25^{8/3})\sqrt{500^2 - 300^2}}{\sqrt{4}}$$

$$Q = \frac{(871)(495.7)(400)}{2}$$

$$Q = 86,350,000 \text{ cu. ft/day}$$

How to calculate the weight of gas in a pipeline

Rule. Find the volume of the pipe in cubic feet and multiply by the weight of the gas per cubic feet. To find the latter, multiply the absolute pressure of the gas times 3 and divide by 1000.

The basis for the latter is that gas with a specific gravity of 0.60 at 70 °F weighs 3.06 lb/cu. ft at 1000 psia. And everything else remaining equal, weight of the gas is proportional to absolute pressure. Thus, to find the weight of gas at, say, 500 psia, multiply 3 × 500/1000 = 1.5 lb (1.53 is more accurate.)

Example. Find the weight of gas in a 1250 ft aerial river crossing where the average pressure reads 625.3 on the gauges; the temperature of the gas is 70 °F, and specific gravity of gas is 0.60.

Solution. Change psig to psia:

625.3 + 14.7 = 640 psia.

Weight of gas per cubic feet:

$3.06 \times \dfrac{640}{1000} = 1.959$ lb/cu. ft.

Volume of 1250 ft of 23¼-in ID pipe = 3685 cu. ft.

Weight of gas = 7217 lb.

This method is fairly accurate; here is the same problem calculated with the formula:

$$W = \frac{(V)(144)(P_{abs})}{RT}$$

where:

W = Weight of gas in lbs
V = Volume of pipe
P_{abs} = Absolute pressure of gas
R = Universal constant 1544/molecular weight of gas, and
T = Temperature of gas in °R

(To find the molecular weight of gas, multiply specific gravity × molecular weight of air, or in this case, $0.6 \times 28.95 = 17.37$.)

$$W = \frac{(3685)(144)(640)}{(1,500 \div 17.37)(70 + 460)}$$

$W = 7209$ lb

Estimate average pressure in gas pipeline using upstream and downstream pressures

To find the average gas pressure in a line, first divide downstream pressure by upstream pressure and look up the value of the factor X in the table shown. Then multiply X times the upstream pressure to get the average pressure.

Example. Find the average pressure in a pipeline if the upstream pressure in 775 psia and the downstream pressure is 420 psia.

$$\frac{P_d}{P_u} = \frac{420}{775} = 0.54$$

Interpolating from the table,

$X = 0.78 + 4/5(0.02) = 0.796$.

$P_{average} = 0.796 \times 775 = 617$ psia

This method is accurate to within 2 or 3 psi. Here is the same problem calculated with the formula:

$$P_{average} = \frac{2}{3}\left(P_u + P_d - \frac{P_u P_d}{P_u + P_d}\right) = \frac{2}{3}\left(1195 - \frac{325,500}{1195}\right)$$

$P_{average} = 615$ psia

P_d/P_u	X
.10	.67
.15	.68
.20	.69
.25	.70
.30	.71
.35	.725
.40	.74
.45	.76
.50	.78
.55	.80
.60	.82
.65	.84
.70	.86
.75	.88
.80	.90
.85	.925
.90	.95
.95	.975
1.00	1.00

Chart for determining viscosity of natural gas

From Carlton Beal, "The Viscosity of Air, Water, Natural Gas, Crude Oil and Its Associated Gases at Oil Field Temperatures and Pressures," *Petroleum Development and Technology*, AIME, 1946.

Flow of gas

For gas flow problems as encountered in oil field production operations, the Fanning equation for pressure drop may be used. A modified form of this equation employing units commonly used in oil field practice is[1-6]:

$$W = \frac{FL(\text{MCFD})^2 STZ}{20,000 \, d^5 P_{av}}$$

and for determining the friction factor F:

$$R_W = \frac{20.14(\text{MCFD})S}{d\mu}$$

where:

P = Pressure drop, psia
F = Friction factor, dimensionless
L = Length of pipe, ft
MCFD = Gas flow at standard condition
S = Specific gravity of gas
T = Absolute temperature (°F + 460)
P_{avg} = Average flowing pressure, psia
d = Internal pipe diameter, inches
Re = Reynolds number, dimensionless, and
μ = Viscosity, centipoises.

In actual practice, empirical flow formulas are used by many to solve the gas flow problems of field production operations. The Weymouth formula is the one most frequently used since results obtained by its use agree quite closely with actual values. Recent modification of the formula, by including the compressibility factor, Z, made the formula applicable for calculation of high-pressure flow problems.

The modified formula is as follows:

$$Q = 433.45 \frac{T_s}{P_s} \times d^{2.667} \times \left(\frac{P_1^2 - P_2^2}{LSTZ}\right)^{1/2}$$

where:

Q_S = Rate of flow of gas in cubic feet per 24 h measured at standard conditions

d = Internal diameter of pipe in inches.

P_1 = Initial pressure, psi absolute

P_2 = Terminal pressure, psi absolute

L = Length of line in miles

S = Specific gravity of flowing gas (air = 1.0)

T = Absolute temperature of flowing gas

Table 1
Values of $d^{2.667}$ for Use in Weymouth's Gas-Flow Formula

(Increments of tenths of an inch with certain standard pipe sizes inserted)

d	$d^{2.667}$	d	$d^{2.667}$	d	$d^{2.667}$	d	$d^{2.667}$
1.0	1.0	6.2	129.9	11.5	673.8	16.9	1881
1.1	1.3	6.3	135.4	11.6	689.5	17.0	1911
1.2	1.6	6.4	141.2	11.7	705.5	17.1	1941
1.3	2.0	6.5	147.1	11.8	721.7	17.2	1971
1.4	2.4	6.6	153.3	11.9	738.1	17.250	1988
1.5	2.9	6.7	159.5	12.000	754.8	17.3	2002
1.6	3.5	6.8	166.0	12.1	771.7	17.4	2033
1.7	4.1	6.9	172.6	12.2	788.8	17.5	2064
1.8	4.8	7.0	179.3	12.3	806.1	17.6	2096
1.9	5.5	7.1	186.2	12.4	823.7	17.7	2128
2.0	6.3	7.2	193.3	12.5	841.6	17.8	2160
2.067	6.9	7.3	200.5	12.6	859.7	17.9	2193
2.1	7.2	7.4	207.9	12.7	878.0	18.0	2225
2.2	8.2	7.5	215.5	12.8	896.5	18.1	2259
2.3	9.2	7.6	223.3	12.9	915.3	18.2	2292
2.4	10.3	7.7	231.2	13.0	934.4	18.3	2326
2.5	11.5	7.8	239.3	13.1	953.7	18.4	2360
2.6	12.8	7.9	247.6	13.2	973.2	18.5	2394
2.7	14.1	7.981	254.5	13.250	983.9	18.6	2429
2.8	15.6	8.0	256.0	13.3	993.0	18.7	2464
2.9	17.1	8.1	264.6	13.4	1013	18.8	2499
3.0	18.7	8.2	273.4	13.5	1033	18.9	2535
3 068	19.9	8.3	282.4	13.6	1054	19.0	2570
3.1	20.4	8.4	291.6	13.7	1075	19.1	2607
3.2	22.2	8.5	300.9	13.8	1096	19.182	2639
3.3	24.1	8.6	310.5	13.9	1117	19.2	2643
3.4	26.1	8.7	320.2	14.0	1138	19.250	2664
3.5	28.2	8.8	330.1	14.1	1160	19.3	2680
3.6	30.4	8.9	340.2	14.2	1182	19.4	2717
3.7	32.7	8.941	344.6	14.3	1205	19.5	2755
3.8	35.2	9.0	350.5	14.4	1227	19.6	2793
3.9	37.7	9.1	361.0	14.5	1250	19.7	2831
4.0	40.3	9.2	371.6	14.6	1273	19.8	2869
4.026	41.0	9.3	382.5	14.7	1297	19.9	2908
4.1	43.1	9.4	393.6	14.8	1321	20.0	2947
4.2	45.9	9.5	404.8	14.9	1344	20.1	2987
4.3	48.9	9.6	416.3	15.0	1368	20.2	3026
4.4	52.0	9.7	428.0	15.1	1393	20.3	3067
4.5	55.2	9.8	439.8	15.2	1418	20.4	3107
4.6	58.5	9.9	451.9	15.250	1431	20.5	3148
4.7	62.0	10.0	464.2	15.3	1443	20.6	3196
4.8	65.6	10.020	467.0	15.4	1468	20.7	3230
4.9	69.3	10.1	476.6	15.5	1494	20.8	3272
5.0	73.1	10.2	489.3	15.6	1519	20.9	3314
5.1	77.1	10.3	502.2	15.7	1545	21.0	3357
5.2	81.2	10.4	515.3	15.8	1572	21.1	3400
5.3	85.4	10.5	528.7	15.9	1599	21.1	3443
5.4	89.8	10.6	542.2	16.0	1625	21.250	3468
5.5	94.3	10.7	555.9	16.1	1653	21.3	3486
5.6	98.9	10.8	569.9	16.2	1680	21.4	3530
5.7	103.7	10.9	584.1	16.3	1708	21.5	3574
5.8	108.6	11.000	598.5	16.4	1736	21.6	3619
5.9	113.7	11.1	613.1	16.5	1765	21.7	3663
6.0	118.9	11.2	627.9	16.6	1793	21.8	3709
6.065	122.4	11.3	643.0	16.7	1822	21.9	3754
6.7	124.2	11.4	658.3	16.8	1851	22 0	3800

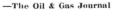

—The Oil & Gas Journal

Standard conditions for measurements:

T_s = Standard absolute temperature
P = Standard pressure, psi absolute, and
Z = Compressibility factor.

Values of expression $d^{2.667}$ for different inside diameter of pipe are given in Table 1.

Another formula quite commonly used is the one determined by F. H. Oliphant:

$$Q_s = 42a\sqrt{\frac{P_1^2 - P_2^2}{L}}$$

where:

Q = Discharge in cu. ft/h
42 = A constant
P_1 = Initial pressure in lb/in^2, absolute
P_2 = Final pressure in lb/in^2, absolute
L = Length of line in miles, and
a = Diameter coefficient.

This formula assumes a specific gravity of gas of 0.6. For any other specific gravity, multiply the final result by

$$\sqrt{\frac{0.6}{\text{sp. gr. gas}}}$$

The values of diameter coefficients for different sizes of pipes are given as:

¼ in	0.0317
½ in	0.1810
¾ in	0.5012
1 in	1.0000
1½ in	2.9300
2 in	5.9200
2½ in	10.37
3 in	16.50
4 in	34.10
5 in	60.00
5⅝ in	81.00
6 in	95.00
8 in	198

10 in	350
12 in	556
16 in	1160
18 in	1570

For pipes greater than 12 in in diameter, the measure is taken from the outside and for pipes of ordinary thickness, the corresponding inside diameters and multipliers are as follows:

Outside	Inside	Multiplier
15 in	14¼ in	863
16 in	15¼ in	1025
18 in	17¼ in	1410
20 in	19¼ in	1860

To determine gas flow rates for specific pressure drops for pipe sizes used in production operations, Tables 3–5 may be used. The tables are calculated by the Weymouth formula for listed inside diameters of standard weight threaded line pipe for the sizes shown.

Tables 3–5 give the hourly rates of flow of 0.70 specific gravity gas, flowing at 60 °F and measured at a standard temperature of 60 °F and at a standard pressure base of 14.4 psi + 4 oz. gauge pressure of 14.65 psi. For specific gravity of gas other than 0.70 and for flowing temperature other than 60 °F, correction can be made by use of the factors shown in Table 2. For instance, to determine the flow of gas of specific gravity 0.06 flowing at the temperature of 20 °F, the value obtained from the gas flow Tables 3–5 would be multiplied by 1.12, the factor obtained from Table 2.

To find from Tables 3–5 the volume of gas delivered through any length L, the volume found from the table should be multiplied by $1/\sqrt{L}$. L must be expressed as a multiple of the length heading the table in use. For instance, if the delivery is to be determined from a 4500 ft line, the volume found for a 1000 ft line in Table 3 should be multiplied by $1\sqrt{4.5}$. At the bottom of each of the two tables in question, multipliers are given to be used as correction factors.

Use of Tables 3–5 may be illustrated by the following example.

Example. What is the delivery of a 2-in gas line, 3600 ft long, with an upstream pressure of 200 psi and a downstream pressure of 50 psi? The specific gravity of gas is 0.07; the flowing temperature, 70 °F.

Table 2
Temperature—Specific Gravity Multipliers

Specific Gravity	FLOWING TEMPERATURE—DEGREES F.												
	0	10	20	30	40	50	60	70	80	90	100	110	129
0.60	1.15	1.14	1.12	1.11	1.10	1.09	1.08	1.07	1.06	1.05	1.04	1.03	1.02
0.70	1.06	1.05	1.04	1.03	1.02	1.01	1.00	.991	.981	.972	.964	.955	.948
0.80	.995	.984	.974	.964	.954	.945	.935	.927	.918	.910	.901	.894	.886
0.90	.938	.928	.918	.909	.899	.891	.882	.874	.965	.858	.850	.842	.835
1.00	.890	.880	.871	.862	.853	.845	.837	.829	.821	.814	.806	.799	.792

Table 3
Volume of Gas Delivered—Thousands of Cubic Feet Per Hour Through 1000- and 2000-ft Lines

0.7 SP. GR. 60° F.

Left section — Gage Pressure 400 to 150 P.S.I.

Up Stream P.S.I.	Down Stream P.S.I.	1000 FT 2	2½	3	4	6	2000 FT 2	2½	3	4	6
400	10	221.5	355.8	635.0	1311.	3910.	156.6	251.6	449.0	926.8	2765.
	50	219.2	352.1	628.4	1297.	3869.	155.0	248.9	444.3	917.2	2736.
	100	213.2	342.5	611.4	1262.	3764.	150.8	242.2	432.3	892.4	2662.
	350	105.6	169.7	302.9	625.2	1865.	74.70	120.0	214.2	442.1	1319.
350	10	194.7	312.7	558.1	1152.	3436.	137.6	221.1	394.6	814.6	2430.
	50	192.0	308.5	550.6	1136.	3390.	135.8	218.1	389.3	803.6	2397.
	100	185.2	297.5	531.0	1096.	3270.	131.0	210.4	375.5	775.1	2312.
	300	98.63	158.4	282.8	583.7	1741.	69.74	112.0	200.0	412.8	1231.
300	10	167.8	269.6	481.2	993.1	2962.	118.7	190.6	340.2	702.3	2095.
	50	164.8	264.6	472.4	975.0	2908.	116.5	187.1	334.0	689.4	2056.
	100	156.8	251.9	449.5	927.9	2768.	110.9	178.1	317.9	656.1	1957.
	250	91.08	146.3	261.1	539.0	1608.	64.40	103.5	184.7	381.1	1137.
250	10	141.0	226.4	404.1	834.2	2488.	99.67	160.1	285.8	589.8	1759.
	50	137.3	220.5	393.6	812.5	2424.	97.08	155.9	278.3	574.5	1714.
	100	127.6	205.0	365.9	755.3	2253.	90.24	145.0	258.7	534.0	1583.
	200	82.84	133.1	237.5	490.3	1462.	58.58	94.09	167.9	346.7	1034.
200	10	114.0	183.2	327.0	674.9	2013.	80.64	129.5	231.2	477.2	1424.
	50	109.5	175.9	313.9	647.9	1933.	77.42	124.4	222.0	458.2	1367.
	100	97.08	155.9	278.3	574.5	1714.	68.65	110.3	196.8	406.3	1212.
	150	73.68	118.4	211.3	436.0	1301.	52.10	83.69	149.4	308.3	919.7
150	10	87.04	139.8	249.6	515.1	1537.	61.55	98.87	176.5	364.2	1086.
	25	85.45	137.3	245.0	505.7	1508.	60.42	97.06	173.2	357.6	1067.
	50	80.98	130.1	232.2	479.3	1430.	57.26	91.98	154.2	338.9	1011.
	100	63.21	101.5	181.2	374.1	1116.	44.70	71.80	128.2	264.5	789·0

Right section — Gage Pressure 100 to 10 P.S.I.

Up Stream P.S.I.	Down Stream P.S.I.	1000 FT 2	2½	3	4	6	2000 FT 2	2½	3	4	6
100	10	59.84	96.12	171.6	354.1	1056.	42.31	67.97	121.3	250.4	746.9
	25	57.50	92.36	164.9	340.3	1015.	40.66	65.31	116.6	240.6	717.7
	50	50.62	81.32	145.1	299.6	893.6	35.79	57.50	102.6	2₁1.8	631.9
	75	38.22	61.39	109.6	226.2	674.6	27.02	43.41	77.5	159.9	477.0
80	5	49.46	79.45	141.8	292.7	873.1	34.97	56.18	100.3	207.0	617.9
	10	48.82	78.43	140.0	288.9	861.8	34.52	55.46	98.98	204.3	609.4
	30	44.50	71.64	127.9	264.0	787.3	31.54	50.66	90.42	186.6	556.7
	55	34.26	55.03	98.23	202.8	604.8	24.23	38.91	69.46	143.4	427.6
60	5	38.45	61.77	110.3	227.6	678.5	27.19	43.68	77.96	160.9	480.0
	10	37.63	60.45	107.9	222.7	664.3	26.61	42.74	76.29	157.5	469.7
	20	35.32	56.73	101.3	209.0	623.5	24.97	40.12	71.61	147.8	440.9
	40	27.17	43.65	77.91	160.8	479.7	19.21	30.86	55.09	113.7	339.2
40	5	27.21	43.71	78.01	161.0	480.3	19.24	30.91	55.16	113.9	339.6
	10	26.03	41.81	74.63	154.1	549.5	18.41	29.57	52.77	108.9	324.9
	15	24.50	39.36	70.26	145.0	432.6	17.33	27.83	49.68	102.5	305.9
	25	20.08	32.26	57.58	118.8	354.5	14.20	22.81	40.71	84.04	250.7
30	5	21.38	34.35	61.30	126.5	377.4	15.12	24.29	43.35	89.48	266.9
	10	19.86	31.90	56.94	117.5	350.6	14.04	22.56	40.26	83.11	247.9
	15	17.81	28.61	51.07	105.4	314.5	12.60	20.23	36.11	74.54	222.4
	20	15.03	24.14	43.09	88.94	265.3	10.63	17.07	30.47	62.89	187.6
20	5	15.21	24.43	43.61	90.00	268.5	10.75	17.28	30.83	63.65	189.8
	10	12.98	20.85	37.22	76.83	229.2	9.180	14.75	26.32	54.33	162.1
10	5	7.923	12.73	22.72	46.89	138.9	5.603	9.000	16.06	33.16	98.90

MULTIPLIERS TO CONVERT 1000 FOOT LINE VALUES TO LENGTHS LISTED:

Length of Line—Feet	200	400	600	800	1000	1200	1400	1600	1800	2000	2200	2400	2600	2800	3000	3200	3400	3600	3800	4000	4200	4400	4600	4800	5000	5200
Multipliers	2.24	1.58	1.29	1.12	1.00	.91	.85	.79	.75	.71	.67	.65	.62	.60	.58	.56	.54	.53	.51	.50	.49	.48	.47	.46	.45	.44

For remarks regarding use of the Table refer to page 78 of source.　　　　Courtesy Armco Steel Corporation.

Table 4
1- and 2½-mile Lines

0.7 SP. GR. 60° F.

Left section — Gage Pressure 400 to 150 P.S.I.

Up Stream P.S.I.	Down Stream P.S.I.	1 MILE 2	2½	3	4	6	2½ MILE 2	2½	3	4	6
400	10	96.39	154.8	276.4	570.4	1701.	60.96	97.92	174.8	360.8	1076.
	50	95.38	153.2	273.5	564.5	1684.	60.32	96.90	173.0	357.0	1065.
	100	92.80	149.1	266.1	549.2	1638.	58.69	94.28	168.3	347.3	1036.
	350	45.98	73.85	131.8	272.1	811.6	29.08	46.71	83.37	172.1	513.3
350	10	84.71	136.1	242.9	501.3	1495.	53.58	86.06	153.6	317.1	945.8
	50	83.57	134.2	239.6	494.6	1475.	52.85	84.90	151.5	312.8	933.0
	100	80.61	129.5	231.1	477.1	1423.	50.98	81.90	146.2	301.7	900.0
	300	42.92	68.95	123.1	254.0	757.7	27.15	43.61	77.84	160.7	479.2
300	10	73.03	117.3	209.4	432.2	1289.	46.19	74.20	132.4	273.4	815.4
	50	71.70	115.2	205.6	424.3	1266.	45.35	72.84	130.0	268.4	800.5
	100	68.23	109.6	195.6	403.8	1204.	43.15	69.32	123.7	255.4	761.8
	250	38.64	63.67	113.6	234.6	699.7	25.07	40.27	71.88	148.4	442.5
250	10	61.34	98.53	175.9	363.0	1083.	38.80	62.32	111.2	229.6	684.8
	50	59.75	95.98	171.3	353.6	1055.	37.79	60.70	108.3	223.6	667.1
	100	55.54	89.21	159.2	328.7	980.4	35.13	56.42	100.7	207.9	620.1
	200	36.05	57.91	103.4	213.4	636.4	22.80	36.63	65.37	134.9	402.5
200	10	49.63	79.72	142.3	293.7	876.1	31.39	50.42	90.00	185.8	554.1
	50	47.65	76.54	136.6	282.0	841.1	30.13	49.41	86.40	178.3	532.0
	100	42.25	67.87	121.1	245.8	745.8	26.72	42.92	76.61	158.1	471.7
	150	32.06	51.51	91.94	189.8	566.0	20.28	32.58	58.14	120.0	358.0
150	10	37.88	60.85	108.6	224.2	668.7	23.96	38.48	68.69	141.8	422.9
	25	37.19	59.74	106.6	220.1	656.5	23.52	37.78	67.44	139.2	415.2
	50	35.24	56.61	101.0	208.6	622.1	22.29	35.80	63.91	131.9	393.5
	100	27.51	44.19	78.87	162.8	485.6	17.40	27.95	49.88	103.0	307.1

Right section — Gage Pressure 100 to 10 P.S.I.

Up Stream P.S.I.	Down Stream P.S.I.	1 MILE 2	2½	3	4	6	2½ MILE 2	2½	3	4	6
100	10	26.04	41.83	74.67	154.1	459.7	16.47	26.46	47.22	97.47	290.7
	25	25.02	40.20	71.75	148.1	441.7	15.83	25.42	45.38	93.66	279.4
	50	22.03	35.39	63.16	130.4	388.9	13.93	22.38	39.95	82.46	246.0
	75	16.63	26.72	47.68	98.4	293.6	10.52	16.90	30.16	62.25	185.7
80	5	21.53	34.58	61.72	127.4	380.0	13.61	21.87	39.03	80.57	240.3
	10	21.25	34.13	60.92	125.7	375.1	13.44	21.59	38.53	79.53	237.2
	30	19.41	31.18	55.65	114.9	342.6	12.28	19.72	35.20	72.65	216.7
	55	14.91	23.95	42.75	88.24	263.2	9.430	15.15	27.04	55.81	166.5
60	5	16.74	26.88	47.98	99.04	285.4	10.58	17.00	30.35	62.64	186.8
	10	16.38	26.31	46.95	96.91	289.1	10.36	16.64	29.70	61.29	182.8
	20	15.37	24.69	44.07	90.96	271.3	9.721	15.62	27.87	57.53	171.6
	40	11.83	19.00	33.91	69.98	208.8	7.479	12.01	21.44	44.26	132.0
40	5	11.84	19.02	33.95	70.08	209.0	7.489	12.03	21.47	44.32	132.2
	10	11.33	18.20	32.48	67.04	200.0	7.165	11.51	20.54	42.40	126.5
	15	10.66	17.13	30.58	63.11	188.3	6.745	10.83	19.34	39.92	119.1
	25	8.740	14.04	25.06	51.72	154.3	5.527	8.879	15.85	32.71	97.57
30	5	9.315	14.95	26.68	55.07	164.3	5.885	9.453	16.87	34.83	103.9
	10	8.643	13.88	24.78	51.15	152.6	5.466	8.781	15.67	32.35	96.49
	15	7.752	12.45	22.23	45.88	136.8	4.903	7.875	14.06	29.02	86.55
	20	6.540	10.51	18.75	38.71	115.5	4.137	6.645	11.86	24.48	73.02
20	5	6.619	10.63	18.98	39.17	116.8	4.186	6.724	12.00	24.77	73.90
	10	5.650	9.076	16.20	33.44	99.74	3.573	5.740	10.25	21.15	63.08
10	5	3.448	5.539	9.886	20.41	60.87	2.181	3.503	6.253	12.91	38.50

MULTIPLIERS TO CONVERT 1 MILE LINE VALUES TO LENGTHS LISTED:

Length of Line—Miles	1.2	1.4	1.6	1.8	2.0	2.2	2.4	2.6	2.8	3.0	3.2	3.4	3.6	3.8	4.0	4.2	4.4	4.6	4.8
Multipliers	.91	.85	.79	.75	.71	.67	.65	.62	.60	.58	.56	.54	.53	.51	~.50	.49	.48	.47	.46

For remarks regarding use of the Table refer to page 78 of source.　　　　Courtesy of Armco Steel Corporation.

Table 5
5- and 10-mile Lines

0.7 SP. GR. 60° F.

GAGE PRESSURE		5 MILE LINE — NOMINAL PIPE SIZE ▼					10 MILE LINE — NOMINAL PIPE SIZE ▼				
Up Stream	Down Stream	2	2½	3	4	6	2	2½	3	4	6
P.S.I.	P.S.I.	Thousands of Cu. Ft. Per Hour					Thousands of Cu. Ft. Per Hour				
400	10	43.10	69.24	123.6	255.1	760.9	30.48	48.96	87.39	180.4	538.0
	50	42.66	68.52	122.3	252.4	753.0	30.16	48.45	86.48	178.5	532.4
	100	41.50	66.67	119.0	245.6	732.6	29.35	47.14	84.14	173.7	518.0
	350	20.56	33.03	58.95	121.7	363.0	14.54	23.35	41.69	86.04	256.7
350	10	37.88	60.86	108.6	224.2	668.8	26.79	43.03	76.81	158.5	472.9
	50	37.37	60.03	107.2	221.2	659.7	26.43	42.45	75.77	156.4	466.5
	100	36.05	57.91	103.4	213.3	636.4	25.49	40.95	73.09	150.9	450.0
	300	19.20	30.84	55.04	113.6	338.9	13.57	21.80	38.92	80.33	239.6
300	10	32.66	52.46	93.64	193.3	576.6	23.09	37.10	66.22	136.7	407.7
	50	32.07	51.51	91.94	189.8	566.0	22.67	36.42	65.01	134.2	400.2
	100	30.51	49.02	87.49	180.6	538.7	21.58	34.66	61.86	127.7	380.9
	250	17.73	28.47	50.82	104.9	312.9	12.53	20.13	35.94	74.18	221.3
250	10	27.43	44.07	78.65	162.3	484.3	19.40	31.16	55.62	114.8	342.4
	50	26.72	42.92	76.61	158.1	471.7	18.89	30.35	54.17	111.8	333.5
	100	24.84	39.90	71.21	147.0	438.5	17.56	28.21	50.36	103.9	310.0
	200	16.12	25.90	46.23	95.41	284.6	11.40	18.31	32.69	67.47	201.2
200	10	22.20	35.65	63.64	131.4	391.8	15.69	25.21	45.00	92.88	277.0
	50	21.31	34.23	61.10	126.1	376.2	15.07	24.20	43.20	89.17	266.0
	100	18.89	30.35	54.17	111.8	333.5	13.36	21.46	38.31	79.07	235.8
	150	14.34	23.03	41.11	84.86	253.1	10.14	16.29	29.07	60.01	179.0
150	10	16.94	27.21	48.57	100.3	299.0	11.98	19.24	34.35	70.89	211.5
	25	16.63	26.72	47.68	98.42	293.6	11.76	18.89	33.72	69.60	207.6
	50	15.76	25.32	45.19	93.27	278.2	11.14	17.90	31.95	65.96	196.7
	100	12.30	19.76	35.27	72.81	217.2	8.699	13.97	24.94	51.48	153.6

0.7 SP. GR. 60° F.

GAGE PRESSURE		5 MILE LINE — NOMINAL PIPE SIZE ▼					10 MILE LINE — NOMINAL PIPE SIZE ▼				
Up Stream	Down Stream	2	2½	3	4	6	2	2½	3	4	6
P.S.I.	P.S.I.	Thousands of Cu. Ft. Per Hour					Thousands of Cu. Ft. Per Hour				
100	10	11.65	18.71	33.39	68.92	205.6	8.235	13.23	23.61	48.74	145.4
	25	11.19	17.98	32.09	66.23	197.5	7.913	12.71	22.69	46.83	139.7
	50	9.852	15.83	28.25	55.31	173.9	6.966	11.19	19.97	41.23	123.0
	75	7.438	11.95	21.32	44.02	131.3	5.259	8.448	15.08	31.12	92.84
80	5	9.626	15.46	27.60	56.97	169.9	6.807	10.93	19.52	40.28	120.2
	10	9.502	15.26	27.24	56.23	167.7	6.719	10.79	19.26	39.76	118.6
	30	8.680	13.94	24.89	51.37	153.2	6.138	9.860	17.60	36.32	108.4
	55	6.668	10.71	19.12	39.46	117.7	4.715	7.574	13.52	27.90	83.23
60	5	7.484	12.02	21.46	44.29	132.1	5.292	8.501	15.17	31.32	93.42
	10	7.324	11.76	21.00	43.34	129.3	5.179	8.319	14.85	30.65	91.42
	20	6.873	11.04	19.71	40.68	121.3	4.861	7.808	13.94	28.77	85.80
	40	5.288	8.495	15.16	31.30	93.36	3.740	6.007	10.72	22.13	66.01
40	5	5.296	8.507	15.18	31.34	93.48	3.745	6.015	10.74	22.16	66.10
	10	5.066	8.138	14.53	29.98	89.43	3.582	5.755	10.27	21.20	63.24
	15	4.769	7.661	13.67	28.22	84.19	3.372	5.417	9.669	19.96	59.53
	25	3.908	6.278	11.21	23.13	69.00	2.764	4.439	7.924	16.36	48.79
30	5	4.161	6.685	11.93	24.63	73.46	2.943	4.727	8.437	17.41	51.94
	10	3.865	6.209	11.08	22.87	68.23	2.733	4.390	7.836	16.17	48.25
	15	3.467	5.569	9.940	20.52	61.20	2.451	3.938	7.029	14.51	43.27
	20	2.925	4.698	8.386	17.31	51.63	2.068	3.322	5.930	12.24	36.51
20	5	2.960	4.755	8.487	17.52	52.25	2.093	3.362	6.001	12.39	36.95
	10	2.527	4.059	7.244	14.95	44.60	1.787	2.870	5.123	10.57	31.54
10	5	1.542	2.477	4.421	9.126	27.22	1.090	1.752	3.126	6.453	19.25

MULTIPLIERS TO CONVERT 10 MILE LINE VALUES TO LENGTHS LISTED.

Length of Line—Miles	5.2	5.4	5.6	5.8	6.0	6.2	6.4	6.6	6.8	7.0	7.2	7.4	7.6	7.8	8.0	8.2	8.4	8.6	8.8	9.0	9.2	9.4	9.6	9.8	Miles—Length of Line
Multipliers	1.39	1.36	1.34	1.31	1.29	1.27	1.25	1.23	1.21	1.20	1.18	1.16	1.15	1.13	1.12	1.10	1.09	1.08	1.07	1.05	1.04	1.03	1.02	1.01	Multipliers

For remarks regarding use of the Table refer to page 78 of source.

Courtesy of Armco Steel Corporation.

From Table 3 the volume for 1000 ft of line, for 0.70 specific gravity gas and 60 °F flowing temperature for 200 and 50 psi up and downstream pressure, respectively, is 109.5 MCF/h. Correction factor for the pipe length, from the same table is 0.53. Correction factor, from Table 2 for specific gravity and flowing temperature is 0.991. Therefore:

Delivery of gas = 109.5 × 0.53 × 0.991 = 57,512 cu. ft/h.

Multiphase flow

As stated at the beginning of this chapter, the problems of the simultaneous flow of oil and gas or of oil, gas, and water through one pipe have become increasingly important in oil field production operations. The problems are complex, because the properties of two or more fluids must be considered and because of the different patterns of fluid flows, depending upon the flow conditions. These patterns, in any given line, may change as the flow conditions change, and they may be coexistent at different points of the line.[3,5,7]

Different investigators recognize different flow patterns and use different nomenclature. Those generally recognized are:

1. Bubble flow—bubbles of gas flow along the upper part of the pipe with about the same velocity as the liquid.

2. Plug flow—the bubbles of gas coalesce into large bubbles and fill out the large part of the cross-sectional area of the pipe.

3. Laminar flow—the gas—liquid interface is relatively smooth with gas flowing in the upper part of the pipe.

4. Wavy flow—same as above except that waves are formed on the surface of the liquid.

5. Slug flow—the tops of some waves reach the top of the pipe. These slugs move with high velocity.

6. Annular flow—the liquid flows along the walls of the pipe, and the gas moves through the center with high velocity.

7. Spray flow—the liquid is dispersed in the gas.

Figure 1. Prediction of solubility from saturation pressure and gravity of crude oil. (Ref. 12.)

The above description of flow patterns has been given to emphasize the first reason why the pressure drop must be higher for the multiphase than the single-phase flow. In the latter the pressure drop is primarily the result of friction. In the multiphase flow, in addition to friction, the energy losses are due to gas accelerating the fluid, due to compression of gas and formation of waves and slugs. These losses are different for different patterns, with patterns changing as a result of changes in flow conditions. The second reason is the fact that with two phases present in the pipe, less cross-sectional area is available for each of the phases. As previously discussed, the pressure drop is inversely proportional to the fifth power of pipe diameter.

The literature on multiphase flow is quite extensive. Ref. 7 gives 72 references on the subject.

Two-phase Horizontal Flow

More than 20 correlations have been developed by different investigators for predicting the pressure drop in the two-phase flow. Under the project, sponsored jointly by the American Petroleum Institute and American Gas Association at the University of Houston, five of these correlations were tested by comparing them with 2620 experimental measurements culled from more than 15,000 available.

The statistical part of the report on the project concludes that of the five correlations tested, the Lockhart—Martinelli correlation shows the best agreement with the experimental data, particularly for pipe sizes commonly used in oil field production operations. The method may be summarized as follows[3,4]:

1. Determine the single-phase pressure drops for the liquid and the gas phase as if each one of these were flowing alone through the pipe. This is done by use of previously given equations.
2. Determine the dimensionless parameter:

$$X = \sqrt{\frac{\sqrt{\Delta P_L}}{\Delta P_G}}$$

where ΔP_L and ΔP_G are single-phase pressure drops for liquid and gas, respectively.

3. The method recognizes four regimens of the two-phase flow as follows:

Flow regimens

Gas	Liquid
Turbulent	Laminar
Turbulent	Turbulent
Laminar	Laminar
Laminar	Turbulent

The type of flow of each of the phases is determined by its Reynolds number (Figure 2).

4. Determine factor Φ from a chart as a function of X for the appropriate flow regimen.
5. The two-phase pressure drop is then:

$$\Delta P_{TP} = \Phi^2 P_L \text{ of } G$$

The single-phase pressure drop of either gas or liquid phase can be used. The results are the same.

For the two-phase flow problems in oil field production operations, certain simplifications can be used if only approximate estimates are desired:

1. Information that would permit determining the volume of oil and gas under flow conditions from equilibrium flash calculation is usually unavailable. The following procedure may be used:

   Gas volume is determined by use of the chart in Figure 1, which shows gas in solution for oil of different

Figure 2. Parameter Φ for Lockhart–Martinelli correlation for two-phase flow. (Ref. 9.)

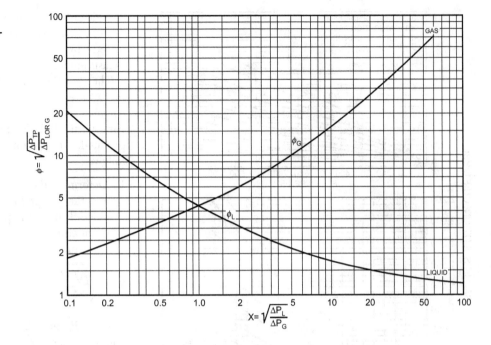

gravities and different saturation pressures (in this case the flowline pressures). If the gas–oil ratio is known for one pressure, the ratio for another pressure can be determined from the chart as follows:

Assume a gas–oil ratio of 600 cu. ft/bbl at 900 psi for a 30°API oil. Determine the gas–oil ratio at 1300 psi. From the chart, at 900 psi, the gas in solution is 205 cu. ft/bbl and at 1300 psi is 300 cu. ft/bbl, an increase of 95 cu. ft/bbl. Therefore, at 1300 psi the gas–oil ratio will be 600–95 = 505 cu. ft/bbl.

Increase in volume of oil because of the increase of the dissolved gas can be obtained from charts for calculation of formation volume of bubble point liquids. However, for approximate estimates in problems involving field flowlines, this step frequently can be omitted without seriously affecting the validity of results.

2. In a majority of cases here under consideration, the flow regimen is turbulent–turbulent. Therefore, the chart in Figure 2 shows the factor for only this regimen.

The procedure is illustrated by the following examples.

Example 1. Determine two-phase pressure drops for different flow ratios and different flow pressures under the following conditions:

Oil	31°API
Viscosity	5 cp at 100 °F
Separator pressure	900 psig
Gas–oil ratio	670 cu. ft/bbl
Flow line	2 in

Wellhead pressure of 1300 psig was assumed, and calculations were made for flow rates of 100 and 50 bbl per day. The single-phase pressure drops were calculated for the gas and liquid phases, each of them flowing alone in the line, according to the Fanning equations.

Calculation of pressure drop for liquid phase:

$$H_f = \frac{f\,L\,(\text{bbl/day})^2\,(\text{lb/gal})}{181{,}916\,D^5}$$

where:

f = Friction factor
L = Length of pipe, ft, and
D = Inside diameter of pipe, inches

Calculation of pressure drop for gas phase.

For the two flow rates and the two pressures, these single-phase pressure drops were calculated to be as follows:

Flow Rate bbl/day	100		50	
Pressure, psia	1300	900	1300	900
ΔP_G psi/100 ft	7.9×10^{-4}	2.0×10^{-3}	2.4×10^{-4}	7.7×10^{-4}
ΔP_1 psi/100 ft	$2. \times 10^{-4}$	2.7×10^{-2}	1.3×10^{-2}	1.3×10^{-2}
Then $\Delta X = \sqrt{\dfrac{\Delta P_L}{\Delta P_G}}$	5.9	3.8	7.4	4.2
From Figure 2 ϕ and the two-phase pressure drop	2.0	2.4	1.9	2.3
Pressure drop $\Delta P_{TP} = \phi_L \times \Delta P_L$	0.12	0.16	0.048	0.069

Figure 3. Determining two-phase flow pressure drops.

With two pressure drops determined for two flow line pressures for each of the flow rates, lines can be drawn showing pressure drop for different flow line pressures for these two rates (Figure 3). From these two lines relationships can be established for pressure drops for other rates and flow line pressures. The procedure is as follows:

Start with, for instance, 800 psia line. For the rate of 50 bpd at 800 psia flow line pressure, the pressure drop from the chart is 0.074 psia/100 ft. From the 50 bpd rate on the scale in the upper part of the chart, go downward to the line representing 0.074 psia pressure drop. This gives one point of the 800 psia line.

For the rate of 100 bpd the chart indicates that at 800 psia flowline pressure, the pressure drop is 0.175 psia/100 ft. From the 100 bpd rate in the upper part of the chart, go downward to the 0.175 psia/100 ft line. This gives the second point of the 800 psia line. The line can now be drawn through these two points. Lines for other pressures can be constructed in a similar way.

The chart is used as follows: For conditions represented by this particular chart, what is the two-phase pressure drop for the rate of 60 bpd and average flow line pressure of 1000 psia?

Start at the rate of 60 bpd on the upper scale of the chart. Go downward to the intersection with the 1000 psia pressure, then to the left, and read the two-phase pressure drop as 0.075 psia/100 ft.

Example 2. If pipe diameter is to be determined for the desired pressure drop and the known length of line, flow rate, and the properties of fluids, then the problem cannot be solved directly. The procedure is as follows:

An estimate is made of the diameter of the two-phase line. The best approach is to determine, for the conditions, the diameter of the line for each of the phases flowing separately. In a majority of cases, the sum of these two diameters will be a good estimate of the diameter for the two-phase line. Calculations are then made for the two-phase pressure drop using the assumed diameter. If this drop is reasonably close to the desired one, the estimate is correct. If not, a new estimate must be made and the calculations repeated.

Inclined two-phase flow

In hilly terrain additional pressure drop can be expected in case of the two-phase flow. Little published information is available on the subject. Ovid Baker suggested the following empirical formula[10]:

$$\Delta P_{TPH} = \Delta P_{TP} + \frac{1.61 N H \rho_L}{144 V_g^{0.7}}$$

where:

$\quad \Delta P_{TPH} =$ Pressure drop, corrected for hilly terrain in psi

$\quad \Delta P_{TP} =$ Calculated two-phase pressure drop for horizontal pipe, psi

$\quad N =$ Number of hills

$\quad H =$ Average height of hills, ft

$\quad \rho_L =$ Density of liquid phase, lb/ft^3, and

$\quad V_g =$ Velocity of gas, ft/s (based on 100% pipe area).

Horizontal three-phase flow

Very little is known about the effect on the pressure drop of addition to the gas–oil system of a third phase, an immiscible liquid, and water. Formation of emulsion results in increased viscosities. Formulas are available for approximate determination of viscosity of emulsion, if viscosity of oil and percent of water content of oil is known. The point is, however, that it is not known what portion of the water is flowing in emulsified form.

From one published reference[11] and from some unpublished data, the following conclusions appear to be well founded:

1. Within the range of water content of less than 10% or more than 90%, the flow mechanism appears to approach that of the two-phase flow.
2. Within the range of water content of from 70% to 90%, the three-phase pressure drop is considerably higher than for the two-phase flow.

For solution of three-phase flow problems of oil field flowlines, the following approach has been used; the oil and water are considered as one phase and the gas as the other phase. Previously given calculation of a two-phase drop is used. The viscosity of the oil-water mixture is determined as follows:

$$\mu_L = \frac{V \times \mu_0 + V_W \times \mu_W}{V_0 + V_W}$$

where: μ_L, μ_0 and μ_W are viscosities of mixture of oil and of water, respectively, and V_0 and V_W represent the corresponding volumes of oil and water.

No test data are available to determine the accuracy of this approach. Until more information is developed on the subject, the method may be considered for approximate estimates.

References

1. *Cameron Hydraulic Data*, 13th ed., Ingersoll-Rand Company, New York, 1962.
2. *Flow of Fluids Through Valves, Fittings and Pipe*, Technical Paper No. 410, Engineering Division, Crane Company, Chicago, 1957.
3. Streeter, Victor L., *Handbook of Fluid Dynamics*, McGraw-Hill Book Company, Inc., New York, 1961.
4. Prof. W. E. Durand's Discussions in *Handbook of the Petroleum Industry*, David T. Day; John Wiley & Sons, Inc., New York.
5. Baker, Ovid, "Simultaneous Flow of Oil and Gas," *The Oil and Gas Journal*, July 26, 1954.
6. Campbell, Dr. John M., "Elements of Field Processing," *The Oil and Gas Journal*, December 9, 1957.
7. Campbell, Dr. John M., "Problems of Multi-phase Pipe Line Flow," in "Flow Calculations in Pipelining," *The Oil and Gas Journal*, November 4, 1959.
8. Duckler, A. E., Wicks, III, Moye, and Cleveland, R. G., "Frictional Pressure Drop in Two-Phase Flow," *AIChE Journal*, January, 1964.
9. Lockhart, R. W., and Martinelli, R. C., "Proposed Correlation of Data for Isothermal Two-Phase, Two-Component Flow in Pipes," *Chemical Engineering Progress* 45: 39–48, (1949).
10. Ovid Baker's discussion of article "How Uphill and Downhill Flow Affect Pressure Drop in Two-Phase Pipe Lines," by Brigham, W. E., Holstein, E. D., and Huntington, R. L.; *The Oil and Gas Journal*, November 11, 1957.
11. Sobocinski, D. P., and Huntington, R. D., "Concurrent Flow of Air, Gas, Oil and Water in a Horizontal Line," *Trans. ASME* 80: 1,252, 1958.
12. Beal, Carlton, "The Viscosity of Air, Water, Natural Gas, Crude Oil and Its Associated Gases at Oil Field Temperatures and Pressures," *Petroleum Development and Technology*, AIME, 1946.
13. Switzer, F. G., *Pump Fax*, Gould Pumps, Inc.
14. *Pump Talk*, The Engineering Department, Dean Brothers Pumps, Inc., Indianapolis.
15. Switzer, F. G., *The Centrifugal Pump*, Goulds Pumps, Inc.
16. Zalis, A. A., "Don't Be Confused by Rotary Pump Performance Curves," *Hydrocarbon Processing and Petroleum Refiner*, Gulf Publishing Company, September, 1961.

Nomograph for calculating Reynolds number for compressible flow friction factor for clean steel and wrought iron pipe

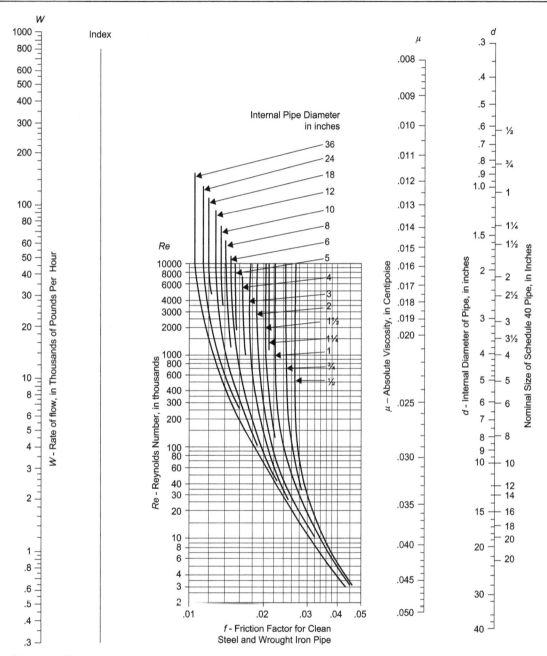

Figure 1. Reynolds number for compressible flow friction factor for clean steel and wrought iron pipe.

Background

The nomograph (Figure 1) permits calculation of the Reynolds number for compressible flow and the corresponding friction factor. It is based on the equation:

$$Re = 6.31\frac{W}{d\mu} = 0.482\frac{q_s \times Sg}{d\mu}$$

where:

W = Flow rate, lb/h
Sg = Specific gravity of gas relative to air

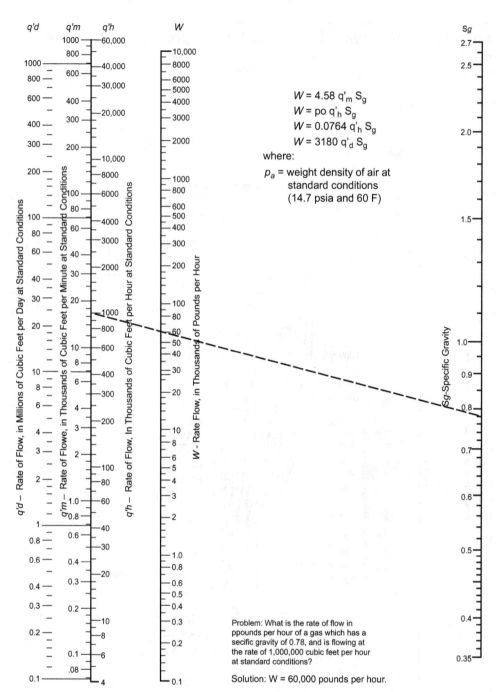

$W = 4.58\ q'_m\ S_g$
$W = po\ q'_h\ S_g$
$W = 0.0764\ q'_h\ S_g$
$W = 3180\ q'_d\ S_g$

where:

p_a = weight density of air at standard conditions (14.7 psia and 60 F)

Problem: What is the rate of flow in ppounds per hour of a gas which has a secific gravity of 0.78, and is flowing at the rate of 1,000,000 cubic feet per hour at standard conditions?

Solution: W = 60,000 pounds per hour.

Figure 2. Equivalent volume and weight flow rates of compressible fluids.

q_s = volumetric flow rate, cu. ft/s (at 14.7 psia and 60 °F), and μ = fluid viscosity, centipoise.

If the flow rate is given in lb/h, the nomograph can be used directly without resorting to Figure 2 to obtain q_s. Figure 2 converts volume flow to weight flow rates if the specific gravity of the fluid is known.

Example. Natural gas at 250 psig and 60 °F with a specific gravity of 0.75 flows through an 8-in schedule 40 clean steel pipe at a rate of 1,200,000 cu. ft/h at standard conditions.

Find the Reynolds number and the friction factor.

At $Sg = 0.75$ obtain $q = 69,000$ from Figure 1, $\mu = 0.011$.

Connect	With	Mark or Read
W = 69,000	μ = 0.011	Index
Index	d = 8 in	Re = 5,000,000
Re = 5,000,000	Horizontally to 8 in	f = 0.014

Source

Flow of Fluids Through Valves, Fittings, and Pipes, Technical Paper No. 410, 3–19, Crane Company, Chicago, (1957).

12: Liquids—General

Determining the viscosity of crude

If the viscosity of a gas-saturated crude oil at the saturation (bubble-point) pressure is known, using this homograph you can quickly estimate the viscosities at higher pressures.

Example. Find the viscosity at 4200 psia for a crude oil when its viscosity is 30 cp at the saturation pressure of 1200 psia. Notice that 4200 psia is 3000 psi above the saturation pressure. Connecting 3000 on the pressure difference scale (left) with 30 on the curved scale for viscosity at the bubble-point pressure, the intersection with the scale on the right at 48 cp is the desired value.

Pressure above bubble point less pressure at bubble point, psia

Abs. viscosity of gas sat'd. crude oil at bubble point pressure, centipoises

Abs. viscosity of crude oil above bubble point, cp

Chart quickly gives API gravity of blends

API gravity of a crude oil blend may be readily estimated from the gravity of the components and their percentage composition. It is a simple procedure to use these curves to find the resulting API gravity of the blend.

Example. If you blend a 14°API pitch (60%) with a 36°API cutting stock (40%), the resulting fuel oil has an API gravity of 22° as read from the nomograph. Calculated result from gravity tables would be 22.1°API.

Results found using this nomograph checked out within 1°API over the range of gravities and percentage of components (1) and (2) in the nomograph below. Estimates from the nomograph are based on the assumption that volumes of blends are additive and that no light components flash off in blending.

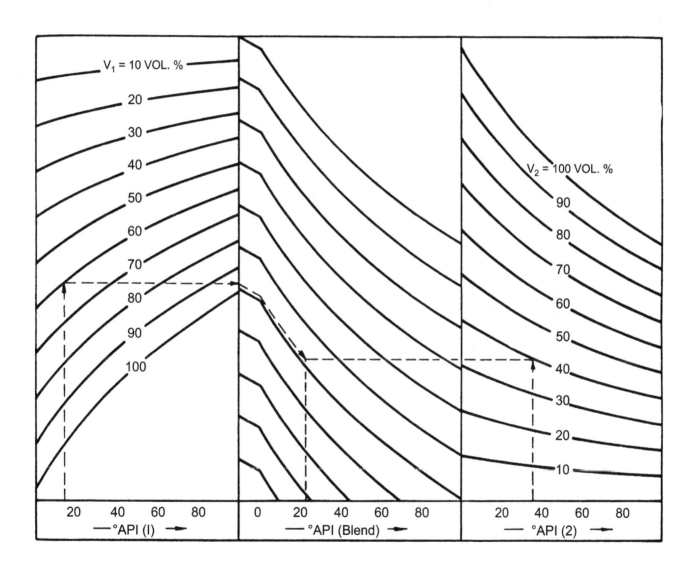

Liquid gravity and density conversion chart

This line chart provides an easy method for converting units of liquid gravity and density. Draw a horizontal line perpendicular to the scale line through a known value and read the equivalent value on any other scale.

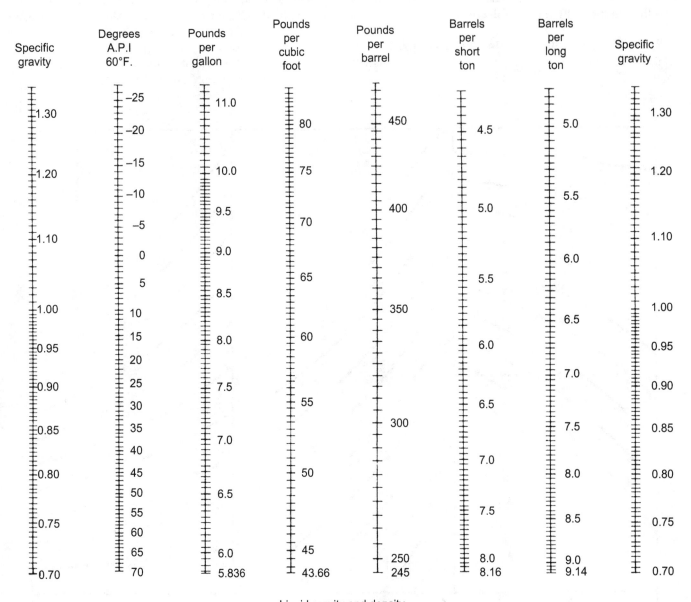

Liquid gravity and density

Nomograph for calculating viscosities of liquid hydrocarbons at high pressure

Lockhart and Lenoir developed a graphical correlation showing the effect of pressure on viscosity of liquid hydrocarbons. This correlation is shown in Figure 1, which is based primarily on data of Hersey and Hopkins,[2] including pure hydrocarbons, lubricating oils, bright stocks, and distillates. Data from Reference 1 have also been included.

To use the nomograph, the characterization factor of Watson, K_w, and the viscosity of the liquid at atmospheric pressure are required.

The accuracy of the correlation decreases as pressure increases.

Example. What is the viscosity of an oil at 5400 psia, if its characterization factor is 11.8 and its viscosity at atmospheric conditions is 90 cP?

Enter 5400 psia in the pressure scale to the viscosity line of 90 and proceed horizontally to the middle reference scale. Follow the curve lines to intersect the vertical line drawn at K_w = 11.8 and read the ratio of viscosity on the extreme left scale at 2.6. The viscosity of the oil is (2.6) (90) = 234 cP.

Source

Lockhart, F. J. and Lenoir, J. M., *Petroleum Refiner*, 40, No. 3, 209 (1961).

References

1. Griest, and others, J. Chem. Physics, 29, 711 (1958).
2. Hersey, M. D., and Hopkins, R. F., *Viscosity of Lubricants under Pressure*, ASME, 1954.

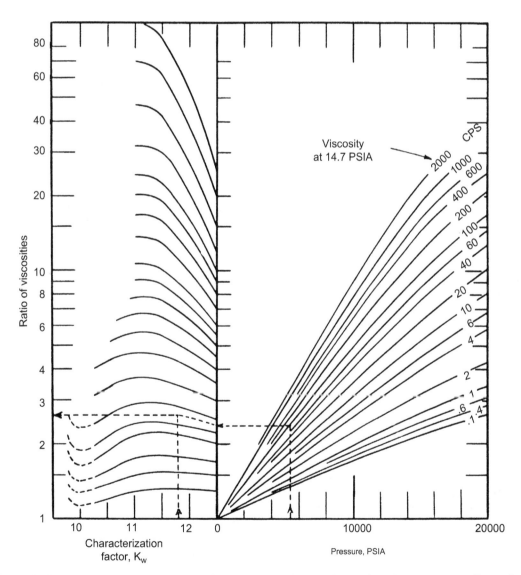

Figure 1. Isothermal effect of pressure on viscosities of liquid hydrocarbons at low reduced temperatures. (Reproduced by permission *Petroleum Refiner*, Vol. 40, No. 3 page 209, copyright 1961, Gulf Publishing Co., Houston.)

Calculate viscosity of a blend

Viscosity is not an additive function of percentage composition. The viscosity of a blend can be calculated with the following equation:

$$V_b = \left[\frac{G_1 + G_2 + G_3 \cdots}{\dfrac{G_1}{\sqrt{V_1}} + \dfrac{G_2}{\sqrt{V_2}} + \dfrac{G_3}{\sqrt{V_3}} \cdots} \right]^2$$

where

V_b = Viscosity of blend,
G = Volume (gals, barrels, etc.),
V = Viscosity in SUS.

Example.

$G_1 = 10{,}000$ bbls	$V_1 = 45$ SUS
$G_2 = 20{,}000$ bbls	$V_2 = 50$ SUS
$G_3 = 5000$ bbls	$V_3 = 37$ SUS

$$V_b = \left[\frac{10000 + 20000 + 5000}{\dfrac{10000}{\sqrt{45}} + \dfrac{20000}{\sqrt{50}} + \dfrac{5000}{\sqrt{37}}} \right]^2,$$

$$V_b = 46.4 \text{ SUS}.$$

Calculate specific gravity of a blend

The specific gravity of a blend may be directly calculated by a ratio of the specific gravity of each component in the blend.

Example. Assume a blend consisting of the following parameters:

Partial specific gravity $= (10{,}000/45{,}000) \times 0.8551$

$= 0.1900.$

Specific gravity $=$ sum of partial blends.

Volume, bbls	Specific Gravity	Partial Specific Gravity
10,000	0.8551	0.1900
15,000	0.8337	0.2779
20,000	0.8481	0.3769
Total 45,000		**0.8448**

Convert viscosity units

Convert SUS to centistokes:

$V = 0.2260 \, V_{SUS} - 135/V_{SUS}$ where $V_{SUS} > 100$.

$V = 0.2260 \, V_{SUS} - 195/V_{SUS}$ where $V_{SUS} \leq 100$.

Convert centistokes to centipoises (cP)
$cP = V_{CS} \times$ density, gm/cm^3

$SUS = cS \times 4.6673$, if $cS > 500$, $SUS = cS \times 4.6347$

$cSt = 2.24(SSF - 184/SSF)$ for $25 < SSF \leq 40$

$cSt = 2.16(SSF - 60/SSF)$ for $SSF > 40$.

Convert specific gravity to API gravity and API gravity to specific gravity

$$API = \frac{141.5}{S_{60}} - 131.5$$

$$S_{60} = \frac{141.5}{API_{60} + 131.5}$$

where $S_{60} =$ specific gravity @ 60 °F

Bulk modulus, psi

API gravity @ 60

Based on Chapter 11.2.1, API MPMS, Compressibility Factors for Hydrocarbons, May 21, 1992.

Calculate bulk modulus

The bulk modulus, K, of a liquid is the reciprocal of its compressibility. The bulk modulus for water is $K = 300,000$ psi. The compressibility factors for most liquids can be found in Chapters 11.2.1 and 11.2.2 of the API Manual of Petroleum Measurement Standards. The following relation, known as the ARCO formula, can be used to calculate the bulk modulus for crude oil.

$$K = 1.286(10^6) + 13.55P - 4.122(10^4)T^{1/2}$$
$$- 4.53(10^3)API - 10.59API^2 + 3.228T \, API,$$

where

P = Average pressure.

Adapted from Fluid Flow Consultants, Inc., Grove, Oklahoma. Reprinted with permission.

Nomograph for calculating viscosity of slurries

This nomogram is based on the Hatschek equation for estimating the viscosity of slurries in an aqueous suspension.

$$\mu = \frac{\mu_w}{1 - x^{0.333}},$$

where

μ = Viscosity of slurry, centipoises,
μ_w = Viscosity of water at the temperature of the slurry, centipoises,
x = Volume fraction of dry solids in the slurry.

The nomogram (Figure 1) shows a scale calibrated in terms of the temperature of the water instead of its viscosity. Thus, the actual water viscosity is not needed.

Example. What is the viscosity of a slurry having a volume fraction of solids of 0.06 at a temperature of 39 °C? Connect $T = 39$ °C with $x = 0.06$ and read $\mu_w = 1.1$ cP on the middle scale.

Source

Davis, D. S., *Brit. Chem. Eng.*, 4, 9, 478 (1959).

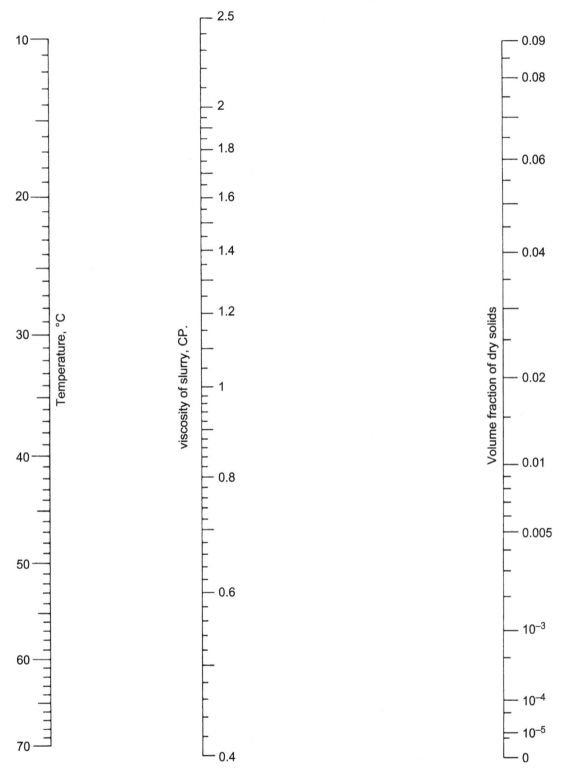

Figure 1. Viscosity of slurries. (Reproduced by permission, *British Chemical Engineering*, vol. 4, page 478, copyright 1959.)

Nomograph for calculating velocity of liquids in pipes

The mean velocity of any liquid flowing in a pipe of circular cross section can be calculated from the following formula:

$$v = 183.3\frac{q}{d^2} = 0.408\frac{Q}{d^2} = 0.0509\frac{W}{d^2\rho}$$

where

v = Average fluid velocity, feet per second,
d = Inside diameter of pipe, inches,
q = Rate of flow, cubic feet per second,
Q = Rate of flow, gallons per minute,
W = Rate of flow, pounds per hour,
ρ = Fluid density, pounds per cubic foot.

The Nomograph for Calculating Velocity of Liquids in Pipes can be used to calculate the liquid velocity when the rate of flow is in cubic feet, gallons, or thousands of lb.

Conversely, knowing the flow rate and velocity, the pipe diameter may be calculated.

Example. What is the velocity of fuel oil at 60 °F flowing through a 2-in schedule 40 pipe at a rate of 45,000 lb/h. The oil density is 56.02 lb/gal.

Connect	With	Mark or Read
W = 45	ρ = 56.02	Q = 100 cu ft/s
Q = 100	d = 2-in schedule 40	v = 10 ft/s

Source

Flow of Fluids through Valves, Fittings, and Pipe, Technical Paper No. 410, 3–7, Crane Company, Chicago, (1957).

Nomograph for calculating velocity of compressible fluids in pipes

The mean velocity of a compressible fluid in a pipe can be computed by the following formula, which is obtained after dividing the rate of flow in appropriate units by the cross-sectional area of the pipe:

$$V = \frac{3.06W\overline{V}}{d^2} = \frac{3/06W}{d^2\rho}$$

where

V = Mean velocity of flow, feet per minute,
W = Rate of flow, in pounds per hour,
V = Specific volume of fluid, cubic feet per pound,
d = Internal diameter of pipe, inches,
ρ = Density of fluid, pounds per cubic foot.

Example. Steam at 600 psig and 850 °F is to flow through a schedule 80 pipe at a rate of 30,000 lb/h. Find the pipe size if the velocity is to be limited to 8000 ft/min.

Connect	With	Mark or Read
850 °F	600 psig line (vertically)	600 psig
600 psig	Horizontally to	$\overline{V}$ = 1.21
$\overline{V}$ = 1.21	W = 30,000	Index
Index	V = 8000	d = 3.7, use 4-in schedule 80 pipe

If a 4-in schedule 80 pipe is used, the actual velocity is found by connecting the Index with 4-in schedule 80 to get V = 7600 ft/min.

Note: If a different fluid is involved, the value of the density (or specific volume) needed to make the calculation can be obtained from Section 9 of this book, under "Density and specific volume of gases and vapors."

Source

Flow of Fluids through Valves, Fittings, and Pipes, Technical Paper No. 410, 3–16, Crane Company, Chicago, (1957).

Nomograph for calculating velocity of liquids in pipes

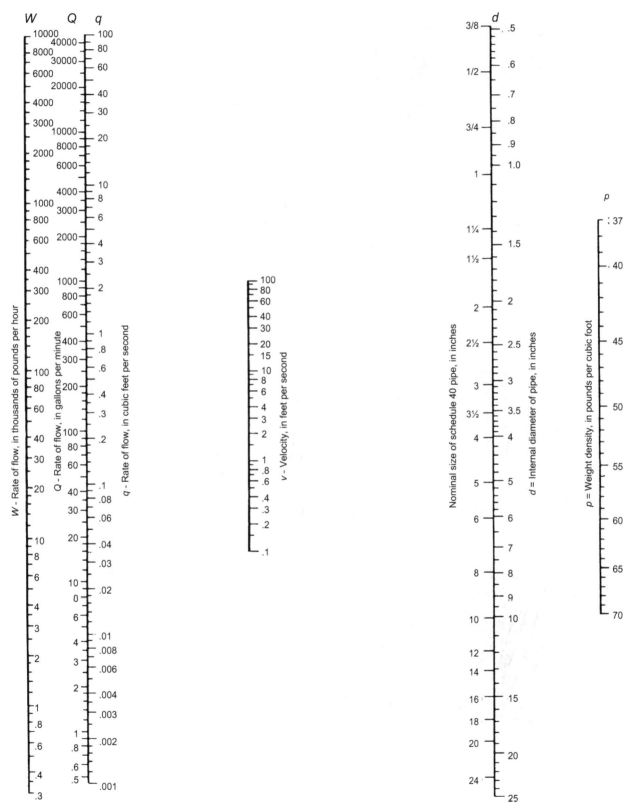

Figure 1. Velocity of liquids in pipe.

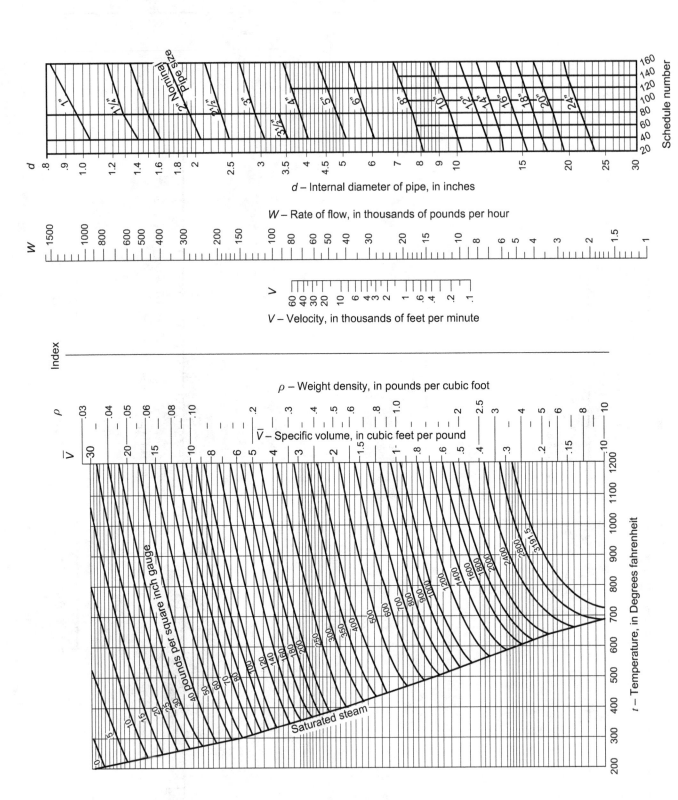

Figure 2. Velocity of compressible fluids in pipes.

Derivation of basic ultrasonic flow equations

The actual measurements performed by ultrasonic flowmeters are of flow velocity, with all other variable effects due to temperature, pressure, density, viscosity, etc., being cancelled via a differential sensing technique.

Multiplication by the cross-sectional pipe area readily enables the output to be conditioned to read in volumetric flow rate scaled to engineering units. Refinement of the technique enables density and other physical properties to be measured as well so that mass flow rates and other parameters pertaining to various process variables can be determined.

Let

D = Pipe diameter,
X = Axial displacement of transducers,
S = Path length between transducers,
θ = Angle of transducers to flow stream,
c = Velocity of sound in the flowing medium,
v = Velocity of flowing medium,
t_1 = Flight time of upstream pulse,
t_2 = Flight time of downstream pulse,

then

$$t_2 = \frac{S}{c + v \cos \theta} \tag{1}$$

$$t_1 = \frac{S}{c - v \cos \theta} \tag{2}$$

To cancel the effects of variations in the velocity of sound c in the medium itself, the circuitry is arranged to compute the reciprocal of each flight time and to subtract one from the other. This difference of the flight time reciprocals leads to the relationship:

$$\frac{1}{t_2} - \frac{1}{t_1} = \frac{2v \cos \theta}{S} \tag{3}$$

Solving for v

$$v = \frac{S}{2 \cos \theta} \left(\frac{1}{t_2} - \frac{1}{t_1} \right) \tag{4}$$

but $\cos \theta = \dfrac{X}{S}$ $\tag{5}$

Therefore, $v = \dfrac{S^2}{2X} \left(\dfrac{1}{t_2} - \dfrac{1}{t_1} \right)$ $\tag{6}$

The circuitry that performs the reciprocal time computation produces an output frequency, since:

$$\frac{1}{t} = f \tag{7}$$

Therefore, $\dfrac{1}{t_2} = f_2$ $\tag{8}$

and $\dfrac{1}{t_1} = f_1$ $\tag{9}$

and $\dfrac{1}{t_2} - \dfrac{1}{t_1} = f_2 - f_1 = \Delta f$ $\tag{10}$

Substituting in Equation 6 we have the basic velocity equation:

$$v = \frac{S^2}{2X} \Delta f \tag{11}$$

or

$$v = k_1 \Delta f \tag{11a}$$

The relationship $k_1 = \dfrac{S^2}{2X}$ $\tag{12}$

is the flow velocity scaling factor and is usually expressed in ft per second or ft per minute. Note that it is a function of the square of the distance between the transducers. This is an important consideration when providing for re-zeroing of the meter calibration each time the transducers are removed and re-inserted.

For volumetric flow, the velocity must be multiplied by the cross-sectional area of the pipe, and the following considerations are involved:

$$A = \pi \left(\frac{D}{2} \right)^2 \tag{13}$$

Volumetric flow

$$V - vA \tag{14}$$

Therefore, volumetric flow

$$V \pi = \frac{S^2 D^2 \Delta f}{8X} \tag{15}$$

The relationship

$$k_2 = \frac{S^2 D^2}{8X} \pi \qquad (16)$$

is the volumetric flow scaling factor and is usually expressed in cubic ft per minute.

Further information on the physical properties of the flowing medium is also acquired in the process of the basic measurement. Specifically, the velocity of sound in the medium itself can be measured independent of the velocity of its motion.

Equation 3 computed the difference of the reciprocals of the upstream and downstream flight times. In the instrument, this is performed in the delta computing channel. Supposing an addition of the reciprocals of the respective upstream and downstream flight times is performed in a separate computing channel, called the sigma channel. The following relationships ensue:

$$\frac{1}{t_2} + \frac{1}{t_1} = \frac{c + v\cos\theta}{S} + \frac{c - v\cos\theta}{S} = \frac{2c}{S} \qquad (17)$$

then $c = \dfrac{S}{2}\sum f \qquad (18)$

Let $k_3 \dfrac{S}{2} \qquad (19)$

be the sound velocity scaling factor.

Then $c = k_3 \sum f \qquad (20)$

The speed of sound c in a medium is a function of its bulk modulus and its density and is expressed by the following equation:

$$c = \sqrt{\frac{B}{\rho}} \qquad (21)$$

where

  B = Bulk modulus of elasticity,
  ρ = Density.

From this, the equation for density may be derived thus

$$c^2 = \frac{B}{\rho} \qquad (22)$$

and $\rho = \dfrac{B}{c^2} \qquad (23)$

Substituting the value of c from Equation 20 into 23 gives

$$\rho = \frac{B}{\left(k_3 \sum f\right)^2} \qquad (24)$$

For a given gas, B is essentially constant.

Therefore, $\dfrac{B}{k_3 2} = k_4 \qquad (24a)$

is a fair representation.

Then $\rho(\text{density}) = k_4 \sum f^{-2} \qquad (25)$

An interesting relationship may be derived at this point, namely, an equation for mass flow. Since

mass flow $= \rho V \qquad (26)$

and since $V = k_1 \Delta f \qquad (11a)$

and $\rho = \dfrac{k_4}{\sum f^2}$

then mass flow $= k_1 k_4 \dfrac{\Delta f}{\sum f^2} \qquad (27)$

or, letting $k_1 k_4 = k_5$

mass flow $= k_5 \dfrac{\Delta f}{\sum f^2} \qquad (28)$

Further, the density of a gas is related by the equation

$$\rho = \frac{P}{T} \qquad (29)$$

where

  P = Pressure,
  T = Temperature (°K).

Therefore, if the pressure is known, the temperature may be extracted from the sigma channel in this way:

From Equation 25, $\dfrac{P}{T} = k_4 \sum f^{-2} \qquad (30)$

and $T = k_4 P^{-1} \sum f^{-2} \qquad (31)$

Let us go back to Equation 24. If the chemical properties of the gas or the ratio of the mixture between two or more gases change, then B is not a constant. If separate transducers are used to measure density, say via the pressure/temperature relationship, then the variation of B can also be measured.

Thus, $\rho = B\left(k_3 \sum f\right)^{-2} \qquad (32)$

and $B = \rho\left(k_3 \sum f\right)^{2} \qquad (33)$

The above relationship holds for liquids or gases. In the case of a gas, we may further say:

$$B = \frac{P}{T}\left(k_3 \sum f\right)^2 \qquad (34)$$

How fast does oil move in a pipeline?

Divide the throughput, in barrels per day, by the square of the nominal diameter, in inches; divide this result by 100 and subtract 1 from the answer. The answer is the speed of the oil column in miles per hour.

Example. A 12-in line has a throughput of 70,000 bpd,

$$70,000 \div 144 = 486,$$

$$486 \div 100 = 4.86,$$

$$4.86 - 1.00 = 3.86.$$

The oil moves about 4 miles per hour.

This rule is rough, but seldom is an exact answer needed for this problem. Greater accuracy can be had by using the true inside diameter of the line in the following formula:

$$V = 0.0081 \times Q/d^2$$

where

$V = $ Speed, mph,
$Q = $ Throughput, bpd,
$d = $ Inside diameter of pipe, in.

Take the above example: If the wall thickness of the pipe is 3/8 in.

$$V = (0.0081)(70,000 \div 144)$$

$$V = 3.94$$

or about 4 miles per hour.

Example. A 20-in pipeline has a throughput of 200,000 barrels per day. How fast does the oil move inside the pipeline?
By the first method,

$$200,000 \div 400 = 500,$$

$$500 \div 100 \div 1 = 4 \text{ miles per hour.}$$

By the second method,

$$V = (.0081)(200,000 \div 19.25^2),$$

$$V = 4.37, \text{ or about 4 miles per hour.}$$

Estimate the volume of a pipeline per linear foot using the inside diameter

To find the volume of a pipeline in gallons per ft, square the inside diameter in inches and multiply by 4%. A more accurate answer may be obtained by adding 2% of this result to the first answer.

Example. Find the volume of a pipeline having an ID of 7.981 in expressed in gallons per ft.

$$\text{Volume} = (0.04) \times (7.981)^2 = 2.548 \text{ gallons per ft.}$$

For greater accuracy, add 2% of 2.548.

$$2.548 + 0.051 = 2.599 \text{ gallon per ft.}$$

Actual volume $= 2.599$ gallons per ft.

To find the volume of the pipeline in barrels per ft, square the inside diameter in inches and divide by 1000. For greater accuracy, subtract 3% of the first answer.

Example. Find the volume of a pipeline with an ID of 7.981 in in barrels per ft.

$$\text{Volume} = \frac{(7.981)^2}{1000} = 0.0637 \text{ barrels per ft.}$$

For greater accuracy, subtract 3% of 0.0637,

$$0.0637 - 0.0019 = 0.0618 \text{ barrels per ft.}$$

Actual volume $= 0.0619$ barrels per ft.

What is the linefill of a given pipe in barrels per mile?

Multiply the inside diameter of the pipe, in inches, by itself, and multiply the result by 5.13; the answer is the linefill in barrels per mile.

Example. Six-in schedule 40 pipe; inside diameter is 6.065 in.

$$6.065 \times 6.065 \times 5.13 = 188.7 \text{ barrels per mile.}$$

If the correct inside diameter is used, the rule gives the correct linefill in standard 42-gallon barrels per mile of pipe.

Estimate leakage amount through small holes in a pipeline

The amount of fluid lost through a small hole in a pipeline can be estimated using the following equation:

$$Q = 0.61A\sqrt{2gh}$$

where

> Q = Flow in cu ft/s,
> A = Cross sectional area, sq ft,
> g = Gravitational constant, ft/s/s,
> h = Head, feet.

Example. Assume the following conditions:

Hole diameter = 0.125 in,

Pressure = 30 psig,

Sp. gr. of fluid = 0.85,

A = 0.0000852 sq ft,

h = 81.5 ft,

g = 32.2,

$Q = 0.61 \times 0.0000852\sqrt{2 \times 32.2 \times 81.5}$,

$Q = 0.002765$ cu ft/s.

Table gives velocity heads for various pipe diameters and different rates of discharge

Rate of Flow in Gallons per Minute	1/8"	1/4"	3/8"	1/2"	3/4"	1"	1¼"	1½"	2"	2½"	3"	4"	5"	6"	Rate of Flow in Gallons per Minute
						(Velocity Heads in Feet)									
1	.488	.148	.043	.017											1
2		.590	.175	.069	.022										2
3			.392	.155	.050	.020									3
4			.698	.276	.090	.035	.012	.006							4
5			1.089	.430	.141	.054	.018	.010							5
10				1.720	.564	.215	.071	.038	.016	.007	.003				10
15					1.265	.488	.160	.087	.036	.015	.007				15
20					2.250	.861	.286	.154	.065	.027	.013				20
25						1.345	.447	.241	.101	.041	.020				25
30						1.933	.643	.346	.146	.060	.029				30
35						2.635	.877	.472	.198	.082	.039				35
40						3.343	1.145	.618	.259	.106	.052	.016			40
45							1.548	.779	.329	.134	.065	.021			45
50							1.787	.963	.406	.166	.080	.026			50
70							3.503	1.888	.795	.326	.157	.050	.020		70
75								2.165	.912	.373	.180	.057	.023		75
100								3.851	1.621	.665	.320	.101	.041	.020	100
120								5.548	2.333	.956	.462	.146	.060	.029	120
125									2.528	1.035	.498	.158	.065	.031	125
150									3.649	1.493	.736	.229	.093	.045	150
175									4.954	2.021	.980	.309	.127	.061	175
200	8"								6.483	2.656	1.281	.406	.166	.080	200
225										3.359	1.618	.514	.209	.101	225
250	.040									4.131	1.992	.633	.259	.125	250
270	.047										2.333	.740	.302	.146	270
275	.048										2.429	.769	.313	.151	275
300	.057										2.883	.912	.373	.180	300
350	.078	10"										1.242	.509	.246	350
400	.102											1.621	.665	.320	400
450	.129	.053										2.053	.839	.406	450
470	.141	.057	12"									2.239	.917	.443	470
475	.144	.059										2.276	.934	.453	475
500	.159	.065	.031									2.535	1.038	.502	500
550	.193	.079	.038										1.256	.606	550
600	.229	.093	.045	14"									1.493	.721	600
650	.269	.109	.053	.026									1.752	.846	650
700	.312	.127	.062	.031									2.035	.980	700
750	.358	.146	.071	.035									2.337	1.123	750
800	.407	.166	.080	.040										1.281	800
850	.462	.187	.090	.045										1.548	850
900	.516	.209	.102	.050	15"									1.618	900
950	.574	.234	.113	.056	.043									1.813	950
1000	.637	.259	.125	.062	.048									1.999	1000
1050	.703	.296	.138	.069	.053									2.202	1050
1100	.771	.313	.152	.075	.058									2.421	1100
1150	.841	.343	.166	.082	.064	16"								2.648	1150
1200	.917	.373	.181	.090	.069									2.880	1200
1250	.995	.406	.196	.097	.075	.062								3.135	1250
1500	1.433	.582	.282	.145	.108	.088									1500
2000	2.547	1.035	.502	.249	.193	.158									2000
2500		1.618	.784	.391	.301	.246	20"	24"							2500
3000		2.329	1.128	.562	.434	.355	.147		30"						3000
3500		3.224	1.540	.766	.590	.502	.199	.096	.051						3500
4000			2.007	1.000	.771	.633	.261	.125	.056						4000
4200			2.209	1.102	.851	.698	.288	.137							4200
4500			2.539	1.265	.975	.799	.330	.158	.065						4500
5000			3.135	1.561	1.204	.985	.407	.195	.079	36"					5000
5500				1.884	1.460	1.196	.493	.227	.096						5500
6000				2.239	1.733	1.421	.586	.281	.115	.056					6000
6500				2.627	2.035	1.669	.690	.329	.134	.065					6500
7000				3.056	2.360	1.936	.797	.384	.156	.076					7000
7200					2.709	2.056	.846	.404	.165	.080					7200
7500					3.047	2.224	.917	.440	.180	.087					7500
8000							1.045	.498	.204	.099					8000
8500							1.177	.564	.230	.112					8500
9000							1.319	.633	.259	.125	42"				9000
9500							1.469	.702	.289	.140					9500
10000							1.630	.779	.319	.154	.083				10000
10500								.858	.352	.169	.092				10500
11000								.943	.387	.186	.101				11000
11500								1.033	.423	.204	.110				11500
12000								1.123	.460	.222	.120				12000
12500								1.218	.500	.240	.130				12500
13000								1.316	.541	.260	.141				13000
13500								1.421	.582	.281	.152				13500

F. G. Switzer "Pump-Fax," Goulds Pumps, Inc.

Viscocities of hydrocarbon liquids

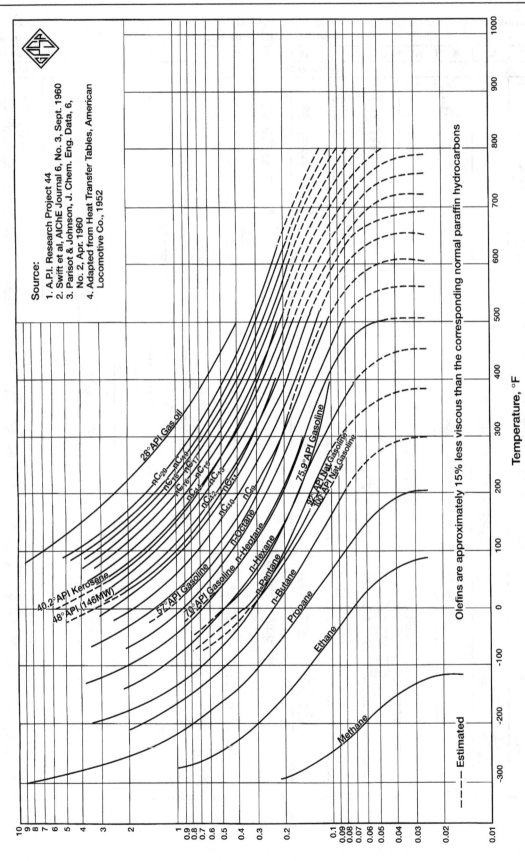

Reprinted with Permission—Gas Prrocessors Suppliers Association, *Tenth Edition Engineering Data Book.*

13: Liquids—Hydraulics

MARINE HOSE DATA

This section presents data on how to calculate the length of hose, and pressure loss, through marine hose.

CALM system

The overall length of the floating hose string is designed as follows:

$$TL = C + RX(1 + E) + B/2 + L + H + M + G \qquad (1)$$

where:

TL = Required overall length
 C = Diameter of buoy
 R = Length of mooring hawser
 E = Max. elongation of mooring hawser
 B = Beam of tanker
 L = Distance between bow and manifold
 H = Maximum (light ballast) freeboard (approx. 0.8 × tanker depth)
 M = Distance between rail and manifold
 G = Height of manifold from deck

SALM system

The required overall length of hose string is designed as follows:

$$\ell_1 < R$$

where:

 ℓ_1 = Distance between hose arm* and floating hose
 R = Length of mooring hawser
TW = $\sqrt{\ell_1 + W^2}$
 W = Maximum water depth at hose arm*

$$TL = HM + RX(1 + E) + TW + (B/2) + L + H \\ + M + G + \ell_1 \qquad (2)$$

where:

 TL = Required overall length of hose string
HM = Max. horizontal movement of buoy
 R = Length of mooring hawser
 E = Max. elongation of mooring hawser
 B = Beam of tanker
 L = Distance between bow and manifold

H = Maximum light (ballast) freeboard (approx. 0.8 × tanker's depth)
M = Distance between rail and manifold
G = of manifold from deck

---

* Hose arm—the horizontal (I_1) or the vertical (W) distance between the underwater hose to swivel connection and the commencement of the main floating line.

Reprinted with permission Bridgestone Engineered Products Co.

Tandem system

The required overall length of hose string is designed as follows:

$$TL = LF + HF + R + (1 + E) + LT + HT + MT + GT$$
(3)

where:

TL = Maximum water depth at hose arm
LF = Distance between manifold and stern of floating storage vessel (FSV)
HF = Max. height of manifold of FSV from sea-surface
R = Length of mooring hawser
E = Max. elongation of mooring hawser
LT = Distance between bow and manifold of off-take-tanker (OTT)
HT = Maximum light (ballast) freeboard of OTT (approx. 0.8 × tanker's depth)
MT = Distance between rail and manifold of OTT
GT = Height of manifold from deck of OTT

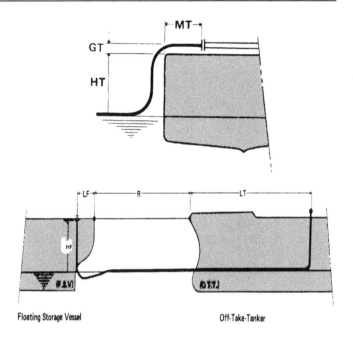

Floating Storage Vessel   Off-Take-Tanker

Multi-point mooring system

The required overall length of hose string is designed as follows:

$$TL = 1.5 \times L + D + H + G + M + \sqrt{(FA/2)^2 + (SW/2)^2}$$
(4)

where:

TL = Required overall length
L = Distance between PLEM and tanker hull
D = Maximum water depth
H = Maximum light (ballast) freeboard (approx. 0.8 × tanker's depth)
M = Distance between rail and manifold
G = Height of manifold from deck
FA = Max. fore and aft excursion of tanker
SW = Max. sideways excursion of tanker

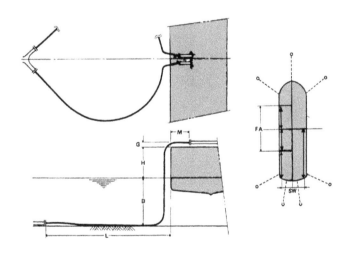

Reprinted with permission Bridgestone Engineered Products Co.

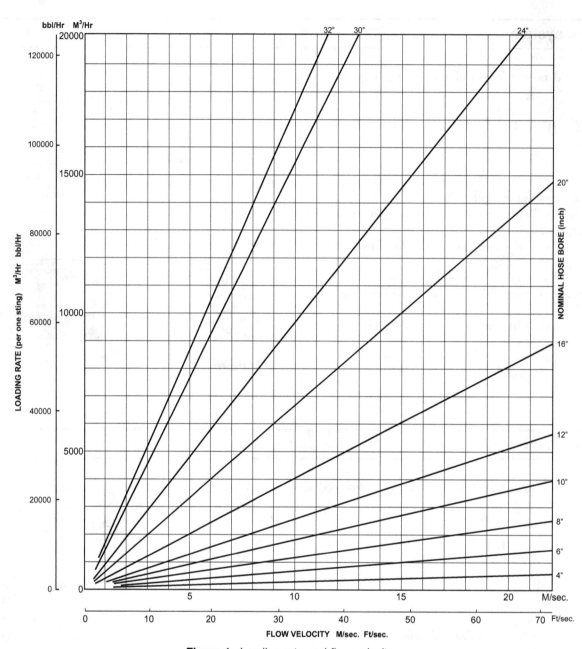

Figure 1. Loading rate and flow velocity.

Pressure loss in hose string

The pressure loss in a hose string is calculated as follows:

$$\Delta P = \frac{\lambda \times L \times V^2 \times \gamma \times 10^{-4}}{2 \times g \times D} \tag{1}$$

where:

ΔP = Pressure loss (Kg/cm^2)
λ = Friction factor
L = Length of hose string (m)
V = Flow velocity (in/sec)
γ = Specific weight of oil (Kg/m^3)
   = $(1000 \times$ specific gravity$)$
g = Acceleration of gravity (in/sec^2)
D = Inside diameter of hose (m)

λ is calculated by using the following Mises' experimental formula:

$$\lambda = 0.0096 + 5.7 \times \sqrt{K/D} + 1.7 \times \sqrt{1/Re} \tag{2}$$

where:

K = Coefficient of wall roughness by Mises' experiment (m)
$K = 0.3 \times 10^{-6}$ for rubber hose
Re = Reynolds number

Re is calculated as follows:

$$Re = \frac{D \times V}{\nu} \tag{3}$$

where:

ν = Kinematic viscosity (m^2/sec)

Input Data:

Viscosity and specific weight of oil depend on temperature, and the values at the operating temperature sometimes need to be converted from values at other given temperatures.

Specific weight can be converted from other units as follows:
Specific weight $(Kg/m^3) = 1000 \times$ specific gravity

$$\text{Specific gravity} = \frac{1415}{1315 + Deg.API} \tag{4}$$

Specific gravity at operating temperature can be calculated as follows:
   S.G. operating = S.G. measured $-$ A $\times$ (Temp. operating $-$ Temp. measured) *Note:* Temp: °C. Coefficient of S.G. change "A" depends on the specific gravity of the oil and is given in Table 1.

Kinematic viscosity is converted from other units as follows:

1 cSt = 1 CP/S.G. (cP is Absolute Viscosity)
1 cSt = 0.245 $\times$ R (Redwood Sec.)
1 cSt = 7.60 $\times$ E (Engler Deg.)
1 cSt = 0.220 $\times$ S (Saybolt Universal Sec.)

Conversion of kinematic viscosity at a different temperature depends on the characteristics of the cargo and can be calculated from data at two different temperatures.

$$\log_{10} \log_{10}(n + 0.6) = \nu - m \times \log T \tag{5}$$

where:

T = oK
ν = cSt

Flow velocity is determined by the internal diameter of the hose and the loading rate as follows:

$$V = \frac{4 \times Q}{3600 \times \Pi \times D^2} \tag{6}$$

where:

V = Flow velocity (m/sec)
Q = Loading rate (m^3/hr)
D = Inside diameter of hose (m)

Reprinted with permission Bridgestone Engineered Products Co.

Table 1
Change of specific gravity of oil

S.G. at meas. temp.	Coeff. "A"	S.G. at meas. temp.	Coeff. "A"
0.640–0.643	0.00095	0.754–0.759	0.00078
0.644–0.648	0.00094	0.760–0.765	0.00077
0.649–0.654	0.00093	0.766–0.771	0.00076
0.655–0.661	0.00092	0.772–0.777	0.00075
0.662–0.667	0.00091	0.778–0.783	0.00074
0.668–0.674	0.00090	0.784–0.790	0.00073
0.675–0.681	0.00089	0.791–0.799	0.00072
0.682–0.688	0.00088	0.800–0.808	0.00071
0.689–0.696	0.00087	0.809–0.818	0.00070
0.697–0.703	0.00086	0.819–0.828	0.00069
0.704–0.711	0.00085	0.829–0.838	0.00068
0.712–0.719	0.00084	0.839–0.852	0.00067
0.720–0.726	0.00083	0.853–0.870	0.00066
0.727–0.734	0.00082	0.871–0.890	0.00065
0.735–0.741	0.00081	0.891–0.970	0.00064
0.742–0.747	0.00080	0.971–1.000	0.00063
0.748–0.753	0.00079		

Example:

Conditions:

Loading rate:	40,000 bbl/hr.
Hose size:	20″ & 16″
Hose length:	100 m for 20″ hose
	50 m for 16″ hose
Operating temp.:	30°C
Specific gravity of oil	34.4° API at 60°F
Kinematic viscosity:	6.90 cSt at 30°C

Calculations:

Loading rate: $Q = 40,000 \times 0.15899 = 6,360 \ \text{m}^3/\text{hr}$

Flow velocity for 20″ hose:

$$V_{20''} = \frac{4 \times 6,360}{3600 \times \Pi \times 0.486^2} = 4.83 \ (\text{m/sec})$$

$$V_{16''} = \frac{4 \times 6,360}{3600 \times \Pi \times 0.380^2} = 5.92 \ (\text{m/sec})$$

Specific weight:

$$S. \, G._{\cdot 60°F} = \frac{141.5}{131.5 + 34.4} = 0.853 \ (\text{at } 15.56°C)$$

$S.G._{30°C} = 0.853 - 0.00066X(30 - 15.56) = 0.843$

Specific weight $= 843 \ (\text{Kg/m}^3)$

Kinematic viscosity:

$\nu = 6.9 \, \text{cSt} = 6.9 \times 10^{-6} \times (\text{m}^2/\text{sec})$

Reynolds number:

$$Re = \frac{D \times V}{\nu}$$

$$Re_{20''} = \frac{0.486 \times 9.52}{6.9 \times 10^{-6}} = 670,539$$

$$Re_{16''} = \frac{0.380 \times 15.6}{6.9 \times 10^{-6}} = 859,130$$

Friction factor:

$$\lambda = 0.0096 + 5.7 \times \sqrt{\frac{K}{D}} + 1.7 \times \sqrt{\frac{1}{Re}}$$

$$\lambda_{20''} = 0.0096 + 5.7 \times \sqrt{\frac{0.3 \times 10^{-6}}{0.486}}$$
$$+ 1.7 \times \sqrt{\frac{1}{670539}} = 0.01615$$

$$\lambda_{16''} = 0.0096 + 5.7 \times \sqrt{\frac{0.3 \times 10^{-6}}{0.380}}$$
$$+ 1.7 \times \sqrt{\frac{1}{859130}} = 0.1650$$

Pressure loss:

$$\Delta P = \frac{\lambda \times L \times V^2 \times \gamma \times 10^{-4}}{2 \times g \times D}$$

$$\Delta P_{20''} = \frac{0.01615 \times 200 \times 9.52^2 \times 843 \times 10^{-4}}{2 \times 9.81 \times 0.486}$$
$$= 2.59 \ (\text{Kg/cm}^2)$$

$$\Delta P_{16''} = \frac{0.01650 \times 50 \times 15.6^2 \times 843 \times 10^{-4}}{2 \times 9.81 \times 0.380}$$
$$= 2.27 \ (\text{Kg/cm}^2)$$

Total pressure loss $= 2.59 + 2.27 = 4.86 \ (\text{Kg/cm}^2)$

Reprinted with permission Bridgestone Engineered Products Co.

Pressure drop calculations for rubber hose

Fluid flow

The general flow formula is as follows:

$$\Delta P = \frac{0.0134 \times f \times S \times L \times Q^2}{d^5}$$

where: ΔP = Pressure drop, psi
  f = Friction factor
  S = Specific gravity
  L = Length of line, ft
  Q = Flow, gpm
  d = Inside diameter of line, in.

$$N_R = \frac{3{,}160 \times Q}{V_C \times d}$$

where: N_R = Reynolds number
  Q = Flow, gpm
  V_C = Fluid viscosity, centistokes
  d = Inside diameter of line, in.

Friction factor, f.

For laminar flow, $f = \dfrac{64}{N_R}$.

Source

Gates Rubber Company.

Examples of pressure drop calculations for rubber hose

Flow rate = Q = 1,000 gpm
Hose ID = d = 6.0 in.
Length = L = 100 ft
Fluid = Gasoline Specific gravity = S = 0.71

Viscosity $= V_C = 0.476$ centipoise/0.71
  $= 0.6704$ centistokes

Reynolds number $= N_R = [3{,}160 \times Q]/[V_C \times d]$
  $= [3{,}160 \times 1{,}000] [0.6704 \times 6]$
  $= 785{,}600$

From Figure 1, f = 0.012 for $N_R = 785{,}600$

P = $[0.0134 \times f \times S \times L \times Q^2]/d^5$
P = $[0.0134 \times 0.012 \times 0.71 \times 100 \times 1{,}000{,}000]/7{,}776$
P = 1.468 psi per 100 ft

Figure 1. Friction factors for rubber hose. Reproduced with permission of Gates Rubber Company.

As a rule of thumb, loss for each set of couplings equals 5% of the loss per 100 ft of hose. For two sets of couplings,

Pressure loss $= 2 \times 0.05 \times 1.468 = 0.15$ psig

Typical formulas used for calculating pressure drop and flow rates for pipelines

Shell/MIT

$P = (0.241\ fsQ^2)/d^5$

For viscous flow,

$f = 0.00207 \times (1/r)$

For turbulent flow,

$f = 0.0018 + 0.00662 \times (1/r)^{0.355}$
$r = 0.0119Q/dv$

Viscous flow,

$r = 0.1$ to 0.135

Turbulent flow.
R > 0.4
Indeterminate region, r = 0.135 to 0.4

Benjamin Miller

$$B = (0.1692)\left(d^5 P/s\right)^{0.5} \left(\log(d^3 sP/z^2) + 4.35\right)$$

T. R. Aude

$$B = 0.871 \times K\left(\frac{1}{z^{0.104}}\right)\left(\frac{\left[(P^{0.552} \times d^{2.656})\right]}{s^{0.448}}\right)$$

$$P = \left[\frac{Bz^{0.104} s^{0.448}}{0.871\, Kd^{2.656}}\right]^{1.812}$$

Hazen-Williams

$$Q = \frac{0.148 \times d^{2.63} \times d^{0.54}}{s^{0.54}}$$

Where:

Q = Barrels per day
B = Barrels per hour
P = Pressure loss, psi per mile
d = Pipe inside diameter, inches
z = Absolute viscosity, centipoises
s = Specific gravity
K = Efficiency factor (Aude's formula)
f = Friction factor
r = Reynolds number modified Re/7,742
C = Hazen-Williams coefficient (use 120 for new steel pipe; 10 for ordinary 15-year-old steel pipe; and 150 for fiberglass pipe)
SUS = Saybolt Universal Seconds
v = Kinetic viscosity, centistokes

Examples

Shell/MIT. Find the pressure loss for a crude oil pipeline where the flow rate is 30,000 barrels per day with the following conditions:

Pipe: 10.750″ × 0.307″ w.t.
Specific gravity = 0.85
Viscosity = 80 SUS

Convert viscosity (80 SUS) to centistokes

$V_{cs} = 0.226\, V_{SUS} - 135/V_{SUS}$
$V = 0.226(80) - 135/80$
$V = 16.39\ cS$
$r = 0.0119Q/dv$
$r = (0.0119 \times 30{,}000)/(10.136 \times 16.39)$
$r = 2.148$

$f = 0.0018 + 0.00662 \times (1/r)^{0.355}$
$f = 0.0018 + 0.00662 \times (1/2.148)^{0.355}$
$f = 0.006846$

$P = (0.241 fsQ^2)/d^5$
$P = (0.241 \times p.006846 \times 0.85 \times 30000^2)/10.136^5$
$P = 11.798$ psi/mile

T. R. Aude. Find the pressure loss, psi/mile, when pumping gasoline with the following conditions

B = 2,000 barrels per hour
Pipe = 8.625″ × 0.322″ w.t.
z = 0.59 centipoise
s = 0.733
K = 0.94 (pipe roughness/efficiency factor)

$P = ((Bz^{0.104}\, s^{0.448})/(0.871\, Kd^{2.656}))^{1.812}$
$P = ((2000 \times 0.59^{0.104} \times 0.733^{0.448})/$
$\quad (0.871 \times 0.94 \times 7.981^{2.656}))^{1.812}$
$P = 44.1$ psi/mile

Note: Aude's formula is based on data taken from 6″ and 8″ lines and no extension data are available.

Benjamin Miller. Find the pressure loss, psi/mile, when pumping gasoline with the following conditions:

B = 4,167 barrels per hour
Pipe = 16″ × 0.3125″ w.t.
z = 0.59 centipoise
s = 0.733
d = 15.375″

$B = (0.169176)(d^5 P/s)^{0.5}(\log(d^3 sP/z^2) + 4.35)$

Let P = 6.4 psi/mile for first try:

$4167 = (0.1691676)(15.375^5 6.4/0.733)^{0.5}$
$\quad (\log(15.375^3 \times 0.733 \times 6.4)/(0.59^2) + 4.35)$
$4167 = 0.1691676 \times 2738 \times (4.69 + 4.35)$
$4167 \neq 4188$

Let P = 6.34 psi/mile for second try:

$4{,}167 = 0.169167 \times 2{,}726 \times (4.68 + 4.35)$
$4{,}167 = 4{,}167$

Therefore, $P = 6.34$ psi/mile.

Hazen-Williams. Typically used for water system hydraulic calculations.) Find the pressure drop, psi/mile, in a water system where the flow rate is 20,000 barrels per day.

The pipe is 4.5-in. $\times$ 0.226-in. wt new steel. Since the flowing medium is water, $s = 1.0$.

$20{,}000 = (0.148 \times 4{,}048^{2.63} \times 120 \times P^{0.54})$

$$P^{0.54} = 20{,}000/702.23$$
$$= 28.48$$
$$P = 493 \text{ psi/mile}$$

All of the preceding examples have assumed no difference in elevation. Where a difference in elevation exists, this must be considered in determining the total pressure required for a given flow rate.

Hydraulic gradients

Some of the pressure drop formulas used to calculate pressure drop and flow rates for pipelines are shown earlier in this section. In order to use these formulas, it is first necessary to determine the characteristics of the fluid to be transported—primarily the specific gravity, viscosity, and either the flow rate or allowable pressure drop.

Before the system design is completed, it is necessary to know the elevation of the terrain to be traversed by the pipeline. Once the elevation of the system is known and a pipe size has been selected, the pressure loss data may be used to calculate or plot a hydraulic gradient. A hydraulic gradient is the height that a column of fluid will rise to if small tubes were installed along the pipeline as shown in Figure 1. Feet of head may be converted to pressure by dividing by 2.31 and multiplying by the specific gravity. Conversely, pressure is converted to feet of head by multiplying pressure by 2.31 and dividing by the specific gravity of the fluid.

The slope of the hydraulic gradient will represent the amount of pressure drop per unit of linear distance. The pressure drop per unit of length is uniform as long as there is no change in pipe size or specific gravity, and therefore the

hydraulic gradient will be a straight line with a slope proportional to the system flow rate.

Since the pipeline elevation profile is usually plotted in feet, it is helpful to plot the hydraulic gradient in feet rather than pressure. The hydraulic gradient may be plotted by converting pressure at the origin and end points to feet using the specific gravity of the fluid in the pipeline. The end point in this case will be the end of the pipeline if there are no intermediate booster stations. If the system is equipped with intermediate booster stations, then the suction of the booster station in the first segment will be the end point for that segment of the line. Don't forget to consider the suction pressure requirement for the booster station. These two points are then connected with a straight line, which will represent the pressure (in feet) at any intermediate point along the pipeline. The pressure at any point may be determined by taking the difference between the gradient and pipeline elevation and converting feet to psi.

Figure 2 shows a typical elevation profile plot with two hydraulic gradients—one for no booster station and one for an intermediate booster station and no change in elevation. In Figure 3 there is an elevation change, and for a hydraulically balanced system, it was necessary to locate the booster station at a point other than the geographic midpoint of the line. The location of the booster station in this example corresponds to the hydraulic mid-point of the system. A hydraulically balanced system is one in which all stations develop the same differential head. In Figure 2 where there was no elevation change, the booster station location is at the geographic mid-point of the system, which also corresponds to the hydraulic mid-point of the system. This is not meant to imply that a system must absolutely be hydraulically balanced. A booster station may be located off the hydraulic center; however, its differential head requirement will differ from that of the origin station and system throughput capacity will be affected. The amount of difference in head required, for a booster station which is located off hydraulic center, will

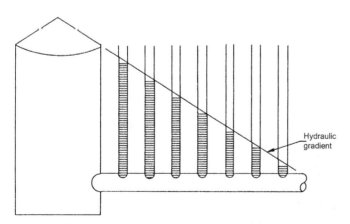

Figure 1. Schematic for hydraulic gradient.

Hydraulic gradient

Figure 2. Profile with no elevation change.

Figure 3. Profile with elevation change.

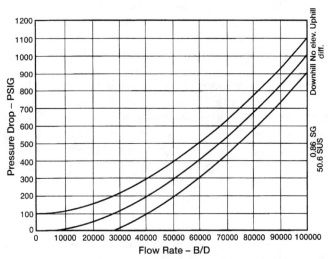

Figure 4. System curve. 30 mi., 12.750″ × 0.250″ pipe.

Figure 5. Pipeline elevation and test gradients and operating profile.

depend upon the amount of system elevation change and back pressure maintained at the end of the line.

The total system friction loss may be divided equally among an optimum number of booster stations. The discharge pressure at each station may be calculated with the following equation:

$$Pdp = (Pt + S + (N - 1) \times Ps + Pr)/N$$

where:

Pdp = Station discharge pressure
Pt = Total system friction loss
S = System static pressure
N = Number of booster stations
Ps = Station suction pressure
Pr = Residual or back pressure at end of line

In Figure 4, three system curves are shown to illustrate how elevation will affect the capacity of a system. With a constant input pressure of 750 psig, the capacity will be 77,500 barrels per day pumping uphill (272 ft); 84,000 barrels per day with no elevation change; and 90,000 barrels per day pumping downhill (272 ft). Note that where there is a difference in elevation, the amount of displacement of the curves, 100 psi in this case, is uniform.

The hydraulic gradient may also be used to analyze a system to assure that it is operated within the pressure limits determined by the hydrostatic test and the pipe design parameters. Figure 5 shows the test and maximum allowable operating gradients for a system. The hydrostatic test point was located at MP 150, and the test pressure was 1,800 psig, which in this case corresponds to 90% of the specified minimum yield strength (SMYS) of the pipe. Water was used as the test medium, and therefore 1,800 psig equates to 4,157 ft of head. The test gradient is drawn as a straight line across the profile from 8,157 ft (4,157 ft of head + 4,000 ft of pipeline elevation). This system may be operated at 80% of the test pressure, since this represents 72% of the SMYS of the pipe. See Section 3 for details on pipe design and limits on maximum allowable operating pressure.

The 80% gradient (still based on water) is drawn as a straight line across the profile from 7,325 ft. The maximum allowable operating profile is now developed by taking the difference between the 80% of test pressure gradient, in feet, and the pipeline elevation and converting this difference to feet of head for the fluid to be transported. For this example, a specific gravity of 0.5 was used. At MP 150, the difference between pipeline elevation and the 80% of test gradient is 3,325 ft of head (water), and for a specific gravity of 0.5, this will convert to 6,650 ft of head. Therefore, the maximum operating pressure in ft of head for a fluid with a specific gravity of 0.5 will be 10,650ft (6,650ft of head + 4,000 ft pipeline elevation) at MP 150.

The maximum allowable operating profile will not be a straight line but will be a displaced mirror image of the pipeline elevation profile. The remainder of the maximum allowable operating profile may be developed by following this procedure at other points along the pipeline. When fluids with different specific gravities are transported, an operating profile will need to be developed for each specific gravity. If the line is hydrostatically tested in several segments rather than in one continuous length, an operating profile will need to be developed for each of the test segments.

Once the maximum allowable operating profile has been established, the hydraulic gradient may be superimposed. If the hydraulic gradient crosses the maximum allowable operating profile, the pipe will be overpressured in the segments where the gradient crosses the profile.

Figure 6 illustrates the effect hills have on the operation of a system. In this case, a hill is located near the end of the line and, due to its elevation, it will control the system capacity, since adequate back pressure must be maintained (if the vapor pressure of the fluid is above atmospheric pressure) at the end of the line to assure the gradient will "clear" the hill. If no back pressure is maintained at the end of the line, the gradient will intersect the side of the hill and, theoretically, a siphon action will occur if the elevation difference is not too great. In reality, the system flow rate will change of its own

Figure 6. Pipeline elevation profile and hydraulic gradients.

accord to cause the gradient to intersect the top of the hill. The flow from that point to the end of the line will cascade or run away from the flow on the other side of the hill. If the vapor pressure of the fluid is very low, this condition can probably be tolerated. If segregated batches of material are handled or if the fluid vapor pressure is above atmospheric pressure, adequate back pressure must be maintained to keep the batches from mixing or forming vapor. Inadequate back pressure will cause difficulties in getting the flow in versus flow out to balance on a short-term basis as would be required for a leak detection system.

Two conditions are illustrated in Figure 6 to show the effect the flow rate will have on the amount of back pressure required to clear the hill. Some safety factor should be included when determining the amount of pressure to maintain in the system—especially when handling high vapor pressure materials. The flowing temperature will also need to be considered in making this determination.

For this example, the fluid has a specific gravity of 0.5 and its vapor pressure requires a minimum of 200 psig in the system. Therefore, it will be necessary to maintain 200 psig (924 ft) at the top of the hill located at MP 140. All gradients must be drawn through this point. If the specific gravity changes, the control point will move up or down depending upon the direction of the specific gravity change. The gradient connecting MP 0 to the end of the line passes through the control point and intersects the end of the line at 6,300 ft or 498 psig. The second condition, which assumes a booster station at MP 78, illustrates what happens to the back pressure requirement when the flow rate changes. The gradient from the discharge of the booster station must also pass through the control point, and since the flow rate is greater, the slope of the gradient is steeper and intersects the end of the line at 5,800 ft or 390 psig. It follows that as the system flow rate changes, it will be necessary to adjust the amount of back pressure. If the back pressure is permanently set to the higher value, system capacity will be penalized. See Section 16: Instrumentation, for a discussion of control systems.

Equivalent lengths

Some pipeline systems are not made up of one single uniform pipe size. As the pressure in a system decreases, it may be desirable to use pipe with a thinner wall thickness to reduce the investment cost, or where deliveries are made along the pipeline, it is sometimes practical to use a smaller diameter pipe for the lower flow rate. Conversely, where injections are made into a system, it may be necessary to use a larger diameter pipe for the increased flow rate.

The capacity of a system may be increased by installing loop lines. Loop lines are two or more pipelines that are operated in parallel. Loop lines are not practical where segregated batches are pumped, since there will be a problem in getting the batches to arrive simultaneously. Also, if there are injections or deliveries along such a system, it will be almost impossible to balance the amount injected or delivered between the two lines. The most practical method for operating loops where segregated batches are handled is to operate each line separately. A product system might, for example, use one line for distillates and the other line for gasoline. If the gasoline volume is too large for one line, the excess could be batched through the other system. One other caution on loops is that the lengths of the legs may not be exactly the same.

Hydraulic systems are analogous to electrical circuits as shown in Figure 1, and this sometimes helps in analyzing a pipeline system. In a series electrical circuit, the same current flows through all of the circuit elements, and the voltage drops across the elements are additive. This same logic may be applied to a pipeline system with segments of different pipe sizes connected in a series. The pressure drops through the various segments are additive, and the sum of the pressure drops will equal the total system pressure drop.

A loop system is analogous to a parallel electrical circuit as shown in Figure 1. Since a common pressure is applied to all the legs of the loop system, the pressure drop across each leg of a loop system must be the same and the flow will divide in proportion to the friction loss in each leg.

An examination of the Shell/MIT flow formula reveals that the friction factor is dependent upon flow rate, pipe ID, and viscosity. Since the flow through each leg of a loop system will change as the total system flow changes, this will have an impact on the calculated equivalent length factor. For best accuracy, the equivalent length factor should be determined as near as possible to the expected operating parameters.

Figure 1. Series and loop system analogy.

$$V = IR1 + IR2$$

Series System

$$V = (I1 + I2)Rt$$

Loop System

Note: Hydraulic equivalents shown in ( )
P = pressure, Q = flow, f = friction
V = voltage, I = current, R = resistance

Series systems

Figure 1 shows a system curve for a pipeline system 30 miles in length and consisting of 15 miles of 12.750-in. OD × 0.250-in. pipe in series with 15 miles of 10.750-in. OD × 0.250-in. pipe. System friction curves were developed for each pipe size and for the combination of the two sizes. Inspection shows that the curve for the combination is the sum of the 12.750-in. OD curve and the 10.750-in. OD curve. Pressure drop is additive on lines of constant flow for different pipe sizes connected in series.

The step of developing the individual curves for 12.750-in. OD and 10.750-in. OD pipe may be eliminated by calculating the equivalent length of 12.750-in. OD or 10.750-in. OD pipe. A system friction loss curve may then be calculated using the equivalent length of one pipe size or the other. For this example, the fluid is crude oil with a specific gravity of 0.86 and a viscosity of 50.6 SUS.

At a flow rate of 40,000 barrels per day, the pressure drop through 12.750-in. OD × 0.250-in. pipe is 6.618 psi/mile, and the pressure drop through 10.750-in. OD × 0.250-in. pipe is 15.469 psi/mile. The ratio of these two pressure drops is 6.618/15.469 or 0.428. This ratio times the length of the 12.750-in. OD pipe equals the equivalent amount of 10.750-in. OD pipe that will be required to yield the same pressure drop as 15 miles of 12.750-in. OD pipe. Therefore, the equivalent length of 10.750-in. OD pipe is 15 + (0.428 × 15) or 21.41 equivalent miles of 10.750-in. OD pipe. If it is

desired to use 12.750-in. OD pipe as the base size, the same procedure may be followed except that the pressure ratio will be reversed and the reciprocal of 0.428 or 2.3364 multiplied by 15 miles of 10.750-in. OD pipe to obtain 35.046 equivalent miles of 12.750-in. OD pipe. The total equivalent length will be 15 miles + 35.046 miles, or 50.046 equivalent miles of 12.750-in. OD pipe.

This may be checked as follows:

At a flow rate of 40,000 b/d, the pressure drop through 15 miles of 10.750-in. OD is

$15.469 \times 15 = 232$ psig

The pressure drop through 15 miles of 12.750-in. OD is

$6.618 \times 15 = 99$ psig

The total pressure drop will be

$232 + 99 = 331$ psig

The pressure drop through 21.42 equivalent miles of 10.750-in. OD is

$21.42 \times 15.469 = 331$ psig

The pressure drop through 50.046 equivalent miles of 12.750-in. OD is

$50.046 \times 6.618 = 331$ psig

Figure 1. System curve. 15 mi., 12.750″ × 0.250″ pipe in series with 15 mi., 10.750″ × 0.250″ pipe.

Looped systems

The ratio of pressure drops may also be used for calculating the equivalent lengths of pipeline loops; however, the procedure is slightly different from that used for series systems.

Figure 1 shows the friction curves for a looped system, which is shown schematically in Figure 2. For reference, individual curves are shown for each pipe size and the combination of the two lines in parallel. The total friction loss for the system may be determined graphically using a system of curves such as shown in Figure 1, or the ratio of pressure drops may be used to determine an equivalent length for the looped system.

To use the ratio of pressure drops to convert the loop to a single line of equivalent size, determine the pressure drop through the larger size pipe (if this is to be the base size) using a chosen flow rate. In this example, a flow rate of 60,000 b/d of crude oil with a specific gravity of 0.86 and a viscosity of 50.6 SUS is used. The pressure drop through the 12.750-in. OD pipe at 60,000 b/d is 13.54 psi/mile. Next, find a flow rate through the 10.750-in. OD pipe that will yield the same pressure drop of 13.54 psi/mile. The required flow rate through the 10.750-in. OD pipe is 37,100 b/d for a drop of 13.54 psi/mile. The sum of these two rates is 97,100 b/d. Use this rate, 97,100 b/d, and determine the pressure drop/mile through the 12.750-in. OD pipe. The pressure drop/mile is 31.91 psi/mile. The factor for converting to an equivalent length of 12.750-in. OD × 0.250-in. pipe is 13.54/31.91 or 0.4243. The equivalent length of the loop combination will be 15 miles × 0.4243 or 6.3645 equivalent miles of 12.750-in. OD × 0.250-in. pipe.

Qc = Qa + Qb = 80,000 B/D

Figure 2. Looped pipelines.

If it is desired to convert the loop combination to a base diameter of 10.750-in. OD, select a flow rate and determine the pressure drop in the 10.750-in. OD leg. At 30,000 b/d, the pressure drop in the 10.750-in. OD leg will be 9.32 psi/mile. Determine the flow rate through the 12.750-in. OD leg to produce a pressure drop of 9.32 psi/mile. The rate will be 48,560 b/d. The sum of these two rates is 78,560 b/d. This rate through the 10.750-in. OD leg will produce a pressure drop of 51.29 psi/mile. The equivalent length factor for 10.750-in. OD pipe is 9.32/51.29 = 0.1817. The equivalent length will be 0.1817 × 15 or 2.7257 equivalent miles of 10.750-in. OD pipe.

This may be checked as follows:

The pressure drop through the loop combination at a flow rate of 80,000 b/d is

9.62 psi/mile × 15 miles = 144 pisg

The flow through the 10.750-in. OD leg will be 30,550 b/d and 49,450 b/d through the 12.750-in. OD leg.

The pressure drop for 6.3645 equivalent miles of 12.750-in. OD pipe at a flow rate of 80,000 b/d is

6.3645 equivalent miles × 22.584 psi/mile = 144 pisg

The pressure drop for 2.7257 equivalent miles of 10.750-in. OD pipe at a flow rate of 80,000 b/d is

2.7257 equivalent miles × 52.982 psi/mile = 144 pisg

The ratio-of-pressures method will work in conjunction with any of the flow formulas listed in this section.

Figure 1. Friction curves. 10.750″ × 0.250″ and 12.750″ × 0.250″ and looped combination.

Calculate pressure loss in annular sections

$$P = 0.000027 \, flpv^2/M$$

where:

f = Friction factor
l = Length of section, ft
p = Density of fluid, #/ft^3
v = Average velocity, fps
M = Mean hydraulic radius, ft
P = Pressure loss of length, psi
u = Absolute viscosity, ft/lb-sec. (0.000672* centipose)
R$_n$ = Reynolds number
Q = Volume, cu. ft per second
e/d = Relative roughness
v = $Q/0.7854d_A^2$
M = (area/wetted perimeter) = $(d_1 - d_2)/4$

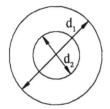

Example:

Throughput 60,000 barrels per day = 6.44 fps
Specific gravity 0.68
Viscosity 0.50 centipoise
$d_1 = 12.615''$
$d_2 = 7.000''$
$l = 2,000$ ft
$d_A = (d_1^2 - d_2^2)^{0.5} = (12.615^2 - 7.0^2)^{0.5}$
$M = (d_1 - d_2)/4 = (1.051 - 0.583)/4 = 0.117$ ft.
v = 6.44 fps
$p = 62.4 \times 0.68 = 42.432 \#/ft^3$
$u = 0.000672 \times 0.50 = 0.000336$ ft.lb.-sec.
Rn = $(4 \times 0.1178 \times 6.44 \times 42.432) \div 0.0000336 = 383,200$
e/d = 0.00017
f = 0.015
P = $(0.000027 \times 0.015 \times 2000 \times 42.432 \times 6.44^2) \div 0.1178$
P = 12.1 psi

Calculate pressure and temperature loss for viscous crudes ≥ 1,000 cP

When calculating pressure and temperature losses for highly viscous Newtonian crude oils, accurate flowing temperature data are required. Usually, crudes of this nature are flowing in the laminar region, and a change of a few degrees in temperature can change the viscosity by as much as 50% or more and thus cause a significant change in the friction factor.

Figure 1 illustates a typical pressure and temperature profile for an uninsulated pipeline.

The data typically required for determining a pressure/temperature profile for a viscous crude oil pipeline is shown in Table 1.

Once the data have been tabulated, a computer program such as BATCH[1] can be used to calculate the temperature/pressure profile of the pipeline. The program performs a rigorous energy balance, i.e., static head, friction, and kinetic mechanical energy, plus friction, compression, and stored and transferred heat are included. Thermal resistances of the inside film, pipe, any insulation, and the outside environment (soil, air, or water) are calculated.

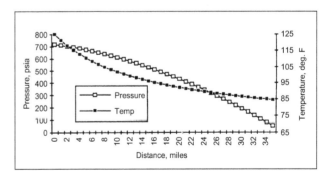

Figure 1. Uninsulated pipeline.

$$P_2 = P_1 - s\Delta x \left[\frac{C_Q Q^2 f}{d^5} + \frac{C_H(e_2 - e_1)}{\Delta x} \right]$$

where:

P_1 = Pressure at the upstream end of the calculation segment

Table 1
Data required for pressure/temperature loss calculation

Length	Heat transfer coefficient (fluid)	Inlet temperature
Size, OD	Insulation thickness	Inlet pressure
Wall thickness	Thermal conductivity of insul.	Throughput
Pipe wall roughness	Thermal conductivity of soil	Sp. gr. at two temp.
Thermal conductivity of pipe	Pipe burial depth	Viscosity at two temp.
Upstream elevation	Soil temperature	Specific heat of liquid at two temp.
Downstream elevation	Vapor pressure at two temp.	Thermal conductivity of liquid at two temp.
Number of calculation segments	Soil K factor	

P_2 = Pressure at the downstream end of the calculation segment

Δx = Length of the calculation segment

f = Darcy–Weisbach friction factor

d = Inside diameter of the pipe

s = Liquid-specific gravity

e_1 = Elevation at the upstream end of the calculation segment

e_2 = Elevation at the downstream end of the calculation segment

C_Q = Flow conversion constant

C_H = Head conversion constant

An energy balance is performed on each calculation segment and its surroundings, resulting in the next equation:

$$t_2 = [wc_{p1}t_1 - UA(LMTD) + w(h_1 - h_2)/J] \div (wC_{p2})$$

where:

t_1 = Flowing temp. at the upstream end of the calculation segment

t_2 = Flowing temp. at the downstream end of the calculation segment

h_1 = Head at the upstream end of the calculation segment

h_2 = Head at the downstream end of the calculation segment

C_{p1} = Heat capacity at constant pressure evaluated at t_1

C_{p2} = Heat capacity at constant pressure evaluated at t_2

w = Total mass flow rate

U = Overall heat transfer coefficient

A = Surface area of the pipe in calculation segment

LMTD = Log mean temp. difference between the average flowing temp. and the average ambient temp.

J = Mechanical equivalent of heat

Example:

Pipeline length	35 miles
Size	14.0000″
Wall thickness	0.250″
Pipe wall roughness	0.0018″
Thermal conductivity (pipe)	29 Btu/hr-ft-F

Upstream elevation	1,500 feet
Downstream elevation	1,650 feet
Length	35 miles
Number of calculation segments	35
Heat transfer coefficient (calculated by program Btu/hr-ft2-F)	
Insulation thickness	0″ (Case 1)
Insulation thermal conductivity	0 Btu/hr-ft-F
Burial depth	36″
Soil temperature	50°F
Soil K factor	1.2
Inlet temperature	125°F
Inlet pressure	calculate
Outlet pressure	50 psia
Throughput	8,000 bpd

Liquid properties:

	Value	Temp.
Specific gravity	0.95	77
	0.90	210
Viscosity, cP	3260	80
	90	180
Specific heat	0.5	60
Btu/Lbm-F	0.5	80
Thermal conductivity	0.2	60
Btu/hr-ft-F	0.22	100
Vapor pressure	0.4	100
psia	0.3	70

The pressure/temperature profile for an uninsulated line, based on the above conditions, is shown in Figure 1 and in Table 3.

If the pipeline crosses areas where the parameters such as soil thermal conductivity, pipe size, etc. change, BATCH will allow the system to be segmented into additional pipes (up to 99 pipes) with each pipe being configured with its own input data.

Assume that 2″ of urethane foam insulation is added to the pipeline previously described. The new pressure/temperature profile is shown in Figure 2 and in Table 4.

Typical soil K factors are shown in Table 2. BATCH may also be used to calculate pump requirements and heater duty.

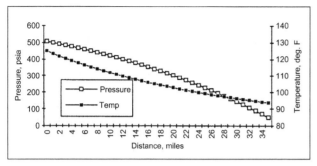

Figure 2. Insulated pipeline—2″ polyurethane.

Table 2
Soil K factors Btu/hr/ft²/°F

Soil	Condition	Cover, In.	K
Sandy	Dry	24	0.25 to 0.40
Sandy	Moist	24	0.50 to 0.60
Sandy	Soaked	24	1.10 to 1.30
Clay	Dry	24	0.20 to 0.30
Clay	Moist	24	0.40 to 0.50
Clay	Moist to wet	24	0.60 to 0.90
Sandy	Dry	8	0.60 to 0.70
Sandy	Moist to wet	8	1.20 to 2.40
River	Water	60	2.00 to 2.50
Air	Exposed	—	2.0 (approx.)

Table 3
Uninsulated pipe

Distance miles	Pressure psia	Temp. deg. F
0	718	125
1	710.59	121.3
2	702.57	118.1
3	693.79	115.3
4	684.25	112.8
5	673.93	110.5
6	662.83	108.4
7	650.95	106.5
8	638.3	104.8
9	624.89	103.3
10	610.73	101.9
11	595.82	100.6
12	580.18	99.3
13	563.81	98.2
14	546.74	97.2
15	528.97	96.2
16	510.51	95.2
17	491.38	94.4
18	471.6	93.6
19	451.18	92.8
20	430.13	92.1
21	408.46	91.4
22	386.21	90.7
23	363.37	90.1
24	339.97	89.5
25	316.02	88.9
26	291.55	88.4
27	266.55	87.9
28	241.06	87.4
29	215.09	86.9
30	188.65	86.5
31	161.76	86
32	134.43	85.6
33	106.68	85.2
34	78.54	84.9
35	50	84.5

Table 4
Insulated pipe

Distance miles	Pressure psia	Temp. deg. F
0	506.76	125
1	499.69	123.4
2	492.32	121.9
3	484.64	120.5
4	476.65	119.1
5	468.33	117.8
6	459.68	116.5
7	450.7	115.3
8	441.38	114.1
9	431.71	112.9
10	421.68	111.8
11	411.3	110.7
12	400.56	109.7
13	389.45	108.7
14	377.98	107.8
15	366.13	106.8
16	353.9	105.9
17	341.3	105.1
18	328.32	104.3
19	314.96	103.5
20	301.22	102.7
21	287.1	101.9
22	272.59	101.2
23	257.71	100.5
24	242.44	99.8
25	226.8	99.2
26	210.77	98.5
27	194.38	97.9
28	177.6	97.3
29	160.46	96.7
30	142.95	96.2
31	125.08	95.6
32	106.84	95.1
33	88.25	94.6
34	69.3	94.1
35	50	93.6

Reprinted with permission: Fluid Flow Consultants, Grove, Oklahoma

Determine batch injection rate as per enclosure

In the operation of a pipeline system, it may be necessary to inject a specific volume into a passing block of liquid. When it is desirable to maintain the same flow rate downstream of the injection and inject at the minimum rate or into most of the passing block, the following equations can be used to determine the upstream injection rates.

$$Q_U = \frac{B + H_O Q_D + Q - \left[(B + H_O Q_D + Q)^2 - 4 \times H_O \times B \times Q\right]^{0.5}}{2 \times H_O}$$

$$Q_I = Q_D - Q_U$$

Example:

$Q = 15,000$ bbls for injection

$Q_I = $ Injection rate, bph

$B = 73,500$ bbls

$Q_D = 15,600$ bph (to be held constant)

$H_O = 0.5$ hours

$$QU = \frac{73,500 + 0.5 \times 15,600 + 15,000 - \left[(73,5000 + 0.5 \times 15,600 - 15,000)^2 - 4 \times 0.5 \times 73,500 \times 15,600\right]^{0.5}}{2 \times 0.5}$$

$Q_U = 12,751$ bph

$Q_I = 15,600 - 12,751 = 2,849$ bph

where:

$Q_D = $ Downstream rate that is to be held constant, bph
$Q_I = $ Injection rate to be determined, bph
$Q_U = $ Upstream pumping rate to be determined, bph
$Q = $ Volume to be injected, barrels
$B = $ Volume of passing block to be injected into, barrels
$H_O = $ Block outage in hours. This is the time that the head and tail end of the block interface is passing the injection location, during which time injection is not allowed.

PRESSURE LOSS THROUGH VALVES AND FITTINGS

"K" FACTOR TABLE—SHEET 1 of 4

Representative Resistance Coefficients (K) for Valves and Fittings

("K" is based on use of schedule pipe as listed on page 2-10)

PIPE FRICTION DATA FOR CLEAN COMMERCIAL STEEL PIPE WITH FLOW IN ZONE OF COMPLETE TURBULENCE

Nominal Size	½"	¾"	1"	1¼"	1½"	2"	2½, 3"	4"	5"	6"	8-10"	12-16"	18-24"
Friction Factor (f_T)	.027	.025	.023	.022	.021	.019	.018	.017	.016	.015	.014	.013	.012

FORMULAS FOR CALCULATING "K" FACTORS* FOR VALVES AND FITTINGS WITH REDUCED PORT
(Ref: Pages 2-11 and 3-4)

- **Formula 1**

$$K_2 = \frac{0.8 \left(\sin \frac{\theta}{2}\right)(1 - \beta^2)}{\beta^4} = \frac{K_1}{\beta^4}$$

- **Formula 2**

$$K_2 = \frac{0.5 (1 - \beta^2) \sqrt{\sin \frac{\theta}{2}}}{\beta^4} = \frac{K_1}{\beta^4}$$

- **Formula 3**

$$K_2 = \frac{2.6 \left(\sin \frac{\theta}{2}\right)(1 - \beta^2)^2}{\beta^4} = \frac{K_1}{\beta^4}$$

- **Formula 4**

$$K_2 = \frac{(1 - \beta^2)^2}{\beta^4} = \frac{K_1}{\beta^4}$$

- **Formula 5**

$$K_2 = \frac{K_1}{\beta^4} + \text{Formula 1} + \text{Formula 3}$$

$$K_2 = \frac{K_1 + \sin \frac{\theta}{2}[0.8 (1 - \beta^2) + 2.6 (1 - \beta^2)^2]}{\beta^4}$$

- **Formula 6**

$$K_2 = \frac{K_1}{\beta^4} + \text{Formula 2} + \text{Formula 4}$$

$$K_2 = \frac{K_1 + 0.5 \sqrt{\sin \frac{\theta}{2}} (1 - \beta^2) + (1 - \beta^2)^2}{\beta^4}$$

- **Formula 7**

$$K_2 = \frac{K_1}{\beta^4} + \beta (\text{Formula 2} + \text{Formula 4}) \text{ when } \theta = 180°$$

$$K_2 = \frac{K_1 + \beta \left[0.5 (1 - \beta^2) + (1 - \beta^2)^2\right]}{\beta^4}$$

$$\beta = \frac{d_1}{d_2}$$

$$\beta^2 = \left(\frac{d_1}{d_2}\right)^2 = \frac{a_1}{a_2}$$

Subscript 1 defines dimensions and coefficients with reference to the smaller diameter.
Subscript 2 refers to the larger diameter.

*Use "K" furnished by valve or fitting supplier when available.

SUDDEN AND GRADUAL CONTRACTION	SUDDEN AND GRADUAL ENLARGEMENT
If: $\theta \lessgtr 45°$ .... K_2 = Formula 1	If: $\theta \lessgtr 45°$ .... K_2 = Formula 3
$45° < \theta \lessgtr 180°$...$K_2$ = Formula 2	$45° < \theta \lessgtr 180°$...$K_2$ = Formula 4

Reprinted by permission and courtesy of Crane Co. from Technical Report 410

"K" FACTOR TABLE—SHEET 2 of 4

Representative Resistance Coefficients (K) for Valves and Fittings

(for formulas and friction data, see page A-26)

("K" is based on use of schedule pipe as listed on page 2-10)

GATE VALVES
Wedge Disc, Double Disc, or Plug Type

If: $\beta = 1$, $\theta = 0$ . . . . $K_1 = 8\,f_T$

$\beta < 1$ and $\theta \leqslant 45°$ . . . . $K_2 =$ Formula 5

$\beta < 1$ and $45° < \theta \leqslant 180°$ . . . $K_2 =$ Formula 6

SWING CHECK VALVES

$K = 100\,f_T$ $K = 50\,f_T$

Minimum pipe velocity   Minimum pipe velocity
(fps) for full disc lift   (fps) for full disc lift

$= 35\,\sqrt{\overline{V}}$ $= 60\,\sqrt{\overline{V}}$ except

U/L listed $= 100\,\sqrt{\overline{V}}$

GLOBE AND ANGLE VALVES

If: $\beta = 1$ . . . $K_1 = 340\,f_T$

If: $\beta = 1$ . . . $K_1 = 55\,f_T$

If: $\beta = 1$ . . . $K_1 = 150\,f_T$ If: $\beta = 1$ . . . $K_1 = 55\,f_T$

All globe and angle valves,
whether reduced seat or throttled,
If: $\beta < 1$ . . . $K_2 =$ Formula 7

LIFT CHECK VALVES

If: $\beta = 1$ . . . $K_1 = 600\,f_T$

$\beta < 1$ . . . $K_2 =$ Formula 7

Minimum pipe velocity (fps) for full disc lift

$= 40\,\beta^2\,\sqrt{\overline{V}}$

If: $\beta = 1$ . . . $K_1 = 55\,f_T$

$\beta < 1$ . . . $K_2 =$ Formula 7

Minimum pipe velocity (fps) for full disc lift

$= 140\,\beta^2\,\sqrt{\overline{V}}$

TILTING DISC CHECK VALVES

	$\alpha = 5°$	$\alpha = 15°$
Sizes 2 to 8" . . . K =	$40\,f_T$	$120\,f_T$
Sizes 10 to 14" . . . K =	$30\,f_T$	$90\,f_T$
Sizes 16 to 48" . . . K =	$20\,f_T$	$60\,f_T$
Minimum pipe velocity (fps) for full disc lift =	$80\,\sqrt{\overline{V}}$	$30\,\sqrt{\overline{V}}$

Reprinted by permission and courtesy of Crane Co. from Technical Report 410

Representative Resistance Coefficients (K) for Valves and Fittings

(for formulas and friction data, see page A-26)

("K" is based on use of scheduled pipe as listed on page 2-10)

STOP-CHECK VALVES (Globe and Angle Types)	FOOT VALVES WITH STRAINER

STOP-CHECK VALVES
(Globe and Angle Types)

If:
$\beta = 1 \ldots K_1 = 400\,f_T$
$\beta < 1 \ldots K_2 = $ Formula 7

Minimum pipe velocity
for full disc lift
$= 55\,\beta^2\,\sqrt{V}$

If:
$\beta = 1 \ldots K_1 = 200\,f_T$
$\beta < 1 \ldots K_2 = $ Formula 7

Minimum pipe velocity
for full disc lift
$= 75\,\beta^2\,\sqrt{V}$

If:
$\beta = 1 \ldots K_1 = 300\,f_T$
$\beta < 1 \ldots K_2 = $ Formula 7

If:
$\beta = 1 \ldots K_1 = 350\,f_T$
$\beta < 1 \ldots K_2 = $ Formula 7

Minimum pipe velocity (fps) for full disc lift
$= 60\,\beta^2\,\sqrt{V}$

If:
$\beta = 1 \ldots K_1 = 55\,f_T$
$\beta < 1 \ldots K_2 = $ Formula 7

If:
$\beta = 1 \ldots K_1 = 55\,f_T$
$\beta < 1 \ldots K_2 = $ Formula 7

Minimum pipe velocity (fps) for full disc lift
$= 140\,\beta^2\,\sqrt{V}$

FOOT VALVES WITH STRAINER

Poppet Disc   **Hinged Disc**

$K = 420\,f_T$ $K = 75\,f_T$

Minimum pipe velocity
(fps) for full disc lift
$= 15\,\sqrt{V}$

Minimum pipe velocity
(fps) for full disc lift
$= 35\,\sqrt{V}$

BALL VALVES

If: $\beta = 1, \theta = 0 \ldots\ldots\ldots\ldots K_1 = 3\,f_T$
$\beta < 1$ and $\theta \lesssim 45° \ldots\ldots K_2 = $ Formula 5
$\beta < 1$ and $45° < \theta \lesssim 180° \ldots K_2 = $ Formula 6

BUTTERFLY VALVES

Sizes 2 to 8" $\ldots K = 45\,f_T$
Sizes 10 to 14" $\ldots K = 35\,f_T$
Sizes 16 to 24" $\ldots K = 25\,f_T$

"K" FACTOR TABLE— SHEET 4 of 4

Representative Resistance Coefficients (K) for Valves and Fittings

(for formulas and friction data, see page A-26)

("K" is based on use of schedule pipe as listed on page 2-10)

PLUG VALVES AND COCKS

Straight-Way

3-Way

View X—X

If: $\beta = 1$,
$K_1 = 18 f_T$

If: $\beta = 1$,
$K_1 = 30 f_T$

If: $\beta = 1$,
$K_1 = 90 f_T$

If: $\beta < 1 \ldots K_2 = $ Formula 6

MITRE BENDS

$\propto$	K
0°	$2 f_T$
15°	$4 f_T$
30°	$8 f_T$
45°	$15 f_T$
60°	$25 f_T$
75°	$40 f_T$
90°	$60 f_T$

90° PIPE BENDS AND FLANGED OR BUTT-WELDING 90° ELBOWS

r/d	K	r/d	K
1	$20 f_T$	8	$24 f_T$
1.5	$14 f_T$	10	$30 f_T$
2	$12 f_T$	12	$34 f_T$
3	$12 f_T$	14	$38 f_T$
4	$14 f_T$	16	$42 f_T$
6	$17 f_T$	20	$50 f_T$

The resistance coefficient, K_B, for pipe bends other than 90° may be determined as follows:

$$K_B = (n - 1) \left(0.25 \ \pi \ f_T \frac{r}{d} + 0.5 \ K \right) + K$$

n = number of 90° bends
K = resistance coefficient for one 90° bend (per table)

CLOSE PATTERN RETURN BENDS

$$K = 50 f_T$$

STANDARD ELBOWS

90°

45°

$K = 30 f_T$ $K = 16 f_T$

STANDARD TEES

Flow thru run . . . . $K = 20 f_T$
Flow thru branch . . . . $K = 60 f_T$

PIPE ENTRANCE

Inward Projecting

$K = 0.78$

r/d	K
0.00*	0.5
0.02	0.28
0.04	0.24
0.06	0.15
0.10	0.09
0.15 & up	0.04

*Sharp-edged

Flush

For K, see table

PIPE EXIT

Projecting   **Sharp-Edged**   **Rounded**

$K = 1.0$ $K = 1.0$ $K = 1.0$

Reprinted by permission and courtesy of Crane Co. from Technical Report 410

Equivalent Lengths *L* and *L/D* and Resistance Coefficient *K*

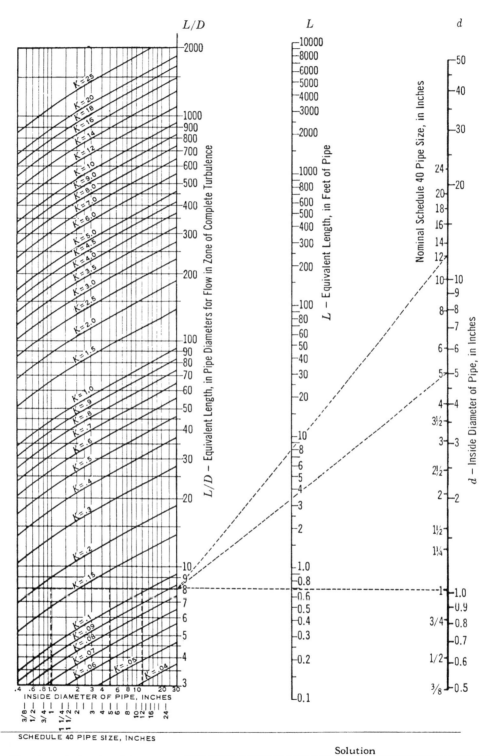

Problem: Find the equivalent length in pipe diameters and feet of Schedule 40 clean commercial steel pipe, and the resistance factor *K* for 1, 5, and 12-inch fully-opened gate valves with flow in zone of complete turbulence.

Solution

Valve Size	1″	5″	12″	Refer to
Equivalent length, pipe diameters	8	8	8	Page A-27
Equivalent length, feet of Sched. 40 pipe	0.7	3.4	7.9	Dotted lines
Resist. factor *K*, based on Sched. 40 pipe	0.18	0.13	0.10	on chart.

Reprinted by permission and courtesy of Crane Co. from Technical Report 410.

Equivalents of Resistance Coefficient K
And Flow Coefficient C_v

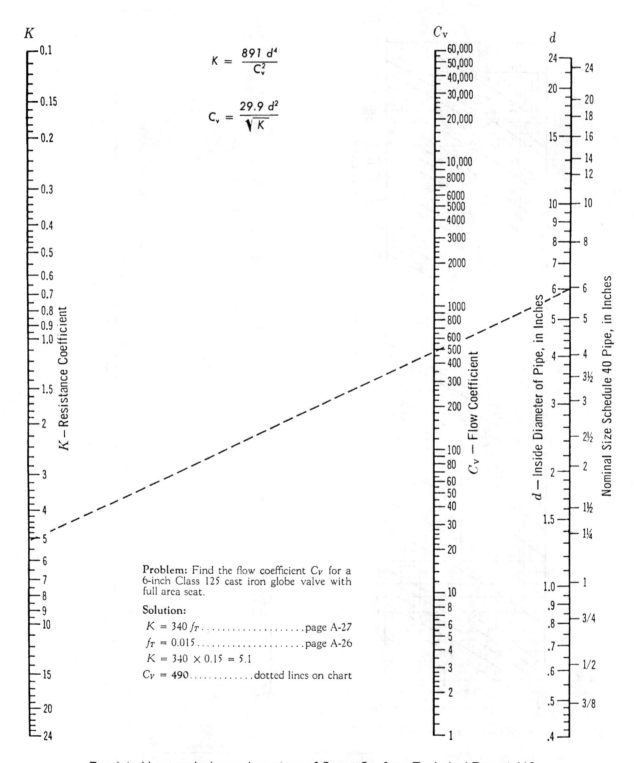

$$K = \frac{891\ d^4}{C_v^2}$$

$$C_v = \frac{29.9\ d^2}{\sqrt{K}}$$

K — Resistance Coefficient

C_v — Flow Coefficient

d — Inside Diameter of Pipe, in Inches

Nominal Size Schedule 40 Pipe, in Inches

Problem: Find the flow coefficient C_v for a 6-inch Class 125 cast iron globe valve with full area seat.

Solution:

$K = 340\ f_T$ . . . . page A-27

$f_T = 0.015$ . . . . page A-26

$K = 340 \times 0.15 = 5.1$

$C_v = 490$ . . . . dotted lines on chart

Reprinted by permission and courtesy of Crane Co. from Technical Report 410.

Nomograph for calculating Reynolds number for flow of liquids and friction factor for clean steel and wrought iron pipe

The nomograph (Figure 1) permits calculation of the Reynolds number for liquids and the corresponding friction factor for clean steel and wrought iron pipe.

The Reynolds number is defined as:

$$Re = 22,700\frac{q\rho}{d\mu} = 50.6\frac{Q\rho}{d\mu} = 6.31\frac{W}{d\mu} = \frac{dv\rho}{\mu a}$$

where: Re = Reynolds number
d = Inside pipe diameter, in.
q = Volumetric flow rate, cu. ft/sec

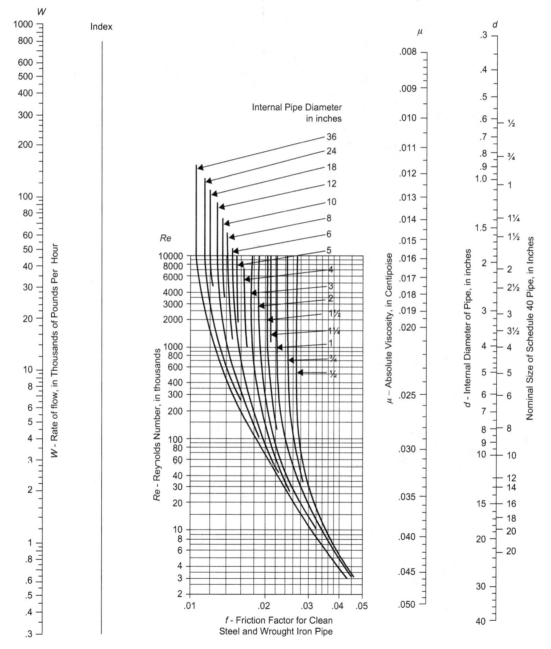

Figure 1. Reynolds number for liquid flow friction factor for clean steel and wrought iron pipe. (Reproduced by permission, *Tech. Paper, 410,* Crane Co., copyright 1957).

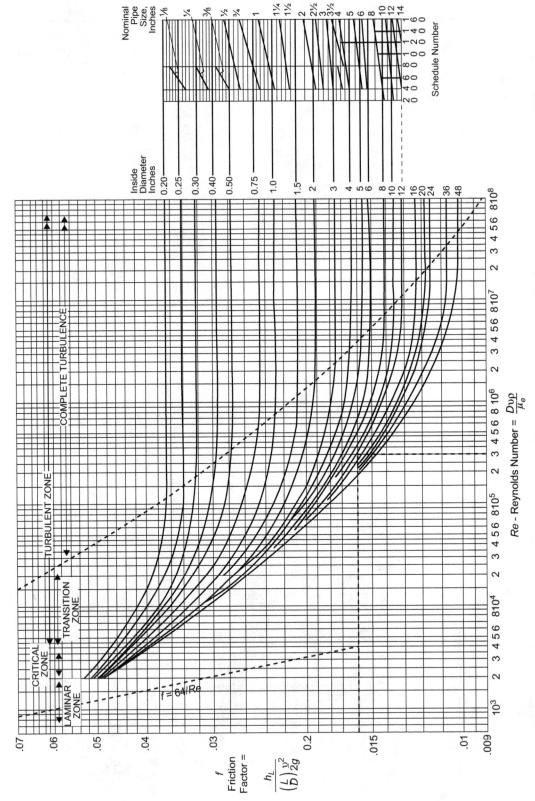

Figure 2. Friction factors for clean commercial steel and wrought iron pipe. (Reproduced by permission, *Tech. Paper, 410*, Crane Co., 1957.)

Q = Volumetric flow rate, gal/min
ρ = Fluid density, lb/cubic ft
μ = Fluid viscosity, centipoise
μa = Fluid viscosity, lb/(ft)(sec)
v = Average fluid velocity, ft/sec
w = Mass rate of flow, lb/hr

Example. Given:

water at 200°F
    d = 4-in. schedule 40 steel pipe
    Q = 415 gal/min
    ρ = 60.13 lb/cubic ft
    μ = 0.3 centipoises

Obtain the Reynolds number and the friction factor.

Connect	With	Mark or Read
Q = 415	ρ = 60.13	W = 200,000
W = 200,000	d = 4-in. schedule 40	Index
Index mark	μ = 0.3	Re = 1,000,000

At Re = 1,000,000 read f = 0.017 from the graph on the center of the nomograph.

Figure 2 is an enlarged version of the friction factor chart, based on Moody's data[1] for commercial steel and wrought iron. This chart can be used instead of the smaller version shown in Figure 1.

Reference

1. Moody, L. F., *Transactions, American Society of Mechanical Engineers*, 66, 671 (1944).

Nomograph for calculating pressure drop of liquids in lines for turbulent flow

The nomograph (Figure 1) is based on a modified version of the Fanning equation. It gives the pressure drop per 100 ft of pipe for liquids flowing through a circular pipe. The basic equation is:

$$P_{100} = \frac{0.0216\ f\ \rho Q^2}{d^2} = 4{,}350\frac{f\rho q^2}{d^2} = 0.000336\frac{fW^2}{d^2\rho}$$

$$= 0.1294\frac{fv^2\rho}{d}$$

where:

    P_{100} = Pressure loss per 100 ft of pipe, psi
    f = Friction factor (4 × f' in Fanning equation)
    ρ = Fluid density, lb/cu. ft
    d = Tube internal diameter, in.
    W = Flow rate, lb/hr
    Q = Flow rate, gal/min
    q = Flow rate, cu. ft/sec

The Moody[1] friction factor and relative roughness generalized charts (Figures 2 and 3) are also included for use with the nomogram.

Example. The estimated Reynolds number is 100,000 when water is flowing at a rate of 100 gal/min and a temperature of 70°F Find the pressure drop per 100 ft of 2-in. diameter pipe, the relative roughness from Figure 3 is 0.0009.

At e/d = 0.0009 and Re = 100,000, the friction factor, f, is 0.022 from Figure 2.

Using Figure 1:

Connect	With	Mark or Read
f = 0.022	ρ = 62.3 lb/cu. ft	Intersection at Index 1
Index 1	Q = 100 gal/min	Intersection at Index 2
Index 2	d = 2-in.	P = 8 psi per 100 ft of pipe

Source

Flow of Fluids Through Valves, Fittings and Pipe, Technical Paper No. 410, 3-11, Crane Company, Chicago (1957).
Anon., *British Chemical Engineering*, 2, 3, 152 (1957).

Reference

1. Moody, L. F., *Transactions, American Society of Mechanical Engineers*, 66, 671 (1944).

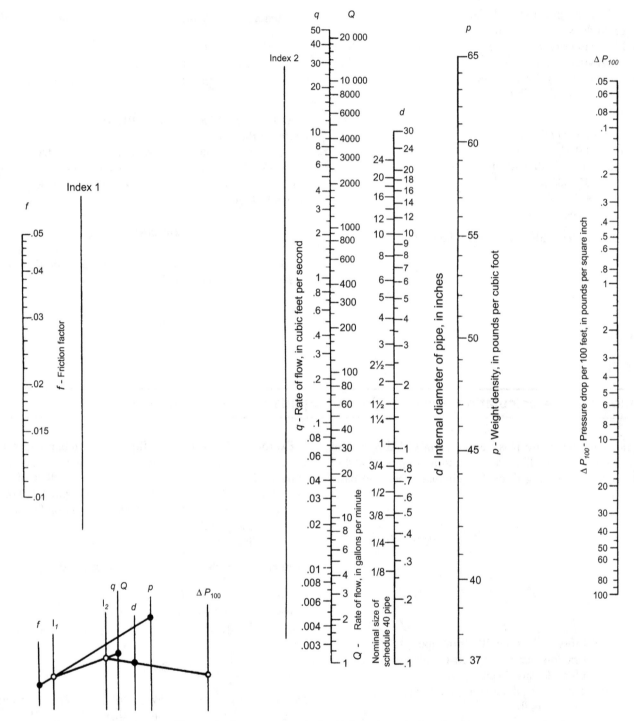

Figure 1. Pressure drop in liquid lines for turbulent flow.

Figure 2. Friction factors for any type of commercial pipe.

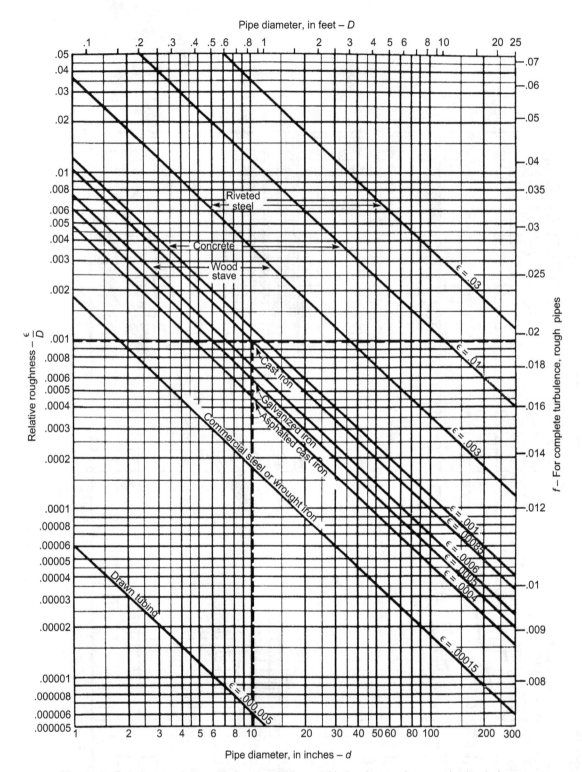

Figure 3. Relative roughness of pipe materials and friction factors for complete turbulence.

Drag-reducing agents

Drag-reducing agents (DRAs) can substantially reduce friction loss in most pipelines when flowing most hydrocarbon liquids.

When a DRA is dissolved in a solvent, it produces a solution that in laminar flow has the same pressure drop, and therefore the same Newtonian viscosity, as the solvent. But when the solution is in turbulent flow, a DRA produces a pressure drop smaller than that which would occur with untreated solvent moving at the same flow rate. The turbulent flow friction of the solution is less than that of the solvent.

DRAs are highly viscoelastic and thixotropic. They may have a viscosity at low shear rates in excess of 3,000,000 cP, with a viscosity 100 to 1,000 times less at higher rates. The viscosity is somewhat affected by temperature and increases with decreasing temperature. DRA-solvent solutions are viscoelastic, time-independent, shear degradable, non-Newtonian fluids. A DRA-solvent solution handles as an ordinary hydrocarbon except in turbulent flow, when the reduced friction becomes evident.

DRAs in present use in oil and products pipelines are themselves hydrocarbons and thus should have no effect on refining processes or refined products. Millions of barrels of crude treated with DRAs on the Alyeska system have been refined without incident. Drag-reduction installations need not involve large-diameter pipelines. The 200,000 b/d increase in the Alyeska system resulting from the injection of some thousands of gallons per day of DRA in this 48-in. line is dramatic.

However, the offshore operator who obtains a 20% increase in capacity in his 8-in. flowline by injecting a few dozen gallons of DRA per day is happy: Without the DRA, he would have to shut in production or loop the flowline. Either alternative would be orders of magnitude more expensive than the injection of a DRA.

As production decays over time, the amount of DRA injected will be reduced, with the eventual result that production will match the untreated flowline capability and operations can proceed as before but with no shut in wells in past history and no sterilized investment in a loop to be carried on the books in future years.

Early applications

The drag-reduction effect as applied to solutions was accidentally discovered by B. A. Toms in the United Kingdom in 1947.[1] It had been known for years that concentrated water—coal slurries and wood pulp—water slurries could produce a friction factor lower than that of water alone. But Dr. Toms's 1948 publication was the first indication of drag reduction in otherwise Newtonian solutions.

The first use of DRAs in the oil industry was to reduce pressure loss during pumping of fluids downhole to fracture tight formations. The quantities of DRA used are proportionally large, but the savings are substantial. Data published on one DRA-kerosene solution with 600 weight ppm of DRA in kerosene showed an 80% drag reduction over untreated kerosene.

The ongoing use of DRAs to debottleneck the Alyeska 48-in. crude line has been described in several published articles.[2] There are many other uses on commercial pipelines—one considerably larger than the Alyeska example—and the list of operators using DRA is growing. Use of DRAs in multistation pipelines is a special application, because the drag-reduction effect does not survive shear stresses raised by centrifugal pumps and appears to be badly degraded going through most positive displacement pumps.

The general rule is that DRAs are always injected on the discharge side of a pump station. If drag reduction is required along a pipeline with N pump stations, the DRA must be injected at N points.

The effectiveness of a DRA depends to a certain extent on the viscosity of the untreated oil and decreases as viscosity increases. It varies slightly with diameter and decreases as diameter increases. The effect varies greatly with flow velocity, which, for a given oil and line diameter, means it varies greatly with Reynolds number. Drag reduction increases as Reynolds number increases. However, dilute solutions are believed to disintegrate at shear rates found in commercial pipelines.

There are some as yet little understood practical aspects of DRA application. For instance, there seems to be some relationship between the physical characteristics of the oil and of the DRA. A certain DRA will produce good results in a given crude yet produce poor results in a seemingly similar oil.

Untreated non-Newtonian oils may produce non-Newtonian DRA—oil solutions, and a particular DRA may increase, rather than decrease, flowing friction. Sometimes an additional, specially compounded additive is required to obtain the desired drag reduction. In addition, there appears to be something in certain DRA—solvent combinations that has to do with the solubility of the DRA in the oil. Test loop studies verify that for these combinations drag reduction appears to increase asymptotically as the distance from the point of injection increases. Specially designed DRA injection nozzles help in these cases.

DRA activity

The precise mechanism describing how a DRA works to reduce friction is not established. It is believed these agents work by directly reducing turbulence or by absorbing and later returning to the flowing stream energy that otherwise would have been wasted in producing the crossflows that comprise turbulence. The absorb-and-return theory has a certain intuitive attractiveness, because one of the characteristics of a viscoelastic fluid is its ability to do this on a physically large scale. Whether the effect exists at the molecular level is unknown.

Pipe liners generally work with the rational flow formula. The friction factor therein, which is a scaling factor applied to velocity head to yield friction loss, is a macro factor in that it lumps all of the friction-producing effects acting in a flowing system into a single numerical value. The friction factor concept does not provide an insight into how a DRA might work. It is necessary to look at the velocity distribution pattern in a pipeline in turbulent flow to understand the bases that support the concept of the f-factor and understand better how drag reduction might take place.

The concept of turbulent flow in pipelines provides that the point flow may be directed in any direction while maintaining net pipe flow in the direction of decreasing pressure. All flow not in the direction of net pipe flow absorbs energy; the more turbulent the flow becomes, the more energy absorbed in these crossflows. In flow in hydraulically rough pipes, the f-factor does not vary with increasing flow, but the unit friction loss increases as the square of the velocity of net pipe flow.

The model of turbulent pipe flow we will use is based on the Universal Law of the Wall, which applies to both smooth and rough pipe flow. It presumes two general modes of flow in a pipe: the wall layer and the turbulent core.

The wall layer is presumed to contain all of the flow where the point velocity varies with the distance of the point from the pipe wall. The turbulent core carries all of the flow where variations in point velocity are random and independent of this distance. The wall layer itself is presumed to be composed of three sublayers: the laminar sublayer, the buffer zone, and the turbulent sublayer.

The laminar sublayer has as its outer boundary the wall of the pipe, and flow in the entire sublayer is laminar. This definition requires that immediately at the wall lies a film of liquid that is not moving.

The increase in point velocity as a point moves away from the wall is a linear function of the distance from the wall and directed parallel to the wall in the direction of pipe flow. There are no crossflows in this sublayer. The contribution of the laminar sublayer to total friction arises in viscous shear generated in the sublayer.

Pipe roughness acts to inhibit laminar flow; therefore, the definition is only statistically true for rough or partially rough pipe.

In the turbulent sublayer, which has as its inner boundary the turbulent core, it is presumed that the velocity varies logarithmically with distance from the wall, but that crossflows in the sublayer may be statistically large and nearly as great as in the core. The contribution of the turbulent sublayer to total friction lies in the turbulent shear generated in this sublayer.

In the buffer zone, which lies between the laminar and turbulent sublayers, variation of point velocity with point position is not established. Ideally, it should be the same as the laminar sublayer at the outer boundary and the same as the turbulent sublayer at the inner boundary. Any number of mathematical functions can be written to approximate these two requirements. Both viscosity and density contribute to shear losses in the buffer zone.

The contribution of the turbulent core to total friction loss lies entirely in turbulent shear; DRAs are presumed not to act on this component of total friction. Therefore, the action of a DRA must take place in the laminar sublayer, the buffer zone, or the turbulent sublayer.

That the action of a DRA increases with increasing flowing velocity would indicate that DRA action could not be primarily in the laminar sublayer. It sometimes would appear that this is so: Liquids treated with a DRA flowing under conditions in which the pipe is hydraulically rough and viscosity presumed to make no contribution to total friction can yield excellent drag reduction. This kind of result would lead to the presumption that a DRA works in the buffer zone.

On the other hand, large doses of a DRA have been observed to keep a liquid in laminar flow at Reynolds numbers much higher than those expected of the same liquid without DRA, which leads to the conclusion that a DRA acts in the laminar sublayer.

This conclusion suggests that a DRA may inhibit the action of crossflows, whether generated in the laminar sublayer or the buffer zone, which move inward into the turbulent sublayer and core and absorb energy that otherwise would go toward producing pipe flow. It appears this is the case.

Reduction in the energy absorbed by crossflows originating at the pipe wall can be accomplished by reducing their number, size, or velocity. The energy in the crossflows may also be absorbed by a DRA and later returned to the flowing stream for reuse, which, in effect, means reducing their effect if not their size, number, or velocity. It is here that theory breaks down; no one really knows which effect predominates, or if either does.

The following is known:

- That DRAs act to reduce turbulence.
- That to do so, they seem to act in the laminar sublayer or the buffer zone by effectively enlarging them or their effect.
- That the net effect is to reduce total friction.
- That the f-factor is less than would be calculated from the pipe liner's usual sources.

Experimental data taken on liquids flowing without a DRA do not translate directly to liquids that have been dosed with a DRA. The rational flow formula still applies, but the usual f-factor relationship does not.[3]

At a given flow rate, the percentage of drag reduction is defined as follows:

$$\%DR = \frac{\Delta Po - \Delta Pi}{Po} \times 100$$

where:
%DR = Percentage drag reduction
Po = Pressure drop of untreated fluid
Pi = Pressure drop of fluid containing drag reducer.

Percentage throughput increase can be estimated as follows:

$$= \left[\left(\frac{1}{1 - \frac{\%DR}{100}} \right)^{0.55} - 1 \right] \times 100$$

where %DR is the same as previously defined.

Given a constant Q (flow rate), elevation 1 = elevation 2, and baseline pressures P_1 and P_2, $DPf = P_2 - P_1$ (without DRA).

The volumetric capacity and/or horsepower of the existing pumps may limit the amount of throughput increase that can be achieved. The installation of additional pump capacity may be required in order to obtain the full benefits of DRA use.

Example

Pipe OD = 10.750 in.
Pipe w.t. = 0.250 in.
Length = 50 miles
Upstream elev. = 500 ft
Downstream elev. = 500 ft

Fluid crude oil
Oil temperature = 85°F
Discharge (source) pressure = 1,440 psig
Residual pressure (sink) = 50 psig
Base throughput (untreated) = 54,000 b/d

Oil properties

Temp.	60°F	85°F	90°F
Visc.	60 SUS	49 SUS	48 SUS
SG	0.8528	0.8423	0.8398
Visc.	8.70 cP	5.99 cP	5.72 cP
Visc.	10.20 cSt	7.11 cSt	6.81 cSt

Desired throughput increase = 20% or 64,800 b/d.

$$20\% = \left[\left(\frac{1}{1 - \frac{\%DR}{100}} \right)^{0.55} - 1 \right] \times 100$$

Solve for %DR
%DR = 28.25%

Using a program such as GENNET M,[4] the calibration factor on the pipe data input screen can be adjusted as follows:

$1 - 0.2825 = 0.7175$. Once this is done, run the program to find the new flow rate of 64,900 b/d.

Contact the DRA vendor to determine the amount of DRA that needs to be added to achieve the desired %DR.

References

1. Toms, B. A., "Some observations on the flow of linear polymer solutions through straight pipe at large Reynolds numbers," *Proceedings of the International Congress on Rheology*, Scheveningen, The Netherlands, vol. 2, pp. 135-141 (1948).
2. Burger, Edward D., Monk, Werner R., and Wahl, Harry A., "Flow increase in the Trans Alaska pipeline using a polymeric drag reducing additive," *Society of Petroleum Engineers*, 9419 (1980).
3. *Oil and Gas Journal*, February 18, 1985.
4. Refer to the multiphase flow calculation example for information on GENNET M (Fluid Flow Consultants, Inc., Broken Arrow, Oklahoma).

How to estimate the rate of liquid discharge from a pipe

A quick and surprisingly accurate method of determining liquid discharge from an open-ended horizontal pipe makes use of the equation:

$$gpm = \frac{2.56 \times d^2}{\sqrt{Y}}$$

where:
gpm = Gallons per minute
d = Diameter of pipe, in.
X = Inches
Y = Inches

Example. From the drawing:

X = 24 in.
Y = 25 in.
d = 2 in.

$$gpm = \frac{(2.56)\ (24)\ 2^2}{\sqrt{25}}$$

$$= \frac{245.76}{5}$$

$$= 49.15$$

Predict subsurface temperature ranges

Ground temperatures are vital to pipeline design for two reasons. First, the heat exchange between the contents of the line and the ground are greatly affected by the differential temperatures of the pipe and the soil. Second, the temperature of the pipe at times of installation will have a considerable effect on stresses, particularly where the subsequent ground temperatures will vary considerably from the installation temperatures.

For example, in the recent construction of a 34-in. line that is to operate at temperatures well over 100°F, care was exercised to limit bends when it was installed at temperatures ranging from 50°F to 70°F and the underground temperature will range from about 50°F to 80°F over a year's time.

Generally speaking, most areas will have somewhat similar curves from 3 ft on down. Of course, with depth, they approach the temperature of underground water, and the summer–winter variation ceases.

To adapt for U.S. work. As a comparison, the following temperature ranges are encountered in winter and summer at depths of 3 ft in various areas in the United States.

California ....51°F to 81°F
Colorado plains ....35°F to 65°F
Texas (Panhandle) ....42°F to 75°F
Texas (Coast) ....60°F to 80°F
Mississippi ....57°F to 80°F

Figure 1. The accompanying curve and table shows the range of underground temperatures you can expect in the prairie regions of Canada. At shallow depths, there is some variation and uncertainty as to temperature because of snow or sleet covering. But from 3 ft down, the temperature stabilizes.

Idaho ....37°F to 56°F
Illinois ....41°F to 56°F
Montana ....33°F to 54°F
Michigan ....38°F to 60°F

Wyoming ....26°F to 65°F

Generally, a curve plotted for any one of the above will closely parallel the curve for Canada.

Sizing pipelines for water flow

This form gives a reasonable degree of accuracy.

$$d^5 = \frac{(gpm)^2(D)}{(10)(H)(100)}$$

where:

    d = Inside diameter of pipe in inches
  gpm = Gallons per minute
    D = Equivalent distance (in ft) from standpipe (or pump) to outlet
    H = Total head lost due to friction in feet

Example. What size line should be run from a water tower to a cooling tower for a minimum of 160 gpm if the equivalent distance (actual distance in ft plus allowances for valves and fittings) is 1,000 ft and the minimum head supplied by the tower will always be at least 40 ft?

Solution.

$$d^5 = \frac{(160)^2(1,000)}{(10)(40)(100)} = 640 \text{ in.}$$

Rather than extract the fifth root of 640, simply raise inside diameters of logical pipe sizes to the fifth power and compare them with 640.

If 3-in. ID pipe were used

$$d^5 = (3)(3)(3)(3)(3) = 243$$

This is obviously too small. Trying 4-in. pipe

$$d^5 = (4)(4)(4)(4)(4) = 1,024$$

How approximate throughput of a line can be estimated from pipe size

Multiply the square of the nominal pipe size by 500.

Example. The throughput of a 12-in. line is approximately 144 by 500 = 72,000 barrels per day.

Naturally, the results of this rule are at best only a very rough approximation, since no account is taken of gravity, viscosity, pressures, or even of spacing between pump stations. Nevertheless, almost all of the oil and products lines now in operation have throughput capacities that will be given by the above rule, with the constant (given as 500) ranging between 300 and 600. The rule does *not* apply to short lines, such as loading lines, or runs within a refinery yard, but only to cross-country pipelines.

Gauge liquid flow where no weir or meter is available

Rule. Measure number of seconds required to fill 5-gallon pail and divide this number into the factor 10,000 to get the answer in barrels per day.

Actually, the factor is 10,286, but it is so close to 10,000 that 10,000 can be used for a rule of thumb number since it is easy to remember.

Example. It requires 100 seconds to fill a 5-gallon bucket. 10,286 divided by 100 equals 102.86 bbl per day, or for practical purposes:

$$10,000 \div 100 = 100 \text{ bpd}$$

Proof: 100 seconds to fill 5-gallon pail equals 20 seconds per gallon equals .05 gallons per second.

$$60 \times 60 \times 24 = 86,400 \text{ seconds in a day}$$

$$86,400 \times .05 = 4,320 \text{ gallons per day}$$

$$4,320 \div 42 \text{ gal per bbl} = 102.86$$

Estimate crude gathering line throughput for a given pipe diameter

To select the proper size line or estimate throughput for crude oil gathering lines, an engineer with a leading oil company has devised this handy rule: **Throughput in barrels per hour is roughly equal to a constant times the square root of pressure drop per mile (psi/mile).** Constants were derived from field data for various diameter lines between 2 and 10 in.

Pipe diameter (in.)	Constant (K)
2	7.8
3	19.2
4	37.8
6	105
8	209
10	385

Example. What size gathering line would be required to move 240 barrels per hour of crude through a 10-mile line using a pump discharge pressure of 400 psi? Assuming no significant changes in head due to elevation and a pressure drop of 40 psi per mile, for the first trial calculation a 4-in. line appears to be the best choice, so try the constant 37.8.

Throughput $(4 - \text{in.})$

$$= 37.8 \sqrt{40} = (37.8)(6.4) = 241 \text{ barrels per hour}$$

So a 4-in. gathering line will be adequate.

How to determine head loss due to friction in ordinary iron pipeline carrying clear water

The form below gives a reasonable degree of accuracy:

$$S = \frac{G^2}{10 \, d^5}$$

where:

 S = Head loss due to friction in ft per 100 ft length of pipeline
 G = Rate of flow in gallons per minute
 d = Nominal diameter of pipe in in.

The exact head loss due to friction in pipelines carrying clear water is given by the Williams and Hazen formula:

$$S = \left[\frac{V}{1.318 \, C \, R^{.63}}\right] 1.85$$

where:

 S = Loss of head due to friction in ft per 100 ft
 V = Average velocity in ft per second
 C = Coefficient depending on the roughness inside the pipe
 R = The mean hydraulic radius = D/4

where:

D is the diameter in ft

The average velocity V is given by Q/a where Q is the rate of flow in ft/sec and a is the cross-sectional area of the pipe.

C is a function of the material and age of the pipe and the type of water. For smooth new iron pipe C = 120. For 15-year-old ordinary pipe C = 100. For 25-year-old ordinary pipe C = 90.

Example. What will be the head loss per 100 ft if a 30-in. line flows 10,000 gpm?

Solution.

$$S = \frac{(10,000)^2}{10\,(30)^5} = \frac{100}{243} = 0.41 \text{ ft}$$

Exact answer using the Williams and Hazen formula:

$$S = 0.34 \text{ ft}$$

Remarks. This rule of thumb could be very satisfactorily used for pipe diameters 3 in. and larger and a discharge greater than 400 gpm.

This rule applies only to wrought and cast iron pipes (20 years service). If new smooth brass or steel pipe is used, a factor should be used = 0.6.

For 30-year-old ordinary pipe factor = 1.2. These are empirical factors.

For pipe diameters 2 in. and less and discharges less than 200 gpm, this rule of thumb does not give satisfactory results.

How to size lines, estimate pressure drop, and estimate optimum station spacing for crude systems

This table will give "horseback" estimates as to optimum line size, pressure drop, and pump station spacing for crude oil of approximately 40°API gravity and 60 SSU viscosity.

The depreciation was assumed to be 6% straight line, return on the investment 8%, and capital and operating costs were based on Mid-Continent averages:

For	Use		
Throughputs (bbl per day)	Pipe Diameter (in. OD)	Pressure Drop (psi per mile)	Pump Station Spacing (miles) (for 800 psi disch. press.)
0 to 2,000	3 ½	—	—
2,000 to 3,000	4 ½	32	25
3,000 to 7,500	6 ⅝	16	50
7,500 to 16,500	8 ⅝	10.5	76
16,500 to 23,500	10 ¾	8.5	94
23,500 to 40,000	12 ¾	7	112

Estimate the optimum working pressures in crude oil transmission lines

The approximate economic optimum working pressure in psi = 1250 − 15 × OD of line.

Example. Estimate the maximum economic optimum working pressure—which is usually station discharge pressure—for a 20-in. crude oil pipeline.

Solution. Approximate maximum working pressure = 1,250 − 15 × 20 = 950 psi.

If the line profile is approximately horizontal, the maximum working pressure occurs at the discharge of a pump station. This method is useful only for approximations.

How to size crude oil and products lines for capacity increases

A new pipeline for crude oil or products service should be sized to carry ⅔ of its ultimate capacity.

Example. Design a pipeline for an initial capacity of 100,000 barrels per day. This could be done by laying a 16-in.

line and spacing 14 pump stations along its length. To increase the capacity by 50% or to 150,000 barrels per day would require adding 14 intermediate booster stations, resulting in excessive operating costs. If the line had been designed so that the 100,000 barrels per day was ⅔ of its ultimate capacity, a 20-in. line with five stations would be installed. To increase the capacity to 150,000 barrels per day would require adding five stations. This is based on the theory that when expanding the capacity of a line, install a loop line if the capacity is to be greater than 50% of the original capacity. If the expansion is to be less, add horsepower. Both of these approaches are based on statistical results rather than economic studies.

How to determine the maximum surge pressure in liquid-filled pipeline when a valve is suddenly closed

The maximum surge pressure that occurs in a liquid pipeline when a valve is suddenly closed may be quickly determined by using the simple rule of thumb: Line pressure equals 0.8 times the weight per cubic foot of liquid times its velocity in feet per second. This compares favorably to the results when using the following equation.

$$P = \frac{12wV\sqrt{\frac{32.2}{w}\left(\frac{1}{\frac{1}{K}+\frac{D}{tE}}\right)}}{144 \times 32.2}$$

where:

P = Surge pressure, psi
w = Weight of liquid, lb./ft^3

V = Velocity change, fps
K = Bulk modulus of the liquid, psi
D = Outside diameter of the pipe, inches
t = Pipe wall thickness, inches
E = Young's modulus of the pipe material, psi

Example. An 8.625-in. × 0.250-in. w.t. products pipeline transports kerosene at 4.5 fps. The weight of the kerosene is 51.12 lb./ft^3, and the API gravity is 41.1 (rel. density = 0.8198). K = 171,000 psi.

Using the simplified equation, P = 0.8 wV = 0.8 × 51.12 × 4.5 = 184 psi

Using the full equation, P = 178 psi; the difference is approximately 3%.

What is the hydrostatic pressure due to a column of liquid H feet in height?

$$P = \frac{H \times 61.41}{API + 131.5}$$

$$P = \frac{20 \times 61.41}{37 + 131.5}$$

$$P = 7.28 \text{ psi}$$

Example. A column of 37° oil is 20 ft high; what is the pressure at the base of the column?

Transient pressure analysis

Transient pressure analysis is the study of a pipeline system's response to a time-dependent upset condition such as a valve closing or a pump shutting down. When a valve is closed downstream of the source pressure, the pressure will increase on the upstream side of the valve while the downstream pressure decreases. The valve closure time will determine how fast the upstream pressure increases and the downstream pressure decreases. The high upstream pressure "surge" will propagate upstream while the low pressure surge moves downstream. If the downstream side is open to a tank, the downstream surge will quickly dissipate due to the surge being absorbed in the tank. Similarly, when a pump shuts down, the pressure downstream of the pump decreases and the pressure upstream of the pump will increase. Other causes of pressure surges include tank changes, manifold changes, or leaks.

Steady-state versus transient conditions

Steady-state means that the pressure gradients and flow rates in the pipeline are not changing. Transient means the gradients and flows are a function of time. The event that causes conditions to be time dependent is the upset condition such as valve closure or pump failure. Transient conditions must be considered when determining if the pipeline will be subjected to abnormally high operating pressures.

Fluid properties

All pressures and flows, whether steady-state or transient, depend on the fluid properties. This example deals with a pipeline filled with batches of motor gasoline and diesel fuel. The program used for the analysis is designed for almost incompressible fluids such as crude oil, products, and water.

The properties required for surge analysis include: specific gravity, viscosity, bulk modulus, and vapor pressure. All properties should be at the same base temperature. Multiple batches can be handled and PVT data will be needed for each batch.

Pipe

The pipe data, such as outside diameter, wall thickness, specified minimum yield strength (SMYS), Poisson's ratio, and Young's modulus of elasticity are required. The length and elevation profile of the pipeline are also required.

Pipe environment

Since an isothermal surge analysis is being performed, the only environmental information required is how the pipeline is anchored. Is the pipe above ground or under water? Does it have expansion loops throughout, or is it just constrained at the source? Multiple pipes can be handled, and each pipe may be constrained differently.

Equipment

The data required for a centrifugal pump includes rated flow, rated head differential, shutoff head differential (PSOH), maximum allowed discharge pressure, minimum allowed suction pressure, discharge pressure trip setting, initial speed, final speed, time it takes to go from initial to final speed (ramp-down time), time when the pump starts to ramp down, and whether the pump is protected by a check valve in the pump discharge line.

The data required for valves includes valve closure time, time or pressure when the valve starts to close, closure curve shape exponent (the exponent is different for gate valves, ball valves, etc.), and the amount of valve closure.

Data required for pressure relief valves include the set pressure and the back-pressure on the relief line.

Initial conditions

Since surge is a time-dependent phenomenon, we need the conditions that exist before any upset occurs. The required initial conditions are: the pressure at the delivery end of the pipeline. The program will establish the initial steady-state flow rate automatically.

Perturbations that cause surges

Any time-dependent change, such as a leak, valve closure, or pump shutdown will cause surges in the system. Cascading upset events, such as internal pumps tripping, may cause other pumps to trip and can be modeled.

The fluid–pipe–equipment interaction

The liquid and the pipe, according to its environment, expand and contract as the surges travel. The pumps, valves, and other equipment change the pressures and/or flows as the surges reach them. The surges travel at the speed of sound in the fluid–pipe system. The fluid itself is what relays the surge energy to the various pipe sections and the equipment. The fluid, pipe, and equipment are inextricably linked, and their behavior must be determined by a model that employs simultaneous fluid–pipe and equipment solution procedures. The Method of Characteristics, which NATASHA PLUS© uses, is a simultaneous technique and therefore models "surge" phenomenon rigorously.

To illustrate the use of the program, the pipeline system shown in Figure 1 is equipped with a valve at the end of the pipeline that is caused to close at time T = 300 seconds.

The valve travels from full open to full closed in 5 seconds. A pressure relief valve is connected to the system and is set to relieve at 275 psig. The system consists of six pipes configured as shown in Table 1.

Simulation

The operation of the system was simulated for 1,200 seconds, and the results are shown graphically in Figure 2. The program outputs a detailed tabular listing showing pressures and flow rates at various locations along the pipeline at selected time steps. A graphics file containing selected node pressures and flow rates is also produced by the program.

Figure 2 was prepared from the graphics file produced by the program. A typical time page report is shown in Table 2. The program also prints a maximum pressure summary which is shown in Table 3. A summary of the relief valve operation is shown in Table 4, indicating that the total volume released was 855 bbls during the simulation.

The pump is not shut down in this simulation because of the action of the relief valve. A review of the maximum pressure summary indicates that the pipeline MOP is exceeded at two locations. In order to prevent overpressure, it will be necessary to make changes to the system such as changing the relief valve setting or slowing the closing time for the valve. NAASHA-PLUS© can then be used to simulate the changes to confirm that pipeline MOP is not exceeded.

Product parameters:

Product	sp. gr.	visc. cSt	bulk modulus psi
Diesel	0.79	2.15	156,495
Gasoline	0.741	0.622	121,505

Initial flow rate is 92,500 BPD
Downstream pressure is 40 psig.

Figure 1. Sample pipeline for transient analysis study.

Table 1
Data Required By The Program

Pipe Design Data						
Pipe No.	Length Mi.	OD in.	W. T. in.	Roughness in.	SMYS psig	Design Fact.
1	2.05	12.75	0.375	0.0018	35000	72%
2	16.3	12.75	0.25	0.0018	52000	72%
3	16.3	12.75	0.25	0.0018	52000	72%
4	15.45	12.75	0.25	0.0018	52000	72%
5	16.53	12.75	0.25	0.0018	52000	72%
6	1.0	12.75	0.25	0.0018	52000	72%

Table 1
Data Required By The Program (continued)

Liquid Properties

Pipe No.	Sp. Gr.	Vis. cSt	Bulk Mod. psi	Vapor Press psia
1	0.741	0.622	0.122	5.7
2	0.79	2.15	0.156	10
3	0.741	0.622	0.122	5.7
4	0.79	2.15	0.156	10
5	0.741	0.622	0.122	5.7
6	0.79	2.15	0.156	10

Initial Conditions

Pipe No.	Flow BPD	Upstream Press psig
1	92500	1398.22
2	92500	1444.19
3	92500	958.58
4	92500	659.75
5	92500	363.1
6	92500	68.76

Pipe No.	Friction Factor	Seg. Length Length, mi.	Press. Drop psi/mi.	Head Drop ft/mi
1	0.014	1.03	21.5907	67.2098
2	0.0157	1.02	23.2816	67.978
3	0.014	0.91	19.4468	60.5359
4	0.0157	1.03	23.2816	67.978
5	0.014	0.92	19.4468	60.5359
6	0.0157	1.00	23.2816	67.978

PUMP PARAMETERS

PUMP AT NODE 7 THAT DISCHARGES INTO PIPE 1

RATED FLOW	=	92,500 bbl/day
ZERO FLOW HEAD DIFFERENTIAL	=	4520.00 ft-head
RATED HEAD DIFFERENTIAL	=	4228.02 ft-head
SUCTION PRESSURE	=	40.00 psig
MAXIMUM DISCHARGE CONTROLLED PRESSURE	=	1440.00 psig
DISCHARGE TRIP PRESSURE	=	1490.00 psig
INITIAL SPEED	=	3565.00 rpm
FINAL SPEED	=	3565.00 rpm
CURVE EXPONENT	=	0.5000

VALVE PARAMETERS

PRESSURE RELIEF VALVE AT NODE 5

MAX PRESS. BEFORE RELIEF	=	275.00 psig
BACK PRESSURE ON RELIEF LINE	=	0.00 psig
VALVE AT NODE 6		
VALVE STARTS TO CLOSE AT T	=	300.00 sec
VALVE CLOSES IN	=	5.00 sec
VALVE STARTS TO CLOSE ABOVE P	=	99999.00 psig
FINAL FRACTIONAL AREA OPEN TO FLOW	=	0.00
CLOSURE SHAPE EXPONENT	=	1.00

Elevation Data

Pipe No.	Point	MP mi.	Elev. ft.	MOP psig
1	1	0	877	1440
	2	1.59	885	1440
	3	2.05	875	1440
2	1	0	875	1440
	2	1.13	890	1440

(table continued on next page)

**Table 1
Data Required By The Program (continued)**

Elevation Data

Pipe No.	Point	MP mi.	Elev. ft.	MOP psig
	3	1.27	860	1440
	4	4.37	940	1440
	5	4.96	890	1440
	6	10.96	990	1440
	7	11.98	1060	1440
	8	13.03	980	1440
	9	16.3	1000	1440
3	1	0	1000	1440
	2	1.71	1065	1440
	3	4.66	1010	1440
	4	6.42	1085	1440
3	5	9.9	930	1440
	6	10.21	1040	1440
	7	11.46	970	1440
	8	14.21	1040	1440
	9	16.3	1071	1440
4	1	0	1071	1440
	2	3.76	1050	1440
	3	5.53	1005	1440
	4	5.92	1020	1440
	5	6.32	1020	1440
	6	15.45	817	1440
5	1	0	817	1440
	2	4.05	770	1440
	3	5.45	875	1440
	4	5.68	810	1440
	5	6.26	915	1440
	6	7.13	805	1440
	7	8.31	930	1440
	8	12.36	755	1440
	9	16.53	746	1440
6	1	0	746	275
	2	0.28	760	275
	3	0.62	762	275
	4	1	762	275

Figure 2. Transient analysis.

Table 2
Typical time page report for T = 0 seconds

Pipe No.	Loc. mi.	Elev Ft.	Up Flow M bbl/day	Down Flow M bbl/day	Head Ft.	Press. psig
1	0	877	93	93	5230	1398
1	1	882	93	93	5161	1374
1	2	875	93	93	5092	1355
2	0	875	93	93	5092	1444
2	1	889	93	93	5023	1416
2	2	880	93	93	4953	1395
2	3	906	93	93	4884	1362
2	4	932	93	93	4815	1330
2	5	892	93	93	4746	1320
2	6	909	93	93	4676	1290
2	7	926	93	93	4607	1261
2	8	943	93	93	4538	1231
2	9	960	93	93	4468	1202
2	10	977	93	93	4399	1172
2	11	1007	93	93	4330	1138
2	12	1041	93	93	4261	1103
2	13	981	93	93	4191	1099
2	14	988	93	93	4122	1074
2	15	994	93	93	4053	1048
2	16	1000	93	93	3984	1022
3	0	1000	93	93	3984	959
3	1	1034	93	93	3929	930
3	2	1063	93	93	3874	903
3	3	1046	93	93	3820	891
3	4	1029	93	93	3765	879
3	5	1012	93	93	3710	867
3	5	1043	93	93	3655	839
3	6	1082	93	93	3600	809
3	7	1048	93	93	3545	802
3	8	1008	93	93	3491	798
3	9	968	93	93	3436	793
3	10	952	93	93	3381	780
3	11	1003	93	93	3326	746
3	12	978	93	93	3271	737
3	13	1001	93	93	3217	712
3	14	1024	93	93	3162	687
3	14	1044	93	93	3107	663
3	15	1058	93	93	3052	641
3	16	1071	93	93	2997	619
4	0	1071	93	93	2997	660
4	1	1065	93	93	2927	638
4	2	1059	93	93	2857	616
4	3	1054	93	93	2787	594
4	4	1041	93	93	2717	574
4	5	1015	93	93	2647	559
4	6	1020	93	93	2577	533
4	7	1000	93	93	2507	516
4	8	977	93	93	2437	500
4	9	954	93	93	2367	484
4	10	932	93	93	2297	468
4	11	909	93	93	2227	452
4	12	886	93	93	2157	435
4	13	863	93	93	2087	419

(table continued on next page)

Pipe No.	Loc. mi.	Elev Ft.	Up Flow M bbl/day	Down Flow M bbl/day	Head Ft.	Press. psig
4	14	840	93	93	2017	403
4	15	817	93	93	1947	387
5	0	817	93	93	1947	363
5	1	806	93	93	1892	349
5	2	796	93	93	1836	334
5	3	785	93	93	1781	320
5	4	774	93	93	1725	305
5	5	811	93	93	1669	276
5	6	858	93	93	1614	243
5	6	894	93	93	1558	213
5	7	828	93	93	1503	217
5	8	925	93	93	1447	168
5	9	892	93	93	1391	160
5	10	853	93	93	1336	155
5	11	813	93	93	1280	150
5	12	773	93	93	1225	145
5	13	754	93	93	1169	133
5	14	752	93	93	1113	116
5	15	750	93	93	1058	99
5	16	748	93	93	1002	82
5	17	746	93	93	947	64
6	0	746	93	93	947	69
6	1	762	93	93	879	40

Table 2
Typical time page report for T = 0 seconds (continued)

Table 3
Maximum pressure summary

Pipe No.	MP mi.	Elev. ft.	Hmax ft.	Pmax psig	MOP psig	Time sec.	
1	0	877	5269.51	1411.07	1440	1199.44	
1	1.03	882.16	5209.78	1390.22	1440	1200.98	
1	2.05	875	5150.04	1373.33	1440	1199.44	
2	0	875	5150.04	1464.14	1440	1199.44	× Exceeds MOP)
2	1.02	888.52	5089.72	1438.85	1440	1200.98	
2	2.04	879.81	5029.4	1421.18	1440	1199.44	
2	3.06	906.1	4969.08	1391.52	1440	1200.98	
2	4.07	932.39	4908.75	1361.85	1440	1199.44	
2	5.09	892.23	4848.44	1354.95	1440	1200.98	
2	6.11	909.21	4788.11	1328.47	1440	1199.44	
2	7.13	926.19	4727.8	1302	1440	1200.98	
2	8.15	943.17	4667.47	1275.53	1440	1199.44	
2	9.17	960.15	4607.16	1249.05	1440	1200.98	
2	10.19	977.13	4546.84	1222.58	1440	1199.44	
2	11.21	1006.9	4486.53	1191.73	1440	1200.98	
2	12.22	1041.33	4426.21	1159.27	1440	1199.44	
2	13.24	981.31	4365.9	1159.18	1440	1200.98	
2	14.26	987.54	4305.58	1136.38	1440	1199.44	
2	15.28	993.77	4245.27	1113.6	1440	1200.98	
2	16.3	1000	4184.95	1090.8	1440	1199.44	
3	0	1000	4184.95	1023.15	1440	1199.44	
3	0.91	1034.42	4137.43	996.82	1440	1200.98	
3	1.81	1063.11	4089.9	972.34	1440	1199.44	
3	2.72	1046.23	4042.38	962.49	1440	1200.98	
3	3.62	1029.35	3994.85	952.65	1440	1199.44	
3	4.53	1012.47	3947.33	942.81	1440	1200.98	
3	5.43	1042.95	3899.8	917.74	1440	1199.44	
3	6.34	1081.54	3852.29	890.08	1440	1200.98	
3	7.24	1048.28	3804.76	885.5	1440	1199.44	
3	8.15	1007.95	3757.25	883.2	1440	1200.98	
3	9.06	967.61	3709.72	880.89	1440	1199.44	
3	9.96	951.68	3662.21	870.74	1440	1200.98	
3	10.87	1003.23	3614.69	838.92	1440	1199.44	
3	11.77	977.95	3567.18	831.77	1440	1200.98	
3	12.68	1001	3519.66	809.1	1440	1199.44	
3	13.58	1024.05	3472.15	786.44	1440	1200.98	
3	14.49	1044.14	3424.63	764.72	1440	1199.44	
3	15.39	1057.57	3377.13	745.15	1440	1200.98	
3	16.3	1071	3329.61	725.57	1440	1199.44	
4	0	1071	3329.61	773.55	1440	1199.44	
4	1.03	1065.25	3268.67	754.64	1440	1200.98	
4	2.06	1059.49	3207.72	735.74	1440	1199.44	
4	3.09	1053.74	3146.78	716.84	1440	1200.98	
4	4.12	1040.85	3085.82	700.38	1440	1199.44	
4	5.15	1014.66	3024.89	688.48	1440	1200.98	
4	6.18	1020	2963.94	665.77	1440	1199.44	
4	7.21	1000.21	2903	651.68	1440	1200.98	
4	8.24	977.31	2842.06	638.65	1440	1199.44	
4	9.27	954.41	2781.13	625.63	1440	1200.98	
4	10.3	931.51	2720.18	612.6	1440	1199.44	
4	11.33	908.61	2659.25	599.57	1440	1200.98	

(table continued on next page)

Table 3
Maximum pressure summary (continued)

Pipe No.	MP mi.	Elev. ft.	Hmax ft.	Pmax psig	MOP psig	Time sec.	
4	12.36	885.7	2598.31	586.55	1440	1199.44	
4	13.39	862.8	2537.39	573.52	1440	1200.98	
4	14.42	839.9	2476.45	560.5	1440	1199.44	
4	15.45	817	2415.53	547.47	1440	1200.98	
5	0	817	2415.53	513.52	1440	1200.98	
5	0.92	806.34	2367.36	501.47	1440	1199.44	
5	1.84	795.69	2319.21	489.42	1440	1200.98	
5	2.76	785.03	2271.05	477.37	1440	1199.44	
5	3.67	774.37	2222.9	465.33	1440	1200.98	
5	4.59	810.63	2174.74	438.21	1440	1199.44	
5	5.51	858.04	2126.59	407.51	1440	1200.98	
5	6.43	893.72	2078.43	380.58	1440	1199.44	
5	7.35	827.95	2030.29	386.24	1440	1200.98	
5	8.27	925.23	1982.13	339.52	1440	1199.44	
5	9.18	892.26	1933.99	334.65	1440	1200.98	
5	10.1	852.58	1885.84	331.93	1440	1199.44	
5	11.02	812.9	1837.7	329.21	1440	1200.98	
5	11.94	773.22	1789.55	326.49	1440	1199.44	
5	12.86	753.93	1741.41	317.22	1440	1200.98	
5	13.78	751.95	1693.26	302.39	1440	1199.44	
5	14.69	749.96	1645.13	287.57	1440	1200.98	
5	15.61	747.98	1596.98	272.73	1440	1199.44	
5	16.53	746	1548.9	257.93	1440	742.37	
6	0	746	1548.9	274.98	275	742.37	
6	1	762	1721.75	328.7	275	460.14	× (Exceeds MOP)

Table 4
Pressure relief valve summary

TOTAL VOLUME RELEASED AT NODE 5 = 855.4117bbl
MAX VOLUME RATE RELEASED AT NODE 5 = 102.958 1000 bbl/day

Tank farm line sizing

By: J. C. Stivers

The designer must locate the tanks, size the tank lines and select pumps. The designer must know the receiving flow rates, discharge flow rates, liquids to be handled along with the properties of the liquid such as viscosity, specific gravity and vapor pressure. The minimum and maximum tank levels must be determined which will set the available volume that can be pumped. Other factors include determining the number of tanks which will depend upon the type of material being stored, tank turnover frequency, system load factor (which takes into consideration system outages) and environmental considerations.

If possible, the tanks should have combination suction and fill lines. To size the lines, it is necessary to consider the fill and discharge flow rates. If a tank farm can receive from a number of small systems and has one out-going pipeline, the size of the tank lines will probably be determined by the outgoing flow rate. If a tank farm is fed by one stream and delivers to a number of smaller systems, the line sizes will probably be determined by the incoming stream rate. It is necessary to check both to confirm this generalization.

The criteria for sizing a tank line based on filling rates is somewhat arbitrary. When there are a number of sources, the operator may specify a maximum pressure of 25 to 50 psig required at the connection. The piping manifold, meters and other equipment must then be sized so the pressure will not exceed 25 to 50 psig at the maximum flow rate. If a metering system is to be used on the incoming stream and a back pressure valve is required for the meters, the tank line size may be sized to produce the minimum back pressure at the maximum flow rate. When the tank farm owner develops the pressure required to fill tanks, an economic analysis should be made to determine line sizes. Large diameter pipelines require less pressure and therefore less power cost but require greater investments than smaller diameter lines.

Determine pipeline size

The equation for determining the line size from the tanks and to fill a pump based on discharge flow rates is as follows:

NPSH available must be $\geq$ NPSH required by the pump

$$\frac{P_a \times 1.135}{sg} - \frac{Pv \times 2.31}{sg} \pm static$$

Where:

P_a = Atmospheric pressure, in HG
P_v = Vapor pressure, psia
sg = specific gravity

Press. (ft of head) − pipe friction loss (psi/mile) × (equiv. feet of suction line × 2.31)/(5280 × sg)

Atmospheric Pressure—The lowest atmospheric pressure for the area should be considered. This is particularly important for tank farms where product is transferred from one tank to another using the pump. This usually occurs when the atmospheric pressure is low

Vapor Pressure—Vapor pressure increases with temperature and therefore reduces the NPSH available. Conversely liquid viscosity decreases with temperature increases and therefore reduces pipe friction losses. Therefore, it becomes necessary to determine the liquid temperature that produces the lowest NPSH available.

Static Pressure—This is the elevation difference between the minimum liquid level in the tank and the centerline of the

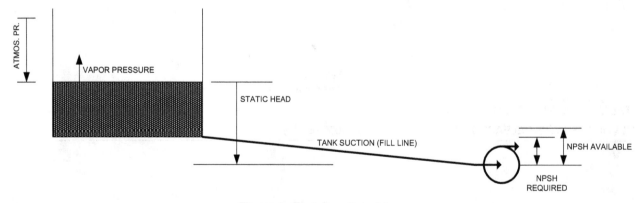

Figure 1. Tank farm line sizing.

pump suction. If the ground elevation in the tank farm changes, the designer can take advantage of the terrain and locate the tanks accordingly.

Pipe Friction Loss—There are many equations used to determine pipe friction loss. See Section 13.

Most tank lines are not designed to be pigged. Some crudes, particularly sour crudes, may cause a build up in the tank line, in turn reducing the pipe i. d. Any buildup will cause the friction factor to be higher and if build up is expected, the designer may wish to increase the pipe size—particularly if NPSH available to the pump is marginal.

NPSH required by pump

The pump manufacturer supplies the NPSH requirement for each pumps. Some positive displacement pump manufacturers may not publish these requirements due to a wide range of operating speeds. They will however, provide requirements for specific conditions. If the NPSH requirements of the pump are not met, the pump will cavitate and a loss of pressure and flow will occur. Mechanical damage may also be possible.

Types of booster pumps

The pumps most typically used include:

Horizontal split case
Vertical can
Vertical inline

A horizontal split case pump may be used in some instances and will be preferred by most maintenance and operations personnel because it is easy to maintain. The NPSH requirement for a horizontal pump may be 7 to 10 feet and the suction nozzle may be 2 to 4 feet above grade. The vertical can pump can have a 0 NPSH requirement with the suction nozzle three or more feet below grade. The NPSH advantage for the can pump can range from 12 to 17 feet which may permit a size reduction of the tank lines. NPSH requirements for the vertical inline pump are similar to the horizontal pump. The advantage of the inline pump is pump and installation costs.

Design considerations

The designer should not automatically assume that a booster pump is required. Most positive displacement pumps have low NPSH requirements and a therefore a booster pump is seldom required unless metering is installed between the tank and mainline pump(s). If the terrain permits the tanks to be located considerably higher than the main pumps, a booster pump may not be required for centrifugal pumps.

If a series of main line pumps are used and the pumping rate is determined by the number of pumps operating, it may be possible to obtain a pump with a lower NPSH requirement for the first pump and thus eliminate the need for a booster pump. This is shown by the following example.

Assume three main line pumps are connected in a series. Pump combinations produce the following flow rates:

No. of Pumps Operating	Flow Rate
3	100%
2	75%
1	50%

NPSH available

Flow Rate	PSH Available
100%	15 feet
75%	20 feet
50%	27 feet

At a flow rate of 100%, it may be necessary to pay a premium for the number one pump to obtain a 15 feet NPSH requirement. However, a minimum cost second unit sized for 100% flow rate may be able to operate at 75% flow rate and 20 feet of NPSH.

Low NPSH requirement booster pumps are normally less efficient than mainline pumps. When this is the case, a booster pump should be selected to develop just enough head to feed the main line or delivery pumps.

If a metering system is required, it would normally be installed between the tanks and mainline pumps. In this case, a booster pump would almost certainly be required to overcome the pressure losses in the metering system. The booster pump then must develop enough head to produce the minimum back pressure required by the meters or the minimum NPSH required by the mainline pumps.

If the booster pump is used to transfer liquid from one tank to another, it will probably develop more head than required. This allows the designer to reduce the size of the transfer header and valves.

Once the minimum atmospheric pressure and vapor pressure of the liquid to be pumped is determined, there are three variables on the left side (NPSH available) and one on the right (NPSH required by the pump). The solution is trial and error and the following procedure is suggested:

- Select a tentative location for the tanks and determine the minimum liquid level elevation.
- Determine the equivalent pipe length of the suction line.
- Locate the booster pump and assume it can operate at a specific NPSH.

This allows the solution of the equation for maximum pipe friction loss. Application of pipe pressure loss formula will give the minimum size pipe that will result in a specific NPSH. Select the next larger standard pipe size and rework the equation to determine the actual minimum NPSH available at the pump. Generally, the cost of the booster pump will decrease as the NPSH requirement increases. Higher NPSH will also allow the designer to select a different style pump. For these reasons, the equation should be solved using the next two larger pipe sizes. Pumps can now be selected for three different pipe sizes. The costs of the three systems should be estimated and an economic analysis made to select the system.

The designer must be aware that a low NPSH pump can pull a vacuum which must be considered in determining pump differential. (Pipe collapse?) Frequently, the required pump discharge pressure is used to specify pump differential head which assumes zero pressure at the pump suction. The following example assumes the booster pump is designed to operate at 7 feet NPSH and 7 feet NPSH is available. The liquid is fuel oil with a vapor pressure of 0.2 psig and 0.85 s.g. and atmospheric pressure is 14.7 psia.

$(2.31/\text{s.g.}) \times (\text{press. avail.} + \text{atmos. press} - \text{vapor press}) = \text{NPSH}$

Press. avail. $= [(\text{s.g.} \times \text{NPSH})/2.31] - \text{atmos. press.} + \text{vapor press.}$

Press. avail. $= [(0.84 \times 7)/2.31] - 14.7 + 0.2 = -11.95$ psig

If a pump discharge pressure of 25 psig is required, the pump differential is $25 - (-11.95) = 36.95$ psig or 102 feet.

Hydraulics calculations for multiphase systems, including networks

The calculation of pressure, temperature, and volumes in multiphase systems (pipelines, wells, and pipe networks), is a complicated function of fluid composition, pressure, temperature, flow rate, fluid properties, pipe properties, angle of inclination (elevation differences), properties of the medium surrounding the pipes, and pipe connectivity. Therefore, for all but the simplest of systems, a computer program is required.

GENNET-M is a powerful Multiphase Flow program from Fluid Flow Consultants, Inc., of Broken Arrow, Okla. GENNET-M handles pipeline, well, and pipe NETWORK models (one or more sources, one or more sinks, and zero or more loops) of up to 500 pipes, has a Liquids Removal ("pigging") option, a pressure-drop calibration option, a viscosity curve input option, and "DP+DT" equipment (pressure and/or temperature change device, i.e., a pump, compressor, heater, cooler, minor loss across a valve or fitting, pressure drop from an IPR curve, etc.). GENNET seeks to solve for the flow rates in each pipe of an arbitrarily connected network subject to constant boundary (source/sink) pressures and/or flows and the resistance (friction and elevation differences) in the pipes.

Figure 1. Diagram of network and pipe data input screen for the example.

Example. The GENNET-M input file name is PLROT.GIN. The text input file is PLROTIN.TXT. The text output file name is PLROTOUT.TXT.

This example (PLROT) is a multiphase (oil, gas, and water) problem with four sources, two sinks, and one loop. The source flows are known (GENNET-M calculates the source pressures), and the sink pressures are known (GENNET-M calculates the sink flows). Each source has a different flow rate and composition (oil API; oil viscosity curve; gas gravity; and CO_2, N_2, and H_2S content). GENNET-M mixes the fluids according to the flow rate in each pipe and then calculates the various fluid properties it needs as it calculates the flowing pressures and temperatures. All source pipes (Pipe001, Pipe010, Pipe004, and Pipe005) are in air exposed to a 10 mph wind. The loop pipes (Pipe002, Pipe003, and Pipe011) and Pipe006 are submerged in water exposed to a 10 mph cross-current. The sink pipes (Pipe008 and Pipe009) and Pipe007 are buried in soil at a depth of 36 in. The complete input data can be found in PLROTIN.TXT, which is shown below.

Input

GENNET-M 5.4.23

GENeral NETwork Multiphase flow analyzer

DATE: 01/30/2004   TIME: 12:13

PROGRAM and its DOCUMENTATION Copyrighted by: Fluid Flow Consultants 1995-2004

— All Rights Reserved —

(Photocopies of Input, Output, User's Manual may be made as required by authorized users of the PROGRAM)

General Data

Title:
Company: Fluid Flow Consultants
Number of Pipes: 11
Fluid Type: Oil
Global Units: English
Pipe Length Units: Feet

PIPE PROPERTIES 1

Pipe Number	Pipe Name	Outside Diameter (in)	Wall Thickness (in)	Incl Angle (Deg)	Rough (in)
1	Pipe010	2.875	0.217	-0.1	0.0018
2	Pipe002	6.625	0.250	0.0	0.0018
3	Pipe004	4.500	0.224	-0.1	0.0018
4	Pipe005	4.500	0.224	-0.1	0.0018
5	Pipe006	6.625	0.250	0.0	0.0018
6	Pipe007	10.750	0.344	0.5	0.0018
7	Pipe008	8.625	0.312	1.8	0.0018
8	Pipe009	8.625	0.312	-0.6	0.0018
9	Pipe001	2.875	0.217	-0.1	0.0018
10	Pipe003	2.875	0.217	0.0	0.0018
11	Pipe011	2.875	0.217	0.0	0.0018

PIPE PROPERTIES 2

Pipe Number	Pipe Name	Thermal Conductivity (Btu/'-h-F)	Length (ft)	Calc Segs	Env	Calibration Factor	Liq Holdup Factor
1	Pipe010	29.00000	4420.0	2	Air	1.00000	1.00000
2	Pipe002	29.00000	4000.0	2	Water	1.00000	1.00000
3	Pipe004	29.00000	6250.0	3	Air	1.00000	1.00000
4	Pipe005	29.00000	10180.0	5	Air	1.00000	1.00000
5	Pipe006	29.00000	4000.0	2	Water	1.00000	1.00000
6	Pipe007	29.00000	24050.0	12	Soil	1.00000	1.00000
7	Pipe008	29.00000	7520.0	4	Soil	1.00000	1.00000
8	Pipe009	29.00000	9200.0	5	Soil	1.00000	1.00000
9	Pipe001	29.00000	2523.0	1	Air	1.00000	1.00000
10	Pipe003	29.00000	2000.0	1	Water	1.00000	1.00000
11	Pipe011	29.00000	2000.0	1	Water	1.00000	1.00000

PIPE CONNECTIVITY

Pipe Number	Upstream Node	Pipe Name	Downstream Node
1	8	Pipe010	1
2	1	Pipe002	2
3	6	Pipe004	3
4	7	Pipe005	3
5	3	Pipe006	2
6	2	Pipe007	4
7	4	Pipe008	9
8	4	Pipe009	10
9	5	Pipe001	1
10	1	Pipe003	11
11	11	Pipe011	2

PIPE ENVIRONMENT

Pipe Number	Pipe Name	Heat Transfer (Btu/ft2-hr-F)	Insul Thick (in)	Insul T.Cond (Btu/'-h-F)	Amb Temp (F)	A/W Vel (mph)	Burial Depth (in)	Soil T.COND (Btu/'-h-F)
1	Pipe010	0.00000	0.00	0.30	60.0	10.0	36.00	0.80
2	Pipe002	0.00000	0.00	0.30	60.0	10.0	36.00	0.80
3	Pipe004	0.00000	0.00	0.30	60.0	10.0	36.00	0.80
4	Pipe005	0.00000	0.00	0.30	60.0	10.0	36.00	0.80
5	Pipe006	0.00000	0.00	0.30	60.0	10.0	36.00	0.80
6	Pipe007	0.00000	0.09	0.30	60.0	10.0	36.00	0.80
7	Pipe008	0.00000	0.09	0.30	60.0	10.0	36.00	0.80
8	Pipe009	0.00000	0.09	0.30	60.0	10.0	36.00	0.80
9	Pipe001	0.00000	0.00	0.30	60.0	10.0	36.00	0.80
10	Pipe003	0.00000	0.00	0.30	60.0	10.0	36.00	0.80
11	Pipe011	0.00000	0.00	0.30	60.0	10.0	36.00	0.80

SOURCE CONDITIONS

Source Name: Source004,   Pipe Number: 1,   Pipe Name: Pipe010
Flowing Temperature	(F) :	124.0
Flowing Pressure	(psig) :	1200.49
Oil Rate	(bbl/day) :	4350.000
Gas/Oil Ratio	(std ft3/bbl) :	448.000
Water Cut	(Volume %) :	8.200
Gas Gravity:		0.695
Separator Pressure	(psig) :	0.000
Separator Temperature	(F) :	60.000
Nitrogen, N2	(mole%) :	2.120
Carbon Dioxide, CO2	(mole%) :	3.210
Hydrogen Sulfide, H2S	(mole%) :	0.082
Oil Gravity	(API) :	24.550
Bub Pt Press	(psig) :	0.000
Water Spec Grav	(FRAC) :	1.060

Source Name: Source002,   Pipe Number: 3,   Pipe Name: Pipe004
Flowing Temperature	(F) :	136.0
Flowing Pressure	(psig) :	721.52
Oil Rate	(bbl/day) :	4400.000
Gas/Oil Ratio	(std ft3/bbl) :	1000.000
Water Cut	(Volume %) :	22.300
Gas Gravity:		0.685
Separator Pressure	(psig) :	0.000
Separator Temperature	(F) :	60.000
Nitrogen, N2	(mole%) :	0.125
Carbon Dioxide, CO2	(mole%) :	1.080
Hydrogen Sulfide, H2S	(mole%) :	0.015
Oil Gravity	(API) :	32.800
Bub Pt Press	(psig) :	0.000
Water Spec Grav	(FRAC) :	1.100

Source Name: Source003,   Pipe Number: 4,   Pipe Name: Pipe005
Flowing Temperature	(F) :	142.0
Flowing Pressure	(psig) :	851.61
Oil Rate	(bbl/day) :	4500.000
Gas/Oil Ratio	(std ft3/bbl) :	1026.000
Water Cut	(Volume %) :	23.800
Gas Gravity:		0.669
Separator Pressure	(psig) :	0.000
Separator Temperature	(F) :	60.000
Nitrogen, N2	(mole%) :	0.095
Carbon Dioxide, CO2	(mole%) :	0.098
Hydrogen Sulfide, H2S	(mole%) :	0.006
Oil Gravity	(API) :	34.600
Bub Pt Press	(psig) :	0.000
Water Spec Grav	(FRAC) :	1.090

Source Name: Source001,   Pipe Number: 9,   Pipe Name: Pipe001
Flowing Temperature	(F) :	128.0
Flowing Pressure	(psig) :	1238.40
Oil Rate	(bbl/day) :	5100.000
Gas/Oil Ratio	(std ft3/bbl) :	750.000
Water Cut	(Volume %) :	6.300
Gas Gravity:		0.723
Separator Pressure	(psig) :	0.000
Separator Temperature	(F) :	60.000
Nitrogen, N2	(mole%) :	1.430
Carbon Dioxide, CO2	(mole%) :	4.230
Hydrogen Sulfide, H2S	(mole%) :	0.065
Oil Gravity	(API) :	28.600
Bub Pt Press	(psig) :	0.000
Water Spec Grav	(FRAC) :	1.060

OIL VISCOSITY

```
Source Name: Source004,     Pipe Name: Pipe010:
   Pipe      Temperature      Viscosity
                 (F)           ( cp  )
-----------  -----------    ---------
      1        100.000        1.62000
      1        210.000        0.88000
Power: Visc(cp) = 0.7154E+02 * Temp(F) ^ -.8225E+00

Source Name: Source002,     Pipe Name: Pipe004:
      3        100.000        1.20000
      3        210.000        0.44500
Power: Visc(cp) = 0.5666E+03 * Temp(F) ^ -.1337E+01

Source Name: Source003,     Pipe Name: Pipe005:
      4        100.000        1.05000
      4        210.000        0.22600
Power: Visc(cp) = 0.1451E+05 * Temp(F) ^ -.2070E+01

Source Name: Source001,     Pipe Name: Pipe001:
      9        100.000        1.45000
      9        210.000        0.95000
Power: Visc(cp) = 0.2001E+02 * Temp(F) ^ -.5699E+00
```

SINK CONDITIONS

Sink Name	Pipe Number	Pipe Name	Flowing Pressure (psig)	Flow Rate (bbl/day)
Sink001	7	Pipe008	100.00	11274.102
Sink002	8	Pipe009	200.00	7075.899

Output

The calculated output can be found in PLROTOUT.TXT.

GENNET - M 5.4.23

GENeral NETwork Multiphase flow analyzer
DATE: 01/30/2004   TIME: 12:12

Pipe No	Pipe Name	Dist (ft)	Flow (bbl/day)	Pressure (psig)	Temp (øF)	Liq Holdup (Bbl)	Vel mix (ft/sec)	Flow Patt
9	Pipe001	0.0	0.510E+04	1238.40	128.00			
9	Pipe001	2523.0	0.510E+04	465.89	109.24	6.9	30.3	Intr

						6.9		
3	Pipe004	0.0	0.440E+04	721.52	136.00			
3	Pipe004	2083.3	0.440E+04	654.40	115.02	13.6	15.3	Intr
3	Pipe004	4166.7	0.440E+04	583.04	99.37	13.1	16.2	Intr
3	Pipe004	6250.0	0.440E+04	504.91	87.90	12.5	17.6	Intr

						39.2		
4	Pipe005	0.0	0.450E+04	851.61	142.00			
4	Pipe005	2036.0	0.450E+04	790.08	120.64	14.4	14.0	Intr
4	Pipe005	4072.0	0.450E+04	726.25	104.37	14.1	14.4	Intr
4	Pipe005	6108.0	0.450E+04	658.77	92.18	13.7	15.1	Intr
4	Pipe005	8144.0	0.450E+04	585.88	83.18	13.1	16.2	Intr
4	Pipe005	10180.0	0.450E+04	504.91	76.64	12.3	17.8	Intr

						67.7		

Pipe No	Pipe Name	Dist (ft)	Flow (bbl/day)	Pressure (psig)	Temp (øF)	Liq Holdup (Bbl)	Vel mix (ft/sec)	Flow Patt
1	Pipe010	0.0	0.435E+04	1200.49	124.00			
1	Pipe010	2210.0	0.435E+04	880.72	106.14	7.9	16.6	Liq
1	Pipe010	4420.0	0.435E+04	465.89	92.76	6.7	21.5	Intr

						14.6		
2	Pipe002	0.0	0.869E+04	465.89	101.70			
2	Pipe002	2000.0	0.869E+04	444.63	60.00	29.9	11.2	Intr
2	Pipe002	4000.0	0.869E+04	423.41	60.00	30.1	11.0	Intr

						60.0		
5	Pipe006	0.0	0.890E+04	504.91	82.17			
5	Pipe006	2000.0	0.890E+04	465.45	60.00	26.5	16.8	Intr
5	Pipe006	4000.0	0.890E+04	423.41	60.00	25.7	17.5	Intr

						52.2		
10	Pipe003	0.0	0.760E+03	465.89	101.70			
10	Pipe003	2000.0	0.760E+03	444.74	60.00	4.8	6.2	Intr

						4.8		
11	Pipe011	0.0	0.760E+03	444.74	60.00			
11	Pipe011	2000.0	0.760E+03	423.41	60.00	4.8	6.0	Intr

						4.8		
6	Pipe007	0.0	0.184E+05	423.41	60.00			
6	Pipe007	2004.2	0.184E+05	409.86	60.00	73.2	11.6	Intr
6	Pipe007	4008.3	0.184E+05	396.04	60.00	72.0	11.9	Intr
6	Pipe007	6012.5	0.184E+05	381.94	60.00	70.7	12.3	Intr

Pipe No	Pipe Name	Dist (ft)	Flow (bbl/day)	Pressure (psig)	Temp (øF)	Liq Holdup (Bbl)	Vel mix (ft/sec)	Flow Patt
6	Pipe007	8016.7	0.184E+05	367.52	60.00	69.3	12.7	Intr
6	Pipe007	10020.8	0.184E+05	352.75	60.00	67.9	13.1	Intr
6	Pipe007	12025.0	0.184E+05	337.60	60.00	66.4	13.7	Intr
6	Pipe007	14029.2	0.184E+05	321.99	60.00	64.9	14.2	Intr
6	Pipe007	16033.3	0.184E+05	305.89	60.00	63.2	14.9	Intr
6	Pipe007	18037.5	0.184E+05	289.21	60.00	61.5	15.6	Intr
6	Pipe007	20041.7	0.184E+05	271.86	60.00	59.6	16.4	Intr
6	Pipe007	22045.8	0.184E+05	253.71	60.00	57.6	17.4	Intr
6	Pipe007	24050.0	0.184E+05	234.60	60.00	55.5	18.6	Intr

						781.7		
7	Pipe008	0.0	0.113E+05	234.60	60.00			
7	Pipe008	1880.0	0.113E+05	206.43	60.00	31.1	19.8	Intr
7	Pipe008	3760.0	0.113E+05	175.85	60.00	28.9	22.5	Intr
7	Pipe008	5640.0	0.113E+05	141.48	60.00	26.3	26.6	Intr
7	Pipe008	7520.0	0.113E+05	99.99	60.00	22.9	33.7	Intr

						109.3		
8	Pipe009	0.0	0.708E+04	234.60	60.00			
8	Pipe009	1840.0	0.708E+04	228.19	60.00	31.0	11.9	Intr
8	Pipe009	3680.0	0.708E+04	221.55	60.00	30.6	12.2	Intr
8	Pipe009	5520.0	0.708E+04	214.65	60.00	30.1	12.6	Intr
8	Pipe009	7360.0	0.708E+04	207.47	60.00	29.6	12.9	Intr
8	Pipe009	9200.0	0.708E+04	199.99	60.00	29.1	13.3	Intr

						150.3		

END

Flow pattern

Some correlations define flow patterns, and others do not. The modified Beggs and Brill-Moody correlations, which GENNET-M uses, define the following patterns:

1. Segregated (usually Stratified or Stratified-Wavy for near horizontal segments; usually Annular for near vertical segments)
2. Intermittent (Slug or Plug)
3. Distributed (Bubble for high liquid loadings; Mist for low liquid loadings)
4. Transition (somewhere between Segregated and Intermittent)

Because any optimization method requires continuity across flow pattern boundaries, three more "transition" zones were added to the Beggs and Brill flow pattern map:

5. D-S (transition zone between Distributed and Segregated)
6. D-T (transition zone between Distributed and Transition)
7. D-I (transition zone between Distributed and Intermittent)

It is important to note that the Beggs and Brill correlations do not necessarily predict the correct flow pattern if the angle of inclination is more than a few degrees from horizontal. If the segment is not close to being horizontal, the flow pattern is used only as a correlating parameter for the holdup.

Adapted from Fluid Flow Consultants, Inc. Grove, Oklahoma; ©1997 Fluid Flow Consultants. Reprinted with permission.

14: Pumps

Centrifugal pumps

Centrifugal pumps are best suited for large volume applications or for smaller volumes when the ratio of volume to pressure is high. The selection of the proper pump will depend on the system throughput, viscosity, specific gravity, and head requirements for a particular application. Where a variety of products with different characteristics are handled, it may be necessary to utilize multiple pumps either in series or parallel, or a series—parallel combination, or pumps with variable speed capabilities to achieve the desired throughput at least cost. Economics will ultimately determine the type and number of pumps to be used.

The first step in selecting centrifugal pumps is to analyze the pipeline system and determine its characteristics such as the pressure required to move the desired flow rate. Initial and future conditions should be evaluated so that pumps with sufficient flexibility to handle both conditions with minor changes may be selected. Refer to Section 13—Liquid hydraulics for data on calculating pipeline pressure drop.

Pump manufacturers publish pump performance maps similar to the one shown in Figure 1 to aid in pump selection. The performance map shows the range of differential head and capacity that will be developed by a family of pumps. Once the system data have been established, a pump may be

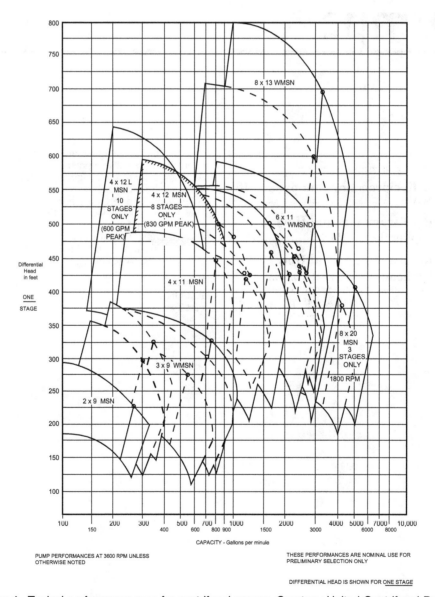

Figure 1. Typical performance map for centrifugal pumps. Courtesy United Centrifugal Pumps.

Figure 2. Typical performance curve for centrifugal pumps. Courtesy United Centrifugal Pumps.

selected from the map. The selection is further refined by referring to a curve such as that shown in Figure 2, which shows a typical pump manufacturer's catalog curve for a specific pump showing the performance for various impeller size, as well as the brake horsepower, NPSH requirement, and pump efficiency. Pump performance curves are plotted in feet and for a specific gravity of 1.0. When pumping a nonviscous liquid, the pump will develop the same head shown on the performance curve; however, the *pressure* will change with specific gravity, as will the horsepower required to drive the pump. Pressure may be calculated by multiplying the pump differential head by the specific gravity and by 0.4329. The horsepower required to drive the pump will also vary directly with the specific gravity of the fluid being pumped.

The viscosity of the fluid will influence the performance of the centrifugal pump. This is discussed in more detail later.

Centrifugal pumps are available in a number of different configurations such as horizontal, vertical, radial split, and axial split casings. The final choice will depend on the system hydraulic conditions (pressure and flow rate), the ease of maintenance desired, and amount of space available for installing the pump. The pressure rating of the case and material selection will be determined by the system requirements and the fluid to be pumped.

It is extremely important that sufficient net positive suction head (NPSH) above the vapor pressure of the liquid being pumped is available for safe and reliable pump operation. NPSH is always specified in feet of head above vapor pressure required at the center line of the impeller. The NPSH required will increase as the capacity of the pump increases, which makes it necessary for the user to make a best effort to furnish the pump manufacturer with an accurate range of flow rate conditions. The pump suction nozzle geometry, pump speed, and the impeller inlet will determine the NPSH that will be required. If too much margin of safety is included in the pump specifications, the pump will probably be oversized to obtain a low NPSH and will not operate at its best efficiency point. Figures 8–10 show three examples of how to calculate available NPSH. As a rule of thumb, the NPSH available should be at least 10% greater than the NPSH required by the pump. It is difficult and costly to correct NPSH problems after the pump is placed in service. NPSH is discussed in detail in *Centrifugal Pumps: Design and Application* by Ross and Lobanoff, Gulf Publishing Co., and it's recommended reading for those

Figure 3. Centrifugal pump performance data.

who are involved in the selection and application of centrifugal pumps.

Table 1 gives some conversion factors and formulas that will be useful in problems dealing with pumps.

All centrifugal pumps, except those pumps that are specifically designed for low suction pressure applications such as tank farm booster pumps, should be protected by a low suction pressure shut down switch. The value of pressure to be used to set the switch will be determined by the amount of NPSH required at the maximum expected operating flow rate, the vapor pressure of the fluid being pumped, and the atmospheric pressure at the pump location. Once the maximum expected operating flow rate is known, the amount of NPSH required may be obtained from the pump performance curve. The NPSH curve is plotted in feet of head, and this may be converted to psi for switch setting purposes as follows:

$$PSIG = NPSH \times 0.433 \times \text{specific gravity} + \text{vapor pressure}$$
$$- \text{atmospheric pressure}$$

Figure 3 shows the performance data for a centrifugal pump that was chosen from the performance map. The impeller diameter was chosen from the curve shown in Figure 2. In this case, an impeller diameter of 11 in. was chosen.

Centrifugal pumps may be operated singly, in series, in parallel, or in a combination of series and parallel. When pumps are operated in series, the heads are additive at a given rate of flow. When pumps are operated in parallel, flow is additive at a given value of head. This is illustrated in Figure 3. When pumps are operated in a series—parallel combination, these same rules will apply to determine the total head developed by the combination of pumps. Figure 4 shows performance data for one pump and for series and parallel operation of two pumps.

Pumps connected in series have the discharge of one pump connected to the suction of the downstream pump. Pumps connected in parallel use a common suction line and common discharge line. (See Figure 4a.) Pumps that are to be operated in parallel must be matched so that both develop the same head at the same flow rate. They should also have constantly rising head curves. Pumps with a head curve that has a hump (one that develops less head at shutoff than at some other point on the head curve) should be avoided since this could possibly preclude the second pump from pumping.

Table 1
Conversion Factors and Formulas[*]

$$\text{gpm} = 0.07 \times \text{boiler hp}$$

$$\text{gpm} = 449 \times \text{cfs}$$

$$\text{gpm} = 0.0292 \times \text{bbl/day}$$

$$\text{gpm} = 0.7 \times \text{bbl/hr}$$

$$\text{gpm} = 4.4 \times \text{cubic meters/hr}$$

$$\text{gpm} = \frac{\text{lb per hr}}{500 \times \text{specific gravity}}$$

$$H = \frac{P \times 2.31}{\text{specific gravity}}$$

$$V = \frac{Q \times 0.321}{A}$$

$$U = \frac{\text{Diameter (in.)} \times N}{229}$$

$$h_v = \frac{V^2}{2g}$$

$$\text{whp} = \frac{Q \times H \times \text{specific gravity}}{3,960}$$

$$\text{bhp} = \frac{Q \times H \times \text{specific gravity}}{3,960 \times e}$$

$$\text{bhp} = \frac{Q \times P}{1,715 \times e}$$

$$T = \frac{\text{bhp} \times 5,250}{N}$$

$$N_s = \frac{N\sqrt{Q}}{H_1^{3/4}} = \frac{N\sqrt{\sqrt{H_1}} \times \sqrt{Q}}{H_1}$$

$$S = \frac{N\sqrt{Q}}{h_{sv}^{3/4}} = \frac{N\sqrt{\sqrt{h_{sv}}} \times \sqrt{Q}}{h_{sv}}$$

$$t_r = \frac{H(\frac{1}{e} = 1)}{780 \times C}$$

where
N = speed in rpm
N_s = specific speed in rpm
S = suction specific speed in rpm
Q = capacity in gpm
P = pressure in psi
H = total head in ft
H_1 = head per stage in ft
h_{sv} = net positive suction head in ft
h_v = velocity head in ft
whp = water horsepower
bhp = brake horsepower
U = peripheral velocity in ft per second
g = 32.16 ft per second (acceleration of gravity)
mgd = million gallons per day
cfs = cubic ft per second
bbl = barrel (42 gallons)
C = specific heat
psi = lb./in.^{2}
gpm = gallons per minute
e = pump efficiency in decimal
D = impeller diameter in in.
V = velocity in ft per second
T = torque in ft-lb.
t = temperature in °F
t_r = temperature rise in °F
A = area in in.^{2}

[*]Courtesy United Centrifugal Pumps.

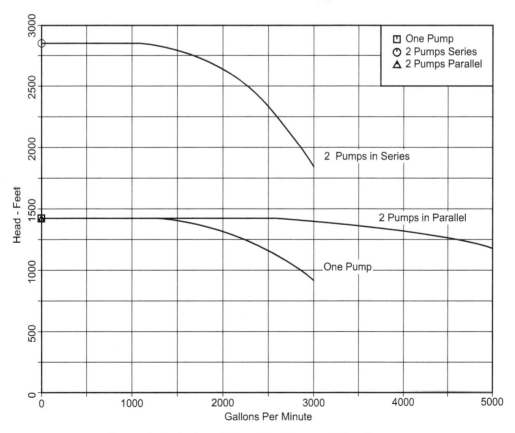

Figure 4. Series/parallel combinations—centrifugal pumps.

Figure 4a. Pumps connected in parallel/in series.

power. If as many as three series units can be justified, the units should be sized as follows:

One unit—full head
One unit—3/4 of the head of the full head unit
One unit—1/2 of the head of the full head unit

This will allow the following pump combinations to be used:

Units	Total head	Units	Total head
½	½	½, 1	½
¾	¾	¾, 1	1¾
1	1	½, ¾, 1	2¼
½, ¾	1¼	¾	

The high cost of energy will in almost all cases justify the additional cost of multiple pumping units or variable speed drivers. The use of multiple units will allow most pumping conditions to be met without throttling and therefore wasting

Increments of ¼ head can be achieved up to the last ½-head increment. This results in maximum flexibility when constant rpm drivers are used.

Figure 5 shows the performance of series versus parallel operation of pumps. In this particular example, series

Figure 5. .Series/parallel combinations with system curve—centrifugal pumps.

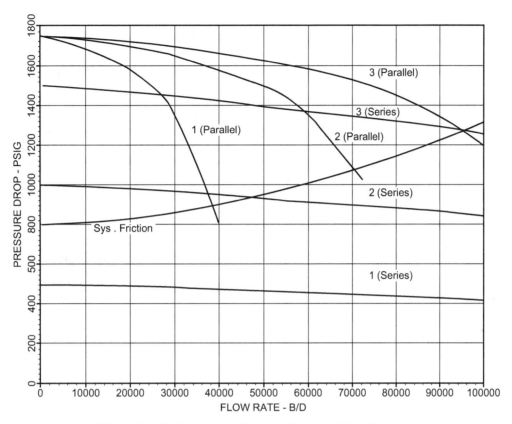

Figure 5a. Series and parallel operation—centrifugal pumps.

operation will provide a higher pumping rate than provided with a parallel operation, and two pumps in parallel will pump only slightly more than one single pump. The maximum flow rate with two pumps in series will be 2594 gpm or 89,439 b/d. The maximum flow rate for one pump will be 1917 gpm or 66,092 b/d.

For system rates between 66,092 b/d and 89,439 b/d, pressure throttling will be necessary to match the pump performance to the system curves, unless the pumps are equipped with variable speed drives. If throttling is necessary, it may be minimized by operating at a high flow rate for part of the day and at a lower rate for the balance of the day. If a system rate of 70,000 b/d is desired with the system shown in Figure 5, one pump operating 20 h per day and two pumps operating 4 h per day would achieve this rate; however, this may not be the most economical method of operation. It will be necessary to calculate and compare the cost of throttling pressure versus the cost to operate the second unit for 4 h. System rates less than 66,092 b/d will require pressure throttling unless the pump speed may be varied. It is not uncommon for pipeline pump stations to contain several pumps, each of which may be sized differently to increase the amount of flow rate flexibility.

The use of multiple pumps affords some protection against throughput reductions due to one or more pumps being out of service.

Parallel pump operation is usually best suited for systems where the residual static pressure is high compared to the line friction loss. Such systems usually have a flat system curve. An example of this is shown in Figure 5a.

For this system, three parallel pumps would offer the most flexibility, since one unit will always be able to overcome the system friction and static and would pump almost as much as two series pumps. One could make the argument that one large pump could be selected to meet the 2 (series) condition shown in Figure 5a and a ½-head pump could be used for the maximum volume condition. This argument is valid as long as the large pump is available. When it is not available, the system will be down, since the ½-head unit will not develop enough head to overcome the static and friction losses. Three pumps operating in parallel will offer the most flexibility for the system shown in Figure 5a.

Changes in system throughputs may require resizing the pump impellers to meet the new conditions. Some types of multistage pumps may be destaged or upstaged by adding or taking out impellers, depending upon the number of stages in

Table 2
Formulas for Recalculating Pump Performance with Impeller Diameter or Speed Change[*]

Q_1, H_1, bhp$_1$, D_1 and N_1 = Initial Capacity, Head, Brake Horsepower, Diameter, and Speed.
Q_2, H_2, bhp$_2$, D_2 and N_2 = New Capacity, Head, Brake Horsepower, Diameter, and Speed.

Diameter Change Only	Speed Change Only	Diameter a Speed Change
$Q_2 = Q_1 = \left(\dfrac{D_2}{D_1}\right)$	$Q_2 = Q_1 = \left(\dfrac{N_2}{N_1}\right)$	$Q_2 = Q_1 = \left(\dfrac{D_2}{D_1} \times \dfrac{N_2}{N_1}\right)$
$H_2 = H_1 = \left(\dfrac{D_2}{D_1}\right)^2$	$H_2 = H_1 = \left(\dfrac{N_2}{N_1}\right)^2$	$H_2 = H_1 = \left(\dfrac{D_2}{D_1} \times \dfrac{N_2}{N_1}\right)^2$
bhp$_2$ = bhp$_1$ $\left(\dfrac{D_2}{D_1}\right)^3$	bhp$_2$ = bhp$_1$ $\left(\dfrac{N_2}{N_1}\right)^3$	bhp$_2$ = bhp$_1$ $\left(\dfrac{D_2}{D_1} \times \dfrac{N_2}{N_1}\right)^3$

[*]Courtesy United Centrifugal Pumps.

operation and the amount of head change required. If it is contemplated that it may be necessary in the future to destage or upstage, this should be specified when the pump is purchased. It is also possible to optimize the efficiency point by chipping the pump volutes and under-filing the pump impeller(s) or adding volute inserts. The pump manufacturer should be contacted for details on how to accomplish this.

The pump affinity laws shown in Table 1 may be used to predict the performance of the pump at different operating conditions. These laws are commonly used to determine new impeller diameters required for nominal changes in flow rates or performance at different speeds. The affinity laws will allow the translation of a point on the performance curves but will not define the operation of the pump in the system. A new pump performance curve may be constructed using these laws and then applied to the system curve to determine the actual operation of the pump.

The system shown in Figure 5, with both pumps configured with 11-in. diameter impellers, will pump a maximum of 89,439 b/d. The system throughput requirements have now changed such that the maximum desired throughput is 87,931 b/d, and it is desired to trim the impellers in one of the pumps to meet this new condition. The new $H-Q$ condition for one pump will be $H_2 = 955$ ft and $Q_2 = 2550$ gpm. In order to establish the new impeller diameter, it will be necessary to translate this point to the existing pump curve to determine H_1 and Q_1. Select flow rates on both sides of the

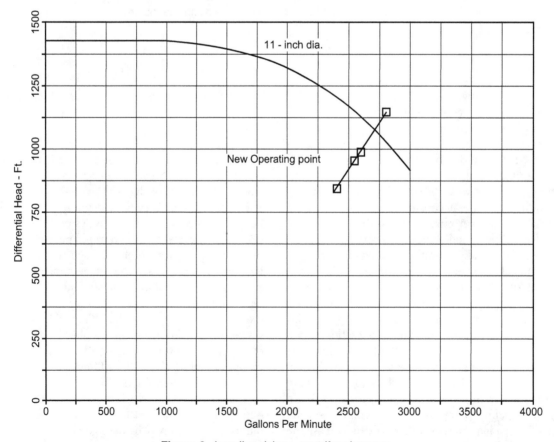

Figure 6. Impeller sizing—centrifugal pumps.

new rate and calculate the corresponding head. These data are plotted, and the resulting line will cross the existing pump head curve as shown in Figure 6. Determine the head and flow rate at the point where this line crosses the existing head curve, and these values will be H_1 and Q_1. For this calculation, flow rates were selected as follows:

Selected Q	Calculated H
2800	1151
2600	993
2400	846

The head was calculated as follows:

$$H = H_2 \times (Q/Q_2)^2$$

This line crosses the existing head curve at 2700 gpm and 1080 ft, and these values will be the values to use for Q_1 and H_1, respectively.

Use the pump affinity law to calculate the new diameter as follows:

$$H_2/H_1 = (D_2/D_1)^2$$

$$D_2 = D_1 \times (H_1/H_2)^{1/2}$$

$$D_2 = 11.0 \times (955/1080)^{1/2}$$

$$D_2 = 10.34 \text{ in}$$

Since the pump affinity laws are not quite exact for a diameter change, it will be necessary to apply a correction factor to the calculated value of D_2 to determine the final diameter. The correction factor is determined by calculating the ratio of the calculated diameter to the original diameter. This value, expressed as a percentage, is then used to enter Figure 7 to determine the actual required diameter as a percentage of the original diameter.

$$D \text{ calc}/D \text{ orig} = 10.34/11 = 94\%$$

Enter the chart with 94% and read 94.5% as the actual required diameter as a percentage of the original diameter.

The new diameter will therefore be $0.945 \times 11.0 = 10.395$ in. The impellers would probably be trimmed to 10 7/16 in.

Once the new diameter is established, the new pump head curve may be calculated by using the affinity laws to calculate new $H-Q$ values.

The pump affinity laws are exact for a speed change. Figure 7a shows the effect of a speed change on the pump performance. The pump manufacturer should be consulted for data on the range of speeds over which the pump may be efficiently operated.

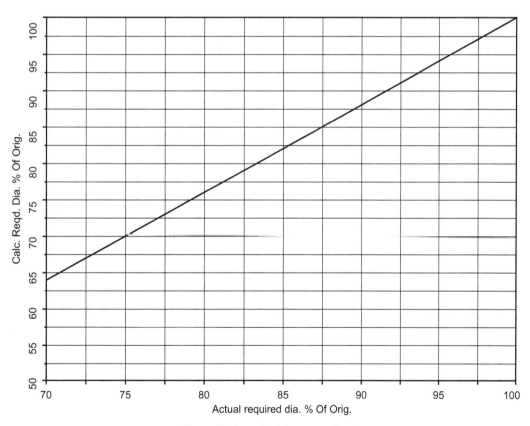

Figure 7. Impeller trim correction.

Figure 7a. Centrifugal pump performance data for speed.

Very viscous liquids will have an impact on the performance of centrifugal pumps. Figure 11 compares pump performance when pumping water to pumping a liquid with a viscosity of 6000 SUS. The effects of viscosity on pump performance may be predicted by the use of Figures 12 and 13. These charts are used as follows. Refer to Figure 11 and select the water capacity at the BEP (best efficiency point). In this example, the water capacity at the BEP is 23,000 gpm. Refer to Figure 12 and enter the chart with 23,000 gpm and move vertically to intersect the desired impeller diameter, 27 in, and then horizontally to intersect the 6000 SUS line. Move vertically to intersect the line representing the water efficiency (85%), then horizontally to read the viscous efficiency 62% (not a correction factor). The previous vertical line is extended to intersect the head correction and capacity correction curves. The capacity correction factor is 93%, and the head correction factor is 95%. The viscous head at shutoff will be the same as the shutoff head for water. The viscous head will be 617 ft at 21,400 gpm. Plot this point and then draw the head-capacity curve through this point. The viscous

Figure 8. NPSH calculation for suction lift.

Figure 9. NPSH calculation for liquid at boiling point.

Figure 10. NPSH calculation for pressured drum.

Figure 11. Effects of viscosity on centrifugal pump performance.

Figure 12. Viscosity correction chart—capacities above 10,000 gpm. Courtesy United Centrifugal Pumps.

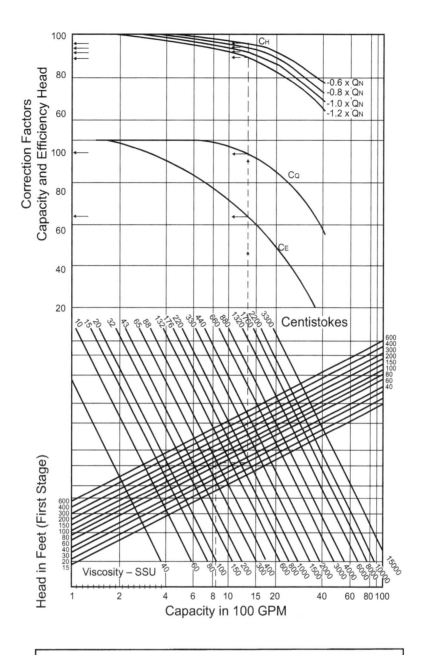

Figure 13. Viscosity correction chart—capacities up to 10,000 gpm. Courtesy United Centrifugal Pumps.

brake horsepower is calculated at the BEP (62% at 21,400 gpm). Draw the viscous brake horsepower curve through this point. The slope of the line will follow the slope of the water brake horsepower curve. The viscous efficiency will be 62% at 21,400 gpm, and the viscous efficiency curve may now be calculated using the viscous brake horsepower, viscous head and capacity, and the viscous specific gravity, which in this example is 0.9.

Figure 13 is used for pumps with capacities up to 10,000 gpm, and Figure 12 is used for pumps with capacities greater than 10,000 gpm.

There may be instances where it is necessary to calculate the pump speed torque requirements to determine if the pump driver will be able to accelerate the pump to operating speed. Example 1 illustrates this calculation procedure, and the results are shown in Figure 14.

Speed torque calculation

The formula for calculating the speed-torque relation for a pump with a given specific gravity and viscosity is:

$$\text{Torque(lb-ft)} = \frac{5250 \times \text{HP}}{\text{RPM}}$$

The following data are required to plot the speed-torque curve:

1. HP @ Shutoff.
2. HP @ Open Valve (use one of the following):
   - 2a. HP @ 120% of peak capacity.
   - 2b. HP rating of driver if greater than HP @ 120% of peak capacity.
   - 2c. Depending on pump application maximum horsepower on performance curve which is greater than driver rated horsepower.

NOTE: Using "2a," or "2b" will cover most of the applications. However on some applications "2c" must be considered.

3. Pump RPM.
4. Torque @ Shutoff.
5. Torque @ "2a," "2b" or "2c". From the above data the torque with the valve closed (using HP @ Shutoff), and the torque with the valve open (using HP @ "2a," "2b" or "2c") at full speed can be computed.

The torque varies as the square of the speed; therefore to obtain the torque at:

3/4 Speed—multiply full speed torque by—0.563
1/2 Speed—multiply full speed torque by—0.250
1/4 Speed—multiply full speed torque by—0.063
1/8 Speed—multiply full speed torque by—0.016

From the figures a speed-torque curve for both closed and open valve, covering the complete speed range, can be plotted.

	Example	
HP @ shutoff = 80	Torque @ Shutoff $= \dfrac{5250 \times 80}{3550} = 118$ lb ft @ 3550 RPM	
HP @ 120% of peak capacity = 150		
Pump RPM = 3550	Torque @ 120% of peak cap. $= \dfrac{5250 \times 150}{3550} = 222$ lb ft @ 3550 RPM	

Speed-RPM	Torque-Valve Open (150 HP)	Torque-Valve Closed (80 HP)
Full—3550	Full—222	Full—118
3/4—2660	0.563—125	0.563—67
1/2—1775	0.250—56	0.250—30
1/4—885	0.063—14	0.063—8
1/8—445	0.016—4	0.016—2
0—0	0.050—11	

Figure 14. Courtesy United Centrifugal Pumps.

Variable Speed or Impeller Trim

The simplest way to change pump flow is to throttle the discharge by the valve. Even in this situation, the impeller diameter cannot be arbitrarily chosen. The diameter should be selected to ensure that the pressure (head) is not excessive and that the power consumed provides the required coverage for the anticipated flow range.

Impeller size selection is straightforward and can be easily found in manufacturer manuals. Figure 1 shows pump performance curves at several impeller trims. Such representation is called a combined curve (versus a single line curve) and is intended to help the pump user select the appropriate impeller diameter for the rated point (and around it) and evaluate efficiency, power, and NPSHR.

Unfortunately, if the conditions of service change to require more flow and head than originally specified, the trimmed impeller size would limit such flexibility. Moving to 100 gpm and 120 ft. of head for a rated point of 70 gpm at 100 ft (as shown in Figure 1) is impossible; the 5.12-in. impeller would not allow it.

Changing the operating conditions by a variable speed drive can be a flexible alternative, as shown in Figure 2. Inefficiently throttling the valve to change the flow is unnecessary, and the entire range of flow and head expands dramatically.

The affinity laws calculate the required pump's speed if a flow range equivalent to that in Figure 1 is desired. Each operating flow-head point scales up or down by simply multiplying the flow by the ratio of the diameter (or speed) and the head by the square of that ratio. This is how Figure 2 is constructed. In our example, several points on the H-Q curve shown in Figure 1 are picked, for example 0, 40, 80, and 120 gpm with heads at these points being 15, 150, 138, and 112 feet.

If we wanted to make the new speeds correspond to exactly the same impeller diameters given in Figure 1 (6, 0.6, 5.0, 4.5, 4.0 and 3.5 in.), we simply calculate the speed ratios corresponding to the ratios of the impeller trim at a given curve to the impeller diameter at the maximum OD.

The main advantage of the VFD with regard to efficiency and energy savings is that efficiency essentially remains the same for a given "scale-down" of speed. Trimming the impeller diameter, while allowing the same modification to the head-flow curve, can be inefficient (compare Figure 1 to Figure 2).

For example, BEP (best efficiency point) efficiency at maximum diameter is 58% and drops to about 43% when the impeller is trimmed from 6 in to 3.5 in. With equivalent reduction of speed from 3500 rpm to 2014 rpm, efficiency essentially remains the same. You can easily calculate the energy saved by this method.

Figure 1.

Figure 2.

Scaling down the H-Q curve by trimming the impeller follows the affinity laws only approximately, and certain correction factors must be applied. Efficiency decreases with a reduction in speed. However, this reduction in efficiency with speed is small.

In general, the higher the power, the more money saved—as long as you properly account for all variables, including initial purchase cost of the VFD, installation and maintenance, and efficiency difference.

Used with permission of Dr Lev. Nelik—Pumping Machinery LLC Atlanta, GA-PumpingMachinery.com. This article originally appeared in the June 2010 issue of Pumps & Systems Magazine.

PULSATION CONTROL FOR RECIPROCATING PUMPS

Suppress the surge in the suction and discharge systems of your positive displacement pumps

V. Larry Beynart, Fluid Energy Controls, Inc.

This article reviews the specific requirements for the selection and sizing of pulsation dampeners to reduce the fluid pressure pulsations generated by positive displacement reciprocating pumps. It includes analyses of the nature of pressure pulsation and explains the advantages of using pulsation dampeners and suction stabilizers. Sizing procedures for dampeners illustrate how their application improves suction and discharge control in reciprocating pumps, reduces pressure pulsations, increases Net Positive Suction Head (NPSH), and makes the service life of pumps longer and more efficient.

Figure 1. The rotary motion of the crank movement is transformed into the reciprocating motion of the piston.

Combating pulsation

The primary purpose of pulsation control for reciprocating pumps is to attenuate or filter out pump-generated pressures that create destructive forces, vibration, and noise in the piping system.

Every reciprocating pump design has inherent, built-in pressure surges. These surges are directly related to the crank-piston arrangement (Figure 1). Fluid passing through a reciprocating pump is subjected to the continuous change in piston velocity as the piston accelerates, decelerates, and stops with each crankshaft revolution. During the suction stroke, the piston moves away from the pump's head, thereby reducing the pressure in the cylinder. Atmospheric pressure, which exists on the surface of a liquid in a tank, forces liquid into the suction pipe and pump chamber. During the discharge stroke, the piston moves toward the pump's head, which increases the pressure in the cylinder and forces the liquid into the discharge pipe. This cycle is repeated at a frequency related to the rotational speed of the crank.

At every stroke of the pump, the inertia of the column of liquid in the discharge and suction lines must be overcome to accelerate the liquid column to maximum velocity. At the end of each stroke, inertia must be overcome and the column decelerated and brought to rest. Figure 2 compares pump pulsation frequencies to pipe span natural frequencies of vibration.

A key benefit of pulsation control is the reduction of fatigue in the pump's liquid end and expendable parts. Reducing the pressure peaks experienced by the piston will in turn reduce the power-end peak loading.

		Frequency, Hz		
RPM	Pump	Duplex Pulse[a]	Triplex Pulse[a]	Quint Pulse[a]
50	0.8	3.2	2.4	4.0
100	1.7	6.8	5.1	8.5
150	2.5	10.0	7.5	12.5
200	3.3	13.2	9.9	16.5

[a]Pulse = pump RPM x number of cyliders/60

(a)

◄──────── 30' ────────►	2",5lb Pipe	6",19 lb Pipe
	2.9 Hz	8.5 Hz
	4.3 Hz	12.5 Hz
	1.9 Hz	5.5 Hz

(b)

Figure 2. (a) Typical pump and pulse frequencies. (b) Typical piping natural frequencies.

The source of fluid pressure pulsations

It is almost mandatory that multiple pump installations have well-designed individual suction and discharge pulsation control equipment. While multiple pumps may be arranged to run at slightly different speeds, it is impossible to prevent them from frequently reaching an "in-phase" condition where all the pump flow or acceleration disturbances occur simultaneously. Having more than one pump can multiply the extent of the pipe vibration caused by such disturbances because the energy involved is similarly increased.

The reduction of the hydraulic pressure pulses achieved by using pulsation control equipment is usually reported in total pulsation pressure "swing" as a percentage of average pressure. This method is used widely within the industry and is often recommended as a standard. Referring to Figure 3, the percent of residual pulsation pressure can be calculated as follows:

Case I: $\Delta P/\text{Ave} \times 100 = 460/1000 \times 100 = 46\%$
Case II: $\Delta P/\text{Ave} \times 100 = 70/100 \times 100 = 7\%$

It is very important that any reference to degree of pulsation should apply to the total excursion in terms of pressure and percentage. In Case I, the total pulsation of 460 psi (46%) implies that the excursion is from 680 psi (32% below the average) to 1140 psi (14% above the average).

Reciprocating pumps introduce into the suction and discharge systems three apparently unrelated pressure disturbances, which are illustrated in Figure 4. These include:

- A low-frequency disturbance at A, based on the rate at maximum flow velocity pressures
- A higher frequency due to maximum acceleration pressure at the beginning of each piston stroke at B
- A pressure disturbance at the point of flow velocity change (valley) at C

Major problems caused by pressure pulsations

Unsteady flow. The most obvious problem caused by pulsation is that flow is not constant, which can lead to problems in processes where a steady flow rate is required, such as applications involving spraying, mixing, or metering. Flow rates can be difficult to measure with some conventional types of flowmeters.

Noise and vibration. Many reciprocating pump installations suffer from problems that can lead to excessive maintenance costs and unreliable operation. Typical examples are noise and vibration in the piping and the pump. Vibration can lead to loss of performance and failure of valves, crossheads, crankshafts, piping, and even pump

barrels. High levels of pulsation can occur when the pulsation energy from the pump interacts with the natural acoustic frequencies of the piping. A reciprocating pump produces pulsation at multiples of the pump speed, and the magnified pulsation in the system is generally worse at multiples of the plunger frequency. Most systems have more than one natural frequency, so problems can occur at differing pump speeds.

Shaking. Shaking forces result from the pulsation and cause mechanical vibration of the piping system. These forces are a function of the amplitude of the pulsation and the cross-sectional area of the piping. When the exciting frequency of the pulsation coincides with a natural frequency of the system, amplification and excessive vibration occur. Amplification factors can be as high as 40 for pulsation resonance and 20 for mechanical resonance. If the mechanical resonance coincides with the acoustic resonance, a combined amplification factor of 800 is possible.

Wear and fatigue. Pulsation can lead to wearing of pump valves and bearings and, if coupled with mechanical vibration, will often lead to loosening of fittings and bolts and thus to leakage. In severe cases, mechanical fatigue can lead to total failure of components and welded joints, particularly when the liquid being pumped is corrosive.

Cavitation. Under certain pumping conditions, areas of low system pressure can occur. If this pressure falls below the liquid's vapor pressure, local boiling of the liquid will occur and bubbles of vapor will form. If the liquid has dissolved gases, they will come out of solution before the liquid boils.

Pressure waves summary

The earlier analysis of the source and nature of fluid pressure pulsations enables us to build the idealized pressure wave pattern (Figure 5). Although identical forces are acting on the liquid in the suction and discharge sides of a pump, the effects of these forces vary greatly. The major difference is that acceleration effects in the suction at B tend to separate the liquid into vapor or "cavitation" pockets that collapse with a high-magnitude pressure pulse. Acceleration on the discharge side of the piston at B, however, tends to compress

Figure 3. Reporting the degree of pulsation and control. (a) Non-dampened waveform from flow variation (Case I). (b) Dampened waveform from flow variation (Case II).

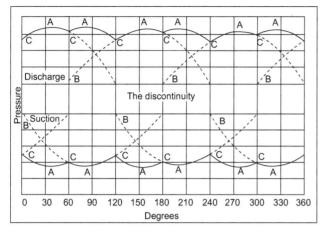

Figure 4. Triplex single-acting pump-points of induced pressure disturbances.

the liquid and create an impulse pressure pulse. This requires separate consideration of the suction and discharge systems when choosing and installing pulsation control and suction stabilizing equipment.

Pulsations in suction systems

The friction losses in a suction system are usually low because of the relatively short length and large diameter of the piping involved. Accordingly, the flow-induced A-type pressures are of low magnitude compared to the accepted percentage of change (Figure 4). For example, if the suction pressure is a static 20 psi, the 23% variation of a triplex single-acting pump would generate a theoretical A-type pressure variation of only 9.2 psi. This is hardly enough energy to set pipes in motion. The same flow variation at 1400 psi working pressure, however, creates more than enough energy—a 644 psi A-type pressure variation in the discharge line.

Even so, the forces of acceleration become overwhelming disturbances in the suction. Pressure pulses of more than 25 psi are often encountered in pumps—even in systems with short suction pipes. A small amount of dampening of the flow-induced pressure can reduce pulsation to a negligible amount, leaving the 25-psi C-type acceleration pulsations. These remaining pulsations can create damage by enabling cavitation (Figure 4).

Cavitation. In practice, cavitation occurs at a pressure slightly higher than the vapor pressure of the liquid. Because vapor pressure is a function of temperature, a system may run without cavitation during the winter months but be subject to cavitation problems during the summer when temperatures are higher. Cavitation is connected with the presence of microscopic gas nuclei, which are in the solid

materials bonded in the liquid. It is because of these nuclei that liquids cannot withstand tension. Without them, it is estimated that water could transmit a tension of approximately 150,000 psi.

During the beginning stages of cavitation, the nuclei give rise to the formation of gas bubbles, which are swept along to an area where the pressure is higher. The bubbles then collapse, producing high-intensity pressure waves. The formation of successive bubbles quickly follows. This repeating cycle lasts for only a few milliseconds, but during this time the localized pressures can be as high as 60,000 psi and temperatures can rise as much as 1500 °F. Cavitation can occur at a pump's inlets, inside the barrels, at the draft tubes of turbines and on propellers, mixing blades, Venturi meters, and other places where changes in fluid velocity occur. The cavitation effects in rotary positive displacement pumps are similar to those in rotodynamic pumps. The usual effects are wear of surfaces in contact due to erosion, slow increase in clearances, reduction in performance, noise, and vibration. Cavitation damage in reciprocating pumps can happen very quickly and be catastrophic. A cavitating 60 hp pump can break a major component every 350 h.

Suction control. Suction control is an essential step in the design of any pump system and is equally important for reciprocating and centrifugal pumps. This process requires the following major steps:

1. Compare Net Positive Suction Head available (NPSHa) to Net Positive Suction Head required (NPSHr). (NPSH is a feature of the system layout and liquid properties. It is often referred to simply as NPSHa or NPSH available. Pump manufacturers usually quote NPSHr. This is a function of the pump's performance and is related to the pump's susceptibility to cavitation. To calculate the NPSHa to the pump, it is necessary to take into account the friction losses of the suction piping. NPSHa is then the difference between the total pressure on the inlet side of the pump and the sum of the vapor pressure of the liquid and the friction losses of the suction pipes.)
2. NPSHa is greater than NPSHr, the general suction conditions are under control, and their improvement (with respect to NPSHa) by installation of a suction pulsation dampener is not required. In this situation, the installation of a pulsation dampener would be required only for pressure fluctuation and noise reduction purposes.
3. If NPSHa is less than NPSHr and the piping rearrangement of a suction system is impossible or insufficient, installation of a pulsation dampener before the pump's inlet is essential.
4. Choose the appropriate dampener design based on steps 1–3. Using this process will help you achieve maximum reduction of pressure fluctuation.

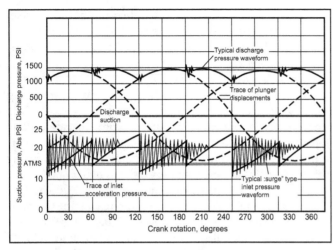

Figure 5. Idealized triplex single-acting pump pressure waveform showing the effect of surge-like acceleration pressures.

Pulsations in discharge systems

The same forces at work on the suction side also affect discharge systems, but the pressure at A due to flow-induced pulsations becomes overwhelming at 460 psi (Figure 4). The 25 psi contribution from acceleration at C is a small percentage (2.5%) of the total discharge pressure. One exception is when the pump is delivering into a low-friction, high-pressure system such as a short vertical discharge system (Figure 6a). An example is a mine dewatering application, where the pump is delivering to an already pressurized system—a pressurized pipeline—through a short connecting pipe. Another exception is when the pump is delivering to already pressurized systems such as hydraulic press accumulators (Figure 6c). In those two cases, the acceleration pressures can become the overwhelming disturbance, particularly if the piping system is relatively long compared to a suction system (but considerably shorter than a "pipeline").

Control of pulsation with dampeners

Successful control of damaging pulsation requires careful selection of pulsation dampeners and proper location of those dampeners in a piping system. The earlier pressure pulsation analysis showed that dampeners will reduce the level of pulsation and consequently pressure fluctuations. With lower pipe vibration and noise, wear at the pump, risk of cavitation, and metal fatigue will be less as well. (There are several OSHA standards dedicated to the control of industrial noise in the workplace. One of the most practical and inexpensive ways to meet OSHA noise level requirements and achieve the overall performance improvement of reciprocating pumps is installation of pulsation dampeners on suction and discharge systems.)

Figure 6. Dynamic differences in discharge systems. Vertical discharge. Horizontal discharge. Pressurized systems.

Dampener types. After evaluating the causes and types of pressure pulsation effects in reciprocating pumps, we can review which available fluid pressure pulsation control equipment is best for specific conditions.

Hydropneumatic pulsation dampeners are gas-type energy-absorbing design that uses a gas-filled bladder to absorb the "A" flow on peaks and to give back that flow on the "lows," thereby reducing flow variations beyond the pump and consequently reducing the flow-induced pressure pulsations. This is the most efficient, low-cost, and widely used type of pulsation control equipment.

This discussion is limited to the two major designs of bladder-type hydropneumatic pulsation dampeners: appendage and flow-through.

Appendage dampeners. Figure 7 shows a typical appendage design suction pulsation dampener. It consists of a homogenous shell and a nitrogen-filled flexible rubber bladder. The fluid port can be threaded (NPT or SAE) or flanged. This design is recommended for installation on pumping systems where suction is flooded. The dampener should be installed as close as possible to the pump inlet, and the nitrogen in the bladder should be precharged to 60% of the pump's inlet pressure. Figure 8 shows a large capacity version of this design.

The same type of suction pulsation dampener (150 psi) was installed on a 120 gpm and 30 psi Wheatley triplex pump with 3″ piston bore, 3-1/2″ stroke. Another two discharge 1500 psi pulsation dampeners were used with a 263 gpm and 642 psi Wheatley quintuplex pump with 3-3/8″ bore, 4-1/2″ stroke.

Flow-through dampeners. This design should be installed on a pump's suction system when inlet pressure (NPSH) is very low and the fluid can contain entrained gases such as air, carbon dioxide, sulfur dioxide, etc. Flow-through units have significant advantages over the appendage design for a few reasons. First, the inlet and outlet diameters of the dampener are equal to the pipe diameter. Second, the in-line installation enables reduction of the inlet fluid velocity into the pump. In addition, the bladder response time is faster than an appendage dampener, due to the minimum distance between the shell inlet and pipe centerline. The flow-through design can also be used to improve the NPSHa of systems with low NPSH because it acts as a suction stabilizer.

Figure 9 shows an actual installation of a discharge flow-through 2.5 gallon capacity dampener. This carbon steel, 3000 psi, 3″ flanged unit was installed on the discharge system of an existing triplex single-acting reciprocating pump with 2.88″ piston bore diameter and 5″ stroke—one of seven such units. The owner, Ormet Aluminum Mill Products Corporation, is a Hannibal Rolling Mill Division, Hannibal, Ohio. Currently, the company is scheduled to install another two pulsation dampeners of the same type. The installation of pulsation dampeners for the remaining pumps is scheduled for sometime in 2000.

Figure 7. Appendage type 60 in³ 1500 psi stainless steel pulsation dampener with threaded fluid port.

Figure 8. Typical large capacity appendage type flanged pulsation dampener.

The abilities of pulsation dampeners

Hydropneumatic bladder pulsation dampeners are able to prevent the generation of the most destructive low-frequency pulses such as those generated by the pump's rotary motion and the combination of flow from each of the pump's cylinders. (This is basically rpm multiplied by the number of cylinders.) Accordingly, based on a maximum of 500 rpm for most small pumps, the maximum frequency involved should not be more than about 50 Hz.

Performance comparisons. Figure 10 compares the performance of the two types of pulsation dampeners on a typical pump. In this particular case, installation of hydropneumatic dampeners offers overall improved performance.

An analysis of reciprocating pump pulsations is based on the premise that pressure or pressure variation (pulsation) due to flow velocity (flow-velocity variation) is a function of the square of such velocity. And the system must present enough resistance to flow—due to pipe friction, bends,

Figure 9. Flow-through pulsation dampener. Courtesy of Ormet Aluminum Mills Products Corporation, a Hannibal Rolling Mill Division.

Figure 10. (a) No dampener installed. (b) Appendage type hydropneumatic dampener. (c) flow-through type dampener. Top chart: downstream of dampener. Bottom charts: upstream of dampener.

fittings, chokes—to enable the generation of a pressure or pulsation in the first place. The magnitude of that pressure is directly related to the resistance. The analysis is also based on the premise that the acceleration of the liquid at the beginning of an increase in velocity (start of stroke) generates a pressure pulse, and the magnitude is directly related to the length of pipe connected to the pump suction or discharge. It is also inversely related to the diameter of the pipe and directly related to the rate of change of velocity.

Because the inherent flow variations of a reciprocating pump are of the same intensity (but out of phase) in both the suction and discharge, for many years it was believed that a pulsation dampener of the same size as the discharge should be used on the suction, and many such installations can be found. However, from the standpoint of flow-induced pulsations alone, it is logical that a smaller dampener can be used on the suction side. For example, a typical triplex single-acting pump has a flow variation of about 23% in both the suction and the discharge. If all of this flow variation were converted to pressure pulsations, the pulsations would be a maximum of 46% of the average pump pressure. In our example, the maximum change in pressure at the discharge would be about 460 psi at 1000 psi average pressure. If there is a 50 psi suction pressure, the change in pressure there would be only 23 psi; that is, there would be about 20 times less energy available to set the piping in motion.

In addition, the acceleration due to velocity variations is the same on the suction and discharge sides. In the suction system, however, such disturbances usually overwhelm the disturbances due to flow variation, and conversely, in the discharge system the flow-induced pressure pulses usually overwhelm the acceleration pressure pulses.

There is evidence that only a flow disturbance or an acceleration disturbance (not both) can exist during any suction or discharge stroke of a pump. One will overwhelm and reduce the effect of the other. A plausible explanation is the theory that in pumps of greater than about 300 fpm piston speed, a disturbance of either type will momentarily reduce the volumetric efficiency to such an extent that it cannot recover fast enough to enable the flow rate to reach an adequate value to affect another disturbance in the short period of time (about 3 ms) between the disturbances. There

is no such worry with slow speed pumps because few if any disturbances are generated.

It is possible for a pump with high inherent flow variations to actually deliver an almost pressure-pulseless liquid into an open-ended (in effect), short, relatively large-diameter horizontal discharge pipe. Some typical cases are pumps with short connecting pipes feeding a large-diameter already-pressurized system such as an existing pipeline or hydraulic press supply. If the discharge pipe is vertical, as in a mine dewatering system, the acceleration of the liquid column due to gravity becomes the predominant generator of pressure pulses. This great dynamic and hydraulic difference is shown in Figure 6, where the pump "sees" the same discharge pressure under three different physical arrangements. Energy-absorbing or hydropneumatic pulsation dampeners may not solve pulsation problems in cases (a) and (c) of Figure 6.

Application of a large-capacity, flange type of surge suppressors should be carefully considered (Figure 8).

Advantages of bladder type pulsation dampeners

- Positive sealing between the gas and fluid chambers, such as the bladder, serves as a permanent wall between them
- Zero leakage between both chambers due to bladder barrier
- Quick pressure response times for pressure control pump pulsation and shock-dampening applications are facilitated by the light weight of the bladder
- Simplicity of design reduces overall maintenance and operating cost of pumping system

Sizing considerations

Suction pulsation dampeners and suction stabilizers. The following sizing procedure will serve two purposes:

- Evaluation of the pump suction system and subsequent sizing of the pulsation dampener and suction stabilizer with respect to the reduction of pressure fluctuation
- Improvement of pump suction conditions represented by NPSHa

Sample data

Piping

- The maximum flow of 200 gpm originates from the storage tank.
- The tank is open to atmosphere.
- The length of the 8″ cast iron suction pipe is 106 ft.
- The difference of elevation between the storage tank and centerline of pump suction inlet is 23 ft.
- The fittings are: (1) 8″-90° elbows, (1) 8″-45° elbows, (4) 8″ Tees.

Pump (4 each)
- Single action triplex
- Max speed = 45 rpm
- Flow rate = 200 gpm (800 gpm for all 4 pumps)
- Min NPSH = 23 ft. of column of water
- Bore = 9.0 in
- Stroke = 6.0 in

Fluid
- Anaerobically digested municipal sludge conditioned with 10% lime and 5% ferric chloride.
- Fluid temperature = 100 °F

Analysis. In order to determinate the net positive suction head available (NPSHa), the acceleration head and the friction losses along the suction piping must be calculated. Acceleration head can be calculated as follows:

$$h_a = \frac{LVnC}{Kg}$$

where

h_a = Acceleration head, ft. of fluid,
L = Length of suction line, ft,
V = Velocity in the suction line, fps,
n = Pump speed, rpm,
C = Pump type constant, for triplex single-acting pump
C = 0.066,
K = 1.4 for hot water,
g = 32.2 ft/s^2.

Pipe friction loss factor

$$K_1 = f\frac{L}{D} = \frac{0.020 \times 106}{0.336} = 6.31,$$

Valve and fitting friction loss factor

$$K_2 = (4 \text{ valves}) \times 1.8 + (1{-}90° \text{ elbow}) \times 1.5$$
$$+ (1{-}45° \text{ elbow}) \times 0.8$$
$$= 9.50$$

Total valve and fitting frictional loss factor

$$K_t = 15.81$$

Total pipe frictional losses

$$h_t = \frac{K_t V^2}{2g} = \frac{(15.81) \times (5.12)^2}{2 \times 32.2} = 6.44 \text{ ft of w.c.}$$

Acceleration head

$$h_a = \frac{(106) \times (5.12)^2 \times 45 \times 0.066}{1.4 \times 32.2} = 34.96 \text{ ft of w.c.}$$

The maximum static elevation head hs = 23.0 ft. of water column.

Net Positive Suction Head Required NPSHr = 9.87 psig @ 45 rpm = 22.8 ft of w.c., say 23 ft of w.c.

$$P_{\text{tank}} = 0 \text{ psig} = 0 \text{ ft of w.c.}$$

Net Positive Suction Head Available

$$\text{NPSHa} = h_s + P_{\text{tank}} - h_a + h_f$$
$$= (23 + 0) - (34.96 + 6.44) = -18.40 \text{ ft of w.c.}$$

Since NPSHr = 23.0 ft of w.c., the NPSHa = 18.40 ft of w.c. and is less than NPSHr. The actual suction head is not adequate to operate the pump at the 23 ft of w.c. NPSH requirement.

This analysis shows the necessity of installing the suction stabilizer to improve suction conditions. Such stabilizers can reduce the acceleration head, thereby increasing the NPSHa. The unit will also eliminate the acceleration head by minimizing the mass of fluid in the suction pipeline that must be accelerated. The suction stabilizer installed at the pump inlet will reduce the fluid column that must be accelerated to the fluid between the bladder and the pump suction manifold. This fluid mass is normally very small and requires a low-pressure fluctuation. The losses correspond to the acceleration effects and are about 67.5% of the operating pressure.

Acceleration losses

$$h_a = (0.075) \times (23 \text{ ft of w.c.}) = 1.73 \text{ ft of w.c.}$$

NPSHa for the system with the suction stabilizer is

$$\text{NPSHa} = (23.0 + 0) - (1.73 + 6.44) = 14.84 \text{ ft of w.c.}$$

As can be seen from the earlier analysis, the installation of a suction stabilizer will effectively reduce the acceleration head and provide a much greater NPSHa to the pump.

This will prevent local cavitation and increase pump efficiency. To provide the required Net Positive Suction Head in our example, the sludge conditioning tank should be raised only 8.0 ft (versus to 18.4 feet) to compensate for the friction losses between the tank and pump suction manifold.

The actual volume of pulsation dampener/suction stabilizer can be calculated using the following sizing procedure:

$$Volume = 10 \times displaced\ flow/revolution$$

$$Displaced\ flow = (0.7854 \times B^2 \times stroke \times no.\ pistons$$
$$\times action)/231\ gallons$$

where

B = Pump cylinder bore, in.
stroke = Pump piston stroke, in.
no. piston = 1 for simplex, 2 for duplex, 3 for triplex.
suction = 1 for single acting pump, 2 for double acting pump.

By installing the pump data in the formula, we get the required actual volume of pulsation dampener/suction stabilizer.

$$Volume = 10 \times (0.7854 \times 9^2 \times 6 \times 3 \times 1)/1 \times 231$$

$$= 49.6\ gallons$$

The 50-gallon suction stabilizer installed at the pump inlet will reduce the pressure fluctuation on the inlet pipeline to 1.73 ft of w.c., thereby providing an NPSHa of 33.24 ft, greater than without a suction stabilizer.

Sizing of discharge pulsation dampeners

In order to achieve optimum results from the dampener, its size (capacity) must be properly calculated. The following simple formula can be applied for sizing a pulsation dampener for either a simplex, duplex, or triplex single-acting pump.

$$Size\ of\ dampener = plunger\ area(sq\ in)$$
$$\times plunger\ stroke(in) \times 50$$
$$\times pump\ type\ factor$$

Notes:

1. 50 = pressure factor for about 65% pulsation control
2. pump type factor = 0.25 for duplex single-acting pump

Example. Determine the size of a pulsation dampener to control pulsations caused by a duplex single-acting plunger chemical metering pump handling 50% sodium chloride.

Data

- Plunger diameter $(D) = 1\frac{3}{4}$ in
- Plunger stroke $(L) = 3$ in
- Max. pressure = 150 psig
- Pump type factor = 0.25

Size of pulsation dampener = plunger area (sq in) × plunger stroke (in) × 50 × pump type factor = $\pi D^2/4 \times 3 \times 50 \times 0.25 = 90.2$ cu in. Therefore, a 0.5-gallon size dampener should be selected.

Material and design specification guidelines for pulsation dampeners and suction stabilizers

The next step is proper specification of the dampener's material of construction with respect to working conditions. The diaphragm or bladder and all other seals in the dampener shall be EPR (ethylene propylene) compound compatible with 50% sodium hydroxide (NaOH). The metal parts of the dampener shall be of 316 stainless steel. The selected dampener in this example shall dampen pulsations within 5%–10% of the mean line pressure.

Conclusion

The following factors must be thoroughly considered during the selection process of pulsation dampeners.

1. Maximum design pressure rating of the dampener must be within the operating pressure limit of the system.
2. Liquid port size of the dampener should be equal to the discharge port of the pump or equal to the discharge line size.
3. Dampener should be charged with a neutral gas (usually nitrogen) to approximately 60% of the mean system pressure.
4. Appendage type dampeners: pipe or nipple connecting the dampener to the line should be as short as possible. The diameter of connecting pipe should not be smaller than the size of the dampener liquid port.
5. The dampener should be installed as close as possible to the pump discharge port.
6. A periodic check of nitrogen gas pressure in the dampener should be made to make sure the dampener is in operable condition.
7. The dampener's rubber and metal parts must be compatible with the pumped liquid.

V. Larry Beynart is an Application Engineer with Fluid Energy Controls, Inc., a Los Angeles manufacturer of hydropneumatic accumulators and pulsation dampeners. He holds a Bachelor of Science Degree in Mechanical Engineering and Hydraulic Certification from Fluid Power Society.

Rotary pumps on pipeline services

James R. Brennan
Imo Pump, Monroe, NC

Abstract

Pipelines handling liquids in excess of about 500 SSU viscosity will normally use rotary, positive displacement pumps both for optimum efficiency and, frequently, lower initial cost. This section provides some insight into the application of rotary pumps to some of the unique requirements facing users of pipelines in crude oil transport, electric utility power generation, and general industrial needs. Coverage includes pumping system designs, flow control on variable flow systems, serial pump operation on multistation pipelines, and key components that help ensure reliable pump operation and maintainability.

Pipelines are used to transport a broad range of liquids, including crude oils, fuel oils, and refined products such as gasoline, as well as water, specialty chemicals, and others. Lines can be as short as several dozen feet to hundreds or even thousands of miles long. They can cross state or country boundaries. The modern world relies upon safe, efficient operation of all manner of pipelines, both liquid and gas. Oil producers, pipeline transport companies, refineries, and power plants are probably the largest operators of pipelines.

Types of pipeline services

Pipeline services can be broadly categorized as loading/ unloading, local transport, and long-distance transport. Examples of load/unload include pumping a relatively short distance from or to rail cars, tank trucks, barges, or tanker ships. These services are typically intermittent, perhaps 8 or fewer hours per day, perhaps not every day of the week. Differential pump pressures are usually low, less than 125 psid. The combination of low power and intermittent use makes pump operating efficiencies less important than in other services.

Local transport would be defined as a single pumping station installation with line lengths less than 50 miles. Design operating pressures can reach 1450 psig or more. Local or national regulation may dictate maximum pressures. These pipelines can be intermittent or continuous duty services and can range from modest power levels to many hundreds of horsepower or more.

Long-distance transport is usually more than 50 miles and includes multiple pumping stations along the length of the pipeline. These pipelines normally operate around the clock at partial or full capacity. Controls and instrumentation tend to be more sophisticated and automated. Long-distance pipelines will frequently operate at variable station flow rates, depending on product demand at the end terminal. Pumping stations will normally have three or more pumps with one in standby mode (three half-capacity pumps). If pipeline capacity is to be increased over time, more pumps may be added to each station.

Pumping stations

Pumping stations for liquid pipelines can be as simple as a single pump located adjacent to a supply tank pumping liquid down the pipeline to a receiving tank (Figure 1). The supply and receiving tank level switches can control the pumping cycle by starting or stopping the pump at appropriate minimum and maximum tank liquid levels. Longer pipelines needing intermediate pumping stations can be handled in one of two basic arrangements. The first is to use storage tank(s) at each station (Figure 2). The upstream station pumps to the downstream station's tanks. The downstream station pump(s) take suction from these tanks and send the liquid farther downstream. This is a simple,

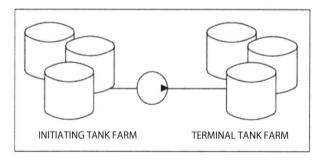

Figure 1. Simple pipeline system.

Figure 2. Multistation, tank system.

Figure 3. Multistation, tight system.

reliable method of long-distance transport. Pumps are started or stopped depending on liquid level in tanks. This has several disadvantages. The cost of storage tanks, dykes, and tank condition monitoring are not small. There is also product in "inventory" in the tanks that does not help a business's cash flow. The alternate arrangement is a "tight" system, leaving the pumping stations in series with each station's pump(s) taking suction from the pipeline (Figure 3). This has the advantage of eliminating the need for storage tanks but requires control over each pumping station in the pipeline such that exactly the same flow rate is going through each station—more on that later.

Types of pumps applied

Most pipeline pumps are centrifugal units. When the product being transported is always a low viscosity liquid like water, gasoline, diesel oil, or very light crude oil, centrifugal pumps are cost-effective, reliable, and efficient. However, as the liquid viscosity increases, the frictional losses within a centrifugal pump quickly reduce pumping efficiency dramatically (Figure 4). For this reason, rotary, positive displacement pumps are often used when products such as heavy crude oil, bunker fuels (no. 6 fuel oil), low sulfur fuels, asphalt, Orimulsion®;* (a manufactured boiler fuel emulsion of 30% water and 70% bitumen), and similar need to be transported. The energy costs of operating a pipeline can represent close to 50% of total operating costs, so pump operating efficiencies play a critical role in profitability. Table 1 illustrates differential annual operating costs for two real examples of centrifugal pumps versus rotary pumps.

Although there are a great many types of rotary pumps, the most common on pipeline services are two- and three-screw pumps, gear pumps (internal and external), and vane pumps for low pressure services. Two- and three-screw pumps are used for higher pressure services, especially at medium to high flow rates. Table 2 shows the approximate maximum flow and pressure capability for each type, not necessarily achievable concurrently.

Gear and vane pumps (Figure 5) tend to have the lower first cost and may or may not have the best lifecycle cost. They

*Orimulsion is a registered trademark of Petroleos de Venezuela S. A.

Figure 4. Rotary versus centrifugal efficiency.

Table 1
Operating Costs Comparison

Parameter	Centrifugal	Rotary
Type	6 stage	3 screw
rpm	3500	1750
Cst	200	200
psid	765	765
gpm	410	410
bhp	395	229
Efficiency %	45	81
Annual energy cost at $0.07/kw h	$180,449	$104,615

Table 2
Rotary Pump Flow and Pressure Ranges

Type	Maximum gpm	Maximum psid
Gear	1500	250
Vane	1000	250
2 Screw	5500	1500
3 Screw	3400	2000

will normally operate at relatively low speeds, requiring speed reducers on all but the very smallest sizes. Multiple-screw pumps (Figure 6) move liquid axially rather than in a radial direction, so fluid velocities are relatively low even when

Figure 5. Gear (top) and vane pump.

directly driven at four- and six-pole motor speeds (1800 and 1200 rpm at 60 Hz, 1500 and 1000 rpm at 50 Hz). Efficiencies remain high over a much broader range of differential pressure than centrifugal pumps. Figure 7 shows 12,000 hp of

Figure 7. Fifteen twin-screw pumps for SUBSEA pipeline.

installed two-screw pumps (15 pumps in all) that are used to load *Orimulsion* tankers 8 miles offshore. Figure 8 illustrates a diesel engine–driven three-screw crude oil pipeline pump operating in Guatemala at 1400 psig.

Specified operating requirements

Obviously, pump and driver specifications will be needed to acquire equipment that will perform to requirements. Unless your organization has these talents in house, an engineering firm will be needed to explore the pipeline requirements. As

Figure 6. Multiple-screw pipeline pumps.

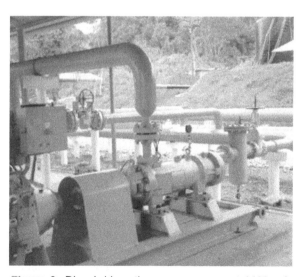

Figure 8. Diesel-driven three-screw pump at 1400 psi.

it relates to rotary pumps, at a minimum the following will need to be defined:

1. Number of pumping stations
2. Number of pumps per station
3. Possible range of
   - Pump inlet pressures
   - Pump discharge pressures
   - Liquid temperature range
   - Liquid viscosity range
   - Per pump flow rate range
   - Particulate content
   - Upset conditions
   - Static conditions
4. Driver requirements
5. Pump instrumentation
6. Pump leak detection

Particulate content can range from unknown (take your chances) to specific weight percentage of sediment, sand, etc. For liquids containing abrasives, pump life is inversely proportional to something between the square and the cube of pump speed. Operating larger pumps at lower speeds will increase initial cost but can save on maintenance costs over many years. Another proven technique to extend pump life in abrasive environments is to select pumps with effectively more stages than needed just for pressure capability. For example, every wrap in a screw pump acts as a pressure stage, similar to a multistage centrifugal pump. More stages reduce the pressure rise per stage, reduce the slip flow (volumetric inefficiency), and spread out the wear. The result is extended pump service life and, frequently, higher average efficiency over time.

Product leak detection could be a critical element in both pumping station safety and environmental contamination. Unless there are observant attendants on site at all times, automatic alarm or shutdown should be considered in the event of product leakage from any source. Rotary pumps most commonly use mechanical seals to contain pumped product within the pump casing. A mechanical seal will wear and can fail catastrophically, and product under pressure can leak large volumes in short time periods. Many pumps used for pipeline services can be equipped with various seal-leak detection schemes to alert/alarm or shut down in the event of excessive seal leakage.

Driver selection depends on available power sources. In remote areas, diesel engines are usually chosen. They provide some range of speed and, therefore, flow control of the pumps. By far, the most common drive is a fixed speed electric motor, although variable frequency drives are becoming more popular. They can improve station efficiency by avoiding the necessity of drawing power to pump flow that is to be bypassed around the station (wasted energy). Spacer couplings are highly recommended between the pump and its drive. They allow access to the pump shaft-end seal and bearing, as well as the motor shaft-end bearing.

Upset conditions and static conditions need to be assessed during pipeline design. Most rotary, high-pressure pumps cannot tolerate anything close to discharge pressure at their inlet (suction) port. In a tight system, if an upstream pumping station is started and the downstream station is not started in very close synchronization, the downstream station may see excessive pressure on its inlet side. A low pressure set inlet side relief valve may be necessary, together with an overspill tank and appropriate instrumentation.

Instrumentation tends to be a very owner-specific issue. Some systems have only a few pressure gauges. Others include bearing temperature detectors, vibration detection, inlet and outlet pressure transducers, and temperature transducers, all sufficient to get a good picture of what is happening at a pumping station, sometimes from many hundreds of miles away.

Pump heating and/or cooling

If the product being pumped is so viscous that heating it before pumping is necessary, then it will probably be prudent or essential to provide a means of preheating the pumps so they are not faced with sudden introduction of hot liquid into cold pumps. Depending on available utilities, pumps can be jacketed for use of heating mediums such as steam or hot oil. Heat tracing with electric thermal wire is also an option. The outside of the pump will need to be insulated to contain the heat. Heating and insulation systems must take into account pump maintainability if the heating or insulating system must be removed to service the pump. Be sure to avoid applying heat to timing gear cases or pump bearings. In some instances, these pump components may actually need cooling to operate for extended periods.

Pump protection

Like most other pumps, rotary pumps are vulnerable to premature wear or failure from too little or too much inlet pressure, too high a discharge pressure, excessive inlet temperature, dry running, and ingestion of foreign material (weld rod, wire pieces from cleaning brushes, etc.). In addition, if the liquid pumped is inherently "dirty," like many crude oils and bunker fuels, then pump wear will be a normal part of operating pipeline pumps and one should prepare for that eventuality.

Systems need to be instrumented to at least alarm if any of these parameters approach a limiting value. If pumps are unattended most of the time, then automatic shutdown is the only protection from catastrophic pump failure. Use of inlet strainers is highly desirable, provided they are instrumented to detect excessive pressure drop and cleaned when needed.

Without regular monitoring and cleaning, strainers eventually will accumulate debris to the point that the pressure loss of liquid going through the strainer will allow pump inlet pressure to fall below the minimum allowable. Worse still, that pressure drop can cause the strainer element to collapse. On collapse, all the accumulated debris, as well as the strainer element, will arrive at the pump inlet. Nothing good will come of this.

As with any positive displacement pump, a pressure relief device is essential to safe operation. The "shutoff head" of a positive displacement pump can be enormous, far above the pressure rating of any system component. A relief valve is normally installed at each pump discharge with the valve outlet connected back to the supply tank (preferred) or to the pump inlet piping as far from the pump as practical.

System considerations

When starting a rotary pipeline pump with the pipeline already full, the system must provide for the time required to accelerate the liquid from zero velocity to final velocity. There may be 50 miles of product that needs to get to 10 ft/sec. This is normally accomplished with an electric-operated valve around each pump that bypasses discharge flow either back to the supply tank (preferred) or back to pump suction. The valve is open on pump startup and closed gradually as the flow velocity in the pipeline builds. If bypassing back to the pump inlet piping, go as far away from the pumps as practical. This will maximize the mass of liquid available to absorb the driver power draw that is not yet going down the pipeline while the bypass is not fully closed.

Flow control can be achieved in several ways. Variable-frequency drives are becoming more affordable in larger sizes. Because rotary pump flow is directly proportional to speed, control is fairly straightforward. In tight systems with fixed-speed drivers, pumping station variable flow is achieved first by how many of the station's pumps are in operation and second by bypassing flow around the station to control how much total flow leaves the station. Figure 9 is a simplified schematic of such a system. The bypass valve senses station inlet pressure. It bypasses whatever flow is necessary to

control station inlet pressure at set point. If the station inlet pressure rises, less flow will be bypassed and more flow will therefore leave the station.

Maintenance

Maintenance for rotary pumps is not very different from any other pump. Hopefully, the installation provided both space to access the pump for service as well as some shelter from the elements. Because rotary pumps have relatively close internal running clearances, be especially careful to achieve and maintain good pump-to-motor shaft alignment. Avoid pipe strain at all costs for the same reason. If pumping station downtime is a serious matter (usually), then have an installed standby pump ready to take over, as well as critical spare parts. If a pump "goes down," try to identify the cause before starting another one. Sadly, too many times a series of failures is allowed to occur before enough time and effort is expended on root cause analysis.

Many pipeline pumps have service parts that need to be removed from both ends of the pump. Spacer couplings go a long way in easing the drive end service. Keeping walls or objects away from the pump end opposite the driver is usually necessary to avoid the need to disconnect piping and remove pumps to a bench area.

Some pumps are, in fact, better maintained on a bench in a maintenance area, where space to perform the work is more available, components can be more easily assessed for their condition, and new parts can be installed more correctly.

Conclusions

Rotary, positive displacement pumps can play key roles in reliable, efficient, effective transport of liquids over short or very long distances. They are an optimum choice where efficiency is a strong driver of pump selection and operation. There is an extensive history of successful applications of rotary pumps to pipeline transport.

Source

This tutorial was presented at the Pump Users Expo 200, Louisville, KY, September 2000.

James R. Brennan is market services manager with Imo Pump, Monroe, NC. He is a 1973 graduate of Drexel University in Philadelphia, a member of the Society of Petroleum Engineers, SPE, and has more than 30 years of service with Imo Pump.

Adapted from Imo Pump, Monroe, NC. ©Imo Pump 2000. Reprinted with permission.

Figure 9. Tight series pump operation.

KEY CENTRIFUGAL PUMP PARAMETERS AND HOW THEY IMPACT YOUR APPLICATIONS—PART 1

Pumps and Systems: They Go Together

By Doug Kriebel, PE, Kriebel Engineered Equipment

The purpose of this article is to familiarize the reader with centrifugal pump manufacturers' performance curves, show how a pump system head curve is constructed, and demonstrate how to analyze the effect of changes in the system.

A pumping system operates where the pump curve and the system resistance curve intersect. This important fact is one of the most overlooked concepts when designing or troubleshooting a pumping system.

The pump curve has a head/capacity portion, which indicates the change in flow with respect to head at the pump. It is generated by tests performed by the pump manufacturer. We will review the key facts that this curve offers.

The system resistance curve is the change in flow with respect to head of the *system*. It must be plotted by the user based upon the conditions of service. These include physical layout, process conditions, and fluid characteristics.

Anyone operating a pumping system must know where the pump is operating. Otherwise, it is like driving a car and not knowing how fast it's going.

The pump curve

The pump curve consists of a head/capacity curve (curve A of Figure 1) that shows the total differential head (H is expressed in feet of pumped liquid). This is the difference between the total discharge head (Hd) at the pump discharge and the total suction head (Hs) at the pump inlet at any given flow rate (Q), expressed in gpm (gallons per minute).

$$H = Hd - Hs \qquad (1)$$

All pressure is in feet of liquid.

At this point, it's important to revisit some pressure units:

psi = pounds/in^2
psia = pound/in^2 absolute
psig = pound/in^2 gauge
psia = psig + barometric pressure

The conversion from psi to feet of liquid is to divide the pressure, in psi, by its density:

$$(lb/in^2)(lb/ft^2) \times (144\,in^2/ft^2) = feet\ of\ pumped\ fluid$$

For water with a specific gravity (SG) of 1, you substitute its density (62.4 lb/ft^3) into the above equation and the result is:

psi $\times$ 2.31/SG = feet of liquid (SG is the ratio of the density of a liquid to water at 60 °F). See Figure 2 for illustration of feet, psia/psig/SG.

The pump curve also includes the pump efficiency (in percent—curve B), BHP (brake horsepower—curve C), and NPSHr (net positive suction head required—curve D) at any

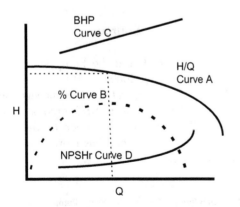

Figure 1. Performance curve data.

Figure 2. Pressure equivalent.

flow rate. This curve is plotted for a constant speed (rpm) and a given impeller diameter (or series of diameters).

Figure 1 is a single line curve; Figure 3 is an actual manufacturer's performance curve showing the same information but expressed as a series of curves for different impeller diameters (from $12 - 5/8''$ to $15 - 15/16''$). The pump impeller can be trimmed to any dimension between the maximum and minimum diameters.

The H, NPSHr, efficiency, and BHP all vary with flow rate, Q. The best efficiency point (BEP) is the point on the curve where the efficiency is at a maximum. For Figure 3, it's 82% at 1300 gpm and 100′ TDH.

BHP can be read from the curve at any flow rate; however, it is usually calculated from the efficiency, head, and flow using Equation 2. Pump curves are based on a specific gravity of 1.0. Other liquids' specific gravity must be considered.

$$BHP = Q \times H \times SG/(3960 \times \% \text{ eff}) \qquad (2)$$

The NPSHr is read from the bottom curve from flows, and the NPSHr in feet is read on the lower right. The following examples apply to Figure 3.

Example 1. For an impeller diameter of $15 - 15/16''$ (maximum), what are the head, efficiency, NPSHr, and BHP at 900 gpm? (Answer: 110′; 76.5%, 5′ NPSHr, and 32 BHP).

Example 2. What are the head, efficiency, NPSHr, and BHP at 1600 gpm? (Answer: 85′, 78%, 7.5′ NPSHr, and 44 BHP).

In a pumping system, the capacity of the pump moves along the pump curve based upon the differential head across the pump. For instance, for Figure 3, with the maximum impeller diameter, if the flow is 900 gpm and you want to increase it, you reduce the head to 85′, and the flow will increase to 1600 gpm. It's by controlling the pump differential head that you control the flow of the pump. If you wanted 1300 gpm, you would have to throttle the head back to 100′.

The slopes of pump curves vary depending on the pump-specific speed (Ns). How they vary depends on Ns and is shown in Figure 4. The Ns is a dimensionless number calculated as:

$$Ns = \text{rpm} \times (Q^{1/2})/H^{3/4} \text{ all at BEP} \qquad (3)$$

The Ns for Figure 3 is 1311.

NPSHr is net positive suction head required by the pump—based on test.

NPSHa is net positive suction head available by the system—must be calculated.

$$NPSHa = P1 + hs - hvp - hfs \qquad (4)$$

P1 = Pressure of suction vessel (atmospheric in open tanks)
hvp = Vapor pressure of liquid at its pumping temperature, t

Figure 3. Manufacturer's performance curve.

shape of typical head-capacity curves for variuos specific speeds.

Figure 4. Head versus capacity versus Ns.

NPSHa = P1 +hs - hvp - hfs (Equation 4)

H= (P₂-P₁)×2.31/Sp.g + (hd-hs) + (hfd + hfs) + (V²d-V²s)/2gc

(Equation 6)

Figure 5. Atypical pump system.

hs = Static suction head, the vertical distance between the eye of the first stage impeller centerline and the suction liquid level. (This may be a positive or negative number, depending on whether the suction liquid is above the pump suction eye or below it, as on a "lift.")

hsf = Friction and entrance losses on the suction side

Pump curves are generated by manufacturers and show a range of impeller diameters from minimum to maximum size for the casing size selected. These are based on a single speed (rpm). Head/capacity relationship changes with speed or impeller diameter per the affinity laws. Affinity laws are discussed in all the references. Other information often found is suction specific speed (Nss), which is a dimensionless number calculated at the BEP as

$$Nss = rmp \times (gpm_{eye})^{1/2}/NPSHr^{3/4} \quad (5)$$

(*Note:* The gpm is per impeller eye, so for a double suction impeller, the flow is half the flow of single suction impeller.)

The Nss in Figure 3 is 10,800.

Nss is a function of NPSHr. To reduce NPSHr, the impeller eye is modified. If the eye is modified too much, it causes a reduction in the low flow capabilities of the pump and limits the low end point of operation.

System head curves (Figure 5)

Step one in constructing a system/head curve is to consider the fluid energy requirement.

At any point in a pumping system, there are four components making up its total energy (head in feet):

- Pressure
- Velocity
- Elevation
- Heat (ignored, since pumping is usually at a constant temperature across the pump)

The difference in energy levels (differential head, H) between the suction (Hs) head and discharge (Hd) head of a pump can be expressed using Bernoulli's equation and adding the friction loss.

The friction loss is the loss in energy due to resistance to flow as it moves through a system. This resistance is due to viscous shear stresses within the liquid and turbulence at the walls of the pipe and fittings. It varies with liquid properties (viscosity, solids entrained); size of the pipe (velocity); roughness of the pipe's internals; length of travel; and losses at fittings, control valves, and the entrance and exits.

$$H \text{ system} = \text{Hd} - \text{Hs} = (P_2 - P_1) + (\text{hd} - \text{hs}) + (V_d^2 - V_s^2)/2g_c + (\text{hfs} + \text{hfd}) \quad (6)$$

where

P_1 and P_2 = The entrance and exit pressures,
hd and hs = The elevations from the pump impeller eye to the entrance and exit liquid levels,
V_s and V_d = The entrance and exit velocities (velocities are given in ft/sec. in this article),
hfs and hfd = The frictional losses in the suction and discharge pipe system, expressed in feet.

To simplify:

Hf = hfs + hfd is the total friction loss
(at each flow rate) through the system (7)

h = hd − hs is the difference in elevation form
discharge level to suction level (8)

hv = $V_d^2 - V_s^2/2g_c$ is the velocity head component (9)

Velocity head is often ignored, since in most cases the velocity is low (in tanks maintained at constant level, it's 0). However, if the discharge is coming out through a nozzle or other orifice, its velocity could be high and, therefore, significant if the other components are low. This should not be confused with entrance and exit head losses, which are accounted for in the Hf calculations.

Velocity head (hv = $V_2/2g_c$) is a confusing concept to many—it's the head (distance) required for a liquid to fall in order to reach a velocity, v. It costs energy to move (accelerate) from some low velocity to a higher velocity.

All components of Equation 6 must be consistent and usually in feet of liquid. Refer to Figure 5 for the following example.

For the pressure component

$$P_2 - P_1 = \frac{(5 \text{ psig}-0 \text{ psig}) \times \frac{2.31 \text{ ft}}{\text{psi}}}{1.0} = 11.5'$$

It could have been

$$29.7 \text{ psia} - 14.7 \text{ psia} = 5 \text{ psia} \times \frac{2.31}{1.0} = 11.5$$

Use either; just be consistent. For open tanks there is 0 psig, 14.7 psia (at sea level).

For the static head component

$$h = \text{hd} - \text{hs} = 60 - 10 = 50'$$

Note that this is from liquid level to liquid level. This layout was for illustration; most industrial applications do not have the discharge pipe located beneath the liquid level because it would back siphon when the pump stops. If the pipe goes overhead and empties above the liquid line (or more often uses a vacuum breaker), use the centerline elevation of the discharge pipe at the exit for the hd. The pressure and static head components are constant with flow and plot as a straight line.

For the velocity head components

Assume the tanks are either being filled/emptied at the pumping rate or they are large enough that the downward and upward velocities are very low, or they are the same diameter and the velocity heads cancel. Therefore, the velocity head component is 0.

The friction component

Hf includes pipe losses and minor losses, as follows.

Pipe losses

There are several ways to calculate these losses:

1. Published data such as the Crane Handbook, the Cameron Hydraulic Data Book, or KSB's Centrifugal Pump Design.
2. Using the Hazen Williams formula:

$$\text{Hf} = [3.022(v)^{1.85}L]/[C^{1.85} \times D^{1.165}] \quad (10)$$

where
  L = Length of pipe in feet
  v = Velocity ft/sec
  D = Pipe diameter in feet
  C = Constant for various pipe materials

This is a simple way to go. The disadvantage is that it assumes constant NRe (Reynolds Number) in the turbulent range and a viscosity of 1.13 cs (31.5 SSU), which is water at 60 °F. This can yield errors of 20% or more for the range of 32−212 °F.

3. Using the Darcy formula:

$$\text{Hf} = fLv^2/2Dg_c \quad (11)$$

where
  f = the friction factor based on the Reynolds number (NRe)

$$\text{NRe} = D \times v/\text{vis} \quad (12)$$

where
  vis = Viscosity in ft^2/s

In laminar flow(less than NRe = 2000)$f = 64/\text{NRe}$ (13)

In turbulent flow (NRe greater than 4000), f varies with NRe and the roughness of the pipe. This is expressed as "relative roughness" = e/D; where e is absolute roughness (in feet) found in tables and f is found in a Moody Friction Factor Chart using NRe and e/D.

If the range of NRe is between 2000 and 4000 (the transition zone), use turbulent flow design.

4. Using computerized flow simulators. (You need to know limits and assumptions upon which the program is based.)

Minor losses

These include obstructions in the line, changes in direction, and changes in size (flow area). The two most common methods of calculating these losses are

1. *Equivalent lengths:* uses tables to convert each valve or fitting into an "equivalent length," which is then added to the pipe length in *L*, and Hf is calculated in Darcy (Equation 11 above).
2. *Loss coefficients:* uses published resistance coefficients (*K*) in the formula: $hf = Kv^2/2g_c$. *K* values for many types of fittings are published. Loss coefficients are the way to determine entrance and exit losses and losses due to sudden reductions or enlargements.

Control valve losses

These losses are calculated using the Cv for the valve:

Cv = gpm pressure drop(psi)

The Hf increases as flow increases and is a parabolic curve. The Hf for the system is calculated using one of the above methods. Assume the suction pipe is 10″ diameter, 20′ long, with two elbows and one gate valve. The discharge pipe is 8″, 1000′ long, with eight elbows, one gate valve, and one check valve. The calculations yield Hf values of

1. Hf = 40′ @ 400 gpm
2. Hf = 12′ @ 800 gpm
3. Hf = 27′ @ 1200 gpm
4. Hf = 37′ @ 1400 gpm

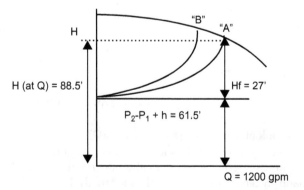

Figure 6. System head curve.

With the system head calculated at several flows (the pressure and height components are not affected by flow), we can plot the results (Figure 6).

The total differential head at 1200 gpm is $H = 11.5' + 50' + 27' = 88.5'$.

The pump's head capacity curve is plotted on the same graph. Where they intersect is the operating point ("A"). Note the pressure and static head components (66.50) are constant with flow, while the Hf component increases as flow increases.

If a control valve in the discharge piping were used, the system head curve could be throttled back to a higher head and therefore lower flow ("B").

What can go wrong?

A centrifugal pump will run where the system head curve tells it to run. The point where the centrifugal pump performance curve and system head curve intersect is the operating point. It's important to understand where the operating point is at all times and to avoid the dangerous areas of pump operation. What are some of the dangerous areas to consider?

Horsepower. A pump must always have a driver with more hp than it consumes. With most centrifugal pumps, as the pump flow increases, the BHP increases to a maximum, and there must always be enough hp available from the driver to supply the needs of the pump. This means as TDH falls, the flow goes up and so does the BHP. Undersizing overloads the driver.

NPSH. As the pump flow increases, the NPSHr increases. As flow increases, NPSHa decreases (hfs increases with flow). There must always be more NPSHa than NPSHr or the pump cavitates. Cavitation is where the pressure at a given point in the pump, prior to entering the impeller eye, falls below the vapor pressure of the liquid pumped, turning it to vapor. As the pressure increases through the impeller, the bubble of vapor collapses. This causes capacity reduction, noise, vibration, pitting of the impeller and, eventually, failure of a component.

No flow. It is possible for the TDH to rise to a point where there is no *flow*. This is called *shutoff* and means no flow is going through the pump. At this point, the temperature will rise in the pump because the inefficiencies of the pump are causing heat to go into the pumpage. This raises the temperature and will eventually cause the pumpage to vaporize, and the close clearances such as ring fits and mechanical seal faces will not receive lubrication. Eventually the seals fail or the pump seizes. High temperatures also cause the pump to distort and change clearances. Failure of some component will occur.

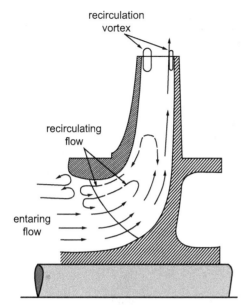

recirculation
vortex

recirculating
flow

entaring
flow

Low-volume Recirculation

Figure 7. Low flow recirculation.

Low flow. It is possible for the TDH to rise to a point of low flow where damage other than no flow occurs. Such circumstances include:

- Low flow temperature rise—As the flow decreases, the efficiency falls and there is more BHP needed to drive the pump than is needed to move the liquid through the TDH. The excess BHP goes into the pump as heat. The flow is lower (less to carry away the heat), and therefore the temperature rise is greater as flow drops. It is possible to reach a point where the temperature rises to the point that it reduces the NPSHa below the NPSHr and flashing occurs as above. This point is the "thermal minimum flow" and is a function of not only the pump, but also the system characteristics.
- Low flow recirculation/separation damage—As flow decreases, the liquid begins to recirculate in both the impeller eye and impeller discharge (at the volute cut-water or diffuser). This recirculation will cause separation, eddy currents, and turbulence that will cause the same noise, vibration, and damage as cavitation (Figure 7). The minimum flow required to prevent low flow recirculation separation damage is called the "continuous minimum flow." It is specified by the pump manufacturer based on the impeller eye geometry and Nss.
- Radial or axial thrust damage—All pumps use bearings and balancing devices to prevent the rotating parts from contacting the stationary parts. These are pressure and flow related, and they may limit the flow of the pump to a minimum or maximum. The pump manufacturer must make known the limits based on the pump design.

References

1. Carver Pump Company. *Performance Curves*, Muscatine, IA.
2. Lindeburg, PE, Michael. *Mechanical Engineering Reference Manual*, 8th *Ed.*, Professional Publications, Belmont, CA (1990).
3. Karassik, Igor. *Pump Handbook*, McGraw-Hill, New York, (1976).
4. Westaway CR. *Cameron Hydraulic Data Book*, 16th *Ed.*, Ingersoll-Rand, Woodcliff Lake, NJ (1984).

Doug Kriebel has presented pump seminars and workshops throughout his 30-year industrial career. He has extensive experience with pumping applications in a wide range of industries including positions in the electric utility industry, as well as with a chemical equipment manufacturer and major pump OEM. Mr Kriebel is currently president of a pump and equipment manufacturer's representative/distributor organization. He holds a B.S. degree in chemical engineering. He is an active member of AIChE and is a registered Professional Engineer in Pennsylvania.

KEY CENTRIFUGAL PUMP PARAMETERS AND HOW THEY IMPACT YOUR APPLICATIONS—PART 2

Pumps and Systems: They Go Together

By Doug Kriebel, PE, Kriebel Engineered Equipment

Editor's note: This article is the second installment of a feature that first appeared in the August 2000 issue of Pumps & Systems Magazine. Part 1 explained the commonly applied parameters of head/capacity, NPSH and brake horsepower. This month's conclusion examines what can go wrong in a system and how to fix common problems. If you've lost track of your August issue, you can read the full text of Part 1 online at www.pump-zone.com.

How to examine a system

Each system must be analyzed to determine what happens to the pump at not only design, but also every other possible point of operation—start-up, off peak, unusual modes, "best case," and "worst case." The following examples illustrate how a system change can create either low flow, high flow, or NPSH problems. Refer to Figure 1, which shows a pump system. We will first look at the suction side to determine the NPSHa. The suction side is 20' of 5" pipe with two elbows and a gate valve. At 300 gpm the hfs is 1'. The pump selected is shown in Figure 2.

Example 1. The tank is open to atmosphere, 0 psig, 34' water = P_1; pumping water at 70 °F, hvp = 0.36 psi = 0.8'; from hs1 = 10' then:

The NPSHa = $P1 + hs - hvp - hfs = 34 + 10 - 0.8 - 1$

$$= 42.2'$$

Referring to the pump curve Figure 2, at 300 gpm, the NPSHr = 23' so we have adequate NPSHa.

P₁=50 psig
liquid level

P₁=0 psig

hd = 45'
h = 35'

liquid level, hs1

hs = 10'

hs = -10'

liquid level, hs2

NPSHa = P1+hs-hvp-hfs   Equation 4

H = (P₂-P₁)x2.31/Sp.g + (hd-hs) + Hf + (V_d²-V_s²)/2gc   Equation 6

Figure 1. Pump system for Examples 1–6.

Example 2. All conditions the same except the tank is drawn down to level $hs_2 = -10'$

Then the NPSHa = $34 - 10 - 0.8 - 1 = 22.2'$ and the NPSHa is less than the NPSHr and the pump cavitates.

Example 3. All conditions the same as Example 1 except the water is now 190 °F; hvp = 9.34 psi = $9.34 \times 2.31/0.966 = 22.3'$.

Then NPSHa = $34 + 10 - 22.3 - 1 = 22.7'$ and the pump cavitates.

Example 4. Same as Example 3 except from hs2 = $-10'$ Then NPSHa = $34 - 10 - 22.3 - 1 = 0.7'$ and the pump cavitates.

Example 5. Conditions same as any above except the suction pipe is installed as 3" instead of 5": hfs = 10' (instead of 1'). All the above NPSHa would be reduced by 90'!

The lesson to be learned on the *suction side:* Have as much static suction head as possible (hs), locate the pump as close as possible to the suction vessel (with a minimum suction side frictional pressure drop (hfs)), and be careful of liquids near their vapor pressure (hvp).

Let's go further and use the Example 1 conditions and look at the discharge side: Assume 150' of 4" pipe with six elbows, one gate valve, and one check valve: hfd = 14' @ 300 gpm.

Example 6. Assume the levels in both suction tank and discharge tank remain constant with suction at hs1 then:

$$H = (P_1 - P_2) + (hd - hs) + (V22 - V21)/2gc + (hfd + hfs)$$
$$= (50 \text{ psig} - 0 \text{ psig})2.31/1.0 + (45' - 10') + 0 + (14' + 1')$$
$$= 115.5' + 35' + 15' = 165.5'$$

The pump selected is shown in Figure 3. The operating point is "A".

What happens if the levels don't remain constant and the suction tank pulls down from level hs1 = +10' to level hs2 = -10' and the discharge tank fills up an equal 20'?

The "h" component (hd - hs) would go up from 35' to 75' and the "P" + "h" components would rise from 150.5 to (115.5 + 75) 190.5'. This would cause the pump to "walk" back on the curve to some lower flow, point B.

All this means is that the pump will pump a lower flow rate. This is not necessarily a bad thing as long as it meets the process needs and is above the minimum flow for the pump. If it does go below minimum flow, damage will occur.

Figure 2. The selected pump, Example 1.

If the pressure ($P2$) is increased to 75 psig, the head goes up $(75 - 0)2.31/1.0 = 173$ plus $h = 35'$ to $H = 208$, which is above the shutoff of the pump ($205'$). The pump would be running with no flow. This could be an upset condition that

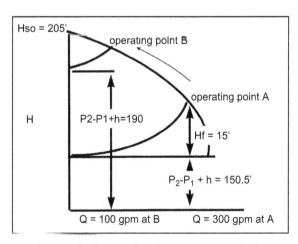

Figure 3. System head curve—Example 6.

causes overpressure to the discharge tank. It is also a situation where an inline filter is installed into an existing installation and the pressure drop is underestimated or the filter is not maintained (cleaned). This example cost an East Coast utility company more than $45,000 in damage to a pump when maintenance installed a filter to protect some heat exchangers without taking into account the pressure drop when fouled!

Example 7. More often the pump is selected for a "worst case" condition, (sometimes called the "design condition"). This example uses the same system as above except use 75 psig for the design $P2$, assuming "worst case" situation (Figure 4, Point A). This accounts for an upset, overpressure situation in the discharge tank or other circumstance.

However, when the system actually operates at 50 psig, the pump runs out to point B. This could overload the motor, since more BHP is required. In addition, the increased flow may cause the NPSHr to exceed the NPSHa (which decreases due to increased hfs). This may cause the pump to cavitate.

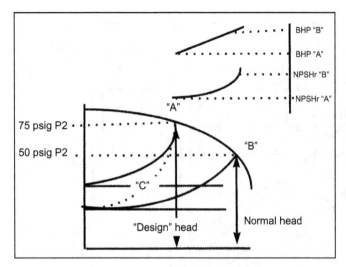

Figure 4. Example 7—changes in head and throttling.

Most systems that are designed for higher $P2$ than normal have a control valve that will increase the hfd and "walk" the pump back on the curve to the correct operating point (back to "A" via line C).

A major chemical company using sealless pumps and no control valves (they were handling toxic material and did not want to risk leaks) designed a system for a specific H based on a $P2$ set by the process. After startup, which went without a hitch, the reactor pressure ($P2$) was decreased to improve yield. Without control valves, the pump ran out on the curve and destroyed itself by cavitation. It took $70,000 in repairs until someone ran the calculations to determine how the system change caused the pump failures!

A similar situation occurs if excess safety factor is utilized (Figure 5). If the actual design point is calculated as 300 gpm at 165′ (point A) and the pump is selected for 345 gpm and

200′ head (point B) and the calculations are correct, the pump will run out to point C. This would pump more than required. So the pump would be throttled back to point D to pump the 300 gpm. This means $215 - 165 = 50'$ f head is throttled out, which costs 5.2 BHP. Since you pay for BHP, this is wasted energy. In this case it would amount to $3100 per year at a rate of $.085/kW h for service 24 h × 360 days. Higher heads or flows would magnify the costs.

There is nothing wrong with safety factors, and it is always better safe than sorry, but after installation, check the hydraulics. If gross over-sizing was the case, trimming the impeller will save considerable money in operations.

Example 8 (variable speed operation). An economical alternative to throttling with a valve is to use variable speed to change the pump curve to meet variable demands of a system head curve. This is illustrated in Figure 6. The system head curve is plotted, a pump is picked to meet the worst case condition, and reduced speed is used to ride down the system curve to meet varying flows at reduced pressure. In some cases this can be very economical when compared to the wasted BHP of throttling control.

If the normal point is the same as above (300 gpm at 165′ point A on Figure 6) and with safety factors (345 gpm at 200′ point B) the pump operates on the uppermost curve for 3500 rpm, where it intersects the system head curve at point C. By slowing down to 3200 rpm, it goes through point A without having to waste energy by throttling the head. Any point can be reached on the system head curve by reducing speed. This is a valuable method of operating when flows or heads have large changes in magnitude.

The danger comes if you reduce the pump speed to a point where the system head curve causes the pump to operate at a point of too low flow or no flow (i.e. beneath the system head curve, such as 2700 rpm). It is possible to

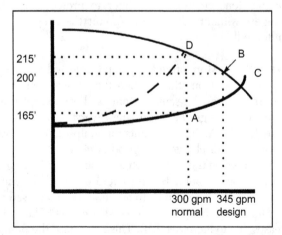

Figure 5. System head curve: design/normal/runout/throttle.

Figure 6. Example 8—variable speed operation.

operate too high flow where BHP or NPSH becomes a factor, or at a speed where other damage could occur. Examples of other speed-related damages are if the pump reaches a critical speed, a speed where auxiliary equipment no longer functions (a shaft driven lube oil pump may not provide proper lubrication at too low a speed, or a motor fan motor may turn too slowly to provide adequate cooling).

Example 9 (parallel pumping applications). It's not uncommon to use two or more pumps in parallel. The flows are additive at a constant head (Figure 7). (The flow, Q, at head H for one pump at A is added to a second pump at the same head to get twice the flow, $2Q$, at the intersection of the two pump curves, the system head curve, B. It's important to draw the system head curve to make sure that if one pump operates alone it does not run out to point which will overload the motor or exceed the NPSH requirements. Point B is the operating point of two pumps in parallel, and point C is for one pump at runout (estimated here as $1.2Q$). The one pump runout must be checked to provide sufficient BHP, NPSHa, submergence against vortexing, and mechanical design for the higher flow.

This is more common in vertical cooling water service where several pumps are used in parallel and a control valve is not used on each pump. If one pump trips, the system allows the remaining pump(s) to run out.

Another potential for parallel pump problems is where two dissimilar pumps (dissimilar curves) are used (Figure 8). The pump curve for pump A is added to pump B at constant head. The system head is drawn. Q1 is the flow with only pump A running, Q2 is the flow with only B running, Q3 is the flow with both A + B running. *Note*: This is not $Q1 + Q2$. If the system is throttled back (by any means), the system head could cause the pump with the lower head to run below a minimum flow. Pump A is shut off at point C.

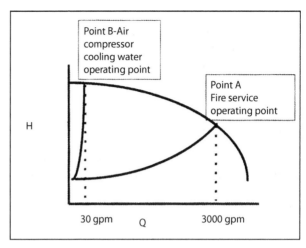

Figure 8. Parallel pumps.

Example 10 (a pump for a given service is used for an alternate service). Figure 9 is the performance curve for a horizontal split case pump put in as a fire protection pump that is to run at 3000 gpm. The plant engineer lost his air compressor cooling water pump and decided to use the fire pump to supply cooling water for the compressor inner and after coolers. (If you need 300 gpm, a 3000 gpm pump will certainly do!)

The pump had six failures of the radial bearings. The maintenance group replaced them, then redesigned the pump and added angular contact back-to-back bearings in place of the double row bearings. They then added oil lubrication to the existing grease bearings and were about to change to oil mist when someone finally asked why the fire pump was running when there was no fire!

Point A is the design point, but point B is where it operated to supply the air compressor's cooling water. This was too far back on the curve and caused high radial loads that were the cause of failure. There was no way any bearing or lubrication upgrade would enable this pump to operate at point B.

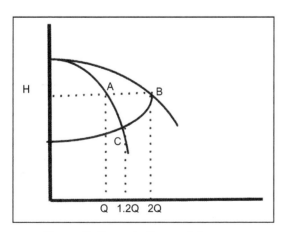

Figure 7. Example 9—parallel pumps.

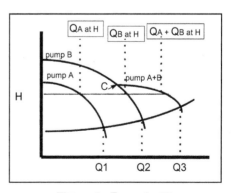

Figure 9. Example 10.

In conclusion, the purpose of this article is to provide operations and maintenance people who do not normally do process calculations with the tools to analyze the pump systems they have to operate. This will enable them to operate the systems more economically and to determine the "real" root cause of failures. Failures are mechanical in nature, but the cause of the mechanical failure is a "system" incompatibility failure. In today's business world, systems are often designed by process engineers who are not available when the system becomes operational, and the system may not have been installed, operated, or maintained the way it was intended to be, or the correct method of operation is no longer the way it was originally designed.

References

1. Carver Pump Company. *Performance Curves*, Muscatine, IA.
2. Lindeburg, PE, Michael. *Mechanical Engineering Reference Manual*, 8th *Ed.*, Professional Publications, Belmont, CA (1990).
3. Karassik, Igor. *Pump Handbook*, McGraw-Hill, New York, (1976).
4. Westaway CR. *Cameron Hydraulic Data Book*, 16th *Ed.*, Ingersoll-Rand, Woodcliff Lake, NJ (1984).

Nomenclature/symbols

H = Total differential head
Hd = Total discharge head
Hs = Total suction head
SG = Specific gravity
Q = Flowrate
NPSH = Net positive suction head (r required, a available)
BHP = Brake horsepower

Ns = Specific speed
Nss = Suction specific speed
P = Pressure (1 suction, 2 discharge)
hvp = Vapor pressure of liquid
hs = Static suction head
hd = Static discharge head
h = (hd − hs)
hfs = Friction losses, suction side
hfd = Friction losses, discharge side
Hf = (hfs + hfd)
vd = Discharge velocity
vs = Suction velocity
v = Velocity (ft/sec.)
hv = Celocity head
g_c = Gravitational constant (32.2 ft/s^2)
L = Pipe length in feet
D = Pipe diameter in feet
C = Hazen Williams constant for pipe materials
NRe = Reynolds number
f = Friction factor
vis = Viscosity
e = Absolute roughness
K = Resistance coefficient for pipe fittings

Doug Kriebel has presented pump seminars and workshops throughout his 30-year industrial career. He has extensive experience with pumping applications in a wide range of industries, including positions in the electric utility industry, as well as with a chemical equipment manufacturer and major pump OEM. Mr Kriebel is currently president of a pump and equipment manufacturers' representative/distributor organization. He holds a B.S. degree in chemical engineering, is an active member of AIChE, and is a registered Professional Engineer in Pennsylvania.

Estimate the discharge of a centrifugal pump at various speeds

Within the design limits of a pump, the discharge will be approximately proportionate to the speed.

$$Q_2 = Q_1 \frac{N_2}{N_1}$$

where

Q_1 is the gallons per minute at a particular speed N_1, and Q_2 is the gallons per minute at N_2.

Example. If a pump delivers 350 gpm at 1200 rpm, what will it deliver at 1800 rpm?

$$Q_2 = 350 \times \frac{1,800}{1,200} = 525 \text{ gpm}$$

Similarly, for a given pump, design head varies as the square of the impeller speed and power as the cube of the speed.

How to estimate the head for an average centrifugal pump

By slightly modifying the basic pump formula:

$$H = \left(\frac{DN}{1,840\,K_u}\right)^2$$

it is possible to find a means of estimating the head a pump will develop for each impeller at various speeds. The terms are defined as follows:

H = Head, ft
D = Diameter of the impeller, in
N = rpm of the impeller
K_u = Pump design coefficient

The values of this expression may vary from 0.95 to 1.15. For estimating purposes, substitute 1900 for 1840 K_u. The formula becomes:

$$H = \left(\frac{DN}{1900}\right)^2 \times (\text{the number of impellers})$$

Example. What head will be developed by a well-designed centrifugal pump with four 11.75-in. impellers turning at 1200 rpm?

$$H = 4 \times \left(\frac{11.75 \times 1,200}{1,900}\right)^2 = 4 \times 7.4^2 = 220 \text{ ft of head}$$

Find the reciprocating pump capacity

Multiply the diameter of the plunger by itself, multiply this by the stroke, then by the number of strokes per minute, and finally by 0.11; the result is the capacity in barrels per day.

$$C = B^2 \times S \times N \times 0.11$$

Example. Four-in. bore, 6-in. stroke, 80 strokes per minute.

$$4 \times 4 \times 6 \times 80 \times 0.11 = 845 \text{ barrels per day}$$

Note. The rule is approximate only; it allows about 5% for "slip," which is actually somewhat variable. Also note that it is applicable only to single-acting pumps; for a double-acting piston type pump, N must be twice the rpm, and a reduction must be made for the area of the piston rod on the crank end of the pump.

How to estimate the hp required to pump at a given rate at a desired discharge pressure

Ignoring effects of temperature, viscosity, etc.

$$\text{hp} = \text{psi} \times \frac{\text{barrels per day}}{1,000} \times 0.022$$

The factor 0.022 is arrived at by using a pump efficiency of 80% and a gear efficiency of 97%.

Example. What horsepower would be required to pump 15,000 barrels per day with an 800 psi discharge?

$$\text{hp} = 800 \times 15 \times 0.022$$
$$= 264 \text{ horsepower}$$

The factor 0.022 was derived from the formula:

$$\text{hp} = \frac{(\text{psi})\ (\text{bpd})}{58,800 \times \text{pump efficiency} \times \text{gear efficiency}}$$

Nomograph for determining reciprocating pump capacity

The theoretical capacity of reciprocating pumps can be calculated by the following equation:

$$C = 6/55\, D^2\, L\, N$$

where

C = Theoretical pump capacity (delivery), gal/min
D = Diameter of cylinder, ft
L = Length of stroke, ft
N = Number of strokes/min

The nomograph (Figure 1) is based on this equation. The diameter and length-of-stroke scales are calibrated in in units instead of ft. The nomograph also gives the delivery in British imperial gallons.

Example. What is the theoretical capacity of a reciprocating pump having the following characteristics?

$D = 3$ in
$L = 3$ in
$N = 60$ strokes/min

Connect	With	Mark or Read
$D = 3$ in	$L = 3$ in	Reference scale
Ref. Point	$N = 60$	$C = 5$ gal/min

Source

Macdonald, J. O. S., *British Chemical Engineering*, 1, 6, 301 (1956).

Figure 1. Theoretical capacity of reciprocating pumps.

Nomograph for determining specific speed of pumps

The specific speed is a criterion used in the selection of pumps for a given service. It can be used to (a) avoid cavitation troubles, (b) select the most economic pump for a given process layout, and (c) assist in redesigning process conditions.

For single-stage, side-entering pumps, the NPSH is defined by the following equation:[1,2]

$$S = N\frac{Q^{1/2}}{H^{3/4}} \tag{1}$$

where

S = Specific speed, no units
Q = Pump capacity
N = Pump speed, rev/min
H = Net positive suction head above the liquid vapor pressure (NPSH) measured at the pump suction flange, ft of fluid

H (in ft of liquid) is given by

$$H = h_a \pm h_h - h_p - h_v \tag{2}$$

where

h_a = Ambient pressure at the liquid surface
h_h = Static head between liquid surfaces
h_p = Combined entrance, piping and fitting friction loss
h_v = Proper pressure at pumping temperature

The specific speed of centrifugal pumps may vary between 1100 and 15,000. Values above 12,000 are not considered practical and, where possible, the use of a specific speed below 8500 is recommended. Specific speed limits for each type of pump are usually available from manufacturers of pumps.

The nomograph (Figure 1) solves Equation 1.

The specific speed is also defined in terms of a dimensionless number, using consistent units, by the following equation:

$$S\frac{Q^{0.5}N}{(\Delta H \times g_c)^{0.75}} \tag{3}$$

where

g_c = gravitational conversion constant, 32.2 (lb mass/lb force) (ft/s^2)

The range for this equation is 0.03–0.87. Results of Equation 1 can be converted to these units by dividing by 17,200.

Example. The following data are given for a single-suction reboiler pump:

Q = 400 gal/min
N = 3500 rev/min

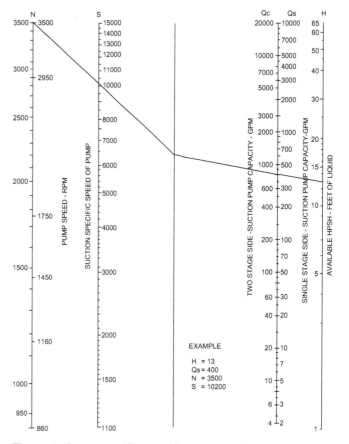

Figure 1. Pump-specific speed from net positive suction head. Reproduced by permission, *Petroleum Refiner*, Vol 36, No. 10, copyright 1957.

The reboiler liquid level is 18 ft above the pump, and friction from entrance, pipe, and fittings loss add up to 5 ft of liquid.

For a reboiler, the liquid is at its boiling point, so $h_a = h_v$, using Equation 2, $H = 13$ ft.

Connect	With	Mark or Read
NPSH = 13	Q = 400 gal/min	Reference line
Ref. line pivot	N = 3500	S = 10,200

Source

Jacobs, J. K., *Petroleum Refiner*, 36, 10, 145 (1957).

Reference

1. *Hydraulic Institute Standards*, 10th Edition, New York.
2. Brown, et al., *Unit Operations*, 190, John Wiley and Sons, New York (1950).

Nomograph for determining horsepower requirement of pumps

In the selection of pumps it is necessary to calculate the horsepower requirement to handle a specified delivery rate of fluid against the specific head. The formula for the horsepower required by a pump is as follows:

$$hp = \frac{G \times H \times \text{specific gravity}}{3,960 \times e}$$

where:

hp = Pump horsepower
G = Flow rate, gal/min
H = Total dynamic head (change in head), ft
e = Pump efficiency

Figure 1 solves this equation.

Example. Given:

$G = 200$ U.S. gal/min
$H = 60$ ft
Specific gravity = 1.1
$e = 0.80$

What is the required horse power?

Connect	With	Mark or Read
$e = 0.80$	$H = 60$ ft	C
C	$G = 200$ gal/min	hp = 0.78 horsepower

Figure 1. Pump horsepower.

Source

Macdonald, J. O. S., *British Chemical Engineering*, 1, 6, 290 (1956).

How to select motors for field-gathering pumps

The following simple formulas will provide a reasonably close approximation of the motor hp necessary to drive duplex double-acting pumps normally used in the field.

Rule.

1. $hp = \dfrac{\text{bpd} \times \text{psi}}{48,000}$

2. $hp = \dfrac{\text{bpd} \times \text{psi}}{2,000}$

3. $hp = \dfrac{\text{gpm} \times \text{psi}}{1,400}$

Motors selected with these formulas will fit delivery rates used in the formula if they are within capacity of the pump. If anticipated rate is below pump capacity, a savings may be made in first cost and power bills by selecting a motor corresponding to the actual rate rather than to full pump capacity.

It should be noted that net bpd values are to be used in Formula 1. That is, if 600 barrels are to be pumped each day, but it is desired to accomplish this in only 8 h, the rate to be used in the formula is 1800 bpd.

Reciprocating pumps

A. How a reciprocating pump works

A reciprocating pump is a positive displacement mechanism with liquid discharge pressure being limited only by the strength of the structural parts. Liquid volume or capacity delivered is constant regardless of pressure and is varied only by speed changes.

Characteristics of a GASO reciprocating pump are (1) positive displacement of liquid, (2) high pulsations caused by the sinusoidal motion of the piston, (3) high volumetric efficiency, and (4) low pump maintenance cost.

B. Plunger or piston rod load

Plunger or piston "rod load" is an important power and design consideration for reciprocating pumps. Rod load is the force caused by the liquid pressure acting on the face of the piston or plunger. This load is transmitted directly to the power frame assembly and is normally the limiting factor in determining maximum discharge pressure ratings. This load is directly proportional to the pump gauge discharge pressure and proportional to the square of the plunger or piston diameter.

Occasionally, allowable liquid end pressures limit the allowable rod load to a value below the design rod load. It is important that liquid end pressures do not exceed published limits.

C. Calculations of volumetric efficiency

Volumetric efficiency (E_v) is defined as the ratio of plunger or piston displacement to liquid displacement. The volumetric efficiency calculation depends upon the internal configuration of each individual liquid body, the piston size, and the compressibility of the liquid being pumped.

D. Tools for liquid pulsation control, inlet and discharge

Pulsation control tools (PCT, often referred to as "dampeners" or "stabilizers") are used on the inlet and discharge piping to protect the pumping mechanism and associated piping by reducing the high pulsations within the liquid caused by the motions of the slider-crank mechanism. A properly located and charged pulsation control tool may reduce the length of pipe used in the acceleration head equation to a value of 5–15 nominal pipe diameters. Figure 1 is a suggested piping system for power pumps. The pulsation control tools are specially required to compensate for inadequately designed or old/adapted supply and discharge systems.

E. Acceleration head

Whenever a column of liquid is accelerated or decelerated, pressure surges exist. This condition is found on the inlet side

RECIPROCATING PUMPS FLOW CHARACTERISTICS

DUPLEX DOUBLE-ACTING

Average Flow — 100%
Maximum Flow — 100% + 24%
Minimum Flow — 100% − 22%
Total Flow Var. — 46%

TRIPLEX SINGLE-ACTING

Average Flow — 100%
Maximum Flow — 100% + 6%
Minimum Flow — 100% − 17%
Total Flow Var. — 23%

Figure 1. Suggested piping system for power pumps.

NOTES

1) Centerline of pump suction (inlet) to be slightly higher than centerline of suction (inlet) valve–so any air in suction system promptly reaches pump and is expelled.

2) Slope by-pass line so low point drain will fully empty relief and choke valves and all liquid in by–pass circuit.

3) Do not locate piping or other equiment in front of or above pump liquid end–preventing servicing.Refer to pump manufacturer for minimum

4) Locate charging pump at point shown–if a charging pump is necessary. (As for volatile fluid, for example.)

5) If desired, a two-way motor-operated by-pass valve may be used rather than manual type. It should be designed to open automatically while starting or stopping.

6) Feed line and by-pass line should feed liquid into tank belo minium liquid level.

7) To remove piping strain and vibration, a flexible hose, expansion joint, or swivel joint pair should be positioned to minimize effects of piping thermal expansion, contraction, and piping weight.

8) Suction and discharge piping must be supported and anchored.

9) To protect suction system against hazard of discharge pressure entry (as when pump is idle), a small relief valve is often connected here.

10) All system components must have adequate pressure ratings for operating, starting, and upset conditions. In order to reduce potential hazards, particular attection is recommended for the surge condition that will result downstream of the relief valve when normal discharge is blocked.

QUINTUPLEX SINGLE-ACTING

Average Flow– 100%
Maximum Flow– 100% + 2%
Mnimum Flow – 100% - 5%
Total Flow Var – 7%

SEPTULEX SINGLE-ACTING

Average Flow– 100%
Maximum Flow– 100% + 1. 2%
Mnimum Flow – 100% - 2.6%
Total Flow Var – 3.8%

NONUPLEX SINGLE-ACTING

Average Flow– 100%
Maximum Flow– 100% + 0. 6%
Mnimum Flow – 100% - $1.5%
Total Flow Var – 2.1%

of the pump as well as the discharge side. Not only can the surges cause vibration in the inlet line, but they can restrict and impede the flow of liquid and cause incomplete filling of the inlet valve chamber. The magnitude of the surges and how they will react in the system is impossible to predict without an extremely complex and costly analysis of the system. Since the behavior of the natural frequencies in the system is not easily predictable, as much of the surge as possible must be eliminated at the source. Proper installation of an inlet pulsation control PCT will absorb a large percentage of the surge before it can travel into the system. The function of the PCT is to absorb the "peak" of the surge and feed it back at the low part of the cycle. The best position for the PCT is in the liquid supply line as close to the pump as possible or attached to the blind flange side of the pump inlet. In either location, the surges will be dampened and harmful vibrations reduced.

Required Formulae and Definitions

Acceleration head

$$h_a = \frac{LVNC}{Kg} \qquad V = \frac{GPM}{(2.45)\,(ID)^2}$$

where

h_a = Acceleration head (in feet)
L = Length of liquid supply line (in line)
V = Average velocity in liquid supply line (in fps)
N = Pump speed (revolutions per minute)
C = Constant depending on the type of pump
  C = 0.200 for simplex double-acting
    = 0.200 for duplex single-acting
    = 0.115 for duplex double-acting
    = 0.066 for triplex single- or double-acting
    = 0.040 for quintuplex single- or double-acting
    = 0.028 for septuplex, single- or double-acting
    = 0.022 for nonuplex, single- or double-acting
K = Liquid compressibility factor
  K = 2.5 For relatively compressible liquids (ethane, hot oil)
  K = 2.0 For most other hydrocarbons
  K = 1.5 For amine, glycol, and water
  K = 1.4 For liquids with almost no compressibility (hot water)
g = Gravitational constant = 32.2 ft/s^2
d = Inside diameter of pipe (inches)

Stroke

One complete unidirectional motion of piston or plunger. Stroke length is expressed in inches.

Pump capacity (Q)

The capacity of a reciprocating pump is the total volume throughput per unit of time at suction conditions. It includes both liquid and any dissolved or entrained gases at the stated operating conditions. The standard unit of pump capacity is the US gallon per minute.

Pump displacement (D)

The displacement of a reciprocating pump is the volume swept by all pistons or plungers per unit time. Deduction for piston rod volume is made on double-acting piston type pumps when calculating displacement. The standard unit of pump displacement is the U.S. gallon per minute. For single-acting pumps:

$$D = \frac{ASnm}{231}$$

For double-acting piston pumps:

$$D - \frac{(2A-a)\ snm}{231}$$

where

A = Plunger or piston area, square inch
a = Piston rod cross-sectional area, square inch (double-acting pumps)
s = Stroke length, inch
n = RPM of crankshaft
m = Number of pistons or plungers

Plunger or piston speed (v)

The plunger or piston speed is the average speed of the plunger or piston. It is expressed in feet per minute.

$$v = \frac{ns}{6}$$

Pressures

The standard unit of pressure is the pound force per square inch.

Discharge pressure (Pd)—The liquid pressure at the centerline of the pump discharge port.

Suction pressure (Ps)—The liquid pressure at the centerline of the suction port.

Differential pressure (Ptd)—The difference between the liquid discharge pressure and suction pressure.

Net positive suction head required (NPSHR)—The amount of suction pressure, over vapor pressure, required by the pump to obtain satisfactory volumetric efficiency and prevent excessive cavitation.

The pump manufacturer determines (by test) the net positive suction head required by the pump at the specified operating conditions.

NPSHR is related to losses in the suction valves of the pump and frictional losses in the pump suction manifold and pumping chambers. Required NPSHR does not include system acceleration head, which is a system-related factor.

Slip (S)

Slip of a reciprocating pump is the loss of capacity, expressed as a fraction or percent of displacement, due to leaks past the valves (including the back-flow through the valves caused by delayed closing) and past double-acting pistons. Slip does not include fluid compressibility or leaks from the liquid end.

Power (P)

Pump power input (Pi)—The mechanical power delivered to a pump input shaft, at the specified operating conditions. Input horsepower may be calculated as follows:

$$Pi = \frac{Q \times Ptd}{1714 \times \eta\, p}$$

Pump power output (Po)—The hydraulic power imparted to the liquid by the pump, at the specified operating conditions. Output horsepower may be calculated as follows:

$$Po = \frac{Q \times Ptd}{1714}$$

The standard unit for power is the horsepower.

Efficiencies (*n*)

Pump efficiency (ηp) (also called pump mechanical efficiency)—The ratio of the pump power output to the pump power input.

$$\eta p = \frac{Po}{Pi}$$

Volumetric efficiency (ηv)—The ratio of the pump capacity to displacement.

$$\eta v = \frac{Q}{D}$$

Plunger load (single-acting pump)

The computed axial hydraulic load, acting upon one plunger during the discharge portion of the stroke, is the plunger load. It is the product of plunger area and the gauge discharge pressure. It is expressed in pounds force.

Piston rod load (double-acting pump)

The computed axial hydraulic load, acting upon one piston rod during the forward stroke (toward head end), is the piston rod load.

It is the product of piston area and discharge pressure, less the product of net piston area (rod area deducted) and suction pressure. It is expressed in pounds force.

Liquid pressure $\left(\dfrac{\text{pounds}}{\text{square inch}}\right)$ or (psi)

= **Cylinder area** (square inches)

= $(3.1416) \times (\text{Radius(inches)})2$

= $\dfrac{(3.1416) \times (\text{diameter (inches)})^2}{4}$

Cylinder force (pounds) − pressure (psi)

× area(square inches)

Cylinder speed or average liquid velocity through

piping (feet/second) = $\dfrac{\text{flow rate (gpm)}}{2.448 \times (\text{inside diameter (inch)})^2}$

Reciprocating pump displacement (gpm)

= $\dfrac{\text{rpm} \times \text{displacement (cubic in/revolution)}}{231}$

Pump input horsepower – see horsepower calculations

Shaft torque (foot-pounds) = $\dfrac{\text{horsepower} \times 5252}{\text{shaft speed (rpm)}}$

Electric motor speed (rpm) = $\dfrac{120 \times \text{frequency (HZ)}}{\text{number of poles}}$

Three phase motor horsepower (output)

= $\dfrac{1.73 \times \text{amperes} \times \text{volts} \times \text{efficiency} \times \text{power factor}}{746}$

Static head of liquid (feet) = $\dfrac{2.31 \times \text{static pressure (psig)}}{\text{specific gravity}}$

Velocity head of liquid (feet) = $\dfrac{\text{liquid velocity}^2}{g = 32.2\,\text{ft/sec}^2}$

Absolute viscosity (centipoise)

= specific gravity × Kinematic viscosity (centistrokes)

Kinematic viscosity (centistrokes)

= 0.22 × Saybolt viscosity (ssu)

= $\dfrac{180}{\text{Saybolt viscosity (ssu)}}$

Absolute pressure (psia)

= Local atmospheric pressure + gauge pressure (psig)

Gallon per revolution

= $\dfrac{\text{Area of plunger (sq in)} \times \text{length of stroke (in)} \times \text{number of plungers}}{231}$

Barrels per day = gal/rev × pump speed (rpm) × 34.3

Example: Find the volumetric efficiency of a reciprocating pump with the following conditions:

Type of pump	3 in diam plunger × 5 in stroke triplex
Liquid pumped	Water
Suction pressure	Zero psig
Discharge pressure	1785 psig
Pumping temperature	140 F
c	127.42 cu in
d	35.343 cu in
S	0.02

Find βt from Table of Water Compressibility (Table 1).

Table 1
Water Compressibility. Compressibility Factor $\beta t \times 10^{-6}$ = Contraction in Unit Volume Per Psi Pressure; Compressibility from 14.7 psia, 32°F–212°F and from Saturation Pressure Above 212°F

Pressure	Temperature																		
	0°C	20°C	40°C	60°C	80°C	100°C	120°C	140°C	160°C	180°C	200°C	220°C	240°C	260°C	280°C	300°C	320°C	340°C	360°C
psia	32°F	63°F	104°F	140°F	176°F	212°F	248°F	284°F	320°F	356°F	392°F	428°F	464°F	500°F	536°F	572°F	608°F	644°F	680°F
200	3.12	3.06	3.06	3.12	3.23	3.40	3.66	4.00	4.47	5.11	6.00	7.27							
400	3.11	3.05	3.05	3.11	3.22	3.39	3.64	3.99	4.45	5.09	5.97	7.21							
600	3.10	3.05	3.05	3.10	3.21	3.39	3.63	3.97	4.44	5.07	5.93	7.15	8.95						
800	3.10	3.04	3.04	3.09	3.21	3.38	3.62	3.96	4.42	5.04	5.90	7.10	8.85	11.6					
1000	3.09	3.03	3.03	3.09	3.20	3.37	3.61	3.95	4.40	5.02	5.87	7.05	8.76	11.4	16.0				
1200	3.08	3.02	3.02	3.08	3.19	3.36	3.60	3.94	4.39	5.00	5.84	7.00	8.68	11.2	15.4				
1400	3.07	3.01	3.01	3.07	3.18	3.35	3.59	3.92	4.37	4.98	5.81	6.95	8.61	11.1	15.1	23.0			
1600	3.06	3.00	3.00	3.06	3.17	3.34	3.58	3.91	4.35	4.96	5.78	6.91	8.53	10.9	14.8	21.9			
1800	3.05	2.99	3.00	3.05	3.16	3.33	3.57	3.90	4.34	4.94	5.75	6.87	8.47	10.8	14.6	21.2	36.9		
2000	3.04	2.99	2.99	3.04	3.15	3.32	3.56	3.88	4.32	4.91	5.72	6.83	8.40	10.7	14.3	20.7	34.7		
2200	3.03	2.98	2.98	3.04	3.14	3.31	3.55	3.87	4.31	4.89	5.69	6.78	8.33	10.6	14.1	20.2	32.9	86.4	
2400	3.02	2.97	2.97	3.03	3.14	3.30	3.54	3.85	4.29	4.87	5.66	6.74	8.26	10.5	13.9	19.8	31.6	69.1	
2600	3.01	2.96	2.96	3.02	3.13	3.29	3.53	3.85	4.28	4.85	5.63	6.70	8.20	10.4	13.7	19.4	30.5	61.7	
2800	3.00	2.95	2.96	3.01	3.12	3.28	3.52	3.83	4.26	4.83	5.61	6.66	8.14	10.3	13.5	19.0	29.6	57.2	238.2
3000	3.00	2.94	2.95	3.00	3.11	3.28	3.51	3.82	4.25	4.81	5.58	6.62	8.08	10.2	13.4	18.6	28.7	53.8	193.4
3200	2.99	2.94	2.94	3.00	3.10	3.27	3.50	3.81	4.23	4.79	5.55	6.58	8.02	10.1	13.2	18.3	27.9	51.0	161.0
3400	2.98	2.93	2.93	2.99	3.09	3.26	3.49	3.80	4.22	4.78	5.53	6.54	7.96	9.98	13.0	17.9	27.1	48.6	138.1
3600	2.97	2.92	2.93	2.98	3.09	3.25	3.48	3.79	4.20	4.76	5.50	6.51	7.90	9.89	12.9	17.6	26.4	45.4	122.4
3800	2.96	2.91	2.92	2.97	3.08	3.24	3.47	178	4.19	4.74	5.47	6.47	7.84	9.79	12.7	17.3	25.8	44.5	110.8
4000	2.95	2.90	2.91	2.97	3.07	3.23	3.46	3.76	4.17	4.72	5.45	6.43	7.78	9.70	12.5	17.1	25.2	42.8	101.5
4200	2.95	2.90	2.90	2.96	3.06	3.22	3.45	3.75	4.16	4.70	5.42	6.40	7.73	9.62	12.4	16.8	24.6	41.3	93.9
4400	2.94	2.89	2.90	2.95	3.05	3.21	3.44	3.74	4.14	4.68	5.40	6.36	7.68	9.53	12.2	16.5	24.1	40.0	87.6
4600	2.93	2.83	2.89	2.94	3.05	3.20	3.43	3.73	4.13	4.66	5.37	6.32	7.62	9.44	12.1	16.3	23.6	38.8	82.3
4800	2.92	2.87	2.88	2.94	3.04	3.20	3.42	3.72	4.12	4.64	5.35	6.29	7.57	9.36	12.0	16.0	23.2	37.6	77.7
5000	2.91	2.87	2.87	2.93	3.03	3.10	3.41	3.71	4.10	4.63	5.32	6.25	7.52	9.28	11.8	15.8	22.7	36.6	73.9
5200	2.90	2.85	2.87	2.92	3.02	3.18	3.40	3.69	4.09	4.61	5.30	6.22	7.47	9.19	11.7	15.6	22.3	35.6	70.3
5400	2.90	2.85	2.86	2.91	3.01	3.17	3.39	3.68	4.07	4.59	5.27	6.19	7.41	9.12	11.6	15.3	21.9	34.6	66.9

$\beta t = 0.00000305$ at 140°F and 1800 psia. Calculate volumetric efficiency:

$$\text{Vol. Eff.} = \frac{1 - \left[P_{td\beta t}\left(1 + \dfrac{c}{d}\right)\right]}{1 - P_{td\beta t}} - S$$

$$= \frac{1 - \left[(1785 - 0)(0.00000305)\left[1 + \dfrac{127.42}{35.343}\right]\right]}{1 - (1785 - 0)\,(0.00000305)}$$

$$-\ 0.02$$

$$= 0.96026 = 96 \text{ percent}$$

Specific gravity (at 60°F) $= \dfrac{141.5}{131.5 + \text{API gravity (degree)}}$

Bolt clamp load (lb) $= 0.75 \times$ proof strength (psi) $\times$ tensile stress area (in^2)

Bolt torque (ft – lb) $= \dfrac{0.2(\text{or } 0.15)}{12}$ $\times$ nominal diameter in inches $\times$ bolt clamp load (lb)

0.2 for dry
0.15 for lubricated, plating, and hardened washers

Calculating volumetric efficiency for water

The volumetric efficiency of a reciprocating pump, based on capacity at suction conditions, using table of water compressibility, shall be calculated as follows:

$$\text{Vol. Eff.} = \frac{1 - \text{Ptd}\beta t\left(1 + \dfrac{c}{d}\right)}{1 - \text{Ptd}\beta t} - S$$

where

$\beta t =$ Compressibility factor at temperature t (degrees Fahrenheit or centigrade). (See Tables 1 and 2).
$c =$ Liquid chamber volume in the passages of chamber between valves when plunger is at the end of discharge stroke in cubic inches
$d =$ Volume displacement per plunger in cubic inches
$\text{Ptd} =$ Discharge pressure minus suction pressure in psi
$S =$ Slip, expressed in decimal value

Calculating volumetric efficiency for hydrocarbons

The volumetric efficiency of a reciprocating pump based on capacity at suction conditions, using compressibility factors for hydrocarbons, shall be calculated as follows:

$$\text{Vol. Eff.} = 1 - \left[S - \frac{c}{d}\left(1 - \frac{\rho d}{\rho s}\right) \right]$$

where

$c =$ Fluid chamber volume in the passages of chamber between valves, when plunger is at the end of discharge strike, in cubic inches
$d =$ Volume displacement per plunger, in cubic inches
$P =$ Pressure in psia ($P_s =$ suction pressure in psia;
$P_d =$ discharge pressure in psia)
$P_c =$ Critical pressure of liquid in psia
$P_r =$ Reduced pressure

$$\frac{\text{Actual pressure in psia}}{\text{Critical pressure in psia}} = \frac{P}{P_C}$$

$P_{rs} =$ Reduced suction pressure $= P_S/P_c$
$P_{rd} =$ Reduced discharge pressure $= P_d/P_c$
$S =$ Slip expressed in decimal value
$t =$ Temperature, in degrees Rankine
$\quad =$ Degrees F $+ 460$ ($t_s =$ suction temperature in degrees Rankine; $t_d =$ discharge temperature in degrees Rankine)

$T_c =$ Critical temperature of liquid, in degrees Rankine (See Table 3)
$T_r =$ Reduced temperature

$$= \frac{\text{actual temp. in degrees Rankine}}{\text{critical temp. in degrees Rankine}}$$

$$= \frac{t}{T_C} \text{ (See Figure 2)}$$

$T_{rs} =$ Reduced suction temperature

$$= \frac{t_s}{T_C}$$

$T_{rd} =$ Reduced discharge temperature

$$= \frac{t_d}{T_C}$$

Vol. Eff. $=$ Volumetric efficiency expressed in decimal value.

$$= \frac{1}{1} \times \cdots \times 62.4 = \text{density of liquid in lb per cu ft}$$

$s =$ Density in lb per cu ft at suction pressure

Table 2
Water Compressibility
Compressibility Factor $\beta t \times 10^{-6} =$ Contraction in Unit Volume Per Psi Pressure
Compressibility from 14.7 psia at 68°F and 212°F and from Saturation Pressure at 392°F

Pressure	Temperature			Pressure	Temperature		
	20°C	100°C	200°C		20°C	100°C	200°C
psia	38°F	212°F	392°F	psia	38°F	212°F	392°F
6000	2.84	3.14	5.20	22,000	2.61	2.42	3.75
7000	2.82	3.10	5.09	23,000	2.59	2.38	3.68
8000	2.80	3.05	4.97	24,000	2.58	2.33	3.61
9000	2.78	3.01	4.87	25,000	2.57	2.29	3.55
10,000	2.76	2.96	4.76	26,000	2.56	2.24	3.49
11,000	2.75	2.92	4.66	27,000	2.55	2.20	3.43
12,000	2.73	2.87	4.57	28,000	2.55	2.15	3.37
13,000	2.71	2.83	4.47	29,000	2.54	2.11	3.31
14,000	2.70	2.78	4.38	30,000	2.53	2.06	3.26
15,000	2.69	2.74	4.29	31,000	2.52	2.02	3.21
16,000	2.67	2.69	4.21	32,000	2.51	1.97	3.16
17,000	2.66	2.65	4.13	33,000	2.50	1.93	3.11
18,000	2.65	2.60	4.05	34,000	2.49	1.88	3.07
19,000	2.64	2.56	3.97	35,000	2.49	1.84	3.03
20,000	2.63	2.51	3.89	36,000	2.48	1.79	2.99
21,000	2.62	2.47	3.82				

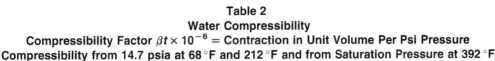

d = Density in lb per cu ft at discharge pressure = Expansion factor of liquid

$\dfrac{1}{\omega}$ = Characteristic constant in grams per cubic centimeter for any one liquid that is established by density measurements and the corresponding values of (See Table 3).

$$\rho_s = \frac{\rho 1}{\omega 1} \times \omega \times 62.4 = 4.803 \times 0.1048 \times 62.4$$
$$= 31.4 \text{ lb per cu ft}$$

Find density at discharge pressure:

$$T_{rd} = \frac{t_d}{T_c} = \frac{460 + 80}{666} = 0.81$$

Table 3

Carbon Atoms	Name	T_c Degrees Rankine	P_c lb per sq in	pl/wl grams per cc
1	Methane	343	673	3.679
2	Ethane	550	717	4.429
3	Propane	666	642	4.803
4	Butane	766	544	5.002
5	Pentane	847	482	5.128
6	Hexane	915	433	5.216
7	Heptane	972	394	5.285
8	Octane	1025	362	5.340
9	Nonane	1073	332	5.382
10	Decane	1114	308	5.414
12	Dodecane	1185	272	5.459
14	Tetradecane	1248	244	5.483

Example. Find volumetric efficiency of the previous reciprocating pump example with the following new conditions:

Type of pump	3 inch dia plunger × 5 inch stroke triplex
Liquid pumped	Propane
Suction temperature	70 °F
Discharge temperature	80 °F
Suction pressure	242 psig
Discharge pressure	1911 psig

Find density at suction pressure:

$$T_{rs} = \frac{t_s}{T_c} = \frac{460 + 70}{666} = 0.795$$

$$P_{rs} = \frac{P_s}{P_c} = \frac{257}{642} = 0.4$$

$$\frac{\rho 1}{\omega 1} = 4.803 \text{(From Table 3, propane)}$$

$$\omega = 0.1048 \text{(From Figure 2)}$$

$$P_{rd} = \frac{P_d}{P_c} = \frac{1926}{642} = 3.0$$

$$\omega = 0.1089 \text{(From Figure 2)}$$

$$\rho d = \frac{\rho 1}{\omega 1} \times \omega \times 62.4 = 4.803 \times 0.1089 \times 62.4$$
$$= 32.4 \text{ lb per cu ft}$$

Therefore

$$\text{Vol. Eff.} = 1 - \left[S - \frac{c}{d} \left(1 - \frac{\rho d}{\rho s} \right) \right]$$
$$= 1 - \left[0.02 - \frac{127.42}{35.343} \left(1 - \frac{32.64}{31.4} \right) \right] = 0.8376$$
$$= 83.76\%$$

Reprinted with permission—Goulds Pumps.

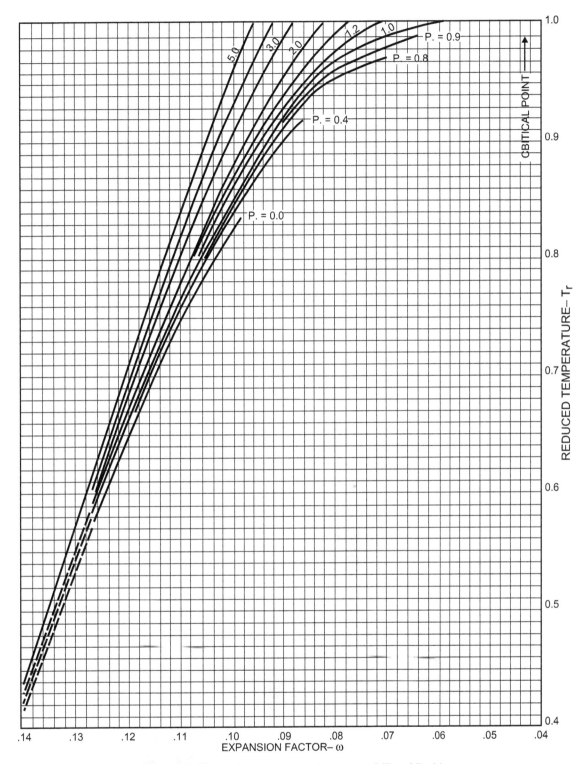

Figure 2. Thermal expansion and compressibility of liquids.

Understanding the basics of rotary screw pumps

...operating principles, application factors, and performance characteristics

Albert A. Zalis, Vice President Engineering, Warren Pumps, Inc., IMO Industries, Inc.

Basically only two categories of pumps need be considered when a pumping application is evaluated: centrifugal pumps and positive-displacement pumps. Centrifugal pumps depend on rotational speed to establish head and capacity characteristics and are capable of moderately high discharge heads and very high pumping capacities. Centrifugal pump design configurations include single and multistage, enclosed and open impeller, single and double suction, and horizontal and vertical.

When the application of a centrifugal pump is considered, in addition to the normal requirements of head and capacity, the pumped medium fluid characteristics must be known in detail:

- Abrasiveness
- Consistency and viscosity
- System head/capacity variations

- Net positive suction head available (NPSHA)
- Corrosiveness
- Required construction materials

Although centrifugal pumps successfully handle all types of slurries and viscous fluids, a suitable relationship between head, capacity, and fluid characteristics must exist. A simple end suction centrifugal pump with an open impeller can be used for handling both clear water and slurries; see Figure 1. Typical centrifugal pump performance curves are shown in Figure 2. Centrifugal pump performance can be evaluated using the affinity laws: With the impeller diameter held constant, capacity Q varies as the ratio of the change in speed N:

$$\frac{Q_2}{Q_1} = \frac{N_2}{N_1}$$

Figure 1. The end suction centrifugal pump is a general-purpose pump that is widely used for pumping a broad range of liquids. Seal may be flushed with clean liquid when corrosive or abrasive slurries are pumped.

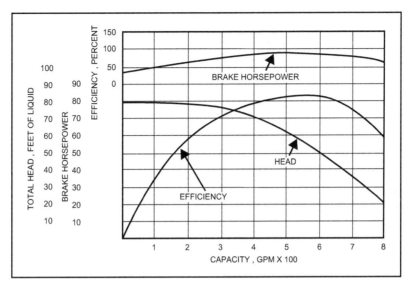

Figure 2. Typical performance curves for a centrifugal pump plot the brake horsepower, efficiency, and head (discharge pressure) as functions of capacity. The plots of horsepower and head are not straight lines, a fact that increases the difficulty of accurately predicting performance.

Head H varies as the square of the speed change ratio:

$$\frac{H_2}{H_1} = \left(\frac{N_2}{N_1}\right)^2$$

Brake horsepower (bhp) varies as the cube of the speed change ratio:

$$\frac{bhp_2}{bhp_1} = \left(\frac{N_2}{N_1}\right)^3$$

These laws make it simple to correct for varying speeds. However, these relationships assume that efficiency remains constant. Tests have shown that the correction for head, with regard to speed, is not precisely correct when head is reduced more than the amount calculated to result from a speed increase. Also, a speed increase yields a reduction in efficiency exceeding that calculated using the brake horsepower and head formula. However, the deviation of values is not great, and the affinity laws are a valuable tool when used with good engineering judgment.

Positive-displacement pumps include several pump designs. The major classifications are reciprocating and rotary pumps. Reciprocating pumps include power and steam driven, single and double acting, and plunger and piston.

Reciprocating pumps have high head capability, easily handle all types of slurries, and have high volumetric efficiency, permitting low-capacity/high-head relationships.

When the application of a reciprocating pump is considered in addition to the usual head and capacity data, the following conditions related to the fluid must be known:

- NPSHA
- Allowable pulsation levels
- Corrosiveness or required construction materials

Rotary pumps cover a broad range, including screw, gear, vane, and radial piston types.

Screw pumps can be further classified as single (progressive cavity), two screw (timed), and three screw (untimed). Gear pumps include external and internal gear designs.

Rotary pumps operate in generally the same pressure ranges as centrifugal pumps, except at lower capacities. Rotary pumps are not particularly suitable for handling severely abrasive fluids, except for the single-screw progressive cavity pump, which performs well in pumping abrasive slurries. Rotary pumps are generally capable of handling a wide range of fluid viscosities, at medium pressures, over a broad range of flow rates.

The remainder of this article is confined to a discussion of screw pumps, primarily the two-screw timed pump that utilizes timing gears to drive the screws and prevent driving contact between the intermeshing screws. This pump uses bearings to support the shafts and absorb some or all of the radial load, giving it a broad field of application suitable for pumping corrosive and noncorrosive fluids, lubricating and nonlubricating fluids, clean fluids and slurries, and Newtonian and non-Newtonian fluids.

Relatively high capacities, for positive-displacement pumps, can be handled by screw pumps. The primary limitation is the production of screws having the displacement required to produce a high discharge rate.

Performance characteristics of a screw pump are functions of screw design, provided it is used with properly designed inlet and outlet ports. In a particular pump size, performance characteristics can be radically changed by altering the screw

Figure 3. Four basic dimensions establish the pump displacement characteristics. One of the flanks must have an epicycloidal profile to establish a seal with the mating screw. The mating flank may have an epicycloidal curve or any form of convex curve that will achieve a sealed running mesh.

design. When a screw pump performs poorly in a specific application, the application parameters should be rechecked and the screw design examined.

The capacity of a screw pump is a function of four critical screw dimensions that control pump displacement (Figure 3). For the pump to deliver maximum displacement efficiently, the profiles of the intermeshing threads must be given careful attention. The outside diameter of the screw and the root diameter are cylindrical. Clearances of these dimensions average 0.0015 in. per in. of diameter. The flanks (thread sides) are profiled and must be carefully controlled to obtain good pumping efficiency. One of the two flanks must have a profile defined as an epicycloid. The other flank may have an epicycloidal curve or any form of convex curve that will form a continuous seal when it is mated with the meshing screw. The total axial or flank clearance normally equals the total diametrical clearance between the screw outside diameter and the casing bore.

There must be small clearances between the screws and between the screws and casing. These clearances allow a small amount of leakage from the pump outlet cavity to the pump inlet cavity. This leakage, identified as slip, is a function of pump outlet pressure, fluid viscosity, and fluid characteristics. If the screws are designed with epicycloidal curves on both flanks, the slip will be 70% greater than if the design provides for one epicycloidal curve and one convex curve. However, for certain applications the double epicycloid has distinct advantages.

In a pump design utilizing four pumping screws, the arrangement usually allows the fluid to enter the screws on each end and discharge in the center (Figure 4). When the screws are all the same diameter, axial thrust forces are equal

Figure 4. In a two-shaft, four-screw pump, the material enters at each end and discharges in the center. End thrust forces are balanced. Radial forces are functions of outlet pressure and screw dimensions and must be carried by the bearings at each end.

Figure 5. Pumps with two screws require timing gears to maintain constant relationship between screws. Axial forces are normally counterbalanced by hydraulic loading. The gear set may be product lubricated (a) or provided with an external lubrication system (b).

and opposite and therefore cancel each other. Although other arrangements for balancing radial forces are available, they are, at best, only partly effective and sometimes introduce other problems. Radial forces in screw pumps are functions of outlet pressure and screw dimensions. If the epicycloid/convex screw flank design is used, proper location of the profiles for the two screws on each shaft will partially balance radial loads.

Timed screw pumps are available in which only two pumping screws are used (Figure 5). In these pumps, strategic location of the flank profiles has no effect on radial forces. Axial forces are of appreciable magnitude and are normally counterbalanced hydraulically.

Screw pumps are capable of discharge pressures from atmospheric to more than 300 psig. However, the same screw design cannot be used throughout this range of pressures. In a centrifugal pump, high discharge heads are achieved either by high speeds or, more often, by staging a number of impellers to operate in series; each impeller carries a proportionate share of the discharge head.

Discharge head in a screw pump is regulated by controlling the number of screw turns. The minimum number of screw turns that can be used is two. Fewer than two turns results in erratic performance and excessive slip. Unlike the centrifugal pump impeller, each screw turn does not generate a proportionate share of the outlet pressure. The additional screw turns assist in reducing slip, thereby increasing volumetric efficiency.

"...lower volumetric efficiency could result in a higher mechanical efficiency."

A higher volumetric efficiency can mean higher mechanical efficiency, and with today's power costs it would appear desirable to use screws with many turns to achieve a high volumetric efficiency. This is not necessarily true. The greater the number of screw turns, the longer the screw (screw length equals the number of turns multiplied by the screw pitch). Screw length has a proportional bearing on the horsepower required to drive the pump at zero outlet pressure.

If a low-viscosity fluid is being pumped, this horsepower will be of small magnitude. However, the horsepower

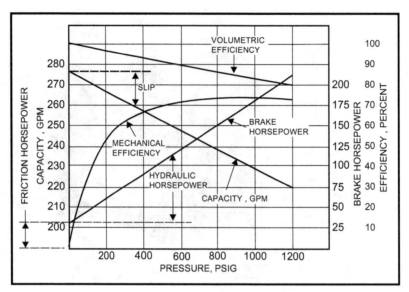

Figure 6. In contrast to performance curves for a centrifugal pump, the screw pump performance curves for capacity and brake horsepower are normally straight lines. Capacity, brake horsepower, mechanical efficiency, and volumetric efficiency are plotted as functions of discharge pressure.

requirement increases rapidly as viscosity increases. There-fore, a slightly lower volumetric efficiency could result in a higher mechanical efficiency. Mechanical efficiency controls power costs; consequently, the application engineer, when selecting a screw pump, must consider both slip and friction horsepower to determine the best arrangement.

The remaining parts of a screw pump are subject to conventional mechanical design: Shafts must be designed for bending and torsional loads, and bearings must match the center distance between the two parallel shafts. Timing gears should be designed in accordance with American Gear Manufacturers Association standards. If spur gears are used, both shafts should be rigidly located axially. If herringbone gears are used to reduce noise level, the driven shaft should be allowed to float axially and be positioned by the apex of the gear. Velocities through the fluid inlet passages should be kept low, normally not more than 5 fps for average viscosities of 150 to 1000 SSU and decreased as the viscosity increases.

The performance characteristics of a screw pump are similar to those of other positive-displacement pumps. Capacity, horsepower, and efficiency are plotted as functions of discharge pressure. For centrifugal pumps, horsepower, efficiency, and discharge pressure are plotted as functions of capacity. With centrifugal pumps, plots of horsepower and head against capacity are not straight lines. Therefore, accurately predicting performance is difficult, and more testing and modeling are required. For positive-displacement pumps, plots of capacity and brake horsepower against pressure normally result in straight lines when coordinate

paper is used. However, there are instances in which the capacity versus pressure line is slightly curved. This charac-teristic usually results from excessive clearance in the pump permitting deflection and eccentric operation of the rotating screw.

Performance conditions can be manipulated to reflect operation at speeds or pressures other than those shown in Figure 6, using these fundamental relationships:

Friction horsepower is the power required to overcome losses including packing seal friction, timing gear inefficien-cies, bearing friction, and pumped fluid viscous shear. All of the losses are so small when compared to the viscous drag of the pumped fluid that it can be safely assumed that friction horsepower is entirely dependent on fluid viscosity and pump speed. Even pump rotor clearances, within the limits of efficient pump operation, have such a minor influence on the power consumed that extremely sensitive instrumentation is required for taking measurements.

Figure 7 shows the relationship between friction horse-power and viscosity at 1200 rpm. If, for example, it is necessary to determine friction horsepower fhp at 1800 rpm, the following relationship exists:

$$fhp_2 = fhp_1 \left(\frac{1800}{1200}\right)^{1.5}$$

With today's programmable calculators and computers, the curves shown in Figure 7 may be easier to work with if they are reduced to mathematical formulas. Because the equation will take the form $x = ay^n$, by means of simultaneous equations the friction horsepower equation becomes

$$fhp_2 = 2.9541 \times (\text{viscosity in SSU})^{0.42065}$$

Hydraulic horsepower is the measure of the total hydraulic work done by the pump. This term should not be confused with water horsepower or work horsepower, the measure of the useful work done by the pump.

These terms can be defined by the following formulas:

$$\text{hydraulic horsepower} = \frac{(\text{displaced gpm})(\text{psi})}{1714}$$

$$\text{water horsepower} = \frac{(\text{displaced gpm})(\text{psi})}{1714}$$

Displaced gpm is the capacity at zero psi discharge pressure. Delivered gpm is the capacity at the rated or required discharge pressure, or the displaced gpm minus the slip gpm.

Referring to Figure 6, the brake horsepower is the sum of the friction horsepower and the hydraulic horsepower. This approach yields a fairly accurate brake horsepower through a wide range of operating conditions.

Slip is the loss of capacity from the higher pressure area through the internal clearances. Slip is a function of viscosity and differential pressure. Figure 7 is a slip/viscosity curve for 500 psi operating pressure. Slip varies in direct proportion to the change in pressure. Similarly, with the friction horsepower curve, slip can be expressed by the following formula:

$$\text{slip, gpm} = 294.95 \times (\text{viscosity in SSU})^{-0.4660}$$

Mechanical efficiency is a calculated quantity that until recently received very little attention. The energy crisis has made it more important. Mechanical efficiency is calculated by

$$\text{mechanical efficiency} = \frac{(\text{displaced gpm})(\text{psi})}{1714(\text{bhp})} \times 100$$

Volumetric efficiency is an infrequently used measure of internal slip and can be calculated by

$$\text{volumetric efficiency} = \frac{\text{delivered gpm}}{\text{displaced gpm}} \times 100$$

It may be well to review a suggested approach to the pumping of non-Newtonian fluids. One fairly common application for screw pumps is pumping pseudoplastic polymers of various types. A pseudoplastic fluid is one whose viscosity decreases as the shear rate in the pump increases. Data on pseudoplastic fluids are usually presented as a curve plotting apparent viscosity versus shear rate in reciprocal seconds. The determination of shear rate in a positive-displacement pump is a widely discussed topic. It is difficult, if not impossible, to arrive at one formula that is applicable to

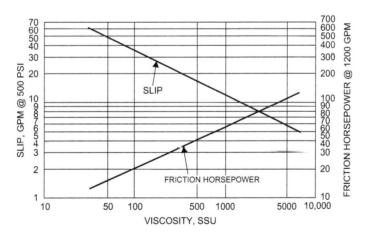

Figure 7. Slip is a loss of capacity from the higher pressure area to the lower pressure area and is a function of viscosity and pressure differential. Slip varies in direct proportion to the change in pressure. At a specific pressure as the viscosity increases, slip decreases, while the friction horsepower increases.

all pumps. For the timed screw pump, successful application has been made using the following formula:

$$\text{shear rate, } s^{-1} = \frac{\pi(\text{rotor diameter})(\text{rpm})}{(\text{radial clearance})(60)}$$

where

rotor diameter = Pumping screw diameter, inches
radial clearance = Clearance between screw diameter and casing bore, inches

This formula gives consideration to only the area of maximum shear in the pump. Experience has shown that refining this formula to include all shear surfaces yields values that are grossly incorrect. The shear rate calculated from the formula will permit a satisfactory pump selection; however, care should be taken to consider the shear rate in the inlet piping. This relatively small quantity results in a much higher viscosity and thereby influences the inlet pressure requirements.

Sometimes it becomes necessary to take a surplus pump and apply it to a new application. This often happens in chemical plants where processes are frequently subject to change. Earlier discussions dealt in some detail with screw profile variations, importance of screw turns, slip, and friction horsepower. All these factors must be considered for the new application. Performance characteristics can be changed radically by installation of a new rotating element in an existing casing.

This article has not covered all there is to learn about screw pumps and their applications but has discussed the more important areas on which there is little, if any, literature published.

Reprinted from **PLANT ENGINEERING** © CAHNERS PUBLISHING COMPANY.

How to evaluate VFD speed on hydraulics

Few would dispute that variable frequency drives (VFDs) save energy, but the exact amount depends on the system. Hydraulically speaking, the main difference between a variable frequency (speed) drive and a discharge valve is that a VFD only changes a *pump* curve, while a valve only changes a *system* curve. A pump operates at the intersection between its H-Q curve and a system curve, and a change in either moves the operating point to a new intersection.

Consider a centrifugal pump operating at 4000 gpm and producing 300 ft of head at 1800 rpm. According to its curve (Figure 1), pump efficiency is 82 percent. This happens to be at the best efficiency point (BEP) point, but, in general, the same logic would apply to any operating point.

Suppose we want to reduce the flow by 50 percent, to 2000 gpm. By closing the discharge valve, we would change a system curve, which would intersect the *same pump curve* at 370 ft of head, 61 percent efficiency. Such control process is simple, quick and easy. However, it has other issues, which we will review. Hydraulically, the system in Figure 1 is referred to as *predominantly friction* (no static head). Such system is ideal for VFD application for flow control. The primary reason is at the heart of the affinity laws, which state

Figure 1. Flow control by throttling discharge valve.

Figure 2. Flow control by VFD speed reduction, friction system head example.

Figure 3. Flow control by VFD, static system head example.

Table 1
Water Compressibility Comparison of Valve Throttling, VFD (Friction) and VFD (Static Head) Examples

	Throtting	VFD (Friction)	VFD (Static)
rpm	1800	900	1650
Flow, qpm	2000	2000	2000
Head, feet	370	75	300
Eff. %	61%	82%	62%
Power, hp	306	46	244
Energy cost, $/KW-h	0.10	0.10	0.10
Energy used per year, $	$200,193	$30,187	$159,701
BEP flow, qpm	4000	2000	3667
Operation %BEP	50%	100%	55%

that pump flow changes directly with the speed ratio, head as a square and power as cube. Therefore, the relationship between head and power as cube. Therefore, the relationship between head reduced as a square of a speed ratio $(900/1800)^2$, i.e. $0.25 \times 300 = 75$ ft. Notice that efficiency does not change as the entire efficiency curve "slides" to the left and remains at a peak of 82 percent.

What would happen if a pump operates against constant (static, or mostly static) head, as in a lift station application? Consider Figure 3.

In Figure 3, 300 ft is a constant static head, with friction losses assumed to be negligible in comparison. A range of RPM variation is significantly reduced, so the entire pump $H-Q$ curve does not drop below the contact $H = 300$ ft. By applying affinity laws, we will find RPM = 1650, a new pump curve $(H-Q)$ intersects a constant 300-ft head line at desired 2000 gpm and efficiency at that point is 62 percent.

Table 1 compares these three cases.

Observations

- For *friction-dominated* systems (long pipes, flow transfer cases), a VFD saves a substantial amount of energy, and operates pumps reliably due to close proximity to BEP flow (100 percent in example shown)
- For *static-dominated* systems (injection against constant pressure, lifting against constant head), the energy savings are substantially less, and the pump surprisingly

operates substantially *off-BEP* position, not significantly different from a valved flow control case
- For cases where both systems are present, an in-between scenario would result VFD drives are expensive devices and require certain knowledge of controls, proper electrical and electronic maintenance and care. Flow control by VFD requires programming and training and is more complex than simple control by throttling a valve. These are some downsides to VFDs, but they offer the convenience of remote control (as do some valves), energy savings when operated in friction-dominated systems and more reliable pump operation. Operating closer to the BEP means less loads, better seal life, lower vibrations and improved suction recirculation conditions.

The user must have a detailed understanding of a system's hydraulics, unique $H-Q$ curves and flow control requirements to make sure the VFD investment is justified. Organizations implementing VFDs must be prepared to deal with the added complexity but also appreciate their usefulness. The anticipated benefits of a VFD hinge on an organization's efforts to train users to operate and maintain such systems. For organizations leaving these issues to chance, the anticipated benefits can quickly become headaches.

Reprinted with permission—Dr. Lev Nelik, P. E. APICS Pumping Machinery, LLC, Atlanta, GA www.PumpingMachinery.com and Pumps & Systems March 2008 issue.

Progressive cavity pumps

Progressive Cavity Pump Parts

1—Rotor
2—Stator (usually elastomer material)
3—Wettable body parts
4—Pump discharge flange (outlet)
5—Pump suction flange (inlet)
7—Drain plug to drain suction housing
8—Universal joint
9—Universal joint seal
10—Pump packing
11—Connecting rod
15—Inspection port

They are especially suited for tough, nasty, multiphase fluids with gas and solids in suspension, at relatively moderate temperatures. PC pumps are ideal for shear sensitive liquids. A progressive cavity (PC) pump is a positive displacement pump but does not rely on valves for operation. They are capable of producing reasonably accurate repeatable flow and the output is if fairly independent of head. Operating without valves, the PC delivers its output free of pulsations. The PC pump will not air lock and because it does not rely on valves as a conventional positive displacement pump, the NPSHr is low.

The stator is made of an elastomer and the rotor can be made of carbon or stainless steel. Carbon steel rotors are usually coated for added wear resistance. PC pumps have a single shaft that requires sealing.

This type of pump transfers fluid by means of the progress, through the pump, of a sequence of small, fixed shape, discrete cavities, as its rotor is turned. This leads to the volumetric flow rate being proportional to the rotation rate (bidirectionally) and to low levels of shearing being applied to the pumped fluid. Hence these pumps have application in fluid metering and pumping of viscous or shear sensitive materials. It should be noted that the cavities taper down toward their ends and overlap with their neighbors, so that, in general, no flow pulsing is caused by the arrival of cavities at the outlet, other than that caused by compression of the fluid or pump components.

Capacity — 70 Durometer Hardness — — 55 Durometer Hardness' Data Based on Water @ 68'F Horsepower

* Std. Nitrile EPDM and Fluoroelastomer - 70 Duro.
  Std Natural Rubber - 55 Duro.

Differential Pressure (PSI)**

PC pump difficulties

Care must be exercised in selecting the stator material to be sure that it is compatible with the fluid being pumped. Erosion is another difficulty. When erosion is a problem rotor material and/or rotor coating that is harder than the base metal must be selected to withstand erosion. The pump should not be allowed to run dry as this may heat the stator above its maximum operating temperature. In certain instances, a temperature probe may be inserted into the stator and used to shut the pump down before a temperature that can damage the stator is reached. Flow switches may also be used to detect "no flow" and shut the pump down.

The pumping rate can be varied over a range by controlling the speed of rotation of the rotor. A variable frequency drive is well suited for use with at PC pump. When selecting the VFD, the application should be carefully specified so that the driver will supply the required torque at start up as the start up torque is greater than the running torque.

PUMP CURVE ACCURACY

George Stephens, Engineered Software, Inc.

Part One

More than 150 pump manufacturers present their pump performance data electronically. These digital pump catalogs can be found online, within stand-alone pump selection software, and through select hydraulic analysis software packages. Thousands of engineers worldwide use these catalogs to evaluate and select the pumps that best meet their needs.

Based on experience in digitizing more than 135,000 pump curves for manufacturers throughout the industry, Part One outlines how digital pump curves are created, as well as some of the accuracy issues that may arise from the methods of curve digitization. It covers pump performance curves and the accuracy ramifications from different digitization methods. As the industry moves toward electronic pump catalogs, end users should know how accurate they are and how they are generated.

Pump Selection Programs

The main function of a pump selection program is the ability to digitally reproduce the manufacturer's pump performance data accurately and to create a selection list of pumps that meet the customer's criteria. All pump manufacturers have "paper curves" representing the performance of their pumps,

and most also have these curves available in pdf format. These curves reproduce the manufacturer's bench test data and are the most accurate depictions of the pump's performance.

However, for a software program to understand and use this pump performance data, the data must first be converted into a form that can be read by the program. This is referred to as digitization of the pump curves. There are various methods of digitizing pump performance data, but occasionally, depending on the method used, some vital information can get lost in the translation. The accuracy of this translation can be critical to the pump selection and evaluation process.

To better understand the performance of the digitized curves, consider an example of a typical published pump curve from a manufacturer. Figure 1 shows a detailed curve that includes the flow versus head performance curves as well as isometric curves for efficiency, power, and net positive suction head required (NPSHR). These curves accurately represent the test performance of the indicated pump in accordance with the *ANSI/HI 1.6-2000 Centrifugal Pump Tests* standard.

Figure 1.

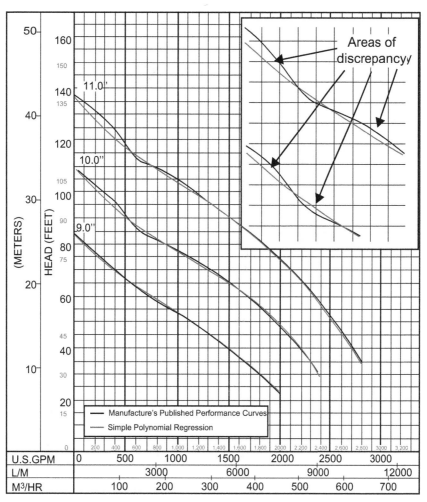

Figure 2.

If a pump were required for system design conditions of 800 U.S. gallons per minute and 35 feet of head and this particular pump were being evaluated, the system designer could feel confident that this pump with a 10-inch impeller would successfully fit the bill. The pump selection and evaluation software must accurately reproduce the manufacturer's published pump curve to aid the customer's evaluation of the pump for the application. The key to generating accurate digital curves is the method with which the pump performance data are saved and displayed.

All forms of pump curve digitization involve obtaining data points from the pump manufacturers. These points can either be extracted from the published paper curves or gathered directly from the manufacturers.

One simple method is to take several points along the performance curve and perform a polynomial expression describing the performance curve, which can be saved and used in a pump selection program.

The original data pints are then discarded, and the pump performance data are now represented solely by a curve drawn from that equation. Depending on the shape of the original published curve, this polynomial expression may or may not be an accurate representation.

A second, more accurate method is to collect as many data points as necessary to accurately fit the shape of the published performance curve and use these saved data points to represent the original published curve in the pump selection program. Storing the original data points is an integral step in maintaining the pump curve's accuracy, and interpolation between those data points is the most accurate method of properly sizing a pump.

When it comes to displaying the pump curve in the software, a simple polynomial regression curve can have a tendency to miss the original pump curve at critical points. On the other hand, if a more robust numerical analysis is performed, such as a cubic spline regression, then a curve can be drawn that precisely represents the original published curve.

Figure 3.

Polynomial Regression

A closer look at some actual pump curve data illustrates the problems associated with the simple polynomial regression method of curve digitization. Figure 2 depicts flow versus head performance curves, which are the primary search parameters for sizing a pump. The black curves are the actual tested performance curves provided by a pump manufacturer. Overlaid in pink are the performance curves generated by a pump selection tool using a simple polynomial regression. The inset is a close-up view of the anomalies in this case study.

Notice that the pump performance curves with anomalies, such as inflection points, are not well represented by the polynomial regression. The polynomial regression has a tendency to smooth out the anomalies such as those shown in Figure 2. The consequences of this curve characterization can be minor, or they can be significant. In this case, if an end user were sizing for a design point of 300 U.S. gallons per minute, at a total head of 126 feet, then this pump might not even be considered for evaluation.

A more detrimental scenario could occur if the polynomial regression curve rose significantly above the actual curve. In this case, the program may suggest a viable pump that might not meet the design point criteria. In addition, if the pump performs in areas where the polynomial regression curve deviates from the manufacturer's published performance curve, an end user will get erroneous energy usage figures and operating costs.

Cubic Spline Regression

Figure 3 depicts the same manufacturer's performance curves in black. The green overlaid curves are generated by a pump selection tool that uses a cubic spline regression analysis. This method forces the curves to pass through every collected data point. It then smoothes the curves from point to point to provide clean, accurate pump performance curve displays. Flow and head are never significantly overstated or understated at any point in the curve range. End users can use these displays with confidence when performing pump evaluations.

Figure 4a.

Even the best numerical analysis techniques cannot predict anomalies between data points. If the performance curve to be digitized has these types of dips or abnormalities, it is imperative that enough data points are collected in and around that range. At a minimum, data points should be collected at the edges and at the peak (or valley) of the anomaly. Ideally, a couple more points should be collected between them. This allows the selection tool to display the most accurate representation of the original performance data for the valuation and selection process to proceed without any false selections.

Part Two

More than 150 pump manufacturers present their pump performance data via electronic means. These digital pump catalogs can be found online, within stand-alone pump selection software, and through select hydraulic analysis software packages. Thousands of engineers worldwide use these catalogs to evaluate and select the pumps that best meet their needs. Part one of this article detailed the digitization of pump performance curves and the accuracy ramifications from different digitization methods. In this part, the digitization of the pump power and efficiency curves will be explored.

To recap Part One, the main function of a pump selection program is the ability to digitally reproduce the manufacturer's pump performance data accurately and to create a selection list of pumps that meet the customer's criteria.

Figure 1 shows a detailed manufacturer's pump curve. This an example of what the manufacturer might have available in a paper catalog or in PDF format. It includes the curves for efficiency, power, and NPSHR. These curves accurately represent the performance of the indicated pump in accordance with the *ANSI/HI 1.6-2000 Pump Tests* standard.

For a software program to understand and use this pump performance data, the data must first be converted into a form that can be read by the program. This digitization process must accurately reflect the manufacturer's published pump curve to aid the customer in evaluating the pump for an application. Part One revealed the shortcomings of using a simple polynomial regression analysis to store and display the manufacturer's performance curves. Areas occur in which the digitized curve deviated significantly from the manufacturer's curve. It was also shown that a cubic spline regression analysis rendered a much more precise curve.

Power and Efficiency Curves

In addition to the standard flow versus head performance curves, many manufacturers also include curves for pump power and pump efficiency. It must be stressed that pump efficiency and pump power are function of each other. The formulas relating power and efficiency are

$$bhp = \frac{Q \times H \times \rho}{247,000 \times \eta_P} \text{ and } \eta_P = \frac{Q \times H \times \rho}{247,000 \times bhp},$$

where

bhp = brake horsepower,
η_P = pump efficiency (expressed as a decimal),
Q = flow rate (gpm),
H = total head (ft.), and
ρ = fluid density (lb./ft.^{2}).

Power is calculated with efficiency as a parameter and vice versa. To reproduce accurate and trustworthy results, one set of data must be gathered from the pump manufacturer, and the other set of data must be calculated. When generating curves, it is *not* accurate for a software program to try to collect and reproduce both sets of data from the manufacturer. This would leave open the possibility of errors. An end

Figure 4b.

Figure 5a.

user might select a flow, head, and efficiency from the graph and then perform a hand calculation of the corresponding power to find that the calculated power does not match up with the power shown on the graph.

For this reason, only one set of empirical data, efficiency *or* power, should be taken from the manufacturer. The other should be calculated. Also, the efficiency data reported by most pump manufacturers typically has much more granularity than the power data. That is to say, there is typically a wider range of efficiency values over the range of pump operation than the range of power values. For this reason, it is usually more accurate to digitize the manufacturer's efficiency data, and then calculate the power data, than the other way around.

Example One

To see the effect that digitizing can have on power and efficiency curve generation, consider the curves in Figures 4a

and 4b. In this example, the pump selection program attempted to digitize the isometric power curves and then calculate isometric efficiency curves from the power. Notice that the software rendered efficiency curves in Figure 4a are well off the mark, particularly in and around the 70%-and-up operating range. Not only are the isometric efficiency curves misaligned, but the percentages clearly do not match up with the published curve. The pump's overall efficiency is overstated by up to 3%.

If a customer were to download these performance curves (Figure 4a in black) from the pump manufacturer's technical literature and then pull up the performance curves (Figure 4a in pink) from the pump selection software, it would be virtually impossible to tell that the curves are from the same pump.

Figure 5b.

Example Two

Now consider the curves in Figures 5a and 5b. In this case, the pump selection properly digitized the efficiency curves from the manufacturer using a cubic spline regression analysis. The isometric efficiency curves are dead accurate at every point in the graph, and the efficiency is never overstated. In Figure 5b, the power curves also track extremely well with the manufacturer's, given that they are calculated power curves, and more importantly, the endpoints at the run out flow rates are accurate.

Conclusion

Using a pump selection program that accurately represents the manufacturer's pump performance is a necessity.

Inaccuracies in curve digitization and rendering can lead to improper pump and motor selections as well as faulty operating cost calculations. Users should be aware of the digitization techniques employed by their pump selection software and the information in these articles should aid them in making an informed decision when choosing a software package.

Used with permission of Engineered Software, Inc. Lacey, Washington. These two articles originally appeared in December 2011 and January 2012 issues of Pumps & Systems Magazine.

15: Measurement

Multiphase flow meter

Principle of operation

The major parts of the TopFlow meter are the Venturi insert and the electrodes inside the Venturi throat. The flow rates of oil, water, and gas are calculated based on the measurements performed by the electrodes and the measurement of the differential pressure across the Venturi inlet.

The TopFlow meter utilizes a capacitance sensor for oil-continuous liquid mixtures, where the capacitance caused by the dielectric constant of the fluid mixture is measured in the throat of the Venturi. Similarly, it utilizes a four-electrode conductance sensor for water-continuous liquid mixtures, where the conductance is caused by the conductivity of the fluid mixture. All the necessary electrodes are incorporated within a Venturi that is somewhat modified. Advanced data processing is utilized to have continuous readings of the flow rates of oil, water, and gas.

The Venturi has a minimum of pressure drop. Nearly all multiphase meter vendors use a Venturi in one way or the other, and the reason is that this is probably the most reliable device for multiphase measurement because it measures the mass flow rate independent of what happens. One objection is the relatively limited turndown ratio that is specified, especially for single-phase measurements within fiscal uncertainties. However, because most wells produce in the high GVF range, where the total mass flow rate does not vary much, there is no general problem connected to the use of this device. Especially when installing one meter each single well and not having to deal with varying flow rates, the Venturi is very suitable for multiphase measurement. Erosion will not give any major difficulties, because the TopFlow meter uses a Venturi in terms of an insert into the pipe. This means that the insert itself can be made of a wear-resistant material like high-quality steel or even ceramics in cases when erosion is considered to be a problem. In extreme cases, due to the simplicity of the TopFlow meter, the insert itself can easily be replaced. However, in the latter case, the other parts of the production pipeline, such as bends and valves, would also be eroded to destruction and would have to be replaced. A Venturi is able to give a good picture of the dynamics of the flow (i.e. it is not sensitive to the flow regime effects for measuring the instant mass flow rate accurately). Modern differential pressure transmitters are also able to follow the fast variations through high-frequency response.

The most important flow regime effects are mainly due to the ratio between gas and liquid. One is the longitudinal effect along the axis of the flow conduit, represented by the slug regime, where longer or shorter liquid plugs are separated by longer or shorter gas pockets. One challenge with this is that the liquid plug is not entirely liquid, but contains varying amounts of small gas bubbles. In addition, the gas pockets are not entirely gas but contain larger or smaller amounts of liquid. In general, the liquid and gas are also flowing with different velocities. It is of very high importance to employ sensors with fast time response to follow the rapid changes in slug flow. It is likewise important to measure the velocity often, because there is neither a constant, distinct gas velocity, nor a corresponding constant liquid velocity. Thereby, the velocity "components" are measured in real time and the slip between liquid and gas is handled automatically, without the need for theoretical or empirical slip models.

Another flow regime aspect is the radial effect, concerning the distribution of the liquid and gas over the cross-section of the flow conduit. In general, this plays a part in all multiphase flow conditions, but especially at high gas volume fractions, where the slip can be significant, the gas tends to gather in the center of the cross section, along the axis of the flow conduit. This is because of the lower flow resistance in the turbulent zone in this center. In plain words, the gas races through this

Figure 1. TopFlow Sys Multiphase Flow Meter.

channel, whereas the liquid is pushed aside and flows in contact with the pipe wall. The velocity difference is quite substantial and increases at increasing gas volume fraction. However, a part of the liquid flows as droplets following the gas phase. Taking these matters into consideration, it is obvious that the main source of uncertainty in multiphase flow rate measurement can be found in the complexity of the flow itself and not necessarily in the type of sensor used for the purpose. The crucial matter is therefore to choose appropriate technology and develop good physical models and reliable interpretation tools to solve the difficult tasks.

Adapted from FlowSys AS, Norway. Reprinted with permission.

Pipeline flow measurement—the new influences

E. L. Upp, Flow Measurement Consultant

Introduction

There are many considerations that influence our choice of methods and equipment in the measurement of liquids and gases in pipelines. A review of these problems is timely, since we have changing influences that are affecting our choices to accomplish the job of flow measurement. These are accuracy, electronic equipment, economics, fluid characteristics, standards and governmental influences, and testing or proving practices. All of these factors have had importance to some degree in the past; however, their relative importance to flow measurement and more importantly to a pipeline's overall operation has changed considerably as we have seen the prices of the fluids handled make dramatic changes along with supply and demand fluctuations of considerable magnitude versus a steady growth pattern of just a few years back.

Accuracy

Today's demand—better flow measurement accuracy. In analyzing the economics of the flow measurement industry, we find the customers are no longer in an expansion mode. With the turndown in their economies, they are more concerned with improving efficiencies in their present installations. Better measurement accuracy is one way to achieve this. The users are asking the manufacturer to improve his present devices as well as develop new meters and new technology to do the job "better." No one can fault these desires. However, measurement is a very conservative, "tradition-following" market. Industry metering standards are slow to change, and in the past it has been difficult to get the users to adopt new meters and products.

In the gas industry, contracts are written by lawyers who are not measurement experts and who tend to follow a contract form that was successful in the past. Manufacturers tried to switch users to electronic readout devices some 20 years ago. Electronic computers and flow recorders have been available to industry all this time, though admittedly not of today's quality. Yet today in the United States—for pipeline large-volume natural gas measurement—less than 2% of the stations installed utilize electronic computer controls and readouts. The industry has been more successful in applying electronic digital readout devices to liquid metering in conjunction with the turbine meter and is beginning to see production fields being automated for both liquid and gas measurement.

If the industry had a new meter today that offered absolute accuracy; infinite rangeability; met all the environmental requirements, codes, and governmental regulations; could be installed and operated by present personnel with minimal maintenance; was priced low compared to present devices on the market—it would still be difficult to sell it in sufficient quantities in the marketplace to make it profitable to manufacturers. The companies would at best be willing to try the meter for a year and compare it with present meters. Then and only then would they add it to their accepted list. The manufacturers are asking the user to change his cash register. The user has to be convinced that he is on solid legal, engineering, and procedural grounds.

Costs of accuracy. Many companies are taking a hard look at all capital expenditures in this time of high interest rates. Management people feel that the high price of the fluids they are handling allows a larger investment in order to get accurate measurement. Simple expenditure of capital does not eliminate the fact that normally the more expensive the device, the "better answer" we will get. However, careful consideration must be given to what the "better answer" will mean to our operations.

Quite often the flow measurement industry is called upon to solve flow measurement problems. Sometimes the end result is not exactly what the users had in mind. A new seller of natural gas or fluids that has a meter reading low by 1% would be delighted to use the new "more accurate meter." However, a purchaser who was just as interested in his balance sheet and who had his flow figure raised by 1% would not be too happy with a decision to use this "more accurate meter."

Many times, users have evaluated electronic flow computers, and when compared with the existing meter, will read differently. There are good, solid technical arguments why the computer is more accurate, but there are also political and economic reasons why customers cannot use more accuracy "if it lowers my billing." When two parties and a contract are involved, you must keep in mind that you cannot have a more accurate metering device that will raise measurement for both sides of the equation. As manufacturers, the only path that can be followed is to sell the most accurate metering device they know how to make and to let the customer work with the political and economic consequences. These influences certainly affect the industry's ability to bring out new products and have them accepted based on accuracy.

Their place in flow measurement. There is no denying that the high technology age is upon us. This basically refers to electronic calculating and data handling systems as far as the pipeline flow measurement problems are concerned. These tools have been available for several decades; however, we are finally seeing the beginning of the significant use of electronics to replace our previous mechanical methods of converting meter readings to custody transfer volumes. Basically, this is due to the availability of low-cost, reliable electronic devices that will meet the field conditions where flow measurement takes place and a growing awareness of the user on the requirements to successfully apply these devices. We have moved through stages of applications from "flow computers can do anything," to "flow computers do not operate in the field," to the stage we are in now where properly applied flow computers are a viable consideration for accomplishing the job we have done in the past, plus additional timely uses of the flow data to improve operations. However, this decision is not made without a serious look at the economic consequences.

Economics of flow measurement

Accuracy versus cost. As the cost of the handled fluids goes up, there is a demand by all parties involved in its purchase and sale to "reduce the tolerances," "get more accurate measurement," or "get more sophisticated equipment to get a better answer." Capital investment required to upgrade measurement is getting a hard look by pipeline management, as are all of their expenditures. So, our desire for better measurement is tempered by available capital and the resulting savings to be expected. The definition of "more accurate answers" of various measuring devices, both in terms of theoretical and practical use, must be reviewed to improve our ability to get the best measurement for the dollars invested.

If you are producing a well at 0.25 MMcfd, one level of equipment and procedure investment is justified, whereas a station delivering 25 or 250 MMcfd may require another. Similar thinking applies to liquid measurement. Physical limitations such as available power, cost of space to install equipment, and effect on the environment must also be considered. On offshore platforms, where space costs a premium, meters that can be installed with a minimum of piping have been used because of the realized savings. A large portion of the new metering devices' electronic readouts require good quality and dependable power. In some remote locations where commercial power is not available, the cost of getting the proper power may exceed the cost of the metering. In this case, the cost of power may dictate the use of mechanical or pneumatic readout devices or some local generated source such as solar or thermoelectric. These power sources are normally low-wattage devices, which may restrict the capabilities of the electronics or limit the choices of devices.

Personnel requirements. A very definite consideration affecting the economics of flow measurement is the personnel required to accomplish a given job. This may mean at least a retraining, if not a replacing of present measurement people. Generally, they will be in an upgraded position that requires both electronic and mechanical skills along with a solid understanding of flow measurement. Otherwise, you may end up with an electronics-oriented person chasing the details of proper programs and round-off procedures of the measurement readout system while not recognizing that the basic meter may be improperly applied or functioning and the total system is in error. The reverse can be true if a mechanically-oriented person is lacking in electronic knowledge. Fortunately, in the past few years most new workers have had good exposure to electronic devices and are more comfortable with electronic flow approaches than some of our older personnel. This problem exists at all levels of our company's operations and to a large degree will be the deciding factor of the final success of our system. It has a definite economic impact on our choice of changing our flow measurement systems.

Improving present flow measurement. Since most pipelines are in a nonexpansion mode, if not a reduction in throughput, operating people are spending a large portion of their efforts in improving efficiencies of their systems. This includes much closer attention to the efficiencies of compressor or pump stations. We are not talking about large percentage reductions in efficiencies being significant cost savings.

As an example, some gas pipelines have as much as 80% of their total cost of operation coming from fuel costs, so even a 5% reduction relates to a reduction in operation cost of 4%. To study these efficiencies for real results requires very accurate flow measurement including individual measurement on each compressor or pump and engine. This is a case where the use of flow measurement to reduce costs is the driving factor rather than the custody transfer service that we generally worry about.

Fluid characteristics

A changing problem. For years we have had the luxury of handling fairly simple fluids with well-defined and/or accepted correction factors to reduce fluids at flowing conditions to fluids at base conditions. We now are being faced with pipelines handling "dense fluids" such as carbon dioxide and ethylene and liquids such as nonstabilized crudes, light hydrocarbons operating at their vapor pressures, and refrigerated LPG. In these cases, the choice of the basic meter is less of a concern than the proper preparation of the fluid prior to passing through the meter, and the proper factors for making corrections to base conditions.

Sometimes we are trapped by our previous practices established with well-behaving fluids and can find large errors when these practices are applied to ill-defined fluids. In these cases the problems are not in the basic flow meter, but in the flow meter system application of the correcting factors based on inadequate knowledge. The proper answer to these problems has not been worked out to give the accuracy of flow measurement we desire, but with the recognition of the problem, we have been able to minimize some of the errors while continuing to seek new flow meters or methods of correcting for the fluid characteristics to solve the problem.

Standards and governmental effects on flow measurement

Outside influences on flow measurement. Industry practices, standards, and contracts, along with the ever increasing governmental requirements, must be considered to complete the picture of present day pipeline flow measurement. It is the belief of some of the bodies that measurement can be accomplished by legislation.

All orifice plates in some countries, for example, must be stamped by a government inspector, which makes them acceptable and correct. The manufacturer or user then can adopt the attitude that if the standards body or government finds it correct, it is not his worry and he really does not need to know much about flow measurement. Fundamentally, this has a serious flaw, since it removes the one item effective in obtaining accuracy—personal commitment and responsibility to accomplish the job. The manufacturers and users in countries where this is acceptable practice have passed the responsibility back to the government. They have reduced the caring as to whether or not their metering is accurate or

properly installed and whether or not it maintains this accuracy in long-term service.

As we get further into the governmental agencies controlling our business, we have seen that the cost of doing the job is of secondary importance to meeting DOEE, EPA, or OSHA standards, as an example. Unfortunately, the governmental bodies are looking at getting further in the measurement of fluids business with every passing day. Experience tells us that present economics of fluid measurement will be less of a controlling factor as this continues to occur.

Testing or proving flow meters

Present practices. The term "proving" in flow measurement means a throughput test and comes to us from the liquid measurement people who developed the mechanical displacement prover (now termed the "pipe prover") for proving turbine and positive displacement meters on crude oil and petroleum products. Prior to this development, we had critical flow and low flow provers used on PD meters on gas, but these had relatively limited use. On the other hand, "proving head meters" was never done. We substituted mechanical inspection of the primary devices and calibration of the secondary devices for the throughput test.

Proving is the checking of the throughput of a meter versus a standard determination of volume from another device. The purpose of all inspection and calibration or proving is to establish a meter's performance accuracy to some tolerance. This tolerance is dependent on the basic accuracy of the standard used plus the conditions of the test. The allowable tolerance is dependent on the value and quantity of the product handled. Because of this, some metering installations cannot justify a few extra dollars in cost. In those cases, we continue to use inspection and calibration to determine the meter's accuracy.

This proves that meters can be quite expensive in terms of permanently installed provers. The decision is based on "cost considerations" and is balanced by the degree of "accuracy needed" and operations and maintenance money available to be spent in obtaining and maintaining the "accuracy." Subsequently, here, are several influences that affect our measurement testing and proving: present practice, standards, accuracy desired, and economics of the measurement job.

Improvement of present practices. Present practices will be continued in concept; however, equipment development has improved our tools such as air deadweight devices for differential pressure tests and precision test gauges for static pressure for gas meters. Liquid meters have improved equipment available for sampling and fluid characteristic testing.

The standards for flow measurement have had a great deal of study done recently to improve the accuracy and application range with such programs as the American Petroleum Institute work on the fluid characteristics, the development of the small provers for liquids, and the basic flow coefficient research on the orifice by the American Petroleum Institute and the Gas Processors Association. This work of upgrading standards to improve accuracy is continuing at a rapid pace. The accuracy requirements for all flow measurement is improved because of the high cost of the measured fluids, and we anticipate that this demand will increase.

Summary

The business of pipeline flow measurement has assumed a more important role in the overall pipeline operation picture because the economics of the job to be done requires: more accuracy; better knowledge of the fluids being handled; wider application of electronics to accomplish the basic flow measurement plus other uses of these data to improve operations; the adaptation to changing standards and governmental rules; and the use of new equipment and techniques to determine the accuracy of a flow meter. The people who are responsible for flow measurement have a challenge to respond to these changing influences to improve their company's flow measurement.

Liquid measurement orifice plate flange taps

Sequence of calculations

1. Calculate the S value.

$$S = \frac{W_m}{D^2 \left(2831 + 0.0567 T_f\right) \sqrt{G_f h_m}}$$

2. Calculate Reynolds number.

$$Re_D = \frac{4.42 W_m}{D_\mu}$$

3. Iterate to find uncorrected beta ratio.

$$S = 0.598\beta^2 + 0.01\beta^3 + 0.518/\beta^{6.425}$$

$$fx = 0.598\beta^2 + 0.01\beta^3 + 0.518\beta^{6.425} - S$$

$$f^1x = 1.196\beta + 0.03\beta^2 + 3.3282\beta^{5.425}$$

Start with $\beta = 0.5$ and calculate fx/f^1x. If greater than 0.0001, subtract fx/f^1x from current value of β and repeat until the error is less than 0.00001.

4. Calculate Reynolds number correction factor. If $R_D < 350\beta^{1.3}$:

$$F_c = \left[A - \left(A \left(0.5^{(\beta^{25} - 0.12)} \right) \sqrt{1 - \frac{700\beta^{1.3}}{1200}} \right) \right]$$

$$\times \left[\frac{\sin \left(90 \sqrt{\frac{R_D}{350\beta^{1.3}}} \right)}{\beta^{0.15}} \right] + 0.997$$

where $A = (\beta - 0.2)^2 + (8.7e^{3.9\beta})(700\beta^{1.3})^{-0.89}$
When $\beta - 2 > 0$

$$(\beta - 0.2)^2 = 0$$

If $350^{1.3} < R_D < 700\beta^{1.3}$:

Table 1
Table of Equal Thermal Expansion Temperatures, °F, for Various Orifice Plate Materials

Type 316 or Type 304	Aluminum	Copper	Type 430	2% CRMO	5% CRMO	Bronze	Steel	Monel
	−298					−360		
	−230	−378				−273		
−325	−178	−266				−216		
−226	−131	−194				−163		−281
−152	−85	−132				−111		−191
−86	−41	−72				−60		−108
−24	+3	−15				−8		−33
+38	+47	+42	+24	+28	+27	+43	+31	+38
+101	+91	+99	+112	+108	+110	+95	+106	+104
+158			+202	+184	+191	+149	+181	+168
213			290	260	272	200	254	230
266			374	331	352	249	324	290
318			456	400	426	297	392	348
367			531	468	493	345	457	405
414			604	534	560	393	520	461
463			677	597	628	440	542	516
512			749	660	696		644	571
560			817	720	763		705	626
607			884	780	827		765	680
655			952	842	888		823	733
703			1020	899	948		879	785
750			1087	956	1009		934	835
796			1155	1011	1071		988	886
840			1223	1065	1133		1041	935

Use the appropriate factor for the orifice plate material.

HP 41CV Data Sheet

```
* * * * * * * * * * *        ENTER DIFF RANGE INS H2O              2.3391
                                    100.0000   ***
   ORIFICE LIQD            ENTER PIPE ID INS         ENTER REQD  ORIFICE INS
   FLANGE    TAPS                   4.0260   ***             2.3750   ***
                            ENTER VISCOSITY cP         CALC    BETA
* * * * * * * * * * *               0.8000   ***       RATIO
                            CALC   BETA                        0.5899   ***
ENTER FLOW  PPH            RATIO                       CALC   FLOW
   100,000.0000   ***              0.5810   ***       PPH
ENTER TEMP DEG F          CALC ORIFICE                     103,700.4948   ***
       150.0000   ***       BORE  INS
ENTER SG                         2.3391   ***         * * * * * * * * * * *
       0.9800   ***        * * * * * * * * * * *
```

HP 41CV Register Loading

SIZE 080		
R00= 0.000000	R26= 0.056700	R53= "DIFF R"
R01= 0.000000	R27= 2,831.000000	R54= "ANGE I"
R02= 0.000000	R28= 0.000000	R55= "******"
R03= 0.000000	R29= 0.000000	R56= "CALC "
R04= 0.000000	R30= 72.450000	R57= "ENTER "
R05= 0.000000	R31= 46.470000	R58= 44.000000
R06= 0.000000	R32= 57.484900	R59= 0.000000
R07= 0.000000	R33= 57.505100	R60= "NS H2O"
R08= 0.000000	R34= 57.520000	R61= "PIPE I"
R09= 0.000000	R35= 57.535460	R62= "D INS "
R10= 0.000000	R36= 57.616200	R63= "VISCOS"
R11= 0.000000	R37= 57.636400	R64= "ITY cP"
R12= 0.000000	R38= 0.000000	R65= "BETA "
R13= 0.000000	R39= 56.656600	R66= "RATIO "
R14= 0.000000	R40= 71.676869	R67= "RIFICE"
R15= 0.000000	R41= 57.707273	R68= " BORE "
R16= 0.000000	R42= 0.000000	R69= "INS "
R17= 0.000000	R43= 56.656600	R70= "REQD "
R18= 0.000000	R44= 56.484900	R71= "CALC O"
R19= 1,200.000000	R45= "E LIQD"	R72= "ORIFIC"
R20= 3.328200	R46= "FLANGE"	R73= "E INS "
R21= 5.425000	R47= " TAPS"	R74= 0.000000
R22= 1.196000	R48= "FLOW "	R75= 0.000000
R23= 0.000010	R49= "PPH "	R76= 0.000000
R24= 0.518040	R50= "TEMP D"	R77= 0.000000
R25= 6.425000	R51= "EG F "	R78= 0.000000
	R52= "SG "	R79= 0.000000

HP 41CV Listing

SUBROUTINES
```
01+LBL "FT LIQ"
02 GTO A
```
PRINT LINE OF STARS
```
03+LBL a
04 FC? 55
05 RTN
06 55.55
07 ADV
08 SF 12
09 XEQ b
10 ADV
11 RTN
```
PRINT LINES
```
12+LBL d
13 SF 12
14 1
```
```
15 ST+ 58
16 RCL IND 58
17+LBL b
18 STO 59
19 CLA
20+LBL c
21 ARCL IND 59
22 RCL 59
23 FRC
24 100
25 *
26 STO 59
27 X#0?
28 GTO c
29 FC? 55
30 GTO 01
31 PRA
32 CF 12
33 RTN
34+LBL 01
```
```
35 CF 12
36 AVIEW
37 RTN
```
PRINT & STORE VARIABLES
```
38+LBL e
39 1
40 ST+ 58
41 RCL IND 58
42 X=0?
43 GTO d
44 XEQ b
45 1
46 ST+ 01
47 RCL IND 01
48 TONE 9
49 FS? 00
50 GTO 01
51 STOP
```

52 STO IND 01
53 .5
54 STO 12
55 RCL IND 01
56◆LBL 01
57 XEQ J
58 GTO e

RE-ENTER VARIABLES

59◆LBL "REENTER"
60 ADV
61 1
62 +
63 STO 01
64 RCL IND 01
65 XEQ J
66 STOP
67 XEQ J
68 STO IND 01
69 ADV
70 SF 00
71 SF 02
72 STOP
73 GTO 02

BYPASS PRINTER

74◆LBL J
75 FC? 55
76 GTO 01
77 PRX
78 RTN
79◆LBL 01
80 VIEW X
81 RTN

CALC S

82◆LBL G
83 RCL 12
84 PSE
85 X↑2
86 .598
87 *
88 RCL 12
89 3
90 Y↑X
91 .01
92 *
93 +
94 RCL 12
95 RCL 25
96 Y↑X
97 RCL 24
98 *
99 +
100◆RTN

ITERATE β (ENDS AT 218)

101◆LBL D
102 RCL 12
103 RCL 22
104 *
105 RCL 12
106 X↑2
107 .03
108 *
109 +
110 RCL 12
111 RCL 21
112 Y↑X
113 RCL 20
114 *
115 +
116 1/X

117 XEQ G
118 RCL 11
119 RCL 10
120 /
121 -
122 *
123 ST- 12
124 ABS
125 RCL 23
126 X<>Y
127 X>Y?
128 GTO D

CALC F_c WHEN $R_D \leq 700\beta^{1.3}$

129 RCL 12
130 1.3
131 Y↑X
132 700
133 *
134 STO 18
135 STO 15
136 2
137 /
138 STO 14
139 RCL 09
140 /
141 1/X
142 SQRT
143 90
144 *
145 SIN
146 RCL 12
147 .15
148 Y↑X
149 /
150 STO 17
151 RCL 09
152 RCL 14
153 X>Y?
154 GTO I
155 RCL 17
156 ABS
157 SQRT
158 STO 17

CALC F_c WHEN $R_D > 700\beta^{1.3}$

159 RCL 09
160 RCL 18
161 X>Y?
162 GTO I
163 1
164 STO 17
165 RCL 09
166 STO 18
167 STO 15
168 RCL 09
169 RCL 19
170 X>Y?
171 GTO I
172 STO 18

CALC F_c WHEN $R_D = 1200$

173◆LBL I
174 0
175 RCL 12
176 .2
177 -
178 X>Y?
179 0
180 X↑2

181 RCL 12
182 3.9
183 *
184 E↑X
185 8.7
186 *
187 RCL 15
188 -.89
189 Y↑X
190 *
191 +
192 STO 28
193 RCL 14
194 RCL 18
195 /
196 RCL 12
197 SQRT
198 SQRT
199 .12
200 -
201 Y↑X
202 *
203 1
204 RCL 18
205 RCL 19
206 /
207 -
208 SQRT
209 *
210 CHS
211 RCL 28
212 +
213 RCL 17
214 *
215 .997
216 +
217 STO 10
218 RTN

MAIN PROGRAM ENTER DATA

219◆LBL A
220 CF 02
221 1
222 STO 01
223 STO 10
224 29
225 STO 58
226 XEQ a
227 XEQ d
228 XEQ d
229 XEQ a
230 XEQ e
231 CF 00

CALC S

232◆LBL 02
233 39
234 STO 58
235 RCL 02
236 RCL 06
237 X↑2
238 /
239 RCL 27
240 RCL 26
241 RCL 03
242 *
243 +
244 /
245 RCL 04
246 RCL 05
247 *
248 SQRT

249 /
250 STO 11

CALC R_D

251 4.42
252 RCL 02
253 *
254 RCL 06
255 /
256 RCL 07
257 /
258 STO 09

CALC β

259 XEQ D
260 XEQ D
261 XEQ D
262 RCL 12
263 XEQ J

CALC d

264 XEQ d
265 RCL 06
266 RCL 12
267 *
268 STO 29
269 GTO 01

ENTER REQD BORI

270◆LBL "EVEN SZ"
271 7
272 STO 01
273 40
274 STO 58
275 XEQ e

CALC β

276 RCL 08
277 RCL 06
278 /
279 STO 12
280 XEQ J
281 XEQ G
282 STO 16

CALC FLOW

283 XEQ d
284 RCL 02
285 RCL 11
286 /
287 RCL 16
288 *
289 STO 29
290◆LBL 01
291 XEQ J
292 FS? 00
293 STOP
294 XEQ a
295 RCL 29
296 VIEW X
297 BEEP
298 ADV
299 STOP
300 .END.

$$F_c = \left[A - \left(A \left(0.5^{(\beta^{.25} - 0.12)} \right) \sqrt{\frac{1 - 700\beta^{1.3}}{1200}} \right) \right]$$

$$\times \left[\sqrt{\frac{\sin \left(90 \sqrt{\frac{R_D}{350\beta^{1.3}}} \right)}{\beta^{0.15}}} \right] + 0.997$$

If $700\beta^{1.3} < R_D < 1200$:

$$F_c = \left[A - \left(A \frac{(350\beta^{1.3})^{(\beta^{.25} - 0.12)}}{R_D} \sqrt{1 - \frac{R_D}{1200}} \right) \right] + 0.997$$

When $R_D > 1200$:

$$F_c = (A + 0.997)$$

5. Divide original S by F_c for viscosity correction.
6. Repeat step 3 for new β for corrected S.
7. Repeat step 4 for final F_c for corrected β.
8. Repeat step 5 for final S.
9. Repeat step 3 for final β.
10. Calculate orifice bore.

$$d = \beta D \text{ in}$$

11. Calculate flow rate for an even-sized orifice plate.

$$\beta = \frac{d}{D}$$

$$S = 0.598\beta^2 + 0.01\beta^3 + 0.518\beta^{6.425}.$$

$$\text{new flow} = \frac{\text{new } S}{\text{original } S} \times \text{original flow, lb/h}$$

Nomenclature

A = Partial calculation of F_c dimensionless
D = Inside diameter of pipe, in
d = Diameter of orifice bore, in
F_c = Reynolds number correction factor, dimensionless
G_f = Specific gravity of liquid at flowing temperature, dimensionless
h_m = Maximum differential range, in of water
R_D = Reynolds number, dimensionless
S = Beta ratio index, dimensionless
T_f = Flowing temperature, °F
W_m = Maximum rate of flow, lb/hr
μ = Absolute viscosity, centipoises
β = Beta ratio d/D, dimensionless

Reference

12. Spink, L. K., *The Principles and Practice of Flowmeter Engineering*, The Foxboro Company.

Vixcosity Corrections for Concentric Orifices

Mass measurement light hydrocarbons

Streams that transport mixed natural gas liquids require the use of mass measurement to accurately account for the volume of the components which make up the mix. When natural gas liquid components are mixed, nonideal mixing occurs. Compressibility factors are difficult to determine for mixes. Mass measurement eliminates the effects of nonideal mixing and the need for compressibility factors.

Certain precautions must be noted in using the mass measurement method. The density and volume must be measured at the same flowing conditions. An accurate composite sample of the flowing stream must be collected and analyzed with care. On systems where there are a number of input points that will be compared to an output point, it may be desirable to have all composite samples analyzed at the same laboratory.

The following data are needed for the calculation of quantities of natural gas liquid components:

- Metered volume at operating conditions
- Meter factor, if applicable
- Density at flowing conditions
- Composite sample and sample analysis for metered volume
- Component lb mass for each component in the stream
- Density—lb mass per U.S. barrel for each component in the stream

Sample calculation

Metered volume = 25,000 barrels
Meter factor = 1.0015
Density = 0.480

Composite sample analysis

Component	mol %
Carbon dioxide	0.08
Methane	1.90
Ethane	41.30
Propane	38.00
Isobutane	5.00
Normal butane	9.30
Isopentane	1.50
Normal pentane	1.10
Hexanes+	1.82
Total	100.00

See Table 1 for the details of the calculation.

Table 1
Mass Measurement–Light Hydrocarbons

Component	Mol % 1	Mol Wt 2	Mol % × Mol Wt 3	Wt Frac. of Comp. 4	# Mass Comp. 5	Density #/bbl 6	U.S. bbl 7
CO_2	0.08	44.010	3.5208	0.000852	3588	286.4358	13
C_1	1.90	16.043	30.4817	0.007381	31,063	105.0000	296
C_2	41.30	30.070	1241.891	0.300739	1,265,556	124.7232	10,147
C_3	38.00	44.097	1675.686	0.405787	1,707,617	177.5256	9619
IC_4	5.00	58.123	290.615	0.070375	296,153	197.0934	1503
NC_4	9.30	58.123	540.5439	0.130899	550,844	204.4980	2694
IC_5	1.50	72.150	108.225	0.026208	110,287	218.7444	504
NC_5	1.10	72.150	79.365	0.019219	80,877	220.9914	366
C_6^+	1.82	87.436	159.1335	0.038536	162,166	244.2300	664
	100.00		4129.461		4,208,151		25,805

Total lb mass = 25,000 × 1.0015 × 0.480 × 8.337 × 42 = 4,208,151.
Column 1—From Sample Analysis.
Column 2—From Table of Hydrocarbon Properties (Figure 1).
Column 3—Column 1 × Column 2.
Column 4—Mol % × Mol Wt divided by sum of Column 3.
Column 5—Column 4 × total lb mass.
Column 6—From Table of Hydrocarbon Properties (Figure 1).
Column 7—Column 5 divided by Column 6.
Reference: GPA Standard 8173-83, Method for Converting Natural Gas Liquids and Vapors to Equivalent Liquid Volumes. GPA Standard 8182-82, Tentative Standard for the Mass Measurement of Natural Gas Liquids.

Pipeline measurement of supercritical carbon dioxide

G. W. Marsden, Shell Pipeline Corporation and **D. G. Wolter,** Shell Oil Company

Introduction

The Shell Companies have been involved in carbon dioxide (CO_2) systems used for enhanced oil recovery for a number of years. Pilot programs existed in the late 1970s using recovered combustion generated carbon dioxide. In 1981, a plan was in place for a large-scale commercial pipeline transportation system connecting carbon-dioxide producing fields in southwestern Colorado and the oil fields of the Permian Basin in West Texas (see Figure 1). The Cortez Pipeline Company was set up as the ownership company and Shell Pipeline Corporation was established as the operator.

The 810-km (502 mi) long, 760-mm (30-in) diameter Cortez pipeline was designed to operate under supercritical conditions of 9000 to 19,000 kPa (1300–2700 psig), 286 to 322 K (55–120 °F), and with the nominal stream composition shown in Table 1. Design flow rates were 4.6–13.83 kmol/s (330–1000 MMscfd). Line fill was completed in May 1984 and since then, the system has been in continuous operation. There are 10 receipt meters and nine delivery meters currently in operation, each with a custody transfer measurement facility.

Table 1
Stream Composition

Major Components	m^3/m^3	% Vol
CO_2 minimum	0.95	95
Nitrogen, maximum	0.04	4
Methane	0.01–0.05	1–5
Contaminants		
Hydrogen sulfide	0.00002	0.002
Water	0.48 g per m^3 at standard conditions	30 lb/MMscf

Definitions

"Carbon dioxide," as referred to here and as used in the pipeline industry, is a CO_2-rich mixture normally having nitrogen and methane as principal volumetric contaminants. A minimum quality of 95% carbon dioxide purity is used by many pipeline carriers, while Cortez experience thus far is circa 98% CO_2 purity. The measured quantity is the total of all mixture constituents.

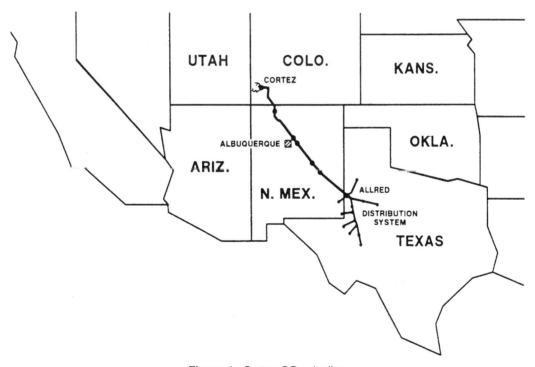

Figure 1. Cortez CO_2 pipeline.

Another term used by liquid pipeline operators is "custody transfer." Liquid pipeline operators typically do not own the fluid being transported, but instead accept custody at pipeline origin (receipt) and transfer that custody back to the owner or other party at the pipeline destination (delivery).

Thus, pipelines have custody transfer measurement facilities, which are composed of generally accepted, commercially available, and accurate measurement equipment. Additionally, the operation of this equipment is accomplished in a uniform manner acceptable to all parties to the transaction: buyers, transporters, and sellers. The *API Manual of Petroleum Measurement Standards* is widely used in the design and operation of these facilities.

These brief definitions provide insight into the selection of custody transfer measurement methods and equipment. Now, let us return to the CO_2-rich fluid we want to address.

Physical properties

CO_2, in its pure form, has a critical point of 304 K (88 °F) and 7415 kPa (1071 psia). Within practical limits, the characteristics of pipeline-quality CO_2-rich mixtures will approximate those of pure CO_2. Refer to the pressure enthalpy diagram for CO_2 (Figure 2), which illustrates the area of normal measurement for Cortez Pipeline.

The most economical pipeline operation is in the supercritical single-phase area. Fortunately, this operating area also coincides with our preferred measurement envelope. There is no pressure letdown for the delivery; in practice, the receiving party prefers the fluid with a pressure level similar to the pipeline operating pressure. This leads to measurement systems where pressure conditions are quite similar at both the inlet to and the outlet from the pipeline.

Product temperature does vary from the inlet measurement facility, which is downstream of compression equipment, to the delivery stations where the CO_2 stream approaches the ambient ground temperature. Future increases in throughput will require additional pumping/compression equipment, which will tend to create measurement operating temperatures that are similar at both pipeline inlet and outlet.

The upshot of these temperature and pressure conditions is the measurement envelope shown in Figure 2. This envelope is considerably more confined during normal measurement operations than is the pipeline operating envelope. The operating envelope includes the changes in hydraulic conditions that occur along the route of the pipeline.

Looking again at the P–E diagram (Figure 2), observe that the density of the CO_2-rich stream is in the range of 640–960 kg/m³ (40–60 lb/ft³). Additionally, note that the prime determinant of a change in density will be a change in temperature. This sensitivity of change in density with temperature is of concern in the design of the measurement facilities, and it has led to special attention being given to temperature loss/gain between elements of the measuring system. There is an approximate 0.9% volume change for each °C (0.5%/°F).

The change of density with pressure change in the measurement envelope is considerably less prominent, being on the order of one tenth as severe as temperature.

Figure 2. Carbon dioxide pressure enthalpy diagram.

Measurement methods in use today

The commercial planning and development of the Cortez carbon dioxide project preceded the engineering activity, and the supply and sales contract negotiations were well advanced before custody transfer flow measurement methods were investigated. The resulting contracts, based on existing natural gas pipeline systems, dictated the use of MMscfd volumetric units for gas phase equivalent volume of carbon dioxide.

Custody transfer measurement of carbon dioxide in large volumes at high pressures presented significant technological hurdles in the methods, performance, and materials areas. First, the probable stream compositions and the variations of the source field carbon dioxide were uncertain. This uncertainty of composition, and thus the physical properties, led to the selection of mass measurement techniques that would provide flexibility and operability when product conditions changed. Shell's decade-long experience with mass measurement of variable mix hydrocarbon streams and with supercritical streams suggested mass measurement as one of the strongest contenders. Mass measurement, however, would not directly provide the volumetric quantities that had been established in the preceding business negotiations.

The selection of meter systems to meet the above requirements resulted in two systems that are used today in the carbon dioxide transportation industry. The orifice meter, common in the natural gas industry, and the turbine meter, similar to those used in liquid measurement, were the dominant choices. As with most other choices, there are advantages and disadvantages (see Table 2) to each method. Both are operated as mass meter systems with subsequent conversion to volumetric terms for contract compliance and accounting requirements.

Fortunately, Shell Pipeline had several years of operating experience with mass measurement of petrochemical products, i.e. ethane and ethylene, as background in the orifice measurement system. The orifice meter was selected based on analysis of the factors shown in Table 2. The units conversion process, however, required additional facilities and methods for routinely and rapidly determining the composition of the streams, the calculation of molecular weight for the equivalent gas, and the conversion to volume at the selected reference pressure and temperature conditions. The resulting methods and facilities in service today are discussed below.

Orifice measurement practices

The measurement system that provided basic simplicity, reliability, wide acceptance, and the capability of handling variable mix streams without breaking new frontiers in measurement methods was resolved to be the orifice flow measuring element with online density meter and microprocessor flow computer. This basic assembly of three proven

Table 2
Carbon Dioxide Pipeline Metering Options

	Orifice Meter	Turbine Meter
Advantages	Industry acceptance	Accuracy $\pm 0.5\%$
	Proven equipment and operations	Rangeability $\pm 10:1$
	Least sensitive to density determination accuracy	
Disadvantages	Limited rangeability	Performance unknown with no experience in CO_2
	Accuracy $\pm 1\%$	No experience with proving in CO_2
	Costly installation	

devices provides real time-mass flow output. The orifice is the key to the success of the measurement system.

The orifice plate has a 50-year record of widespread use in the measurement of all manner of fluids. It is a static device generally inert to the measured fluid conditions, and calibration consists of simple dimensional measurement and conformance to specified physical tolerances. This, along with the periodic calibration of the associated differential pressure transmitter, completes the verification of this part of the measurement system. The orifice meter is one of the few measurement devices accepted for custody transfer that does not require in-service proving.

The second element, the density meter, has been selected to be an external unit that is easily isolated from the flowing stream for calibration, inspection, and maintenance. The relative insensitivity of CO_2 density to small changes in pressure in the primary flow to the orifice meters permits locating the density meter upstream of the meter manifold, thereby serving several meters. All installations are insulated to assure that pipeline density is maintained in the density meter.

The microprocessor flow computer, or third element of the system, is essential to achieving the advantages of integrated mass flow. This advantage comes from the ability of the computer to make computations in essentially "real time." How this advantage comes about is seen upon examination of the flow equation.

Orifice flow equation

The *API Manual of Petroleum Measurement Standards*, Chapter 14, Section 3 (also API 2530, AGA Report No. 3,

ANSI/API 2530-1985, and GPA 8185-85) provides the widely accepted formulation for mass flow. Mass flow is considered by this source to be the fundamental orifice meter flow equation.

Equation 38 of this publication, "Equation for Mass Flow Rate when Density of Fluid Flowing is Known", is shown in Figure. 3.

The terms used in Figure. 3 are:

q_m is the mass flow rate.

K and Y, the flow coefficient and the expansion factor, are the empirical elements of the orifice equation, based upon the work reported by Buckingham and Bean in 1934.

The flow coefficient, K, is in part dependent upon the viscosity of the fluid. Pipeline CO_2 viscosity ranges from approximately 0.05 to 0.08 mPa's (0.05–0.08 cP) in the measurement envelope.

The expansion factor, Y, is a function of the ratio of the orifice differential pressure to static pressure and to the specific heat ratio of the CO_2. The specific heat ratio ranges from 2.9 to 4.0 for CO_2 in the measurement envelope.

ρ_{fl} is the density output of the density meter.

ΔP is the differential pressure across the orifice plate.

Two correction factors, F_a and F_{pwl}, from Appendix E of Chapter 14, Section 3 are also utilized. F_a provides for the thermal expansion of the orifice plate, and F_{pwl} provides for local gravitational correction for deadweight testers.

Note that errors in density are taken at half their real value by integrating the density value with the orifice differential before taking the square root.

Figure 4 illustrates the accumulation of individual tolerances to arrive at the mass measurement system uncertainty tolerance. Individual element tolerances are modified by an

---

14.3.5.4.3   "Equation for Mass Flow Rate when Density of Fluid Flowing is Known"

$$q_m = \frac{\pi}{4}[2g_c]^{0.5} K Y_1 d^2 [\rho_{fl}\Delta P]^{0.5}$$

Correction Factors in Accordance with Appendix E

E.3: Orifice Thermal Expansion Factor

$$F_a = 1 + [0.0000185(t_f - t_{meas})]$$

E.7: Local Gravitational Correction for Deadweight Tester

$$F_{pwl} = \left[\frac{g}{g_o}\right]^{0.5}$$

Figure 3. CO_2-rich mixture orifice measurement equation. From the *API Manual of Petroleum Measurement Standards*, 2nd Edition, September 1985.

VARIABLE	TOLER-ANCE %	EFFECT	SQ OF TOL & EFF	SUM OF SQUARES	SQ. RT. OF SUM %
Basic Orifice, F(B)	0.50	1.00	.2500	.2500	.5000
Diffntl Press, H	0.50	0.50	.0625	.3125	.5590
Density	1.00	0.50	.2500	.5625	.7500
Processor Error	0.01	1.00	.0001	.5626	.7500
Reynolds, F(R)	0.03	1.00	.0009	.5635	.7506
Expansion, Y	0.02	1.00	.0004	.5639	.7509
Orifice, F(A)	0.05	1.00	.0025	.5664	.7523

Individual factors with tolerance as above will provide system measurement with tolerance of (plus or minus)   0.75%

Figure 4. Orifice meter system tolerance (root-sum-square method).

effect factor, then the square root of the sum of the squares of these products is computed. The resulting overall measurement uncertainty for the Cortez mass measurement system is ±0.75%.

Pipeline operation

Full Cortez pipeline operation is accomplished using the mass flow output of the measurement system. This system also forms the basis for all custody transfer measurements in volumetric units.

It is necessary to introduce a mass-to-volume conversion step to satisfy the volumetric contracts. This conversion is accomplished using an analysis of the composite stream with a gas–liquid chromatograph maintained at each measurement area. The CO_2, N_2, and CH_4 composition of the CO_2-rich stream is determined for each measurement ticket period. An off-line equation of state computer program then calculates the appropriate standard volumetric units. A custody transfer measurement ticket (see Figure 5) is produced to document all pertinent data, including mass units, three-component composition, and the standard volumetric units.

The equation of state computer program utilized by Cortez Pipeline for this conversion was developed specifically for CO_2-rich mixtures. It is based upon the International Union of Pure and Applied Chemists (IUPAC) publication, "CO_2 International Thermodynamic Tables of the Fluid State," Volume III, compiled by Angus, Armstrong, and DeReuch. The method of extended corresponding states is utilized for the stated properties of the CO_2-rich mixture.

Turbine metering

The authors do not have firsthand experience or knowledge with turbine metering of carbon dioxide and therefore will not delve into the details of the application. In 1981, turbine metering was considered as one of the two major alternatives and rejected for a variety of reasons. Interestingly, other companies did select turbine metering, and experience is now being accumulated on large-scale carbon dioxide pipelines.

Turbine meters were not selected for Cortez or subsequent Shell systems, because the experience with mechanical displacement provers at such high pressure with compressible fluids was nonexistent. Likewise, the application of liquid or gas turbine meters in supercritical carbon dioxide service was a complete unknown.

The volumetric accuracy of turbine meters was attractive and no doubt led others to select them. However, the direct influence of density in the mass flow calculation is a disadvantage (as compared to orifice metering where density enters the equation as the square root of density and effectively halves the sensitivity of mass flow accuracy to density determination errors) (See Figure 6).

The proving of turbine meters was a concern and has been a significant commissioning problem on systems utilizing turbine meters. High-pressure provers in the 900 to 1500 ANSI flange ratings were scarce and expensive. The seals, design, and materials were considered potential problems. Significant progress has been made by those utilizing the equipment. In the long term, turbine meters and provers will likely be more widely applied as seal systems are

```
FIELD LOC        CO2 DENVER UNIT STATION                 DELIVERY
LEASE/COMP
FOR ACCT OF
CREDIT
ACCTG CODE  74222502 01   TICKET NO. 0004  ACCTG MO/YR  05 85
TICKET FROM DATE 04 01 85      TICKET TO DATE 05 01 85
METER NO.    1
CLOSE(MLBS)        226998
OPEN(MLBS)         186118
SUBTOTAL           40880
TOTAL ALL METERS (MT)           40,880  SAMPLE NO. 4
COMPOSITION,%MOL         CO2 98.350   CH4 .136   N2 1.514
SPEC. VOL(SCF/LB)     8.6568048
VOL = SPEC VOL X MT =        353,890   MSCF AT 14.650 PSIA/60F
OPEN   7:00 A M
CLOSE  7:00 A M              ISSUED BY:                 DATE:
COMMENTS:
```

Figure 5. Cortez Pipeline Company. CO_2 meter ticket.

```
          FOR ORIFICE METERING:

MASS = (CONSTANT) (DENSITY * DIFFERENTIAL) 0.5


        FOR TURBINE OR VORTEX METERING:

         MASS = (VOLUME) (DENSITY)
```

Figure 6. Mass flow calculation for orifice metering.

demonstrated to be reliable and as the other operational aspects are demonstrated to be manageable.

Potential measurement methods

Two additional meter types appear attractive and are coming into use in carbon dioxide service for various applications. Cortez Pipeline now utilizes the vortex shedding meter for line integrity/leak detection purposes, and the coriolis meter is attractive due to its simplicity and apparent accuracy.

The vortex shedding meter offers advantages in higher rangeability, requiring fewer meter runs than orifice meters, and thus lower installation costs. We believe the rangeability to be between 10 and 20 to 1 and the off-the-shelf accuracy to be 1% or better. Operating personnel have made online comparisons with adjacent, though several miles distant, orifice metering systems, and observe agreement within less than 1% for extended periods. No detailed studies have been made, however.

Currently, no custody transfer applications of the vortex shedding meter are known, although consideration is becoming more common. Application will first occur on low-volume service between agreeable parties or at the opportunity that assures a proving system will be available. (Proving via master meter and master prover techniques is one possibility that has been tested on other similar products.) Once the proving methods are applied, it is expected that accuracies better than 0.5% mass basis would be achievable.

Application of the coriolis meter lags the vortex shedding meter since it is a newer development. The expected accuracy of better than 1% and rangeability of 10 to 1 are attractive. The coriolis meter also provides direct mass measurement, which removes the need for a separate density determination and should significantly reduce both the installed and operating costs. The meter currently lacks widespread industry acceptance for custody transfer measurement applications, and the high-pressure operation of carbon dioxide systems currently limits selection to the smaller meter sizes. No custody transfer metering applications are known at this time.

Future application of the coriolis meter in carbon dioxide service will be much the same as described earlier for the vortex shedding meter. Industry experience, acceptance, application of proving methods, and availability in larger sizes with higher pressure ratings will open the routes for increased use.

The future

Carbon dioxide measurement technology is advancing rapidly, and the major difficulties in accuracy and operation are now identified. One accuracy enhancement that is progressing is the proprietary development of a flow computer with an accurate equation of state for carbon dioxide density determination from on-line/real time pressure and temperature inputs and keypad, or possibly analog, input of stream composition.

At this time, the Shell equation of state is a proprietary unit and is being used in its mainframe form for conversion of mass to volumetric units on Cortez Pipeline. Industry acceptance and verification in field practice has been excellent. The equation of state is a modification of the IUPAC equation for pure carbon dioxide and incorporates the effects of nitrogen and methane impurities over a range of 80–100% carbon dioxide. The equation of state is considered to be accurate to ±0.2% for density determination in the supercritical regions. The program is being applied to a flow computer with density update calculations every 1–2 min. Pressure and temperature are real-time analog inputs, while composition will be manually input via the keypad on a periodic basis dependent upon compositional stability of the fluid at each location. The planned operation region is shown in Table 3.

It is recognized that the National Bureau of Standards and others are pursuing similar equation-of-state expressions. Early investigation, however, determined that an industry-accepted equation of state was in the distant future and the earliest implementation could be achieved on a proprietary development with industry acceptance to follow. This decision was also supported from the view that numerous inhouse

Table 3
Region of Operation for CO$_2$ Equation of State Flow Computer

Fluid	Single Phase CO$_2$		
		m^3/m^3	Vol %
Composition	CO$_2$	0.8−1.0	80−100
	Methane	0−0.02	0−2
	Nitrogen	0−0.20	0−20
Pressure	3,400−20,500 kPa absolute (500−3000 psia)		
Temperature	273.16−333.16 K (32−140 °F)		

uses could be made of the computers where outside acceptance was not required. These applications included process operations at the carbon dioxide production, separation, and reinjection plants where density meter operation is a considerable burden and optimized accuracy is not justified. Likewise, the computer is attractive for comparison against critical density meters and for line integrity/leak detection flow measurement facilities. We are hoping to implement some of these applications in 1986.

In addition to the equation of state, it is our intent to remain flexible and consider the use of the potential meter types discussed earlier. The meter manifolds on the Cortez system have been designed with flexibility for other meter types and proving operations if these become sufficiently attractive.

The technology of carbon dioxide pipeline measurement is gaining uniformity via conferences such as this and joint industry development projects such as the equation-of-state work at NBS. In the commercial area, it is desirable that similar efforts be directed toward the use of mass measurement for operation and custody transfer purposes and to do away with the processes necessary for conversion to volumetric units. Mass units are obtained with simpler systems, at a lower cost, and more accurately. We should all work toward converting the contracts and agreements to mass units.

Source

Reprinted from *Oil & Gas Journal*, November 3, 1986 by permission and courtesy of Perm Well Publishing Co.; G. W. Marsden, Shell Pipeline Corporation; and D. G. Wolter, Shell Oil Company.

GAS MEASUREMENT

Master meter proving orifice meters in dense phase ethylene

James E. Gallagher, Shell Oil Company, Houston

Abstract

Shell Pipeline Corporation, cognizant of the importance of highly accurate orifice measurement for polymer-grade ethylene, embarked on a developmental program to improve the overall uncertainty of orifice metering. The program's goal was to successfully apply master meter proving techniques to dense phase ethylene orifice meter facilities operated on a mass basis.

This paper presents the techniques and results of mass proving concentric, square-edged, flange-tapped orifice meters. All data were obtained on commercially operated metering facilities that are in compliance with ANSI 2530 requirements.

Results indicate an overall orifice meter uncertainty tolerance of approximately ±0.31% is attainable when orifice meters are mass proved in situ using master meter proving techniques.

Goal of development program

During the past decade, increasing commercial activity for ethylene demanded improvement of measurement technology. Significant measurement uncertainties were being experienced on ethylene pipeline systems. In response, a developmental program to improve the overall uncertainty of orifice metering was initiated by Shell Pipeline Corporation.

The program's goal was to develop an economical method for proving ethylene orifice meters under actual operating conditions. Master meter proving on a mass basis was selected as the only viable approach.

Pipeline transportation and measurement systems

An ethylene distribution system consists of a pipeline network and salt dome storage facility linking producers and consumers. Since producers and consumers are not equipped with on-site storage facilities, the pipeline is designed with maximum flexibility to satisfy the continually changing demands on the operations.

Shell's ethylene systems are operated in the dense phase fluid region due to lower transportation costs. The ethylene meter stations operate in two regions depicted in Figure 1—the dense-phase fluid and single-phase gas regions.

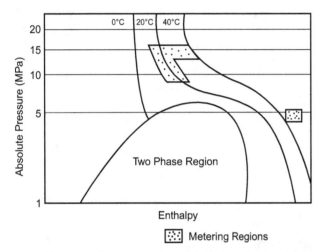

Figure 1. Ethylene pressure-enthalpy diagram.

The quantity of ethylene measured by concentric, square-edged, flange-tapped orifice meters, shown in Figure 3, is calculated through the combined use of the mass orifice equation and API's ethylene density equation. In the United States, the applicable standards are API, Chapter 14, Section 3—Orifice Metering of Natural Gas and Other Related Hydrocarbon Fluids, Second Edition, September 1985 (ANSI 2530), and API Chapter 11.3.2.1—Ethylene Density.

In Shell, the concentric square-edged, flange-tapped orifice predominates as the primary element. Most installations are equipped with a dual chamber orifice fitting to allow retrieval of the orifice plate under pressure. For the most part, the orifice plates are sized to maintain a 0.3 to 0.6 beta ratio. Solid state electronic flow computers and transmitters are installed to maximize metering accuracy. Parallel meter runs and stacked "dp" transmitters maximize rangeability while maintaining a high degree of accuracy.

Historically, the industry has presumed that compliance with the reference standard piping conditions infers accuracy. No consideration was given to deviation from the empirical equation's criteria upon initial installation or

Velocity Profile
- **Obstructions, Deposits, Etc.**
- **Piping Configuration**
- **Pre & Post Run Lengths**
- **Pipe Condition**
  - Wall Roughness
  - Roundness
- **Plate Condition**
  - Concentric
  - Edge Sharpness
  - Nicks or Burrs
  - Flatness
  - Roughness
- **D.P. Sensing Taps**
  - Burrs

Operating Conditions
- **Pulsating Flow**
- **Widely Varying Flow**

Fluid Properties
- **Density**
  - Equation of State Uncertainty
  - Transmitters' Sensitivity
- **Expansion Factor**
- **Viscosity**

Instrumentation
- **Ambient Effects**
- **Static Pressure on "DP" Sensing**
- **Calibration Accuracy**

Figure 2. Influence quantities on orifice meters.

Figure 3. Ethylene orifice meter.

further deviation over the life of the facility. The major weakness of orifice metering has been the inability to prove the meter under actual operating conditions.

In 1982, the Industry Ethylene Measurement Project conducted at Mt. Belvieu, Texas, demonstrated the feasibility of gas turbine meters, conventional piston pipe provers, and API's equation of state for custody transfer. The project also demonstrated the turbine meter and pipe prover approach produced a lower uncertainty than orifice meters. The ability to accurately prove the turbine meter, shown in Figure 4, on a mass basis under actual operating conditions eliminated the reliance on reference conditions.

However, installation costs could not justify this technique for most custody transfer installations. As a result, the majority of Shell's ethylene custody transfers continues to be based on the traditional orifice meter.

Figure 4. Ethylene turbine meter.

Development program

Development of the master meter proving technique was conducted on commercially operating ethylene pipelines by field personnel. Orifice meters associated with these systems were proven under actual operating conditions.

Fluid properties. The thermodynamic and physical properties of polymer-grade ethylene have been the subject of intensive investigations. Comprehensive treatises on the density of pure ethylene have been published by two highly reputable sources—API and the NBS. The pipeline industry in the United States presently utilizes API.

Chapter 11.3.2–Ethylene Density (API 2565). Ethylene densities are calculated using measured temperatures and pressures. The accuracy of the calculated density depends on the accuracy of the equation of state, the accuracy of the measuring devices, and the sensitivity of the calculated density to small temperature and pressure variations.

Portable proving facility. A truck-mounted 900 ANSI proving system was developed to allow maximum utilization of capital investment. The truck is equipped with a small volume prover, master turbine meter, and associated instrumentation, as shown in Figures 5 and 6.

Figure 5. Master meter method—mechanical.

Figure 6. Master meter method—electronic.

The small volume prover, SVP, operates on the principle of repeatable displacement of a known volume of fluid in the calibrated section. The mass of the fluid in the prover is determined through the combination of the prover's base volume, the fluid's pressure and temperature, and the API ethylene density equation.

The SVP was selected based on the following criteria:

- Minimize capital investment by using portable system
- Unidirectional flow design
- Honed and calibrated pipe section
- Dynamic piston seal leak detection
- Single fiber optic piston position detector
- Control of launching pressure wave
- Double chronometry pulse interpolation

Principal parts include an outer housing, a honed inner measuring cylinder, a piston displacer with an Invar extension rod, and an automatic hydraulic system. The hydraulic system returns the piston displacer to the upstream launch position after each proving run and actuates the prover bypass valve to allow the fluid to bypass the prover while the piston is being returned to the launch position.

The pressure wave that occurs when launching the prover's piston is monitored by a high-speed recorder, which was added during initial testing. The differential pressure, or pressure wave, across the prover is recorded to assure the piston does not act as a compressor or brake. The hydraulic system controls the piston to maintain a constant velocity through the calibrated section.

A gas turbine meter specially modified by the manufacturer for ethylene service was selected for the master turbine meter. The meter is equipped with a Lexan rotor, ball bearings, external lubricator, and conventional preamplified electronic pulse output. The master meter has a stated linearity of ±0.25% over a 5:1 turndown. The truck is equipped with two master turbine meters to cover normal operating flow rates—10 mm (4 in) and 50 mm (2 in) turbine meters.

The SVP is controlled by a microprocessor-based computer. The computer monitors repeatability, piston position, and prover bypass valve position; performs dual chronometric pulse interpolation; and calculates/prints a hard copy report for each successful proving.

The computer receives input from the prover's temperature and pressure transmitters, the turbine meter's pulse train, the fiber optic gate detector, and various hydraulic controls.

Two microprocessor-based flow computers are used to calculate the turbine meter's mass flow. The first computer calculates the API ethylene density based on the turbine meter's pressure and temperature transmitters. The second computer calculates the turbine's mass flow based on the turbine meter's pulse input and contains an internal mass proving counter.

Solid state temperature transmitters, equipped with high-accuracy 100 ohm platinum RTD sensors, are installed 4–5 pipe diameters downstream of the turbine meter and at the exit of the prover. The calibrated accuracy is ±0.08 °C (0.15 °F).

Solid state pressure transmitters, equipped with a sealed capacitance sensing element, are installed 4–5 pipe diameters downstream of the turbine meter and at the inlet to the prover. The calibrated accuracy is ±6.89 kPa (1.0 psi).

Solid state differential pressure (dp) transmitters, equipped with a sealed capacitance sensing element, are installed across the prover and the piston's seal. The calibrated accuracy is ±12 Pa (0.05 in H_2O).

A high-speed chart recorder monitors the prover's "dp" transmitters:

- To assure that the prover does not induce a pressure wave greater than 6.89 kPa (1 psi) while the piston is in the calibrated section.
- To assure that the piston's seal does not leak in the dynamic state.

A solid state counter, equipped to receive a contact closure from the orifice's computer and send a dry contact closure output to the turbine meter's flow computer, is installed to start and stop the master meter proving run. This device triggers the running start/stop master meter proving.

Master meter proving principle. The proving is based on the concept that a "master meter" can act as a transfer standard between the primary standard and another meter.

Meter proving by the master meter method requires two distinct steps:

- Mass prove the master turbine meter on site under operating conditions.
- Master meter prove the orifice meter on a mass basis using the master turbine meter.

To assure the devices are experiencing identical conditions, the orifice meter, turbine meter, and SVP are placed in series, as shown in Figure 5, during the entire calibration. Before proving the "master meter," the flow and/or density conditions through the orifice meter, turbine meter, and SVP must approach steady state conditions.

In the "master meter" proving method, the turbine meter is calibrated on a mass basis under actual operating conditions by the SVP. Once the turbine meter has been proved, a comparison of the 2 m mass outputs forms the basis for determining the orifice meter factor as shown in Figure 6.

Proving procedures

In order to assure accuracy, the operator must be certain the system approaches steady state conditions (flow rate, density, pressure, temperature) and that no leakage is occurring between the orifice, turbine, and the prover. If steady state conditions do not exist, accuracy and/or repeatability will be extremely difficult to obtain. The orifice meter's instrumentation is calibrated prior to proving.

Preparation.

1.1 Initiate flow through the SVP. Check for leakage on all critical valves, relief valves, and piston displacer seal.
1.2 Calibrate all instrumentation associated with the prover truck (pressures, temperatures, flow computers).
1.3 Check operation of assembled instrumentation by performing preliminary turbine meter proving runs.

1.4 If necessary, adjust hydraulic system to assure the prover does not act as a compressor or a brake. Adjust the hydraulic system if the piston displacer's differential pressure exceeds ±6.89 kPa (1 psi).

1.5 Allow ethylene to circulate through the prover truck until the outlet temperature is stabilized.

Proving run.

2.1 Initiate the SVP proving sequence by commanding the prover computer/controller to obtain a VMFt. (Acceptable repeatability for the turbine meter is five consecutive runs in which the volumetric meter factor, VMFt, does not exceed a tolerance of 0.05%.) The computer will print a proving report when this criterion has been met.

2.2 Calculate the turbine meter's mass meter factor, MMFt.

2.3 Clear the turbine meter flow computer's internal mass prover counter.

2.4 Reset and program the master meter gate simulator to allow the master meter proving to occur for a minimum of 1 h. Record the turbine meter's mass flow rate. (The simulator initiates a master meter proving after receiving five dry contacts from the orifice computer. The simulator stops the proving after approximately 1 h. Example—for flow at 50,000 lb/h, the gates would be set at 5 and 505.)

2.5 When the master meter run is complete, record the turbine meter's mass counts.

2.6 Calculate the orifice's mass meter factor (MMFo).

2.7 Repeat steps 2.1 thru 2.6 until two consecutive master meter runs satisfy repeatability criteria. (Acceptable repeatability for orifice meter is two consecutive runs in which the MMFo does not exceed a tolerance of 0.10%.)

2.8 Average the results of the master meter runs. Enter the average MMFo into the orifice flow computer for use in continuing operations.

Figure 7 is a typical orifice proving report.

Analysis of data

Analysis of the data is contained in this section, along with the calculation methods for reduction of the data.

Proving calculations. The prover base volume, PBV, is stated at 101.325 kPa (14.696 psia) and 15.6 °C (60.0 °F). Therefore, correction for the effect of operating pressure and temperature shall be applied to determine the prover's volume at operating conditions. These corrections are as follows:

Correction for Effect of Pressure on Steel Prover (Cpsp)

$$\text{Cpsp} = 1.0000 \, (\text{since the prover is equipped with a double wall design})$$

Correction for Effect of Temperature on Steel Prover (Ctsp)

$$\text{Ctsp} = \left[1 + 0.000024 \left(T_f - 15.6\right)\right] \text{ SI Units}$$

$$\text{Ctsp} = \left[1 + 0.0000134 \left(T_f - 60\right)\right] \text{ IP Units}$$

Figure 7. Shell Pipeline Corporation orifice master meter proving report version 86.2A.

The calibrated volume of the prover at operating conditions is:

$$PVtp = [PBV*Cpsp*Ctsp]$$

The mass of fluid in the prover's calibrated section at operating conditions is:

$$PMtp = [PVtp*DENp]$$

In the "master meter" proving method, the turbine meter is calibrated on a mass basis under actual operating conditions. The master meter's Mass Meter Factor, MMFt, can be calculated using the following equations:

$$TVtp = [(Proving\ Run\ Pulses)/(K\ Factor)]$$

$$TMtp = [TVtp*DENt]$$

$$VMFt = [PVtp/TVtp]$$

$$MMFt = [PMtp/TMtp]$$

The average mass meter factor, MMFt, for the turbine meter is calculated along with the repeatability for five consecutive runs. Acceptable repeatability is based on a maximum tolerance of 0.05% for the turbine's volumetric meter factor, VMFt. The master turbine meter is reproved at the start of each master meter proving run. This mass factor is applied forward.

In the "master meter" proving method, the orifice meter is proved on a mass basis under actual operating conditions. The average mass meter factor for the orifice meter, MMFo, is calculated along with the repeatability for two consecutive runs. Acceptable repeatability is based on a maximum deviation of 0.15%. An MMFo is calculated for each master meter proving run using the following equation:

$$MMFo = [(Turbine\ Mass)*(MMFt)]/[(Orifice\ Mass)]$$

Discharge coefficients. When an orifice proving report was calculated, the empirically derived mass discharge coefficient was compared with two empirical predictions—the Stolz and Buckingham equations. The Mass Discharge Coefficient was calculated from the master meter proving results using the following formula:

$$Mass\ Cd = [(MMFo)*(Buckinghams\ Cd)]$$

Influence quantities. Orifice meter applications are based on steady flow of a homogeneous, single-phase, Newtonian fluid with a velocity profile that follows the reference standard piping conditions. Departures from these reference piping conditions, the effect of the dynamic properties of the fluid, instrumentation effects, and operating conditions are termed influence quantities. Influence quantities are summarized in Figure 2.

Repeatability and reproducibility. Repeatability is the ability to obtain the same result when measuring the same quantity several times in succession. The turbine meter's VMFt has a required repeatability range of 0.05%. The orifice meter's MMFo has a required repeatability range of 0.10%. Repeatability has been expressed as a percentage on each orifice proving report.

Reproducibility is the ability to obtain the same result when measuring the same quantity under different conditions (i.e. time, observer, operating conditions, etc.). With respect to the turbine meter, reproducibility is not significant, since the meter is proved immediately prior to proving an orifice meter. The orifice meter's MMFo reproducibility was demonstrated over a time period of several days at different flow rates and was generally within 0.10%.

Uncertainty of orifice calibrations. The generally accepted statistical method of the quadrature sum (square root of the sum of the squares) has been applied in deriving the overall measurement accuracy.

In calculating the uncertainties, the following assumptions were made:

- The universe is homogeneous (statistically controlled) with respect to MMFo.
- The distribution is approximately Normal or Gaussian.
- The sample is random.
- A 95% confidence limit is required.

The PBV was established to within ±0.031% by water draw calibration using NBS certified standards.

When proved separately as a volume meter, the turbine meter's VMFt demonstrated a standard deviation of ±0.025% for five consecutive proving runs. An assigned uncertainty of ±0.05%, or two standard deviations, was used in calculating the overall uncertainty.

The flow computer has a stated uncertainty of 0.10%. The density of ethylene was determined using API Chapter 11.3.2.1—Ethylene Density, using measured pressures and temperatures. The uncertainty of the equation of state is stated as ±0.20%.

Additional uncertainties when calculating density using an equation of state, no matter how accurate the equation of state might be, are errors resulting from the uncertainties of the measured temperature and pressure. Transmitter uncertainties of ±0.08 °C (0.15 °F) and ±6.89 kPa (1 psi) at metering conditions of 8 MPa (1160 psia) and 20 °C (68 °F) would result in a calculated density uncertainty of ±0.15%.

Following is a summary of the master meter uncertainty equation:

	Tolerance (%)	Effect	Square of tol & eff
Prover calibration	0.03	1.0000	0.0009
Turbine repeatability	0.05	1.0000	0.0025
Turbine computer	0.10	1.0000	0.0100
Density computer	0.10	1.0000	0.0100
Equation of state	0.20	1.0000	0.0400
Density sensing	0.15	1.0000	0.0225
		Sum	0.0861

Square root of sum ±0.2934%

When the orifice meters were proved on a mass basis, they demonstrated a standard deviation of ±0.05% for two consecutive proving runs. An assigned uncertainty of ±0.10%, or two standard deviations, was used in calculating the overall uncertainty.

The following is a summary of the overall orifice meter uncertainty equation as a result of proving on a mass basis using the master meter method:

	Tolerance %	Effect	Square of tol & eff
Master turbine	0.2934	1.0000	0.0861
Master meter proving	0.10	1.0000	0.0100
		Sum	0.0961

Square root of sum ±0.3100%

The uncertainties calculated are indicative at the time of proving and do not reflect additional uncertainties associated with measurements occurring beyond the proving period.

Many factors contribute an element of uncertainty to the flow rate indicated by the flow meter in relation to the actual or true flow. Orifice metering systems are made of an assembly of parts originating from different manufacturers. To determine the uncertainty with which the total assembly can measure, the tolerances on the different components and the way they affect the metering accuracy have to be considered. Since the orifice meter has been proved in situ, the practical working formula for the uncertainty can be expressed as follows:

$$\% = \left(\left[(\text{orifice calibration})^2 \right] \pm 0.25 \left[(\text{dp error})^2 \right] \right.$$
$$\left. + 0.25 \left[(\text{density sensing error})^2 \right]^{0.5} \right)$$

Conclusions

The results of the orifice calibrations were rather interesting. Installations that were not in compliance with ANSI 2530

requirements were discovered, as well as human errors associated with flow computer programming.

The master meter method identified and measured errors associated with the following problems:

- Deformed orifice plates
- Eccentrically mounted orifice plates
- Orifice plates installed backwards
- Improperly programmed computers
- Effect of compressor pulsations.

Additional information was no doubt masked by the influence quantities. From an application standpoint, identifying the source of error is not as important as measuring and correcting for the error.

Based upon the results, the majority of orifice meters have indicated a general understatement of measured quantities prior to proving.

In summary, our original goal was achieved—to develop an economically viable method for proving ethylene orifice meters under actual operating conditions. The master meter proving method has demonstrated to be a reliable measurement tool in identifying the quantifying measurement problems.

References

1. *Orifice Metering of Natural Gas and Other Related Hydrocarbon Fluids*, API MPMS Chapter 14.3, Second Edition, September 1985 (AGA Report No. 3), (GPA 8185-85), (ANSUAPI 2530).
2. *Ethylene Density*, API MPMS Chapter 11.3.2.1, 1974.
3. *Shell Flow Meter Engineering Handbook*, Second Edition, McGraw-Hill Book Company (UK) Limited, 1985.
4. Miller, R. W., *Flow Measurement Engineering Handbook*, McGraw-Hill Book Company, 1983.
5. Stolz, J., *A Universal Equation for the Calculation of Discharge Coefficients of Orifice Plates*, Flow Measurements of Fluids, IMEKO Conference on Flow Measurement of Fluids, Groningen, The Netherlands, 1978.
6. Kolb, R. H., *Ethylene Mass Measurement System with Prover*, The Second Petroleum Measurement Seminar, 1980.
7. Pfrehm, R. H., "Improved Turbine Meter System Measures Ethylene Accurately," *Oil & Gas Journal*, April, 1981.
8. Bellinga, H, Kooi, A., Hoeks, C. P., van Laak, F A. L., and Orbons, P. J., "Piston Prover Used to Calibrate Gas Meters," *Oil & Gas Journal*, August, 1985.
9. Boncher, F. S., Shipman, L. M., and Yen, L. C., *Hydrocarbon Processing*, 54 (2), 89 (1975).
10. *ASTM Manual on Presentation of Data and Control Chart Analysis*, ASTM Special Technical Publication 15D, 1976.
11. Measurement Uncertainty for Fluid Flow in Closed Conduits, ANSI/ASME MFC-2M-1983.

Notation

β = ratio of orifice plate bore to internal meter diameter
Cd = orifice discharge coefficient
Cpsp = correction for effect of pressure on prover
Ctsp = correction for effect of temperature on prover
d = orifice plate bore, in (mm)
dp = differential pressure across orifice plate, in H_2O @ 60 (Pa)
D = meter internal diameter, in (mm)
DENo = density of fluid at orifice, lbm/ft^3 (kg/m^3)
DENp = density of fluid at prover, lbm/ft^3 (kg/m^3)
DENt = density of fluid at turbine, lbm/ft^3 (kg/m^3)
F_a = orifice thermal expansion factor
Fb = basic orifice factor
Fdp = correction for static pressure on dp transmitter
Fr = Reynolds number factor
IP = Inch-Pound System of Units
MMFt = mass meter factor for turbine
MMFo = mass meter factor for orifice
PBV = prover base volume, ft^3 (m^3)
PMtp = prover mass, lbm (kg)

PVtp = prover volume, ft^3 (m^3)
q_m = mass flow rate, lbm/s (kg/s)
Q_m = mass flow rate, lbm/hr (kg/hr)
Rd = orifice bore Reynolds number
R_D = pipe Reynolds number
SI = International System of Units
SVP = small volume prover
T_f = flowing temperature, °F (°C)
TMtp = turbine meter's mass, lbm (kg)
TVtp = turbine meter's volume, ft^3 (m^3)
VMFt = volumetric meter factor for turbine meter
Y2 = expansion factor based on downstream static pressure

Source

Reprinted from *Oil & Gas Journal*, December 15, 1986 by permission and courtesy of Perm Well Publishing Co. and James E. Gallagher, Shell Oil Company.

Gas or vapor flow measurement—orifice plate flange taps

Sequence of Calculations

1. Calculate $K_0\beta^2$ value.

$$K_0\beta^2 = \frac{W_m}{D^2\left(358.5 + .007T_f\right)\sqrt{\gamma_f h_m}}$$

2. Iterate to find uncorrected beta ratio.

$$K_0\beta^2 = \beta^2\left[\left(0.5993 + \frac{0.007}{D}\right) + \left(0.364 + \frac{0.0076}{\sqrt{D}}\right)\beta^4\right]$$
$$+ \left[0.4\left(1.6 - \frac{1}{D}\right)^5\right.$$
$$\times \left(0.07 + \frac{0.5}{D} - \beta\right)^{5/2}\left(0.009 + \frac{0.034}{D}\right)$$
$$\times (0.5 - \beta)^{3/2} + \left(\frac{65}{D^2} + 3\right)(\beta - 0.7)^{5/2}\right]$$
$$\div \left[1 + 0.000015\left(830 - 5000\beta + 9000\beta^2\right.\right.$$
$$\left.\left. - 4200\beta^2 + \frac{530}{\sqrt{D}}\right)\right]$$

Any negative factor, such as $(\beta - 0.7)^{5/2}$ when β is less than 0.7, shall be taken as equal to zero, and thus the whole term $(65/D^2 + 3)(\beta - 0.7)^{5/2}$ will drop out of the equation.

Let fx equal the equation in step 2—$K_0\beta^2$, and let:

$$f^1x = 2\left(\left(0.5993 + \frac{0.007}{D}\right)\right.$$
$$+ 4\left(1.6 - \frac{1}{D}\right)^5\left(0.07 + \frac{5}{D} - \beta\right)^{5/2}$$
$$- \left(0.009 + \frac{0.034}{D}\right)(0.5 - \beta)^{3/2}$$
$$+ \left(\frac{65}{D^2} + 3\right)(\beta - 0.7)^{5/2}\beta + 6\left(0.364 + \frac{0.076}{\sqrt{D}}\right)\beta^6\right)$$

Start with $\beta = 0.7$ and calculate fx/f^1x. If greater than 0.000015, subtract fx/f^1x from current value of β and repeat until the error is less than 0.000015.

3. Calculate the expansion factor.

$$Y = 1 - \frac{(0.41 + 0.35\beta^4)h_m}{39.6kP_f}$$

4. Divide $K_0\beta^2$ by Y.
5. Calculate the Reynolds number correction factor.

$$F_r = 1 + \left(830 - 5000\beta + 9000\beta^3 - 4200\beta^3 + \frac{530}{\sqrt{D}}\right)$$
$$\times \frac{(D\beta)^2 \mu_{cp}}{4.42 W_m}$$

6. Divide $K_0\beta^2$ by F_r.
7. Repeat step 2 to find corrected beta ratio.
8. Repeat steps 3, 4, 5, and 6 using corrected beta ratio.
9. Repeat step 2 for final beta ratio.
10. Calculate orifice bore.

$$d = BD \text{ in}$$

11. Calculate flow rate for an even-sized orifice bore.

$$\beta = \frac{d}{D}$$

Calculate $K_0\beta^2$ using the equation in step 2.

$$\text{new flow} = \frac{\text{new } K_0\beta^2}{\text{original } K_0\beta^2} \times \text{original flow, lb/h}$$

Notation

D = inside diameter of pipe, in
d = bore diameter of primary device, in
F_r = Reynolds number correction factor for steam, vapor, or gas, dimensionless
h_m = maximum differential or range of differential gauge, in of water, dry calibrated
K_o = coefficient of discharge, including velocity of approach at a hypothetical Reynolds number of infinity, dimensionless
k = ratio of specific heats C_p/C_r, dimensionless
P_f = flowing pressure, psia
T_f = flowing temperature, °F
W_m = maximum rate of flow, lb/h
Y = expansion factor, dimensionless
β = diameter ratio d/D, dimensionless
γ_f = specific weight of flowing conditions, lb/ft^3
μ_{cp} = absolute viscosity, centipoises

Source

Instrument Engineering Programs, Gulf Publishing Company, Houston, Texas.

Reference

1. Spink, L. K., *The Principles and Practice of Flowmeter Engineering*, The Foxboro Company.

Properties of gas and vapors

Table 1
Physical Properties

| No. | Compound | Formula | Critical Constants | | Molecular Weight (M) | Ratio of Specific Heats C_p/C_v at 14.7 psia and 60 °F (k) | Heat of Vaporization, 14.696 psia at Boiling Point, Btu/lb (*NBP* ΔHV) |
			Pressure, psia (P_c)	Temperature, °R (T_c)			
1	Methane	CH_4	667.8	343.4	16.043	1.311	219.22
2	Ethane	C_2H_6	707.8	550.0	30.070	1.193	210.41
3	Propane	C_3H_8	616.3	666.0	44.097	1.136	183.05
4	*n*-Butane	C_4H_{10}	550.7	765.0	58.124	1.094	165.65
5	Isobutane	C_4H_{10}	529.1	735.0	58.124	1.097	157.53
6	*n*-Pentane	C_5H_{12}	488.6	845.7	72.151	1.074	153.59
7	Isopentane	C_5H_{12}	490.4	829.1	72.151	1.075	147.13
8	Neopentane	C_5H_{12}	464.0	781.0	72.151	1.074	135.58
9	*n*-Hexane	C_6H_{14}	436.9	913.7	86.178	1.062	143,95
10	2-Methylpentane	C_6H_{14}	436.6	895.8	86.178	1.063	138.67
11	3-Methylpentane	C_6H_{14}	453.1	908.3	86.178	1.064	140.09
12	Neohexane	C_6H_{14}	446.8	860.1	86.178	1.064	131.24
13	2,3-Dimethylbutane	C_6H_{14}	453.5	900.3	86.178	1.065	136.08
14	*n*-Heptane	C_7H_{16}	396.8	972.8	100.205	1.054	136.01
15	2-Methylhexane	C_7H_{16}	396.5	955.0	100.205	1.054	131.59
16	3-Methylhexane	C_7H_{16}	408.1	963.8	100.205	1.054	132.11
17	3-Ethylpentane	C_7H_{16}	419.3	973.5	100.205	1.054	132.83
18	2,2-Dimethylpentane	C_7H_{16}	402.2	937.2	100.205	1.053	125.13
19	2,4-DimethylperUane	C_7H_{16}	396.9	936.0	100.205	1.053	126.58
20	3,3-Dimethylpentane	C_7H_{16}	427.2	965.9	100.205	1.053	127.21
21	Triptane	C_7H_{16}	428.4	956.4	100.205	1.055	124.21
22	*n*-Octane	C_8H_{18}	360.6	1024.2	114.232	1.046	129.53
23	Diisobutyl	C_8H_{18}	360.6	990.4	114.232	1.049	122.80
24	Isooctane	C_8H_{18}	372.4	979.5	114.232	1.049	116.71
25	*n*-Nonane	C_9H_{20}	332.0	1070.7	128.259	1.042	123.76
26	*n*-Decane	$C_{10}H_{22}$	304.0	1112.1	142.286	1.038	118.68
27	Cyclopentane	C_5H_{10}	653.8	921.5	70.135	1.117	167.34
28	Methylcyclopentane	C_6H_{12}	548.9	959.4	84.162	1.085	147.83
29	Cyclohexane	C_6H_{12}	591.0	996.7	84.162	1.089	153.00
30	Methylcyclohexane	C_7H_{14}	503.5	1030.3	98.189	1.068	136.30
31	Ethylene	C_2H_4	729.8	506.6	28.054	1.243	207.57
32	Propene	C_3H_6	669.0	656.9	42.081	1.154	188.18
33	1-Butene	C_4H_8	583.0	755.6	56.108	1.111	167.94
34	*cis*-2-Butene	C_4H_8	610.0	784.4	56.108	1.121	178.91
35	*trans*-2-Butene	C_4H_8	595.0	771.9	56.108	1.107	174.39
36	Isobutene	C_4H_8	580.0	752.6	56.108	1.106	169.48
37	1-Pentene	C_5H_{10}	590.0	836.9	70.135	1.084	154.46
38	1,2-Butadiene	C_4H_6	653.0	799.0	54.092	1.118	181.00
39	1,5-Butadiene	C_4H_6	628.0	766.0	54.092	1.120	174.00
40	Isoprene	C_5H_8	558.4	872.0	68.119	1.089	153.00
41	Acetylene	C_2H_2	890.4	555.3	26.038	1.238	—
42	Benzene	C_6H_6	710.4	1012.2	78.114	1.117	169.31
43	Toluene	C_7H_8	595.9	1065.6	92.141	1.091	154.84
44	Ethylbenzene	C_8H_{10}	523.5	1111.2	106.168	1.072	144.00
45	*o*-Xylene	C_8H_{10}	541.4	1135.0	106.168	1.069	149.10
46	*m*-Xylene	C_8H_{10}	513.6	1111.0	106.168	1.072	147.20
47	*p*-Xylene	C_8H_{10}	509.2	1109.6	106.168	1.072	144.52

Table 1
Physical Properties (continued)

| No. | Compound | Formula | Critical Constants | | Molecular Weight (M) | Ratio of Specific Heats C_p/C_v at 14.7 psia and 60 °F (k) | Heat of Vaporization, 14.696 psia at Boiling Point, Btu/lb (*NBP* ΔHV) |
			Pressure, psia (P_c)	Temperature, °R (T_c)			
48	Styrene	C_8H_8	580.0	1166.0	104.152	1.076	151.00
49	Isopropylbenzene	C_9H_{12}	465.4	1136.4	120.195	1.060	134.30
50	Methyl alcohol	CH_4O	1174.2	923.0	32.042	1.237	473.0
51	Ethyl alcohol	C_2H_6O	925.3	929.6	46.069	1.149	367.0
52	Carbon monoxide	CO	507.0	240.0	28.010	1.403	92.7
53	Carbon dioxide	CO_2	1071.0	547.9	44.010	1.293	238.2
54	Hydrogen sulfide	H_2S	1306.0	672.7	34.076	1.323	235.6
55	Sulfur dioxide	SO_2	1145.1	775.5	64.059	1.272	166.7
56	Ammonia	NH_3	1636.1	730.3	17.031	1.304	587.2
57	Air	N_2O_2	547.0	238.7	28.964	1.395	92.0
58	Hydrogen	H_2	188.1	60.2	2.016	1.408	193.9
59	Oxygen	O_2	736.9	278.9	31.999	1.401	91.6
60	Nitrogen	N_2	493.0	227.4	28.013	1.402	87.8
61	Chloride	$C1_2$	1118.4	751.0	70.906	1.330	123.8
62	Water	H_2O	3208.0	1165.6	18.015	1.335	970.3
63	Helium	He	—	—	4.003	1.660	—
64	Hydrogen chloride	HCl	1198.0	584.5	36.461	1.410	185.5

Note: C_p/C_v calculated by the formula $k = C_p/(C_p - 1.1987)/M$.

Sequence of calculations

1. Mixture values are calculated by the mixture law, which assumes that the property of the mixture is the sum of the partial fractions.
2. The compressibility factor is iterated from the Redlich–Kwong equation:

$$fx = Z^3 - Z^2$$
$$+ BP_f \left[\left(\left(\frac{A^2}{B} \right) - BP_f - 1 \right) Z - \left(\frac{A^2}{B} \right) BP_f \right]^2$$

$$f^1x = 3Z^2 - 2Z + BP_f \left(\left(\frac{A^2}{B} \right) - BP_f - 1 \right)$$

where $A = \dfrac{0.42748 T_c^{2.5}}{P_c T_f^{2.5}}$

$$BP_f = 0.0866403 \frac{T_c P_f}{P_c T_f}$$

$$\frac{A^2}{B} = 4.93396 \left(\frac{T_c}{T_f} \right)^{1.5}$$

Start with $Z = 1$ and calculate fx/f^1x. If greater than 0.00001, subtract fx/f^1x from the current value of Z and repeat until the error is less than 0.00001.

3. Calculate the specific weight:

$$\gamma 3_f = \frac{P_f M}{T_f 1.314 Z} \text{ lb/ft}^3$$

Notation

A, B = partial calculations for Z
M = molecular weight, dimensionless
P_f = flowing pressure, atmospheres
P_c = critical pressure, atmospheres
T_f = flowing temperature, °K
T_c = critical temperature, °K
Z = compressibility, dimensionless
γ_f = specific weight, lb/ft³
Note: These units do not correspond with the calculator input units. The program converts as necessary.

Source

Instrument Engineering Programs, Gulf Publishing Company, Houston.

Determine required orifice diameter for any required differential when the present orifice and differential are known in gas measurement

Procedure. Draw a straight line between the values of the present and required differentials on scales "A" and "C." From the intersection of this line on scale "B", draw second line to connect with the size of the present orifice diameter on scale "E." The size of the required orifice will be found at the intersection of his second line with scale "D."

Estimate the temperature drop across a regulator

The temperature drop that will be experienced when gas is passed through a regulator may be estimated as 1° drop for each atmosphere (15 psi) of pressure drop.

Example. Determine the temperature drop that will be experienced when reducing a gas from 750 to 600 psig.

Solution

$$750 - 600 \text{ psig} = 150 \text{ psig}$$

$$\frac{150}{15} = 10 \text{ atmospheres}$$

$$10 \times 1°/\text{atm} = 10° \text{ of drop}$$

Estimate natural gas flow rates

This chart provides a good estimate of natural gas flow rates through 4-in meter runs equipped with flange taps. No calculations are required to obtain a flow rate. Orifice plate size and differential and static pressures from the orifice meter chart are used to make the estimate.

To use the Gas Flow Rate Estimator, enter the chart at the proper differential value at the indicated lower horizontal scale. Project this value vertically upward to the appropriate static pressure line, and then horizontally to the proper plate size line. Drop vertically from the plate size line to the lowermost horizontal rate scale at the base of the chart to obtain the estimated flow rate.

How to estimate the average pressure differential on the remaining meter runs of a parallel system when one or more runs are shut off

On gas systems where inferential meters are installed in parallel, at times it may be necessary to shut down one or more of the runs temporarily for repair or maintenance work. When this occurs, it is important to know beforehand what the pressure differential on the remaining runs will be so that the meters will not be over-ranged. The equation is:

Average differential for runs remaining in service (A)

$$= \left[\frac{(\text{No. of runs in service})\sqrt{(\text{Average differential})}}{(\text{No. of runs remaining in service})}\right]^2 .$$

Example. A measurement engineer wishes to shut off 1 m run on a 4-m parallel system to inspect the orifice plate. The average differential on the four runs is 49 in and his meters have a 100-in range. He wishes to determine whether shutting down 1 m will over-range the three remaining meters.

$$(A) = \left(\frac{4 \times \sqrt{49}}{3}\right)^3 = (9.3)^2 = 86.5 \text{ in}$$

Since the average differential for the remaining runs is 86.5 in, the operator may safely shut the meter down as planned to inspect its orifice plate.

Sizing a gas metering run

Multiply the absolute pressure by the desired differential pressure (the differential in inches of water should not be greater than the absolute pressure in psia). Take the square root of the product. Divide this figure into the maximum flow rate in cubic ft/hr to obtain the approximate coefficient. Then divide the maximum coefficient by 250 and take the square root to determine the approximate inside diameter of the orifice in inches. The metering run diameter should not be less than one-third larger than the orifice diameter.

Example. Find the orifice diameter and meter run for a line designed to carry a maximum flow of 2 MMcf/h at a pressure of 600 psia. Use a differential pressure of 50 in of water.

$$\sqrt{600 \times 50} = \sqrt{30,000} = 173.2$$

$$\frac{2,000,000}{173.2} = 11,540$$

$$\frac{11,540}{250} = 46.2$$

$$\sqrt{46.2} = 6.8 \text{ in, approximate orifice diameter}$$

Meter run should not be less than 10 in in diameter.

List of typical specifications for domestic and commercial natural gas

1. **Heating value:** Not less than 950 Btus per std cubic ft.
2. **Purity:** The gas to be delivered shall be free from dust, gums, crude oils, and hydrocarbons liquefiable at temperatures in excess of 15 °F at 800 psig.
3. **Sulfur content:** The gas shall contain not more than one grain of hydrogen sulfide per 100 cu ft of gas, and not more than 20 grains of total sulfur per 100 cu ft of gas.
4. **CO_2 content:** The gas shall contain not more than 2% carbon dioxide.
5. **Moisture content:** The gas shall contain not more than 4 lb of water vapor per MMcf of gas measured at a pressure of 14.4 psi and a temperature of 60 °F.
6. **Temperature:** Maximum temperature at point of delivery shall be 120 °F.

Determine the number of purges for sample cylinders

To determine the proper number of times a sample cylinder should be pressured to ensure removal of the original gas, multiply the reciprocal of each filling pressure (in atmospheres) by each succeeding reciprocal. The final filling will determine what fraction of the original gas is still in the cylinder.

Example. The problem is to fill a gas sample for analysis to obtain an accuracy of 0.05%. Source pressure is 450 psia.

Solution. First filling to line pressure and emptying

$$\frac{1}{450/15} = \frac{1}{30} \times 100 = 3.33\%$$

Second filling to line pressure and emptying:

$$\frac{1}{30} \times \frac{1}{30} \times 100 = 0.11\%$$

Third filling to line pressure:

$$\frac{1}{30} \times \frac{1}{30} \times \frac{1}{30} \times 100 = 0.0037\%$$

Find the British thermal units (Btu) when the specific gravity of a pipeline gas is known

Use the graph below prepared from the Headlee formula:
    Btu @ 30 in Hg and 60° saturated = 1545 (specific gravity) + 140 − 16.33 (% N_2)

Example. Assume a gas having a gravity of 0.670, which contains 1.0% nitrogen. From the chart, read up from 0.670 gravity on the baseline to the intercept with the curve marked 1% nitrogen. Reading horizontally across, the total Btu of the gas is 1160 Btu at 30 in, 60° saturated.

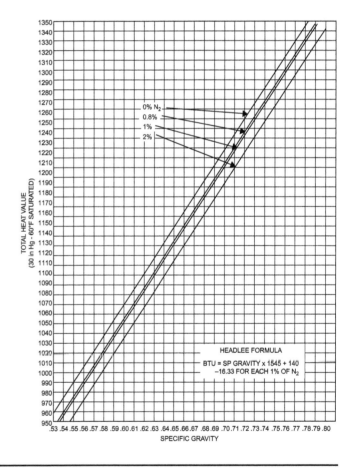

Estimate for variations in measurement factors

Orifice measurement

1. Each 10° change in flowing temperature will cause a 1% change in flow calculated.

2. Each 0.01 change in specific gravity of the gas measured will result in a 1% change in passing flow calculated.

3. Each 10-psi error in static pressure will make the following error in flow calculated:
   at 100 psia—5%
   at 500 psia—1%
   at 1000 psia—0.5%
4. Each 0.1 in of differential pressure error will make the following error in flow calculated.

at 5 in differential—1%
at 25 in differential—0.2%
at 50 in differential—0.1%
at 100 in differential—0.05%

5. Supercompressibility affects volume measured approximately 0.5% per 100 psi pressure in flowing temperature range of 60–90 °F.

Rules of measurement of gas by orifice meter

Where the static pressure remains constant, to reduce the flow by one half, lower the differential reading to one fourth. For example, if the differential is reading 64 in, to reduce the flow by one half, reduce the differential to 16 in.

For better measurement, keep the beta ratio between 0.2 and 0.6.

Headers on meter stations should be twice the internal diameter of the meter runs they manifold together.

Measurement stations should be installed on the suction side of a compressor station rather than the discharge side where possible, to reduce effect of pulsation.

Use the wine size orifice plate in meter runs that are manifolded together where possible; recordings can then be used to check against the other, since their differentials should be approximately the same.

Make quick calculation of the gas flow through orifice meters

On meter runs up to and including 10-in, this can be accomplished by multiplying the coefficient for a 1-in orifice times the actual orifice installed squared times the square root of the extension (the product of the differential in inch of water times the static pressure in lb/in^2 absolute). AGA-3 coefficients for a 1-in orifice in 2, 3, 4, 6, 8, and 10 in standard orifice meter runs using flange taps should be committed to

memory. Base conditions would be for a 0.60 specific gravity 60 °F, and 14.73 psia pressure base. The answer would be in standard cubic ft/hr or day depending upon the coefficient used.

AGA Committee Report Number 3 coefficients at the conditions shown are:

2 in meter run—1 in orifice—271.4 cfh, 0.6G, 60 °F, and 14.73 psia
3 in meter run—1 in orifice—262.8
4 in meter run—1 in orifice—260.8
6 in meter run—1 in orifice—259.3
8 in meter run—1 in orifice—258.7
10 in meter run—1 in orifice—258.5 (heavy)

Example. Find the hourly gas flow through an 8-in meter run having a 2.00-in orifice when the differential pen is recording 16 in of water and the static pressure is 900 lb/in^2 absolute.

Solution. From memory the coefficient for an 8-in meter run, 1-in orifice is 258.7. Multiplying 258.7 times 4 (or 2 squared) times 120 (the square root of the product of 900 times 16) equals 124,176 cfh, the approximate flow in cubic ft of gas per hour at 0.6 gravity, 60 °F temperature, and 14.73 psia base pressure.

How to measure high pressure gas

Accurate gas measurement is only obtained through a good maintenance and operating schedule

T. J. Lewis, Measurement Engineer, Columbia Gulf Transmission Co., Houston

The rapidly increasing value of natural gas, along with rising operating costs, has made accurate gas measurement vital for the natural gas industry.

Accurate gas measurement is obtained only through an efficient maintenance and operating schedule, adhered to rigidly, after a properly designed measuring station has been

installed. Any error in the measurement of gas at high pressures and large volumes magnifies itself, so that there can be either a considerable gain or loss in revenue.

Field measurement at high pressure is defined as the measurement of gas at a specific point in a gas producing area at a pressure higher than that of contract base pressure.

Contracts

The gas purchase contract is of prime importance in determining the requirements for gas measurement at a given location.

Information such as gas analysis, gas/oil ratios, specific gravity determinations, flowing temperatures, pressures, etc., should be gathered before a gas purchase contract is written. This collection of data should be used to eliminate any assumptions regarding the requirements for performing accurate measurement. It also provides information for the engineering and accounting functions, so that they are made aware of their responsibilities and limitations in helping provide good measurement.

The contract should also state as clearly as possible all parameters of the gas measurement operation to prevent possible disputes by having a clear understanding between the buyer and the seller.

Design and planning

In the planning and designing of a natural gas measuring station, many items of equal importance must be considered if accurate measurement is to be obtained.

The design and fabrication of the orifice meter runs must meet the specifications of the AGA Gas Measurement Committee Report No. 3, latest revision. In addition to these requirements, consideration must be given to safety, volume and quality of gas, term of contract, flexibility of station design, and physical location of the station. The most important point in this whole procedure is remembering that the measuring station is similar to a large cash register with money flowing both ways for the company.

A meter tube should be designed to allow easy internal inspection. Shutoff valves should be placed on each end of the meter, so that it may be depressured independently of any other piping. The instrumentation and/or control piping should be located so and supported to furnish maximum protection against accidental breakage.

Meter runs, piping, etc., should be designed as a Class III area facility. If possible, a properly vented walk-in type meter building should be used, with the total installation laid out with the safety of the operating personnel firmly in mind (Figure 1). As periodic orifice plate inspections are necessary, provisions should be made so that these inspections can be made without interruption of the gas flow. This can be done by installing a bypass line or by using a senior type orifice fitting.

The flow and temperature recording instruments should be located in a walk-in type meter building. This arrangement provides a year-round working area for chart changing, meter tests, maintenance work, etc.

Installation

The measuring or metering components in a measuring station are divided into two elements: primary and auxiliary.

The *primary element* of an orifice-type measuring station consists of the meter tube, the orifice plate fitting, the orifice plate, pressure taps, and straightening vanes. While the

Figure 1. Meter runs, piping, etc., should be designed as a Class III area facility. If possible, a properly vented walk-in type meter building should be used, with the total installation laid out with the safety of the operating personnel firmly in mind.

primary element is being fabricated, its components should be checked against the specifications listed in AGA Gas Measurement Committee Report No. 3. Any discrepancies found can then be corrected before the meter tube is finally assembled and shipped to the field.

When the meter tube is being installed in the field, the orifice plate is checked to make sure it is centered in the meter tube, and that it meets AGA Report No. 3 specifications. Meter tubes are normally installed in a horizontal or vertical plane. Pressure taps in a vertical plane are preferred, since this allows the gauge lines to be installed from the top of the orifice fitting. The pressure taps located on the bottom of the orifice fittings then can be used for installing drips or drains.

The *auxiliary element* consists of the meters (recording devices) used for measuring the static pressure, differential pressure, and temperature of the gas in the meter tube. Meters are placed as close as possible to the orifice or flange fitting. Gauge lines are connected from the meter to the orifice fitting in a way that will best provide a drain to the meter tube. Liquids trapped in a gauge line can cause significant errors in the recorded pressures.

There are two principal types of orifice meter gauges: the mercury manometer-type and the bellows-type meters. Each of these meters is suitable for field measurement, with the bellows-type meter becoming the favorite for wet gas measurement.

Installation of these elements in a measuring station should be under the supervision of experienced measurement personnel.

Operation and maintenance

After a measuring station has been properly installed, accurate measurement is obtainable only by conscientious care and maintenance by the field measurement technicians.

Any discrepancy noticed on the chart should be explained on the chart, and a correction should be made immediately. Since immediate correction is not always possible, a report should then accompany the chart describing the condition of the meter and when the correction was completed. These reports, along with test and inspection reports, are necessary for the calculation of accurate volumes.

The accuracy of recording instruments depends on the men who test and operate the measuring station equipment. Good test equipment, properly maintained, is absolutely necessary in checking the calibration of temperature and pressure recording instruments and the quality of the gas being measured. The frequency of meter tests, gas quality tests, and meter tube inspections depends on the individual company and its contractual agreements with other companies.

Monthly calibration tests and inspections on the flow and temperature recorders are normally adequate for accurate

Figure 2. Relative effect of various factors on calculated flow volume (also Figure 3).

Figure 3. The above (also Figure 2) shows what happens to volume calculations when there are small errors in static pressure, differential pressure, temperature, and specific gravity measurements.

measurement. In special cases, however, a weekly inspection may be required. Monthly calibration tests and inspections are vital to accurate measurement practices. Figures 2 and 3 show what happens to volume calculations when there are

Figure 4. Charts are shown being changed, and clocks wound and adjusted. Each measuring station has its own flow pattern, which is recognizable by experienced chart changers.

small errors in static pressure, differential pressure, temperature, and specific gravity measurements.

Since external tests and inspections do not reveal the internal condition of a meter tube, the meter tube and its components must be inspected internally on an annual basis for possible buildups and internal corrosion. This inspection ensures continuing accuracy.

Chart changing

Charts must be changed on time, and clocks must be kept wound and adjusted (Figure 4).

Each measuring station has its own flow pattern, which is recognizable by experienced chart changers. Any change in this flow pattern indicates a possible error, which should be corrected if possible. If the correction cannot be made by the chart changer, he should notify the field measurement personnel who are qualified to perform required adjustment and/or repairs.

Each chart must be properly dated and identified along with field information necessary for calculating the required volumes. Automatic chart changers have become increasingly popular, especially in remote areas such as offshore Louisiana.

Periodic pulsation tests should be made at each measuring station, particularly where a measuring station is located on the suction or discharge side of a compressor station. In this instance, pulsation dampeners must be installed between the compressor and the measuring station. Errors caused by

pulsation can be very large, but they will not necessarily show in calibration tests.

A comprehensive maintenance program, faithfully followed, can detect and prevent costly errors before they occur. Leaking valves, gauge lines, etc., will cause errors in an increase or decrease in company revenue. Such leaks should be corrected as soon as possible.

Wet gas measurement

The problems encountered in measuring wet gas are considerably different from dry gas measurement.

Wet gas contains entrained liquids consisting of water and hydrocarbon condensates that can cause hydrate formations in the orifice plate fitting, flow recorders, and instrument lines. These hydrates are formed with the aid of free water under certain pressure and temperature conditions. Formation of hydrates can be reduced or eliminated in several different ways.

One successful solution to prevent hydrate formation is to use drip or condensate bottles in instrument and gauge lines. Infrared gas heaters can be used in meter buildings for heating gauge lines and flow recorders. This type of heater has become very popular lately, since it does not produce an open flame.

Errors caused by hydrate formations in the meter tube and around the orifice plate are common in wet gas measurement. The formation of these hydrates can be prevented by continuously injecting methanol into the gas stream ahead of the meter run. Experience and experimentation will determine the amount of methanol to be injected.

The use of methanol-containing inhibitors is not recommended as the inhibitor often solidifies in the meter tubes.

Four ways to calculate orifice flow in field

Rogers G. Thompson, Measurement Technician, Texas Eastern Transmission Corp., Shreveport, Louisiana

Several methods are available for orifice flow field calculations.

The particular method used is determined by such factors as company policy, advance warning, time available, individual preference, accuracy required, and the information and equipment available at the time of need. A good rule to follow is to calculate as precise a volume as possible at all times.

Any volume calculated manually will be considered a field calculation.

Most companies now use some type of computer system, and rarely will a manually calculated volume match a computer volume. Field calculations are usually made under conditions that eliminate the use of adding machines and calculators. Accuracies can vary from an extremely

Table 1
Combined Specific Gravity and Temperature Correction
Factors—Orifice Meters Combined Factor $= \left(\sqrt{\frac{1}{G}} \sqrt{\frac{520}{460°F}} \right)$

Flow Temp. °F	Specific Gravity								
	.590	**.591**	**.592**	**.593**	**.594**	**.595**	**.596**	**.597**	**.598**
35	1.3343	1.3332	1.3321	1.3309	1.3298	1.3287	1.3276	1.3264	1.3254
36	1.3330	1.3319	1.3208	1.3296	1.3285	1.3274	1.3263	1.3251	1.3241
37	1.3317	1.3306	1.3295	1.3283	1.3272	1.3261	1.3250	1.3238	1.3228
38	1.3303	1.3292	1.3280	1.3269	1.3258	1.3247	1.3235	1.3224	1.3214
39	1.3290	1.3279	1.3267	1.3256	1.3245	1.3234	1.3222	1.3211	1.3201
40	1.3277	1.3266	1.3254	1.3243	1.3232	1.3221	1.3209	1.3198	1.3188
41	1.3264	1.3253	1.3241	1.3230	1.3219	1.3208	1.3197	1.3185	1.3175
42	1.3251	1.3240	1.3228	1.3217	1.3206	1.3195	1.3184	1.3172	1.3162
43	1.3238	1.3227	1.3215	1.3204	1.3193	1.3182	1.3171	1.3159	1.3149
44	1.3223	1.3212	1.3201	1.3190	1.3179	1.3168	1.3156	1.3145	1.3135
45	1.3210	1.3199	1.3188	1.3177	1.3166	1.3155	1.3143	1.3132	1.3122
46	1.3197	1.3186	1.3175	1.3164	1.3153	1.3142	1.3130	1.3119	1.3109
47	1.3184	1.3173	1.3162	1.3151	1.3140	1.3129	1.3118	1.3106	1.3096
48	1.3171	1.3160	1.3149	1.3138	1.3127	1.3116	1.3105	1.3093	1.3083
49	1.3158	1.3147	1.3136	1.3125	1.3114	1.3103	1.3092	1.3080	1.3070
50	1.3147	1.3135	1.3124	1.3113	1.3102	1.3091	1.3080	1.3069	1.3059
51	1.3134	1.3122	1.3111	1.3100	1.3089	1.3078	1.3067	1.3056	1.3046
52	1.3121	1.3109	1.3098	1.3087	1.3076	1.3065	1.3054	1.3043	1.3033
53	1.3108	1.3096	1.3085	1.3074	1.3063	1.3052	1.3041	1.3030	1.3020
54	1.3095	1.3083	1.3072	1.3061	1.3050	1.3039	1.3028	1.3017	1.3007
55	1.3081	1.3070	1.3059	1.3048	1.3037	1.3026	1.3015	1.3004	1.2994
56	1.3070	1.3059	1.3048	1.3037	1.3026	1.3015	1.3004	1.2992	1.2982
57	1.3057	1.3046	1.3035	1.3024	1.3013	1.3002	1.2991	1.2980	1.2970
58	1.3044	1.3033	1.3022	1.3011	1.3000	1.2989	1.2978	1.2967	1.2957
59	1.3032	1.3021	1.3010	1.2999	1.2988	1.2977	1.2966	1.2955	1.2945
60	1.3019	1.3008	1.2997	1.2986	1.2975	1.2964	1.2953	1.2942	1.2932
61	1.3006	1.2995	1.2984	1.2973	1.2962	1.2951	1.2940	1.2929	1.2919
62	1.2994	1.2983	1.2972	1.2961	1.2950	1.2939	1.2928	1.2917	1.2907
63	1.2981	1.2970	1.2959	1.2948	1.2937	1.2926	1.2915	1.2904	1.2894
64	1.2970	1.2959	1.2948	1.2937	1.2926	1.2915	1.2904	1.2893	1.2883
65	1.2957	1.2946	1.2935	1.2924	1.2913	1.2902	1.2891	1.2880	1.2870
66	1.2945	1.2934	1.2923	1.2912	1.2901	1.2890	1.2879	1.2868	1.2858
67	1.2932	1.2921	1.2910	1.2899	1.2888	1.2877	1.2866	1.2855	1.2845
68	1.2920	1.2909	1.2898	1.2887	1.2876	1.2865	1.2855	1.2844	1.2834
69	1.2908	1.2897	1.2887	1.2876	1.2865	1.2854	1.2843	1.2832	1.2822
70	1.2895	1.2884	1.2874	1.2863	1.2852	1.2841	1.2830	1.2819	1.2809
71	1.2884	1.2873	1.2862	1.2851	1.2840	1.2829	1.2818	1.2807	1.2798
72	1.2872	1.2861	1.2850	1.2839	1.2828	1.2818	1.2807	1.2796	1.2786
73	1.2859	1.2848	1.2837	1.2826	1.2815	1.2805	1.2794	1.2783	1.2773
74	1.2847	1.2836	1.2825	1.2815	1.2804	1.2793	1.2782	1.2771	1.2761
75	1.2835	1.2825	1.2814	1.2803	1.2792	1.2781	1.2770	1.2760	1.2750
76	1.2824	1.2813	1.2802	1.2791	1.2780	1.2770	1.2759	1.2748	1.2738
77	1.2811	1.2800	1.2789	1.2778	1.2767	1.2757	1.2746	1.2735	1.2725
78	1.2799	1.2788	1.2777	1.2767	1.2756	1.2745	1.2734	1.2723	1.2713
79	1.2787	1.2776	1.2766	1.2755	1.2744	1.2733	1.2722	1.2712	1.2702

accurate spot check to a very basic calculation. External factors usually determine what procedure to use.

Uses for the calculations determine the degree of accuracy required for each particular instance. Since time is usually at a premium, it would be absurd to calculate to the cubic ft when calculating to the Mcf would suffice. A good example is automatic control of a compressor station where rates are given in terms of MMcf.

Field calculations are used for such things as determining purge gas volumes, relief valve losses, set point determination, and sizing of plates. In general, volume, rate, differential, and plate size are the form variables that are determined by field calculations.

This discussion will be based on the AGA Committee Report No. 3 as it is the most commonly used. It is also the most complicated due to the large number of corrections applied.

Four categories have been devised to aid in discussing field calculations: the AGA formula, graphs, computers, and the quick method.

AGA formula

In making field calculations, a good rule to follow is—the simpler, the better. Detailed calculations are difficult to perform under usual field conditions.

Another good rule to follow is cool and check. Always allow some time to elapse between calculating and checking. This can be from 5 min to 5 h. To have someone else check your work is even better. Following these rules will eliminate most errors.

The gas flow equation as stated in the AGA Committee Report No. 3 is:

$$Qh = Fb\ Fr\ Y\ Fm\ Fpb\ Ftb\ Fm\ Fa\ Fl\ Ftf\ Fg\ Fpv\sqrt{h_wpf}$$

Obviously, the most accurate method of volume calculation would use all of the above factors. Most field calculations will use some modified form of the AGA equation.

In order to present a logical discussion, these correction factors are broken down into three groups—major, minor, and intermediate. This grouping is based upon the effect each factor has on the volume.

The major factors are the basic orifice factor Fb and the specific gravity factor Fg. These two factors have to be included in all computations.

The intermediate factors are the pressure base correction Fpb, flowing temperature correction Ftf, and super-compressibility correction Fpv. The value of these factors is variable. Usually, they are determined by operating conditions at each specific location or are set by contract as in the case of pressure base.

Supercompressibility can become a large correction with transmission companies and producers where gas is measured at elevated pressures. For field calculation purposes, intermediate factors will be considered as either major or minor.

The Fr, Y, Fm, Ftb, Fa, and Fl corrections are minor factors for field calculation because of their minimal effect upon the final volume. Most of these factors are usually ignored.

Condensed formula

As previously stated, the most accurate method of field calculation is the AGA formula using all corrections. However, this is usually not possible because of insufficient time, advance warning, or operating information.

If a little work is done prior to the time of use, the formula can be condensed to three or four calculations. However, some operating information must be available. The static pressure, differential, run and plate size, specific gravity, and gas composition must either be known or assumed.

Assumptions should be based on knowledge of normal operating procedures and characteristics of similar M&R stations. As an example, the inert composition of gas is assumed to be 0% nitrogen and 0% carbon dioxide until an analysis is performed. This procedure permits volume calculations until more definite information is available.

By using known and assumed data, an hourly coefficient composed of Fb, Fr, Y, Fm, Ftb, Fpb, Fa, and Fl can be computed. The only corrections left to be made are Fg, Ftf, and Fpv. By giving some thought, time, and prior work, the formula can be reduced to:

$$Qh = C\ Fg\ Ftf\ Fg\sqrt{h_wpf}$$

Tables are available that further reduce the amount of calculation that has to be done. Combined specific gravity and temperature tables are available from different sources. Also available are extension values for $\sqrt{h_wpf}$ or the individual values of $\sqrt{h_w}$ and $\sqrt{pf}$. These tables eliminate what is probably the hardest part of a field calculation.

When following these guidelines, calculations become fairly simple. One note of caution: always make computations in a legible and logical order. This makes everything fairly simple to check, because many errors are caused by illogical layout of calculations as any other factor.

From this point on, all methods are an adaptation of the AGA Report No. 3 equation.

Flow graphs

Typically, the next most accurate method is the company-developed flow graphs.

Although the graphs necessarily have to be based on average data, they are tailored to each particular company's

Figure 1. Example of orifice meter flow graph used by Texas Eastern.

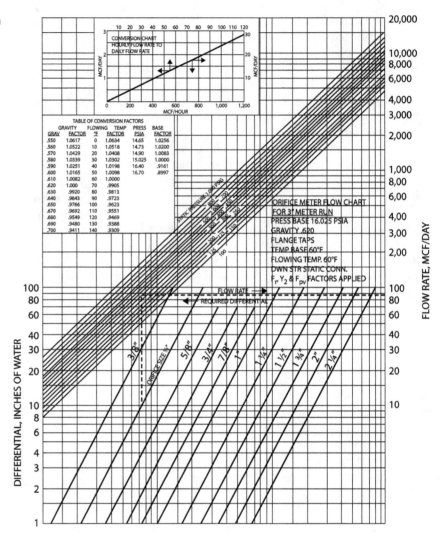

operating conditions. Thus, you get more accurate results for your particular requirements. Corrections such as gravity can be made to the final figure if desired, but usually are not.

The graph has the advantage of giving considerable information quickly. All plate size and differential combinations are shown at a glance. Daily and hourly flow rates are also given. This is the fastest and easiest of all methods to use, and accuracy is usually good enough to use under most conditions. No calculations are necessary under most conditions.

Most paper graphs are bulky, requiring a straight edge and space to use. Humidity will destroy them. However, replacements are usually available upon request. Oddly enough, this is probably the least used of all methods.

Most companies use the AGA equation to develop their graphs. The newer factors Fa and Fl are almost certainly not included in current data. Most factors are based on a particular company's average data.

Some factors are given a value of 1.000 and therefore ignored. Supercompressibility corrections are usually by average specific gravity and inert composition.

Computers

Flow computers and slide rules are probably the most widely used method of field calculation. This is due to their compactness and simplicity of operation. Accuracy is generally in the acceptable range for field calculations.

Most manufacturers of measurement equipment furnish some type of flow computer either free or at a nominal cost. Construction usually is of paper and plastic.

All of these flow computers are based on the AGA Report No. 3 flow equation. Most of the minor corrections are not used, with the omissions listed in the instructions. Since computers are designed for use throughout the industry, they must be kept standard. Consequently, they will not fit a particular set of conditions without making additional corrections.

Most flow computers incorporate Fb, Fg, Ftf, Ftb, and sometimes a generalized form of Fpv. The minor factors are usually ignored. The pressure base is usually set at 14.7 or 14.73, but some computers do incorporate a pressure base correction feature.

The biggest single drawback to computers is the fact that they must be general in scope and not slanted to one particular method of calculating volumes. Corrections are kept too general to fully correct for a specific set of conditions. There must be enough use of a computer for a person to maintain speed and accuracy of operation. If not, the instruction booklet will have to be studied each time the computer is used.

Flow computers are compact, easy to carry and store, and are very quick to use. As long as their limits are recognized, they can do an excellent job.

Quick method

Primarily, the quick method uses the major factors Fb and Fg only.

Inclusion of the intermediate factors is desirable if sufficient information is available. These intermediate corrections are made in a general way only. For example, if the pressure base is 15.025, subtract 2% of the volume. Algebraically, add 1% of the volume for each 10° variation from the temperature base. For each 100 psi of static pressure, add 8% of the volume.

This procedure will suffice for a lot of field calculation requirements. Keep in mind that this method is a rough calculation and is not intended to be used as a precise quantity.

Some mention must be made of the extension that probably causes more trouble than any other part of a volume calculation. In determining average differential and pressure, spot readings are the easiest to get and the most commonly used method. Sometimes predetermined values are used. Quite often, differential and pressure are averaged over several hours. When averaging, care should be exercised for differentials under 25 in, especially for differentials under 10 in. Due to the square root effect, errors are magnified in these ranges.

Extension books are available either in the form $\sqrt{h_w pf}$ or the individual $\sqrt{h_w}$ and $\sqrt{pf}$ tables. Although the combined tables are bulky, the individual tables can be limited to a few small sheets. Use of these tables will improve accuracy and speed.

Practical maintenance tips for positive displacement meters

Here are the most common malfunction causes and points to watch on the widely used meters

J. A. Wilson, Meter Shop Manager, Northern District, Skelly Oil Co., Pampa, Texas

Thorough understanding of a meter's function is necessary to properly repair any mechanism within it.

There is a definite reason for each meter part, and each part has a definite relation to the other. The alignment of any one part with another has been precisely established for each type and size of meter.

The functions of a positive displacement meter are similar to a four-cylinder engine. For one revolution of the tangent, each chamber must fill and exhaust once.

One revolution of the tangent is 360°. Therefore, to make a smooth running meter with uniform displacement, one of the chambers must start to exhaust every 90°. Since the valves control the time at which the chamber starts to fill and exhaust, they must be coordinated, one with the other, so that this 90° relationship is maintained.

Similarly, the diaphragms must also be connected to assure this relationship, at least in their extreme positions. Both valves and diaphragms must be connected to coordinate both 90° relationships.

The following are practical tips for PD meter repair and testing.

PDM repairs

The repairman cannot always carry parts for every type meter. Also, testers who only do an occasional repair job are not as efficient as men performing repairs daily. Therefore, most major repairs and all major overhauls are more efficiently handled in the meter shop.

Operating conditions in the field vary and are not always the best. Some of the operating problems in the field are gas saturated with water and hydrocarbons, iron oxide, sour gas, and lubrication.

Shop repair. Meter shops, no matter how well equipped, cannot perform satisfactory meter repair without qualified and dedicated personnel.

When an experienced man disassembles a meter for repair, he inspects each part as he tears it down, considering each as a possible cause of the malfunction. When the problem is found, if possible, it is corrected before proceeding.

Figure 1. Low pressure flow provers for PD meters can be used in the field or shop. The principle is essentially that of the orifice meter, except the orifice discs are calibrated for different flow rates instead of using various coefficients. Shown above is nomenclature for the Rockwell prover.

Some of the most common causes of malfunction and points to watch on the more widely used meters are:

- Exhaust port gaskets should be inspected to make sure the seal is good. Some meters have metal spacers under the seal and must be in place when reassembled.
- The clearance of the flag rods at the top of the meter should be checked. After a meter has been time checked and balanced, it will sometimes run rough. Look for a bright spot right over the top of the flag rods in the top cover. This is not a common problem, but is difficult to locate, and a few strokes with a file will take care of it.
- The reassembly of the tangent gear between the valve gears of meters, as well as other gear-driven valves, should be carefully centered. There is usually a small amount of side play as well as end play on the tangent shaft bracket. A little extra care here will contribute to a quiet operating meter.
- The body seal O-rings on the later models should be placed carefully in the slot when the top cover is put into place.

Flag arms. Because new materials are being used in meters and the new aluminum-case meters are designed differently, there have been changes in some repair procedures. Some meters use an aluminum, long flag arm with a brass bushing cast in the center, so the flag arm can be soldered to the flag rod.

Most repairmen use a wet sponge to rapidly cool a solder joint on flag rods when flag arms are attached, but because of the different expansion coefficients of brass and aluminum, the repairman must use extra precaution in cooling these solder connections.

Tests have proven that cooling the aluminum part of the flag arms too quickly will cause fractures in the aluminum surrounding the brass bushing. Most of these fractured flag arms will go undetected and later break down in operation. However, if the repairman cools the solder joints by easing a wet sponge slowly up the flag rod, the joint is cooled from the inside out and no damage is done.

Valves and seats. Careful grinding of the valves and valve seats is very important. Number 80 grit may be used to clean up bad valves and seats, finishing with number 120 grit. Most repairmen use the figure eight motion in grinding, letting the weight of the glass block control the pressure on the valve seat. Use a very light touch on the valves. Where it is practical, both valve seats should be ground together. Grinding paper that is self adhesive is certainly recommended over grinding paper that has to be cemented to a glass block.

There are thousands of ironcase meters that have been in service for a great number of years. The valve seats on these meters should be carefully checked to make sure they are still bonded to the meter body. Remove and reseal the seats if there is any doubt about the strength of the bond. Use an old

Crank — Adjustable Tangent

Partition, Table, Case and channels

Grommet-Type Seals

Valve

Flag-Rods — Duramic Diaphragms

Figure 2. Thousands of ironcase PD meters, such as the American 25B and 35B shown above, have been in service for many years. Valve seats should be carefully checked to make sure they are still bonded to the meter body.

grinding block to true up the bottom of the valve seats and buff the meter case clean to ensure a good bond.

Clean all surplus cement from the intersurfaces of the valve seat opening. Be careful not to use screws that are longer than standard in valve guides that screw into the valve seat. An overlength screw can jack the valve seat off the body and break the seal.

Watch the clearance on the new plastic parts when they are installed in old style meters. The raised reinforcing rib on some of these parts may pose clearance problems in many meters. Soldering acid will dissolve some plastic, so if you must use acid core solder, be careful to keep it off the plastic.

Balance and time check. Take plenty of time to balance and time the valves and diaphragms to ensure a smooth operating meter. Use care in setting the tangent in proper relation to the valves. Time spent here will be regained manyfold when the meter is calibrated.

Most shops have a utility-type vacuum sweeper. This is an excellent means of rotating a meter before the top cover is installed. This gives the repairman a good chance to watch the valve action and make a final check on timing and balance, gauge, and depth of the undergear driving arm in relation to the tangent. Make sure the tangent wrist is engaged, but do not touch the short flag arm.

Installing a long pointer on the index driver dog will help proof test the meter, since the circumference of the circle will be much longer than the test-hand on the index. It is a good practice to rotate the meter at least two rotations of the test-hand after the index is installed. This gives a final check to make certain the meter is operating smoothly. A vacuum cleaner can also be used to good advantage here.

Pan-type diaphragms can be rebuilt with a minimum amount of investment in jigs and other equipment. Many parts, such as crank brackets and flag arms that have pressed-in bushings, are easily repaired. Rebuilding parts can save up to 75% of the cost of a new item.

When it comes to repairing meters, "about right" is not good enough. The meter repairman who wants to make his company money will strive to make all checks and adjustments as accurately as possible.

PDM testing

Positive displacement meters may be tested by any of the following methods: (1) bell type meter prover, (2) low pressure flow prover, (3) critical flow prover, and (4) test meter. All of these methods, when correctly used under proper conditions, will give similar results.

Bell prover. Meter shop personnel are more familiar with the bell prover than with the other three types.

When a bell of sufficient size is available, large-capacity meters can be rapidly calibrated in the shop. The use of a slope gauge will certainly reveal if the repairman has assembled a well-balanced, timed meter with a minimum of internal friction.

Bell provers should be in a temperature-controlled area, and all meters should be at room temperature before testing. Normally both a 20% rate and capacity rate are run when proof testing flow rate on the meter. To test small domestic meters, 2 in of water pressure on the bell is used instead of the 1 1/2 in used.

Low pressure flow prover. The low pressure flow prover is used both in the field and in the shop. It is very compact and portable, consisting of two pieces of aluminum pipe with straightening vanes in the upstream sections.

The principle is essentially that of the orifice meter, except the orifice discs are calibrated for different flow rates instead of using various coefficients. Many shops use the low pressure flow prover to test large capacity meters when a bell prover of sufficient size is not available. One or two blowers may be used to supply the necessary volume of air, or low pressure gas may be used where safe.

Critical flow prover. Critical flow is a condition occurring in the gas flow when the downstream pressure is approximately one half the upstream pressure.

The critical flow prover consists of a piece of 2-in, extra heavy pipe approximately 18-in long. One end is threaded to screw into a test connection placed as close to the meter outlet as possible. The other end has a special orifice plate holder. The critical flow prover can be used to test positive displacement meters operating on pressure in excess of 15 psi by passing gas or air through the meter and the prover and discharging it into the air.

The orifice plates of a critical flow prover are stamped on the downstream side with the time (in seconds) required for one cubic ft of air to pass through the orifice at critical flow conditions. Plates used in critical flow provers have rounded edges on the upstream side, whereas orifice plates in a low pressure flow prover are sharp on the upstream edge.

It is vitally important that both the critical flow and low pressure orifice plate be handled with care to prevent damage, which would change the time stamped on the plate.

The stopwatch, which is important in timing prover tests, should be serviced and checked at regular intervals. Such as many pieces of mechanical equipment, the stopwatch will operate for a long time before it completely gives out, and there is a very good possibility that it will be operating inaccurately during this time. Remember that tests taken with a defective timer are inaccurate.

Test meter method. Test meters have become popular in recent years with larger companies and contract testers. Continued development of rotary displacement has made the test meter a practical method for field use.

Although the shop tester normally runs only two rates— check rate and capacity rate—on a bell prover, most testers run three rates on the critical flow, low pressure, and test meter methods. These rates are usually about 20, 50, and 100%.

Testing frequency is a much-discussed subject. Sales through a large-capacity meter represent a great deal of money; thus, it is important to regularly check the meter's accuracy. Some of the larger meters should be tested at 90-day intervals. Others of intermediate size could perhaps operate for two, five, or seven years without being tested and at the end of the scheduled time, simply be changed out.

Many factors affect a testing and change-out schedule. The meter's mileage, its type of service, the type and amount of gas the meter must measure, and the length of time the meter is in service, are important factors in setting up a testing and change-out schedule. Only after all of these factors have been considered can an adequate testing and change-out program be established.

Sizing headers for meter stations

Calculate the velocity of gas in meter headers for proper sizing to achieve good measurement

Glerm E. Peck, Senior Engineer, Pipe Line Division, Bechtel Corp., San Francisco

Meter and regulator station design for natural gas transmission systems continues to grow more complex and demand more of the design engineer. Basic elements of today's station remain about the same, but higher pressures and ever-increasing capacities have replaced the positive displacement meter with multi-tube orifice meters and automatic chart changers. Buildings are often prefabricated structures specifically designed to house telemetering transmitters and supervisory and communications equipment in addition to basic recording instruments. The multi-run setting may have a computer giving the dispatcher instantaneous rate and totalized flow for the station. Electrical equipment may cost several times that of the entire station of a few years ago.

There is a sparsity of published criteria for sizing of orifice meter headers in the multirun station. A discussion of factors that affect the sizing of the header and arrangement of component parts for satisfactory operating conditions should be in order.

We shall assume that the orifice tube and its component parts are designed in accordance with AGA Gas Measurement Report No. 3, that piping meets all the requirements of USASI B31.8 for pressure, materials, and class of construction, that all valves and fittings meet appropriate codes or standards, and that the differential pressure across the orifice plate will not exceed 100 in of water.

The design goals may be listed as follows:

1. Commercially available equipment will be used that is specified to the closest reproducible economic tolerances.
2. The arrangement of the equipment shall conform to accepted practice for good measurement and allow fullest use of the available meter tubes.
3. Fittings and connections to the header shall minimize the flow turbulence in so far as is practical.

Meeting the above goals allows considerable latitude in the choice of components according to the judgment and experience of the designer without contradicting code, contract, or economic requirements.

In sizing the header, a number of factors must be considered, namely:

- The number of orifice tubes that will eventually be connected to the header, or the ultimate flow capacity.
- The setting layout, or arrangement, which governs the resistance flow paths the gas must take from entering the inlet header and dividing through the meter tubes before recombining in the outlet header and exiting.
- The type of connections which the tubes make with the header.

Station layout

There are as many opinions of what constitutes good meter station layout as there are engineers working with gas measurement. Four of the more common arrangements are illustrated in Figures 1–4 and will be discussed relative to the flow-resistance paths offered.

Header connections

Not many years ago, the header was a common piece of pipe with "orange-peel" ends and notched-miter nipple connections. The next step was a shaped nipple reinforced by a welding saddle or pad. Higher pressures brought forged steel tees into use. On larger-sized tees, the spacing between runs or tubes depended on the measurement for the tee and required a number of girth welds. With the introduction of the extruded nozzle header, the limitation of tees was eliminated for a generally more economical header.

The radius of curvature or flaring of the nozzle as extruded has superior flow characteristics over the reduced branch of a forged tee, and both are vast improvements over the square cut opening for a shaped nipple. Some designers have gone a step further and reduced the 90° header nozzle by the use of a 45° or 60° lateral nipple to the axis of the header. This reduces the turbulence entering or leaving the header and could reduce the diameter of the header. This method can be much more costly for fabrication, is suspect for structural integrity at higher pressures and extreme ambient conditions, and goes beyond the configurations used in establishing the proportions, beta ratios, and coefficients for the orifice meter.

Runs per header

The number of runs, or meter tubes, for a given header is limited by the velocity criterion selected, the operating pressure, the size of fittings obtainable, and the relationship of the peak load to the normal load or capacity for which the meter setting is designed. The operating pressure has an important bearing on the number of orifice tubes used with

Figure 1. The flow-resistant paths are identical for this layout and each of the two runs will furnish its full capacity with near identical differential readings on the meters. However, when additional capacity is required, an additional run will have a higher or longer resistance-flow path and will furnish a lower differential reading and capacity as compared with the first two runs.

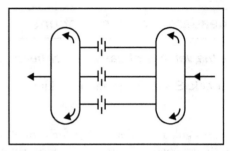

Figure 2. In this arrangement, the middle tube has the lower resistance-flow path compared with the outer tubes and will affect the meter reading accordingly. The difference between meter readings will be more noticeable if the headers are undersized.

Figure 3. The "donut" header has been frequently used when large headers could be utilized. At higher pressures and capacities, this layout places a high economic burden on the station and requires more space, which can be a problem when a limited site is available.

Figure 4. This layout has been a favorite for many since the advent of the extruded or drawn nozzle header and because the resistance-flow paths are as nearly identical as is possible using commercially available equipment. It is limited to three or 4 m tubes before the header begins to be "out-sized."

a given header, because the capacity of the tube is proportional to the square root of the pressure while the capacity (velocity) of the header is proportional to the pressure in atmospheres. At lower pressures the number of runs might be limited to three, while at higher pressures the limit would be four runs. Size of fittings is generally less limiting than other factors unless the meter tube is greater than 12 in in diameter. For the very large tube, the extruded nozzle header is required, since reducing branch tees are not generally available. The relationship of the peak load to the sizing of the header is discussed under the velocity sizing criterion.

Velocity sizing

The following velocity criterion is suggested for sizing of headers for the sales orifice meter setting (expressed at ft per minute):

Target	Normal Limit	Maximum Limit
1500	2000	2500

The relationship of the peak hour load to the average hour for the peak day designed for the station determines the range of velocity used in sizing the header. For example, the peak hour which is 10% of the peak day and one that is only 1% higher than the average hour for the peak day obviously demands that different limits be used for economic justification. The 10% peak hour suggests that for higher fluctuation in demand, a higher velocity be allowed in the header for the short time involved.

Keeping the velocity as low as possible within this criterion will provide the basis for good measurement and minimize the difference between adjacent meter readings. The way that gas enters and leaves the header connections to the meter tube is important in considering the velocity criterion that is chosen.

Sizing procedure. The normal procedure for sizing the station follows:

1. Take the peak day forecast or contract load and calculate the appropriate peak hour figure—provided the peak hour

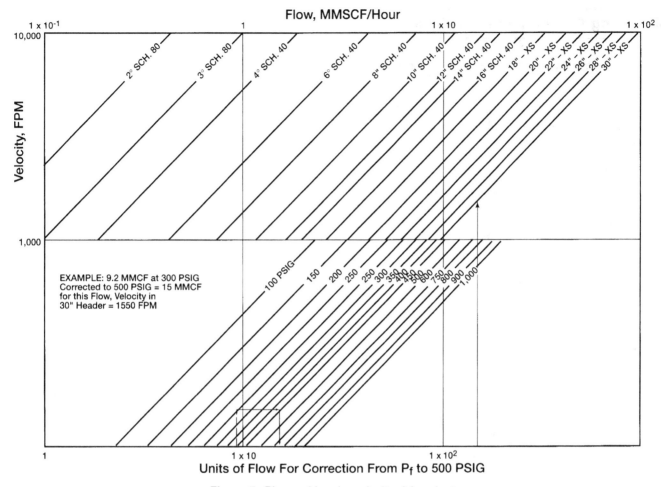

Figure 5. Pipe and header velocity sizing chart.

figure has not been fixed as a part of the forecast or contract. (Suggested period: 10 years, minimum).

2. Divide the peak hour by the number of parallel tubes desired.

3. Use a flow rule or chart to size the orifice tube. Check the initial minimum hour against the minimum beta ratio and differential desirable, since this may affect the size and number of tubes.

4. Use a chart, such as Figure 5, to calculate the velocity in the header for the proper criterion. Figure 5 has been found useful since the capacity can be traced up to the desirable velocity and the nearest appropriate size of header chosen.

Figure 5 was prepared for 500 psig or 514.65 psia at a temperature of 60 °F in order that the sizing of pipe or headers be made against velocity and capacity in Mscf/h on the upper part of the chart. The lower part of the chart is used to correct the flow at other pressures to the proportionate flow at 500 psig. Since the sizing procedure is an approximate one, it is felt that corrections for temperatures other than the standard condition and for compressibility are not necessary. Use of Figure 5 or a chart prepared to your own physical conditions will quickly and accurately allow sizing of headers for the orifice meter setting within the present state of the art.

Measuring flow of high-viscosity liquids

By Peter B. Jakubenas, Manager of Installed Product Services FMC Technologies Measurement Solutions, Houston, Texas

Measuring custody transfer of high-viscosity fluids has always presented challenges, and as the world increasingly relies on heavy crudes with various amounts of wax, bitumen, and sand, the challenge increases. Crude oils typically flow at rates up to 50,000 barrels per hour (BPH) at export terminals, up to 1500 BPH at ship and truck loading facilities and to 7000 BPH in pipelines.

In addition to heavy crude, other challenging high-viscosity products include heavy fuel oils, In addition to heavy crude, other challenging high-viscosity products include heavy fuel oil, bunker fuels, lubricating oils, grease components, and asphalts. In most cases, flow requires elevated temperatures—up to 400 °F (205 °C)—and flow meters must be reliable in this environment.

Viscosity, specific gravity and temperature

A viscous hydrocarbon is considered to be any liquid hydrocarbon that requires special treatment or equipment in order to be handled or stored.

Viscosity of water and liquid petroleum products

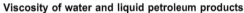

Viscosity (resistance to flow) is related to specific gravity (density), although the relationship between the two is not direct, as the table below shows.

In addition, each hydrocarbon has a characteristic viscosity versus temperature curve making it possible to infer viscosity for known stable hydrocarbon mixtures using a curve developed in the laboratory. In cases where mixtures of hydrocarbons vary over time, viscosity can be measured using one of the many on-line viscometers available.

The figure in this article shows viscosity—temperature curves for the same fluids. In all cases, viscosity decreases with increasing temperatures, but the decline may be more or less dramatic.

Viscosity of water and liquid petroleum products

1. Ethane (C_2H_6)
2. Propane (C_3H_8)
3. Butane (C_4H_{10})
4. Natural Gasoline
5. Gasoline
6. Water
7. Kerosene
8. Distillate
9. 48 Deg. API Crude
10. 40 Deg. API Crude
11. 40 Deg. API Crude
12. 32.6 Deg. API Crude
13. Salt Creek Crude
14. Fuel 3 (Max.)
15. Fuel 5 (Min.)
16. SAE 10 Lube (100 V. I.)
17. SAE 30 Lube (100 V. I.)
18. Fuel 5 (Max.) or Fuel 6 (Min.)
19. SAE 70 Lube (100 V. I.)
20. Bunker C Fuel (Max.) and M. C. Residum
21. Asphalt

Measurement alternatives

Measuring flow of fluids with viscosities above 100 cP requires special consideration by those charged with designing, operating, and maintaining the equipment. The sliding vane positive displacement has been the first choice

Typical viscosity and specific gravity of petroleum products*

Typical Viscosity in cP** @ degrees F (degrees C)							S.G. @ degrees F (degrees C)	
Product	30 (−1)	60 (15)	100(38)	150 (66)	300(150)	400 (205)	60 (15)	50 (66)
Ethane LPG	0.07	0.05	0.03				38	0
Propane LPG	0.14	0.12	0.09	0.07			0.51	0.41
Butane LPG	0.20	0.18	0.15	0.13			0.56	0.49
Gasoline	0.83	0.63	0.49	0.38			0.75	0.70
Water	1.8	1.2	0.7	0.4	0.2	0.1	1.00	0.95
Kerosene	3.5	2.2	1.7	0.9	0.4	0.2	0.82	0.77
Jet fuel	3.5	2.2	1.7	0.9	0.4	0.2	0.82	0.77
48 API crude	3.5	2.7	1.7	1.1			0.79	0.74
40 API crude	10	7	4	2			0.81	0.77
35.6 API crude	25	16	6	3			0.85	0.80
32.6 API crude	42	21	9	5			0.86	0.81
Fuel 3 (max.)		10	5	3	1	0.5	0.90	0.85
Fuel 5 (min.)		16	7	4	1	0.6	0.97	0.92
Fuel 6 (min.)		820	150	43	5	2	0.99	0.94
SAE 10 Lube		68	29	11	2	1	0.88	0.83
SAE 30 Lube		450	105	31	4	2	0.90	0.85
SAE 70 Lube			450	95	8	3	0.92	0.87
Bunker C			1500	290	16	5	1.01	0.96
Asphalt				>3000	80	19	0	0

*Data made available from Crane Co.
**Centipoise (cP) = Specific Gravity × Centistokes (cSt).

for measurement since the 1930s; however, in the last 10−15 years, other technologies have emerged that can offer improved cost/benefit ratios, depending on the viscosity and temperature of the fluid and the required accuracy of the meter. Most custody transfer flow meters fall into one of four categories:

- Positive displacement (PD) meters
- Helical turbine meters
- Coriolis meters
- Liquid ultrasonic meters

The positive displacement meter is the only direct measuring alternative, in that every molecule of fluid passes through the meter. The other technologies infer total flow by measuring velocity within a separate flow conduit, where changes in the physical properties of the fluid or of the conduit itself increase the uncertainty of the measurement.

This disadvantage can be overcome by proving the meters on a regular basis and monitoring fluid characteristics to be sure the meter readout has not gone beyond acceptable tolerances. Inferential meters can also be influenced by upstream (and to a lesser extent downstream) flow conditions, such as elbows, strainer debris, or partially open valves.

In defining appropriate flow meters for given conditions, the total turndown range is the range of flow rates over which a meter will perform within a specified linearity, usually ±0.15%.

There are two components of the total turndown range: the viscosity turndown range (which depends largely on the type of meter). Each of the four technologies noted above has a characteristic turndown range, defined as the ratio of the maximum flow divided by the minimum flow rate. The higher the turndown range, the greater the range of flow rates over which the meter will be linear.

The total turndown range is equal to the flow turndown range multiplied by the viscosity turndown range. If meters are operated within a narrow band of flow rate and viscosity and proved at flow rates and viscosities close to these, linearity becomes less of a factor, and meter performance becomes more repeatable. Meter linearity is less of an issue with large parcels such as pipeline and ship loading/unloading and applications, but becomes more important in truck loading, rail-car loading, bunkering, and other small-parcel applications.

Most of the very high-viscosity refined products are handled at temperatures above ambient to facilitate pumping, transportation, and metering, so it is important to specify measuring component materials that can withstand the elevated temperatures provided by steam, hot oil, or electric heat-traced and recirculation lines. In cases where it is uneconomical to heat large quantities of high-viscosity crudes, they are treated with diluents to enable transportation and metering at near-ambient temperatures.

Meter selection and sizing

This section lists the advantages and disadvantages of the 4 m types used for high-viscosity fluids and shows graphs of suitability depending on viscosity and flow rate.

Positive displacement meters

Advantages

- High accuracy over a wide range of viscosities and flow rates up to 2000 cSt with proper clearances
- Extremely good repeatability on high-viscosity fluids, very low slippage, long life if there is little or no abrasive material in the fluid
- Low pressure drop
- Functions without external power
- Special construction available for high viscosities and temperatures
- Can register near zero flow rate
- Flow conditioning not required
- Measures directly, not an inferential device

Disadvantages

- More moving parts leads to increased maintenance compared to other meters
- May become damaged by flow surges and gas slugs
- Chance of corrosion and erosion from chemicals and abrasive materials
- Derated flow rate capacity for high viscosities and temperatures
- Relatively high cost for large meter sizes, since all fluid must pass through the meter.

Helical turbine meters

Advantages

- Higher turndown range on high-viscosity crudes than conventional turbine meters
- Very good repeatability
- Reduced susceptibility to fouling, abrasives, and deposits
- Less sensitive to viscosity changes
- Lower pressure drop than conventional turbine meters
- Available in large sizes, providing good value for high flow rates

Disadvantages

- Requires flow conditioning
- Back pressure required
- Requires pulse interpolation due to low-resolution pulses
- An inferential device

Coriolis meters

Advantages

- Low maintenance, minimally affected y abrasives and corrosives
- Not susceptible to damage by gas slugging
- Registers near zero flow rate
- Minimally affected by viscosity changes
- Flow conditioning not normally required

Disadvantages

- Sensitivity to installation conditions, including shock, vibration, pulsations, and effects of adjacent parallel runs
- Deposits can affect accuracy
- Difficult to prove due to time lag of the pulse output

- Requires periodic 43-zeroing
- Needs back pressure control
- High pressure drop that drastically increases with viscosity
- An inferential device

Liquid ultrasonic meters

Advantages

- High accuracy
- Wide dynamic flow range
- Negligible pressure drop
- Nonintrusive and no moving parts, making this the least influenced by abrasive materials
- Diagnostic capabilities
- No need for upstream strainer
- Self diagnostic capabilities

Disadvantages

- Flow conditioning recommended
- Susceptible to fouling or deposits
- Sensitive to installed conditions
- Sampling and microprocessor-based output contributes to difficulty in proving
- High cost for small sizes
- Maximum size limited by proving capability
- Back pressure required
- An inferential device

Operational issues

Air removal. All of the meter types discussed require the removal on entrained air and/or vapor which may be difficult with most viscous liquids. As viscosity increases, so does the time required to separate fine bubbles of air or vapor from the liquid. A vapor removal system may be required and a number of such systems are available.

Proving the meters. For custody transfer applications, all of the inferential meter types described above must be periodically proved at conditions close to operating temperature, viscosity, gravity, flow rate, and pressure. For high flow rates, in situ pipe provers are recommended. For lower flow rates, portable proving equipment or master meters may be used.

Although all proving methods may be used with high-viscosity fluids, pipe provers have an inherent advantage over tank provers because the displacer seal against the pipe wall ensures that all fluid is measured. When a tank prover is used, the operator should be alert to high-viscosity fluid clinging to the tank wall because this will affect the measurement. The amount of clingage will vary with the temperature and it is difficult to control temperature within a tank.

Pipe provers used in above ambient temperatures should be insulated and possibly even steam or electrically heat traced. Make sure that the maximum operating temperatures of the elastomer prover displacer sphere and the internal prover coatings are sufficient to handle fluid temperatures.

Readout and control equipment. Since most high viscosity liquids are handled at line temperatures far above the 60 °F reference temperature, accurate electronic temperature compensation of integrated volumes is essential, and we recommend electronic flow computers. For large systems where several meters operate in parallel, we recommend supervisory computers capable of automatic proving plus temperature monitoring and control.

For small stand-alone PD meters deployed where no power is available, mechanical wheel and disk temperature and gravity compensators (ATG and ATC) are still available.

Special operational and maintenance considerations. For viscous hydrocarbons heated above ambient, it is essential carefully to monitor and control temperature because viscosity drastically increases with a drop in temperature. If the temperature of the system decreases enough that the pumps cannot sustain flow, pump damage can result. Removing solidified product from the pipelines and metering equipment can be very difficult and expensive.

If systems are used intermittently, it may be more practical to evacuate or flush them with a low-viscosity solvent between uses. Upon restarting, slowly fill the system and completely evacuate air or vapor before resuming flow, since the meters and other components can be damaged by gaseous slugs.

If the temperature control system fails and temperatures rise too high, serious damage to valves, meters, and provers can occur. In addition, overheating the fluid can cause coking or chemical changes—and some overheated petroleum liquids may ignite on exposure to air.

Hydraulic considerations are also important since the pressure drop in the measuring system will dramatically increase with an increase in fluid viscosity. Pump sizing assumptions become very important. For example, consider a crude oil with 12 API gravity and viscosity of 300 cP @ 180 °F and 6500 cP @ 60 °F. the maximum flow rate desired is 6000 BPH, but will this really be achieved when the product is heated? This needs to be confirmed. If this flow rate can be achieved, what is the maximum flow rate that is hydraulically possible when the viscosity falls to 6500 cP?

Conclusions

The figure below shows a composite of the four technologies. While there are many choices of flow meters for high-viscosity liquids, no single metering technology is best for all applications.

For abrasive-free liquids, positive displacement meters can handle the widest range of flow rates and viscosities and will have a very long life. They have a track record of infield service of over 70 years, and some meters are still in operation after 30 years on heavy, waxy crudes. For applications where future liquid viscosities and flow rates are uncertain, positive displacement meters are the most flexible.

Helical turbine meters offer the lowest cost alternative for high flow rates on high-viscosity, abrasive-free fluids.

Coriolis meters perform well within their limited flow and viscosity range although they have high pressure drops.

Ultrasonic meters only become economically attractive for high flow rates and are the best at accommodating abrasive fluids.

Selecting the proper ancillary equipment is essential to meter performance and equipment operation and maintenance of high-viscosity flow meters may require special techniques to each application.

References

1. API Manual of Petroleum Measurement Standards Chapter 1 "Vocabulary" provides a comprehensive list of definitions for many of the terms used in this article.
2. API Manual of Petroleum Measurement Standards Chapter 6, "Metering Viscous Hydrocarbons" provides a comprehensive guide for the design, installation, operation, and proving of meters and their auxiliary equipment used in metering viscous hydrocarbons.

Reprinted with permission *Pipeline & Gas Journal* July 2007 issue.

Matching the flow meter to the application

Selection Strategies for Systems Designers

By **Greg Livelli**

Two significant specifications for a flow meter are accuracy and turndown (or rangeability). The accuracy of all flow-meters to some degree depends on the circumstances of their application. These effects include pressure, temperature, fluid dynamic influences, and external influences. Pressure and temperature cause dimensional and physical changes. Fluid flow and pipe roughness can shift meter performance. Mass meters and displacement meters are the least affected, while turbine, vortex, and ultrasonic meters are the most affected. The upshot is that a traceable and complete cali-bration in a lab is no guarantee that the same performance and accuracies will be obtained or maintained in the field.

A flow meter's turndown is the ratio of the maximum to the minimum flow rate capable of being measured at a specified accuracy, providing a measure of rangeability. High values of turndown can be irrelevant because typical processing flow velocities often fall within relatively narrow ranges. For example, most practical liquid pipeline flow rates in the processing industries range from 0.5 to 12 ft/s, a turndown ration of 24-to-1. Lower rates are difficult to accurately measure and higher rates result in high pressure drops, pumping energy costs, and erosion (if solids are present). In the case of pipelines carrying gases in the processing indus-tries, the practical flow velocities range from 15 to 200 ft/s—a turndown ratio of about 13-to-1. Many actual applications have flow rate ranges well within these extremes.

Instrument engineers commonly apply volumetric flow-meters to measure liquids. When the process fluid is a gas or steam, measuring mass flow becomes especially important.

Figure 2. The Venturi DP element minimizes pressure loss.

Because these fluids are compressible, their density changes with pressure and temperature. Volumetric flow measure-ment is often meaningless in these cases. However, volu-metric flow meters work if the application also includes measurement of temperatures and pressures at the meter or these values are known quantities. Simple calculations based on the three variables can infer mass flow rates.

Differential-pressure flow meters

Flow meters based on differential pressure insert an element within the piped flow. The unrecoverable pressure loss across the element—typically an orifice plate, Venturi, nozzle, pitot tube, or wedge—is a measure of the volumetric flow rate.

Flow meters based on differential pressure represent a popular choice in the processing industries, constituting nearly 30% of the installations. They have good application flexibility since they can measure liquid, gas, and steam flows and are suitable for extreme temperatures and pressures with moderate pressure losses. Accuracy ranges from 1 to 5%. Compensation techniques can improve accuracy to 0.5–1.5%.

On the other hand, DP flow meters are expensive to install and have limited rangeability (turndown) compared to other types. They require a separate transmitter to compute the volumetric flow rate and develop a standard signal. Changes in density, pressure, and viscosity can significantly affect

Figure 1. The orifice plate is a popular DP element.

Figure 3. The nozzle has Venturi-like properties.

Figure 5. The wedge DP element handles tough, dirty fluids.

accuracy of DP flow meters. And while they have no moving parts, maintenance can be intensive.

Orifice plate. Instrument engineers can employ a variety of elements to create the differential pressure, with the orifice plate being the most common. Orifice plates are inexpensive and available in a variety of materials. The turndown ratio, however, is less than 5-to-1, and accuracy is moderate— 2–4% of full scale. Maintenance of good accuracy requires a sharp edge to the upstream side. This edge will wear and degrade over time. Pressure loss for orifice plates is high relative to other DP elements.

Venturi. This element finds use primarily in water and wastewater applications. It handles dirty fluids and does not require upstream flow profiling. Pressure loss is minimal, making it a good choice when little pressure head is available.

Rangeability, while better than orifice plates, is less than 6-to-1, with accuracy of ±1–2% of full scale. Viscosity effects on accuracy are high. Flow must be turbulent (Reynolds numbers > 10,000). The Venturi has limited acceptance within the processing industries.

Nozzle. Characteristics of nozzle elements mimic those of the Venturi. Nozzles come in three types: ISA 1932 nozzle (common outside of the United States); the long radius nozzle; and the Venturi nozzle.

Pitot tube. This element is another low-cost DP element used to measure fluid flow, especially air flow in ventilation and HVAC systems. Pitots commonly measure the speed of airplanes. This element works by converting the kinetic energy of the flow velocity into potential energy (pressure). Engineers can easily insert the pitot tube into existing piping,

Figure 4. The pitot tube converts fluid kinetic energy into pressure.

Figure 6. Electromagnetic flow meters require conductive fluids.

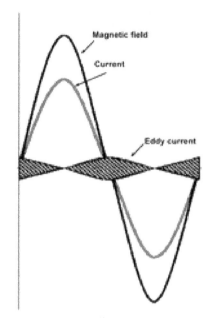

Figure 7. AC magmeters have good signal-to-noise ratios, but require frequent calibrations to eliminate zero drift.

minimizing installation costs. It cannot handle dirty flow and has limited rangeability.

Such as an insertion meter, the pitot tube makes a measurement at a point within the pipeline or ductwork. An exception is the annubar pitot tube, which contains multiple orifices. This element sees the dynamic pressure across the velocity profile, providing an averaging effect.

Wedge. This element consists of a V-shaped restriction molded into the meter body. In profile, the wedge looks similar to a segmental orifice plate to the incoming fluid, but it is more expensive than an orifice plate. This basic meter has been on the market for over 40 years, demonstrating its ability to handle tough, dirty fluids. The slanted faces of the wedge provide self-scouring action and minimize damage from impact with secondary phases. Rangeability of 8-to-1 is relatively high for a DP element and accuracies are possible to ±0.5 percent of full scale.

The wedge flow meter, however, is not approved for flow measurement by the American Gas Association (*www.aga. org*) or the American Petroleum Institute (*www.api.org*). Despite this, it is a popular choice in oil and gas applications, especially in production fields. For difficult fluids, it can be equipped with a pair of remote seals that effectively isolate the metered fluid from the DP transmitter without affecting accuracy.

Electromagnetic flow meters

Electromagnetic flow meters (magmeters) are a popular choice among instrument engineers, making up about 20% of flow meter installations. Faraday's law says that a conductor moving through a magnetic field produces an electric signal. In this case, the fluid is the conductor and electromagnetic coils surrounding the meter body generate the magnetic field.

Magmeters have no moving parts. They offer wide rangeability and an unobstructed flow path. Magmeters are ideal

Figure 8. Coriolis mass flow meters are ideal, but relatively costly.

Figure 9. Thermal mass flow meters excel at measuring low flow rates.

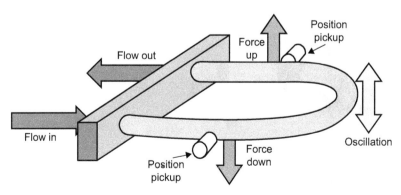

Figure 10. Vortex meters have no moving parts, but may be affected by vibration.

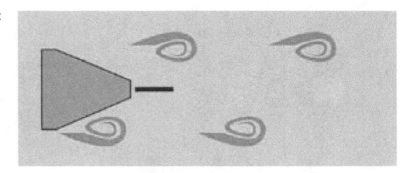

for slurries, and perform well with corrosive and erosive fluids. Engineers can pick a nonconductive liner material for the flow tube—Teflon, Tefzel, rubbers, ceramic, etc.—to suit the fluid measured. Magmeters require minimum lengths of straight pipe to condition the flow profile. On the downside, the fluid must be conductive, which rules out most petroleum-based flows. Magmeters also have physical pressure and temperature limits.

Magmeters have rangeability up to 1000-to-1, depending on the maximum tolerable measurement error. For flow velocities of 0.5–50 ft/s, accuracies are usually stated as a percent of rate, for lower velocities, accuracies are stated as a percent of span. Typical velocities measured range from three to 15 ft/s for water and clean chemicals, three to six ft/s for abrasive fluids, and six to 12 ft/s for coatings and liquids with entrained air.

The voltage developed across the electrodes is in the millivolt range and should be carried via shielded cable to a nearby converter. The converter near the flow tube boosts the signal to a standard (four to 20 mA) or frequency output (zero to 10,000 Hz) for transmission to a display or controller.

The coils producing the magnetic field may be excited by AC, DC, or pulsed DC. DC excitation accounts for most magmeter installations today. With AC excitation, line voltage powers the magnetic coils, and the resulting signal has a sine wave shape of line frequency whose amplitude is a linear function of flow velocity. AC designs produce a high coil current (typically 3.2 amp) and a good signal-to-noise ratio. They are relatively insensitive to media noise. However, they consume much power and can be plagued with zero drift caused by eddy current noise. To compensate, operators stop the flow and set the transmitter output to zero. Frequent operations to eliminate eddy current noise are sometimes necessary to maintain accuracy, which is about ±1 percent of rate.

With DC excitation designs, a lower frequency DC pulse excites the magnetic coils. The converter reads both the flow and noise signals during a pulse. In between pulses, however, it sees only noise, permitting noise cancellation after each cycle. Therefore, zero drift is not a problem for pulsed-DC designs. These magmeters offer 0.15–0.5% accuracy and low power consumption. The low coil current (0.1–0.5 amp) comes with smaller signal and low signal-to-noise ratio. Low frequency designs have longer response times.

A newer DC design develops a stronger flow signal by increasing the DC coil excitation current (0.7 to five amp).

Figure 11. Turbine meters are well-accepted in the processing industries.

Figure 12. Doppler ultrasonic flowmeters work best on slurries and dirty liquids.

This design improves the signal-to-noise ratio, but increases power consumption and cost. Flow tubes must generally be less than 20 in in diameter. Another new design employs dual excitation, pulsing the coils at about seven Hz for zero stability and also at 70 Hz for a stronger signal and improved response times (0.1 s). But, this design raises costs and applies only to flow tubes smaller than 16 in in diameter. Eddy current instability increases proportionally as sensor size increases.

Mass flow meters

Mass flow meters directly measure mass flow rate, as compared to other flow meter types, which measure flow rate based on volumetric principles. Two primary types of mass flow meters, Coriolis and thermal mass flow meters, account for about 18% of installations.

Coriolis flow meters represent the ideal flow meter. They can be applied to virtually any fluid—liquids, gasses, and slurries. They are immune to changes in viscosity, pressure, temperature, or density of the piped fluid and are unaffected by flow profile disturbances. They offer high accuracy (±0.1%) and rangeability (100-to-1), and provide an additional measurement of fluid density. However, their purchase and installation costs are relatively high, as well as their sensitivity to vibration. In addition, they come in a limited selection of sizes, rarely over two inches.

In these meters, the fluid measured runs through one or two vibrating tubes. Tube designs take on a multitude of shapes, depending on the manufacturer. Coriolis forces resulting from the vibration and flowing fluid act on the tubes, causing them to twist or deform. Sensors pick up the change in tube shapes. Sensor signals may be directly related to mass flow rate.

The thickness of the tubing walls varies considerably from design to design. But, even the sturdiest tubing will be thinner than the process piping. Long, bent tubes twist more easily than do short, straight tubes, and so they will generate stronger signals under the same conditions. Curved-tube designs provide wider rangeability (up to 200-to-1), while straight-tube meters are limited to 30-to-1 to 50-to-1, with lower accuracy. Straight-tube meters are more immune to pipeline stresses and vibration.

Thermal mass flow meters add heat to the flow stream to make measurements. One or more temperature sensors

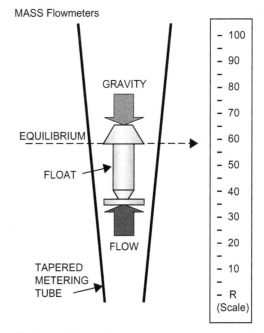

MASS Flowmeters

GRAVITY

EQUILIBRIUM

FLOAT

FLOW

TAPERED
METERING
TUBE

- 100
- 90
- 80
- 70
- 60
- 50
- 40
- 30
- 20
- 10
- R
(Scale)

Figure 13. Transit-time ultrasonic flow meters work best on clean fluids.

Figure 14. Positive-displacement flow meters are unaffected by poor flow profiles.

measure how quickly this heat dissipates. One technique maintains a heated sensor at a constant temperature, measuring the current required. Another way is to measure the temperature difference between a sensor in the flow stream and a heated sensor outside the stream. Higher flow rates result in increased cooling. The cooling effects are a function of mass flow.

These meters are compact, and offer wide rangeability (50-to-1) and good repeatability. They have a fast response time, especially for gasses, and excel at measuring low mass flow rates. They assume a constant specific heat and thermal conductivity of the fluid. Changes in density cause calibration shift, and coating of the sensor can cause drift. Additionally, they measure a point within the pipe diameter, so the flow profile must be known. Accuracy levels typically range from 1 to 3%. However, manufacturers expect to improve their accuracy, which will widen their application.

Vortex vs. Swirl Meters	
Vortex	Swirl
Lower cost	Minimal inlet/outlet straight piping
All cast construction	Better low-flow capability
Water and flanged meter bodies	Flanged bodies only
Higher maximum flow capability	Better rangeability on more viscous liquids
Measures liquids (viscosity < 8 cp) gases and steam	Measures liquids (viscosity <30 cp) gases and stream
Typical sizes ½ to 12 inches	Sizes ½ to 16 inches flanged
316L SS or Hastelloy	316L SS
Accuracy +/– 0.75% of rate for liquids and +/– 1.0% of rate for gases/stream	Higher accuracy +/–0.50% of rate for liquids, gases and stream

Vortex/swirl flow meters

Place a bluff body within a fluid pipeline and the stream will alternately produce whirling vortices on either side of the body. The spacing between the vortices remains constant, regardless of the fluid velocity. Within a pipeline, the vortex effect attenuates a few pipe diameters downstream from the bluff body. Typically, piezoelectric or capacitance sensors count the pressure oscillations caused by the vortices. The vortex-shedding frequency provides a measure of fluid velocity.

Such as a vortex meter, swirl meters detect pressure pulsations, whose frequency depends on the flow rate. Vanes within the meter cause the incoming fluid to take the shape of a rotating helical coil of relatively low-pressure zones. At low flow rates the low pressure swirls or coils are far apart. As flow rates increase, the coils become compressed, coming closer together. Detectors within the meter measure the frequency of the swirling low-pressure zones as they pass through the meter. A deswirler at the meter's exit eliminates any effects on downstream instruments.

Engineers can apply these meters to liquid, gas, or steam, but should avoid dirty, corrosive, or abrasive fluids. The meters have no moving parts, have wide rangeability, and are immune to pressure, temperature, or density changes. These meters, however, lose effectiveness as viscosity increases. They are also affected by vibration or pulsation.

Swirl meters are less sensitive to flow profiles than vortex meters. While vortex meters may require a minimum straight-pipe run of 15 or more pipe diameters upstream and five pipe diameters downstream, swirl meters require only three diameters upstream and one diameter downstream. Both may be installed horizontally or vertically, but must be kept full of fluid during operation. About half of all vortex meter installations require "necking down" oversized process piping by concentric reducers and expanders.

Turbine flow meters

A piped fluid that acts on turbine vanes will rotate the turbine with a rotational rate proportional to flow rate. Turbine meters make up about eight percent of the installed base. Rangeability can reach 100-to-1 if the meter measures the rate of a single fluid at constant conditions. Accuracy is high and may top ±0.1%.

Turbine meters are suitable for extreme temperatures and pressures and can be applied to liquids and gasses. They install easily and are well accepted in processing applications, especially the oil and gas industries. On the other hand, turbine meters suit only low viscosities and are sensitive to flow profile and vibration. Having moving parts, they usually require frequent calibration. Partially open valves upstream from a turbine meter can cause significant errors. Typically, turbine meters require flow upstream straightening vanes.

Additional considerations include:

- Possible flashing or cavitations from pressure drops. Flashing causes the meter to read high while cavitations result in rotor damage. The downstream pressure must equal 1.25 times the vapor pressure plus 2 times the pressure drop.
- Small amounts of air entrainment (<100 mg/1) will make the meter read high, while large quantities can destroy the rotor.
- Solids entrained in the fluid can damage the meter. In these cases, a flushing strainer or filter must be installed upstream of the flow meter.

Ultrasonic flow meters

Doppler flow meters, limited to liquids, send an ultrasonic beam into the fluid and look at reflections from discontinuities such as entrained particles or bubbles. Reflections from the moving objects will experience a frequency shift that is a measure of the flow rate. Transit-time flow meters have beam transmitters and receivers separated by a known distance. They look at the differences in time of flight for ultrasonic beams moving with and against the flow. Doppler meters work best on slurries or dirty liquids, while transit-time meters work best on clean fluids.

These flow meters have no moving parts and offer an unobstructed passage for fluid flow. Ultrasonic flow meters have wide rangeability, and some designs can simply be strapped on to the outside of pipeline. Typical accuracy of these clamp-on flow meters is about five percent. Generally,

clamp-on units will not work well if the pipe has a lining or if the pipe material is stainless steel, PVC, or fiberglass. In addition, the inside of the pipe must be free of sound-absorbing material, such as dirty grease or scale. Multibeam, transit-time units in direct contact with the fluid, however, now have sufficient accuracy to be used for custody transfer of gases and liquids.

Published data on ultrasonic flow meters is limited. They have a relatively high initial cost and performance is sensitive to multiple installation factors. Deposit buildup on the pipe walls can significantly affect accuracy, as can extreme vibrations.

Variable-area flow meters

Commonly known as rotameters, these devices provide practical solutions for many flowmetering applications. They are simple and inexpensive, performing as well or better than many high-tech flow meters on the market. Every rotameter basically consists of two components—a tapered metering tube made of glass, metal, or plastic and a float that rides within the tube. These components come in a wide variety of shapes, sizes, weights, and materials of construction, making them adaptable to a wide range of application needs. They are easy to install and maintain, but must be mounted vertically and plumb. Some designs can handle high pressures and viscosities. Today, most manufacturers have taken all the factors and data involved in sizing a rotameter and incorporated them in a sophisticated software program.

Accuracy, however, is relatively low (±2%) and depends on precise knowledge of the fluid and process. Rotameters are susceptible to vibration and plugging by solids and are affected by density and temperature changes. They primarily apply to flow rates below 200 gpm and pipe sizes less than three inches.

Target flow meters

This meter inserts a physical target within the fluid flow. The moving fluid deflects a force bar attached to the target. The deflection caused by the force depends on the target area, as well as the fluid density and velocity. These flow meters have no moving parts such as bearings and can measure fluid flow rates for any line size above 0.5 in. Engineers can adapt the flow meter to different fluids or flow rate ranges by changing the target size and material. Rangeability is about 15-to-1. One disadvantage is that the target meter's calibration must be verified in the field.

Positive-displacement flow meters

Precision-fitted rotors measure fluid flow in positive-displacement flow meters. The volumes between the rotors are known and fixed. The rotor rotation varies directly with the volume of the fluid being displaced. Counters monitor the number of rotations and a transmitter converts the counts to volumetric flow rate. Principal types include:

- Single or multiple reciprocating piston meters.
- Oval-gear meters with synchronized, close-fitting teeth.
- Movable nutating discs mounted on a concentric sphere located in spherical side-walled chambers.
- Rotary vanes creating two or more compartments inside the meter's housing.

These flow meters are applicable to a wide range of nonabrasive fluids, including those with high viscosities.

Examples include heating oils, lubrication oils, polymer additives, ink, and many more. They are unaffected by poor flow profiles. Accuracy may be up to ±0.1% of full scale with a rangeability of 70-to-1 or better. Moreover, they require no power and can handle high pressures. Positive-displacement flow meters cannot handle solids, entrapped air in liquids, or entrained liquids in gases. They are expensive to install and maintain, having many moving parts. Pressure drop across the meters is high.

Reprinted with permission *Flow Control* August 2007 issue.

Use liquid ultrasonic meters for custody transfer

By **P. Syrnyk and D. Seiler**, Daniel Division, Emerson Process Management Houston, Texas

Using liquid ultrasonic meters for liquid petroleum applications such as custody transfer or allocation measurement is gaining global acceptance by the hydrocarbon processing industry. Ultrasonic technology is well established, but using it for custody transfer and allocation measurement is relatively new. End users often try to utilize the same measurement practices that apply to turbine technology.

While the two approaches share some similarities—such as the need for flow conditioning and upstream and downstream piping requirements—there are also differences. Several in situ proving methods can be used to successfully field calibrate a liquid ultrasonic meter.

Transit time ultrasonic flow meters. This device uses the transit times of the signal between two transducers to determine the fluid velocity. The transducer transmits ultrasonic pulses with the flow and against the flow to

a corresponding receiver (Figure 1). Each transducer alternates as a transmitter and a receiver.

Consider stationary fluid in a full meter spool. In theory, it will take precisely the same amount of time for a pulse to travel through the fluid in each direction since the speed of sound is constant within the fluid. If fluid is flowing through the pipe, then a pulse traveling with the flow traverses the pipe faster than the pulse traveling against the flow. The resulting time difference is proportional to the fluid velocity passing through the meter spool. Single and multiple acoustic paths can be used to measure fluid velocity. Multipath meters tend to be more accurate since they collect velocity information at several points in the flow profile.

The transit time of the ultrasonic signal is measured and used with other variables to calculate flow. Although the ultrasonic signal is traveling in a straight line, it is traveling at an angle, Θ, to the pipe axis (Figure 2). Equations 1 and 2

Figure 1. Transit-time principle for ultrasonic flow meters.

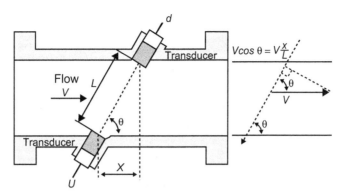

Figure 2. Variables for calculating transit time.

define the flow rate between two transducers located at positions u (upstream) and d (downstream):

$$t_{ud} = \frac{L}{C + V_i \cos \Theta} \qquad (1)$$

$$t_{du} = \frac{L}{C + V_i \cos \Theta} \qquad (2)$$

Solving Equations 3 and 4 simultaneously yields the following results for V_i and C (notice that V_x/L can be substituted for "$V \cos \Theta$" to simplify Equation 4):

$$V_i = \frac{L}{2 \cos \Theta} - \frac{t_{du} - t_{ud}}{(t_{ud})(t_{du})} = \frac{L^2}{2X} \frac{t_{du} - t_{ud}}{(t_{ud})(t_{du})} \qquad (3)$$

$$C = \frac{L}{2} \frac{t_{du} + t_{ud}}{(t_{ud})(t_{du})} \qquad (4)$$

where:

t_{ud} = transit time from transducer u to d
t_{du} = transit time from transducer d to u
L = path length between transducer faces u and d
X = axial length between transducer faces
C = velocity of sound in the liquid in still condition
V_i = mean chord velocity of the flowing liquid
Θ = acoustic transmission angle.

Since the equations are valid for fluid flowing in either direction, the method is inherently bidirectional. Also, the speed of sound term through the medium drops out of the velocity equation. Consequently, velocity is determined from the transit times through the predetermined distances, and is independent of factors such as temperature, pressure, and fluid composition.

Field proving. In custody transfer or allocation measurement applications, it is common practice to periodically prove liquid meters in the field. This is done to reduce uncertainty and maintain accuracy for fiscal measurement, establish meter factors at working conditions, and determine meter factors for different fluids. The measurement

application or the customer contract generally determines the frequency at which it is proved.

Meters in a pipeline that transports several liquids may be proved at every product change. The American Petroleum Institute provides guidelines and requirements. There are four basic requirements for successful proving:

1. The meter should be proved under the normal conditions in which it is expected to operate. Flow variables, including flow rate, pressure, and temperature must be stable and unchanging.
2. Adequate prover capacity is required to provide proving runs of sufficient duration. Larger volumes and longer run times will improve repeatability performance. Upstream flow conditions can affect the run time and volume required.
3. A sufficient number of runs are required to establish a valid proving. If the run-to-run variation is small, then an adequate meter can be found by averaging a few runs. As the run-to-run variation increases, the average of more runs is required for an adequate meter factor.
4. Results should be traceable to the National Institute of Standards and Technology (NIST). The meter prover calibration must be traceable to NIST-calibrated test measures.

Proving principle. Displacement prover systems operate on the displacement principle. This is the repeatable displacement of a known volume of liquid from a calibrated section of pipe between two detectors (Figure 3).

Liquid displacement is achieved by a sphere or a piston traveling through the pipe. A corresponding volume of liquid is simultaneously measured by a meter installed in series with the prover. A meter that is being proved on a continuous-flow basis must be connected at the time of proof to a proving counter. The counter is started and stopped when the displacing device actuates the two detectors at either end of the calibrated section of the prover.

A displacement-type prover must be installed so that the flow of the liquid stream through a meter being proved will

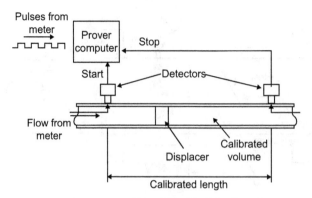

Figure 3. Proving a meter by displacement.

fully pass through the prover. When flow through the meter and prover has stabilized, the proving sequence is initiated. The displacer then travels through the pipe, while maintaining a seal between it and the pipe wall. Each of the detector switches is individually actuated upon displacer passage.

A high-resolution pulse counter connected to the meter output is gated on and off by the detectors. In the event the meter being proven cannot generate a minimum of 10,000 whole meter pulses between detector switches, some form of pulse interpolation must be used to allow discrimination at a rate higher than 1 part per 10,000.

Types of provers. A displacement prover may utilize either a sphere or a piston as a displacer and may be either unidirectional or bidirectional. Liquid always flows in the same direction through a unidirectional prover which has a mechanism that returns the displacer to its initial position in the prover.

A bidirectional prover utilizes a flow reversal mechanism (such as a four-way valve) that causes the displacer to alternate its direction of passage through the prover pipe. Unidirectional proving runs consist of single trips of the displacer past the detectors in the same direction. Bidirectional proving runs consist of "round trips" of the displacer (a complete cycle past the detectors in both directions). A typical bidirectional prover is depicted in Figure 4; a typical unidirectional "small volume" prover is depicted in Figure 5.

Small-volume displacement provers are defined as those with a calibrated volume that permits a minimum of 10,000 pulses to be collected by the meter under test. Its attributes include smaller size and weight, shorter proving cycles, and extremely wide rangeability (in excess of 1000:1).

Meter characteristics. Flow meter characteristics are important in developing a proving methodology. For example, both positive displacement and turbine meters have mechanical elements that rotate as fluid passes through the meter. It is convenient to use electrical sensors to detect the passing of gear teeth or rotor blades. These pulses are totaled to determine the flow volume through the meter.

Figure 4. A bidirectional ball prover.

Figure 5. A small volume prover.

Since these mechanical elements rotate as a result of the flow passing through them, it tends to average out eddies associated with turbulent flow. The frequency output, representing pulses per unit volume, is quite steady and is considered to be a uniform pulse output. These meters exhibit good proving repeatability with fairly short volumes and run times.

An ultrasonic meter has no moving parts. It determines the average velocity over one or more acoustic paths. The actual measurement made is the transit time of an ultrasonic pulse over the path in both the upstream and downstream directions. From this, an average velocity is calculated for each path. The path averages are then combined to form an average velocity for the whole meter, and the volume flow through the meter can be determined based on the cross-sectional area of the meter.

Since an ultrasonic meter has no moving parts, there is nothing to rotate to provide pulse counts. Pulse generation is done by means of a digital-to-frequency converter. Since the frequency is generated as the result of a measurement, it always lags the flow by the update period.

Also, since the meter is sampling the flow at several locations, turbulence in the flow will be measured. The standard deviation of the path velocity measurements is typically 2–5%. This implies a peak-to-peak variance of 12–30%. The turbulence is noise in the flow measurement that averages to zero. As a result, these meters have varying frequency output and are said to have a nonuniform pulse output. In proving ultrasonic meters, volumes and run times must be sufficient to average the turbulence to an acceptably low value.

Regardless of the calibration method, repeatability is an important criterion used to determine the acceptability of proving results. Ultrasonic meter performance verification can be ascertained by conventional means and to a consistent level. Table 1 provides guidance in determining the consecutive number of proving runs that may be required to obtain a meter uncertainty factor of ±0.027% at a 95% confidence level.

Any of the number of runs chosen from this table will produce results that verify a meter's performance to ±0.027% uncertainty. There is no difference between the traditional 5 consecutive runs that repeat within 0.05% and 10 consecutive runs that repeat within 0.12%, they both demonstrate the exact same uncertainty. The operator in the field will attempt to prove the meter in the smallest number of runs possible.

How proving volume has a direct effect on repeatability is shown in Figure 6. The example shows a turbine and ultrasonic meter being proven simultaneously with a small-volume displacement prover. Each of the three proving cycles could be as short as 0.5 s, depending upon flow rate. As a result of the small proving volume, it can be seen that the turbine meter, with its uniform pulse output, displays excellent repeatability (0.011%) over the 3 runs, while the ultrasonic meter does not—it shows a repeatability of 0.23%.

This does not mean that the turbine meter is more accurate or repeatable than the ultrasonic meter since both meters would provide excellent results over a longer time period. An ultrasonic meter proven using a prover with a larger volume, which effectively allows more time for the nonuniform pulse

Table 1
Number of Runs and Repeatability Needed to Achieve ±0.027% Uncertainty in Proving an Ultrasonic Flow meter

$$\text{Repeatability} = \frac{\text{Highcounts} - \text{lowcounts}}{\text{lowcounts}} \times 100$$

Runs	Repeatability	Uncertainty
3	0.02%	±0.027%
4	0.03%	±0.027%
5	0.05%	±0.027%
6	0.06%	±0.027%
7	0.08%	±0.027%
8	0.09%	±0.027%
9	0.10%	±0.027%
10	0.12%	±0.027%
11	0.13%	±0.027%
12	0.14%	±0.027%
13	0.15%	±0.027%
14	0.16%	±0.027%
15	0.17%	±0.027%
16	0.18%	±0.027%
17	0.19%	±0.027%
18	0.20%	±0.027%
19	0.21%	±0.027%
20	0.22%	±0.027%

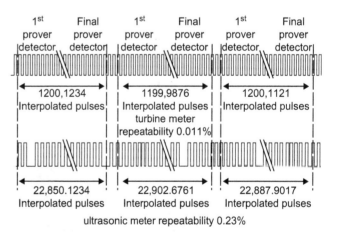

Figure 6. Typical results for turbine and ultrasonic meters using small volume provers.

variations to average out, is depicted in Figure 7. The important point is that increasing proving volumes improves repeatability.

Increasing proving volume. It can be seen that repeatability has improved significantly from 0.23% in

Figure 7. Typical proving results for ultrasonic meters using a larger proving volume.

Figure 6 over three consecutive runs to 0.017% in Figure 7 over two consecutive runs. This is a direct result of increasing the proving volume, thus allowing additional time for the meter to average the signal output. Increasing the proving volume can be accomplished by using a larger-volume prover or by using another method that is being adapted by many end users—the prover-master method (Figure 8).

Master meter proving is an accepted and recognized method of proving flow meters. However, master meters must also be calibrated on a regular basis to minimize errors. Calibrating a master meter in any situation other than the actual operating conditions can introduce uncertainties that must be compensated for. Changes in fluid composition, viscosity, density, and pressure can affect a meter's performance as can installation.

Using a prover-master meter combination installed in series with the meter under test eliminates all of these uncertainties. The master meter is first calibrated by the meter prover in situ under the actual operating conditions and using the actual fluid being measured. Once accomplished, then the master meter is used to calibrate the ultrasonic meter using the master meter calibration

methodology. The advantage of this method is that it allows much larger volumes since the volume can be as large as necessary, thus allowing an acceptable level of repeatability.

Existing provers, whether ball-type or small-volume can be fitted with master meters in the field if it is determined that the existing prover does not have sufficient volume to provide acceptable repeatability. An actual calibration curve on a 6-in ultrasonic meter that was calibrated using a bidirectional ball prover with a volume in accordance with API recommendations, is shown in Figure 9. The same 6-in meter calibrated using the prover-master meter method, is depicted in Figure 10.

A small-volume prover was used to calibrate an 8-in turbine meter, which was then used as the master meter to calibrate the 6-in ultrasonic meter. The results are virtually identical to those above, illustrating the viability of both methods and the reproducibility of the meter itself.

Figure 9. Liquid ultrasonic meter 6-in calibration chart—using Ball prover with ample volume.

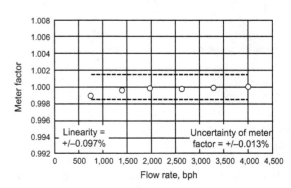

Figure 8. Small volume prover with turbine master meter prover.

Figure 10. Liquid ultrasonic meter 6-in calibration chart—using in situ master-meter small volume.

Summary. Liquid ultrasonic meters are a relatively new technology and they continue to gain acceptance in a variety of applications. Proving a meter per API guidelines requires a ±0.027% meter uncertainty factor with a 95% confidence level. A key to meeting this requirement is having sufficient proving volume, which has a direct impact on repeatability. Generally, the longer the proving cycle, the larger the proving volume and the more repeatable the results.

Despite the relatively recent use of ultrasonic technology in this application, oil and gas companies with a good understanding of the issues are successfully proving liquid ultrasonic meters in the field on a regular basis.

Bibliography

1. API Manual of Petroleum Measurement Standards (MPMS), Chapter 5—Metering, Section 8- Measurement of Liquid Hydrocarbons by Ultrasonic Flow meters Using Transit Time Technology
2. API Manual of Petroleum Measurement Standards (MPMS), Chapter 4—Proving Systems, Section 2—Displacement Provers
3. API Manual of Petroleum Measurement Standards (MPMS), Chapter 4—Proving Systems, Section 5—Master-Meter Prover

Reprinted with permission − Hydracarbon Processing − August 2007 issue.

Handling entrained gas

By **Tim Patton**—Director of Measurement Technology for Micro Motion a division of Emerson Process Management

Gas and liquid coexist in many processes. This mixing of dissimilar states can manifest itself in many ways and with many names—transient, two-phase, flashing, and slugs, to name a few. In a primarily liquid process, it is perhaps best known as entrained gas. By any name, entrained gas is often a troublesome challenge to accurate and reliable flow measurement.

Entrained gas is present in many liquid flow measurement applications, such as the manufacture of film emulsions, beer (wort), glucose, liquid fertilizers, soaps and lotions, hair-care products, phenols, dairy products, polyethylene, and ethylene oxide. Flow measurement performance in such environments can be unpredictable, depending upon the application, as measurement accuracy can be affected by many factors, such as the positioning of the measuring devices or the characteristics of the fluid being measured.

Characterizing entrained gas

To understand how recent technologies have largely overcome problems related to entrained gas, it is important to understand the different categories of entrainment: slug flow, bubble flow, and Empty-Full-Empty (EFE).

Figure 1. Types of entrained gas.

Slug flow is the occurrence of periodic large, coalesced gas bubbles in a liquid process. Slugs are usually not intentional and can be disruptive to flow and density measurement. Slug flow is commonly found in long-distance piping that does not fill properly. It may occur as a result of leaks in pump suction or tank agitation, allowing air to be pulled into a line.

Bubble flow is continuous distribution of gas bubbles (usually air) in a liquid process. The gas could be intentionally injected or could be caused by faulty equipment somewhere in the piping. Bubble flow is commonly found in high-viscosity liquids, such as toothpaste or peanut butter. It can also be present in products that are agitated or pumped at high speeds, as well as in environments where cavitating pumps or broken pump seals are a factor in the process. Most processes with bubble flow have less than five percent entrained gas, though in rare cases the value can be higher.

Empty-full-empty (EFE): In the case of EFE, piping starts empty, fills with liquid, remains full for the duration of the batch, and then empties upon completion of the batch. This batching is commonly found with railcar and tanker truck loading. Because the piping and the meter must be emptied between each fill, measurement can be a challenge when using a flow meter that is configured for only liquid processes.

A subset of empty-full-empty batching is a condition referred to as multiphase flow—processes with multiple density phases with some degree of separation and a mixing layer between them. The width of the mixing layer, dispersion of bubbles, and pipe orientation are all large contributors to the flow structure. Multiphase flow is commonly found in gas injections and well-head testing applications where multiple fluids coexist in the pipe—namely oil, water, and gas.

The design challenge

The biggest design challenge for Coriolis flow measurement is that gas entrainment in a pipeline is actually a continuum offering an infinite variety of conditions. To determine how these conditions affect the performance of the meter, development engineers at Emerson Process Management (*www.emersonprocess.com*) used flow meters with clear tubing to observe what actually occurred inside the meter. These observations, along with simplification of gas entrainment into bubble, slug, and EFE, enabled the engineers to develop a flow meter sensor and electronics that significantly expand the meter's capabilities under entrained-gas conditions.

These investigations concluded that four key elements played a role in the entrained-gas performance for Coriolis meters—the signal processing speed, processing algorithms, sensor design, and meter stability independent of environmental changes.

Improving slug flow & EFE performance

Emerson's Micro Motion Coriolis flow meters feature multivariable digital technology. The accuracy and stability of the Coriolis signals are considerably improved by digitally processing the signal from the flow sensor. Digital signal processing also delivers online monitoring and diagnostics data, providing predictive process and flow meter health information.

The characteristics of gas entrainment in slug flow and EFE batches cause frequent and large disturbances within the Coriolis meter that require the signal processing to be able to control and interpret a chaotic vibration. By increasing the processing speed in the Coriolis electronics, the meter is able to measure all these variables in synchronization with the disturbance. The improved signal processing and drive control optimize performance for transient events.

These developments have helped Micro Motion significantly reduce the flow rate error during the entrained-gas state from 20% with traditional technologies to just 1% using the company's latest-generation ELITE flow meter with enhanced core processor.

Overcoming bubble-flow conditions

Under bubble-flow conditions, the Coriolis sensor design and sensor stability dictate measurement performance. Sensor design is important because the critical bit of information is the relative difference in motion of the bubbles and fluid. Sensor stability is important because sensor vibration during bubble flow can be noisy and can cause the sensor to couple to the environment. With added stability, Coriolis flow meters can ensure that noise introduced by bubbles causes minimal flow measurement errors. Measurement is further compromised when bubbles change the compressibility of the fluid. As the level of gas entrainment increases, so does

Figure 2. Accuracy graph showing performance of new Coriolis flowmetering technology under entrained gas conditions.

the measurement error. Tests show that for a given amount of entrained gas, the measurement error becomes more pronounced as the meter drive frequency is increased. By operating at lower frequencies, Coriolis flow meters can minimize these errors.

Final results

By addressing the four key design elements identified during its development research for entrained-gas applications—processing speed, signal processing, sensor stability, and sensor design—Micro Motion believes that many measurement applications once deemed impossible are now well within today's measurement capabilities. Regarding gas/liquid mixture applications specifically—whether the combination of gas and liquid flow is by design, by flaw, or as a consequence of day-to-day operations—Micro Motion is confident that it has developed methods that are capable of providing measurement accuracy and minimizing waste under such conditions.

Tim Patten *is the director of measurement technology for Micro Motion, a division of Emerson Process Management. Tim earned a master's degree in Mechanical Engineering from Worcester Polytechnic Institute, and a bachelor's degree.*

Coriolis flow meter

The Coriolis flow meter market has been one of the fastest growing flow meter markets over the past five years, and despite their higher initial purchase price, many users find them a good investment when cost of ownership is considered. As such, process plants are increasingly selecting Coriolis meters to replace DP devices, and their use is also growing in the oil and gas industry.

One of the most important features of Coriolis flow meters is that they measure mass flow. While in many cases volumetric flow is sufficient, it is also desirable at times to measure mass flow. Many products are sold by weight rather than by volume, and in these cases it is often desirable to measure mass flow. Chemical reactions are often based on mass rather than volume, so mass flow measurement is often required in the chemical industry.

While both volumetric and mass flow apply to liquids and ases, mass flow is especially appropriate for measuring gases. This is because gases are much more affected by temperature and pressure than are liquids. Pressure has minimal effect on liquids in terms of compressibility, and is often ignored in making volumetric measurements of flow. The effects of temperature on liquids are also often disregarded, except where high accuracy is desired.

Coriolis flow meters are the most accurate flow meters made. While many magnetic flow meters have accuracies in the range of 0.5%, many Coriolis flow meters achieve accuracy of 0.1%. It is the high accuracy of Coriolis flow meters that is one of the major reasons for their extremely rapid growth over the past five years. Companies that need flow meters for custody transfer, or want highly accurate measurements in terms of mass, have a good reason to select Coriolis flow meters.

What is the Coriolis Effect?

Coriolis flow meters are named after Gaspard Gustave de Coriolis, a French mathematician and engineer. In 1835, he wrote a paper in which he described the behavior of objects in a rotating frame of reference. While this is sometimes called the Coriolis force, it is more accurately called the Coriolis effect, since it is not the result of a force acting directly on the object, but rather the perceived motion of a body moving in a straight line over a rotating body or frame of reference.

One common example given to illustrate the Coriolis effect is that of a ball propelled through the air a long distance in a straight line from the North Pole toward a target on the equator. By the time the ball arrives at the equator, it will not land at its apparent target, because the earth will have sufficiently rotated underneath the moving ball so that it will land some distance away from the perceived target on the equator. From the perspective of the person standing where the ball is thrown," the ball will appear to have curved.

A similar example is someone throwing a ball from the center of a rotating merry-go-round. If she throws the ball at a horse on the edge of the merry-go-round, assuming the merry-go-round is rotating at a rapid pace, she will not hit the horse she is aiming at, but perhaps the one after it, since the horse she aimed at will have moved by the time the ball reaches the edge of the merry-go-round.

Figure 1. A diagram showing the location of the inlet and outlet sensors in a Coriolis flow tube. *Image courtesy of Micro Motion.*

A diagram showing the location of the inlet and outlet sensors in a Coriolis flow tube.

Image courtesy of Micro Motion.

Figure 1 showing the inlet and outlet sensors in a Coriolis flow tube. Image courtesy of MicroMotion

In both these cases, there is no force acting to push the ball in a curved direction. Instead, its motion will appear to be curved from the perspective of the ball-thrower because the frame of reference is moving underneath the ball. This is why the Coriolis "force" is more appropriately called the Coriolis effect. It refers to the apparent effect on the motion of an object passing over a rotating frame of reference when viewed from the perspective of the point of origin of that moving object.

Do Coriolis flow meters embody the Coriolis Effect?

If Coriolis flow meters embody the Coriolis effect, it is logical to assume that they must have rotating tubes. Yet, it is not clear why fluid passing through rotating tubes would have a Coriolis effect on the motion of the tubes, since the effect of the fluid is simply the result of inertial motion on the tubes. However, in the Coriolis effect, there is no force exerted on the ball thrown on the earth or the merry-go-round apart from the force exerted in throwing the ball. The ball simply appears to curve when viewed from the perspective of the rotating body.

Proponents of the Coriolis theory also cite a similar example when trying to explain the Coriolis effect. They point to a hose full of water that is held in someone's hand. When no water is running through the hose, it remains still. But as water runs through the hose, the hose starts to twist and turn, due to the inertial effect of the running water on the hose. This is very much such as the action of fluid on Coriolis tubes, but it is not clear why this is an example of the Coriolis effect. Of course rapidly flowing fluid through a narrow flexible object like a hose will cause it to twist and turn, due to the inertia of the fluid. If the pipes that carry many of the liquids

in a plant were flexible instead of rigid, and the flow velocity was high enough, they would twist and turn as well. But, what does this have to do with the Coriolis effect?

How Coriolis flow meters actually work

Instead of having a rotating frame of reference, Coriolis flow meters work on the principle that the inertia created by fluid flowing through an oscillating tube causes the tube to twist in proportion to mass flow rate. Many Coriolis meters have two tubes. They are made to vibrate in opposition to each other by means of a magnetic coil. Sensors in the form of magnet and coil assemblies are mounted on the inlet and the outlet of both flow tubes. As the coils move through the magnetic field created by the magnet, they create a voltage in the form of a sine wave. These sine waves are the key to measuring mass flow.

Under no-flow conditions, the inlet and outlet sine waves are in phase with each other. When fluid is moving through the tubes, the tubes twist in proportion to mass flow rate. The amount of this twist is detected by the inlet and outlet sensors, based on a phase shift (time difference) that occurs in the sine waves formed by the two sensors. The mass flow rate is derived from the difference in phase shift in the sine waves formed by the inlet and outlet sensors.

Coriolis flow meters or inertial mass flow meters?

It is difficult to see the relation between the Coriolis effect and the inertial principle that Coriolis flow meters actually operate on. However, this does not cast doubt on their effectiveness or undermine their theory of operation. It simply calls into question whether they are appropriately named, and what connection the inventors of this meter saw between the Coriolis effect and the inertial operation of Coriolis meters. Perhaps they saw the Coriolis effect as a type of metaphor or analogue to the inertial force exerted by fluid flowing through a vibrating tube.

Coriolis meters do have a kind of "mystique" about them that no other flow meter has. It may be due in part to the difficulty people have in understanding how they operate. Also, manufacturers typically enclose the sensors inside a container, concealing their inner workings. No doubt, part of this mystique is that they are the most accurate flow meter and they do the measurement with no moving parts, apart from the vibration of the tubes.

Recent developments

Regardless of what one calls the principles that form the basis of Coriolis flow meters, there is no denying that these flow meters are playing an increasingly important role in the flow meter market today. One important development is the introduction of large-line-size Coriolis flow meters. Formerly, only Rheonik (now part of **GE Measurement & Control Solutions**) produced Coriolis flow meters for line sizes above six inches. Following Rheonik's introduction of large-line Coriolis flow meters, **Micro Motion**, **Endress+Hauser**, and **KROHNE** have all entered this market. While all Coriolis flow meters above six inches are large and somewhat unwieldy, they have been selling well enough to have a significant impact on the market.

Despite their hefty price tag, 8-, 10-, and 12-inch Coriolis meters, many of which are designed for custody-transfer applications, are selling well in the oil and gas industry. Most recently, Endress+Hauser has introduced a 14-inch Coriolis flow meter designed for custody-transfer applications. End-users are willing to pay a higher price for the high accuracy and reliability of Coriolis flow meters when measuring high-value products such as crude oil. Likewise, end-users pay a high initial price for ultrasonic flow meters for custody transfer of natural gas, because they want the high accuracy and reliability of ultrasonic meters. In both these cases, life-cycle costs outweigh initial purchase price considerations.

While most of the large-line Coriolis flow meters use the traditional bent-tube design, KROHNE has introduced a large-line Coriolis meter using its straight-tube Coriolis design. While this meter is somewhat longer than the bent-tube meters, it is less bulky because the straight-tube design requires less space. This may prove to be a strategic advantage to KROHNE in the competition for the large-line-size Coriolis flow meter market.

Used with permission of Flow Research, Inc., Wakefield, MA 01880.

Jesse Yoder, Ph.D., is president of Flow Research Inc. in Wakefield, Mass., a company he founded in 1998. He has 24 years of experience as an analyst and writer in process control. Dr. Yoder specializes in flow meters and other field devices, including pressure and temperature products. He has written over 150 journal articles on instrumentation topics. Dr. Yoder can be reached at jesse@flowresearch.com.

Proving liquid ultrasonic flow meters for custody transfer measurement

Raymond J. Kalivoda, FMC Technologies

Introduction

Liquid Ultrasonic Flow Meters (LUFMs) continue to gain popularity in petroleum measurement with the promise of high accuracy and low maintenance. These are favorable features, but because of the high volume and value of petroleum products buyers and sellers must have a high level of confidence in the measurement. This assurance in Custody Transfer measurement is gained by adherence to the standards, procedures, and methods that define the measurement process.

There are two fundamental ways that petroleum products are measured—statically or dynamically. The measured volume determined by either method must be validated at operating conditions and traced to a fundamental standard. In a static system, the product is transferred to a tank or similar container. After the transaction is completed, the volume can be measured and validated by a suitable method. In a dynamic system PD, turbine or ultrasonic meters provide instantaneous information on the rate and volume of the transfer. As with static methods, the measurement results must be validated by a suitable method.

A key difference between a static and dynamic measurement is time. A static measurement is similar to a bank transaction, where the cash draw can be pulled and verified at any time. When a static transaction is complete, there is time to review the measured volume, and if an error is suspected, the volume can be re-measured. With a dynamic system the measurement must be right the first time; there is no means to re-measure the volume.

The proving system and process are essential in either static or dynamic measurement systems to validate the accuracy of volume measured. There are various worldwide standards and regulations that define the measurement system requirements, but they are fundamentally the same. Each is traceable to a national standard which in turn is traceable to the international standard. In dynamic measurement systems, this link is the prover volume. There are gravimetric and volumetric methods where the prover volume is traceable the International Bureau of Legal Metrology (BIML). The technical arm of the bureau is OIML (International Organization for Legal Metrology). This is an intergovernmental organization that deals with all aspects of legal metrology. It has 59 member states and 50 corresponding members. In this way, all the countries are tied to the same weights and measurement standards, which facilitates world trade.

The various measurement standards either directly or indirectly define the accuracy requirements over a range of operating conditions which include: flow rate, viscosity, temperature, pressure, piping configurations, etc. It is therefore important in the proving process to define both the operating conditions and the accuracy requirements because they are interdependent. A change in operating conditions can affect the meter's accuracy. As a general rule, the meter must be proven at operating conditions to validate its accuracy with the highest degree of confidence. The greater the change in operating conditions, the more uncertainty is introduced into the measurement.

The object of all custody transfer meters is the same, highly accurate measurement, but the procedures to achieve the results may vary with the different meter technologies. This paper will focus on the standards, procedures, and methods used to prove ultrasonic meters.

Custody Transfer Requirements

Custody Transfer refers to the fiscal measurement used to determine the quantity and financial value of a petroleum product transaction (delivery). The custody transfer requirements can be of two types:

Legal, which is defined by Weights & Measures (W&M) in the country or jurisdiction in which the sale is conducted. The various W&M codes and regulations control the wholesale and retail trade requirements to facilitate fair trade. The regulations and accuracy requirements widely vary between countries and commodities, but they all have one common characteristic—traceability. There is always a procedure that defines the validation process where the duty meter is compared to a standard that is traceable to the legal metrology agency of the respective region. The meters for wholesale and retail trade are normally smaller (4 in and under) to handle volume flow rates of under 2800 l/m (750 gpm). These applications are outside the range of current liquid ultrasonic meters and therefore will not be addressed in this paper.

Contract, is a written agreement between buyers and sellers that defines the measurement requirements. These are large-volume sales between operating companies where refined products and crude oils are transported by marine, pipeline, or rail. Since even a small error in measurement can amount to a large financial difference, custody transfer measurement must be at the highest level of accuracy possible. Because of the critical nature of these measurements, petroleum companies around the world have developed and adopted standards to meet the industry's needs.

A typical contract may define a specific measurement standard such as the American Petroleum Industry (API) Petroleum Measurement Standards or an OIML standard that is tied to the International Standards Organization (ISO). These standards include all the required equipment and details on the process to achieve an acceptable level of measurement.

Specifying a standard in a contract eliminates the need to enumerate many details concerning the equipment and measurement process that are common industry practice. For example, API standards are based on "best practice." They define the proper application parameters of a specific flow meter from field experience. OIML standards are "performance based." They set quantitative conditions that any custody transfer meter must meet. To gain approval, a specific size and type of meter must be tested over the operating range by a notified body (qualified testing agency) that issues a "pattern test report." The "pattern test report" is then submitted to the respective weights and measures authority as part of the application for approval.

Even though the standards may appear to widely vary, they have two fundamental points in common: (1) They all strive to maintain a minimum measurement error for a specific application and (2) They require traceability to national/international metrology standards by proving the meter. In terms of accuracy, which is expressed as uncertainty, a good API system would typically have a $\pm0.1\%$ extended uncertainty. OIML requires a minimum system uncertainty of $\pm0.3\%$ over a wide operating range. For large volume transfers, a better accuracy is required, so the systems are proven over a narrower range of operating conditions and typically achieve a $\pm0.1\%$ extended uncertainty, the same as a system using API standards.

Measurement Definitions

The terminology associated with measurement can be confusing, but is important in understanding the measurement process. Two good sources for measurement terminology are:

- **API** Manual of Petroleum Measurement Standards Chapter 1—"Vocabulary" and Chapter 13—"Statistical Aspects of Measuring and Sampling"
- **VIM** "International Vocabulary of Basic and General Terms in Metrology," commonly known as VIM, which is accepted by the international community of legal metrology

Included below is a summary of the key terminology used in evaluating a meter's performance. Definitions were selected from both sources. As expected the terminology is similar because accurate measurement is essentially the same worldwide.

Accuracy [API 1.0]: the ability of the measuring instrument to indicate values closely approximating the true value of the quantity measured.

Accuracy of measurement [VIM 3.5]: closeness of the agreement between the result of a measurement and the value of the measurand.

Measurand [API 1.0] is a physical quantity that has been or is to be measured.

Repeatability (of results of measurements) [VIM 3.6]: closeness of the agreement between the results of successive measurements of the same measurand carried out under the same conditions of measurement which include the same: measurement procedure, observer, measuring instrument, conditions and location over a short period of time.

API [1.0] adds "…the ability of a meter and prover system to repeat its registered volume during a series of consecutive proving runs under constant operating conditions."

Reproducibility (of results of measurements) [VIM 3.7]: closeness of the agreement between the results of measurements of the same measurand carried out under changed conditions of measurement. A valid statement of reproducibility requires specification of the conditions changed. The changed conditions may include: principle of measurement, method of measurement, observer, measuring instrument, reference standard, location, conditions of use and time. Results here are usually understood to be corrected results.

API [1.0] adds "…the ability of a meter and prover system to reproduce results over a long period of time in service where the range of pressure temperature, flow rate, and physical properties of the metered liquid is negligibly small."

Linearity of a meter [API 1.0]: The ideal accuracy curve of volume meters is a straight line denoting a constant meter factor. Meter linearity is expressed as the total range of deviation of the accuracy curve from a straight line between the minimum and maximum recommended flow rates.

Bias [API 13] is any influence on a result that produces an incorrect approximation of the true value of the variable being measured. Bias is the result of systemic error.

Error (of measurement) [VIM 3.10] is the result of a measurement minus the value of the measurand. Since the value of the measurand cannot be determined, in practice a conventional value is used.

Types of Error [API 13 edited]: The difference between the measured quantity and the true value of the quantity includes all the errors associated with the measurement process the person taking the measurement, the process, the instruments and changes in conditions during the period of measurement. In evaluating a system there are three (3) types of errors:

Spurious error [API] is a gross mistake or blunder that must be identified and eliminated. There are statistical methods for testing for outliners (e.g. Dixon Outline Test) but data should not be discarded unless there are sound reasons for its elimination.

Random error [API] is a variation at constant conditions, normally evenly distributed about a mean, which can be statistically analyzed and the uncertainty defined within a given confidence limit.

Random error [VIM 3.13]: result of a measurement minus the mean that would result from an infinite number of measurements of the same measurand carried out under repeatability conditions. Because only a finite number of measurements can be made, it is possible to determine only an estimate of random error.

Systemic error [API] is one that remains constant as a positive or negative bias resulting in over or under estimating the true value of the measurement. In many liquid measurement applications, systemic error may make a larger contribution to the overall uncertainty than random error. There are two general classes of systemic error:

Constant systemic error is a bias that is particular to an installation and operating conditions. These errors include hydraulic and zero calibration effects.

Variable systemic error is a bias that varies with time and includes bearing wear or changes in meter tolerances.

Systematic error [VIM 3.14]: mean that would result from an infinite number of measurements of the same measurand carried out under repeatability conditions minus the value of the measurand.

Uncertainty of a Measurement [API 13]: the true value of a measurement cannot be determined, but a valid estimator can be obtained by the statistical analysis. The range or interval within which the true value can be expected to lie is the uncertainty range.

Confidence Limit [API 13 edited] It is normally impossible to absolutely set a range of uncertainty. It is more practical to indicate a degree of confidence on this range. This degree of confidence indicates the probability that the uncertainty range contains the "true value." The most common practice is to use a 95% confidence level. This level implies that there is a 95% probability (19 chances in 20) that the "true value" lies within the stated range of uncertainty. The 95% confidence level is recommended for all commercial applications in petroleum measurement. It should also be noted that this is the same confidence level used by legal metrology worldwide.

Uncertainty Analysis: encompasses the statistical analysis of the data derived from a measurement process to determine uncertainty range and confidence limits. It is the highly-mathematical and rigorous process defined in The Guide to the Expression of Uncertainty in Measurement or GUM, as it is commonly called. GUM is the result of ISO/TAG/WG 3 and printed in 1995 by ISO; the terminology is defined by VIM. It was fully supported by OIML and BIPM all worldwide legal metrology. In general the analytical method for calculating uncertainty and confidence limit of 95% is fully agreed upon throughout the world.

It is the uncertainty limits that are set by W&M that are used in the respective countries for the purpose of legal trade. For custody transfer transactions these limits are set by contract. Most times the uncertainty is not precisely stated, but based on precedence. An expected level of accuracy has been established over many transactions based on a specific measurement practice. In the US and many other parts of the world, the best measurement practice for pipeline or marine transfers of larger petroleum volumes is based on proving the custody transfer meter in situ with a displacement-type prover. Typically, a custody meter is proven at the start of a batch once the maximum transfer flow rate is established. Since the product is not changing and the variation in flow rate is limited (e.g. ±10%), the measurement is at optimum accuracy if the measurement process is sound.

For the purpose of this paper;

$$\text{Total Uncertainity} = \text{Random Uncertainity} + \text{System Uncertainity}$$

By viewing uncertainty in this form, it is easier to understand the measurement process and identify the potential factors that can affect uncertainty. There are a number of factors that can affect the measurement, but we need only be concerned with the key factors which include installation conditions, flow rate, viscosity, and temperature for most petroleum products, and also pressure for the lighter products like LPGs. It also should be noted from a review of the definitions that random errors are evenly distributed and will cancel out with a large enough sample. Systemic errors, on the other hand, "remain[s] constant as a positive or negative bias resulting in over or under estimating the true value of the measurement." Since systemic error or measurement bias is a business risk, identifying and reducing these errors is essential. Proving has always been, and remains, fundamental to custody transfer measurement because it is the only sure method to determine and correct systemic errors.

Unique Problems in Proving Liquid Ultrasonic Meters

Proving, as described, is highly important to the custody transfer measurement and in detail is defined in the respective measurement standards. These standards, especially API "Best Practice" standards, were developed around the available technology. PD meters were the only custody transfer meters until the early 1970s when API 5.3 "Measurement of Liquid Hydrocarbons by Turbine Meters" was issued. Coriolis meters were introduced for petroleum service in the early 1990s, but API 5.6 "Measurement of Liquid Hydrocarbons by Coriolis Meters" was not issued until 2002. API 5.8 "Measurement of Liquid Hydrocarbons by Ultrasonic Flow meters Using Transit Time Technology" was issued in 2005.

Each of these meters uses fundamentally different technologies to determine the custody transfer measurement. The measurement technology can dictate the application and proving technology. Since the PD and turbine meters have been used for custody transfer measurement over a long period, the proving technology and procedures have been developed around these technologies.

There is a fundamental difference between PD and turbine meter technology and Coriolis and Ultrasonic meter technology that affects proving. The PD and turbine meters produce a discreet volume output for every volume unit measured. For example, a PD meter with electronic output or a turbine meter can produce 1000 pulses/barrel. Therefore, each time a barrel passes through the measuring chamber a pulse will be transmitted. Coriolis and Ultrasonic meters use a sampling technique. They sample the flow for a period of time, typically 50–500 ms, and then the microprocessor transmits the correct number of pulses for the volume that passed through the meter. The measurement by this method is accurate, but the microprocessor-based flow meters have two unique problems in regard to proving: (1) **sample time** required to achieve an accurate measurement and (2) **time delay** in the pulse per unit volume output. There is a further difference between Coriolis and Ultrasonic meter in the sampling process. This paper only focuses on the issues affecting ultrasonic meters.

Sample Time

Proving LUFMs can be difficult because of their unique measurement method. A fluid flow stream is a complex flow field with numerous eddies, nonaxial components, and changing flow profiles. Unlike PD and turbine flow meters which integrate the flow field by mechanical convergence, ultrasonic meters measure all the numerous eddies and nonaxial velocity components in a turbulent flow field. To provide an accurate measurement, the ultrasonic meter must take a number of readings (or snapshots) of the fluid velocity at multiple locations along the product's flow path. The time it takes to acquire a sufficient number of samples for an accurate measurement depends on the flow regime and meter processing time, but, in general, the longer the sample time, the better the measurement. These sample times are in the milliseconds which don't effect the batch measurement, but become a problem with proving. Proving technology, as indicated, is based on discreet pulses for PD or turbine meters. Ultrasonic meters normally require larger prove volumes for the same flow rates than traditional custody transfer meters. However, with proper provers and sound proving practice, highly accurate measurement can be achieved.

Time Delay

Because of the sampling method used in ultrasonic meters, the output at any instant of time will represent a volume that has already passed thought the meter. This lag poses no problem under steady flow conditions. But, if the flow rate is changed, which is typical with displacement provers, this lag can cause poor run repeatability and even bias error in the meter factor. This problem was identified during the development of the API ultrasonic meter standard 5.8. In 2003, API formed a task group and funded testing to determine the affect of this pulse delay on meter proving. The group was headed by Kenneth Elliott, who reported the preliminary finding at the 2004 North Sea Measurement Workshop in paper 5.4, "API's Microprocessor Based Flow meter Testing Program." There were a number of interesting findings, but the most important results are that: (1) the potential meter factor bias is proportional to the time delay and (2) no accurate factor, based on available data, could be used to correct this systemic error. The practical application of this data is to reduce the meter sample time, and when possible, minimize flow variations during proving and use larger-volume displacement provers.

Proving Ultrasonic Meters

The problems with proving ultrasonic meters can be resolved by either or both of the two methods:

1. Increase the number of runs. The data scatter for an ultrasonic meter is greater than a conventional meter, therefore it may not repeat in 0.05% in five consecutive runs as required for custody transfer measurement. A more general statement would state the repeatability in terms of uncertainty, which is ±0.027% at 95% confidence level. This more general statement allows a wider repeatability test tolerance. For example, a repeatability range of 0.12% in 10 consecutive runs or a repeatability of ±0.17% in 15 consecutive runs both meet ±0.027% at a 95% confidence level. Table 1 is an abridged version of the number of runs at different repeatability ranges to achieve ±0.027% uncertainty.
2. Increase the "prove volume" to allow a sufficient number of samples. This will also increase the time it takes to complete each prover run, which reduces the time lag to a relatively insignificant uncertainty. As a rule, an SVP will not have sufficient "prove volume" or "prove cycle time" for proving an ultrasonic meter unless it is

Table 1
Runs vs. Repeatability

Runs	Repeatability	Uncertainty[*]
3	0.02%	0.027%
5	0.05%	0.027%
8	0.09%	0.027%
10	0.12%	0.027%
15	0.17%	0.027%
20	0.22%	0.027%

[*]Uncertainty of Meter Factor at 95% confidence level.

Table 2
Prover Volume

Meter Size	5 runs 0.05%		8 runs 0.09%	10 runs 0.12%
	Prover Size (barrels)			
	TM	Ultrasonic Meters		
4″	5	33	15	10
6″	12	73	34	22
8″	20	130	60	40
10″	24	203	94	62
12″	48	293	135	89
16″	100	521	241	158

oversized and the change in flow rate during the prove is minimal. In the ultrasonic meter standard, API proves suggested "prover volume" for different size meters. Table 2 shows these suggested prover volumes compared to the typical "prove volume" for turbine meters.

Proving Systems

Proving systems can be divided into:

Direct Proving, where the custody transfer meter is proven against:

- Displacement Provers—Conventional or Small Volume.
- Tank Provers—Volumetric or Gravimetric. This method is only suitable for PD meters and Coriolis mass meters and will not be included in this paper.

Indirect Proving, where the custody transfer meter is proven against:

- Master Meter Provers—Transfer Standard Proven, on-site or laboratory-proven.

Conventional displacement provers are well-established in this sector of the petroleum measurement industry. SVPs, because of their size and flow range, have recently gained the industry's interest. Master meters, which were rarely used on large volume custody transfer systems, are being evaluated because of technical and economic advantages.

Proving systems and proving is costly. On large pipelines and marine loading/unloading facilities, the flowline is normally divided into a number of meter runs, each handling a percentage of the flow. Metering systems are designed in this manner to facilitate proving and increase reliability. There is a technical and economic limit on the size of a prover for a given system. For example, it may be more economical on a 12,000 m^3/h (75,000 bph) measurement system to use 4–12″ flow meters and a 6.5 m^3 (40 bbl) prover than 3–16″ meters with a 15 m^3 (95 bbl) prover. Also, by adding an additional meter run, the system's reliability can be increased because the full flow could be handled by any of the 3 m.

In economic terms, design of custody transfer measurement systems becomes a cost versus risk relationship. Table 3 provides the potential cost of systemic measurement errors at different levels of uncertainty. The table clearly illustrates that the cost of a systemic error on a large system can be a substantial risk. A more subtle implication of this table is that using a number of smaller meters is less risky than using a larger meter.

Proving has always been, and remains, fundamental to custody transfer measurement. Choosing the proper prover and method is critical. Below is a review of each method prover with its advantages and disadvantages.

Displacement Provers

Displacement provers are the fundamental device used to verify the performance of large-volume custody transfer

Table 3
Systemic Uncertainty Risk*

Systemic Uncertainty ($\pm U_s$)	Daily Value (US $)			Annual Value (US $ millions)		
	Barrels per hour					
	10,000	50,000	100,000	10,000	50,000	100,000
0.1%	$11,520	$57,600	$115,200	$4	$21	$42
0.2%	$23,040	$115,200	$230,400	$8	$42	$84
0.3%	$34,560	$172,800	$345,600	$13	$63	$126
0.4%	$46,080	$230,400	$460,800	$17	$84	$168
0.5%	$57,600	$288,000	$576,000	$21	$105	$210

*Based on crude oil @ US $60/bbl and 80% duty cycle.

Figure 1. Uni-directional prover.

meters ($\geq 4''$ PD) and all types of turbine meters. There are various methods to calibrate displacement provers, but all are traceable to a primary metrology standard recognized by BIML for international trade. Displacement provers can be built into a measurement system or a portable prover can be connected into the system. This allows the meter to be validated at the operating conditions.

Conventional Displacement Provers. These are the most common provers in the petroleum transportation market because of their accuracy and reliability. There are two types of conventional displacement provers—uni-directional (Figure 1) and bi-directional (Figure 2). The displacer, which is either a sphere or piston, is installed in a coated pipe of a specific length. The difference between an uni-directional and bi-directional prover is how the displacer is launched and retrieved. The pipe is also fitted with detector switches at two or more points in the pipe. When activated by the displacer, the detector switches define a precise volume which is known as the "prove volume." There are API and other standards that define methods to calibrate this "prove volume" to a high degree of accuracy which is traceable to a national/international metrology standard. When a conventional displacement prover is connected in series with a meter, the displacer is launched and travels at the same speed as the liquid flowing in the pipe.

The switches control the high-speed pulse stream transmitted by the meter. The first detector switch gates an electronic prover counter "on" to collect meter pulses and the second detector switch gates the electronic prover counter

Figure 2. Bi-directional prover.

"off". The total number of pulses on the electronic prover counter represents the volume recorded by the meter. The indicated meter volume is then compared to the precise volume between the prover detector switches. It is possible to calibrate the "prover volume" to an expanded uncertainty of ±0.02%.

To achieve the required level of accuracy for a custody transfer measurement, the meter must generate at least 10,000 m pulses between the detector switches. The 10,000 pulse minimum is required because there can be ±1 pulse uncertainty at either detector switch, so the resolution is ±2 pulses/10,000 pulses or ±0.02%. An uni-directional-type prover normally collects a minimum of 10,000 pulses per proving run. The proving run for a bi-directional prover is both the forward and reverse directions, so a minimum of 20,000 pulses are collected.

Advantages of Conventional Displacement Provers.
Ultrasonic meters can be successfully proven with the properly-sized ball-type displacement prover because of the large "prove volume."

The "time delay" of the ultrasonic meter has little effect on ball provers because of the large "prove volume."

It is relatively inexpensive to increase the "prover volume" of ball provers.

If a ball prover is on site with a smaller than required "prover volume," the meter can still be successfully proven by increasing the number of prover runs. Increasing the number of runs compensates for the smaller "prove volume" provided that the required number is not excessive, typically over 20. Normally, doubling the number of runs (e.g. 5 runs to 10

runs) with a ball prover is sufficient to achieve a custody transfer uncertainty level of ±0.027%.

Ball provers are accepted technology that many users know how to operate and maintain. Ball provers cost 20–50% less than SVP-type displacement provers.

Disadvantages of Conventional Displacement Provers.
Size and limited operating range are the key disadvantages. Bi-directional provers have an operating range which is limited to 10:1 (10–100% of max flow) for well-lubricated products and 5:1 (20–100% of max flow) for dry products. The operating range of uni-directional provers is twice that of bi-directional provers.

Small Volume Type Displacement Provers.
SVPs (Figure 3) are displacement provers that have a volume between detector switches which does not permit collecting 10,000 m pulses. To achieve the required level of proving uncertainty, pulse interpolation techniques are used. These techniques divide the initial and final pulse by the use of chronometry. This allows the volume between detector switches to be substantially less than a conventional prover.

Advantages of Small Volume Provers.
Size and flow range are the key advantages of SVP's. They have a much smaller footprint than comparable ball provers and have a 100:1 (1–100%) flow range.

The small "prover volume" in the SVP makes it easier to clean and reduces the contamination between products when used for proving meters on multi-products.

Figure 3. Small volume prover.

SVPs are preferred for offshore platforms because of their size and flow range. The limitation for platforms is flow rate at 17,500 bph for a 12″ turbine and 12,500 bph for a 16″ PD or 10″ ultrasonic meter (2).

They are ideal as portable provers because of size and flow range. The limitation is the weight of the prover. The largest practical size has a 40 gal "prove volume," which is limited to proving an 8″ turbine meter (7200 bph) or a 10″ PD (4800 bph) or 6″ ultrasonic meter (2) (4000 bph).

SVPs are also becoming popular at pipeline terminals that are being automated, which is the goal of operators. Though in many cases, this excludes ultrasonic meters until the proving problems are resolved.

Disadvantages of Small Volume Provers. The small "prove volume" is a disadvantage for proving ultrasonic meters because the "time delay" and "sample time," as previously explained, are the major disadvantages of SVP's. In most, if not all cases, issues are demonstrated in the API Microprocessor Based Flow meter.

Testing programs severely limit the use of SVPs for proving ultrasonic meters. SVPs are also much more complex than pipe provers, and they are more susceptible to damage from fluids contaminated by debris, particularly abrasives. One method that is developed in the "Master Meter Proving" section is Transfer Meter Proving.

Master Meter Proving Systems

A master meter is termed an indirect meter proving method as compared to a displacement prover, which is a direct proving method. API 4.5 "Master Meters Provers" only recognizes the use of PD meters or turbine meters as master meters. Currently, there is an API working group that is adding Coriolis Flow Meters and Liquid Ultrasonic Flow Meters to this standard. A revision to the standard is expected for balloting by this spring. Until the ballot results are reviewed it is difficult to predict when the revised standard will be completed, but there is not any apparent technical reason why these meters would not be added to the standard.

Since the master meter proving is a secondary proving method, it has the highest uncertainty of all meter-proving methods. As previously defined, the total or extended uncertainty of a proving system is the combination of random and systemic uncertainty. The expanded uncertainty for proving a master meter with an SVP can be expressed as:

$$U_T = U_R + \sqrt{U_{MM}^2 + U_{SVP}^2}$$

where:

U_T—Total uncertainty of a typical prove
U_R—Random uncertainty which is the repeatability of the meter runs

U_S—Systemic uncertainty which is the combined uncertainty of the master meter and SVP
U_{MM}—Master meter uncertainty (systematic)
U_{SVP}—SVP volume uncertainty (systematic)

Systemic error, as defined, is one that remains constant as a positive or negative bias in the measurement. Since it can make a large contribution to the overall uncertainty (U_T), it is imperative to determine and correct for this bias by proving the meter. The wider the variation in installation and operating conditions, the greater the potential systemic error and associated uncertainty.

Two separate sequences of events are necessary in master meter proving. First, the master meter must be calibrated using a prover that is traceable to a national/international standard. The master meter can then be used to calibrate the field-operating meter.

Factors That Contribute to Uncertainty in Master Meter Proving. To best understand the uncertainty in master meter proving it is helpful to distinguish between the Constant Systemic Errors introduced by installation and operating conditions and Variable Systemic Errors that vary with time, which include bearing wear, changes in meter tolerances, strainers, and piping conditions. In general, ultrasonic meters are more sensitive to installation and operating conditions than PD and turbine meters, but less sensitive to the Variable System Errors because there are no moving parts.

Constant Systemic Error Contributions. **Installation Effects**—refers to anything in the measurement system set-up that can cause a systemic change in the meter factor. Certain metering technologies like PD and Coriolis meters are only slightly affected by installation conditions whereas velocity meters such as turbine and ultrasonic meters can be highly affected. Multipath ultrasonic meters may have algorithms that detect and compensate for these effects, but it is problematic on how robust the specific meter can handle these anomalies.

Currently with transfer proving, the total uncertainty (U_T) of a liquid custody transfer measurement can be made in the range of $U_T = \pm 0.1\%$ at a 95% confidence level. This is industry norm for most large volume liquid transactions. If a large systemic installation error (e.g. $\pm 0.5\%$) goes undetected for even a small period of time, it increases the business risk for over or under delivery.

Operating Effects—The most important operating conditions that affect the accuracy of liquid measurement are: flow range, viscosity range, and temperature.

Flow Range is the minimum/maximum flow rate at which a meter/measurement system can operate and maintain the stated accuracy. Custody transfer meters are normally specified to operate over a 10:1 (10—100% of maximum flow) flow range with a linearity of ±0.15%. This means that if all the operating conditions are held constant and only the flow rate is varied, then the maximum meter factor variation will be ±0.15% over the flow range. This variation in meter factor does not include any systemic error due to installation conditions or changes in the other operating conditions. The accuracy of the measurement can be improved by reducing the flow range or establishing a separate meter factor for each flow rate. With today's flow computers, a number of meter factors versus flow rates can be programmed and the meter factor will automatically be linearized. At a given flow rate, the measurement uncertainty (U_R) becomes the repeatability of the meter which must be ≤ ±0.027% for custody transfer measurement.

Viscosity Range of petroleum products can vary from less than 0.1 cP for LPG's to over 1000 cP for heavy crude oils (see Table 4). All meters are sensitive to viscosity, but each metering technology is affected differently. Because of the slippage principle, PD meters are affected on low viscosity products (<2 cP) whereas turbine and ultrasonic meters are sensitive to high viscosity products. Multi-path ultrasonic meters may have algorithms that compensate for viscosity, but on high viscosity crude oils, the other factors such as particles and entrained gaws may also affect the measurement accuracy.

Temperature affects the measurement in several ways: (1) the viscosity can change as illustrated in Table 4; (2) changes in liquid volume as defined by the volumetric coefficient of expansion; (3) changes in meter tolerances that affect the MF; and (4) a change in the prover volume. There are correction factors for the volumetric expansion of liquids and prover volumes, but master meter factors must be established for specific temperatures within the operating range. Since the repeatability of a meter during proving is based on a gross volume (nontemperature compensated volume), it is highly important to maintain a minimum temperature difference between the meter being proved and the prover. For example, a 2 °F change in temperature between the meter under test and the prover will impact meter factor and custody volume by 0.1% across the entire product range.

Variable Systemic Error Contributions. The variable systemic errors are associated with the stability of the master meter over time. Wear in a mechanical meter can cause a shift in the factor due to bearing wear or changes in critical tolerances. There can also be waxing or other coatings that similarly cause shifts in the calibration factor. Debris caught in the strainer basket and/or on flow conditioning elements can have a significant effect on the master meter's calibration.

The only way to confidently detect and correct for these systematic bias is by periodically reproving the meter and comparing the proving results. This raises the question, how often does the master meter need reproving? Because of the variation in master meters, service conditions and acceptable level of uncertainty (risk), it is impractical to establish a fixed time. The answer to this question can only be gained from careful analysis of the data. API 13.2 provides statistical methods to evaluate and track proving data, but a practical approach would be:

- Establish the baseline meter factor for the master meter on each test fluid under consideration. This would consist of ideally six runs at different times and conditions over which the master meter will operate.
- Develop a control chart based on the uncertainty tolerance that is acceptable. Typically for custody transfer measurement of $U_T = ±0.1\%$ the current MMF should be within 0.02% of the last meter factor and the baseline meter factor. The maximum allowance from the baseline should be 0.04%.
- The time between provings can be increased based on the master meter factor's stability.

Typically, a good master meter should remain within the stated tolerances for at least one year or 200 h of service. The precise criteria, though, is based on the system and the agreement of the parties affected by the measurement.

Transfer (Master Meter) Proving. With "Transfer Proving," the master meter is proven on site, which eliminates systemic errors due to installation effects. It also provides the ability to prove over the complete operating range, and if the conditions change or if measurement is challenged, the master meter can be easily reproven.

There is an added uncertainty of the master meter to the total uncertainty budget, but it can be precisely calculated. In practice, the uncertainty added by a master meter should be far less than the bias caused by a change in conditions.

Table 4
Viscosity of Selected Products

Product	API Gravity	Viscosity in cP 60 °F (15 °C)	100 °F (38 °C)	150 °F (66 °C)
Propane—LPG	145	0.12	0.09	0.07
Gasoline	56	0.63	0.49	0.38
Kerosene	40	2.2	1.7	0.9
Light crude	48	2.7	1.7	1.1
Light crude	32	21	9	5
Medium crude	25	1442	243	93
Heavy crude	16	3440[*]	574	230[*]
Heavy crude	10	5100[*]	1294	520[*]

[*]Estimated.

Currently, this method is used to prove terminal load rack meters at large terminals. Normally, a PD master meter is installed in series with one of the line meters and validated using a volumetric can prover. The other delivery meters on the same product are then calibrated against the master meter with the product being pumped directly into the tank truck compartment. This is far more efficient than proving each meter with a volumetric tank and pumping the product back to storage.

The same method, if applied to proving ultrasonic meters, would eliminate both the "time delay" and "sample time" issues. The master meter, whether it is a PD or turbine meter, can be proven with a conventional or SVP and then used to calibrate the ultrasonic meter. Since the master meter is not limited by "prove volume," an adequate volume can be run through the master meter to eliminate both the "time delay" and "sample time" issues.

Laboratory Calibrated Master Meter. Laboratory calibrated master meters have the highest level of uncertainty because of the bias introduced by installation and operating conditions. All meters have some sensitivity to installation and operating conditions and the resulting bias must be quantified for custody transfer measurement. The difficulty of determining this bias depends on the total uncertainty (U_T) required (level of acceptable risk) and the degree of variation in the installation and operating conditions. Obviously, a lower level of uncertainty (e.g. $U_T = \pm 0.1\%$) with a wide variation in operating conditions is significantly more difficult to determine than a higher level of uncertainty (e.g. $U_T = \pm 0.5\%$) over a limited flow range, with one product at a stable operating temperature. In either case, the analytical methods used to determine uncertainty analysis are dependent on test data for the specific type of meter and operating conditions. Specific data is required for each type of meter, brand, meter model, and size to develop an analytical model. The calibration factors developed by the model should then be validated in the field by proving the master meter in situ over the operating conditions. As more data is gathered, the confidence in the measurement will increase and the factors affecting the measurement can be identified and improved to reduce the total uncertainty (U_T).

Laboratory-proven meters are not normally used for large volume custody transfers because of the limited knowledge of the installation and operating effects. However, even today the economical advantages may far outweigh the uncertainty risk. In the long run, improving the diagnostic capabilities of master meters and defining the installation conditions will lower the uncertainty to an acceptable level.

Choosing a Master Meter for an Ultrasonic Application. The use of master meters for large custody transfer applications (6″ and larger meter, over 4000 bph) is not common practice because of the increased risk with large volume transfers. Normally, the meters are directly proven with a displacement prover, either a Conventional Displacement Prover or more recently an SVP. The latter provers are becoming more popular because of their smaller space requirement and recent improvements in reliability. The use of SVP to prove ultrasonic meters is difficult because of the time delay and sampling requirements. On-site or portable conventional provers may be too small for the respective ultrasonic meter. Master meters for these applications may offer a technically sound alternative.

For the purpose of this paper, it will be assumed that Coriolis Flow Meters (CFMs) and Liquid Ultrasonic Flow Meters will be added to API Standard 4.5—Master Meter Provers. Since CFMs are applied at low flow rates, they have limited application as a master meter for ultrasonic applications. PD meters, conventional turbine meters, helical turbine meters, and ultrasonic meters are possible selections.

Traditionally, PD meters were applied for high viscosity products and turbine meters for low viscosity products. This rule-of-thumb has a sound technical foundation. The PD direct measurement principal is based on slippage theory. That is, the volume through-put is segmented and directly measured in the measurement chamber, except for a small amount that slips though the clearances. This slippage is a constant for a given product, flow rate, and temperature. By proving the meter, the bias due to slippage can be identified and compensated with a calibration factor. As the product viscosity increases, the slippage decreases. At a given viscosity (about 10 cP for a 10″ PD meter) the slip is essentially "zero." Furthermore, the meter factor remains constant for all higher viscosity products and over a wide flow range.

PD meters directly measure volumetric flow and as such are not highly affected by installation conditions. Coriolis meters are likewise not affected by installation conditions as long as a stable "zero" can be established and maintained. Turbine and ultrasonic meters measure velocity and are therefore affected by the dynamic flow conditions established by the installation conditions.

Turbine meters are far more sensitive to viscosity. A conventional multibladed turbine meter is normally applied to liquid viscosity of two times the meter's size in inches and a helical turbine meter to 10 times the meter's size in inches. In either case, as the viscosity increases, the minimum flow rate must be increased, limiting the meter's overall flow range.

Ultrasonic meters, such as turbine meters, are sensitive to viscosity effects. They are not independent of these effects as some suppliers claim, but they can be designed to compensate for viscosity profile effects.

Conclusion

When properly proven, liquid ultrasonic flow meters can provide custody transfer accuracy. Conventional uni-directional and bi-directional provers provide the best direct proving method for custody transfer meters. A direct method has the least uncertainty because it eliminates the added uncertainty of a master meter.

It is difficult to directly prove a liquid ultrasonic meter with an SVP because of its small "prove volume". The time lag, with changes in flow rate, becomes a significant percentage of the run time per pass and can cause a bias error. The small "prove volume" also limits the number of samples which affects repeatability. An SVP can be used in a transfer (Master Meter) proving system. In these applications, a PD or turbine would normally be used for the master meter.

Master meter proving can be used to calibrate a liquid ultrasonic meter, but it does add uncertainty to the measurement. The transfer master meter proving method is preferred because it eliminates the bias caused by installation effects and allows the master meter to be calibrated at operating conditions. The master meter can also be easily reproven if there is any doubt about the measurement.

The greatest uncertainty is introduced when the laboratory proven master meter method is used. In this method, the systemic bias caused by the installation effects is difficult to estimate. Similar operating conditions are often difficult to simulate in a laboratory and any differences in operating conditions add to the expanded uncertainty budget. When the operating conditions change, the master meter will need to be recalibrated at these new conditions.

Liquid ultrasonic meters can accurately be proven by a number of different methods. The best proving method depends on properly assessing the "risk cost" of a measurement error compared to the cost of the measurement system. In most high-volume custody transfer applications, the proving system with the least uncertainty and corresponding lowest risk is the best.

References

1. Barry N. Taylor and Chris E Kuyatt. NIST Technical Note 1297, 1994 Edition, Guidelines for Evaluating and Expressing the Uncertainty of NIST Measurement Results, Appendix D, Terminology from International Vocabulary of Basic and General Terms in Metrology (abbreviated VIM), second edition ISO (International Organization for Standardization, Geneva, Switzerland, 1993).
2. American Petroleum Institute (API) Manual of Petroleum Measurement Chapter 1 – Vocabulary, second edition, July 1994.
3. OIML R 117 (Organisation Internationale De Metrologie Legale, French), International Recommendation, Measuring systems for liquids other than water, edition 1995 (E).
4. API Manual of Petroleum Measurement Chapter 4 - Proving Systems, Section 2 – Displacement Provers, third edition, September 2003.
5. API Manual of Petroleum Measurement Chapter 4 - Proving Systems, Section 5 – Master- Meter Provers, second edition, May 2000.
6. API Manual of Petroleum Measurement Chapter 5 - Metering, Section 8 – Measurement of Liquid Hydrocarbons by Ultrasonic Flow Meters Using Transit Time Technology, first edition, February 2005.
7. API Manual of Petroleum Measurement Chapter 13 – Statistical Aspects of Measuring and Sampling, Section 1 – Statistical Concepts and Procedures in Measurement, first edition, June 1985.
8. Kenneth Elliott.APIs Microprocessor based Flowmeter Testing Program, paper 5.4, North Sea Measurement Workshop in 2004.

This paper was originally written and presented at the 7th South Asia Hydrocarbon Flow Measurement Workshop, 2008 by Raymond Kalivoda, Petroleum Measurement Marketing Manager, FMC Technologies, Erie, Pennsylvania USA.

Used with Permission of FMC Technologies – Erie, Pennsylvania USA.

V-Cone® flow meter

The McCrometer V-Cone® flow meter is a patented technology that accurately measures flow over a wide range of Reynolds numbers, under all kinds of conditions and for a variety of fluids. It operates on the same physical principle as other differential pressure-type flow meters, using the theorem of conservation of energy in fluid flow through a pipe. The V-Cone's remarkable performance characteristics, however, are the result of its unique design. It features a centrally-located cone inside the tube. The cone interacts with the fluid flow, reshaping the fluids velocity profile and creating a region of lower pressure immediately downstream of itself. The pressure difference, exhibited between the static line pressure and the low pressure created downstream of the cone, can be measured via two pressure sensing taps. One tap is placed slightly upstream of the cone, the other is located in the downstream face of the cone itself. The pressure difference can then be incorporated into a derivation of the Bernoulli equation to determine the fluid flow rate. The

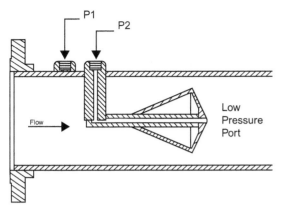

Figure 1. V-Cone flow meter.

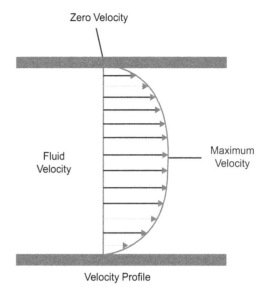

Figure 2. Velocity profile.

cone's central position in the line optimizes the velocity profile of the flow at the point of measurement, assuring highly accurate, reliable flow measurement regardless of the condition of the flow upstream of the meter.

The V-Cone flow meter is a differential pressure type flow meter. Basic theories behind differential pressure type flow meters have existed for over a century. The principal theory among these is Bernoulli's theorem for the conservation of energy in a closed pipe. This states that for a constant flow, the pressure in a pipe is inversely proportional to the square of the velocity in the pipe. Simply, the pressure decreases as the velocity increases. For instance, as the fluid approaches the V-Cone flow meter, it will have a pressure of P_1. As the fluid velocity increases at the constricted area of the V-Cone, the pressure drops to P_2, as shown in Figure 1.

Both P_1 and P_2 are measured at the V-Cone flow meter's taps using a variety of differential pressure transducers. The Δp created by a V-Cone flow meter will increase and decrease exponentially with the flow velocity. As the constriction takes up more of the pipe cross-sectional area, more differential pressure will be created at the same flow rate. The beta ratio equals the flow area at the largest cross section of the cone (converted to an equivalent diameter) divided by the meter's inside diameter.

The V-Cone flow meter is similar to other differential pressure (Δp) meters in the equations of flow that it uses. V-Cone flow meter geometry, however, is quite different from traditional Δp meters. The V-Cone constricts the flow by positioning a cone in the center of the pipe.

This forces the flow in the center of the pipe to flow around the cone. This geometry presents many advantages over the traditional concentric Δp meter. The actual shape of the cone has been continuously evaluated and tested for over 10 years to provide the best performance under differing circumstances.

One must understand the idea of a flow profile in a pipe to understand the performance of the V-Cone flow meter. If the flow in a long pipe is not subject to any obstructions or disturbances, it is well-developed flow. If a line passes across

the diameter of this well-developed flow, the velocity at each point on that line would be different. The velocity would be zero at the wall of the pipe, maximum at the center of the pipe, and zero again at the opposite wall. This is due to friction at the pipe walls that slows the fluid as it passes. Since the cone is suspended in the center of the pipe, it directly interacts with the "high velocity core" of the flow. The cone forces the high velocity core to mix with the lower velocity flows closer to the pipe walls. Other Δp meters have centrally located openings and do not interact with this high velocity core. This is an important advantage to the V-Cone at lower flow rates. As the flow rate decreases, the V-Cone continues to interact with the highest velocity in the pipe. Other Δp meters may lose their useful Δp signal at flows where the V-Cone can still produce one.

The pipe flow profile in actual installations is rarely ideal. There are many installations where a flow meter exists in flow that is not well developed. Practically, any changes to the piping, such as elbows, valves, reductions, expansions, pumps, and tees can disturb well-developed flow. Trying to measure disturbed flow can create substantial errors for other flow meter technologies. The V-Cone flow meter overcomes this by reshaping the velocity profile upstream of the cone. This is a benefit derived from the cone's contoured shape and position in the line. As the flow approaches the cone, the flow profile "flattens" toward the shape of a well-developed profile.

High Accuracy

The V-Cone flow meter primary element can be accurate to ±0.5% of reading and the Wafer-Cone® can be accurate to ±1.0%. The level of accuracy is dependent to a degree on application parameters and secondary instrumentation.

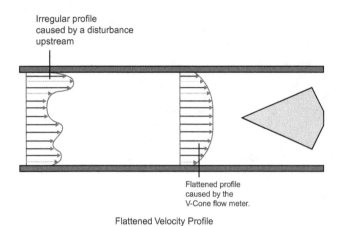

Flattened Velocity Profile

Figure 3. Flattened velocity profile.

Double Elbow and
V-Cone Flow Meter

Figure 5. Double elbow and V-Cone flow meter.

Repeatability

The V-Cone flow meter and the Wafer-Cone flow meter primary element exhibits excellent repeatability of ±0.1% or better.

Turndown

The turndown of the V-Cone flow meter can reach far beyond traditional Δp meters. A typical turndown for a V-Cone flow meter is 10 to 1. Greater turndowns are attainable. Flows with Reynolds numbers as low as 8000 will produce a linear signal. Lower Reynolds number ranges are measurable and are repeatable by applying a curve fit to the measured Δp, derived from calibration over a specific Reynolds number range.

Since the V-Cone flow meter can flatten the velocity profile, it can function much closer to upstream disturbances than other Δp meters. The recommended installation for the V-Cone flow meter is zero to three diameters of straight run upstream and zero to one diameters downstream. This can be

a major benefit to users with larger, more expensive line sizes or users which have small run lengths. McCrometer conducted performance tests of the V-Cone downstream of a single 90° elbow and two close coupled 90° elbows out of plane. These tests show that the V-Cone flow meter can be installed adjacent to either single elbows or two elbows out of plane without sacrificing accuracy.

Upstream and Downstream Minimum Straight Pipe Run Requirements For Gas Metering at Reynolds Numbers > 200,000 For $\beta \geq 0.7$, add 1D			
Size Range	**Obstruction**	**Upstream**	**Downstream**
All sizes	1 Elbow	1D	1D
	2 Elbows	1D	1D
	Tees	1D	1D
	Butterfly valve (control valve)	Not the preferred position	Valve downstream 1D
	Butterfly valve (Shutoff valve)	2D	1D
	Full port ball valve (Shutoff)	1D	1D
	Heat exchanger (depends on type)	1D	0D
	Expander (0.67D to D) over a length of 2.5D	2D	1D
	Reducer (3D to D) over a length of 3.5D	0D	0D

The meter and the adjoining pipe should have equal IDs.

Single Elbow and
V-Cone flow meter

Figure 4. Single elbow and V-Cone flow meter.

Long Term Performance

The contoured shape of the cone constricts the flow without impacting the flow against an abrupt surface. A boundary layer forms along the cone and directs the fluid away from the beta edge. This means the beta edge will not be as subject to the usual wear by unclean fluids, as is the case with an orifice plate. The beta ratio will then remain unchanged and the calibration of the meter will be accurate for a much longer time.

Signal Stability

Every Δp meter has a "signal bounce." This means that even in steady flow, the signal generated by the primary element will fluctuate a certain amount. On a typical orifice plate, the vortices that form just after the plate are long. These long vortices create a high amplitude, low frequency signal from the orifice plate. This could disturb the Δp readings from the meter. The V-Cone forms very short vortices as the flow passes the cone. These short vortices create a low amplitude, high frequency signal. This translates into a signal with high stability from the V-Cone. Representative signals from a V-Cone flow meter and from a typical orifice plate are shown in Figure 6.

Signal Stability

Figure 6. Signal stability.

Precision Tube
V-Cone Flow Meter

Figure 7. Precision tube V-Cone flow meter.

Wafer-Cone Flow Meter

Figure 8. Wafer-Cone flow meter.

Low Permanent Pressure Loss

Without the impact of an abrupt surface, the permanent pressure loss is lower than a typical orifice plate meter. Also, the signal stability of the V-Cone flow meter allows the recommended full scale Δp signal to be lower for the V-Cone flow meter than other Δp meters. This will lower the permanent pressure loss.

Sizing

The unique geometry of the V-Cone flow meter allows for a wide range of beta ratios. Standard beta ratios range from 0.45, 0.55, 0.65, 0.75, and 0.80.

No Areas of Stagnation

The "swept through" design of the cone does not allow for areas of stagnation where debris, condensation, or particles from the fluid could accumulate.

Mixing

The short vortices described earlier mix the fluid thoroughly just downstream of the cone. The V-Cone flow meter is currently used in many applications as a static mixer where instant and complete mixing is necessary.

Application Data

The user must provide application parameters so that the appropriate V-Cone flow meter may be selected. McCrometer has an extensive meter performance database of fluid properties which can be utilized for sizing purposes.

Flow Calculations Nomenclature (All references to "V-Cone" are to the V-Cone flow meter.)

Symbol	Description	English Units	Metric Units
α	Material thermal expansion α or α_{cone}, α_{pipe} (alpha)	$^\circ R^{-1}$	$^\circ R^{-1}$
β	Beta ratio	—	—
C_p	Flow meter coefficient	—	—
d	Cone outside diameter	In	mm
D	Pipe inside diameter	In	mm
ΔP	Differential pressure (dp)	In WC	Mbar
ΔP_{max}	Maximum differential pressure on sizing	Refer to note 4	Refer to note 4
F_a	Material thermal expansion factor	—	—
K	Gas isentropic exponent	—	—
K_1	Flow constant	$\sqrt{\dfrac{lbm \times ft^3}{s^2 \times in\ WC}}$	$\sqrt{\dfrac{kg \times m^3}{s^2 \times mbar}}$
K_2	Simplified liquid flow constant	See note 4	See note 4
μ	Viscosity	cP	cP
P	Operating pressure	psiA	barA
P_b	Base pressure	psiA	barA
Q	Actual volume flow	ACFS	m^3/s
Q_{max}	Maximum flow rate on sizing	See note 4	See note 4
Q_{STD}	Standard gas volume flow	SCFS	Nm^3/s
Re	Reynolds number	—	—
ρ	Flowing density (rho)	lbm/ft^3	kg/m^3
ρ_{water}	Water density	$62.3663\ lbm/ft^3$	$999.012\ kg/m^3$
S_g	Specific gravity of the gas	—	—
T	Operating temperature	$^\circ R$	K
T_b	Base temperature	$^\circ R$	K
T_d	Deviation from standard temperature ($^\circ R$)	$T_d = T - 527.67$	$T_d = \frac{9}{5}T - 527.65$
U_1	Unit conversion	0.0360912 psiA/inWC	0.001 barA/mBar
U_2	Unit conversion	144 in^2/ft^2	1,000,000 mm^2/m^2
U_3	Unit conversion	167.213 lbm/s^2 ft in WC	100 kg/ms @ mbar
U_4		124.0137 cP ft s/lbm	1
U_5	Unit conversion	2.6988 $^\circ R\ lbm/ft^3$ psiA	348.338 K kg/m^3 barA
V	Velocity	ft/s	m/s
Y	Gas expansion factor	—	—
Z	Gas compressibility	—	—
Z_b	Base gas compressibility	—	—

General Flow Calculations

3.2.1	V-Cone Beta Ratio	$\beta = \sqrt{1 - \frac{d^2}{D^2}}$	β from sizing report
3.2.2	Flow constant	$k_1 = \frac{\Pi\sqrt{2 \times U_3}}{4 \times U_2} \times \frac{D^2 \times \beta^2}{\sqrt{1 - \beta^4}}$	
3.2.3	Material thermal expansion factor	$F_a = 1 + 2 \times \alpha \times T_d$	See note 1
3.2.4	Material thermal expansion factor if cone and main pipe are made of different materials	$F_a = \frac{D^2 - d^2}{[(1 - \alpha_{pipe} \times T_d) \times D]^2 - [(1 - \alpha_{cone} \times T_d) \times d^2]^2}$	See note 1
3.2.5	Pipeline velocity	$v = \frac{4 \times U_2 \times Q}{\Pi \times D^2}$	
3.2.6	Reynolds number	$\text{Re} = U_4 \times \frac{v \times D \times \rho}{\mu}$	A dimensionless number that can be used to correlate meter calibration in different fluids
3.2.7	V-cone gas expansion factor	$Y = 1 - (0.649 + 0.696 \times \beta^4)\frac{U_1 \times \Delta P}{k \times P}$	For liquids, $Y = 1$
3.2.8	Wafer gas expansion factor	$Y = 1 - (0.755 + 0.678 \times \beta^8)\frac{U_1 \times \Delta P}{k \times P}$	For liquids, $Y = 1$
3.2.9	Liquid density	$\rho = \rho_{water} \times S_L$	
3.2.10	Gas density	$\rho = U_5 \times \frac{S_g \times P}{Z \times T}$	
3.2.11	Actual volume flowrate	$Q = F_a \times C_D \times Y \times k_1 \times \sqrt{\frac{\Delta P}{\rho}}$	See notes 2, 3 & 5
3.2.12	Standard gas volume flowrate	$Q_{STD} = Q \times \left(\frac{P \times T_b \times Z_b}{P_b \times T \times Z}\right)$	Converts actual flow to standard flow at base conditions

Simplified Liquid Calculation

3.3.1	Simplified Liquid Flow Constant	$k_2 = \frac{Q_{max}}{\sqrt{\Delta P_{max}}}$	See note 4
3.3.2	Simplified liquid flow rate	$Q = k_2\sqrt{\Delta P}$	See note 4

Notes:

1. Material Thermal Expansion—The thermal expansion equations correct for dimensional changes which occur as the operating temperature deviates from the base temperature of 68 °F (see 3.2.3 and 3.2.4) The F_a factor can be excluded from the flow equation if the operating temperature is <100 °F, <559.67 °R, <37.78 °C, <310.93 K. If the F_a factor is significant and the operating temperature is stable, then a constant F_a value can be used. If the F_a factor is significant and the temperature varies, then an F_a factor should be calculated prior to every flow calculation.

2. Discharge Coefficient—Discharge coefficients can be implemented in the flow equations via several different methods. Typical methods are average C_D, C_D look up table, or C_D fitted data. If a C_D look up table or fitted data is utilized, additional calculations must be made based on the Reynolds number [see example process 3(d) and 5(b)].

3. Liquid—Typical Calculation Process
   a. Given: D, β, ρ, C_D, and input of ΔP Calculate: 3.2.2, 3.2.11
   b. Given: D, β, ρ, C_D, and input of ΔP, T Calculate: 3.2.2, 3.2.3 or 3.2.4 if req., 3.2.11
   c. Given: D, β, S_l, C_D, and input of ΔP, T Calculate: 3.2.2, 3.2.3 or 3.2.4 if req., 3.2.9, 3.2.11
   d. Given: D, β, μ, ρ, C_D look up, and input of ΔP Calculate: initially set $C_D = 0.8$, 3.2.2, 3.2.3 or 3.2.4 if required 3.2.11

Iterate until flowrate is <0.01% different from last calculation

Iterate until flowrate is <0.01% different from last calculation

4. Simplified Liquid Calculation—The simplified liquid calculation can be used if the operating temperature is stable and the C_D is constant. The simplified flow constant (k_2) can be calculated from Equation 3.3.1 using the V-Cone Flow Meter Application Sizing sheet. The flow rate can then be calculated using Equation 3.3.2. Units of measure will be the same as those listed on the V-Cone Flow Meter Application Sizing sheet.
5. Gases and steam—Typical Calculation Process: Given: D, β, μ, S_g, Z, k, C_D, and inputs of ΔP, P, T Calculate: 3.2.2, 3.2.3 or 3.2.4 if req., 3.2.7 or 3.2.8, 3.2.10, 3.2.11 b. Given:

D, β, μ, S_g, Z, k, C_D look up, and inputs of ΔP, P, T Calculate: initially set $C_D = 0.8$, 3.2.2, 3.2.3 or 3.2.4 if req., 3.2.7 or 3.2.8, 3.2.10, 3.2.11.

Fluid Properties—Fluid properties such as velocity, compressibility, and isentropic exponent vary with temperature and to some extent pressure. The viscosity in the calculations above could affect the selected C_D value, the compressibility directly affects the density and the isentropic exponent effects the Y factor, although to a small degree. The instrumentation industry uses many different approaches to calculate flow. McCrometer application engineering and the customer must determine which fluid properties are calculated at each set of flow conditions and which properties are constant.

Used with permission of McCrometer, Inc., Hemet, California.

16: Instrumentation

Types of control systems

Early pipeline systems depended upon manual control and the use of tanks to maintain a pipeline system in balance. On long pipelines that required multiple pump stations, a surge tank was used to catch excess flow or make up the deficiency. Engine speed was varied to maintain the level in the surge tank at an acceptable level. This type of operation was fine as long as one grade of product was being pumped but was not satisfactory for batching different products unless there was a tank available for each grade of product handled.

It was obvious that to build a tight line system, one that would not require surge tankage, some kind of automatic control system was needed to keep the booster stations in balance. Before the widespread use of centrifugal pumps, engine speed (for reciprocating pumps) was controlled to maintain the suction and discharge pressure within limits by sensing discharge and suction pressure and directing a control signal to either the engine governor or a control valve. When control valves are used with reciprocating pumps, they are connected to recycle fluid from the discharge of the pump to the pump section.

When centrifugal pumps came into use, the same type of control system was used except that the final control element was a control valve installed in the pump discharge line as shown in Figure 1. The control system at origin pump stations normally controls discharge pressure or system flow rate. Intermediate booster stations require a control system that

will control suction or discharge pressure and in some cases, motor load. An examination of the system line friction and pump head curve will show that simultaneous control of suction and discharge pressure and high flow rate is not required since the pipeline system characteristics will not permit the coexistence of these events except in case of extreme system upset. As will be seen later, the control system is configured so that only one controller will be in control at a given time.

There are several modes of control available to the pipeline engineer. Mode of control is the manner in which control systems make corrections relative to the deviation of the controlled variable. The modes of control commonly used by pipeliners include:

- Proportional position
- Proportional plus reset
- Proportional-speed floating

Other modes such as proportional plus reset plus derivative are more commonly used in the process industries. This mode of control may, however, be used in pipeline measurement stations where it is necessary to control the temperature of a heater used to heat the product being pumped or measured.

In the proportional-position mode, there is a constant linear relation between the controlled value and the position

Figure 1. Typical control scheme for a pipeline booster station.

of the final control element. With this type of control system, the amount of correction will be the same regardless of the amount of deviation of the controlled variable. A proportional-position control system will not necessarily return the controlled variable to the exact value of the set point when an upset occurs. This may be satisfactory for some types of control systems, but not for pipeline pump station control.

The offset deviation noted in the proportional-position system can be overcome by adding reset to the controller. The reset action will continue to correct valve position until no deviation occurs between the controlled variable and controller setpoint.

Proportional-speed floating causes the final control element to change at a rate proportional to the amount of deviation between the setpoint and the measured variable. Correction to the final control element will continue until there is no deviation between the setpoint and the measured variable. The amount of the correction signal is proportional to the amount of deviation between the setpoint and the measured variable. This mode of control will produce an exact correction for any system load change.

Proportional-position and proportional-speed floating are the control modes most commonly used for pipeline control.

Figure 1 shows a typical control scheme for a single unit booster station. A minimum control system will include a suction pressure and discharge pressure controller. Other controllers such as motor current controllers and a ramp generator may be connected in the control loop.

The suction pressure controller is set to maintain the suction pressure above the low suction pressure shut down setting while the discharge pressure controller limits the station discharge pressure to a value that will not exceed the maximum allowable operating pressure. The output from each of the controllers is fed to a signal selector that will select either the high or low signal, depending upon the control action selected, and direct this signal to the control valve. Some control systems are designed so that a selector module may not be required, but they will include a means for selecting the proper signal. In a properly designed system, the transfer of control from one controller to another will be smooth and bumpless.

A ramp generator is used for ramping the control valve to a new position. The control valve may be closed when the station is at rest, and upon station startup the ramp generator will cause the valve to open slowly until one of the controllers is in control of the system or until the pump is "floating"— that is, operating at a condition where neither suction nor discharge pressure control is required. It may also be used to ramp the valve partially closed prior to unit shut down to minimize system upset. This feature is desirable in a multi-unit booster station since when one unit is shut down, the remaining operating units will be operating momentarily at a no-flow condition. This condition will continue to exist until the system pressure decays to a value equal to the discharge

pressure of the remaining units. Ramp speed should be adjustable so that it can be fine-tuned to the system requirements.

The control valve action, i.e., valve opens on increasing signal or valve closes on decreasing signal, will determine how the controller action is set. The suction pressure controller will be set opposite of the discharge controller.

Although not shown, the control system is usually equipped with a manual/automatic switch so that the control valve may be controlled in case one of the controllers fails or during system maintenance.

The controllers may also be equipped with remote setpoint control so that controller setpoint may be varied from a remote location to accommodate a changed condition in the system.

Figure 2 shows a control system for a multi-unit booster station. The control systems for single- and multi-unit stations are identical except that in a multi-unit station the station pressure transmitters are located so that one control system will suffice regardless of the number of units installed. It should also be noted that one control valve serves both units.

The control valve may be operated electrically, pneumatically, or hydraulically. Hydraulic systems offer the advantage of being able to operate the valve at a high rate of speed at virtually any system load. Most hydraulic operators are equipped with dual cylinders and may be equipped with quad cylinders where the system loads are high. The selection of control valves is covered in Section 7.

Variable speed drives or controls are becoming an economically attractive choice for the operation of pumps. The use of these devices will virtually eliminate throttling excess head and the attendant energy costs. Few systems are operated at a constant rate, and therefore some throttling is inevitable. Also, systems transporting multiple grades will require throttling to keep the system in balance as the batches move through the system.

Most control systems used for pipeline control are made up of standard control system elements. Some applications may, however, have unique characteristics that will require a system that is just a little bit different from a standard control system. Most of these types of systems can be put together using standard modules, or a programmable logic controller may be used.

In Section 13—Liquid Hydraulics, the discussion on hydraulic gradients illustrated a case where a hill required maintaining back pressure at the end of the line. It was shown that the amount of back pressure required varied with the system flow rate. A conventional back pressure controller and control valve installed at the end of the line and set for the worst case (maximum back pressure) would work; however, it would not be economical since more pressure than required would be maintained at the higher flow rates. The pressure at the hill could be telemetered to the end of the line, where it could be used as the measured variable for the controller,

SDP = Sta. Disch. Pr.
SFR = Sta. Flow. Rate.
SSP = Sta. Suc. Pr.
RTU = Remote Terminal Unit

Figure 2. Multi-unit booster station.

The setpoint would be set to the amount of pressure required at the hill which in this case is 200 psig. This type of control system would not penalize system economics since it would always maintain the required 200 psig at the hill. However, it would require the installation and maintenance of a telemetry link.

Another option is to use a relationship controller that would not require telemetry of the hill pressure. The relationship controller would use the system flow rate to drive the setpoint of a pressure controller. The controller would need to be equipped with limit stops so that the setpoint could only be changed between the upper and lower limits. High system flow rates would drive the controller setpoint to the minimum allowable value of back pressure and low system flow rates would drive the controller to the allowable maximum limit.

Intermediate system flow rates would drive the setpoint to an intermediate pressure setting.

With today's technology and hardware, the control engineer is limited only by his imagination in solving simple as well as complex control problems. Programmable logic controllers (PLC) and self-tuning controllers are being used more and more in the control of pipelines and pipeline stations. Control systems are being incorporated as a part of the PLC. With the addition of a modem and a suitable communication protocol, the PLC may also be used as the remote terminal unit (RTU) for the supervisory control and data acquisition system (SCADA).

The following article, "Developments in Pipeline Instrumentation," by Dr. R. A. Furness, is an excellent discussion on the state-of-the-art in pipeline instrumentation.

DEVELOPMENTS IN PIPELINE INSTRUMENTATION

R. A. Furness, Center for Flow Measurement, Department of Fluid Engineering and Instrumentation, Cranfield Institute of Technology

Abstract

Pipelines are an integral part of the world's economy and literally billions of pounds (£) worth (lb is a unit of weight) of fluids are moved each year in pipelines of varying lengths and diameters. As the cost of some of these fluids and the price of moving them has increased, so the need to measure the flows more accurately and control and operate the line more effectively has arisen. Instrumentation and control equipment have developed steadily in the past decade but not as fast as the computers and microprocessors that are now a part of most large-scale pipeline systems. It is the interfacing of the new generation of digital and sometimes "intelligent" instrumentation with smaller and more powerful computers that has led to a quiet but rapid revolution in pipeline monitoring and control. This paper looks at the more significant developments from the many that have appeared in the past few years and attempts to project future trends in the industry for the next decade.

Introduction

Our need to transport fluids from the point of production to the area of end use has led in the past 25 years to a large increase in the number of pipelines being designed and constructed. Many of these carry hazardous or toxic products, often through areas of high environmental sensitivity or close to centers of population. The cost of oil production rose sharply to more than $35 per barrel before the recent sudden fall, and the cost of manufactured chemicals and industrial gases has also risen in the same period. With the need to safeguard the cost of the transported fluid and also the population and the environment close to the pipeline, on-line monitoring and in some cases 24-hour surveillance is now becoming routine in many systems throughout the world.

When one sees maps of the pipeline network in some parts of the world, the number and length of pipelines is staggering. At the recent Aberdeen conference, one speaker gave just two of the many statistics that can be produced to show how vast the network is. One line runs from central Canada eastwards to the Toronto/Montreal area. This line alone has 54 million square meters of pipe surface to survey. Also in 1984 some 6 billion barrels of oil and 4.5 billion barrels of manufactured liquids were transported in pipelines in the USA. The actual world volume of pipeline fluids is many times this amount and the value immense.

Pipelines are not new, indeed almost 50% of the world's total network is 30 or more years old. These have been subject to corrosion and wear, and some have carried fluids for which they were not designed. Others have been operated outside design limits. Monitoring and control systems therefore are not just required for measurement purposes but also for integrity surveillance and on-line fault diagnostics. Some of the many developments that can be applied in modern day line operation are briefly reviewed, and some design pointers for their integration into complete systems are given.

Flow measurements

Flow measurement is the most important process variable in the operation and control of pipelines. Simple line flow balances are used to check for discrepancies on hourly, daily, or weekly bases. This may identify that a problem exists, but flowmeters at the end of a line will not identify the location of the problem. For this, pressure transducers along the pipeline will be required together with the flowmeters to establish where the pressure gradient has deviated from that predicted from the basic equations of fluid flow. These equations use pressure/flow values at inlet and outlet to confirm the line is in balance. Such a simple system is described in more detail later in this paper.

Most flowmeters sold are of the differential pressure type, and of these the predominant meter is the sharp-edged orifice plate. Many hundreds if not thousands of these are installed on pipelines around the world to measure gas, liquid, and mixed fluid streams. With the arrival of the modern flow computer these types of flowmeter have almost received a new lease on life. One fact not appreciated, though, is the total measurement uncertainty of such devices. Frequently we read performance claims of $\pm0.5\%$ and better for these pipeline metering systems. This is very difficult to believe when it must be remembered there is an uncertainty of $\pm0.6\%$ on the data used to derive the flow coefficient in the basic calculation. When all the other variables that can affect the uncertainty of measurement are included, such as fluid composition changes, ambient and process temperature and pressure variations, errors in temperature and pressure measurement, A/D conversion and computational errors, to name but a few, then it is unreasonable to expect uncertainties much better than $\pm1\%$ FSD. If wide flow rates are also experienced, then the uncertainty of the low flow determination could be $\pm5\%$ FSD or worse. This makes their use as accurate pipeline flow devices somewhat questionable, although they are very common in the metering of natural gas in this country.

Turbine flowmeters also find widespread use in pipeline applications, particularly hydrocarbons. They all suffer from bearing limitations, however, and this means they require regular calibration to ensure good performance. Recent developments have included twin rotor designs, where measurements of the speed of both rotors simultaneously has led to the meter becoming self diagnostic. These meters can now detect their own bearing wear and also deviations from initial calibration due to blade damage. The same on-board microprocessors can also check the integrity of the data generated by the turbine meter, as well as providing alarm output for loss of signal and other problems.

Two different designs have been applied in pipeline systems with good success. The first design, from Quantum Dynamics and shown in Figure 1, has two rotors with two independent sets of bearings. The shaft and slave rotor ride on one set of bearings with the measurement rotor upstream of the slave motor riding on a second set of bearings on the rotating shaft.

Relative motion between the indicating turbine and the shaft therefore remains close to zero. This plus a drag-free RF pick-off results in an almost frictionless flowmeter. Should the shaft or measuring motor bearings begin to wear, this will result in a change in the ratio of the rotor speeds for a fixed flow rate. Thus, bearing wear can be detected on-line and action taken. This design of meter is also different from all other turbines because of the slender center body and axially short blades. Because friction is very low, the fluid does not have to be accelerated past the rotor to obtain sufficient torque to overcome retarding forces. Thus, pressure drop is also low and pressure and temperature

Figure 1. Twin rotor turbine meter (courtesy Quantum Dynamics Inc).

measurements can be carried out at the point of volume flow rate determination. They have been shown to be outstanding meters for leak detection purposes on gas lines.

The Auto Adjust turbo-meter from Rockwell International is also self-checking but is in addition self-adjusting. The design shown in Figure 2 is quite different from the Quantum meter. The rotors are next to each other with the measuring rotor downstream of the main rotor. Constant performance is obtained because the sensor rotor detects deviations in blade angle of fluid leaving the main rotor. With a very small angle, the sense rotor runs at a much lower speed than the upstream rotor. Performance is then given by the difference in the speed of the two rotors rather than from a single rotor. When bearings begin to wear, the main rotor will spin slightly slower and so the exit angle will also change. This is sensed by the second rotor, which adjusts speed by an equal amount. Bearing wear is therefore automatically adjusted out. These

meters have been installed in many gas transmission lines in North America with generally very good performance. The Quantum meter, on the other hand, is a newcomer to pipeline applications, but personal firsthand experiences in a gas pipeline were favorable.

Pipelines are generally operated on material balances (mass), but most flowmeters are volumetric devices. This means that for pure fluids measurements of temperature and pressure plus an appropriate equation of state are required to enable density to be determined. For mixtures of gas or liquids or indeed as an alternative to the above method, on-line densitometers are used. The product of volume flow and density gives mass flow. Until recently pipeline mass flow-meters were not available, but recently a number of Coriolis type direct mass meters have appeared. These have found slow but steady acceptance in industrial applications and are beginning to find applications in pipeline monitoring. One such design is shown in Figure 3, and recently four other designs were launched at the INTERKAMA trade show in West Germany. This type of meter is applicable to both liquids and gases, and some have been used on liquid/liquid and gas/liquid mixtures but with mixed success. They are currently undergoing rigorous evaluation and analysis in many chemical, process, and small pipeline applications, and indications are that they are standing up well in some applications but not others. Accuracy is around ±0.5% of reading or better, and they do not need additional measurements to determine flowing density. Indeed, by monitoring the output signal, they can be used to indicate density as well as mass flow. The output is in digital and/or analog form making them compatible with many types of flow computer.

Direct mass measurement has one main advantage over volume measurement, since an integration of the signal over a period of time will indicate the pipeline fluid inventory directly. The disadvantage for pipeline operators is the current size range that is available, the largest being 150 mm bore. As most major pipelines are in the range of 500–2,000 mm bore, these Coriolis meters must be installed in parallel to meet the larger applications. Mass metering is relatively

Figure 2. Auto adjust gas pipeline meter (courtesy Rockwell International, Inc.).

Figure 3. Micro motion direct mass flowmeter (courtesy Brooks Instruments).

new, and the next decade should see a steady increase in sales and acceptance of these meters.

The ideal flowmeter is one that is installed on the outside of the line but gives the performance of the best flowmeters installed inside the line. Ultrasonic time-of-flight meters are slowly developing toward these apparently conflicting but demanding criteria. Multiple beam systems have been installed on many pipelines, the most notable of which is the Alaskan oil pipeline. Some 23 meters, each 4 ft in diameter, look for gross imbalances along sections of this important installation. They permit the early detection of product spillage and also indicate the section of line giving the problem. This system has been operating since the late 1970s with millions of barrels of fluid being shipped through the line in that time. Experience of these meters has been very favorable in this application.

Ultrasonic meters, like most modern types of meter, were initially oversold, but the main problem has been to convince users that the two types available, the Doppler and transit time meters, are completely different animals that both use ultrasonics to infer flow. When ultrasonic meters are mentioned, people tend to think of Doppler types. Most of these are relatively crude flow indicating devices. The modern clamp-on transit time meter can indicate flow to around ±2% of reading or better depending on design, compared to the average Doppler meter, which is a ±5% FSD device. The advantage of ultrasonic methods is the negligible headloss and the ability to install other portable or dedicated systems without line shutdown. It is surprising that more have not been installed in pipeline systems, but as discussed later in this paper, this is an example of the conservatism often shown in the application of new technology.

Flow measurement has really been transformed by the application of digital techniques to on-line signal analysis. Much more data than just flow rate is now available, and with the latest systems, operators are taking advantage of the new generation of micro and flow computers to control and run their pipelines. It is unusual these days not to install a local flow computer with each flowmeter, with temperature, pressure, and differential pressure signals being fed into the machine for local "number crunching."

Proving devices

The problem with all flowmeters is that they are affected to some degree by the fluid they are metering. For the least uncertainty of measurement, the flowmeter, whatever type is chosen, should be calibrated on-line under actual operating conditions. In the past this was performed with large and expensive pieces of hardware called ball provers. These had to be large to provide adequate volume within the calibrated section of pipe to ensure sufficient resolution and accuracy for the flowmeter they were intended to calibrate.

Figure 4. Modern compact meter prover (courtesy Brooks Instruments).

The latest generation of proving devices, called compact provers, are essentially time/volume calibrators that eliminate the ±1 count error present on their "big brothers." It is physically much smaller but with the same or better performance. There are currently 10 different versions on the market, of which five are available in the UK. The prover, one design of which is shown in Figure 4, uses pneumatics to launch a piston, which is then driven by the fluid through a calibrated chamber, displaying a known volume. The key to their excellent performance is the data acquisition system illustrated in the lower part of Figure 4. As the piston passes through the calibrated section, pulses from a high frequency internal clock are counted to determine the time, A, to displace the calibrated volume D. Once the counting has started, a second internal timer is gated to respond to the first input pulse from the flowmeter. This timer then records the time interval B to collect C whole flowmeter pulses.

During the proving run there were then C × A/B meter pulses. The meter factor is found by dividing this count by the volume displaced:

$$K = \frac{A}{B} \times \frac{C}{D}$$

Some designs are uni-directional and some bi-directional. In the uni-directional types pneumatics or hydraulics launch and retrieve the piston; in the bi-directional designs the process fluid is used to drive the piston. Other versions incorporate double-barrelled liners to remove the need for pressure compensation and also reduce the effect of temperature variations.

In one proving run there may be as few as 100 pulses collected, so timing error must be minimized. If the pulse output is frequency modulated, from a positive displacement bulk meter, for example, then these few pulses may not represent a true statistical sample of the pulse train and multiple passes may be required. Alternatively, pulse interpolation techniques may be required, several of which are applicable to give the accuracy required. These are currently the subject of intense activity in standards committees who are producing a new general code of practice on meter provers.

The compact prover is better suited to offshore platform applications and to field use on operating pipelines. Being portable, it can be easily coupled in series with the meter to be proved. A suitable choice of materials allows proving on a vast range of fluids to be carried out. Operational experience has generally been favorable, and several direct comparison tests with ball provers have shown the new generation of compact provers to be superior in most respects. However, their acceptance in custody transfer applications has been slower than had been expected, even though several countries have certified some designs for Weights and Measures approval. Some international calibration centers are now using this technology for routine traceable calibrations, and confidence is growing as operational experience is accumulated.

By far the most economic method of measuring flows in pipelines is to install pairs of flowmeters in custom-designed metering stations. By monitoring the ratio of the outputs, a measure of wear or drift is obtained. Provided the outputs of the meters remain within preset tight limits, uncertainty of flow rate measurement is low. The disadvantage of this on pipeline operation is that there are two meters to be maintained and checked, and should faults develop, the pipeline may have to be shut down more frequently, unless by-pass loops are provided when the metering station is designed. Turbine meters and PD meters with pulse outputs give the best performance, particularly if they are regularly proved with either a dedicated or portable prover of the type described.

For large bore lines, particularly gas lines, the proving of flow meters is very expensive if it can be performed at all. The majority of installations rely on orifice plates built to the standards. Recently a new high-pressure gas proving station was completed by British Gas at Bishop Auckland. Here turbine and other meters can be calibrated in actual operating conditions on natural gas. Such an installation has previously only been available in Holland, where gas calibration standards have been developed to a high level of confidence over many years of patient data collection. With the new British Gas facility we should see the ability to cross check our own standards against international standards.

Valves

Valves are an important part of any pipeline system. They come in many shapes and sizes and perform numerous functions. Some act as primary flow regulators, some as check valves for flow direction control, and some as pressure reducers for safety purposes. Developments have centered in two main areas, materials and actuators (both simple and intelligent). One of the development sources is again the North Sea, where the demands for better reliability, better in-field performance, and reduced cost have forced many manufacturers to look critically at designs. As one of the most expensive components in the valve is the body, effort has been concentrated in this area, both in terms of new materials to meet more demanding specifications and of quality and price. These last two items are critical to the survival of many companies, and in a market sector that is growing slowly or even declining, as suggested by one leading instrument journal, valve body cost savings, however small, may mean the difference between life and death.

Slimline butterfly valves have recently been introduced by a number of manufacturers to meet the demand for lower cost reliable devices. These can be installed between existing pipeline flanges. They can be manufactured in a wide range of materials with special seal materials chosen for the most demanding of duties. With suitable actuators, even valves as large as 60 in. in diameter do not take up much space.

Other designs of control valve provide straight-through flow patterns to minimize flow turbulence, vibration, and noise. Should a pipeline be fitted with acoustic line break detectors (see section following), line noise will be important in setting the sensitivity of the detectors, and hence the type of valve fitted will be important to system operation. Other operational problems such as "hunting" in globe valves have prompted the designers to work more closely with the operators to define full specifications. Pipeline operators often run their systems with rapid transients present. Demand changes of both a positive and negative value can subject the control valves to very severe step changes in flow and pressure, and initially many conventional valves could not withstand such treatment.

This is where intelligent sensors on pressure relief valves have come into their own. They are able to give better speed of response and meet some stringent operating criteria. Pressure may also rise gradually due to a steady rise in line temperature, as well as the rapid transient caused by block valve closure or pump failure/surge. These transients travel at sonic velocity and even tenths of a second's advantage in operating the valves can affect safety on the line. Microprocessor-based detectors fitted to the valves can be used to sense line conditions and provide suitable inputs to the valve. In some advanced designs manufactured in the United States, it is possible to program intelligence into the unit to begin closure/opening based on measurements of an anticipated condition such as pump surge further down the line. If the signals from one part of the line can be transmitted to other sectors, corrective action can be taken before safety blowoff valves are activated, releasing product into the environment.

Figure 5. Intelligent pressure relief valve.

A pipeline pressure relief system based on this principle is shown in Figure 5. A strain gauge sensor fitted on the suction side of the compressor is monitored by a microprocessor that linearizes and conditions the signal. The microprocessor operates a pair of solenoid valves acting as the pilot mode with the line pressure acting as the actuating medium. By switching solenoids, the relief valve can be opened, closed, or programmed to act as a regulator at the pressure trapped in the valve body. Such a system can provide adequate response to step changes on the order of 70 bar in 3 milliseconds. In this system the status of the pipeline pumps is provided as an additional input to the microprocessor to trigger action before the pressure surge occurs. Tests on large lines have shown such intelligence to be vital in the safe operation of systems when 5,000 hp pumps were suddenly lost.

Line block valves can also be actuated by output signals from line break detectors. This is important in isolating problem sections of line carrying toxic products. The wave detector will sense problems in seconds, and the line can be closed to minimize the amount of toxic product released. That section of the line may well depressure, but the release effects will be reduced. Such systems are fitted as anti-terrorist devices on lines in problem areas of the world such as the Middle East and Central Africa, where all too often pipelines are political targets.

The choice of materials for valve bodies and the styles of valve available is now very wide. A survey in *Processing* (January 1986) listed 15 different types supplied by 110 companies. These range from the cryogenic applications up to line temperatures in excess of 700°C. Selecting the correct valve for the duty is now the biggest problem facing the user, and where it is necessary to take outputs into computers for safety and control purposes, the choice becomes more difficult. In such a competitive market, developments are bound to come thick and fast, and it will be interesting to watch the development of the next generation of intelligent pipeline valves.

Acoustic line break detectors

Problems with the pipeline are frequently accompanied by the generation of noise, either from problem valves (see above) or line breaks. If the flow becomes too high from a surge or a leak occurs through corrosion or third-party intervention, then there are instances where vibrations in excess of 20 kHz are produced. These frequencies are in the ultrasonic range and can be detected with suitable transducers. Such devices have been made portable so that pipeline crews can clamp them at any point along the line to check for noise. By noting the signal strength, the source of the problem can be pinpointed. British Gas has used this principle to develop one of two types of pipeline monitoring tools that are described later.

A similar technique, though based on a different principle, is the acoustic "wavealert" monitor, though more correctly called a negative pressure wave detector. The central element is a piezo-electric sensor that gives an output when stressed. When a line rupture or leak occurs, there is a sudden drop in line pressure at the source of the problem, followed by line repressurization a few milliseconds later. The rarefaction wave generated by these events moves away from the leak in both directions at the speed of sound. The pipewalls act as wave guides enabling this wave to travel for great distances, attenuating in amplitude as it does so. Sensors placed at known points along the line can detect the passage of this pressure transient. From the triggering of internal timing circuits and a knowledge of the prevailing acoustic velocity for the line, the leak location can be calculated.

Such devices are particularly useful for detecting large breaks in lines very rapidly. The transient wave moves at 1 mile per second in liquids and 1 mile in 5 seconds in gases.

Figure 6. Negative pressure wave "wavealert" monitor (courtesy Spectra-Tek).

If the detectors are positioned at intervals of 2 to 5 miles in critical areas, then the break is detected in a few seconds. If the output signal is fed into a computer, then this can be used to trigger block valves to isolate the problem section of the line very easily. Figure 6 shows one further development where the instrument can be made to cancel line noise and use the full measuring capacity of the sensor for signal detection. There is, however, the practical problem in setting the background threshold correctly. Pipeline noise is generated by bends, valves, pumps, regulators, etc., and it is crucial to be able to remove this background noise to avoid unnecessary line shutdowns when the line noise exceeds the detector setting. The Reynolds number of the flow has been shown to be the important parameter in installations both in the United States and the United Kingdom. Experience has also shown that the installation of these devices is critical to reliable performance. It is important to have the sensor as close as possible to the line to avoid attenuation effects and improve sensitivity. Both this and the background line noise will influence the number of "false alarms" from the instrument.

ICI has been experimenting with strain gauge-type sensors on their cross country ethylene line, and these are showing more promise. These experiments will be closely monitored in the next few years to see whether we can improve the reliability and sensivity of line break detectors.

"Smart" pressure sensors

One area where the impact of microprocessors has been felt is in the development of smart sensors, particularly pressure. Many physical effects respond directly or indirectly to pressure, and as a result, the first generation of "smart" process transducers appeared on the market some years ago. Surprisingly, however, market penetration has been slow, mainly, I suspect, because hand-picked conventional pressure and differential pressure cells perform almost as well over the limited range often required. There is virtually no limit to the amount of intelligence and diagnostics that can be incorporated into pressure transducers, but so far only those that are reasonably linear or that have well defined characteristics have been commercially exploited. The need for linearity is now no longer there because all signal processing the digital and algorithms can be written to cope with any signal/pressure curve, provided it is repeatable. Even hysteresis can be taken account of.

It is in the communications area that the full potential of intelligent transducers is likely to be felt. Hand held field communicators enable the user to rerange, configure, and display process output values without the need to be close to the transmitters. On-line reranging is particularly useful when the pressure transducer is located in a very inaccessible place. It is possible to use two-way communication links to examine the outputs from a variety of intelligent transducers located at remote outstations, but the cost of such a feature should be carefully considered as this is likely to be expensive. Intelligent transducers are more expensive than the commercial instrumentation, but in certain applications cost is not a consideration.

The general features of the new generation of smart sensors, in addition to those mentioned above, include:

- On-line temperature compensation
- Built in-diagnostics
- Claimed accuracies of ±0.1% of span

Leading companies in the field include Honeywell, Bailey, and SattControl, but other established companies are working on similar devices. Most of these sensors use the properties of silicon plus the addition of micro electronics to give the performance. Silicon is an elastic drift-free material, and pressure sensing elements are diffused onto it. A cavity is usually etched on the reverse side to form the diaphragm. Where the line fluid is very aggressive, a second insulating diaphragm is required to protect the silicon from corrosion. Generally speaking, the cost of these sensors is coming down, and it appears we can look forward to many more instruments appearing in the more difficult applications. Provided we can overcome the general conservatism about applying new technology, this is one section of the instrumentation industry where microelectronics will have a great future influence, particularly when used with the computer and real-time models. This will give the pipeline operator more confidence and more information on both the condition of the instrument and the accuracy of the measurement. Interfacing problems should also reduce.

Another type of pressure transducer that looks promising for future applications in pipelines is the vibrating wire sensor. One such instrument from Foxboro has shown considerable reduction in size and manufacturing costs from other sensors. A tungsten wire encapsulated inside a cell is vibrated at its natural frequency. As pressure is applied to the silicon barrier diaphragms, this changes the tension in the wire and so alters the resonant frequency. This is sensed and amplified and gives the transducer excellent resolution characteristics. Pressure and temperature compensation is accomplished within the cell, and although the instrument is not strictly a "smart" transducer, it does possess basic characteristics that should be attractive to pipeline applications.

Densitometers

Continuous high accuracy on-line density measurement has become increasingly important in pipeline operations. The composition of natural gas and hydrocarbon streams are different from each production well, and assumptions about constant composition cannot be made. In fiscal measurement and where the energy content of the stream is required, a knowledge of density is essential, particularly for gases.

Recently great strides in the development of rugged on-line densitometers have been made with the introduction of the vibrating element transducer. This type of instrument is fast becoming the industry standard for pipeline applications. One such instrument is shown in Figure 7. The operation is very simple, being based on the spring mass principle. A flat plate, cylinder, or fork is maintained at resonance by an

Figure 7. Pipeline densitometer (courtesy Solartron Inst and Control and Instrumentation).

electronic circuit. A fluid flowing over or through the vibrating element adds to the vibrating mass and causes a decrease in natural frequency. The higher the fluid density, the lower the frequency and vice versa. By calibration using fluids of precisely known density, the relationship between frequency and density for each element can be determined very accurately. The characteristic equation for most vibrating element densitometers is of the form:

$$\rho = At^2 + Bt + C$$

where A, B, and C are calibration constants, t is the period of oscillation, and ρ the fluid density. Once the constants have been determined by the laboratory tests, the characteristic equation can be programmed into the system computer for on-line calculation of density. In some cases it may be possible to check the performance *in situ* by passing a fluid of known density through the transducer. This will establish if there has been any drift from the original manufacturer's calibration and will help to minimize the uncertainty of the operating system. The latest generation of instruments can approach accuracies of ±0.1% of reading under ideal conditions and are typically ±0.2% of reading under actual operating conditions.

When used in conjunction with orifice meters, for example, the flow through the densitometer is maintained by taking fluid from a tapping downstream of the orifice and feeding this back at the downstream tapping of the plate. This method, known as pressure recovery, is very popular in pipeline applications, and a typical installation is shown in Figure 8. Other methods include pumping a small percentage of the stream flow through a flowmeter and densitometer before feeding back into the main stream.

Figure 8. Standard installation for pipeline density measurement (courtesy IP manual section 6.2 Nov 1983).

There are other types of fluid densitometer such as nuclear or force balance, but these do not find the same widespread use on pipelines as the vibrating element devices. The choice of transducer will depend on the application, since fluid properties, flow rate, sensor element design, and materials of construction all influence the type of densitometer used. For example, if the fluid contains solid particles or is of a "dirty" nature, then straight pipe vibrating cylinder elements are more likely to give better long-term performance. The actual elements are manufactured in a wide range of materials. Ni-Span-C is used for high accuracy metering, and this material also has a low temperature coefficient. Stainless steel, on the other hand, has better corrosion characteristics and is cheaper to use. Materials such as hastelloy C-276 offer excellent resistance to the most corrosive of fluids.

It is probable that the use of these instruments will increase as general experience on pipelines has been favorable. They must, however, be operated in single-phase fluids, as the presence of either droplets in gas lines or vapor bubbles in liquid lines will produce erroneous outputs of unknown magnitudes. The straight tube cylinder-type designs could be used in low second-phase concentration applications if the densitometer is installed vertically with upward flow. This will ensure closer approximation to homogeneous flow. Experience has shown, however, that two-phase flows are significantly more troublesome to handle, and the output of the transducer should be used with some caution.

Pipeline samplers

Oil lines, particularly crude oil lines, often contain significant amounts of water. This has to be monitored and allowed for when the oil is being metered for custody transfer or taxation purposes. It is a growing and important area of pipeline technology to install on-line samplers for the measurement of water content, and work is also under way to use statistical methods to estimate how many samples are representative of the composition of the flowing stream. There are currently two different basic methods being used on pipelines to monitor the water content. These are the capacitance method and grab samplers.

The first method is based on the fact that the difference in the relative permittivities of water and oil makes a capacitive transducer sensitive to the water content flowing between two electrodes. In the past 3 years, commercial instruments using this method have appeared and have attracted a great deal of interest, particularly in North Sea applications where water content could be a problem. They are basically also non-intrusive sensors with the potential for in-line installation. There are two basic designs, by-pass and main line types. By placing electrodes in a by-pass line from the main flow, a measure of water content in the main stream can be made. The accuracy of these meters, however, is highly dependent on the extent to which the flow in the by-pass loop is representative of the main stream. With the main line type, this source of error is reduced, but the problem of flow regime effects still remains. Both types are also affected by the excitation frequency, the salinity of the water content, the water droplet size, the thickness of the isolation material between electrode and fluid, and the homogeneity of the electric field.

Field experience has shown that meaningful measurements can only be made if the flow regime is constant and the best results are obtained when the water is evenly dispersed and in droplet form in the liquid. Such criteria often do not exist, with slugs of water passing down the pipeline at irregular intervals. One such commercial instrument is shown in Figure 9. The important feature of the design is the short distance between the electrodes, which results in high sensitivity. The electrodes are intrusive and therefore liable to either coating or blockage effects if the oil is particularly dirty.

Grab samplers are devices that take small volumes of fluid from the line at regular intervals to build up a representative sample of the mixture over a period of time. An international standard is currently being reassessed in the light of operating experiences by many oil producers. The current practice states that 10,000 "grabs" of around 1 cc each should be taken, and there are guidelines laid down for the sampling

Figure 9. Capacitance-type water content transducer (coutesy Endress and Hauser).

Figure 10. Fast loop pipeline sampling system.

frequency and other criteria to be considered. Work at Cranfield and elsewhere has been evaluating this standard, and some of the limitations have been identified. As with the capacitance devices discussed previously, these may be installed either in the main stream or in a by-pass loop. Often signals from flowmeters in the line are used to set the sampling frequency, and the use of jet mixers, bends, pumps, and other devices that produce a more homogeneous stream is desirable to improve performance and accuracy.

Another variation, the fast loop sampler, is used for continuous monitoring of water content, and such a system is shown in Figure 10. The main advantage of the fast loop sampler is that flow is continuously extracted, and the sampler itself seems to have a smaller perturbing effect on the stream characteristics. A great deal of work is currently under way to examine the effect that the presence of the sampling probe has on the fluid dynamics of the stream. Early work has indicated that the flow streamlines around the probe head do have an influence on the droplet distribution and hence the oil/water ratio. This is developing technology, and we should see increased activity in this area as some of the older North Sea fields begin to produce greater concentrations of water.

Pipeline monitoring systems

On-line inspection techniques have recently been developed to a very high level of sophistication. British Gas has spent some 12 years and more than £100 M producing intelligent monitoring tools and the data analysis software to inspect line sizes from 8-in. to 42-in. bore. There is a demand worldwide for their expertise, and with a potential 1 million miles plus of line to survey, they should be busy for years to come! These inspection tools are commonly called "pigs" but should be distinguished from the pipeline pigs used, for cleaning, dewaxing, batching, filling, and dewatering of pipelines. I feel the term *monitoring tool* should be used, because these devices are very advanced indeed.

Two types of monitoring tools have been developed, one based on magnetic sensors to detect metal loss and the second based on ultrasonics to sense defects and cracks in pipe walls. The two used together give a complete picture of the mechanical state of the pipeline. They are not continuous monitors like pressure, temperature, and flow transducers but are useful in supplementing the on-line measurements. If the lines are inspected at regular intervals, then changes can be identified and possible trouble spots pinpointed. The data can provide a very useful historical record to be used when maintenance, cleaning, or line repairs are needed.

The magnetic pipeline inspection vehicle operates by magnetizing a short section of line as it travels down the pipe, driven by the process fluid. If a section of pipe has suffered metal loss, it will cause a local distortion of the magnetic field. The data are recorded on 14 channel tape systems, and onboard software reduces the amount of data saved to only that where significant signals are noted. The information from the clean or undamaged pipe is not saved. Once the inspection run is complete, the data are analyzed to identify the problem areas. The first offshore run was completed in early 1986 when the Rough and Morecambe gas field lines were surveyed. This run took around 5 hours. The method can be used to survey sections of line around 300 km long, although a more typical single pass would be around 80 km (50 miles) (*Monitor*, No. 2, 1986).

The ultrasonic system is virtually identical except for the transducers and detectors. The sensors are ultrasonic, operating in the megahertz region. They are housed in wheels that press against the pipe walls. They are pulsed from an on-board electronic control system, and the reflected waves are stored in the same way as in the magnetic system. Both inspection vehicles travel at around 3 meters per second. If a line problem is detected by an imbalance between flowmeters, for example, the location could be confirmed by the acoustic detectors on board the inspection vehicle. This could be arranged to alarm when the detector output reaches a maximum, and the precise location can be given by radio transmitters mounted on board.

Other problems that can be identified include corrosion, fatigue, ground movement (causing pipe buckling), and third-party damage. The inspection center at Cramlington has already undertaken a number of overseas surveys and has experience in gas, crude oil, kerosene, and gasoline lines. The impressive part of these pieces of equipment is the engineering. They have to negotiate bends, valves, crossovers, etc. and still protect the relatively fragile electronic recording equipment from damage. They have been used both in on-shore and off-shore lines with equal success. This is one area of pipeline instrumentation development that I will be watching over the next 2 years as new on-line vehicles are developed for pipe coat-less monitoring.

Computer systems

It is in computer systems that the greatest amount of data can be gathered, analyzed, and acted upon in the shortest amount of time. Most operators now have some form of computer-based monitoring system employing either commercially available or custom-developed software to run the system. Programs can calculate the inventory of the line at any point in time and compare this with line measurements to provide independent cross checking of pipeline measurement data. These data can be transmitted by telemetry, radio, or telephone links to a central computer that continuously monitors the "health" of the line. The central operations room is now more likely to contain VDUs rather than banks of process indicators and enables the engineers to have rapid and accurate updates of conditions along the entire line or network. By changing programs and subroutines, a large number of tasks and functions can be performed easily, simultaneously, and cost effectively.

The many functions that can be performed by computer-based systems include:

- Leak detection and location
- Batch tracking of fluids
- Pig tracking
- On-line flow compensation
- Real-time pipeline modeling
- Instrument data and malfunction checking
- Inventory and material balance accounting

Such systems have very rapid responses but are dependent on the accuracy of the input data to be effective. Most computer-based systems consist of two major elements, namely:

1. A supervisory computer plus associated software
2. A number of independent pipeline monitoring stations plus data transmission equipment for communication with the computer

The complexity of the software, the number of input variables, and the pipeline instrumentation selected all influence the size and the choice of computer. Some systems are capable of running on the compact powerful micros such as the IBM AT or XT, while the more complex pipeline networks will require a mainframe machine, possibly communicating with satellite micros installed close to the measurement stations. The overall design time varies with each system but should not be underestimated, especially if SCADA systems are being developed in house (see section following).

At the most basic level the computer system can take flowmeter inputs to perform gross balances along the pipeline. Such a method is very useful in identifying small but consistent loss of product from corrosion pits in the pipe wall. This will arise if the integrated flowmeter outputs diverge with time when the flow in the line is maintained constant. If the line flow rate varies with time, then imbalances are more difficult to detect, since the meters at each end may have different characteristics or their outputs may vary nonlinearly with flow rate. A loss of product will be identified simply as the difference between the steady state inventory of the system and the instantaneous inlet and outlet flows. Mathematically, this is:

$$\Delta V = V_{in} - V_{out} - V_1$$

where ΔV is the leakage volume, V_{in} and V_{out} are the metered inlet and outlet flows, and V_1 is the inventory of the pipeline. This last term can be calculated as the average of the integrated flowmeter signals but is dependent on variations in temperature, pressure, and many other variables. It is the use of computers that enables this last term to be calculated on line and as a function of pipeline elevation, pipe material characteristics, process variations, pipeline fitting loss data among the many variables. Temperature and pressure variations change both the density of the fluid and the actual dimensions of the line. Both will give rise to sources of errors in pipeline inventory if compensation is not included in the

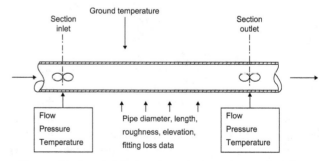

Figure 11. Input data for computer inventory calculation.

Figure 12. Examples of modern computer-based pipeline (courtesy Oil and Gas Journal, 1986).

calculations. Figure 11 shows typical input data requirements for simple on-line inventory calculations.

At the other extreme, computer-based systems can control entire networks with almost no need for the operators to intervene unless serious problems arise. Figure 12, taken from *Oil and Gas Journal*, March 3, 1986, shows two examples from the many hundreds that exist of computer-based monitoring. The benefits from the addition of real-time pipeline models cannot be overemphasized. Even when operating transients occur, some commercially available programs enable problems to be detected. An example is shown in Figure 13. Here a pipeline leak was induced following pump failure and restart, and although the line was operating under highly transient conditions, a material imbalance was detected. Real-time modeling also allows density profiles to be calculated along the line. It is fairly common practice to have amounts of different fluids in the same pipeline separated by a pig or sphere. From the pressure drop/flow characteristics, the real time can track the progress of these batches of fluids along the pipe.

Figure 13. Example of transient real time modelling of pipelines (data from Ref 9).

SCADA systems

Real time computer modeling of pipelines is a technique that uses the full data gathering capabilities of modern digital systems and the computational power of computers to give accurate "snapshots" of the line. These enable the process operator to act quickly when problems develop. The whole system is under the control of the SCADA (Supervisory Control And Data Acquisition) system. This suite of programs polls the data stations on the line in turn, processes the data, controls the running of any pipeline model, and activates alarm and other monitoring routines. Most modern SCADA systems now include checking routines to determine the validity of the data before any calculations are made. It has been found that the use of continuous quality and cross checking of data improves the sensitivity of the system. In addition to these basic functions, some of the more advanced SCADA systems might include predictive models to analyze "what if" operating scenarios, include automatic startup and shutdown procedures, or provide optimization routines for least-cost operating strategies.

There are a number of companies supplying standard SCADA packages, but the majority of systems use custom written master routines to enable the commercially available programs to be used. Each pipeline is different, in line characteristics, instrumentation installed, fluid properties, and need of the user to name but four requirements. Experience has shown that the adaptation of standard software modules to specific needs takes much longer than is first estimated and also gives rise to the greatest number of problems when pipeline monitoring and control systems are being commissioned.

It is very important for the software specialist to communicate directly with the instrument engineer who specified the hardware, to ensure proper integration of the various parts of the system. All too frequently these two key members of the design team do not meet until the project is well under way.

Cathodic protection

Protecting a pipeline from corrosion is an important and expensive part of building and running a pipeline system safely. Every year, operators spend millions of pounds on pipe coatings, gas detectors, and sacrificial anodes to lengthen the effective life of pipelines by reducing the effects of corrosion. Whenever a steel structure of any type comes into contact with the soil and/or water there is almost certainly going to be corrosion. The structure, in our case the pipeline, divides itself into anodes and cathodes with corrosion taking place at the anode. This chemical process is accompanied by the flow of electric current from one point on the line to another via the ground or water. There are two basic preventative methods, sacrificial anodes and impressed currents (more commonly called cathodic protection). This second method is the more common and preferred method for long pipelines to identify the problem spots.

If a voltage is applied to a section of line, a profile of the current flow can be drawn up by measuring the current flow at intermediate points. This gradient will show areas where current is escaping, indicating a pipe defect. Recently, techniques using AC current have been perfected where the magnitude of the current can be sensed inductively from above ground.

This method can be used by pipeline crews walking the line, and compact lightweight handheld units store data on the mechanical condition of the line in lengths as short as 5 meters. On long cross-country lines the profile can be plotted at intervals of a few kilometers, and from historical cross-comparisons, line changes can be identified.

The object of cathodic protection is to prevent current flow from the pipe. The operating principle is to apply a voltage in the opposite direction to the natural potential of the pipe. The pipeline voltage is made negative with respect to earth so that in theory the voltage and current flowing from the cathodic protection system balance the corrosion current that would otherwise flow. Recent developments include the use of solar panels and converters so that energy from the sun is used to run the system. One of the largest examples of this new technique is the Sarir to Tobruk in Libya, where a 32-in. diameter, 319-mile pipeline is protected by such an installation.

Remote station monitoring is easily accomplished with the data being telemetered back to central control rooms many miles from the location of the actual installation. Continuous automatic control is often performed by means of thyristors. Such continuous monitoring is now routine on new lines that are being constructed, and many older lines are being fitted with some form of cathodic protection system. In most cases it is probably more important for the older lines to be surveyed than the newer ones, especially where different fluids have been transported over the lifetime of the line.

System design guidelines

The aims and requirements of a project should be specified as fully as possible as early as possible when designing a pipeline monitoring system. Based on personal experience, the following points should be borne in mind. They are not in any special order but do form a useful checklist when starting system development.

1. Make the system as simple as is practical but do not underspecify the needs for the sake of simplicity.
2. Select instrumentation and pipeline equipment based on performance and not economic grounds. It is better to

install fewer high-quality pieces of equipment than numerous poor ones.

3. Equipment compatibility is important. Use transducers, interface modules, and other pieces or hardware that have standard communications protocol. Custom-designed systems invariably put up development costs and do lead to endless sources of trouble and frustration.

4. When calibration and maintenance are due on the system, do not put them off. Wherever possible use computer software to cross check data to aid in the identification of problems. This will help to reduce operating and maintenance costs.

5. Where system availability is important, install devices that are self checking, self diagnosing, or dual systems where appropriate. The initial cost may be higher but the benefits may be more economical in the longer term.

6. Seek independent references, user experience, or validation of the equipment chosen. Most pieces of hardware perform differently in real applications to the specification published under ideal conditions.

Future trends

Future trends are in one respect easy to assess, but it is very difficult to estimate where we will be in 10 years' time. There is no doubt that the application of computer-based technology is going to proceed at an ever-increasing speed to both the instrumentation and pipeline monitoring systems. There will be reliable and rugged intelligent flow, pressure, valve, and other hardware components appearing, and once the initial testing over the application has occurred, the second-generation systems will give the desired performance, if previous history is anything to go on.

By far the biggest problem facing the whole spectrum of control and instrumentation is to develop standard protocols to facilitate communication between instruments. The situation at present means that most, if not all, monitoring computers must have a multitude of interface boxes scattered through the system to enable some elements of the system to "talk" to others. It is sincerely hoped that manufacturers will develop microcomputers that are able to communicate directly with each other, with full-scale industrial process monitoring and control systems and with the individual process sensors and associated controllers. The more elements there are in the system, the greater the potential of line downtime. Some moves are being made in this area, but it could be some time before we see commonality of process protocol.

The most rapid area of advancement has been in the microchip. Oil pipeline operators are among the many people who have been watching the development of 32-bit machines. The Intel 80386 chip for example, now contains more than 275,000 transistors in one piece of silicon. This device has an operating speed of 4 million instructions per second and has the capability to address over 4 gigabytes of main memory. One other manufacturer is working on a computer that will perform 10 billion calculations per second, more than 10 times faster than any other computer on the current market. Such machines will permit easier real-time modeling of pipelines and will present the line operator with many extra pieces of data on the condition of the line. Of course, complete new suites of software will be required for these fifth generation machines.

Conclusion

In such a short space, the paper cannot do justice or even mention the vast number of recent instrumentation developments that have had an impact on pipeline operation and control. The technology of measurement and control is proceeding at an ever-increasing pace, and it is now quite difficult to keep up with developments in one area. Pipeline instrumentation covers a very wide spectrum from pressure and surge control to intelligent instrumentation and advanced software. Certainly our "friend" the microchip has totally transformed our thinking in the past decade, and the poor pipeline operator now needs specialized help in specifying a system to meet his needs. There are many options open: control using PLCs at sources, control using software in the process computer, and manual override if required. With the development of expert systems, systems can now be programmed to almost preempt problems before they arise and take appropriate corrective action. The flexibility in the industry is vast, but one thing is sure, you cannot control unless you measure accurately, and all too frequently the primary elements of a system are under-specified compared to the rest of the system. It is of very little use, for example, to pressure and temperature compensate transducer outputs when the basic measurement uncertainty is poor.

Source

Measurement | Control, Vol. 20, February 1987.

References

Adams, R. B., "Buckeye modernizes computer operations and SCADA system," *Oil and Gas Journal*, 11, 55–59, August 1986.

Anderson, C. C., "The Ekofisk pipeline control system," *Pipes and Pipelines International*, 30, No. 3, 21–25, 1985.

Baker, R. C. and Hayes, E. R., "Multiphase measurement problems and techniques for crude oil production systems," *Petroleum Review*, 18–22, November 1985.

Chester, J. R. and Phillips, R. D., "The best pressure relief valves: simple or intelligent," *In Tech*, 51–55, January 1985.

Covington, M. T, "Transient models permit quick leak identification," *Int Pipeline Industry*, August 1979.

Davenport, S. T, "Forecast of major international pipeline projects for the 1986–2000 period," *Pipes and Pipelines International*, Vol. 31, No. 1, 9–14, 1986.

Furness, R. A., "Modern pipeline monitoring techniques—Parts 1 and 2," *Pipes and Pipelines International*, Vol. 30, No. 3, 7–11, and Vol. 30, No. 5, 14–18, May and September 1985.

Kershaw, C., "Pig tracking and location—Parts 1 and 2," *Pipes and Pipelines International*, Vol. 31, No. 2, 9–15 and Vol. 31, No. 3, 13–16, 1986.

Stewart, T L., "Operating experience using a computer model for pipeline leak detector," *Journal of Pipelines*, Vol. 3, 233–237, 1983.

Taffe, P., "Intelligent pigs crack pipe corrosion problems," *Processing*, 19–20, September 1986.

Tinham, B., "Micros and sensors make smart transmitters," *Control and Instrumentation*, 55–57, September 1986.

Williams, R. I., "Fundamentals of pipeline instrumentation, automation and supervisory control," ISA *Transactions*, Vol. 21, No. 1, 45–54, 1982.

CHOOSING THE RIGHT TECHNOLOGY FOR INTEGRATED SCADA COMMUNICATIONS

Service was consolidated for both current and projected computer environments

Vernon J. Sterba, Director, System and Telecommunications, Enron Gas Pipeline Group, Group Technical Services, Houston

Consolidation of overlapping telecommunication systems for Enron subsidiaries Florida Gas Transmission, Houston Pipeline, Northern Natural Gas, and Transwestern Pipeline eliminated redundant facilities and integrated information and communication required for upcoming open transportation services.

A wide-area communications study group (WAC) was formed to define and document operations/SCADA communications required for both current and projected computer environments. A functional and technical design would be developed and documented, as well as supporting costs, an implementation timetable, and equipment specifications.

The WAC project was to meet the following requirements:

- Pipeline monitoring and control (SCADA)
- Electronic flow measurement/custody transfer (EFM)
- Real-time pipeline modeling and optimization
- Compressor automation

Since installing an advanced, integrated communication system required a considerable capital investment, the WAC team chose not to take a "SCADA-only" view, since this could miss opportunities to effectively develop the existing communications network.

Mainframe data and voice communications were considered for consolidation. This later became a key factor in technical decision making. The project emphasis was to identify where a cost-effective solution could simultaneously meet SCADA and other voice and data requirements.

The WAC team consisted of Enron members representing both organizational and technical experience. In addition, a telecommunications specialist served as project leader and expediter for formulating project methodology and imparting a "third party" perspective to team discussions.

WAC methodology

The main sequence of tasks was as follows:

- Define existing communication systems
- Define communication requirements
- Analyze adequacy of existing systems
- Identify and evaluate technology alternatives
- Establish communication "direction"
- Develop functional and technical design
- Develop project implementation schedule

This approach allowed the team to fulfill project objectives in a scheduled time frame. Next, an inventory of communication circuits and facilities was entered into a PC database. The database included the latitude and longitude of all facilities, type and name of facility, communication media, circuit number (where applicable), and monthly cost of leased communication.

The database was used to develop communication system maps by interfacing with a computer-automated design package. The maps identified redundant communication paths, duplication of facilities, high-cost communication routes, and other details. Without the maps' visual assistance, it is unlikely the team could have quickly developed an understanding of the extensive and complex communication system.

Existing communication facilities included the following technology combinations:

- Multi-drop and point-to-point leased lines. These lines represented a major cost component; many redundant circuits were identified, and some surprising cost anomalies were discovered. For example, the cost of a 40-mi hop from an RTU to the nearest microwave tower could cost more than a 300-mi hop from the RTU to Houston.
- Company-owned 2/6-GHz microwave system. The system provides voice and data communications in Texas inland and Gulf Coast areas and incorporates voice circuits for toll-free calling at 30 field offices along the microwave system.
- Radio telemetry. This system was used by Northern Natural Gas and Transwestern for gathering SCADA data at master control units (MTUs) or central data concentration points in major supply basins. The MTUs, in turn, relayed data to the SCADA master via leased phone lines. Meteor burst technology. The system "bounces" radio waves off the ionized trails of micro-meteorites entering the earth's atmosphere. Northern Natural Gas used this technology, as a prototype, for nine RTU sites.
- VSAT technology. The VSATs (very small aperture terminals) use satellites in orbit for relaying radio signals. Dish antennas of 1.8 to 2.4-meter diameter (6 to 8 ft) were required. In 1986, Transwestern Pipeline installed 11 satellite terminals to eliminate high construction costs involved with leased line service at remote RTU sites.

Analysis of technology

Sending data to the SCADA master required both short- and long-haul communication segments. The segments could be established by separate and independent technologies.

For example, to aggregate data from several RTUs at a central hub, use radio telemetry or local leased phone lines. The long-haul technology (central hub to SCADA master) could use either leased phone line, microwave, or VSAT. This requires a serial byte-oriented, SCADA protocol standard, which had been in place since 1987.

To determine the best communication direction, various technology "mixes" were applied system-wide. This provided comparative technology costs for the entire communication system.

Table 1
Communication guidelines

Short-haul choices
1. Use radio, if available, in a radio hub arrangement.
2. Use leased lines if economically available, connected to a hub or the home office site.
3. Use VSAT only at remote sites when the above options are infeasible.

Long-haul choices
1. Use existing microwave, when available.
2. Use VSAT for sites with multiple applications.
3. Use single-application leased lines when the monthly cost is less than $450; otherwise, use VSAT.

Amortized capital cost and O&M cost, including leased lines, were estimated for each design case. The cost analysis was conducted for both current and projected data density. The technology mixes considered were:

Case	Short-haul	Long-haul
1	radio	leased line
2	radio	microwave
3	radio	VSAT
4	leased line	leased line
5	leased line	microwave
6	leased line	VSAT

In both current and future data density environments, the combination of radio telemetry and long-haul leased lines (Case no. 1) yielded the minimum monthly cost. This was followed closely by radio telemetry and VSAT (Case no. 3). When VSAT was used also as a mainframe connection, Case no. 3 became the minimum cost option.

In all cases, radio telemetry was the most economical technology for the short-haul communication segment. Radio systems bypass local telephone-company leased lines to eliminate escalating costs and service problems found in remote areas. Radio is not always available, due to frequency availability.

For existing facilities, the microwave network was a clear choice for long-haul systems, with no capital investment required and a small incremental cost for adding circuits. However, when considering microwave expansion, the situation changed significantly. In the current data density environment, increased microwave coverage (Cases no. 2 and 5) did not justify capital costs involved, due to insufficient data traffic.

In the projected environment, higher data densities began to support the capital and O&M costs as microwave system capacity was fully utilized. This moved microwave technology from seventh to third most favorable system, compared to other technologies.

Generalizations shown in Table 1 apply to one system's economics and topography and are not intended to suggest the same options should be chosen by other gas pipeline companies.

For example, the short-haul radio option was not viable in some cases. In one case studied, the cost of a high radio tower exceeded the cost of a VSAT system. Frequency availability was a problem as well.

In addition, the long-haul segment was affected by site-specific factors such as data density, leased-line installation and operating costs, and VSAT line-of-sight availability. VSATs require a clear view to the south.

The technical evaluation recommended combining short haul radio telemetry with existing microwave or VSAT for the long-haul segment. Economics were especially favorable when the VSAT could provide mainframe connections, a situation already found at a majority of sites.

Another decision was whether to recommend C-band or Ku-band VSAT technology. The company was using C-band VSATs in the Transwestern Pipeline but at the time had no experience with Ku-band VSATs.

C-band VSAT advantages

- More cost effective for low data rates; C-band hardware is generally 25%–50% less expensive.
- Capabilities meet existing SCADA transmission speed requirements.
- C-band SCADA communication services were available and had been used since 1986, a "known-quantity" advantage.

C-band VSAT disadvantages

- Not amenable to moderate or high-speed data transmission rates, voice, or video.
- Shares the radio wave spectrum with terrestrial microwave services, increasing the possibility of interference, a rare occurrence in rural areas, but a major concern in metropolitan areas.
- A single vendor was the main provider of VSAT transponder capacity.
- Existing C-band satellites have low utilization.
- C-band is older technology. The expected life span of a satellite, both C and Ku-band, is about 7 years, usually set by fuel limitations for keeping the satellite onstation in geostationary orbit.
- Projected shortages of C-band satellite capacity were expected beyond 1991, as existing satellites reach lifespan limits. However, the projected shortage did not develop.
- When conducting the WAC project, there were no launch plans for replacement C-band satellites.

Ku-band advantages

- Growth trend in satellite communications is toward Ku-band, with future satellite launches planned.
- More suppliers are expected in the Ku-band marketplace.
- Ku-band VSATs can be used for high-speed applications such as voice, video, computer-to-computer data communications, and SCADA and mainframe workstations. Local networks can be extended into wide-area networks (WANs).
- Although the value of satellite voice communications is limited by the inherent 0.5-second delay, receive-only video and data communications will become more valuable in the future, if costs decrease.
- Television receive-only (TVRO) can be used to communicate with a dispersed workforce and provides cost-effective training when compared to increasing airfare costs.
- Computer-to-computer data communications will become necessary components as microcomputer installations increase at field locations.

Ku-band disadvantages

- Cost usually exceeds C-band cost for one-application installations.
- Can be affected by heavy rainfall, such as rainfall rates sufficient to restrict driving. This occurs because water droplets absorb the satellite frequency; however, this can be mitigated by proper antenna design and other equipment. Most users report actual outage rates less than statistically predicted.

VSAT decisions

Because a distributed mainframe application was planned, the Ku-band scenario was cost effective at a majority of sites. A secondary decision was made not to add more C-band VSAT sites. Also, because of the economics mentioned earlier, radio became the first choice for the short-haul option. The combined radio/VSAT or radio/microwave became the "radio hub."

Operating costs, for a Ku-band VSAT at each SCADA site, were $375 to $450 per month. The exact cost varied depending on vendor, data traffic costs, and capital amortization period. This estimated cost agreed with other published analyses.

One disadvantage of VSAT technology was the delay introduced by equipment and related store-and-forward message handling. With one vendor, this delay ran as long as 12 seconds, reducing the effective circuit speed.

Do's and don'ts of VSAT technology installation:

- Do install bumper guards around VSATs in compressor station yards.
- Do install a security fence around VSATs in or near public areas and be prepared to install camouflage, such as shrubbery to hide the VSAT from public view.
- Do allow extra time for installing VSATs mounted on building roofs, especially if the building is not company-owned.
- Do use dish heaters where ice storms are common.
- Do install antennas high enough to avoid snow drifts in northern latitudes.
- Don't install dish heaters in snowy regions where ice storms are frequent.
- Don't install VSAT antennas just outside office buildings; this may obstruct view or reflect sunlight into windows.

Because of inherent VSAT delays, special features were mitigated into the SCADA system. A report-by-exception feature was implemented at data concentration (MTU) sites. Delay was reduced also by using single-hop rather than double-hop VSAT technology.

In a single-hop system (Figure 1), the message path is:

- Outbound poll—SCADA master to vendor master earth station (MES) via leased lines.
- Inbound response-VSAT at RTU to MES via space segment; MES to SCADA master via leased lines.

The double-hop system (Figure 2) replaces the SCADA master to an MES leased-line link with another pair of space-segment hops, increasing time delay, but still cost-effective for smaller systems.

A dial-up backup to key radio hubs and other critical VSAT sites was chosen. The SCADA system supports this failure mode by automatically dialing up failed VSAT sites. This is cost effective, since the business line's monthly cost is low relative to other options. Dial-up on-line costs are high, but rarely needed.

Implementation

Because of very close bids and a WAC team recommendation to use multiple VSAT vendors, two vendors were awarded contracts to install Ku-band VSATs across the nationwide 38,000-mi pipeline grid. Initial construction was completed in 1989.

Enron currently has 125 VSAT sites throughout the United States and has changed over to primarily Ku-band technology. Not all sites use the same VSAT options, as shown below:

Ku-band application area	Number of sites
SCADA	104
Mainframe	60
TV RO	63
Other (ASCII/X.25)	14
Total number of sites	125

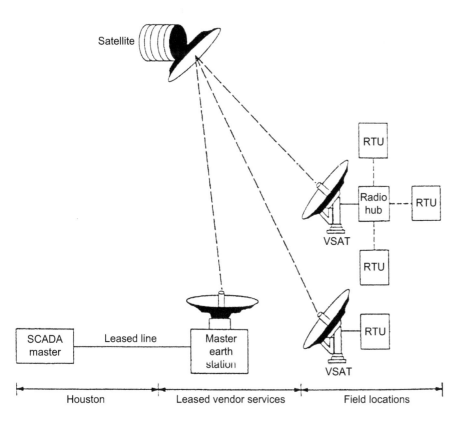

Figure 1. Single-hop VSAT.

Figure 2. Double-hop VSAT.

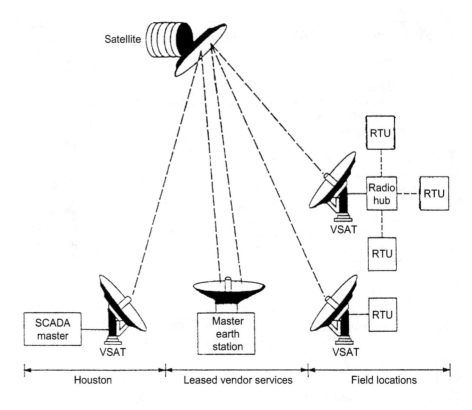

The television receive-only (TVRO) application currently is being installed. The X.25 application permits linkup between field and home office systems, including local area network, electronic mail, and remote access to centralized DEC VAX applications without any dial-up cost. Because many sites are in remote areas, the dial-up costs are often dwarfed by the costs and frustrations of poor quality telephone connections.

Acknowledgment

Based on a paper, "VSAT Communications Technology for SCADA System Communications," presented by the author at the AGA Distribution Conference, Nashville, Tenn., April 1991. The author received the Acker Medal Award 1991–92 for this technical paper, sponsored by Central Hudson Gas and Electric Corp., in recognition of excellent contribution to gas industry.

Literature cited

1. Sharma, R., "VSAT Network Economics: A Comparative Analysis," IEEE Communications, Feb. 1989.
2. Simo, E., "VSATs and the Pipeline Industry," Satellite Communications, Aug. 1988.
3. Coulter, J., "Design and Installation of a Gas Control SCADA Network using Satellite, Microwave and Radio Systems," Entelec News, Vol. 61, no. 2, Aug. 1988.

17: Leak Detection

Pipeline leak detection techniques

R. A. Furness, Cranfield Institute of Technology, Cranfield, Bedford, UK, and **J. D. van Reet,** Scientific Software Intercomp, Houston, Texas

Summary

Pipelines are now an integral part of the world's economic structure, and literally billions of dollars worth of products are now moved annually in pipelines. Both economic and environmental factors are influential in pipeline operation, and therefore integrity monitoring is vitally important in the control and operation of complex systems.

Leak detection systems range from simple, visual line walking and checking to complex arrangements of hardware and software. No one method is universally applicable, and operating requirements dictate which method is the most cost effective. The aim of this paper is to review the basic techniques of leak detection that are currently in use. The advantages and disadvantages of each method are discussed, and some indications of applicability are outlined.

Modern pipeline computer modeling and control is then reviewed in detail. These systems are the most flexible and versatile and are steadily being adopted. The influence of instrument accuracy on system design and performance is discussed, and the basic modeling equations are reviewed.

Introduction

Our need to transport fluids from the point of production to the area of end use has led to a rapid increase in the number of pipelines being designed and constructed. Many of these carry toxic or hazardous products, often close to centers of high population or through areas of high environmental sensitivity. With the need to safeguard these lines, online monitoring is becoming routine and in some cases 24-hour surveillance is mandatory. With the increase in world terrorism, the need for rapid and reliable pipeline measurement and control systems will increase.

The review begins with a summary of causes of leaks and the implications of the failure to detect them. Basic techniques are covered, and the features of each are briefly discussed. The bulk of the paper deals with modern computer-based techniques. Basic flow equations are covered, and the on-line dynamic calculations required are listed together with the input data required to enable the monitoring with minimal downtime. The latest systems are capable of resolving down to 1% of the maximum rated flow with a response time of a few minutes. Practical experience verifies this figure, but the total costs of such a system could be high. The system therefore becomes a compromise among response, performance, and alarm availability, and choice of instrumentation is crucial. The integration of good quality instruments with advanced realtime models seems to be the current trend, and the paper closes with some personal thoughts on future trends.

Causes and economic aspects of leaks

There are four main categories of pipeline failures. These are:

- Pipeline corrosion and wear
- Operation outside design limits
- Unintentional third-party damage
- Intentional damage

Many pipelines are operated for a number of years with no regard to any possible mechanical changes occurring in the line. Some of the products may be corrosive, the pipeline may be left partially full for periods of time, or atmospheric effects may cause external damage. These three reasons are responsible for pipeline corrosion, and this may give rise to corrosion "pits" developing along the line. These are small in nature and could be responsible for material imbalances over a period of time. Very accurate flowmetering can be used to detect this, as discussed in the next section. Abrasive fluids or dust-laden gas streams can give rise to pipeline wear. Again, this is a slow process, but should a weak spot develop (more often than not close to a change in direction or section) then a pipe break may occur very rapidly and totally unexpectedly.

Operation outside design guidelines is more common than is realized, as operators seek to use the line for as many fluids as possible. If the line is designed for a certain maximum temperature and pressure, then operation at higher pressure and/or higher temperature could lead to spontaneous failure. The problem could be compounded if the line has a large but unknown amount of corrosion.

Unintentional third-party damage may occur if excavation or building occurs near buried lines. More often than not the right-of-ways are not clearly marked, and lines are sometimes broken by bulldozers or similar plant machinery, possibly with fatal results. An example occurred in West Virginia in 1984 when a 20-in. natural gas line was punctured by an excavator, and product leaked slowly into a nearby supermarket during the night. The building was totally destroyed early next morning by a massive explosion caused by a staff member lighting a cigarette. Such an occurrence could have been avoided by some form of leak detection on the line.

Intentional damage is unfortunately on the increase, and pipelines carrying flammable or high-value products make ideal targets. Alarm systems linked to block valves can help to minimize the amount of product released as a result of sabotage, so again certain lines are instrumented with the intention of reducing the effects of planned terrorism.

The costs of failure to detect leaks also fall into four main areas:

- Loss of life and property
- Direct cost of lost product and line downtime
- Environmental cleanup costs
- Possible fines and legal suits

These are all self explanatory, with the most costly of these being the last, although any of the four areas could be very expensive. The size of claims can run into many millions of dollars, so the cost of fitting and operating leak detection systems is often insignificant compared with the costs of failure of the line.

It is this background that is causing operators and designers to turn to on-line integrity monitoring systems, and the next section looks at the more basic methods.

Simple leak detection systems

The most basic method of leak detection involves either walking, driving, or flying the pipeline right-of-way to look for evidence of discoloration of vegetation near the line or actually hear or see the leak. Often "unofficial" pipeline monitoring is performed by people living nearby who can inform the operator of a problem with the line.

The most cost-effective way to detect leaks in nonflammable products is to simply add an odorant to the fluid. This requires some care in selection, as frequently the odorant has to be removed before the transported fluid can be used. Organic compounds make the most useful odorizers, especially when the fluid being carried has no natural smell of its own. A good example is carbon monoxide, a highly toxic but odorless gas that is often pumped in large quantities between chemical plants. Chemicals such as mercaptans (rotten egg smell) or Trimethylamine (rotten fish smell) can be added in small quantities to enable any leak to be located by smell. The disadvantage of such a method is that if the leak occurs in an area of no population, the leak will go undetected unless the line is walked regularly by pipeline surveillance crews carrying suitable "sniffer" detectors. Thus, to the apparent low costs of this method have to be added the costs of removing the odorant and maintaining staff to check the line at frequent intervals. The location of a leak is also dependent on prevailing weather conditions. Strong winds may disperse the smell, and atmospheric inversions may give an incorrect location of the leak. The uncertainty of relying on this method alone is high. Nevertheless, it is a useful method if used in conjunction with other techniques.

Simple line flow balances are frequently used to check for gross imbalances over hourly or daily bases. This method may identify that a leak is present, but flowmeters at each end of the line will not identify the leak location. A line pressure measurement system will be required in conjunction with the flowmeters to establish that the pressure gradient has changed from the no-leak situation. The method is useful, however, in identifying the existence of corrosion pits, as the outputs of the flowmeters at each end of the line will consistently diverge if flow in the line is maintained constant. If line flow rate varies with time, then imbalances are more difficult to detect, since the flowmeter outputs may vary nonlinearly with flow rate or may have different flow characteristics from each other.

A loss of product will be identified simply as the difference between the steady state inventory of the system and the instantaneous inlet and outlet flows. Mathematically this is:

$$\Delta V = V_{in} - V_{out} - V_1$$

where: ΔV = Leakage volume

V_{in} = Metered inlet flow
V_{out} = Metered outlet flow
V_1 = Pipeline fluid inventory

This last term can be calculated as the average of the integrated inlet and outlet flows in simple systems, but as will be seen later, the value of this term can be calculated more accurately and easily in real time as a function of several variables.

Another method is based on detecting the noise associated with or generated by a leak. There are many instances where fluid flow can generate vibrations at frequencies in excess of 20 kHz. These frequencies are in the ultrasonic range but can be detected with suitable transducers. The device can be made portable so that pipeline crews can clamp a transducer at any point along the line to check for noise. By noting the signal strength, the source of the leak can be pinpointed.

A similar technique, though based on a different principle, is the acoustic "wavealert" monitor, more correctly called a negative pressure wave detector. This is a piezoelectric sensor that gives an output when dynamically stressed. When a leak occurs, there is a sudden drop in pressure at the leak followed by rapid line repressurization a few milliseconds later. The low pressure wave moves away from the leak in both directions at the speed of sound. The pipe walls act as

Figure 1. Negative pressure wave detector (acoustic monitor).

a waveguide so that this rarefaction wave can travel for great distances, attenuating in amplitude as it does so. Sensors placed at distances along the line can be triggered as the wave passes, and the location of the leak can be calculated from the line conditions and the internal timing devices in the instrument.

Such devices are particularly useful in identifying large breaks in lines very rapidly, since the transient wave typically moves 1 mile in 5 seconds in gases and almost 1 mile per second in liquids. Response is therefore on the order of a few seconds depending on the positioning of the transducers. Figure 1 shows one adaptation where the instrument can be made to cancel line noise and use the full measuring capability of the sensor for signal detection. There is the problem of setting the background threshold correctly, as this may be affected by the location of the instrument in relation to bends, valves, pumps, regulators, etc. Experience has shown that the installation is also critical to reliable performance, and there is also a dependence on the Reynolds number of the flow. Both of these affect the number of "false alarms" from the instrument.

Pig-based monitoring systems

Pipeline pigs are frequently used for pipeline commissioning, cleaning, filling, dewaxing, batching, and more recently pipeline monitoring. This last type of pig can be designed to carry a wide range of surveillance and monitoring equipment and can be used at regular intervals to check internal conditions rather than continuously monitoring the line. Data, however, can be built up over a period of time to provide a history of the line. This information can be used to predict or estimate when maintenance, line cleaning, or repairs are required. If a leak is detected, for example, by flow meter imbalance, the location can be found by using a pig

with acoustic equipment on board. This will alarm when the detection equipment output reaches a maximum, and the precise location of the pig can be confirmed by radio transmitters also mounted on board.

Pigs require tracking because they may become stuck, at a point of debris build-up, for example. Pigging should be carried out at a steady speed, but occasionally the pig may stop and start, particularly in smaller lines. Information on when and where the pig stops is therefore important in interpreting the inspection records. Pig tracking is not new, and many such proprietary systems exist. In the best systems, however, a picture of the line is often programmed in so that outputs from junctions, valves, cross-overs, and other geometries act as an aid to location. Pig tracking can make use of the acoustic methods discussed earlier. When the sealing cups at the front of the pig encounter a weld, vibrational or acoustic signals are generated. Each pipeline therefore has its characteristic sound pattern. When a crack occurs, this pattern changes from the no-leak case, and the location can be found from direct comparison. The technology has become so advanced that information on dents, buckles, ovality, weld penetration, expansion, and pipeline footage can be generated.

The equipment is often simple, consisting of sensor, conditioning, and amplifier circuits and suitable output and recording devices. Such a device developed by British Gas is shown in Figure 2. The range of detection is dependent on the pipeline diameter and the type of pig. Operational data have shown that light pigs in a 200 mm line can be detected at a range of 8 km, increasing to 80 km for a heavy pig in a 900 mm line. As the signals travel at acoustic velocity, this means a signal from a pig at 80 km range will take 190 seconds to be picked up. Such technology is now becoming routine in both offshore gas and onshore liquid lines.

Figure 2. "Intelligent" pipeline monitoring pig. Courtesy British Gas Corp.

Computer-based monitoring systems

It is in computer-based systems that the greatest amount of data can be gathered, processed, analyzed, and acted upon in the shortest period of time. Programs can calculate the inventory of the line at any time and compare this with accurate measurements at any section in the system. The effects of pressure and temperature on line dimensions, for example, can be calculated to provide an accurate estimate of the mass of fluid in the line. Data from a wide range of instruments can be transmitted by telemetry, radio, or phone links to a central computer that monitors the "health" of the line continuously. By changing programs and subroutines, a vast amount of functions and tasks can be accomplished very easily and cost effectively.

The many functions that can be performed by computer-based systems include not only leak detection but also:

- Pig tracking
- Batch tracking of fluids
- Inventory accounting
- On-line flow compensation
- Instrument data and malfunction checking, etc.

Such systems have very rapid response and have the advantage of multiple inputs being required before leaks are declared. Thus, some systems can run for short outage periods with no loss of integrity. They are often complex and costly to install, but once the initial capital investment has been made, running costs are low. The first section in the detailed review of such systems looks at the phenomena that need to be modeled when a leak occurs.

Pipeline leak phenomena

When a leak occurs in a pipeline, the measured pressure downstream of the leak falls, but the pressure at the same location is predicted to rise. The first is not difficult to understand as the line is depressurizing as mass leaves through the leak. The second effect can be explained as follows. The equations presented later in the paper predict pressure based on measured flow or flow based on measured pressure. As mass leaves the system through the leak hole, a reduced flow at the downstream end is compared to the inlet flow. This may not have changed, and so to balance the system, the equations predict a downstream pressure rise. In physical terms the model thinks the line is "packing" and total system inventory is increasing. There is therefore a divergence between measured and modeled pressure.

The same is true of flow changes, but here the inlet flow could increase due to lower pipe flow resistance between meter and leak, while the section outlet flow will fall as mass leaves through the leak instead of passing through the meter. Thus, a real imbalance will result. The model, however, will show an inconsistency since the pressure comparison will indicate line packing and the flow

comparison a line unpacking. If selected pressure and flow imbalance limits are exceeded, then a leak is declared. The magnitude of the leak is predicted from the flow imbalance and the location from the pressure profile imbalance and the flow leak indicators. The impact of instrument accuracy on the predicted location is discussed later in the paper. It is vitally important to good leak sizing and location to have the best pipeline instrumentation possible to minimize uncertainty.

Background philosophy of pipeline modeling

Real-time modeling is a technique that uses the full data-gathering capabilities of modern digital systems and the computational power of small computers to give accurate "snapshots" of the pipeline. The whole system is under the control of a SCADA package of programs, which poll the data stations on the line, process the data, control the running of the transient pipeline model, and activate the alarm and leak location routines. In addition to these basic software modules, more complex systems might include a predictive model to analyze "what if" operating scenarios, provide an optimization routine for least-cost operating strategies, or include a separate man/machine interface for the model system.

The SCADA interface is responsible for acquiring the data from the SCADA system and relating them to the model representation of the line. As a point in the system where two large and independently developed systems join, this can be the source of many problems in the implementation of the real-time modeling. Once the measurement data have been obtained, noise filtering and plausibility checking can be performed prior to running of the model. The model is the mathematical representation of the pipeline and will include such features as elevation data, diameters, valve and pump locations, changes of direction, and the location of cross-overs and junctions. The model provides data on the flow conditions within the line at intervals between seconds and minutes, depending on operational needs.

With the data available from both the measurement system and the pipeline model, the real-time applications modules are run. These are the leak detection and location routines in the context of integrity monitoring. The leak detection module functions by computing the difference between the modeled flows and pressures and the measured values at all points where measurements not already used as boundary conditions are located. Because the model accounts for normal transient operations, these differences will be small under normal conditions. When a leak is present, the differences become larger since the model system does not account for leakage. When these differences exceed preselected values, a leak alarm is declared. Sophisticated voting schemes that require multiple leak indicators to be in alarm for several consecutive time intervals are used to reduce false alarms while maintaining low thresholds. Often a simple pipeline balance of the type discussed earlier is used as a back-up to verify the transient model. Response characteristics are, however, much slower than the real time model.

Once the leak detection module declares a leak, the location routine is activated. The location is calculated from the magnitude and distribution of the leak indicators. As an example, in a straight pipeline with an upstream flow discrepancy and a downstream pressure discrepancy as leak indicators, it is an easy calculation to determine where the leak must be such that the leak flow when added to the modeled flow will produce the additional pressure drop observed at the downstream end. Solutions for pipe networks are more complicated, and unique locations do not always exist. This might be the case with parallel looped lines, for example. In this case all the calculated locations should be checked.

The components of the real-time modeling system work together to reduce the large volumes of raw data from the data acquisition system to a much smaller number of parameters and alarms that are more meaningful to pipeline operations. In the case of integrity monitoring, this means leak events that could not be detected by inspection of the measured data can be found and isolated quickly and reliably.

Basic pipeline modeling equations

The transient pipeline flow model is the heart of a pipeline modeling system. The model computes the state of the pipeline at each time interval for which data are available. The state of the pipeline is defined as a set of pressures, temperatures, flows, and densities that describe the fluids being transported at all points within the system. These quantities are found as the solution to a set of equations that describe the behavior of the pipeline system. These basic equations are the Continuity equation, the Momentum equation, the Energy equation, and an equation of state.

The continuity equation enforces the conservation of mass principle. Simply stated, it requires that the difference in mass flow into and out of a section of pipeline is equal to the rate of change of mass within the section. This can be expressed mathematically by the relation:

$$\frac{d(\rho A)}{dt} + \frac{d(\rho A V)}{dx} = 0$$

The momentum equation describes the force balance on the fluid within a section of pipeline. It requires that any unbalanced forces result in an acceleration of the fluid element. In mathematical form, this is:

$$\frac{dV}{dt} + V \times \frac{dV}{dx} + \frac{1}{\rho} \times \frac{dP}{dx} + g \times \frac{dH}{dx} + \frac{fV|V|}{2 \times D} = 0$$

The energy equation states that the difference in the energy flow into and out of a section equals the rate of change of energy within the section. The equation is:

$$\frac{dV}{dt} + V \times \frac{dT}{dx} + \frac{T}{\rho \times c} \times \frac{dP}{dT} \times \frac{dV}{dT} - \frac{f|V^3|}{2cD} + \frac{4U}{\rho cD}(T - T_g) = 0$$

These three are the basic one-dimensional pipeflow equations and are present in one form or another in all transient pipe models. What is needed to solve them, however, is a relation between the pressure, density, and temperature for the fluid—an equation of state.

The state equation depends on the type of fluid being modeled, as no one equation fully describes the variety of products that are shipped in pipelines. Some of the forms in use include a bulk modulus type of relation of the form:

$$\rho = \rho_o \left[1 + \frac{P - P_o}{B} + \alpha(T - T_o) \right]$$

This is normally used for liquids that can be regarded as incompressible. The bulk modulus B and the thermal expansion coefficient α can be constant or functions of temperature and/or pressure depending on the application. For light hydrocarbon gases a basic equation such as $P = \rho \times R \times Z \times T$ is appropriate, where Z (the compressibility) is a known function of temperature and pressure. For reasonable ranges of temperature and pressure, a function of the form:

$$\frac{1}{Z} - \cong \frac{P}{T^y}$$

may be adequate. For conditions where fluids are transported at or near the critical point, a more sophisticated correlation is required to obtain the required accuracy, but there is still a large uncertainty in the true density under these operating conditions, and they should be avoided wherever possible. Many real-time systems have been installed on lines carrying ethylene, butane, propane, and LNG/LPG products and have used the Benedict–Webb–Rubin correlation as modified by Starling (commonly called the BWRS equation) with reasonable results. Alternatively, tables such as NX-19 or special correlations such as the NBS ethylene equation can be used, but this increases the complexity of the programming.

A further complication arises if the product is not uniform throughout the system. This can occur due to batching of fluids or from varying inlet conditions. The first is more common where different products are shipped in a common line. The properties are essentially discontinuous across the interface of the two fluids but can be considered as uniform within batches. The basic problem here is to keep track of the location of the interface. Systems with continuous variations in inlet conditions occur in both liquid and gas systems. The variations can result from mixing of fluids of slightly different composition or from large variations in supply conditions.

The governing equations presented are non-linear partial differential equations, which are not suitable for machine computation. They have to be solved by implicit or explicit finite difference techniques or the method or characteristics. Of these, the implicit method seems the most appropriate, as the other two methods could give rise to mathematical instabilities if the wrong timestep or distance interval is used.

In order for the transient pipeline model to compute the state along the line at the end of each time interval, a set of initial conditions and a set of boundary conditions are required. The initial conditions specify the state at the beginning of the time interval and are normally the last set of data from the model. A steady state model must be used to generate an initial state when the model is started from rest, a so-called "cold start." In this case a period of time must elapse before the pipeline model truly represents the actual state of the line. This time period allows any transient conditions present and not represented by the steady state model to die out. Generally, less compressible systems will cold start faster than the more compressible ones, but the actual time for the transient model to be activated depends on the application. This may typically be on the order of 30 minutes for a gas pipeline.

The boundary conditions required by the model are taken from measured data along the line. For each point where fluids enter the system, its temperature, fluid type or composition, and either a flow or a pressure is required. For any equipment in the system that affects or controls the line, a suitable boundary condition must also be given. For a gas compressor, for example, either its suction pressure, flow rate, or discharge pressure must be specified. Additional measurements are used by the applications modules, generally by comparing their values to the corresponding model calculations.

Impact of instrument accuracy

The performance of real-time pipeline monitoring systems is limited primarily by the accuracy of the instrumentation installed on the line. To estimate the performance of the leak detection and location routines, it is important to understand the effect of measurement uncertainty on the model and the

real-time applications module. Measurement uncertainty is composed of bias and random components. The first is usually a fixed error between the indicated and true values, but this could change with time as components wear. The second is temporal and possibly spatial fluctuation of the output about its mean value. Fortunately there are techniques that can be employed within the software to largely mitigate the effects of bias, but care should be exercised as this is not always the case.

Leak detection and location both use differences between the measured values and the modeled values to discern leak characteristics. The measured values could contain both bias and/or random errors as discussed. The model values, because they are driven by the measured values as boundary conditions, also include error terms that are less obvious. Because of these errors, the differences in the model and measured values, or leak indicators, will not normally be zero but will fluctuate about some non-zero mean. This mean value is determined by observation during periods when no leak is present and is attributed to bias errors in the measuring system. By subtracting this from the leak indicators, the bias component can be eliminated. The leak detection then will be a function of the measurement precision errors.

Problems with this technique arise when dealing with pipelines whose operations change substantially from time to time. An example would be a liquid line operating intermittently. By monitoring differences in this way, the instruments contributing to the error are not identified. Because fluid flow in a pipeline is governed by highly non-linear relationships, fixed errors in the boundary conditions can cause variable differences in the leak indicators when the pipeline operation changes. As an example, consider a steady state pipeline that is driven by pressure difference between upstream and downstream boundary points. The leak indicator is the difference in the measured and modeled flow. (Note that a transient model would have a flow difference at each end of the pipeline.) The true value of the data for this pipeline is an upstream pressure of 1,000 units with a pressure drop of 100 units for a flow of 100 units. The model of this system is then described by the equation:

$$P1 - P2 = 0.01Q^2$$

If a 1% pressure error is introduced into the upstream pressure, the modeled flow becomes 105 and a difference of 5 units between measured and modeled flow would show up in the leak indicators. As long as the flow stays near this value, the error in the leak indicator will remain nearly constant. For instance, an actual flow of 80 units would result in a modeled flow of 86 units. Thus, the effects of the bias error in the pressure measurement can be substantially mitigated by subtracting 5 units from the leak indicator. This is termed the leak indicator "offset."

Now assume the line is shut down. The 10-unit pressure measurement error causes the flow to compute a flow of 31 units. After applying the offset, the value of the leak indicator still remains at 25. In general, a line that undergoes large and rapid changes in operation will be affected by instrument bias errors. The more common case of lines that operate within relatively narrow bounds, or that change operations slowly so that the offset can be automatically adjusted, will only be affected by the precision error of the measurements.

Measurement errors impact leak detection by limiting the size of leak that can be detected by the monitoring system. The problem comes in finding a threshold value for each alarm in the system. For a simple system that operates within narrow bounds, this can be as simple as the offset previously discussed. The values of the leak indicators (now after the offset has been removed) can be observed during normal operation and the appropriate alarm values set. For more complicated systems, the thresholds can be set in a more rigorous way. The pipeline hydraulic equations can be used to determine the sensitivity of each leak indicator to the measurement at each boundary point. The error, whether bias or precision, of each boundary point can be estimated from knowledge of the transducers and the data-gathering equipment installed on the pipeline. The error in the boundary instruments times the sensitivity of the leak indicator to the boundary point will give the threshold required to prevent normal noise in the measurement value from being interpreted as a leak. Using the root sum square as a result of combining the threshold for each boundary point with the error for the leak indicator's comparison measurement, a threshold for the leak indicator can be calculated online. This "auto tuning" of the leak indicator is found in the more advanced systems commercially available.

As an example, consider the steady state pipeline used in the earlier paragraph. The sensitivity of the modeled flow to the upstream and downstream pressure are:

$$\frac{dQ}{dP1} = -\frac{dQ}{dP2} = \frac{50}{Q}$$

Thus, at a flow of 100 units, the sensitivity of the flow to either pressure would be 0.5, with a decrease in the downstream pressure being equal to an increase in the upstream pressure. For a 1%, or 10-unit error, in the pressure, a 5-unit error in the flow would be expected, which is consistent with the previous results. The threshold required for this system would then be:

$$TL = \left[\left(\frac{50}{Q} \times EP1 \right)^2 + \left(\frac{50}{Q} \times EP2 \right)^2 \right]^{1/2}$$

This shows that the required threshold for the system increases with decreasing flow, with leak detection being impossible at zero flow. The overly stringent requirement at zero flow is due to the simplified model used. Other than that, the results are representative of the manner in which the leak thresholds must be adjusted when large flow variations occur in a pipeline.

Measurement uncertainty affects leak location by increasing the uncertainty in the calculated leak position. Leaks are located in a line by discovering where a leak of a given size would need to be located to best match the observed discrepancies in the leak indicators. Consider our steady state model again. If there is a leak of 20 units halfway down the line so that the flow is 120 units before and 100 after the leak, then the pressure drop would be 122 units. This pressure drop would correspond to a modeled flow of 110. This would result in discrepancies in the leak indicator at both ends of the pipe at 10. The leak location for a pressure–pressure boundary condition is given by:

$$X = L \times \frac{Q2}{Q1 + Q2}$$

For the conditions given above, the correct location of half the pipe length is obtained. If, however, the flow-meters have a 2-unit error such that the upstream leak indicator is 12 and the downstream indicator is 8, the leak is located 40% down the pipe, not halfway. In a 10-mi line this is an error of 100 miles, which is very significant.

When evaluating the effect of measurement uncertainty on leak detection or location, it is useful to compare the uncertainty to the magnitude of the hydraulic events of interest. Thus, a pressure transducer that is 1% accurate over a 1,500 psi span is only 30% accurate when the pressure drop between two closely spaced valve sites is 50 psi. This is because the flow is governed by pressure differences and not absolute pressures. Likewise, a flow-meter that is 2% accurate in comparison to its span is 70% accurate for sizing a 3% leak.

System design aspects and guidelines

The availability of leak alarm uptime depends heavily on the system design and the choice of hardware. Generally the more complex the system, the greater the risk of leak indicator loss, but the more accurate the location of the leaks. Design is therefore a compromise between cost, performance, and reliability. Consider a section of line shown in Figure 3. For the three simple alarms of flow imbalance, pressure imbalance, and acoustic alarms

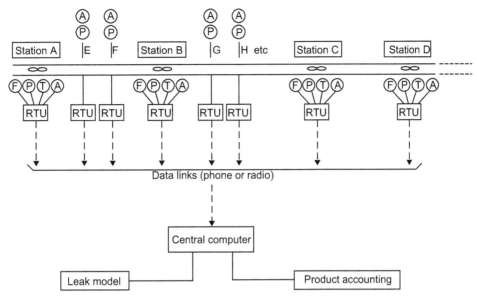

Figure 3. Schematic overview of pipeline computer-based monitoring system.

between stations A and B or B and C, the following components are needed:

Mass flow: 2 flowmeters, 2 pressure sensors, 2 temperature sensors, 2 RTUs, 2 communications links, 1 computer (11 elements)

Pressure: 4 pressure sensors, 4 RTUs, 4 communications links, and 1 computer (13 elements)

Acoustic: 2 acoustic monitors, 2 RTUs, 2 communications links, and 1 computer (7 elements)

If combined hybrid alarms of flow/acoustic, flow/pressure, or pressure/acoustic are used, then the number of components in the chain is increased. By summing the component availabilities for each element, an uptime for each alarm can be estimated. For example, a flowmeter with a failure of once in 3 years with a repair time of 4 hours has an availability factor of 0.99985 (or 99.985%).

From such an analysis, the conclusion can be drawn that the system should be made as simple as possible or instruments should be installed in duplicate to maximize alarm uptime. Section 9 also showed that instruments should be selected on performance and not on economic grounds. It is better to install fewer high-performance instruments than numerous poor ones. Digital conversion also requires attention. Often 12- or 14-bit conversion is required to give the necessary accuracy of data processing, and usually double-precision computation is also required. The use of standard outputs should be made wherever possible. Custom-designed electronics invariably lead to problems. The best guideline, however, is to seek users of pipeline monitoring systems to ask their advice and experience with instruments and system components. Independent validation of all information

should be made wherever possible. Companies that supply such complete systems usually have a client list, and it is worth spending time talking to these clients before the final design specification is fixed.

With regard to instrumentation, flowmeters with the highest accuracy are required for mass balance functions. Suitable types include turbine and displacement meters with pulse outputs. Orifice meters are not really suitable, since the best accuracy that can be obtained from a well-maintained system is around ±1% of full scale. The correct choice of turbine by comparison is ±0.25% of reading or better.

Development of pipeline monitoring systems

The speed of instrumentation development generally is rather frightening. The impact of micro-electronics is still being felt some 10 years after they first appeared, and new and improved transducers with on-board "intelligence" are being sold in increasing numbers. At the same time the size of computers is decreasing and the computating capability is increasing. Software is also advancing rapidly, and the performance of modern flow monitoring systems is becoming dependent on the accuracy of the modeling equations. Equations of state and the behavior of hydrocarbon mixtures are not particularly advanced or well understood, and fundamental research is required before the next advance in this type of technology can proceed.

All of these points indicate that computer-based monitoring systems will become the standard technique of operating and controlling pipelines in the future. Control algorithms can be integrated with the applications modules to produce a semi-intelligent complete integrity monitoring scheme. As experience in the design and operation of such systems grows, they will be applied with increasing confidence.

Future systems will use a combination of the new technology discussed. This could include internal monitoring pigs and advanced pipeline models, both run from a central control room. Thus, the internal and external state of the line could be checked simultaneously. Such technology can enable safer and more economic operation of pipelines to be carried out.

Conclusion

This section cannot do justice in such a short space, to the complex and diverse subject of leak detection. Such systems have been in operation in many forms all over the world, but it is only recently that environmental as well as economic factors have influenced their development. Modern digital systems are transforming operation and design, with many parallel functions being possible with a single system. There is now a clear need to ensure complete integration of all components in the system to guarantee safer and more accurate pipeline management. Instrument selection is critical, as is the need to develop better thermodynamic models for the next generation of systems to become more reliable and accurate.

References

1. Furness, R. A., "Modern Pipe Line Monitoring Techniques," *Pipes and Pipelines International*, May–June, 7–11 and September–October, 14–18, 1985.
2. Mailloux, R. L. and van Reet, J. D., "Real time Transient Flow Modeling Applications," PSIG Annual Meeting, 1983.
3. Covington, M. T., "Pipe Line Rupture Detection and Control," ASME Paper 78-PET-54, 1978.
4. Bernard, G. G., "Energy Balance Derivation," CRC Bethany Internal Report, 1982.
5. Anon, "Pinpointing pigs in pipe lines," *R & D Digest*, 8, 14–15, British Gas Corporation, 1986.
6. Fujimori, N. and Sugaya, S., "A study of a leak detection based on in–out flow difference method," Proceedings of IMEKO Symposium on Flow Measurement and Control, pp. 205–209, Tokyo, Japan, 1979.
7. Stewart, T. L., "Operating experience using a computer model for pipeline leak detection," *Journal of Pipelines*, 3, 233–237, 1983.

Notation

A	cross-sectional area
B	fluid bulk modulus
c	specific heat at constant volume
D	pipeline diameter
EP1	uncertainty in upstream pressure measurement
EP2	uncertainty in downstream pressure measurement
EQ	uncertainty in flow measurement
g	gravitational acceleration
h	elevation
L	pipe length
P	pressure
P_o	reference or base pressure
P_1	section upstream pressure
P_2	section downstream pressure
Q	pipeline flow rate
Q_1	upstream measured vs. modeled flow discrepancy
Q_2	downstream measured vs. modeled flow discrepancy
R	gas constant
T	temperature
T_g	ground temperature
TL	leak detection threshold level
T_o	reference or base temperature
t	time
U	heat transfer coefficient
V	velocity
X	leak location
x	incremental distance along the pipeline
y	Z factor correlation coefficient
Z	gas compressibility factor
α	coefficient of thermal expansion
ρ	density
ρ_o	reference or base density

18: Tanks

Charts give vapor loss from internal floating-roof tanks

S. Sivaraman, Exxon Research & Engineering Co., Florham Park, N.J.

Nomographs, based on the guidelines presented in American Petroleum Institute (API) Publication No. 2519, have been constructed to estimate the average evaporation loss from internal floating-roof tanks.[1] Loss determined from the charts can be used to evaluate the economics of seal conversion and to reconcile refinery, petrochemical plant, and storage terminal losses.

The losses represent average standing losses only. They do not cover losses associated with the movement of product into or out of the tank.

The average standing evaporation loss from an internal floating-roof tank depends on:

- Vapor pressure of the product
- Type and condition of roof seal
- Tank diameter
- Type of fixed roof support

The nomographs (Figures 1-4) can estimate evaporation loss for product true vapor pressures (TVP) ranging from 1.5 to 14 psia, the most commonly used seals for average and tight fit conditions, tank diameters ranging from 50 to 250 ft, welded and bolted designs, and both self-supporting and column-supported fixed roof designs. The charts are purposely limited to tank diameters 250 ft and less, because internal floating-roof tanks are generally below this diameter.

Typical values of the deck fitting loss factors presented as a function of tank diameters in the API Publication 2519 have been used in the preparation of these nomographs. In addition, for the calculations of the evaporation loss for the bolted deck design, a typical deck seam loss factor value of 0.2 has been assumed.

Table 1 gives the proper axis to use for various seal designs and fits.

Table 1
Selection of seal axis

Seal type	Seal axis	
	Average fit	Tight fit
Vapor-mounted primary seal only	H	G
Liquid-mounted primary seal only	F	E
Vapor-mounted primary seal plus secondary seal	D	C
Liquid-mounted primary seal plus secondary seal	B	A

Use of these nomographs is illustrated by the following example.

Example. Determine the evaporation loss for an internal floating roof tank given the following:

- Tank diameter 200 ft

Figure 1. Loss from welded deck, self-supporting fixed roofs.

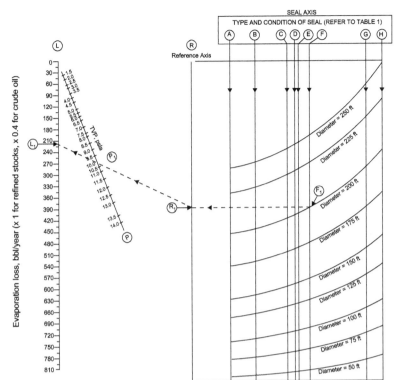

Figure 2. Welded deck, column-supported fixed roofs.

Figure 3. Bolted deck, self-supporting fixed roofs.

- Liquid-mounted primary seal only and an average seal fit
- Product true vapor pressure of 10 psia
- Welded deck with self-supporting fixed roof

Solution.

1. Use Figure 1 for the welded deck and self-supporting fixed roof.

Figure 4. Bolted deck, column-supported fixed roofs.

2. From Table 1 select the seal axis. The seal axis for the example problem is F.
3. Locate the point of intersection F1 between the seal axis F and the tank diameter contour for the 200-ft diameter tank.
4. From the point F1 traverse horizontally to intersect the reference axis R at R1.
5. Locate the true vapor pressure point P1 corresponding to 10 psia on the pressure axis P.
6. Connect the point R1 on the reference axis R and the point P1 on the pressure axis P and extend in to intersect the evaporation loss axis L at L1.

Read the evaporation loss in bbl/year at L1. The average evaporation loss is 188 bbl/year for this example. The same example is shown in Figures 2, 3, and 4 for other deck designs and roof supports.

Source

Oil & Gas Journal, March 9, 1987.

Reference

1. "Evaporation Loss from Internal Floating-Roof Tanks," American Petroleum Institute Publication No. 2519.

Estimating the contents of horizontal cylindrical tanks

Horizontal cylindrical tanks are frequently used for water and fuel storage, and in many cases it is important to be able to gauge these vessels to determine the volume of liquid contained in them. However, it is normally much more difficult to establish a volume-per-inch scale for a horizontal tank than for one in a vertical position. The accompanying nomograph simplifies this problem.

To use the nomograph, it is necessary to gauge the tank and determine the ratio of the depth of liquid in the tank to the tank diameter. After this is found, draw a straight line from the point on the "ratio" scale through the known point on the "diameter of tank" scale and read the intercept on the "gallons per ft of length" scale. From this point, draw a second line

through the known point on the "length of tank" scale and read the intercept on the "gallons (or barrels) in total length" scale.

Example. Find the volume of liquid contained in a horizontal cylindrical tank 7 ft in diameter and 20 ft long when the depth of the liquid is 4 ft, 10.8 in.

The ratio of depth of liquid to tank diameter is:

58.8/84 = 0.70

Connect 0.70 on the ratio scale with 7 ft on the diameter scale and continue the straight line to obtain the intercept 215 on the gallons per ft of length scale. Draw a second line

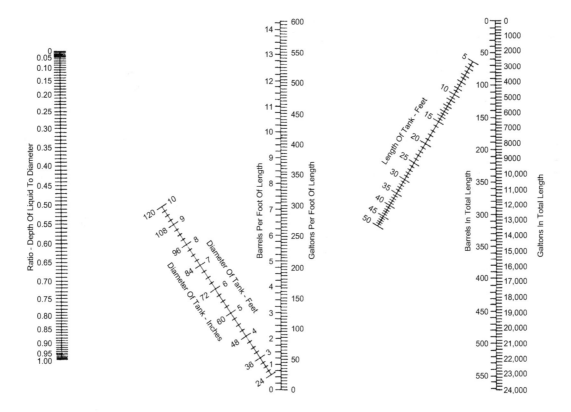

from the point 215 through the 20 on the length of tank scale and continue the line to obtain the intercept 4,300 on the

gallons in total length scale. Total liquid content of the tank is 4,300 on the gallons.

How to gauge a horizontal cylindrical tank

Express the depth in % of the diameter; then the result will be given in % of total capacity of the tank.

Rule 1. For depth up to 30; multiply the square root of the depth by the depth, and then by 0.155.

Example. Liquid depth is 16% of tank diameter

$$16 \times 16 \times 0.155 = 4 \times 16 \times 0.155 = 9.9\%$$

The correct answer is 10.3%; error is about 0.4%.

Rule 2. For depth between 30 and 50; subtract 10 from the depth, multiply by 1.25.

Example. Liquid depth is 44% of tank diameter

$$(44 - 10) \times 1.25 = 34 \times 1.25 = 42.5\%$$

The correct answer is 42.4%.

The maximum error for depths less than 5% may be as great as 10%; gauging in this range is always difficult, and a very small slope can introduce a much larger error than this. When the depth is greater than 50%, apply the same rule to get the volume of the empty space above the fluid, and subtract.

Use nomograph to find tank capacity

This simple nomograph can be used to find the capacity of your vertical cylindrical tanks. Here's how it works:

Draw a straight line from the "height" scale through the "diameter" scale and to the first "capacity, barrels" scale.

Read directly the capacity of the tank in barrels. (*Note:* The "height" scale may be used to indicate the overall height of the tank or the depth of liquid in the tank.)

Draw a second straight line connecting the two "capacity, barrels" scales at the same reading on each scale. Read the capacity of the tank in gallons and cubic ft on the proper scales.

The nomograph was constructed as follows:

1. The "height" scale is based on two log cycles per 10 in. with a range of 1–60 ft.
2. The "capacity, barrels" scale is based on four log cycles per 10 in. with a range of 20–150,000 barrels.
3. The "diameter" scale is based on three log cycles per 10 in. with a range of 4–150 ft.
4. The distance between the height and diameter scales is exactly two-thirds the distance between the height and "capacity, barrels" scale.
5. Determine points to locate the diameter scale from the following equation:

$$\text{Capacity, barrels} = 0.1399 \, (\text{diameter})^2 \, (\text{height}),$$
$$\text{units in ft}$$

6. The "capacity, gallons" scale is based on four log cycles per 10 in. The initial point on the scale is determined as follows:

$$20 \text{ barrels} \times 42 \text{ gallons per barrel} = 840 \text{ gallons}$$

The range of the scale is 900 to 6 million gal.

7. The "capacity, cubic feet" scale is based on four log cycles per 10 in. The initial point on the scale is determined as follows:

$$20 \text{ barrels} \times 5.6146 \text{ cu. ft per barrel} = 112.292 \text{ cu. ft.}$$

The range of the scale is 120 to 800,000 cu. ft.

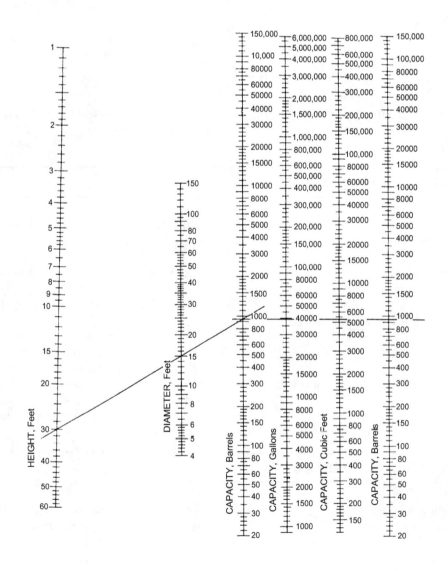

Correct the volume of light fuels from actual temperature to a base of 60°F

To approximate quickly the volume of gasoline or other light liquid fuel at 60°F from a known volume at any temperature in the atmospheric range, use the formula:

$$V_a - V_{60} = 0.0006(T - 60)V_{60}$$

where: V_a = Volume at actual temperature

V_{60} = Volume corrected to 60°F
T_a = Actual temperature of fuel

Example. A tank contains 5,500 gallons of gasoline at 46°F Correct the volume to a base of 60°F.

$(5,500 - V_{60}) = 0.0006(46 - 60)V_{60}$
$(5,500 - V_{60}) = 0.0006(-14)V_{60}$
$5,500 = V_{60} - 0.0084V_{60}$
$5,500 = 0.9916V_{60}$
Volume at 60°F = 5,546.6 gallons

Shrinkage

To approximate the shrinkage or expansion, obtain the difference between the actual volume measured and the corrected volume. In this case:

Shrinkage = 5,546.6 − 5,500 = 46.6 gallons

Volume of liquid in vertical cylindrical tanks

Measure the depth of the liquid and either the diameter or circumference of the tank, then the volume in:

Gallons = 0.0034 d^2h or 0.00034 c^2h
Barrels = 0.000081 d^2h or 0.00082 c^2h
Gallons = 5.88 D^2H or 0.595 C^2H
Barrels = 0.140 D^2H or 0.0142 C^2H

where: d = Diameter, inches

c = Circumference, inches
h = Depth, inches
D = Diameter, feet
C = Circumference, feet
H = Depth, feet

If the circumference is measured on the outside, then three times the thickness of the tank wall should be subtracted before using the formula. Naturally, these rules cannot supplant the results of accurate tank strapping, which take many other factors into account.

Example. How many gallons will a tank 12 ft in diameter and 16 ft high hold when full?

Gallons = 5.88 D^2H
= (5.88) (144) (16)
= 13,548 gallons

Example. How many barrels will a tank 8 ft in diameter and 16 ft high hold when full?

Barrels = 0.140 D^2H
= (0.140)(64)(16)
− 143 barrels

Chart gives tank's vapor formation rate

When sizing the vapor piping for a manifolded expansion-roof tank system, the rate of vapor formation must be known. While the rate of vapor formation can be computed by longhand methods, the calculation is tedious and takes much valuable time.

Example. Determine the rate of formation of vapor in a 140,000 barrel capacity tank when it is filled at the rate of 8,000 barrels per hour.

Solution. Enter the chart on the left at a capacity of 140,000 barrels and draw a straight line through the filling rate of 8,000 barrels per hour on the right. At the intersection with the central scale read the vapor formation rate as 55,000 cu. ft per hour. The vapor piping for this tank would have to be designed for this formation rate if the maximum filling rate anticipated were 8,000 barrels per hour. But if a great filling rate were expected, the vapor formed would have to be computed for the higher rate. The chart could, of course, also be used for this computation.

This chart is based on the following equation:

$$V = \frac{\text{tank capacity, bbl}}{14.3} + \frac{\text{filling rate, bbl per hr}}{0.178}$$

where V = vapor formed, cubic feet per hour

19: Maintenance

How to plan for oil pipeline spills (part 1)

Here are ideas to consider in development of a comprehensive contingency plan

J. D. Sartor, Vice President, and **R. W. Castle,** Senior Project Engineer, Woodward-Clyde Consultants, Environmental Systems Division, San Francisco, California

Federal regulations and permit stipulations require the preparation of written plans and procedures for dealing with accidental spills of materials carried by liquids pipelines. These requirements can be met through the preparation of comprehensive spill contingency plans.

Proper contingency planning involves analysis of the material(s) carried and the environments crossed by the pipeline. Based on company personnel and operating procedures, a response organization and necessary equipment are developed. The planning concept assumes that preplanned emergency actions are appropriate for protection of human life, property, and the environment.

In addition, however, the planning must provide the flexibility to respond to unanticipated situations. The final planning element is the design of a comprehensive training program to ensure the safety and efficiency of the response teams.

Regulatory requirements

Historically, federal requirements pertaining to liquids pipeline construction and operation have contained no specific requirements or guidelines for the preparation of spill contingency plans.

In spite of this, such plans have been required in conjunction with various permits and right-of-way grants. Cases where contingency plans have been stipulated include the Trans Alaska, SOHIO West Coast to Midcontinent, Northern Tier, and the Northwest Alaska Gas pipelines.

The National Environment Policy Act (NEPA) of January 1, 1970 and ensuing amendments have led to federal and state requirements for analysis of the potential for spills, their probable impact, and the degree to which spill impacts may be mitigated through the Environmental Impact Statement— Environmental Impact Report process. Spill contingency is often the major method of spill impact mitigation, and its association with the environmental review process is becoming increasingly common.

Effective July 15, 1980, the Department of Transportation (DOT) amended regulations for the transportation of liquids by pipeline. Title 49 of the Code of Federal Regulations, Part 195 (revised October 1, 1979) describes these regulations and amendments. Subpart—Operation and Maintenance— requires the preparation of written emergency plans and procedures for dealing with accidental release of commodity, operational failures, and natural disasters affecting facilities. The same regulations also require assignment of personnel to emergency functions and the design and implementation of emergency procedures programs. These requirements summarize the essential elements of a spill contingency plan.

Contingency plan objectives

A rapid response capability is essential to adequately control and remove spilled material.

To provide this capability, adequate advance planning is mandatory. Such pre-spill planning should recognize the various types of spill situations that could occur and incorporate a response strategy for each.

These strategies should outline methods to be used and describe the required types and amounts of equipment, materials, and manpower.

In addition to pre-spill planning for anticipated spill types and situations, a comprehensive plan must allow the flexibility to respond effectively to unanticipated spill situations.

Related studies

The development of a spill contingency plan is based on consideration and analysis of a wide variety of factors. Included are geographical elements (location of the spill, drainage characteristics, surface conditions, soil type, accessibility, trafficability), environmental elements (weather conditions, hydrology, etc.), and ecological elements (sensitive and vulnerable areas, rare and endangered species, etc.).

For special environments, additional studies may be required. For example, a series of special studies on reclamation of spill affected areas was required for tundra during the implementation of the Alaska pipeline. These studies included evaluation of plant types that could be used for rehabilitation, effects of spilled oil on native plants, effects of various types of cleanup measures, and residual effects of oil over different weathering periods.

Engineering elements necessary for contingency planning include pipeline pumping and drainage characteristics, valve placements, monitoring equipment, operating procedures, and communications and control systems. Typically much of this information is available through project-related design engineering and environmental studies at an early phase in project implementation.

In some cases, however, all of the required information may not be available and may require generation prior to the completion of the final spill plan. Early versions of spill contingency plans are considered "draft" plans and are submitted to satisfy the requirements of regulatory agencies. The "draft" plans are submitted with the stipulation that final spill plans are prepared and implemented prior to the pipeline becoming operational.

Planning concepts

Pipelines characteristically extend considerable distances and encounter a variety of environmental conditions. The resulting spill planning must respond to these factors and is commonly either overly simplified or so complex that it is unusable in practice. To counter either tendency and yet provide sufficient information in a usable format, it is advantageous to design the contingency plan as a series of volumes, each specific to a particular need. Using this approach, the master volume or General Provisions contains organizational and operational information common to the entire pipeline system. It also forms a basis for personnel training.

The second volume or volumes delineate specific response actions and personnel for specific sections of the pipeline. Commonly the areas covered by these volumes are defined by major drainage patterns. If necessary as a result of drainage basin size or other unusual factors, each of these sections may be divided into further subdivisions until workable size field response plans are achieved.

Typical contents of successive volumes of a pipeline contingency plan are shown in Figure 1. Information pertinent to the entire system is contained only in the general provisions.

An effective contingency plan indicates the actions to be taken, their sequence, and timing in relation to other events.

GENERAL PROVISIONS		DISTRICT PLAN		CONTENGENCY AREA	
100	Introduction	100	Introduction	100	Introduction
101	Back ground	101		101	
102	Policy	102		102	
103	Scope	103	Scope	103	Scope
104	Plan Concept	104	Plan Concept	104	
105	Plan Format	105	Plan Format	105	
200	Contingency Response Organization	200	Contingency Response Organization - District	200	Contingency Response Organization - Area
300	Immediate Response Actions	300	Immediate Response Actions - District	300	Immediate Response Actions - Area
301	Detection	301	Detection	301	Detection
302	Alert Procedures	302	Alert Procedures	302	Alert Procedures
303	Containment	303	Containment	303	Containment
304	Equipment	304	Equipment	304	Equipment
305	Pipeline Operational Actions	305	Pipeline Operational Actions	305	Pipeline Operational Actions
306	Location Cross References	306		306	
400	Contingency Plan	400	Contingency Plan	400	Contingency Plan
401	System Description	401	District Description	401	Area Description
402	Oil Spill Detection	402	Oil Spill Detection	402	
403	Assessment	403	Assessment	403	
404	Containment	404	Containment	404	Containment
405	Exclusion	405	Exclusion	405	Exclusion
406	Cleanup	406	Cleanup	406	Cleanup
407	Disposal	407	Disposal	407	
408	Restoration	408	Restoration	408	
409	Documentation	409	Documentation	409	
500	Environmental Parameters	500	Environmental Parameters	500	Environmental Parameters
501	Biological Environment	501	Biological Environment	501	
502	Meterology	502	Meterology	502	
503	Oceanography	503	Oceanography	503	
504	Hydrology	504	Hydrology	504	
600	Training Program	600	Training Program	600	
700	Response Plan Annexes	700	Response Plan Annexes	700	Response Plan Annexes
701	Governmental Relations	701	Governmental Relations	701	
702	Third Party Contractors	702	Third Party Contractors	702	
703	Logistics	703	Logistics	703	
704	Communications	704	Booms	704	
705	Oil Spill Task Force Directory	705	Skimmers	705	
706	Pipeline Drainage Calculations	706	Sorberts	706	
707	Booms	707	Revegetation	707	
708	Skimmers	708	Wildlife Care & Rehabilitation	708	
709	Sorbents	709	Communications	709	
710	Manpower	710	Oil-On-Water Sensors	710	
711	Revegetation	711	Portable Lightering Systems	711	
712	Oil-On-Water Sensors	712	Public Relations Guidelines	712	
713	Public Relations Guidelines	713	Telephone Addresses - Company	713	
714	Population Centers		Personnel - District	714	
715	Airports & Landing Strips	714		715	
716	Commercial Food & Lodging Facilities	715		716	
717	Academic Institutions & Laboratories	716		717	
718	Oil Spill Cooperatives	717		718	
719	Wildlife Care & Rehabilitation	718	Airports & Landing Strips	719	
720	Lightering Systems	719		720	
		720	Commercial Food & Lodging Facilities		
800	Bibliography				

Figure 1. Contingency plan content.

During or immediately following a spill, the following representative actions should be taken:

a. Verify and locate the spill and notify the spill response team to initiate emergency procedures.
b. Control or limit the amount of material spilled. Exclude or prevent the spread of material to sensitive areas. Evaluate threats to the general public and act accordingly.
c. Determine the extent of the spill and predict its subsequent dispersion as a function of time.
d. Plan and direct overall operations; provide administrative support, liaison with government/local officials, and public information.
e. Remove and clean material from contaminated land areas, subsoils, shorelines, and water surfaces.
f. Determine procedures and site locations for the disposal of contaminated materials.
g. Document all phases of spill incident and subsequent cleanup activities.

Selection of the action to take at different times during a spill depends on: (a) the location and nature of the spill; (b) the quantity and type of material; (c) hydrology, topography, and soil types; (d) ice, weather, and currents. Supervision will start immediately, and documentation must be conducted simultaneously with all other operations.

Spills require prompt action whether they are on land or on water. Detailed response plans must be prepared for immediate action to be taken wherever and whenever spills occur.

In general, the contingency plan describes the logical and sequential order of what-to-do actions. Some actions are taken in all situations. Other actions are contingent on and initiated only by special circumstances. The basic plan is supported by appendices that provide how-to-do-it instructions for basic actions. These appendices also provide information on regulations, sources of equipment, logistics, and general background.

Contingency response

The nature of each individual response organization is dependent on the nature of the pipeline system and the area through which it crosses. In some remote areas such as Alaska, any effective response organization must rely heavily on internal resources. In more populated areas, a spill response organization can be designed more around the use of local resources.

Any response organization must, however, incorporate certain functions. Typically these functions may be grouped as management, advisory, support, and local field response. The management, advisory, and support roles generally perform staff functions and are usually staffed by company personnel. The local field response functions are normally staffed by local operating personnel stationed along the pipeline route. As necessary, these personnel may be supplemented by local contractors.

The assemblage of the above roles is known as the Spill Task Force (Figure 2).

Source

Pipe Line Industry, February 1982.

Figure 2. Spill Task Force organization.

How to plan for oil pipeline spills (part 2)

What to consider for initial and followup actions for spills and task force training

J. D. Sartor, Vice President, and **R. W. Castle,** Senior Project Engineer, Woodward-Clyde Consultants, Environmental Systems Division, San Francisco, California

Within the Spill Task Force, each district uses a two-level organization to provide fast and effective initial and followup response. A typical district organization is shown in Figure 1. The first level of response, the Immediate Response Team (IRT), reacts immediately to any spill within its area of concern. This team takes pre-planned response actions, limiting the spills' impacts until further actions can be activated.

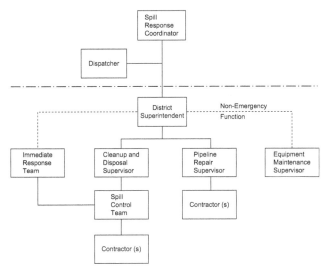

Figure 1. District Response Team organization.

The IRT is trained and equipped to handle minor spills without additional assistance. Should this team be unable to completely control a spill, it will take pre-planned control actions and notify the spill control coordinator. The spill coordinator will then activate company and contractor resources as necessary. The second level, the remainder of the District Response Team, is then activated as necessary to complete the control and recovery of the spill. For sufficiently large spills, the overall Spill Task Force may require activation.

Spills will normally be detected by or reported directly to the pipeline dispatcher, who will ensure proper communications within the Spill Task Force and who will shut down pumping operations if necessary. The dispatcher will ensure that the district superintendent responsible for the area in question is notified and that the appropriate Immediate Response Team has been activated. The dispatcher will then notify the spill coordinator, who will activate additional resources as necessary after consulting with the local area superintendent. Figure 2 illustrates how the contingency response organization functions.

Immediate response

The composition of any individual IRT will vary with the size and nature of the pipeline system. In many cases several company personnel will be available along a specific pipeline segment. With proper training and equipment, however,

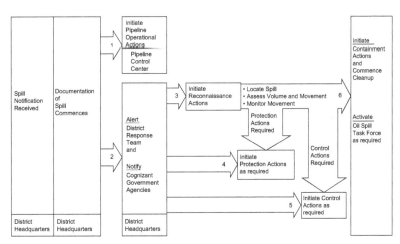

Figure 2. Immediate response action flow chart.

even two- or three-man teams can be constructive in implementing immediate actions.

The duties of the IRT may need to be taken simultaneously. These duties include:

- Verification and location of the leak
- Leak control
- Conducting prescribed initial actions to ensure the safety of the general public, contain the spill, and protect sensitive areas
- Assess the situation and provide initial inputs to the district supervisor and the spill coordinator
- Assume assigned positions in the spill task force when activated

IRT positions can be manned 24 hours per day or on a 24-hour standby basis depending on the nature of the facility and available along-route personnel. Typically, the dispatcher position will be manned on a 24-hour basis. Individuals filling response positions are listed in the contingency plan by name and method of notification.

Immediate response actions

When a spill is detected by monitoring equipment or reported through some other mechanism, the dispatcher immediately notifies the IRT. A typical alert and notification procedure is shown in Figure 3. The primary functions of the IRT are to provide for public safety, contain the spill, and exclude the spill from sensitive or vulnerable areas. For small spills, the IRT may be capable of complete control and cleanup. For larger spills, they provide stop-gap measures until greater resources can be mobilized.

The type of response is dependent on the specific spill area and the nature of the material. The intention of all actions is to locate the leak and begin emergency actions immediately. To facilitate such actions, the contingency plan must provide predesignated control sites and actions, equipment recommendations, and sufficient information on the physical and ecologic characteristics of the threatened area to allow flexible and defendable response actions. Figure 4 depicts a presentation of pre-planned response actions and support information for a typical pipeline segment.

Directions of probable spill movement, access points, and potential control sites are indicated on the pipeline segment map. Supporting information is tabularized and includes predicted maximum spill sizes at selected points, location of equipment storage, recommendations for primary control actions under varying conditions (letter codes refer to a contingency plan appendix that details respective technique implementation), special features of concern during spill control, and general hydrologic characteristics. Ecologic characteristics of the area are provided in an additional supporting appendix.

Flexible response actions

While it is a basic premise in spill planning that preplanned response actions are appropriate, it is also accepted that no two spills are ever identical.

As a result, spill contingency planning must provide the flexibility to respond to unpredicted situations. This can be accomplished through the design of decision guides that equate the types and ranges of environmental parameters to appropriate control and recovery techniques. A representative decision guide for the protection of inland waters is shown in Figure 5. A pre-planned decision making process can be particularly useful in designing cleanup and disposal programs.

Figure 3. Alert and notification procedure.

PIPELINE MILEPOSTS   5.8 to 9.4   COUNTY   County C   MAP REF.   404-5

MAXIMUM SPILL SIZE   River 1: 1,682 bbl (16,148 bbl west of river)
                     Creek A:

NEAREST EQUIPMENT SITE   Site X:

SPILL CONTROL RECOMMENDATIONS

Location	Low Flow	High Flow	Terrestrial	Remarks
CS-1	F,G,H	G,H	–	Private property
CS-2	F,G,H	G,H	–	
CS-3	F,G,H	G,H	–	
CS-4	F,G	–	ABC	
CS-5	F,G,H	I,J,K,L	–	
CS-6	F,G,H	I,J,L	–	Limited access, if oil
CS-7	F,G,H	I,J,L	–	passes this point, con-
CS-8	F,G,H	G,H	–	tain in river at River 1
CS-9	F,G,H	G,H	–	Limited access, private
CS-10	F,G,H	I,J,K	–	property
CS-11	F,G,H	I,J,K	–	
CS-12	F,G,H	I,J	–	
CS-13	F,G,H	I,J	–	
CS-14	F,G,H	I,J	–	Private property

SPECIAL FEATURES/SPILL CONTROL PROCEDURES

1 Stream velocity during moderate to peak flow may preclude effective
   booming. During lower flow periods bury boom skirt on banks and bars

   • River 1 is a major salmon migration and spawning watercourse
   • See Table C-29, Appendix C for an inventory of the Natural Resources
     of River 1.

WATER CROSSINGS/HYDROLOGIC CHARACTERISTICS

   • Depth to water table generally exceeds 75 feet
   • Irrigation ditch (mp 6.0)
   • River 1 (mp 7.0)

Condition	Discharge	Velocity	Max. Width	Max. Depth	Occurrence
Approx. 100 yr. flow	8,800 cfs	8-9 fps	270-350 ft.	9-10 ft.	Dec. - Feb.
Low flow	70 cfs	3 fps	20 ft.±	2 ft.	Jun. - Aug.

   • Unnamed Creek (mp 7.75):   perennial; water quality classification A
   • Irrigation ditch (mp 7.9): underground
   • Creek A (mp 8.6):   perennial; water quality classification A
   • Unnamed Creek (mp 8.75):   perennial; water quality classification A

Figure 4. Typical pipeline segment information and response actions.

Training

Training of all spill response personnel, including simulated spill and equipment drills, is critical. Accordingly, a comprehensive training program is an integral part of the contingency plan. The objectives of the training program are to:

• Maintain the plan as a fully operable and working document
• Inform Task Force members of their respective duties and standard communication procedures
• Allow Task Force members to become familiar with the use of all equipment and to update the plan to reflect the current state of the art, including new equipment and procedures

• Modify the plan in the light of information gained from the field exercises
• Incorporate experience gained from response to spills into the plan

The training program is structured according to the level of responsibility of the participants and may include:

• Classroom instruction of field demonstrations
• Oil spill response drills to test notification, alert, and mobilization procedures
• Drills to practice specific containment measures and general containment and cleanup equipment and techniques

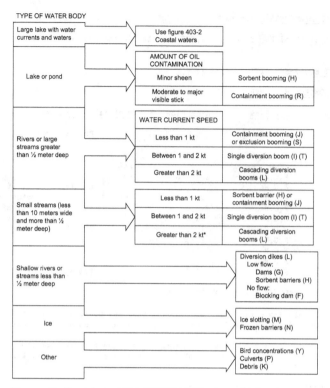

Figure 5. Decision guide for the protection of inland waters.

Conclusion

Accidental pipeline spills have and will continue to occur. Recognizing this fact, regulatory agencies have required emergency spill planning to minimize resulting impacts.

Comprehensive spill contingency planning satisfies these requirements and provides quick and effective response to environmental emergencies. Included in the preparation of such plans are the development of a response organization and response actions for each individual system. Also important in the development of a plan is the inclusion of a training and proficiency program.

(Based on a paper, Contingency Planning for Liquid Pipe Line Spills, presented by the authors as a contribution of the American Society of Mechanical Engineers at the Energy-Resources Conference and Exhibition, January 18–22, 1981, Houston, Texas.)

Source

Pipe Line Industry, March 1982.

20: Economics

Rule of thumb speeds payroll estimates

Hewlett Packard HP41-CV computation provides an estimate of the weekly expense of a "manpower spread" for pipeline construction in a given year

R. D. Green, Energy Center Fellow, University of Pennsylvania

Pipeline construction payroll expenses can be estimated in a rule-of-thumb computation that indexes 1982 survey data to any year of interest when an inflation figure is provided.

An estimate of the weekly payroll expense of a "manpower spread" for pipeline construction in a given year is obtained as follows (Table 1):

$$P(y) = \left(\sum_{j=1}^{15} P_j E_j \right) e^{i(y-1982)}$$

where

P_j = Number of employees in category j,
E_j = Compensation expense per employee in category j,
i = Rate of inflation, and
y = Year for the construction estimate.

The present value (PV) of a stream of payroll expenses is found as follows:

Table 1
Compensation Categories of Manpower

Category* (j)	Position	Weekly compensation** (E$_j$)	Category* (j)	Position	Weekly compensation** (E;)
I	Welding crew foreman	1,750	VII	Field office & overhead spreadman	1,100
II	Welding crew welders	1,670		Stringing foreman	
	Pipe crew foreman			Over-the-ditch, cleaning &	
	Pipe crew stringer bead welder			priming crew foreman	
	Pipe crew hot pass welders		VIII	Testing crew operator	1,050
	Tie-in welders		IX	Field office & overhead asst.	
III	Field office & overhead heavy-duty			spreadman	1,000
	mechanics	1,250	X	R.O.W. clearing & fencing power	
IV	Bending foreman	1,200		saw operators	995
	Bending engineer			Pipe crew skid truck drivers	
V	Field office & overhead fuel truck driver	1,175	XI	Field office & overhead lowboy	
	Field office & overhead grease			truck driver	980
	truck driver			Field office & overhead float	
VI	R.O.W. & fencing dozer operator	1,125		truck driver	
	R.O.W. grading foreman			Pipe crew welder helpers	
	R.O.W. grading dozer operators			Tie-in welder helpers	
	Ditching (earth) foreman		XII	R.O.W. clearing & fencing laborers	960
	Ditching (earth) machine operators			R.O.W. grading laborers	
	Ditching (earth) backhoe operators			Ditching (earth) laborers	
	Road crossing road boring			Stringing truck drivers	
	machine operator			Stringing laborers	
	Road crossing backhoe operator			Bending laborers	
	Stringing crane operator			Pipe crew laborers	
	Stringing sideboom operator			Over-the-ditch, cleaning & priming	
	Bending sideboom operator			crew laborers	
	Pipe crew sideboom operator			Tie-in laborers	
	Pipe crew tack rig operators			Backfilling & clean-up laborers	
	Tie-in foreman			Lower-in laborers	
	Tie-in sideboom operators		XIII	R.O.W. clearing & fencing foreman	900
	Lower in foreman		XIV	Field office & overhead office	800
	Lower in sideboom operators			manager	
	Testing crew foreman			Field office & overhead payroll clerk	
	Backfilling & clean-up foreman		XV	Field office overhead night	
	Backfilling & clean-up dozer operators			watchman	350

$$PV = \sum_{n=1}^{N} \frac{P(y)}{\left(1 + \frac{i}{52}\right)^n}$$

where

N = Number of weeks, and
i = Decimal "interest" rate.

Estimate

If an estimate for P_j is needed, example manpower spreads for 8- and 30-in cross-country pipelines are provided in Table 2. In this case, an estimate of the number of weeks the spreads will be in operation is obtained as follows:

$$PD = \frac{L}{S \times d}$$

where

      PD = Project duration in weeks,
      L = Length of the pipeline in miles,
      S = Average daily spread movement rate in miles per day,
      d = Number of days in the work week (e.g. $d = 6$ if the crews work 6 days per week).

For the manpower spreads of Table 2, S is probably in the range expressed by these inequalities:
1 mile/day $\leq S \leq$ 0.35 miles/day for 8-in size.

Table 2
Typical Manpower Spreads

Category (j)	Numbers of employees in category j for 8- and 30-in pipeline[*]	
	(P_j for 8-in pipeline)	(P_j for 30-in pipeline)
I	1	1
II	6	25
III	2	7
IV	1	3
V	1	5
VI	29	74
VII	3	4
VIII	1	3
IX	1	1
X	14	26
XI	11	36
XII	40	144
XIII	1	2
XIV	2	4
XV	1	1

[*]Cross-country pipeline construction.

Source

Pipe Line Industry, March 1985.

Rule of thumb estimates optimum time to keep construction equipment

Hewlett Packard HP41-CV computation estimates the optimum retention period for construction equipment

R. D. Green, Energy Center Fellow, University of Pennsylvania

Construction firm executives prefer cost-wise decisions concerning whether and when equipment will be replaced. The problem is extensive since, as depicted in Table 1, equipment spreads are both extensive and varied.

Equipment needed to construct a pipeline deteriorates with use, and maintenance costs rise over time. These maintenance costs might include, among other things, (1) funds for a preventive maintenance program, (2) funds to be set aside for costs incurred due to equipment failures, (3) replacement parts, and (4) labor expenses for repairs.

The problem of how to decide when to replace equipment is simplified with a "life cycle analysis" that compares (over an infinite time horizon) any number of alternatives of maintaining or replacing machines and then identifies the optimum choice. A "rule of thumb" approach to life cycle analysis for construction equipment can be performed by making some assumptions to simplify the usually complex calculations. Taking this approach, the assumptions and method are as follows.

Procedure

Let the original cost of the equipment item (e.g. a machine) be K. The problem is to find the number of years, N, that the machine should be kept such that the Net Present Worth (NPW) of operating costs is a minimum. This optimum number of years is called N^* (read "N star"). Suppose that the machine maintenance costs increase geometrically by a fraction, G, over the previous year and that the first year maintenance cost is C dollars. To simplify the equations, assume that all costs are paid as they are incurred; hence all costs are at present value. The relevant discount rate is R.

Table 1
Equipment Spread

Typical spread—equipment type and number for 8-in. and 30-in. cross-country lines	8"	30"		8"	30"
Field Office and Overhead			Sidebooms		3
Cars		2	Tow tractor		1
Pickups	2	7	Clean & prime machine		1
Welding rig	1	1	Coat & wrap machine		1
Fuel truck	1	1	Tar kettles		4
Grease truck	1	1	Winch truck		1
Utility truck	1	1	Flatbed truck		1
Tractor w/lowboy	1	1	Sets, cradles & strongbacks	2	2
Tractor w/float	1	1	Holiday detectors		2
Right of way, clear, burn, and fence			**Lower in**		
Pickups	3	2	Pickups		2
Dozers	2	2	Sidebooms		6
Winch truck		1	Clamshell		1
Flatbed truck	1	1	Dozer		1
Power saws	8	6	Trucks		2
Right of way, grade			Tar kettles		1
Pickup	2	1	Water pump		5
Dozers	4	3	**Tie in**		
Ditch (Earth)			Pickups	2	4
Pickups	2	7	Sidebooms	3	4
Ditching machines	1	2	Welders	1	4
Dozer w/ripper		1	Truck		1
Backhoes	3	4	Clamshell		1
Clamshell		1	Tar kettle		1
Ditch (rock)			Water pump		1
Sidebooms		2	**Testing**		
Wagon drills		2	Special truck	1	1
Compressors		2	Pumps, high pressure		2 sets
Trucks		2	**Utilities**		
Jackhammers		2	Pickup		1
Road crossings			Clamshell		1
Pickup	1	1	Clamps and cutters		1 set
Boring machine	1	1	Welders		2
Sideboom		1	**Backfill and clean up**		
Winch truck		1	Pickups	2	4
Tar kettles		1	Backfiller		1
Back hoe	1	1	Dozers	3	3
Stringing			High lift		1
Pickup	1	1	Farm tractor	1	1
Crane hoist	1	1	Dragline		1
Tractors w/pipe trailers	3	7	Winch truck		1
Sideboom	1	1	Flatbed truck		1
Tow tractors	1	3	**Other*****		
Bending			Base station	1	3
Pickups	1	2	Two-way radios	11	33
Sideboom	1	1	Comfort station/toilet	3	9
Bending machine	1	1	Skids	5000	15000
Transit & level rod		1	Helicopters		
Pipe crew			Internal line-up clamp		
Pickup	1	1	External line-up clamp	5	15
Tractor welders	1	2	Tamper tractor		
Sidebooms		2	External plant coating machine		
Trucks		2	Vacuum-lift unit		
Set inside clamps		1	Laybarge		
Buffing rig	1	1	Motor patrol/grader		
Welding crew			Crane		
Pickup		1	Kitchen		
Welders	3	11	Trailer/sleeping		
Tow tractor		1	Medical station		
Crew bus		1	First aid kits	12	36
Coating crew					
Pickups	1	4			

*Survey data, 1982.
***Source: A Manual of Pipeline Construction, 3rd Ed., Dallas, Energy Communications, 1979, except for "other" category.
***In "Other" category only: 30" data is estimated by multiplying 8" data by three.
Note: For more information on construction spreads, see Cost Estimating Manual For Pipelines And Marine Structures, John S. Page, Houston, Gulf Publishing Co., 1977.

The solution to this problem is to derive an equation that allows a comparison for machines with different lifetimes. This is done by finding the NPW of an infinite stream of constant scale replications. The analysis is performed by using the (algorithm) equations below to find an optimum replacement period for each alternative considered:

Find N^* such that:

$$f(N_i) - f(N_{i+1}) \leq 0 \text{ and } f(N_{i-1}) - f(N_i) > 0$$

where

$$f(N) = K + C\left(\frac{1-\left(\frac{1+G}{1+R}\right)^N}{R-G} \times \frac{(1+R)^N}{(1+R)^N - 1}\right)$$

and

$$f(N+1) = K + C\left(\frac{1-\left(\frac{1+G}{1+R}\right)^{N+1}}{R-G} - \frac{(1+R)^{N+1}}{(1+R)^{N+1} - 1}\right)$$

and where

$K =$ The initial cost of the machine,
$C =$ The first-year maintenance cost,
$G =$ The fraction of increased maintenance costs (e.g. $G = 0.20$ when maintenance costs, C, are rising at 20% per year compounded annually), and
$R =$ The rate (e.g. $R = 0.15$ for a discount rate of 15%).

A stopping rule for operating the algorithm is to halt the computation when the value of N is no longer decreasing (e.g. the computation is over when, as N decreases, an N for which $f(N_i) - f(N_{i+1}) \leq 0$ is finally obtained). An example program and program run is given in the Appendix (written in the language of the Hewlett Packard HP41-CV). Figure 1 is a flowchart that is useful for interpreting the algorithm and translating the program into another computer language.

The procedures for choosing among the technical and managerial options (e.g. alternatives of repairing, replacing, rehabilitating, abandoning, or decreasing the work load of machines) for managing the construction spread are (1) Given i options, use the algorithm to identify the optimum replacement period N^* for each equipment replacement alternative (that is, given i alternatives, find all N^*_i); (2) For each N^*_i, there is an associated NPW_i. Compare these net present worths and identify the least expensive one (e.g. of several N^* values, each with an associated cost and equipment replacement scheme, the least expensive one is the smallest NPW_i); (3) Usually, the equipment replacement plan associated with this minimum NPW_i will be selected and then to implement the decision, that machine will be replaced every N^* years (where the N^*_i of interest is the one

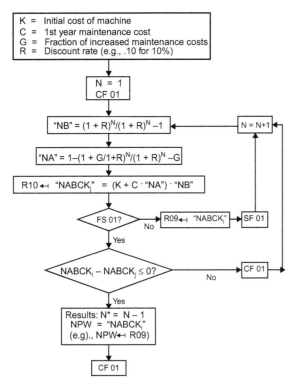

Figure 1. Flow chart for interpreting the algorithm and translating the program into another computer language.

for which NPW_i is a minimum); (4) A simple sensitivity analysis may be performed by reworking items 1 through 3 for various values (e.g. for checking the sensitivity to interest rates, compute N^* for various R values).

"Cutoff rates" may be used to decide when equipment failures are too high despite favorable cost curves for repairs since the pipeline construction firm must be reliable in serving customers, but these cutoff rates should be selected in recognition of the N^* values for the various alternatives to avoid short-circuiting the whole decision process.

The result of the above procedure is a cost-wise strategy for the management of the equipment used in pipeline construction spreads.

If the assumptions of the life cycle analysis are modified to more accurately reflect the rising maintenance costs over time, then a new equation must be derived. If the analytical problem becomes unwieldy, an option is to solve the problem via a simulation; the saddle of the cost curve can then be found using a numerical searching method.

Another perspective is to view N^* as the optimum retention period, e.g. the length of time that a machine should be kept if costs are to be minimized. Costs would be greater if the machine were kept shorter or longer than the ideal duration expressed by N^*.

Source

Pipe Line Industry, August 1985.

Acknowledgments

Prof Zandi, Civil and Urban Engineering Department, University of Pennsylvania; Dr Pollack-Johnson, Department of Mathematics, Oberlin College, Ohio; and Mr V James Joyce, Joyce Western Corp., Andover, NY.

How to estimate construction costs

Two hand-held calculator programs divide pipeline costs into four component areas: right of way, material, labor, and miscellaneous

R. D. Green, Energy Center Fellow, University of Pennsylvania

Pipeline cost models are numerous, and their uncertainty is associated with the degree of generality.

Costs vary with "distances from manufacturers, terrain, locality, labor costs, season of the year, and other variables."[1] Hence, it is vital to the interests of construction firms to maintain historical records of costs and to update these records routinely. A well-developed model may be used as a management tool during construction to identify *either* discrepancies between the model and "reality" or to encourage project superintendents to avoid cost over-runs and outperform previous time and motion constants.

A detailed account of cost estimating has been written by J. S. Page (1977)[2]; the work is similar to the approach taken by contractors seeking to estimate costs of manpower, materials, etc., based on historical time frames and the like. Pipeline cost models that identify cost parameters and the inter-relationship of these parameters have been developed by I. Zandi (1982)[3] and K. Gimm and I. Zandi (1976).[4]

A rule of thumb calculation has also been computed (1964),[5] and it is updated here. Table 1, "Pipeline rule of thumb cost estimate," presents a means of estimating the cost based on costs of pipelines in the United States. The "1964"

Table 1
Pipeline Rule of Thumb Cost Estimate, Cost = $D \times A \times$ Miles of Pipeline

Year	Diameter of Pipe, Inches	$\left(\dfrac{\text{Average Cost US\$}}{\text{Inch Diameter} \times \text{Mile}}\right)$	$\left(\dfrac{\text{Range US\$}}{\text{Inch Diameter} \times \text{Mile}}\right)$		Average % Inflation Since 1964[*]
			Low	High	
1964[**]	10	2059	—	—	—
	16	2849	—	—	—
	20	2632	—	—	—
	24	3325	—	—	—
	30	4085	—	—	—
	36	4260	—	—	—
1980[***]	8	16,920	6,953	24,125	—
	12	18,446	13,882	16,644	—
	16	10,823	9,200	12,368	8.3
	20	14,226	12,039	16,698	10.5
	24	14,738	8,184	28,263	9.3
	30	16,730	13,100	28,756	8.8
	36	25,468	19,087	33,929	11.2

[*]Compounded continuously; from $100\% \times \left(\ln\left(\dfrac{A\ 1980}{A\ 1964}\right)\right) \div 16$.

[**]*Source is Pipeline Rules of Thumb Handbook,*[1] and these figures are based on 1964 data from the Federal Power Commission.

[***]Based on averages of costs for onshore projects from FERC Construction Permit applications as reported in the *International Petroleum Encyclopaedias.*[5]

Table 2
A Real Division of Pipeline Costs

Size (Inches)	ROW	Material	Labor	Misc.	Total
Part I: By Area, % of Total Pipeline Costs Over 10-Year Period:[*]					
8	2.8	37.2	49.2	10.8	100.0
12	5.7	37.3	44.3	11.2	98.5
16	3.9	46.6	38.8	10.5	99.5
20	3.1	46.0	41.8	9.4	100.3
24	3.1	46.3	38.0	11.8	99.2
30	3.7	49.6	37.9	8.5	99.7
36	4.1	48.1	37.2	10.6	100.0
Average	3.8	44.4	41.0	10.4	99.6
Part II: Simplified version of Part I					
8	3.5	37.	49.	10.5	100.0
12	5.7	37.	44.3	10.5	97.5
16	3.5	47.	38.75	10.5	99.75
20	3.5	47.	38.75	10.5	99.75
24	3.5	47.	38.75	10.5	99.75
30	3.5	47.	38.75	10.5	99.75
36	3.5	47.	38.75	10.5	99.75
Average	3.8	44.1	41.0	10.5	99.46
Part III: Version Used in Hand-Calculator Program "CSTA" (D Inches)					
$D \leq 8$	3.5	37	49.	10.5	100.0
$8 < D \leq 12$	5.7	37	44.3	10.5	97.5
$12 < D \leq 36$	3.5	47	38.75	10.5	99.75
Average	—	—	—	—	—

[*]Tabulated from FERC based data presented in the *International Petroleum Encyclopaedias.*[5]

figures are listed for comparison only and are from the (1978) *Pipeline Rules of Thumb Handbook.*[1]

Adjusting for inflation

The pipeline rule of thumb cost estimate is made more useful by including a means for adjusting the "1980" based formula

of Table 1 to other years:
where

$C(y) =$ Cost of the pipeline in year y,
$D =$ Diameter of the pipeline,
$A =$ Average cost of pipeline construction per inch-diameter-mile,
$m =$ Length of pipeline in miles,

i = Continuously compounded rate of inflation (e.g. % inflation from Table 1 divided by 100), and

y = Year for which the cost estimate is desired.

Hand calculator program CST performs this cost calculation when the user inputs 1980-based data from Table 1; an example run is included in the Appendix.

Estimate for cost areas

The cost of a pipeline may be divided into four component cost areas: Right of way (ROW), material, labor, and miscellaneous.

An estimate of any one of these component costs (C_i) is obtained from the following:

$$C_i = a_i C(y),$$

where

a_i = 10-year average proportion of total cost allocated for one of the four areas, and

$C(y)$ = Total cost of the pipeline (or an estimate).

The a_i's are taken from Table 2, Part III, in the program CSTA (see Appendix), according to the size (diameter) of the pipeline given the following divisions:

$D \geq 8,$

$8 < D \leq 12,$

$12 < D \leq 36.$

Appendix—Software for Hewlett Packward HP41-CV

CSTA

Given both the cost and size of a pipe line, this program estimates the division of expenses between four categories: Right of way (ROW), material, labor and miscellaneous. Table 2 forms the basis of the program, and in Part III of the table it is apparent that the program user may elect to use pipe line sizes (diameters) for which no data is supplied. For example, a 10-inch line is viewed as a 12-inch line by the program.

```
PROGRAM "CSTA"
01 ▲ LBL "CSTA"
02 CF 01          Clear flags.
03 CF 02
04 "COST?"        User inputs pipe line construction cost
05 PROMPT         and diameter of line (D).
06 STO 03
07 "D?"
08 PROMPT
09 STO 04
10 8              Is D≤8?, 8<D≤12?, or 12<D≤36?
11 X( )Y
12 X( = Y?
13 SF 01          Set flags accordingly (1, 2, or neither).
14 FS? 01
15 GTO A
16 RCL 04
17 12
18 X( )Y?
19 X( = Y?
20 SF 02
21 FS? 02
22 GTO "AA"
23 ▲ LBL A        ROW cost calculation for D ≠ x
                  such that 8 < x ≤ 12.
24 RCL 03
25 .035
26 *
27 STO 08
28 GTO "AAA"
29 ▲ LBL "AA"     "ROW" cost calculation for D such
                  that 8<D≤12.
30 RCL 03
31 .057
32 *
33 STO 08
34 ▲ LBL "AAA"
35 "R.O.W. = "    output
36 PRA
37 RCL 08
38 PRX
39 FS? 01         "Material" calculation.
40 GTO B
41 FS? 02
42 GTO B
43 RCL 03
44 .47
45 *
```

CST

This program is to be used with the "1980" data of Table 1. The program estimates the cost of a pipe line from the equation,

$$\text{Cost} = D \times A \times \text{Miles} \times \exp\left(\frac{\% \text{ inflation}}{100} \times \# \text{ years}\right).$$

The exponential in this cost equation is merely a means for adjusting the "1980" figures to other years. User input is as follows: "Year" (e.g., 1983; the program also accepts years prior to the base year 1980), "INFLATION" (e.g., 8.3%; this will be compounded continuously), "D" (e.g., 16; diameter of pipe line from Table 1, "1980" category); "A" (e.g.,10,823; average cost from Table 1, "1980" category); "MILES" (e.g., 20.5; length of pipe line in miles).

Program output is as follows: "COST" estimate of pipe line in

U.S. dollars, based on U.S. data (e.g., 4,553,660); "IN YEAR", is the year for which the cost estimate has been computed (e.g., 1983).

```
PROGRAM "CST"
01 ▲ LBL "CST"
02    "YEAR?"     1980 is the base year; the program ac-
                  cepts any year prior to, after or includ-
                  ing 1980.
03 PROMPT
04 1980
05 –
06 STO 00
07 "INFLATION?"   Inflation rate (in percent) is any figure
                  selected by the user; it is continuously
                  compounded.
08 PROMPT
09 100
10 /
11 STO 01
12 "D?"           Diameter of pipe line.
13 PROMPT
14 "A?"           Average cost in units of US$ per (inch
                  diameter × mile).
15 PROMPT
16 *
17 "MILES?"       Length of pipe line in miles.
18 PROMPT
19 *
20 STO 02
21 RCL 01
22 RCL 00
23 *
24 E↑X            Compound factor is exp (rate × time).
25 RCL 02
26 *
27 STO 03
28 "COST = "      Cost estimate of pipe line in U.S. dol-
                  lars for a specified year. The cost esti-
                  mate is for construction and encom-
                  passes an average of material, labor,
                  right of way, and miscellaneous ex-
                  penses, as presented in Table 1.
29 PRA
30 RCL 03
31 PRX
32 "IN YEAR-"
33 PRA
34 RCL 00
35 1980
36 +
37 PRX
38 END
```

EXAMPLE RUN

		XEQ "CST"	After program has been entered, user executes the program and responds to the prompts; after each data entry, the user presses the "RUN" button. User assignable keys may be defined to facilitate rapid program execution.
YEAR?			
	1983	RUN	
INFLATION?			
	8.3	RUN	
D?			
	16	RUN	
A?			
	10,823	RUN	
MILES?			
	20.5	RUN	
COST =			
	4,553,660.	...	
IN YEAR			
	1983	...	

```
46 STO 05
47 GTO C
48 ▲ LBL B
49 RCL 03
50 .37
51 *
52 STO 05
53 ▲ LBL C
54 "MATERIAL = "   output
55 PRA
56 RCL 05
57 PRX
58 FS? 01          "Labor" calculation.
59 GTO D
60 FS? 02
61 GTO E
62 RCL 03
63 .3875
64 *
65 STO 06
66 GTO F
67 ▲ LBL D
68 RCL 03
69 .49
70 *
71 STO 06
72 GTO F
73 ▲ LBL E
74 RCL 03
75 .43
76 *
77 STO 06
78 ▲ LBL F
79 "LABOR = "      output
80 PRA
81 RCL 06
82 PRX
83 "MISC. = "      "Misc." calculation and output.
84 PRA
85 RCL 03
86 .105
87 *
88 STO 07
89 PRX
90 "TOTAL = "      "Total" = ROW + Material + Labor + Misc.
                   Due to assumptions of model (see Table 2), "Total"
                   might be different from input cost.
91 PRA
92 RCL 08
93 RCL 05
94 +
95 RCL 06
96 +
97 RCL 07
98 +
99 PRX            output
100 END
```

EXAMPLE RUN

		XEQ "CSTA"
COST?		
	5,605,238.00	RUN
D?		
	8.0	RUN
R.O.W. =		
	196,183.33	...
MATERIAL =		
	2,073,938.06	...
LABOR =		
	2,746,566.62	...
MISC. =		
	588,549.99	...
TOTAL =		
	5,605,238.00	...

The percentages listed in Table 2 are simple averages of pipeline construction cost data for a specific pipeline size:

$$a_i \times 100 = \sum_{y=1971}^{1980} \frac{A_y}{C_y},$$

where

A_y = Average pipeline cost for a given component cost area (e.g. ROW) in a given year, and

C_y = Average total pipeline cost in a given year (e.g. sum of average costs for a given year in the four categories).

Source

Pipe Line Industry, September 1984.

References

1. *Pipeline Rules of Thumb Handbook*, Gulf Publishing Company, 1978, p. 163.
2. *Cost Estimating Manual For Pipelines and Marine Structures*, Gulf Publishing Company, 1977.
3. "Coal Slurry Pipeline Cost Model," *Journal of Pipelines*, Vol. 2, No. 3, 1982, pp. 53—66.
4. "Freight Pipeline Cost Model," *Transportation of Solid Commodities via Freight Pipeline*, Department of Transportation (DOT/RSPA/DPB-50/78/33—39), 1978.
5. "Pipelining Continues To Set New Records," *International Petroleum Encyclopaedia, 14th ed.*, Petroleum Publishing Co., 1981, pp. 368—373.

Cost estimating strategies for pipelines, stations, and terminals (part 1)

Microcomputers and up-to-date estimating software are more efficient means for cost studies

R. D. Green, Energy Center Fellow, University of Pennsylvania, Folcroft, Pennsylvania

Cost estimating strategies vary in the accuracy of representing cost factor data. Cost factors for pipelines, stations, and terminals are presented in Tables 1 and 2.

In the pipeline industry, the practice has been to assume that the most accurate cost estimates for constructing pipelines, stations, and terminals will come from a very detailed estimating procedure that sums the expected value or best guess for each cost factor involved. Such an assumption ignores more accurate cost-estimating strategies that require many more computations and have therefore traditionally been avoided.

Historically, cost factor information has been organized in files, tables, and ledgers. Estimating has been a lengthy procedure taking days, if not weeks, of work effort by a full staff of estimators. Now, with microcomputers and up-to-date estimating software, we have a more efficient means of storing, retrieving, and handling data. It is time to reconsider the use of more accurate estimating strategies.

By accurate, I mean representing data in a form that identifies what is and is not known about the cost factor.

Current, future estimating practices

Currently, spreadsheet-based cost models for pipelines, stations, and terminals sum the simply represented costs for the cost factors given in Tables 1 and 2. Future construction estimating practices will likely involve two changes from the current practices.

These changes will be a response to the two problems estimators face. One problem is how to speed the computational procedure that yields the final or grand total project cost estimate, and the other is how best to represent cost factor data. This article considers these two problems.

Speeding computational procedures

Estimating computations are often lengthy and redundant. Estimating can be performed quickly either by simplifying the computations (often with an associated loss in accuracy) or by computerization, in which case accuracy need not be sacrificed. Current and future construction estimating procedures involve the use of spreadsheet-based cost models. There are three basic types of spreadsheets:

- *Spreadsheets* are typically large ledger-like sheets of paper upon which are spread the cost factors and costs that go in to the cost estimating procedure in estimating the costs of a construction project.
- *Electronic spreadsheets* are a computerized version of the spreadsheet in which cost factors are listed and the estimator fills in the costs for the appropriate items and the computer automatically tallies the results.
- *Automated electronic spreadsheets* are an automated version of the electronic spreadsheet in the sense that

Table 1
Cost Factors in Pipeline Estimate

Design Engineering
Electrical & mechanical; engineering-supervisor; engineer, drafting; corrosion-engineer, drafting; planning engineering-supervisor; engineer; survey & drawing; aerial photography; stake out; as-built information & drawing; data processing-supervisor; programmer; travel; supplies.

Right of Way, Legal, Environmental & Safety
Acquisition labor; damages labor; permit/legal labor; environmental & safety labor; travel; field office rental; supplies; considerations—existing line; rights, new R/W rights; damages-damage payments, land; purchases, permit payments-permit fees; cost of deposits/bonds; first year rentals/leases.

Construction
Labor—electrical, mechanical; corrosion, inspection labor, expenses—regular, temporary; field office rental, equipment; rentals, supplies; inspection—corrosion coating mill; pipeline, survey, mechanical-pipe mill; pipeline, X-ray, tuboscope, caliper pig; contract —pipeline installation—open area, moderate area, congested; install casing; supplementary work-install valves, purge line, caliper pig (contractor's labor), river crossing, hydrostatic test, gravitometer, station work, buildings, ground bed, rectifier, anodes, environmental requirements, performance bond; extra work—rock removal blasting, air hammer; sand padding; contingencies.

Materials
Instrumentation—transmitters, gravitometers, miscellaneous; mainline pipe (size, grade, & wall), vent & marker, freight charges; casing pipe (size, grade & wall), freight charges; valves (size, series, type); fittings (description, size, series); valve operators; buildings, furnishings & fencing (size, type); coating (pipe size, type); cathodic protection—rectifiers; test stations; insulation joints; anodes (type); seals and insulators; data processing.

Overheads
Indirect labor (managers and above); travel; supplies; interest; fringe benefits.

much of the data, e.g. costs and cost factors, are provided by the computer rather than by having the estimator fill in the blanks with costs. The computer has a database to turn to for cost and material information, thus the estimator need not look up such information in compiling the estimate.

The time and effort required to produce routine estimates may be reduced by using an automated electronic spreadsheet. Given the tremendous data-handling abilities of even low-cost computers vis-a-vis manual computation, cost estimating may be performed using various strategies ranging from the traditional to the more complex.

A fully integrated cost estimating approach enables the estimator to rely on data that have as their basis historical, current, and forecast aspects of the project. Since such an approach is conveniently packaged as a (high-speed) computerized estimating program, the estimator may base estimates on cost factor data in any of the strategies for representing data.

Three estimating strategies

The three basic strategies for obtaining an estimate differ in the way uncertainty is viewed and represented. Each strategy is characterized by the type of estimates that are utilized:

- A point estimate views factors in the estimate as quantifiable without regard to error, range, and

uncertainty (such estimates are rarely viewed as conclusive since managers often ask estimators, "How good are these numbers?").

- A range estimate is generally more sophisticated than the point estimate and may yield, for example, an "expected cost of pipeline construction, ranging as low as…and as high as…"

- A probability estimate is the most sophisticated estimate since it does not ignore uncertainty and is based on an evaluation of probability distributions for (a) historical data, (b) current conditions, and (c) projections of future conditions.

Probability estimates may be depicted in simplified forms without regressing to a range or point estimating strategy. In particular, a probability distribution can be represented with some loss of accuracy as any of several simplified distributions (such as a histogram or a triangular distribution or some smooth distribution that is beta or normal).

One problem with point estimates is that these do not convey an awareness of any uncertainty behind the estimate. Point and range estimates can be misleading. As Hertz describes, the "estimated probability distribution paints the clearest picture of all possible outcomes,"[1] and he gives examples in the area of capital investment projects. Risk analysis is designed to provide more information than conventional techniques that fail, as Hespos and Strassman note, for not taking into account the "skewed factors," and for being "influenced by the subjective aspects of best guesses."[2]

Table 2
Cost Factors in Station and Terminal Estimate

Design Engineering

Electrical & mechanical; engineering supervisor; engineer, drafting; corrosion-engineer, drafting; planning engineering-supervisor; engineer; supplies; civil-engineer, drafting; surveying—property, topograph, aerial, soil borings; stake out/as-built-crew; draftsman; data processing-supervisor; programmer; travel; expenses; supplies.

Right of Way, Legal, Environmental & Safety

Acquisition/damages labor; permit/legal labor; environmental & safety; travel; field office rental; supplies; payments-considerations; damages; land purchase; title searches; permits & fees—first year rental/leases.

Construction

Labor—electrical, mechanical; corrosion, expenses regular & temporary; office & equipment rental; supplies; inspection mechanical; electrical; X-ray; soil testing; concrete testing; contracts; corrosion—anodes; test stations; deep well; groundbed; rectifier; set pole; insulating joint; painting (including sandblasting, primer, all coats); tanks; piping; pipe coating; excavation, machine, backfill; cable, substation—install; transformers; install switchgear; install fuse/circuit breakers; structure work; motors-set motor, install motor; control; starters; instrumentation—install transmitters, recorders, etc.; tank gauges; ticket printers; meter totalizers; electrical cabinets; valve operators; gravitometers; miscellaneous; conduit & wire; lighting—set poles, mount lights; install computer; other—concrete, excavation; modular building; mechanical—welds, valves, cuts & bevels; pipe, civil—excavation, backfill; concrete; stoning; fencing; gates; water well; septic system; building; oil/water separator; seeding; extra work—corrosion, electrical, mechanical, civil; electrical & mechanical—manpower; equipment.

Material

Motors; motor control center; starters; substation and transformers—transformer, substation structure, lightning arrestors, miscellaneous fittings; switchgear; instrumentation-general—gravitometers, transmitters, receivers, power supply, alarm annunciator, recorders, totalizers, ticket printers; switches—pressure, temperature, level; vibration monitor; cabinets; micro pumps; other & miscellaneous; instrumentation—tanks, tank gauges, transmitters, other; lighting—pole, lights; miscellaneous items; circuit breakers; wire & conduit; miscellaneous special equipment; mechanical—pipe (size, grade, & wall), valves (size, series, type); fittings (description, size, series); valve operators; pumps & pump/motor combinations—spare rotating element; meters; prover loop; meter/prover skid; strainers; tanks—large storage tanks, other; communication—radios, other; fire protection—pipe, valves, fittings, fire extinguishers, foam proportioners, foam tank, foam; environmental control equipment—boat, spill boom, vapor recovery system; miscellaneous and special equipment; civil-buildings, building furnishings; oil water separator; structural steel supports; stone; special equipment; corrosion-coating, anodes; insulating joint; test stations; cable; rectifier; other; paint; data processing—computers, CRT displays, printers, disks, RTU; special equipment.

Overheads

Indirect labor—directors & managers of engineering, planning & analysis, construction, right of way, other; travel; supplies; interest; fringe benefits.

Administrative overheads

Administrative labor; travel expenses; supplies.

Indirect costs

Labor; materials, pipe, other.

To illustrate the subjective aspects, note that an estimator typically places the "best point estimate" in the low end if a high outcome is more serious than a low outcome. And the estimator's "best point estimate" is in the high end if a low outcome is more serious than a high outcome. The point estimate is a result of the estimator's judgments about uncertainties, outcomes, and risks or values of the consequences of outcomes. Probability estimates are more useful than point estimates since they more accurately represent what is and is not known about cost factors. There is less need or motivation to distort probability estimates (e.g. these are less vulnerable to subjective aspects).

Although the range estimate conveys the notion that a precise estimate is not available, it does not convey any probability of where the actual cost will fall. Range estimates are often misleading since the actual cost may fall not only within but outside of the range specified. Range estimates are usually defined by two point estimates (e.g. low and high).

The probability estimate conveys the most information and offers a means of expressing the uncertainty of the estimate.

Probability estimates of uncertain parameters are made by converting point forecasts to probabilistic forecasts. This is done generally according to either the historical records of the firm or the "guesstimates" provided by experienced personnel.

For example, where the future is statistically like the past (e.g. no identifiable characteristics such as cycles or trends), probability estimates for future cost factors can be based on historical relative frequency distributions that may have been "refined" by interpolation or smoothing. Concisely stated, the probability estimate conveys information concerning all that is and is not known about the estimated cost.

Therefore, it can visually represent the likeliness that the real value will be, for example, below a given cost level. This is especially important for construction contractors who engage in bidding to perform work on a project.

For the contract-engineering and construction firms, the variability in construction costs must be closely monitored in view of the competitive nature of the construction industry where profits are related to the accuracy of forecasting engineering and construction costs. As for pipeline operating firms that often have work performed by outside contractors, there is a need to pay attention to the variability in construction costs for both large and small projects.

The database ought to reflect the variability in cost factors so that uncertainties for large projects, which can impact the firm's capital position, are more clearly comprehended. Also, for an operating firm that has a large proportion of small construction projects relative to large projects, most of the data gathering opportunities will be missed if the management posture is one of ignoring variability for small projects.

Cost estimating strategies for pipelines, stations, and terminals (part 2)

Developing a cost model for automated electronic spreadsheets may involve three estimating strategies

R. D. Green, Energy Center Fellow, University of Pennsylvania, Folcroft, Pennsylvania

Developing a cost model (e.g. an automated electronic spreadsheet) for pipelines, stations, and terminals, which provides a probability estimate of cost, may involve the use of three estimating strategies. For example, the variability in cost for many cost factors may be so limited that it is feasible to use point estimates. This is often the case for material estimates.

Similarly, range estimates may be used where probability estimates will not add any significant information (e.g. where a probability distribution is approximately "mesa" shaped).

But for the cost factors that exhibit great variability, it is essential to use probability estimates (estimated probability distributions).

Manpower (cost) estimates are often in this category. Whether or not historical data are available, probability estimates can be created. For example, based on the experience of an estimator, a three-point estimate can be obtained. Or if adequate data are available, histograms can be used to estimate the probability density functions (probability estimates). Some commonly used approaches for estimating the probability density functions are:

- The three-point estimate approach (which may involve any of several distributions, e.g. beta or the triangular distribution)
- The histogram approach (which may closely approximate the "smooth curve" of a probability distribution).

Three-point estimate approach

The three-point estimate, e.g. low (L), expected (E), and high (H), can be used to estimate the probability distribution for the cost factor. Given the simplifying assumption that the probability distribution for the cost factor is approximately beta and that the standard deviation equals 1/6 of the range of reasonably possible costs for factors, then the estimated expected value of a given cost factor is

$$C_e = (1/3) \times [2E + (1/2) \times (L + H)].$$

The variance is the square of the standard deviation:

$$s^2 = ((1/6) \times (H - L))^2.$$

Given many cost factors that have been estimated in this way, the expected value of the final or grand total cost is

$$G = \sum_{i=1}^{n} C_{ei}.$$

The variance for G is

$$V_G = \sum_{i=1}^{n} s_i^2.$$

The final or grand total cost may be viewed as a sum of many independent random variables. By invoking the general version of the central limit theorem, the probability distribution of such a sum is approximately normal. Given the expected value G and the variance V_G, it is then straightforward to find the probability that this normal random variable (grand total project cost, G) will be less than a given (e.g. "bid," B) price. This is done by finding the area under the standard normal curve from negative infinity to Z, where

$$Z = (B - G)/\sqrt{V_G}.$$

That is, the estimated probability that the actual grand total cost is less than the given bid price is

$$P = 0.5 + \int_{0}^{x=Z} 1/\sqrt{2\pi} - x^2/e^2 \mathrm{dx}.$$

This equation is solved by invoking Simpson's rule for several intervals and yields the following approximation (if more accuracy is needed, more intervals can be used):

$$p = 0.5 + \frac{\frac{Z}{6}}{3\sqrt{2\pi}} \times \left(1 + \left[\frac{\frac{4}{Z^2}}{e^{72}} \right] + \left[\frac{\frac{2}{Z^2}}{e^{18}} \right] + \left[\frac{\frac{4}{Z^2}}{e^{8}} \right] + \left[\frac{\frac{2}{2Z^2}}{e^{9}} \right] \right. $$
$$\left. + \left[\frac{\frac{4}{25Z^2}}{e^{72}} \right] + \left[\frac{\frac{1}{Z^2}}{e^{2}} \right] \right)$$

Histogram approach

The histogram approach can be a more accurate means of representing the estimated probability distributions for cost factors (rather than using the three-estimate approach) since bimodal and skewed characteristics, etc., can be portrayed. Depending on the mix of estimating strategies and the types of probability estimates involved, the final cost estimate will be obtained in part via a simulation by Monte Carlo sampling. Further details of this approach are beyond the scope of this article.

Observations and conclusion

The advantage of a well-developed electronic spreadsheet is that it can automatically maintain historically based data in the form of, for example, histograms that may be sampled later on to extract information. This supports future efforts to obtain probability estimates for cost factors. The system may accommodate inflation rates that vary among the cost factors. These probability estimates enable the estimator to communicate information about the project risks and the uncertainty of the estimate so that the right risks can be taken by the executive or managerial component of pipeline construction and operating firms.

An implication for the future is related to the current "data policy" of the pipeline organization. In view of the estimating tools that are needed by the pipeline industry, data currently being gathered and preserved ought to reflect the variability of costs for the cost factors. Such data are important to the future of the pipeline organization.

	A	B	C	D	E	F	G
1	CST						
2	DATA:	PIPE DIA.	AVG. COST	RANGE		AVG. % INFLATION	
3		D INCHES	A	LOW TO	HIGH	SINCE 1964	
4		8	$16920	$6953	$24125	8.3	
5		12	$18446	$13882	$16644	8.3	
6		16	$10823	$9200	$12368	8.3	
7		20	$14226	$12039	$16698	10.5	
8		24	$14738	$8184	$28263	9.3	
9		30	$16730	$13100	$28756	8.8	
10		36	$25468	$19087	$33929	11.2	
11	YEAR?	1983	expected	lo	hi		
12	D?	16.00					
13	MILES?	20.50					
14	A	$10823	GIVEN D=	16.00	24.00		
15	INFLATION	8.3					
16	_____	AVERAGE	LOW	HIGH			
17	COST=	$4,553,662.26	$3,870,802.25	$5,203,704.60			
18	_____	_____	_____	_____			
19	CSTA						
20	DATA:						
21	8	2.8	37.2	49.2	10.8	100.0	
22	12	5.7	37.3	44.3	11.2	98.5	
23	16	3.9	46.6	38.8	10.5	99.5	
24	20	3.1	46.0	41.8	9.4	100.3	
25	24	3.1	46.3	38.0	11.8	99.2	
26	30	3.7	49.6	37.9	8.5	99.7	
27	36	4.1	48.1	37.2	10.6	100.0	
28	FOR COST=	$4,553,662.26	$3,870,802.25	$5,203,704.60			
29	FOR D =	16					
30	_____	POINT ESTIMATE	_____				
31	ROW	$178,485.25	GIVEN D=	16	24		
32	MATERIAL	$2,132,669.95					
33	LABOR	$1,775,699.44					
34	MISC.	$480,537.22					
35	TOTAL	$4,567,391.86					
36	_____	RANGE	ESTIMATE	_____			
37		LO	HI				
38	ROW	$151,719.89	$203,964.30				
39	MATERIAL	$1,812,858.13	$2,437,111.89				
40	LABOR	$1,509,418.36	$2,029,183.29				
41	MISC.	$408,476.62	$549,134.66				
42	TOTAL	$3,882,472.98	$5,219,394.13				
43	_____	PROBABILITY	ESTIMATE	_____			
44	B =	$6,000,000			Ce	s square	
45				ROW	$178,270.87	75,818,855	
46				MATERIAL	$2,130,108.30	10,824,798,700	
47				LABOR	$1,773,566.56	7,504,321,729	
48				MISC.	$479,960.03	549,574,535	
49				G = $4,561,905.74	VG =	18,954,513,700	
50						Z =10.445543	
51							
52							0.2315
53						THE PROBABILITY	1.0000
54						OF THE ACTUAL	0.8789
55						GRAND TOTAL COST	0.0047
56						BEING LESS THAN	0.0000
57						A GIVEN COST "B"	0.0000
58						IS APPROXIMATELY	0.0000
59						EQUAL TO "P"	0.0000
60						WHERE	0.5000
61						P =	0.9361

Figure 1. Worksheet example of an electronic spreadsheet.

```
A1 CST                              B40 @LOOKUP(D31,A21:A27,3)*C17/(@LOOKUP(D31,A21:A27,5))   D45 (1/3)*(2*B31+(1/2)*(B38+C38))
A2 DATA:                            B41 @LOOKUP(D31,A21:A27,4)*C17/(@LOOKUP(D31,A21:A27,5))   D46 (1/3)*(2*B32+(1/2)*(B39+C39))
A11 YEAR?                           B42 @SUM(B38,B41)                                          D47 (1/3)*(2*B33+(1/2)*(B40+C40))
A12 D?                              B43 _____PROBABILI                                      D48 (1/3)*(2*B34+(1/2)*(B41+C41))
A13 MILES?                          B44 1E+10                                                  D49 @SUM(D45,D48)
A14 A                               C2 AVG. COST                                              E2 NGE
A15 INFLATION                       C3 A                                                      E3 HIGH
A16 _____                        C4 16920                                                  E4 24125
A17 COST=                           C5 18446                                                  E5 16644
A18 _____                        C6 10823                                                  E6 12368
A19 CSTA                            C7 14226                                                  E7 16698
A20 DATA:                           C8 14738                                                  E8 28263
A21 8                               C9 16730                                                  E9 28756
A22 12                              C10 25468                                                 E10 33929
A23 16                              C11 expected                                              E11 hi
A24 20                              C14 GIVEN D=                                              E14 @IF B12<=24 THEN 24 ELSE ( @IF B12<=30 THEN 30 ELSE 36)
A25 24                              C16 LOW                                                   E21 10.8
A26 30                              C17 B12*B13*(@LOOKUP(D14,B4:B10,2))*@EXP((B15/100)*(B11-1980))  E22 11.2
A27 36                              C18 _____                                              E23 10.5
A28 FOR COST=                       C21 37.2                                                  E24 9.4
A29 FOR D =                         C22 37.3                                                  E25 11.8
A30 _____                        C23 46.6                                                  E26 8.5
A31 ROW                             C24 46                                                    E27 10.6
A32 MATERIAL                        C25 46.3                                                  E31 @IF B29<=24 THEN 24 ELSE ( @IF B29<=30 THEN 30 ELSE 36)
A33 LABOR                           C26 49.6                                                  E43 _____
A34 MISC.                           C27 48.1                                                  E49 VG =
A35 TOTAL                           C28 C17                                                   F2 AVG. % INFLATION
A36 _____                        C30 _____                                          F3 SINCE 1964
A38 ROW                             C31 GIVEN D=                                              F4 8.3
A39 MATERIAL                        C36 ESTIMATE ____                                         F5 8.3
A40 LABOR                           C37 HI                                                    F6 8.3
A41 MISC.                           C38 @LOOKUP(D31,A21:A27,1)*D17/(@LOOKUP(D31,A21:A27,5))   F7 10.5
A42 TOTAL                           C39 @LOOKUP(D31,A21:A27,2)*D17/(@LOOKUP(D31,A21:A27,5))   F8 9.3
A43 _____PROBABILITY ESTIMATE_____  C40 @LOOKUP(D31,A21:A27,3)*D17/(@LOOKUP(D31,A21:A27,5))  F9 8.8
A44 B =                             C41 @LOOKUP(D31,A21:A27,4)*D17/(@LOOKUP(D31,A21:A27,5))   F10 11.2
B2 PIPE DIA.                        C42 @SUM(C38,C41)                                         F21 100
B3 D INCHES                         C43 TY ESTIMATE___                                        F22 98.5
B4 8                                C45 ROW                                                   F23 99.5
B5 12                               C46 MATERIAL                                              F24 100.3
B6 16                               C47 LABOR                                                 F25 99.2
B7 20                               C48 MISC.                                                 F26 99.7
B8 24                               C49 G =                                                   F27 100
B9 30                               D2        RANGE                                           F43 _____
B10 36                              D3   LOW TO                                               F44 s square
B11 1983                            D4 6953                                                   F45 ((C38-B38)/6)*((C38-B38)/6)
B12 16                              D5 13882                                                  F46 ((C39-B39)/6)*((C39-B39)/6)
B13 20.5                            D6 9200                                                   F47 ((C40-B40)/6)*((C40-B40)/6)
B14 @LOOKUP(D14,B4:F10,1)           D7 12039                                                  F48 ((C41-B41)/6)*((C41-B41)/6)
B15 @LOOKUP(D14,B4:F10,4)           D8 8184                                                   F49 @SUM(F45,F48)
B16 AVERAGE                         D9 13100                                                  F50 Z =
B17 B12*B13*B14*@EXP((B15/100)*(B11-1980))  D10 19887                                         F53 THE PROBABILITY
B18 _____                  D11 lo                                                    F54 OF THE ACTUAL
B21 2.8                             D14 @IF B12<=8 THEN 8 ELSE ( @IF B12<=12 THEN 12 ELSE     F55 GRAND TOTAL COST
B22 5.7                                 ( @IF B12<=16 THEN 16 ELSE ( @IF B12<=20 THEN 20 ELSE E14)))  F56 BEING LESS THAN
B23 3.9                             D16 HIGH                                                  F57 A GIVEN COST "B"
B24 3.1                             D17 B12*B13*(@LOOKUP(D14,B4:B10,3))*@EXP((B15/100)        F58 IS APPROXIMATELY
B25 3.1                                 *(B11-1980))                                          F60 WHERE
B26 3.7                             D18 _____                                              F61 P =
B27 4.1                             D21 49.2                                                  G43 _____
B28 B17                             D22 44.3                                                  G50 (B44-D49)/@SQRT(F49)
B29 B12                             D24 41.8                                                  G52 (G50/6)/(3*@SQRT(2*@PI))
B30 POINT ESTIMATE                  D25 38                                                    G53 1/@EXP(0)
B31 @LOOKUP(D31,A21:A27,1)*B28/(@LOOKUP(D31,A21:A27,5))  D26 37.9                             G54 4/@EXP(G50*G50/72)
B33 @LOOKUP(D31,A21:A27,3)*B28/(@LOOKUP(D31,A21:A27,5))  D27 37.2                             G55 2/@EXP(G50*G50/18)
B34 @LOOKUP(D31,A21:A27,4)*B28/(@LOOKUP(D31,A21:A27,5))  D28 D17                              G56 4/@EXP(G50*G50/8)
B35 @SUM(B31,B34)                   D31 @IF B29<=8 THEN 8 ELSE ( @IF B29<=12 THEN 12 ELSE     G57 2/@EXP(2*G50*G50/9)
B36 _____ RANGE    ESTIMATE _____      ( @IF B29<=16 THEN 16 ELSE ( @IF B29<=20 THEN 20 ELSE E31L))  G58 4/@EXP(25*G50*G50/72)
B37 LO                              D36 _                                                     G59 4/@EXP(G50*G50/2)
B38 @LOOKUP(D31,A21:A27,1)*C17/(@LOOKUP(D31,A21:A27,5))  D43 _____                     G60 0.5
B39 @LOOKUP(D31,A21:A27,2)*C17/(@LOOKUP(D31,A21:A27,5))  D44 Ce                               G61 (G52*(@SUM(G53,G59)))+G60
```

Figure 2. Program listing of an electronic spreadsheet.

Example spreadsheet

A simple automated electronic spreadsheet example is provided (with software). It illustrates the use of "lookup tables" as a means of accessing a "built-in" database. In this example, a rule of thumb estimate[3] is used for estimating the probability distribution of pipeline construction costs in several cost areas. An overall distribution is obtained by using the procedure described earlier (three-point estimate approach). The estimated probability distribution for the grand total cost of the project is used in a step to assess the probability that the project cost is less than a given "bid" price. Given relatively few inputs (year of construction, pipeline diameter, and length) and a limited database, this simple automated electronic spreadsheet finds point, range, and probability estimates for several cost areas. The expected grand total cost $G = \$4,561,905.74$ is surrounded by considerable uncertainty. In a full-scale automated electronic spreadsheet, the variance surrounding the cost factors and expected grand total cost will be greatly reduced relative to this simple example.

A possible improvement upon this example is as follows. If more representative distributions for the cost factors are obtained and presented as histograms, then the estimated probability distribution for grand total cost will be obtained through a simulation that performs a summation involving random sampling of histograms (histogram approach).

Overall, the example provides an introductory automated electronic spreadsheet. Given some project-specific information and the historical record held within the program, cost projections are effortlessly produced. A large-scale model would work in a similar way but with more detailed data to represent the many cost factors given in Figures 1 and 2.

Source

Pipe Line Industry, September 1986.

Acknowledgments

Mr B. H. Basavaraj, Facilities Planning Supervisor, Buckeye Pipe Line Co., Emmaus, Pennsylvania; Prof I. Zandi, Civil and Urban Engineering Dept., University of Pennsylvania; Mr Andy Anderson and Mr K. N. Wheelwright, Weyher/Livsey Constructors, Inc., Salt Lake City, Utah.

References

1. Hertz, D. B., "Risk Analysis in Capital Investment," *Harvard Business Review*, Sept.—Oct. 1979, pp. 169—180.
2. Hespos, R. F. and Strassmann, P. A., "Stochastic Decision Trees for The Analysis of Investment Decisions," *Management Science*, August 1965, p. 11.
3. Green, R. D., "Rules of thumb, How to estimate construction costs," *Pipe Line Industry*, Vol. 61, No. 3, Sept. 1984, p. 47—48.

Appendix

A worksheet and program listing for the example spreadsheet discussed above are provided in Figures 1 and 2. The software and computer used is Syncalc on the Atari 800XL. (Note the similarities to VisiCalc and Lotus-123 electronic spreadsheets.) For industrial scale work, a more powerful system such as the Atari 520ST or IBM XT/AT computer offers an ideal system for developing a full-scale automated electronic spreadsheet for pipeline construction estimating. For more information on estimating, see *Cost Estimating Manual for Pipelines and Marine Structures* by John S. Page (Gulf Publishing Co., Houston, 1977).

Economics

The goal of all economic evaluations should be to eliminate all unprofitable projects. Some borderline, and perhaps a few good projects, will be killed along with the bad ones. However, it is much better to pass up several good projects than to build one loser. Every company should develop a standard form for their economic calculations, with "a place for everything and everything in its place." As management becomes familiar with the form, it is easier to evaluate projects and choose among several based on their merits, rather than the presentation.

Components of an economic study

Most economic studies for processing plants will have revenue, expense, tax, and profit items to be developed. These are discussed in the following sections.

Revenues. The selling price for the major product and the quantity to be sold each year must be estimated. If the product is new, this can be a major study. In any case, input is needed from sales, transportation, research, and any other department or individual who can increase the accuracy of the revenue data. The operating costs may be off by 20% without fatal results, but a 20% error in the sales volume or price will have a much larger impact. Further, the operating costs are mostly within the control of the operator, while the sales price and volume can be impacted by circumstances totally beyond the control of the company.

Depending on the product and sales arrangement, the revenues calculation can take from a few man-hours to many man-months of effort. For instance, determining the price for a synthetic fuel from coal can be done in a very short time, based on the cost of service. However, if the price is tied to the price of natural gas or oil, the task becomes very difficult, if not impossible. On the other hand, determining the sales growth and selling price for a new product requires a great deal of analysis, speculation, market research, and luck, but projections can be made.

Most revenue projections should be on a plant net back basis with all transportation, sales expense, etc., deducted. This makes the later evaluation of the plant performance easier.

Operating costs. The operating costs, as developed in the previous section, can be presented as a single line item in the economics form or broken down into several items.

Amortization. Items of expense incurred before start-up can be accumulated and amortized during the first 5 or 10 years of plant operations. The tax laws and corporate policy on what profit to show can influence the number of years for amortization.

Such items as organization expense, unusual prestart-up expenses, and interest during construction, not capitalized, can be amortized. During plant operations this will be treated as an expense, but it will be a non-cash expense similar to depreciation, which is discussed later.

Land lease costs. If there are annual payments for land use, this item of expense is often kept separate from other operating costs, since it does not inflate at the same rate as other costs. If the land is purchased, it cannot be depreciated for tax purposes.

Interest. The interest for each year must be calculated for the type of loan expected. The interest rate can be obtained from the company's financial section, recent loan history, or the literature. The *Wall Street Journal* publishes the weekly interest rates for several borrowings. The company's interest rate will probably be the bank's prime rate, or slightly higher.

The calculation of interest is usually based on the outstanding loan amount. For example, if semiannual payments are expected, the amount of the first year's interest (I) would be

$$I = \left(\frac{i}{2}\right)P + \left(\frac{i}{2}\right)(P - PP), \tag{1}$$

where

i = Annual interest rate,
P = Original principal amount, and
PP = First principal payment.

The equation for any annual payment would be

$$I_y = \sum_{n=1}^{n=N} \left(\frac{i}{N}\right)\left(P_{y,n-1} - PP_{y,n-1}\right), \tag{2}$$

where

I_y = Interest for any year,
$P_{y,n-1}$ = Principal or loan balance during the previous period,
n = Period for loan payment,
N = Total number of payments expected each year,
y = Year during the loan life, and
$PP_{y,n-1}$ = Principal payment made at the end of the previous period.

Debt payments. The method of making principal and interest payments is determined by the terms of the loan agreement.

Loans can be scheduled to be repaid at a fixed amount each period, including principal and interest. This is similar to most home mortgage payments. The size of level payments can be calculated using the formula:

$$I + PP = P \frac{\left(\frac{i}{n}\right)\left(1 + \frac{i}{n}\right)^{n \times y}}{\left[\left(1 + \frac{i}{n}\right)^{n \times y}\right] - 1}. \tag{3}$$

In this type of loan, the principal payment is different each period, so the calculation of interest using Equation 2 must take into account the variable principal payment. Even though both interest and principal are paid together, they must be kept separate. Interest is an expense item for federal income taxes, while principal payments are not.

Alternatively, the debt repayment can be scheduled with uniform principal payments and varying total payments each period. The principal payment can be calculated by the formula:

$$PP_{y,n-1} = PP = \frac{P}{n \times y} \tag{4}$$

In some loans a grace period is allowed for the first 6–12 months of operation, where no principal payments are due. The economics must reflect this.

Depreciation. This is a non-cash cost that accumulates money to rebuild the facilities after the project life is over. There are usually two depreciation rates. One is the corporation's book depreciation, which reflects the expected operating life of the project. The other is the tax depreciation, which is usually the maximum rate allowed by tax laws. To be correct, the tax depreciation should now be called "Accelerated Capital Cost Recovery System" (ACRS).

Economics can be shown using either depreciation rate. If book depreciation is used, a line for deferred income tax must be used. This is a cash-available item that becomes negative when the tax depreciation becomes less than the book depreciation. At the end of the book life, the cumulative deferred income tax should be zero.

The 1984 allowable ACRS tax depreciation is shown in Table 1. Most chemical plants have a 5-year tax life.

Federal income tax. The calculation of the federal income tax depends on the current tax laws. State income taxes are often included in this line item, since many are directly proportional to the federal taxes.

The federal tax laws are complicated and ever-changing. It is recommended that expert tax advice be obtained for most feasibility studies. If no current tax advice is available, many tax manuals are published each year that can be purchased or borrowed from a library.

Table 1
ACRS Schedule in % Per Year of Depreciable Plant

Year	Years for Recovery			
	3	5	10	15
1	25	15	8	5
2	38	22	14	10
3	37	21	12	9
4		21	10	8
5		21	10	7
6			10	7
7			9	6
8			9	6
9			9	6
10			9	6
11–15				6

Current tax laws allow deductions from revenues for sales expenses, operating costs, amortizations, lease costs, interest payments, depreciation, and depletion.

The net income before federal income tax (FIT) is calculated by subtracting all of the deductions from the gross revenues. If there is a net loss for any year, it can be carried forward and deducted the next year from the net income before FIT. If not all of the loss carried forward can be used in the following year, it can be carried forward for a total of 7 years.

At times, tax laws allow a 10% investment tax credit for certain qualifying facilities. Thus, 10% of the qualifying investment can be subtracted from the first 85% of the FIT due. If the taxes paid in the first year are not enough to cover the investment tax credit, it can sometimes be carried forward for several years.

Exhaustible natural deposits and timber might qualify for a cost depletion allowance. Each year, depletion is equal to the property's basis, times the number of units sold that year, divided by the number of recoverable units in the deposit at the first of the year. The basis is the cost of the property, exclusive of the investment that can be depreciated, less depletion taken in previous years.

At times, mining and some oil and gas operations are entitled to an alternate depletion allowance calculated from a percentage of gross sales. A current tax manual should be consulted to get the percentage depletion allowance for the mineral being studied.

Percentage depletion for most minerals is limited to a percentage of the taxable income from the property receiving the depletion allowance. For certain oil and gas producers the percentage depletion can be limited to a percentage of the taxpayer's net income from all sources.

The procedure for a feasibility study is to calculate both the cost and percentage depletion allowances, then choose the larger. Remember, either type reduces the basis. However, percentage depletion can still be deducted after the basis reaches zero.

The cash taxes due are calculated by multiplying the book net income before tax, times the tax rate, then adjusting for deferred tax and tax credits, if any.

Economics form

As stated at the start of this chapter, it is desirable to establish a form for reporting economics. Table 2 shows such a form. By following this form, the procedure can be seen for calculating the entire economics.

As can be seen in Table 2, all revenues less expenses associated with selling are summed in Row 17. All expenses including non-cash expenses such as depreciation, amortization, and depletion are summed in Row 30. The net profit before tax, Row 32, is obtained by subtracting Row 30 from Row 17 and making any inventory adjustment required. Row 34 is the cash taxes that are to be paid unless offset by investment or energy tax credits in Row 36. The deferred income tax is shown in Row 35. The deferred tax decreases the net profit after tax in the early years and increases the net profit after tax in later years. The impact on cash flow is just the other way around, as discussed later. Row 37, profit after tax, is obtained as follows:

  Row 32 (profit before tax)
− Row 34 (income tax)
− Row 35 (deferred tax)
+ Row 36 (tax credits)
= Row 37 (profit after tax)

Row 33, loss carried forward, is subtracted from Row 32 before calculating the income tax.

The cash flow, Row 39, is calculated as follows:

  Row 37 (profit after tax)
+ Row 35 (deferred tax)
+ Row 29 (depletion)
+ Row 28 (amortization)
+ Row 27 (depreciation)
− Row 38 (principal payments)
= Row 39 (cash flow)

If one of the rows is a negative number, such as deferred tax, the proper sign is used to obtain the correct answer. Any required inventory adjustment is also made.

Table 2
Economic Calculation Example

Periods/Year = 1		
Fiscal years	**1992**	**1993**
12 Sales revenue	239,093	248,657
13 Byproduct revenue	17,031	17,713
14 Advertising & marketing	(1638)	(1703)
15 Shipping & freight	(1310)	(1363)
16 Other sales expense	(983)	(1022)
17 Income from sales	252,194	262,282
18 Raw material cost	76,075	78,738
19 Operating labor	26,508	27,701
20 Maintenance labor	23,984	25,063
21 Maintenance material	37,869	39,573
22 Utilities	25,246	26,382
23 Administration & general	9467	9893
24 Unchanging expense	5000	5000
25 Temporary expense	0	0
26 Interest	7614	7188
27 Depreciation	5500	5500
28 Amortization	3639	2139
29 Depletion	12,621	19,541
30 Total cost	233,524	246,719
31 Inventory adjustment	0	0
32 Profit before tax	18,671	15,563
33 Loss carried forward	12,621	4031
34 Income tax	0	7835
35 Deferred tax	2783	(2530)
36 Tax credits	0	7835
37 Profit after tax	15,888	18,093
38 Principal payments	4105	4531
39 Cash flow	36,325	38,212

Economic analysis

Once the cash flow has been calculated, various yardsticks can be applied to determine the desirability of the project. The years for payout (payback) is often used. This is obtained by adding the cash flow each year, from the first year of construction, until the sum of the negative and positive cash flows reaches zero. The payout is usually reported in years to the nearest tenth, i.e. 5.6 years.

Each company has its own cutoff period for payout, depending on the economy, the company position, alternate projects, etc. One company used 5 years as the maximum payout acceptable for several years. As conditions changed, this was reduced to 3 years, then 1 year. It was hard to find projects with 1 year payout after taxes.

Some companies feel they should earn interest on their equity as well as get a payout. To calculate this, the cash flows are discounted back to year 0 using some arbitrary interest (discount) rate, then the positive discounted cash flows are summed each year of operation until they equal the sum of the discounted negative cash flows. This yardstick also works, but consistency must be maintained from one project to another. The discount (interest) rate and payout both become variables that must be set and perhaps changed from time to time depending on economic conditions inside and outside the company.

A simple return on equity can be used as a yardstick. Utilities are required by regulators to set their rates to obtain a given rate of return. This return is usually constant over the years and from project to project, so management must choose among projects on some basis other than profit.

In free enterprise projects, the return on equity will probably vary from year to year, so an average would have to be used. However, use of an average would consider return in later years to be as valuable as income in the first year of operation, which is not true.

The best way to evaluate free enterprise projects is to use discounted cash flow (DCF) rate of return, sometimes called internal rate of return.

This calculation can be expressed as follows:

$$SDCF = \sum_{y=1}^{y=N} \frac{CF_y}{(1+i)^y} = 0, \tag{5}$$

where

> $SDCF$ = Sum of the equity discounted cash flows,
> CF_y = Cash flow, both negative and positive for each year y,
> i = Discount rate in %/100 that results in the sum being zero, and
> N = Total number of years.

Since this calculation uses trial and error to find a discount rate at which all discounted negative and positive cash flows are equal, it can be tedious without a computer. Table 3 demonstrates this calculation for a simple set of cash flows. The DCF rate of return results in a single number that reflects the time value of money. It is certainly one of the yardsticks that should be used to measure projects.

Sensitivity analysis

Once the economic analysis has been completed, the project should be analyzed for unexpected as well as expected impacts on the economics. This is usually done through a set of "what if" calculations that test the project's sensitivity to missed estimates and changing economic environment. As a minimum, the DCF rate of return should be calculated for

Table 3
Discounted Cash Flow Calculations

Discount Rate (%) Year	Discounted Cash Flows			
	0	20	30	25.9
1	−20	−17	−15	−16
2	−20	−14	−12	−12
3	−10	−6	−5	−5
4	10	5	3	4
5	15	6	4	4
6	20	7	4	5
7	25	7	4	5
8	30	7	4	5
9	40	8	4	5
10	50	8	4	5
Totals	+140	+11	−5	00

±10% variations in capital, operating expenses, and sales volume and price.

The impact of loss of tax credits, loss carry forward, and other tax benefits should be studied. Such things as start-up problems that increase the first year's operating costs and reduce the production should also be studied. With a good economic computer program, such as described earlier, all of these sensitivity analyses can be made in an afternoon.

Additional reading on economics can be found in the References.

References

1. Perry, R. H. and Chilton, C. H., *Chemical Engineers Handbook*, Fifth Edition, McGraw-Hill, Inc., New York, section 25, p. 3.
2. Constantinescu, S., "Payback Formula Accounts for Inflation," *Chemical Engineering*, Dec. 13, 1982, p. 109.
3. Horwitz, B. A., "How Does Construction Time Affect Return?" *Chemical Engineering*, Sept. 21, 1981, p. 158.
4. Kasner, E., "Breakeven Analysis Evaluates Investment Alternatives," *Chemical Engineering*, Feb. 26, 1979, pp. 117–118.
5. Cason, R. L., "Go Broke While Showing A Profit," *Hydrocarbon Processing*, July, 1975, pp. 179–190.
6. Reul, R. I., "Which Investment Appraisal Technique Should You Use," *Chemical Engineering*, April 22, 1968, pp. 212–218.
7. Branan, C. and Mills, J., *The Process Engineer's Pocket Handbook, Vol. 3*, Gulf Publishing Co., 1984.

TIME VALUE OF MONEY: CONCEPTS AND FORMULAS

The value of money is not constant but changes with time. A dollar received today is worth more than a dollar received a month from now, for two reasons. First, the dollar received today can be invested immediately and earn interest. Second, the purchasing power of each dollar will decrease during times of inflation, and consequently a dollar received today will likely purchase more than a dollar next month. Therefore, in the capital budgeting and decision-making process for any engineering or financial project, it is important to understand the concepts of *present value, future value,* and *interest.* Present value is the "current worth" of a dollar amount to be received or paid out in the future. Future value is the "future worth" of an amount invested today at some future time. Interest is the *cost* of borrowing money or the *return* from lending money. Interest rates vary with time and are dependent upon both risk and inflation. The rate of interest charged, as well as the rate of return expected, will be higher for any project with considerable risk than for a "safe" investment or project. Similarly, banks and investors will also demand a higher rate of return during periods of monetary inflation, and interest rates will be adjusted to reflect the effects of inflation.

Simple interest versus compound interest

Simple interest is interest that is only earned on the original principal borrowed or loaned. Simple interest is calculated as follows:

$$I = (P)(i)(n),$$

where

> I = Simple interest,
> P = Principal (money) borrowed or loaned,
> i = Interest rate per time period (usually years), and
> n = Number of time periods (usually years).

Compound interest is interest that is earned on both the principal and interest. When interest is compounded, interest is earned each time period on the original principal *and* on interest accumulated from preceding time periods. Exhibits 1 and 2 illustrate the difference between simple and compound interest and show what a dramatic effect interest compounding can have on an investment after a few years.

Exhibit 1
Comparison between $5000 Invested for 3 Years at 12% Simple and 12% Compound Interest

Simple Interest			
End of Year	**Interest Earned**	**Cumulative Interest Earned**	**Investment Balance**
0	$0.00	$0.00	$5000.00
1	$5000.00 × 0.12 = $600.00	$600.00	$5600.00
2	$5000.00 × 0.12 = $600.00	$1200.00	$6200.00
3	$5000.00 × 0.12 = $600.00	$1800.00	$6800.00

Compound Interest (Annual Compounding)			
End of Year	**Interest Earned**	**Cumulative Interest Earned**	**Investment Balance**
0	$0.00	$0.00	$5000.00
1	$5000.00 × 0.12 = $600.00	$600.00	$5600.00
2	$5600.00 × 0.12 = $672.00	$1272.00	$6272.00
3	$6272.00 × 0.12 = $752.74	$2024.64	$7024.64

Exhibit 2
Comparison of Simple versus Compound Interest

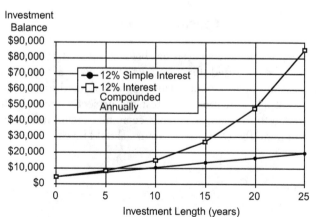

Nominal interest rate versus effective annual interest rate

Almost all interest rates in the financial world now involve compound rates of interest, rather than simple interest. Compound interest rates can be quoted either as a *nominal rate* of interest or as an *effective annual rate* of interest. The nominal rate is the stated annual interest rate compounded periodically (at the stated compounding time interval). The effective annual rate is the rate that produces the same final value as the nominal rate, when compounded only once per year. When the compounding period for a stated nominal rate is one year, then the nominal rate *is the same* as the effective annual rate of interest. The following formula can be used to convert nominal interest rates to effective annual rates:

$$i_{\text{eff}} = \left(1 - \frac{i_{\text{nom}}}{c}\right)^c - 1,$$

where

i_{eff} = Effective annual rate,
i_{nom} = Nominal rate, and
c = Number of compounding periods per year.

Note: Interest rates are expressed as fractions (0.12) rather than percentage (12%).

For *continuous* compounding, natural logarithms may be used to convert to an effective annual rate:

$$i_{\text{eff}} = e^{i_{\text{cont}}} - 1,$$

where

i_{cont} = Continuous compounding rate.

Note: Interest rates are expressed as fractions (0.12) rather than percentage (12%).

Present value of a single cash flow to be received in the future

How much is an amount of money to be received in the future worth today? Stated differently, how much money would you invest today to receive this cash flow in the future? To answer these questions, you need to know the *present value* of the amount to be received in the future. The present value (PV) can easily be calculated by *discounting* the future amount by an appropriate compound interest rate. The interest rate used to determine PV is often referred to as the *discount rate, hurdle rate, or opportunity cost of capital*. It represents the opportunity cost (rate of return) foregone by making this particular investment rather than other alternatives of comparable risk.

The formula for calculating present value is:

$$PV = \frac{FV}{(1 + r^n)},$$

where

PV = Present value,
FV = Future value,
r = Discount rate (per compounding period), and
n = Number of compounding periods.

Example. What is the present value of an investment that guarantees to pay you $100,000 three years from now? The first important step is to understand what risks are involved in this investment, and then choose a discount rate equal to the rate of return on investments of comparable risk. Let's say you decide that 9% effective annual rate of interest (9% compounded annually) is an appropriate discount rate.

$$PV = \frac{\$100,000}{(1 + 0.09)^3} = \$77.218.$$

Thus, you would be willing to invest $77,218 today in this investment to receive $100,000 in 3 years.

Future value of a single investment

Rearranging the relationship between PV and FV, we have a simple formula to allow us to calculate the future value of an amount of money invested today:

$$FV = PV(1 + r)^n$$

where

FV = Future value,
PV = Present value,
R = Rate of return (per compounding period), and
N = Number of compounding periods.

Example. What is the future value of a single $10,000 lump sum invested at a rate of return of 10% compounded monthly (10.47 effective annual rate) for 5 years? Note that monthly compounding is used in this example.

$$FV = \$10,000(1 + 0.10/12)^{60} = \$16,453.$$

The $10,000 investment grows to $16,453 in 5 years.

The importance of cash flow diagrams

Many financial problems are not nearly as simple as the two previous examples, which involved only one *cash outflow* and one *cash inflow*. To help analyze financial problems, it is extremely useful to diagram cash flows, since most problems are considerably more complex than these previous examples and may have multiple, repetitive, or irregular cash flows. The *cash flow diagram* is an important tool that will help you understand an investment project and the timing of its cash flows and help with calculating its PV or FV. The cash flow diagram shown represents the earlier example (in which PV was calculated for an investment that grew to $100,000 after annual compounding at 9% for 3 years).

The cash flow diagram is divided into equal time periods, which correspond to the compounding (or payment) periods of the cash flows. Note that this diagram is also presented from the investor's (lender's) point of view, since the initial investment is a cash outflow (negative) and the final payoff is a cash inflow (positive) to the investor. The diagram could also have been drawn from the borrower's point of view, in which case the PV would be a cash inflow and therefore

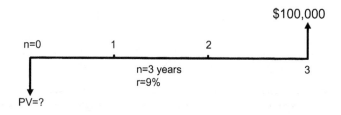

positive. Before drawing a cash flow diagram, you must first choose which viewpoint (lender or borrower) the diagram will represent. It is then extremely important to give each cash flow the correct sign (positive or negative), based upon whichever point of view is chosen. Other examples of cash flow diagrams will be presented in additional sample problems in this chapter.

Analyzing and valuing investments/projects with multiple or irregular cash flows

Many investments and financial projects have multiple, periodic, and irregular cash flows. A cash flow diagram will help in understanding the cash flows and with their analysis. The value of the investment or project (in current dollars) may then be found by discounting the cash flows to determine their present value. Each cash flow must be discounted individually back to time zero and then summed to obtain the overall PV of the investment or project. This *discounted cash flow formula* may be written as

$$PV = \sum_{i=1}^{n} \frac{C_i}{(1+r)^i}$$

where

PV = Present value,
C_i = Cash flow at time i,
n = Number of periods, and
r = Discount rate of return (opportunity cost of capital).

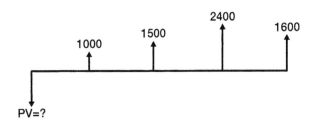

PV=?

Example. How much would you be willing to invest today in a project that will have generated positive annual cash flows of $1,000, $1,500, $2,400, and $1,600, respectively, at the end of each of the first 4 years? (Assume a 10% annual rate of return can be earned on other projects of similar risk.)

Step 1: Cash flow diagram:
Step 2: Discount cash flows into current dollars (present value):

$$PV = \frac{1.000}{(1+.1)^1} + \frac{1,500}{(1+.1)^2} + \frac{2,400}{(1+.1)^3} + \frac{1,600}{(1+.1)^4}$$
$$= \$5,44.74$$

Thus, you should be willing to invest a maximum of $5044.74 in this project.

Perpetuities

A *perpetuity* is an asset that provides a periodic fixed amount of cash flow in perpetuity (forever). Endowments and trust funds are often set up as perpetuities so that they pay out a fixed monetary amount each year to the beneficiary, without depleting the original principal. For valuation purposes, some physical assets such as an oil well, natural gas field, or other long-lived asset can also be treated as a perpetuity, if the asset can provide a steady cash flow stream for many years to come. The cash flow diagram for a perpetuity is represented here:

The value of a perpetuity can be found by this simple formula

$$PV_{perpetuity} = \frac{C}{r}$$

where

C = Cash flow per time period, and
r = Discount rate of return (per time period).

A *growing perpetuity* is a cash flow stream that grows at a constant rate. The formula to value a growing perpetuity expands the basic perpetuity formula to include a growth rate for the cash flow stream:

$$PV = \frac{C_1}{r-g},$$

where

C_1 = Initial cash flow,
r = Discount rate of return (per time period), and
g = Growth rate of cash flows (per time period).

C (equal cash flow each period)

$n = 1$ $n = \infty$

PV

Example. An existing oil well has been producing an average of 10,000 barrels per year for several years. Recent estimates by geologists show that this well's oil field is not large enough to justify additional drilling but can maintain the existing well's production rate for hundreds of years to come. Assume that the recent price of oil has averaged $20 per barrel, prices are historically estimated to grow 3% per year forever, and your discount rate is 10%. Also assume that maintenance costs are insignificant, and the land has no value other than for the oil. What is the current monetary value of this oil field and well, if the owner were to sell it?

Solution:

$$PV = \left(\frac{10,000 \times 20}{0.10 - 0.03} \right) = \$2,857,143.$$

The value of the oil field and well should be $2,857,143 based upon these assumptions.

Future value of a periodic series of investments

Both individual investors and companies often set aside a fixed quantity of money every month, quarter, or year, for some specific purpose in the future. This quantity of money then grows with compound interest until it is withdrawn and used at some future date. A cash flow diagram for a periodic investment is shown.

C (fixed investment each time period)

It can easily be calculated how much these periodic, fixed investments each time period will accumulate with compounding interest to be in the future. In other words, we want to determine the future value (FV), which can be calculated using the following formula:

$$FV = \frac{C}{r}\left[(1+r)^t - 1\right],$$

where

C = Cash flow invested each time period,
r = Fractional compound interest rate (each time period), and
t = Number of time periods (length of investment).

Example. A self-employed contract engineer has no company-paid retirement plan, so he decides to invest $2000 per year in his own individual retirement account (IRA), which will allow the money to accumulate tax-free until it is withdrawn in the future. If he invests this amount every year for 25 years, and the investment earns 14% a year, how much will the investment be worth in 25 years?

Solution:

$$FV = \frac{\$2,000}{0.14}\left[(1+.014)^{25} - 1\right] = \$363,742.$$

The periodic investment of $2000 each year over 25 years will grow to $363,742 if a 14% rate of return is achieved. This example illustrates the "power of compounding." A total of only $50,000 was invested over the 25-year period, yet the future value grew to more than seven times this amount.

Annuities, loans, and leases

An *annuity* is an asset that pays a constant periodic sum of money for a fixed length of time. Bank loans, mortgages, and many leasing contracts are similar to an annuity in that they usually require the borrower to pay back the lender a constant, periodic sum of money (principal and interest) for a fixed length of time. Loans, mortgages, and leases can therefore be considered "reverse" annuities, as shown by the cash flow diagrams. The relationship between the value (PV) of the annuity/loan/lease, its periodic cash flows, and the interest or discount rate is

Cash Flow Diagrams: Annuity

C (level cash flow each period)

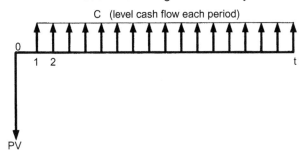

Cash Flow Diagrams: Reverse Annuity

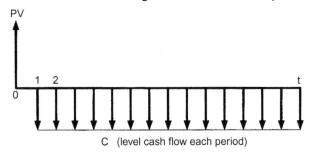

C (level cash flow each period)

$$PV = \frac{C}{r}\left[1 - \frac{1}{(1+r)^t}\right],$$

where

C = Cash flow received/paid each time period,
r = Interest or discount rate (per time period), and
t = Number of time periods (time length of the annuity).

This formula may be used to determine the PV of the annuity, or conversely, to determine the periodic payouts or payments when the PV is known. It should be noted that the future value (FV) at time *t* will be zero. In other words, all the principal of the annuity will have been expended, or all the principal of the loan or lease will have been paid off, at the final time period. Financial calculators are also pre-programmed with this formula and can simplify and expedite annuity, loan, and other cash flow problem calculations.

Example. Oil pipeline equipment is to be purchased and installed on a working pipeline in a remote location. This new equipment will have an operational life of 5 years and will require routine maintenance once per month at a fixed cost of $100. How much would the oil company have to deposit now with a bank at the remote location to set up an annuity to pay the $100 a month maintenance expense for 5 years to a local contractor? Assume annual interest rates are 9%.

Solution:

$$PV = \frac{100}{(0.09/12)}\left[1 - \frac{1}{(1+0.09/12)^{60}}\right] = \$4,817.34.$$

Example. Apex Corp. borrows $200,000 to finance the purchase of tools for a new line of gears it's producing. If the terms of the loan are annual payments over 4 years at 14% annual interest, *how much are the payments*?

$$\$200,000 = \frac{C}{0.14}\left[1 - \frac{1}{(1+0.14)^4}\right]$$

Solution:

$$C = \$68,640.96 \text{ per year.}$$

Gradients (payouts/payments with constant growth rates)

A *gradient* is similar to an annuity, but its periodic payouts increase at a constant rate of growth, as illustrated by its cash flow diagram. A gradient is also similar to a growing perpetuity, except that it has a fixed time length. The cash flow diagram for a gradient loan or lease is simply the inverse of this cash flow diagram, since the loan/lease payments will be negative from the borrower's point of view. The PV of a gradient may be found by

$$PV = \frac{C_1(1 - x^n)}{(1+r)(1-x)} \qquad \text{when } g \neq r$$

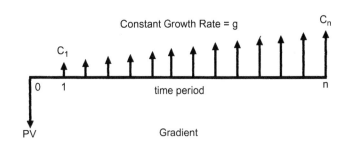

Gradient

or

$$PV = \frac{C_1 n}{(1+r)} \quad \text{when } g = r$$

where

$C_1 =$ Initial cash flow at the end of first time period,
$R =$ Rate of return (discount rate),
$n =$ Number of time periods,
$x = (1+g)/(1+r)$, and
$g =$ Fractional growth rate of cash flow period.

Analyzing complex investments and cash flow problems

No matter how complex a financial project or investment is, the cash flows can always be diagrammed and split apart into separate cash flows if necessary. Using the discounted cash flow (DCF) formula, each cash flow can be discounted individually back to time zero to obtain its present value (PV), and then summed to determine the overall valuation of the project or investment. It may also be possible to simplify things if you can identify a series of cash flows in the complex investment. Cash flow series can be discounted using either annuity or perpetuity formulas, and then added together with the PV of any singular cash flow.

When analyzing any complex cash flow problem, it is important to remember the basic time value of money (TVM) concept (i.e. *the value of money changes with time*). Therefore, if you wish to add or compare any two cash flows, they both must be from the same time period. Cash flows in the future must be discounted to their present value (PV) before they can be added or compared with current cash flows. Alternatively, current cash flows can be compounded into the future (FV) and then added or compared with future cash flows.

Example. An automotive components company has been asked by a large automobile manufacturing company to increase its production rate of components for one particular car model. Sales of this car model have been much higher than expected. In order to increase its production rate, the components company will have to purchase additional specialized machinery, but this machinery will easily fit into its existing facility. The automobile manufacturer has offered to sign a contract to purchase all the additional output that can be produced for 4 years, which will increase annual net revenues by $300,000 without any additional administrative overhead or personnel expense. The added machinery will cost $450,000 and will require maintenance every 6 months at an estimated cost of $20,000. The machinery will be sold at the end of the 4-year period, with an estimated salvage value of $80,000. *If the company requires a minimum rate of return of 20% on all investments, should it sign the contract, purchase the machinery, and increase its production rate?*

(Depreciation and tax effects are not considered in this example.)

Step 1: From the point of view of the automotive components company, the cash flow diagram will be:

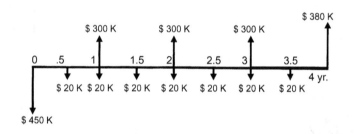

Step 2: Separate the overall cash flow diagram into individual components and series, and then determine the PV of each future cash flow or series of flows by discounting at the required annual rate of return of 20%:

$$PV = \frac{\$20,000}{0.20/2}\left[1 - \frac{1}{(1+0.20/2)^7}\right]$$

$$= \$97,368 \quad (\text{negative})$$

ANNUITY

PV=?

$300k every year

t=0   t=4 yr.

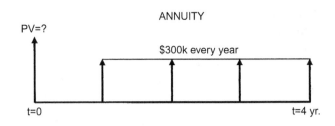

$$PV = \frac{\$30,000}{0.2} \left[1 - \frac{1}{(1+0.2)^4} \right] = \$776,620 \quad \text{(positive)}$$

SINGLE CASH FLOW FROM SALVAGE VALUE

$80K

PV=?

t=0   t=4 yr.

$$PV = \frac{\$80,000}{(1+0.2)^4} = \$38,580 \quad \text{(positive)}$$

Step 3: Sum the PV of all cash flows to obtain the project's overall PV from all *future* cash flows:

$$PV = \$776,620 + \$38,580 - \$97,368$$
$$= \$717,832 \quad \text{(positive)}$$

Step 4: Compare the PV of the future cash flows with the project's *initial* cash flow (initial investment):

The $717,832 PV of the project's future cash inflows (positive) is clearly greater than the $450,000 (negative) initial investment. Thus, this project has a positive *net present value* (NPV), because the PV of the future cash flows are greater than the initial investment. A positive NPV means that the rate of return on the investment is greater than the rate (20%) used to discount the cash flows.

Conclusion: The company should buy the additional machinery and increase its production rate.

DECISION AND EVALUATION CRITERIA FOR INVESTMENTS AND FINANCIAL PROJECTS

How should you decide whether or not to make an investment or go forward with a financial project? If you also have a number of alternative investment options, how do you decide which is the best alternative? How do you evaluate an investment's performance some time period after the investment has been made and the project is up and running? There are a number of analysis techniques that can help with these questions. Four separate methods are presented here, although each method varies in its effectiveness and complexity. There are also additional methods and

techniques that can be found in other engineering economics, finance, and accounting textbooks. However, most modern books and courses in these fields of study recommend the *net present value* (NPV) method as the most effective and accurate technique of evaluating potential investments and financial projects. Accounting measures such as *rate of return* (ROR) on an investment or *return on equity* (ROE) are useful to evaluate a project's financial results over a specific time period or to compare results against competitors.

Payback method

The payback method is a simple technique that determines the number of years before the cash flow from an investment or project "pays back" the initial investment. Obviously, the shorter the length of time it takes a project to pay off the initial investment, the more lucrative the project.

Example. A company wants to add a new product line and is evaluating two types of manufacturing equipment to produce this new line. The first equipment type costs $700,000 and can produce enough product to result in an estimated $225,000 in additional after-tax cash flow per year.

The second type of equipment is more expensive at $900,000, but has a higher output that can bring an estimated $420,000 per year in after-tax cash flows to the company. Which type of equipment should the company buy?

Solution:

Equipment Type 1

Payback period $= \$700,000/\$225,000$ per year

$$= 3.11 \text{ years}$$

Equipment Type 2

Payback period $= \$900,000/\$420,000$ per year

$$= 2.14 \text{ years}$$

Based upon this payback analysis, equipment type 2 should be purchased.

The advantage of the payback method is its simplicity, as shown in the example. However, there are several disadvantages that need to be mentioned. First, the payback method does not evaluate any cash flows that occur after the payback date, and therefore ignores long-term results and salvage values. Secondly, the payback method does not consider the time value of money (TVM) but assumes that each dollar of cash flow received in the second year and beyond is equal to that received the first year. Thirdly, the payback method has no way to take into account and evaluate any risk factors involved with the project.

Accounting rate of return (ROR) method

The *accounting rate of return* (*ROR*) for an existing company or project can be easily calculated directly from the company's or project's quarterly or annual financial accounting statements. If a potential (rather than existing) project is to be evaluated, financial results will have to be projected. Since ROR is an accounting measure, it is calculated for the same time period as the company's or project's income statement. The ROR on an investment is simply the net income generated during this time period, divided by the book value of the investment at the start of the time period.

$$ROR = \frac{\text{Net Income}}{\text{Net Investment(book value)}}$$

Net income is an accounting term defined as "revenues minus expenses plus gains minus losses" (i.e. the "bottom line" profit or loss on the income statement). Net investment is the "book value" of the investment, which is the original investment amount (or purchase price) less any depreciation in value of the investment.

Example. Company XYZ invested $500,000 in plant and equipment 4 years ago. At the end of year 3, the company's financial statements show that the plant and equipment's value had depreciated $100,000 and had a listed book value of $400,000. If company XYZ has a net income of $60,000 during its fourth year of operation, what is its ROR for this time period?

Solution:

$$ROR = \frac{60,000}{400,000} = 0.15(15\%)$$

Like the payback method, the advantage of the accounting ROR is that it is a simple calculation. It is especially simple for an existing company or project when the accounting statements are already completed and don't have to be projected. One disadvantage of ROR is that it doesn't consider TVM when it is used for more than one time period. It also uses accounting incomes rather than cash flows, which are not the same (see Accounting Fundamentals in this chapter). Another disadvantage is that ROR is a very subjective method, because the ROR is likely to vary considerably each time period due to revenue and expense fluctuations, accumulating depreciation, and declining book values. (Some textbooks recommend averaging ROR over a number of time periods to compensate for fluctuations and declining book values, but an "averaged" ROR still does not consider TVM.)

In general, accounting measurements such as ROR are best used for evaluating the performance of an *existing* company over a specific annual or quarterly time period or for comparing one company's performance against another. Net present value (NPV) and internal rate of return (IRR) are superior methods for evaluating *potential* investments and financial projects and in making the final decision whether to go ahead with them.

Internal rate of return (IRR) method

Internal rate of return (IRR) is an investment profitability measure that is closely related to net present value (NPV). The IRR of an investment is that rate of return which, when used to discount an investment's future cash flows, makes the NPV of an investment equal zero. In other words, when the future cash flows of an investment or project are discounted using the IRR, their PV will exactly equal the initial investment amount. Therefore, the IRR is an extremely useful quantity to know when you are evaluating a potential investment project. The IRR tells you the exact rate of return that will be earned on the original investment (or overall project with any additional investments) if the projected future cash flows occur. Similarly, when analyzing past investments, the IRR tells you the exact rate of return that was earned on the overall investment. The IRR decision rule for whether or not to go ahead with any potential investment or project being considered is simple: *If the IRR exceeds your opportunity cost of capital (rate of return that can be earned elsewhere), you should accept the project.*

The following example will make it clear how the IRR of an investment is calculated. It is highly recommended that a cash flow diagram be drawn first before trying to calculate the IRR. Particular attention should be paid to make sure cash inflows are drawn as positive (from the investor's point of view) and cash outflows are drawn as negative. It is also recommended that a financial calculator (or computer) be used to save time; otherwise, an iteration procedure will have to be used to solve for the IRR.

Example. Acme Tool and Die Company is considering a project that will require the purchase of special machinery to produce precision molds for a major customer, Gigantic Motors. The project will last only 4 years, after which Gigantic Motors plans to manufacture its own precision molds. The initial investment for the special machinery is $450,000, and the machinery will also require a partial overhaul in 3 years at an estimated cost of $70,000. Projected after-tax cash flows resulting from the project at the end of each of the 4 years are $140,000, $165,000, $185,000, and $150,000, respectively. These projections are based upon a purchase agreement that Gigantic is prepared to sign, which outlines the quantities of molds it will buy. Expected salvage value from the machinery at the end of the 4-year period is $80,000. Should Acme proceed with this project, if its opportunity cost of capital is 23%?

Step 1: Cash flow diagram:

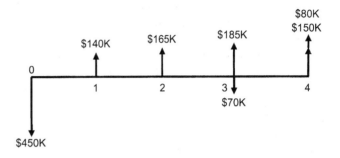

Step 2: Write an equation that sets the original investment equal to the PV of future cash flows, using the discounted cash flow formula. Solve for *r*, which is the IRR. (This is a simple process with a financial calculator or computer but otherwise requires iteration with any other calculator.)

$$450,000 = \frac{140,000}{(1+r)} + \frac{165,000}{(1+r)^2} + \frac{(185,000-70,000)}{(1+r)^3}$$
$$+ \frac{(150,000+80,000)}{(1+r)^4}$$

Solving for r (IRR):

IRR = 0.1537(15.37%)

Step 3: Apply decision rule to results: Since the IRR for this project is 15.37%, and Acme's required rate of return (opportunity cost of capital) is 23%, Acme should not proceed with this project.

IRR has a number of distinct advantages (as does NPV) over both the payback and ROR methods as a decision-making tool for evaluating potential investments. The first advantage is that the IRR calculation takes into account TVM. Secondly, IRR considers all cash flows throughout the life of the project (including salvage values), rather than looking at only one time period's results like accounting ROR does, or the years until the initial investment is paid off like the payback method. Therefore, IRR is an objective criterion, rather than a subjective criterion, for making decisions about investment projects. Thirdly, IRR uses actual cash flows rather than accounting incomes like the ROR method. Finally, since IRR is calculated for the entire life of the project, it provides a basis for evaluating whether the overall risk of the project is justified by the IRR, since the IRR may be compared with the IRR of projects of similar risk and the opportunity cost of capital.

The IRR method has several disadvantages compared to the NPV method, though only one disadvantage is mentioned here for purposes of brevity. Further information about potential problems with the IRR method (compared to NPV) may be obtained from most finance textbooks. One major problem with IRR is the possibility of obtaining multiple rates of return (multiple "roots") when solving for the IRR of an investment. This can occur in the unusual case where cash flows change erratically from positive to negative (in large quantities or for sustained time periods) *more than once* during the life of the investment. The graph in the following example illustrates how multiple IRR roots can occur for investments with these types of cash flows. The second graph is for comparison purposes and typifies the IRR calculation and graph for most investments. (Normally NPV declines with increasing discount rate, thus giving only one IRR "root".) When multiple roots are obtained using the IRR method, it is best to switch to the NPV method and calculate NPV by discounting the cash flows with the opportunity cost of capital.

Example. A project under consideration requires an initial investment/cash flow of $3000 (negative) and then has cash flows of $26,000 (positive) and $30,000 (negative) at the end of the following 2 years. A graph of NPV versus discount rate shows that two different IRR "roots" exist. (IRR is the point on the curve that intersects the horizontal axis at NPV = 0.) The second graph is from the previous example, showing that most "normal" investments have only one IRR root.

The IRR for this example is both 37.1% and 629.6%, as NPV equals zero at these two discount rates.

Illustration of Multiple I.R.R.s
(This Example)

Illustration of Single I.R.R. Root
(Previous Example - Acme Tool & Die)

Net present value (NPV) method

Net present value (NPV), as its name suggests, calculates the net amount that the discounted cash flows of an investment exceed the initial investment. Using the discounted cash flow (DCF) formula, the future cash flows are discounted by the rate of return offered by comparable investment alternatives (i.e. the opportunity cost of capital) and then summed and added to the initial investment amount (which is usually negative).

$$NPV = C_0 + \sum_{i=1}^{n} \frac{C_i}{(1+r)^i}$$

where

C_0 = Initial cash flow (negative for a cash "outflow"),
C_i = Cash flow in time period i,
n = Number of time periods, and
r = Opportunity cost of capital/discount rate.

Though similar to the IRR method, NPV does not calculate an investment's exact rate of return but instead calculates the exact dollar amount that an investment exceeds, or fails to meet, the *expected* rate of return. In other words, if an investment provides a rate of return exactly equal to the opportunity cost of capital, then the NPV of this investment will be zero, because the discounted future cash flows equal the initial investment. Thus, NPV provides an excellent decision criterion for investments. An investment with a positive NPV should be accepted, since it provides a rate of return above the opportunity cost of capital. By the same reasoning, an investment with a negative NPV should be rejected. NPV also does *not* suffer from any of the drawbacks of the payback, accounting ROR, or IRR methods. Because of this, NPV is the method most recommended by financial experts for making investment decisions. IRR is still useful to determine the exact rate of return for an investment, but NPV has none of the problems that IRR may have with unusual investments (such as the "multiple root" problem illustrated previously).

The following example shows how NPV is used to decide between investment alternatives.

Project	C_0	C_1	C_2	C_3	C_4
A	−$125,000	$31,000	$43,000	$48,000	$51,000
B	−$210,000	$71,000	$74,000	$76,000	$79,000

Example. A company must choose between two alternative manufacturing projects, which require different amounts of capital investment and have different cash flow patterns. The company's opportunity cost of capital for projects of similar risk is 15%. Which project should it choose?

Solving with the NPV formula:

$$NPV(A) = -\$4809$$

$$NPV(B) = \$2833$$

Conclusion: The company should choose project B. Despite a higher initial cost than A, Project B earns $2833 more than its required rate of return of 15%. Incidentally, IRR also could have been used, since future cash flows are all positive, and therefore no multiple roots exist. Project A's IRR is 13.23%; Project B's IRR is 15.65%.

SENSITIVITY ANALYSIS

The previous sections of this chapter show how to analyze investments and financial projects. These analysis methods can help quantify whether or not an investment is worthwhile and also help choose between alternative investments. However, no matter what method is used to analyze a financial project, that method and the decision about the project will ultimately rely upon the *projections* of the project's future cash flows. It is, therefore, extremely important that the future cash flows of any financial project or investment under consideration be forecast as accurately and realistically as possible. Unfortunately, no matter how precise and realistic you try to be with your projections of cash flows, the future is never certain, and actual results may vary from what is projected. Thus, it is often useful to perform a *sensitivity analysis* on the investment or financial project.

Sensitivity analysis is a procedure used to describe analytically the effects of uncertainty on one or more of the parameters involved in the analysis of a financial project. The objective of a sensitivity analysis is to provide the decision maker with quantitative information about what financial effects will be caused by variations from what was projected. A sensitivity analysis, once performed, may influence the decision about the financial project, or at least show the

decision maker which parameter is the most critical to the financial success of the project. The NPV method lends itself nicely to sensitivity analysis, since the discount rate is already fixed at the opportunity cost of capital. One parameter at a time can be varied in steps (while the other parameters are held constant) to see how much effect this variance has on NPV. A financial calculator or computer spreadsheet program will greatly expedite the multiple, repetitive calculations involved in this analysis. Plotting the results graphically will help show the sensitivity of NPV to changes in each variable, as illustrated in the following example.

Example. A-1 Engineering Co. is contemplating a new project to manufacture sheet metal parts for the aircraft industry. Analysis of this project shows that a $100,000 investment will be required up front to purchase additional machinery. The project is expected to run for 4 years, resulting in estimated after-tax cash flows of $30,000 per year. Salvage value of the machinery at the end of the 4 years is estimated at $32,000. A-1's opportunity cost of capital is 15%. Perform a sensitivity analysis to determine how variations in these estimates will affect the NPV of the project.

Step 1: Cash flow diagram:

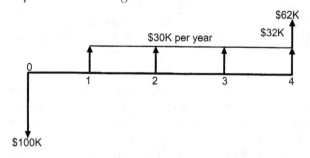

Step 2: Calculate baseline NPV:

$$NPV = -100,000 + \frac{30,000}{(1+0.15)^1} + \frac{30,000}{(1+0.15)^2}$$
$$+ \frac{30,000}{(1+0.15)^3} + \frac{62,000}{(1+0.15)^4}$$
$$= \$3.945.45$$

Step 3: Change salvage value, annual cash flow revenues, and initial investment amount in steps. Calculate NPV for each step change.

Step 4: Plot values to illustrate the NPV sensitivity to changes in each variable.

Conclusion: The graph shows that this project's NPV is equally sensitive to changes in the required initial investment amount (inversely related) and the annual revenues (after-tax cash flow) from the project. This is easily seen from the slope of the lines in the graph. The graph also shows that NPV is not

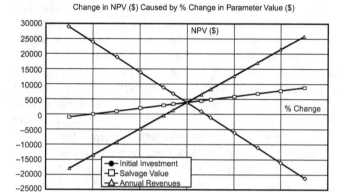

Change in NPV ($) Caused by % Change in Parameter Value ($)

affected nearly as much by changes in salvage value, primarily because the revenues from salvage value occur at the end of the fourth year and are substantially discounted at the 15% cost of capital. In conclusion, this sensitivity analysis shows A-1 Machinery that it cannot accept anything more than a 5% increase in the initial cost of the machinery or a 5% decrease in the annual cash flows. If these parameters change more than this amount, NPV will be less than zero and the project will not be profitable. Salvage value, however, is not nearly as critical, as it will take more than a 20% decrease from its estimated value before the project is unprofitable.

DECISION TREE ANALYSIS OF INVESTMENTS AND FINANCIAL PROJECTS

Financial managers often use *decision trees* to help with the analysis of large projects involving sequential decisions and variable outcomes over time. Decision trees are useful to managers because they graphically portray a large, complicated problem in terms of a series of smaller problems and decision branches. Decision trees reduce abstract thinking about a project by producing a logical diagram that shows the decision options available and the corresponding results of choosing each option. Once all the possible outcomes of a project are diagrammed in a decision tree, objective analysis and decisions about the project can be made.

Decision trees also allow *probability theory* to be used, in conjunction with NPV, to analyze financial projects and investments. It is often possible to estimate the probability of

success or failure for a new venture or project, based upon either historical data or business experience. If the new venture is a success, then the projected cash flows will be considerably higher than what they will be if the project is not successful. The projected cash flows for both of these possible outcomes (success or failure) can then be multiplied by their probability estimates and added to determine an *expected* outcome for the project. This expected outcome is the *average* payoff that can be expected (for multiple projects of the same type) based upon these probability estimates. Of course, the *actual* payoff for a single financial project will either be one or the other (success or failure), and not this average payoff. However, this average expected payoff is a useful number for decision making, since it incorporates the probability of both outcomes.

If there is a substantial time lapse between the time the decision is made and the time of the payoff, then the time value of money must also be considered. The NPV method can be used to discount the expected payoff to its present value and to see if the payoff exceeds the initial investment amount. The standard NPV decision rule applies: *Accept only projects that have positive NPVs.*

The following example illustrates the use of probability estimates (and NPV) in a simple decision tree:

Example. A ticket scalper is considering an investment of $10,000 to purchase tickets to the finals of a major outdoor tennis tournament. He must order the tickets 1 year in advance to get choice seats at list price. He plans to resell the tickets at the gate on the day of the finals, which is on a Sunday. Past experience tells him that he can sell the tickets and double his money, but only if the weather on the day of the event is good. If the weather is bad, he knows from past experience that he will only be able to sell the tickets at an average of 70% of his purchase price, even if the event is canceled and rescheduled for the following day (Monday). Historical data from the *Farmer's Almanac* shows that there is a 20% probability of rain at this time of year. His opportunity cost of capital is 15%. Should the ticket scalper accept this financial project and purchase the tickets?

Step 1: Construct the decision tree (very simple in this example).

Decision Tree

- ☐ = Decision Branch   ◯ = Probability Branch

Step 2: Calculate the expected payoff from purchasing the tickets:

$$
\begin{aligned}
\text{Expected payoff} &= (\text{probability of high demand}) \\
&\quad \times (\text{payoff with high demand}) \\
&\quad + (\text{probability of low demand}) \\
&\quad \times (\text{payoff with low demand}) \\
&= (0.8 \times 20{,}000) + (0.2 \times 7000) \\
&= \$17{,}400
\end{aligned}
$$

Step 3: Calculate NPV of expected payoff:

$$
\text{NPV} = -10{,}000 + \frac{\$17{,}400}{(1+1.5)^1} = \$5130
$$

Remember that the cash flow from selling the tickets occurs 1 year after the purchase. NPV is used to discount the cash flow to its present value and see if it exceeds the initial investment.

Conclusion: The ticket scalper should buy the tickets, because this project has a positive NPV.

The preceding example was relatively simple. Most financial projects in engineering and manufacturing applications are considerably more complex than this example; therefore, it is useful to have a procedure for setting up a decision tree and analyzing a complex project.

Here is a procedure that can be used:

1. Identify the decisions to be made and alternatives available throughout the expected life of the project.
2. Draw the decision tree, with branches drawn first for each alternative at every decision point in the project. Secondly, draw "probability branches" for each possible outcome from each decision alternative.
3. Estimate the probability of each possible outcome. (Probabilities at each "probability branch" should add up to 100%.)
4. Estimate the cash flow (payoff) for each possible outcome.
5. Analyze the alternatives, starting with the most distant decision point and working backwards. Remember the time value of money and, if needed, use the NPV method to discount the expected cash flows to their present values. Determine if the PV of each alternative's expected payoff exceeds the initial investment (i.e. positive NPV). Choose the alternative with the highest NPV.

The following example illustrates the use of this procedure.

Example. Space-Age Products believes there is considerable demand in the marketplace for an electric version of its barbecue grill, which is the company's main product. Currently, all major brands of grills produced by Space-Age and its competitors are either propane gas grills or charcoal grills. Recent innovations by Space-Age have allowed the development of an electric grill that can cook and sear meat as well as any gas or charcoal grill, but without the problems of an open flame or the hassle of purchasing propane gas bottles or bags of charcoal.

Based upon test marketing, Space-Age believes the probability of the demand for its new grill being high in its first year of production will be 60%, with a 40% probability of low demand. If demand is high the first year, Space-Age estimates the probability of high demand in subsequent years to be 80%. If the demand the first year is low, it estimates the probability of high demand in subsequent years at only 30%.

Space-Age cannot decide whether to set up a large production line to produce 50,000 grills a month or to take a more conservative approach with a small production line that can produce 25,000 grills a month. If demand is high, Space-Age estimates it should be able to sell all the grills the large production line can make. However, the large production line puts considerably more investment capital at risk, since it costs $350,000 versus only $200,000 for the small line.

The smaller production line can be expanded with a second production line after the first year (if demand is high) to double the production rate, at a cost of an additional $200,000.

Should Space-Age initially invest in a large or small production line or none at all? The opportunity cost of capital for Space-Age is 15%, and cash flows it projects for each outcome are shown in the decision tree. Cash flows shown at the end of year 2 are the PV of the cash flows of that and all subsequent years.

Steps 1–4 of procedure: (Construct decision tree and label with probabilities and cash flow projections).

Step 5: Analysis of alternatives:

(a) Start the analysis by working backwards from the R/H side of the decision tree. The only decision that Space-Age needs to make at the start of year 2 is deciding

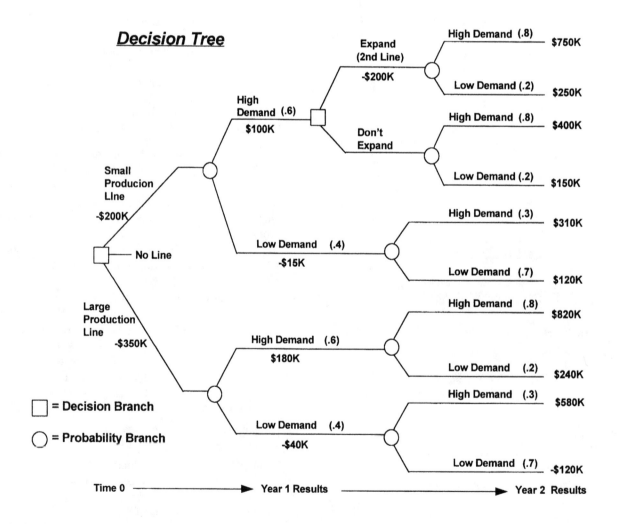

Decision Tree

Small Production Line −$200K

No Line

Large Production Line −$350K

High Demand (.6) $100K

Low Demand (.4) −$15K

High Demand (.6) $180K

Low Demand (.4) −$40K

Expand (2nd Line) −$200K — High Demand (.8) $750K / Low Demand (.2) $250K

Don't Expand — High Demand (.8) $400K / Low Demand (.2) $150K

High Demand (.3) $310K / Low Demand (.7) $120K

High Demand (.8) $820K / Low Demand (.2) $240K

High Demand (.3) $580K / Low Demand (.7) −$120K

☐ = Decision Branch

◯ = Probability Branch

Time 0 ⟶ Year 1 Results ⟶ Year 2 Results

whether or not to expand and install a second production line if demand is high the first year (if the initial decision was for a small production line).

Expected payoff of expanding with a second small production line (at the end of year 2):

$$= (0.8 \times \$750\text{K}) + (0.2 \times \$250\text{K}) = \$650,000$$

Expected payoff without the expansion (at the end of year 2):

$$= (0.8 \times \$400\text{K}) + (0.2 \times \$150\text{K}) = \$350,000$$

To decide if expansion is worthwhile, the NPVs of each alternative (at end of year 1) must be compared:

$$\text{NPV(with 2nd line)} = -\$200\text{K} + \frac{\$650\text{K}}{(1+0.15)^1}$$
$$= \$365,217$$

$$\text{NPV(without 2nd line)} = \$0 + \frac{\$350\text{K}}{(1+.15)^1} = \$304,348$$

These NPV amounts show that Space-Age *should clearly expand with a second small production line* (if its initial decision at time 0 was to open a small production line and the demand was high the first year). Now that this decision is made, the \$365,217 NPV of this expansion decision will be used to work backwards in time and calculate the overall NPV of choosing the small production line.

(b) The expected payoff (at the end of year 2) if the demand is low the first year with the small production line:

$$= (0.3 \times \$310\text{K}) + (0.7 \times \$120\text{K}) = \$177,000$$

This expected payoff must be discounted so that it can be added to the projected cash flow of year 1:

$$\text{PV} = \frac{\$177,000}{(1+0.15)} = \$153,913$$

(c) The expected payoff (at the end of year 1) for the small production line can now be determined:

$$= [0.6 \times (\$100,000 + \$365,217)] + [0.4 \times (-\$15,000 + \$153,913)]$$
$$= \$334,695$$

Notice that the cash flow for year 1 is the sum of the first year's projected cash flow plus the discounted expected payoff from year 2.

(d) Calculate the NPV (at time 0) of the small production line option:

$$\text{NPV} = \$200,000 + \frac{\$334,695}{(1+.15)} = \$91,039$$

(e) Calculate the NPV (at time 0) of the large production line option. Since there are no decisions at the end of year 1, there is no need to determine intermediate NPVs, and the probabilities from each branch of each year can be multiplied through in the NPV calculation as follows:

$$\text{NPV} = -\$350 + \frac{(0.6)(\$180\text{K}) + (0.4)(\$40\text{K})}{(1+0.15)^1}$$
$$+ \frac{(0.6)[(0.8)(\$820\text{K}) + (0.2)(\$240\text{K})]}{(1+0.15)^2}$$
$$+ \frac{(0.4)[(0.3)(\$580\text{K}) + (0.7)(-\$120\text{K})]}{(1+0.15)^2}$$

Solving: NPV = \$76,616

(f) *Conclusion*: Both the NPVs of the large and small production lines are positive, and thus both production lines would be profitable based upon the expected payoffs. However, the NPV of the small production line is higher than the NPV of the large production line (\$91,039 versus \$76,616). Therefore, Space-Age's decision should be to set up a small production line, and then expand with a second line if the demand for electric grills is high in the first year of production.

ACCOUNTING FUNDAMENTALS

Regardless of whether a person is managing the finances of a small engineering project or running a large corporation, it is important to have a basic understanding of accounting. Likewise, it is also impossible to study engineering economics without overlapping into the field of accounting. This section, therefore, provides a brief overview of accounting fundamentals and terminology.

Accounting is the process of measuring, recording, and communicating information about the economic activities of accounting entities (financial projects, partnerships, companies, corporations, etc.). Public corporations, and most other types of companies, issue financial statements from their accounting data on a periodic basis (usually quarterly and yearly). These financial statements are necessary because they provide information to outside investors and creditors and assist a company's own management with decision making. Four primary types of financial statements are normally issued:

- Balance sheet
- Income statement
- Statement of changes in retained earnings
- Statement of cash flows

The *balance sheet* provides information (at one particular instant in time) about financial resources of a company and claims against those resources by creditors and owners. The balance sheet's format reflects the *fundamental accounting equation:*

Assets = Liabilities + Owners Equity

Assets are items of monetary value that a company possesses, such as plant, land, equipment, cash, accounts receivable, and inventory. *Liabilities* are what the company owes its creditors, or conversely, the claims the creditors have against the company's assets if the company were to go bankrupt. Typical liabilities include outstanding loan balances, accounts payable, and income tax payable. *Owner's equity* is the monetary amount of the owner's claim against the company's assets. (Owner's equity is also referred to as stockholder's equity for corporations that issue and sell stock to raise capital.) Owner's equity comes from two sources: (1) the original cash investment the owner used to start the business and (2) the amount of profits made by the company that are not distributed (via *dividends*) back to the owners. These profits left in the company are called *retained earnings*.

There are numerous asset, liability, and owner's equity *accounts* that are used by accountants to keep track of a company's economic resources. These accounts are shown in the following "balance sheet" illustration, which is a typical balance sheet for a small manufacturing corporation.

Notice in this illustration that the balance sheet's total assets "balance" with the sum of liabilities and owner's equity, as required by the fundamental accounting equation. There are also a number of additional *revenue* and *expense* accounts that are maintained but not shown in the balance sheet. The balances of these revenue and expense accounts are closed out (zeroed) when the financial statements are prepared, and their amounts are carried over into the income statement. The *income statement* is the financial statement that reports the profit (or loss) of a company over a specific period of time (quarterly or annually). The income statement reflects another basic and fundamental accounting relationship:

Revenues − Expenses = Profit(or Loss)

A sample "income statement" for our typical small corporation is illustrated.

The income statement simply subtracts the company's expenses from its revenues to arrive at its *net income* for the period. Large companies and corporations usually separate their income statements into several sections, such as income from operations, income or loss from financial activities and discontinuing operations, and income or loss from extraordinary items and changes in accounting principles.

Companies frequently show an expense for depreciation on their income statements. *Depreciation* is an accounting method used to allocate the cost of asset "consumption" over the life of the asset. The value of all assets (building, equipment, patents, etc.) depreciate with time and use, and depreciation provides a method to recover investment capital by expensing the portion of each asset "consumed" during each income statement's time period. Depreciation is a *noncash expense*, since no cash flow actually occurs to pay for this expense. The *book value* of an asset is shown in the balance sheet and is simply the purchase price of the asset minus its accumulated depreciation. There are several methods used to calculate depreciation, but the easiest and most frequently used method is *straight-line depreciation*, which is calculated with the following formula:

$$\text{Depreciation Expense} = \frac{\text{Cost} - \text{Salvage Value}}{\text{Expected Life of Asset}}$$

Straight-line depreciation allocates the same amount of depreciation expense each year throughout the life of the asset. *Salvage value* (residual value) is the amount that a company can sell (or trade in) an asset for at the end of its useful life. "Accelerated" depreciation methods (such as sum-of-the-years'-digits and declining balance methods) are available and may sometimes be used to expense the cost of an asset faster than straight-line depreciation. However, the

XYZ Company
Balance Sheet
As of March 31, 1994

ASSETS

Current Assets:

Cash	$25,000.00	
Accounts Receivable	$268,000.00	
Inventory	$163,000.00	
Total Current Assets		$456,000.00

Property, Plant, & Equipment:

Building	$600,000.00		
less Accumulated Depreciation	($250,000.00)	$350,000.00	
Equipment	$300,000.00		
less Accumulated Depreciation	($125,000.00)	$175,000.00	
Land		$100,000.00	
Total Property, Plant, & Equipment			$625,000.00
Total Assets			**$1,081,000.00**

Liabilities

Current Liabilties:

Accounts Payable	$275,000.00	
Salaries Payable	$195,000.00	
Interest Payable	$13,000.00	
Income Tax Payable	$36,000.00	
Total Current Liabilities		$519,000.00

Long-Term Liabilities:

Notes Payable	$260,000.00	
Total Liabilities		$779,000.00

Owner's (Stockholder's) Equity

Capital Stock	$200,000.00	
Retained Earnings	$102,000.00	
Total Owner's Equity		$302,000.00
Total Liabilities & Owner's Equity		**$1,081,000.00**

use of accelerated depreciation methods is governed by the tax laws applicable to various types of assets. A recent accounting textbook and tax laws can be consulted for further information about the use of accelerated depreciation methods.

The profit (or loss) for the time period of the income statement is carried over as retained earnings and added to (or subtracted from) owner's equity, in the "statement of changes in retained earnings" illustration.

If the company paid out any dividends to its stockholders during this time period, that amount would be shown as a deduction to retained earnings in this financial statement. This new retained earnings amount is also reflected as the retained earnings account balance in the current balance sheet.

The last financial statement is the *statement of cash flows*, which summarizes the cash inflows and outflows of the company (or other accounting entity) for the same time period as the income statement. This statement's purpose is to show where the company has acquired cash (inflows) and to what activities its cash has been utilized (outflows) during this time period. The statement divides and classifies cash flows into three categories: (1) cash provided by or used by operating activities; (2) cash provided by or used by investing activities; and (3) cash provided by or used by financing activities. Cash flow from operations is the cash generated (or lost) from the normal day-to-day operations of a company, such as producing and selling goods or services. Cash flow from investing activities includes selling or purchasing long-term assets and sales or purchases of investments. Cash flow

XYZ Company
Income Statement
For the Quarter Ended March 31, 1994

Sales Revenues		$2,450,000.00
Less: Cost of Goods sold		($1,375,000.00)
Gross Margin		$1,075,000.00
Less: Operating Expenses		
Selling expenses	$295,000.00	
Salaries expense	$526,000.00	
Insurance expense	$32,000.00	
Property Taxes	$27,000.00	
Depreciation, Equipment	$35,000.00	
Depreciation, Building	$70,000.00	($985,000.00)
Income from Operations		$90,000.00
Other Income		
Sale of Assets	$16,000.00	
Interest Revenue	$2,000.00	$18,000.00
		$108,000.00
Less: Interest Expense		($22,000.00)
Income before Taxes		$86,000.00
Less: Income Tax Expense		($34,400.00)
Net Income		*$51,600.00*

XYZ Company
Statement of Changes in Retained Earnings
For the Quarter Ended March 31, 1994

Retained Earnings, Dec. 31, 1993	$50,400.00
Net Income, Quarter 1, 1994	$51,600.00
Retained Earnings, March 31, 1994	$102,000.00

from financing activities covers the issuance of long-term debt or stock to acquire capital, payment of dividends to stockholders, and repayment of principal on long-term debt.

The statement of cash flows complements the income statement, because although income and cash flow are related, their amounts are seldom equal. Cash flow is also a better measure of a company's performance than net income, because net income relies upon arbitrary expense allocations (such as depreciation expense). Net income can also be manipulated by a company (e.g. by reducing R&D spending, reducing inventory levels, or changing accounting methods to make its "bottom line" look better in the short term. Therefore, analysis of a company's cash flows and cash flow trends is frequently the best method to evaluate a company's performance. Additionally, cash flow rather than net income should be used to determine the rate of return earned on the initial investment or to determine the value of a company or financial project. (Value is determined by discounting cash flows at the required rate of return.)

The statement of cash flows can be prepared directly by reporting a company's gross cash inflows and outflows over a time period, or indirectly by adjusting a company's net income to net cash flow for the same time period. Adjustments to net income are required for (1) changes in assets and liability account balances, and (2) the effects of noncash revenues and expenses in the income statement.

For example, in comparing XYZ Company's current balance sheet with the previous quarter, the following information is obtained about current assets and liabilities:

- Cash increased by $10,000
- Accounts receivable decreased by $5000
- Inventory increased by $15,000
- Accounts payable increased by $25,000
- Salaries payable increased by $5000

Additionally, the balance sheets show that the company has paid off $166,600 of long-term debt this quarter. This information from the balance sheets, along with the noncash expenses (depreciation) from the current income statement, is used to prepare the "statement of cash flows." The example demonstrates the value of the statement of cash flows, because it shows precisely the sources and uses of XYZ Company's cash. Additionally, it reconciles the cash account balance from the previous quarter to the amount on the current quarter's balance sheet.

XYZ Company
Statement of Cash Flows
For the Quarter Ended March 31, 1994

Net Income		$51,600.00	
Adjustments to reconcile net income to cash flow from op. activities:			
Decrease in accounts receivable	$5,000.00		
Increase in inventory	($15,000.00)		
Increase in accounts payable	$25,000.00		
Decrease in salaries payable	$5,000.00		
Depreciation expense, equipment	$35,000.00		
Depreciation expense, building	$70,000.00		
Total Adjustments:		$125,000.00	
Net cash flow from operating activities:			$176,600.00
Cash flows from financing activities:			
Retirement of long-term debt (notes payable)	($166,600.00)		
Net cash flow from financing activities:			($166,600.00)
Net Increase in cash:			$10,000.00
Cash account balance, Dec. 31, 1993			$15,000.00
Cash account balance, March 31, 1994			$25,000.00

REFERENCES AND RECOMMENDED READING

Engineering economics

1. Newinan, D. G., *Engineering Economic Analysis*. San Jose, CA: Engineering Press, 1976.
2. Canada, J. R. and White, J. A., *Capital Investment Decision Analysis for Management and Engineering*. Englewood Cliffs, NJ: Prentice-Hall, 1980.
3. Park, W. R., *Cost Engineering Analysis: A Guide to Economic Evaluation of Engineering Projects*, 2nd ed. New York: Wiley, 1984.
4. Taylor, G. A., *Managerial and Engineering Economy: Economic Decision-Making*, 3rd ed. New York: Van Nostrand, 1980.
5. Riggs, J. L., *Engineering Economics*. New York: McGraw-Hill, 1977.
6. Barish, N., *Economic Analysis for Engineering and Managerial Decision Making*. New York: McGraw-Hill, 1962.

Finance

1. Brealey, R. A. and Myers, S. C., *Principles of Corporate Finance*, 3rd ed. New York: McGraw-Hill, 1988.
2. Kroeger, H. E., *Using Discounted Cash Flow Effectively*. Homewood, IL: Dow Jones-Irwin, 1984.

Accounting

1. Chasteen, L. G., Flaherty, R. E., and O'Conner, M. C., *Intermediate Accounting*, 3rd ed. New York: McGraw-Hill. 1989.
2. Eskew, R. K. and Jensen, D. L., *Financial Accounting*, 3rd ed. New York: Random House, 1989.

Additional references

Owner's Manual from any Hewlett-Packard, Texas Instruments, or other make business/financial calculator.

Note: A business or financial calculator is an *absolute must* for those serious about analyzing investments and financial projects on anything more than an occasional basis. In addition to performing simple PV, FV, and mortgage or loan payment calculations, modern financial calculators can instantly calculate NPVs and IRRs (including multiple roots) for the most complex cash flow problems. Many calculators can also tabulate loan amortization schedules and depreciation schedules and perform statistical analysis on data. The owner's manuals from these calculators are excellent resources and give numerous examples of how to solve and analyze various investment problems.

Estimate the cost of a pipeline in the United States (based on 1994 data)

The following cost data are based on average costs for pipelines constructed on and offshore in the United States. Cost data are based on FERC permit filings. These costs are average and consideration will need to be given to the location, terrain, season of the year, length of line, environmental issues, and other variables before using these data.

Example. To find the cost of 100 miles of 36-in diameter pipeline (onshore construction), $27,277 \times 100 \times 36 = \$98,197,200$.

Line size inches	Onshore ($/diam. in/mi)	Offshore ($/diam. in/mi)
8	25,321	
10	81,515	
12	21,160	
16	22,382	
20	54,688	
24	29,243	
26	29,923	
30	40,022	49,664
36	27,277	47,037
42	42,193	
48	26,305	

How to compare the cost of operating an engine on diesel and natural gas

Rule. For a rough estimate of the cost of diesel fuel per hour of operation, multiply: (0.05) (engine hp) (cost per gallon of diesel fuel). To find the cost of natural gas per hour's operation, multiply: (0.008) (engine hp) (cost of natural gas per Mcf).

Example. What would it cost per hour to operate a 500 hp engine on diesel fuel? On natural gas? Diesel fuel costs 10 cents per gallon, and natural gas costs 20 cents per Mcf.

Solution.

Cost/hour for diesel fuel operation = (0.05)(500)(0.10)
$$= \$2.50$$

(A more accurate answer can be obtained by substituting 0.0533 for 0.05. This would make the cost of diesel fuel operation $2.67 per hour.)

Cost/hour for natural gas = (0.008)(500)(0.20) = $0.80

Example. Compare the cost of operating the engine on straight diesel fuel with dual fuel consisting of 90% natural gas and 10% diesel pilot fuel, if the conditions are the same as those just described.

Solution.

Cost per hour = (0.10)(2.50) + (0.90)(0.80) = $0.97

Generally speaking, 1000 cu. ft of natural gas has a Btu content of 1,000,000 and a gallon of diesel fuel has a Btu content of 150,000. Thus 1 Mcf of natural gas is equivalent to 6 2/3 gallons of diesel fuel. To arrive at the figures above, the engine was assumed to require an average of 8000 Btu per brake horsepower hour. This is an arbitrary figure, which is about the average for engines in good mechanical condition operating under high percentage of load. Some pipeline companies are equipped with engines that consume about 7000 Btu/bhp; others run as high as 11,000 Btu/bhp. Condition of the engine and the loading account for most of the difference.

If an engine used an average of 7000 Btu/bhp in developing its full horsepower of say, 1,000, naturally it would use 7,000,000 Btu per hour. This would require 7 Mcf of natural gas or about 46.7 gallons of diesel fuel.

How to estimate energy costs for different pipeline throughputs

The rule of thumb given below will apply to liquid pipelines where the flow is in the turbulent region only. It will serve to estimate energy requirements for pumping through pipelines between increments of +15% and decrements of −15%.

Rule. Multiply the normal throughput in barrels times the cost of energy per barrel. Then multiply the increment or decrement (provided it is within a range of +15% or −15%) times the cost of energy per barrel at normal throughput times 3. Then add or subtract the latter amount from the cost of normal throughput.

Or, restated in simpler form:

Cost of energy per barrel at normal throught = Y

Cost of energy per barrel of increment or decrement within a range of ± 15% = 3Y.

Example. Assume energy requirements for pumping 2000 barrels per hour through a 12-in pipeline are $0.01 per barrel. What are energy requirements for pumping 2200 barrels per hour?

Energy cost for 2,000 b/h:

2000 × 0.01 = $20 per hour

Energy cost for additional 200 b/h:

200 × 0.01(3) = $6 per hour

Energy cost for 2, 200 b/h = $26 per hour

Comparing fuel costs for diesel and electric prime movers

Here's a formula that will give you a quick comparison of fuel costs of a pump station—for diesel engine or electric motor. Just find out which of the expressions below, C_e or C_d, is smaller:

$$C_e = \frac{0.7457}{n} \times E,$$

$$C_d = 0.065D + 0.04,$$

where

- C_e = Power consumption cost of electric motors, cents/bhp-h,
- C_d = Fuel and lube oil cost of diesel engines, cents/bhp-h,
- E = Electric power rate, cents/kW-h,
- D = Diesel fuel cost, cents/gal, and
- n = Efficiency of electric motor.

Example. Let's see which is more economical: running a 1000 bhp unit on diesel or electric driver.

The data are

$n = 95\%$,

$E = 1 \ \text{¢}/\text{kW} - \text{h}$,

$D = 10 \ \text{¢}/\text{gal}$.

So, we have

$C_e = 0.7457/0.95 \times 1 = 0.785$,

$C_d = 0.065 \times 10 + 0.04 = 0.690$.

Then, for a 1000 hp unit the estimated operating cost per hour would be

motor : \$7.85/h,

engine : \$6.90/h.

Keep in mind that this rule deals only with costs of power and fuel consumption. It does not take into consideration such expenses as operating personnel, repairs and renewals, depreciation, overhead, and so on.

Nomograph for calculating scale-up of equipment or plant costs

In correlating the cost of equipment or complete plants as a function of size, it has been shown that very often the cost varies as a fractional power of the ratio of capacities (or equipment sizes), as follows:

$$\left(\frac{C_2}{C_1}\right)^n = \frac{I_2}{I_1},$$

where

- C_1 and C_2 = The capacities (or sizes) of the equipment or plants in question,
- I_1 and I_2 = The cost or fixed investment of the known unit and the new unit, respectively,
- n = The exponent that applies to the particular case at hand.

The nomograph enables one to calculate the capacity ratio raised to the fractional power n and the equipment or plant cost if the size for one capacity level is known.

Example. A rotary drum filter costs \$20,000 when its filtering area is 50 ft^2. Assuming the cost varies as the 0.6 power of the capacity, how much will a 100-ft^2 filter cost?

The capacity ratio is

$$C_2/C_1 = 100/50 = 2.$$

Enter 2 on the capacity ratio scale (A) and align with 0.6 to the intersection point on the middle scale with \$20,000 on the (D) scale and read I2 = \$30,000 on the (E) scale.

Source

Kuong, J. F., *Brit. Chem. Eng.*, 7, No. 7, 546 (1962).

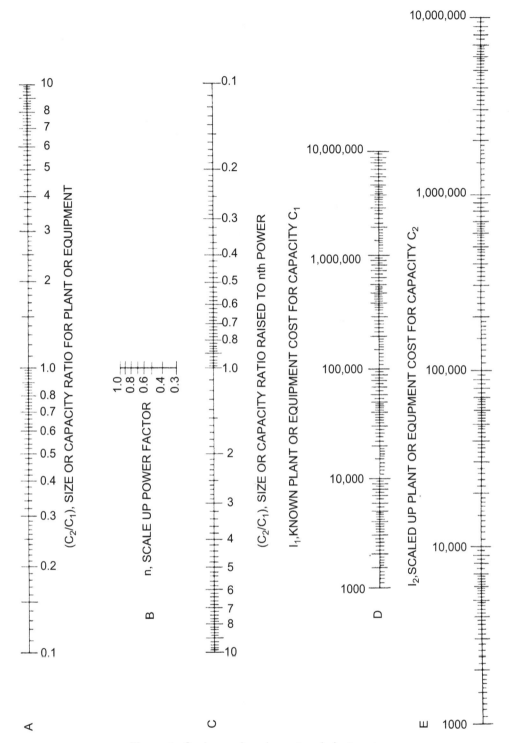

Figure 1. Scale-up of equipment and plant costs.

Nomograph for calculating scale-up of tank costs

The nomograph enables one to estimate the cost of a steel container of given volume when one knows the cost and volume of another or standard container.

The chart is based on the equation[1]

$$C = cV_S^{0.3}V^{0.7},$$

where

    C = Cost of container in question, dollars,
    c = Cost of standard container, dollars per unit volume,
    V_s = Volume of standard container,
    V = Volume of container in question.
V_s and V are expressed in the same units: gallons cubic ft, barrels.

Based on Dickson's suggestions, the chart can be used for drums of the same diameter and internal pressure and for spheres of the same internal pressure.

Example. If a 30,000-barrel tank costs \$36,000, what is the cost of a 15,000-barrel tank? For the standard tank, $c = 36,000/30,000$ or \$1.2 per barrel. Follow the key; connect 30,000 on the V_s scale with 15,000 on the V-scale and note the intersection with the α-axis. Connect this point with 1.2 on the c-scale and read the desired cost of approximately \$22,000 on the C-scale.

Source

Davis, D. S., *Petrol Refiner*, 37, No. 4200 (1958).

Reference

1. Dickson, R. A., *Chem. Eng.*, 54, (8) 122 (1947).

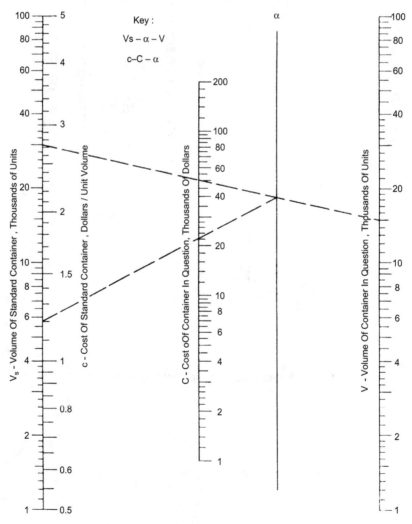

Figure 1. Scale-up of tank costs.

Nomograph for determining sum-of-years depreciation

In estimating annual depreciation of fixed capital assets, the sum-of-the-years-digits depreciation method is commonly used. This method allows for the effect of accelerated depreciation in the earlier years of a project. It is based on the number of years of service remaining each year and on the sum of the arithmetic series of the years from 1 to N. The depreciation factor for each year is calculated by dividing the number of remaining service years by the sum of the digits corresponding to the total service years. The depreciation charge can then be obtained by multiplying the depreciation factor each year by the total fixed asset cost.

The nomograph was constructed from the following equation:

$$D = \frac{N_D}{\sum\limits_{i}^{N} N_i} \times V = \frac{N_D}{1+2+3+\ldots N} \times V,$$

where

$D =$ Annual depreciation, dollars,

$N =$ Number of years of service of the depreciable item,

$N_D =$ Number of years of service remaining each year $(N - 0, N - 1, \ldots N - (N - 1)$ for the first, second and Nth year of depreciation), and

$N_i =$ Digit corresponding to the ith year.

Example. The value of a piece of equipment is $5200. Its salvage value at the end of 10 years is $200. What is the annual depreciation for the first, second, and third years?

Solution. The depreciable amount is $5200 - 200 = $5000. The first, second, and third years correspond to 10, 9, and 8 years of remaining service life, respectively.

Connect	With	Mark or Read
55 for 10 yr	$N_D = 10$ (1st yr)	R
R	$V = 5000$	$D = \$910$

Figure 1. Sum-of-years depreciation.

Repeat the procedure for $N_D = 9$ and 8 and obtain $D = 820$ and 720 for the second and third years of depreciation, respectively.

If the total depreciable value V is larger than the range of the scale, divide by a factor that brings it within the range of the scale, then read the D value, but remember that the result must be multiplied by the same factor.

Source

Kuong, J. F., *Chem. Eng.*, 67, No. 23, 237 (1960).

Nomograph for estimating interest rate of return on investment ("profitability index")

Several methods are used for evaluating the profitability of a given project. A method for estimating profitability that has received considerable emphasis in the past few years is the "interest rate of return," (Figure 1), also known as

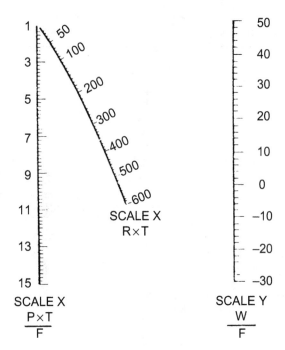

Figure 1. Interest rate of return.

"profitability index" or "discount cash flow" method. This profitability method takes into account the time value of money. Mathematically, interest rate of return may be defined as that rate at which the company's outstanding investment is repaid by proceeds from the project.[1]

The nomograph was developed for the case in which the yearly profits (P) (or the difference between income and costs) are equal through the life of a project. The nomograph is based on the following assumptions:

1. Depreciation method is the sum of the years-digits.
2. Project economic life is equal to depreciable life (T).
3. Construction time is zero, i.e. total investment occurs instantaneously at time zero.
4. There is no salvage value at the end of the project life.
5. There is no start-up expense ($S = 0$).

Where these conditions apply, the nomograph gives the interest rate of return R, as a function of P and T, fixed investment (F), and working capital (W).

Example 1. *Basic Profitability.* For the preliminary evaluation of a project the following estimates are available:

Plant capacity and sales	10,000,000 lb/yr
Average price	$0.20/lb
Unit cost (at capacity)	$0.10/lb
Overhead 15% of price	$0.03/lb

Project life 10 years (T)
Depreciation is sum of the years-digits

Plant investment (F)	$1,000,000
Working capital (W)	$200,000

It may be assumed that no start-up expense is incurred.

Estimate the interest rate of return if the project is undertaken. The reference or base case is not to invest.

First compute P, the yearly before-tax and depreciation profit.

Gross income: $10,000,000 \times 0.20 = \$2,000,000$
Less overhead @ 15%: $10,000,000 \times 0.03 = \$300,000$
Less operating cost: $10,000,000 \times 0.10 = \$1,000,000$
P, profit before tax and depreciation $= \$700,000$

Compute:

$$\frac{P \times T}{F} = \frac{700,000 \times 10}{1,000,000} = 7$$

Compute:

$$\frac{W}{F} = \frac{200,000}{1,000,000} = 0.2$$

Connect	With	Read
$\dfrac{P \times T}{F} = 7$	$\dfrac{W}{F} = 0.2$	$R \times T = 330$

Since $T = 10$, the interest rate of return for the project is 33%/yr.

This is equivalent to saying that the outlay of capital of $1,000,000 in plant facilities plus 200,000 in working capital when invested in the project would generate enough cash to pay back the total investment plus interest at the rate of 33%/yr.

Example 2. *Comparison of alternatives.* Consider the following alternatives for replacing a pump:

- *Alternative A.* Replace existing pump with one of the same type at a cost of $3,000, working capital of $50, and maintenance expense of $300/yr.
- *Alternative B.* Install a new-type pump costing $3500 with working capital of $100 and maintenance costs of $100/yr. Economic and depreciable life for both pumps is 10 years.

The savings per year for using pump B versus pump A are

$$\$300 - \$100 = \$200/yr$$

to justify an increment of investment of $500 ($3500 − $3000). What is the profitability index or interest rate of return on the differential investments?

Calculate:

$$\frac{P \times T}{F} = \frac{200 \times 10}{500} = 4$$

$$\frac{100(W)}{F} = \frac{50}{500} = 0.10$$

Connect	With	Read
$\frac{P \times T}{F} = 4$	0.1 on $\frac{W}{F} = 0.1$ scale	$R \times T = 210$

Since $T = 10$, $R = 21\%/\text{yr}$ = profitability index for the project.

If 21% is considered an adequate rate of return, after consideration of risk, then pump B would be the preferred one.

Source

Greist, W. H., Bates, A. G., and Weaver, J. B., *Ind Eng. Chem.*, 53, 52A (1961).

References

1. Dean, J., *Harvard Bus*. Rev., 32, 120−130 Jan., Feb. (1954).
2. Martin, J. C., in *Chemical Process Economics in Practice*, Hur, J. J., Ed., Reinhold Publishing Corp., New York, 1956.
3. Weaver, J. B., and Reilly, R. J., *Chem. Eng. Progr.*, 52, 405 (1956).

Nomograph for determining break-even point

The nomograph gives the break-even point for a plant as a percent of design capacity when the unit fixed cost, Cf, the sales price, S, and the unit variable cost, Cv, are known. It is based on the following equation:

$$\text{BE} = 100 \cdot \text{Cf}/(S - \text{Cv})$$

The nomograph can also be used to determine the sensitivity of the break-even point to any combination of fixed variable costs or sales price or to determine the maximum fixed cost that can be absorbed for a given break-even point.

Example. Determine the break-even point when:

Design output	10,000 lb/yr
Unit fixed cost	$0.045/lb
Unit variable cost	$0.075/lb
Sales price	$0.200/lb

Calculate $\Delta = 0.200 - 0.075 = 0.125$ or 12.5¢/lb.
Align Cf = 4.5¢ with $\Delta = 12.5$ and obtain BE = 36% of plant capacity.

Source

Kuong, J. F., *Petro/Chem. Engineer*, 35, No. 1, 44 (1963).

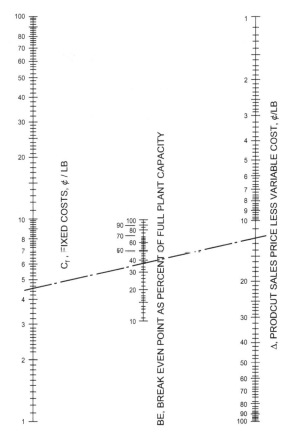

Figure 1. Break-even point.

Chart gives unit cost per brake horsepower of reciprocating compressors with various types of prime movers

Chart shows influence on unit cost of numbers of reciprocating compressor units installed in one station

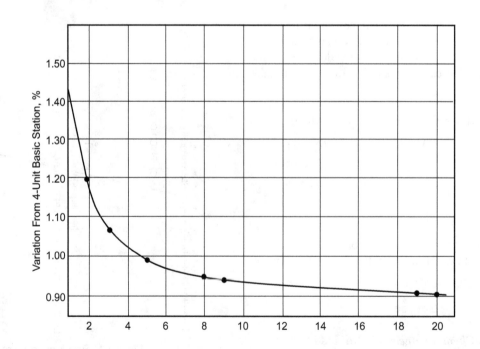

Chart gives unit cost per brake horsepower of centrifugal compressors with various types of prime movers

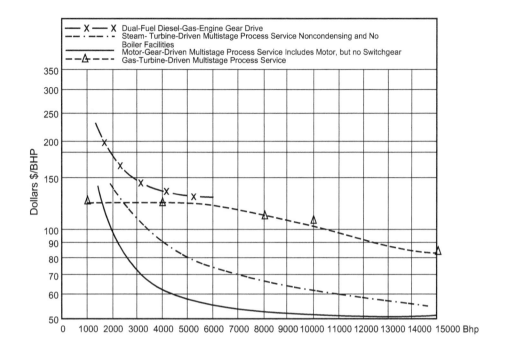

21: Rehabilitation–Risk Evaluation

When does a pipeline need revalidation? The influence of defect growth rates and inspection criteria on an operator's maintenance program

John F. Kiefner and Patrick H. Vieth, Kiefner and Associates, Inc. Worthington, Ohio

Introduction

Buried, high-pressure pipelines are subjected to loads and environmental effects that may cause them to become degraded with the passage of time. All pipeline operators are well aware of this, and the prudent operators have active programs to mitigate deterioration and to remediate defective pipe. The efficiency of an operator's revalidation program can be optimized if the necessary responses can be made when needed, but not until then. The authors have assisted several pipeline operators in establishing timely intervention programs to assure continuing pipeline serviceability. A variety of techniques are used depending on the nature of the pipeline and the perceived problems. Some of the basic techniques are described in this document.

Background

Pipelines may suffer degradation due to a variety of causes, including corrosion, mechanical damage, fatigue, and stress-corrosion cracking. The appropriate remedies for these problems are well known, but other than the routine patrolling of the rights-of-way and monitoring of cathodic-protection potentials and rectifier currents, such remedies are too expensive to be applied on a regular or routine basis. By that we mean that most operators cannot afford to routinely and periodically utilize in-line inspection and/or hydrostatic testing to revalidate their pipelines. Usually, these techniques are invoked when some special circumstances exist. The special circumstances may be the existence of excessive amounts of low pipe-to-soil potential readings, the occurrence of leaks or ruptures, or just an intuitive feeling that it is time to check the condition of a pipeline. Alternatively, as we are finding, more and more operators are coming to depend on more sophisticated models to determine when intervention is needed. The types of models we have used are described.

General concepts

Failure pressure versus defect size. Pipeline integrity is usually defined in terms of pressure-carrying capacity. For a given diameter, wall thickness, and grade of material, one can expect that a sound piece of line pipe will be able to sustain an internal pressure level of at least 100% of its specified minimum yield strength (SMYS). And, that pressure level can readily be calculated by means of the Barlow

formula. For example, if one has a 16-in O.D. by 0.250-in wall thickness X52 pipeline, the pressure level corresponding to 100% of SMYS is 1625 psig. For comparison, the 72% of SMYS maximum operating pressure level is 1170 psig. Thus, one has the expectation that a new piece of this particular pipe has a safety margin against failure of 1625/1170, or 1.39. In fact, if it is free of defects, it will have a failure pressure of at least 1938 psig. The latter is based upon research conducted by the pipeline research Committee of the American Gas Association.[1]

Realistically, not all pipe is defect-free; otherwise there would never be preservice hydrostatic test breaks. But, after a preservice test to a pressure level of 90–100% of SMYS, one can expect a pipeline to perform satisfactorily at an operating stress level up to 72% of SMYS, that is, until or unless it becomes degraded in service by corrosion, mechanical damage, fatigue, or stress-corrosion cracking.

Thanks to extensive research by the previously mentioned Pipeline Research Committee of the American Gas Association (AGA),[1] an experimentally validated model exists for calculating the effects of a longitudinally oriented part-through-the-wall defect on the pressure-carrying capacity of the pipe. For any given piece of pipe, the model can be used to generate the relationships between failure pressure and flaw size. For our example, namely the 16-in O.D. by 0.250-in wall thickness X52 material, the relationships for selected depths of flaws are shown in Figure 1. Each of the nine parallel curves represents the failure-pressure-versus-flaw-length relationship for a particular depth (*d*) to wall thickness (*t*) ratio. The curve that

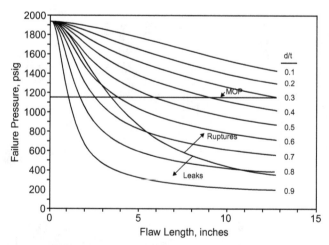

Figure 1. Failure pressure versus flaw size for a material of normal toughness (16 in by 0.250 in, X52, normal toughness).

cuts across the others is the dividing line between leaks and ruptures. Defects with failure at pressures above and to the right of this curve will fail as ruptures, whereas those that have failure pressures below and to the left of this curve will fail as leaks.

A horizontal line is drawn in Figure 1 to represent the maximum operating pressure (MOP) of this pipe material, 1170 psig, corresponding to a hoop stress level of 72% of SMYS. All flaws with length and d/t combinations that lie below this line will fail at pressure levels below the MOP. All flaws that lie above this line could exist in the pipeline in service. For example, the curve representing a d/t of 0.5 crosses the line at a flaw length of 5.8 in. Flaws of that length that are more than halfway through the wall will fail at the MOP; flaws of that length that are less than halfway through will not.

The family of curves shown in Figure 1 is quite useful in the kinds of models we use to determine the time-dependent degradation of a pipeline. Two features of the curves should be kept in mind as the discussion proceeds. First, these curves represent failure pressures that are achievable with steadily increasing pressure over a relatively short period of time (minutes). Because of the phenomenon of time-dependent growth, it is possible to observe failures at pressure levels of 5—10% lower than these curves predict if pressure levels 5—10% below the predicted levels are held long enough.

Secondly, these curves are affected by the ductile-fracture toughness of the material. The curves shown in Figure 1 are characteristic of a garden-variety X52 line-pipe material having a Charpy V-notch upper-shelf energy (full-size specimens) of 25 ft-lb. The set of curves shown in Figure 2 was generated for a hypothetical material of the same pipe geometry and grade but with a Charpy energy of 250 ft-lb (a material of optimum toughness). Note that for a given defect size, the material of optimum toughness has a higher failure pressure. To illustrate this, it is recalled from Figure 1 that a 5.8-in-long flaw halfway through the wall fails at 1170 psig. In contrast, the same 5.8-in-long flaw with a d/t of 0.5 in the very tough material has a failure pressure of 1210 psig. It is noted that blunt flaws such as corrosion-caused metal loss tend to behave as illustrated in Figure 2 rather than Figure 1 regardless of the toughness of the material. This is because the material in a corrosion pit is strained over a large "gauge length", unlike the situation of a crack-like flaw. For the latter, toughness becomes a very important parameter.

Representing the other extreme, Figure 3 presents the failure-pressure-versus-flaw-size relationship for a very low toughness material such as the electric resistance welded (ERW) bond line in an older, low-frequency welded material. Even though the diameter, thickness, and grade are the same as that represented in Figures 1 and 2, the critical flaw sizes are much smaller.

Benefits of hydrostatic testing and in-line inspection

Another useful concept in evaluating pipelines for revalidation arises from the unique failure-pressure-versus-flaw-size relationships of the type described in Figures 1—3, that is, the manner in which those relationships illustrate the benefits of hydrostatic testing and in-line inspection. For illustrative purposes, let us consider the 25 ft-lb material for which Figure 1 gives the failure-pressure-versus-flaw-size relationship. This relationship is replotted in Figure 4, but with additional information. It contains horizontal lines at

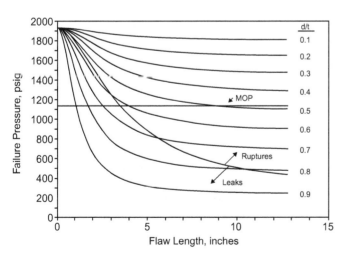

Figure 2. Failure pressure versus flaw size for a material of optimum toughness or for blunt defects (16 in by 0.250 in, X52, optimum toughness or blunt defects).

Figure 3. Failure pressure versus flaw size for a material of low toughness (16 in by 0.250 in, X52, low toughness).

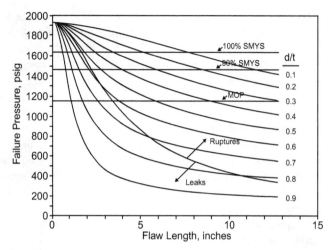

Figure 4. Sizes of flaws located by hydrostatic testing (all types except gouge and dent combinations) (16 in by 0.250 in, X52, normal toughness).

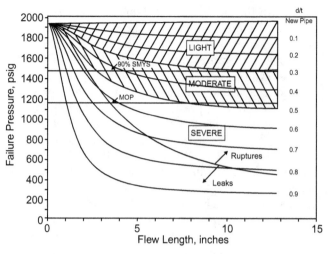

Figure 5. Sizes of flaws located by in-line inspection (corrosion) (16 in by 0.250 in, X52, blunt defects).

1170 psig (maximum operating pressure), 1463 psig (90% of SMYS), and 1625 psig (100% of SMYS). As illustrated previously, no flaws having failure pressures below 1170 psig can exist if the pipeline is operating at 72% of SMYS; otherwise, they would fail (i.e. their particular lengths and d/t values would lie below the horizontal line at 1170 psig). Similarly, if the pressure in the pipeline is raised to 1463 psig (90% of SMYS), any flaws with length and d/t combinations lying above the line at 1170 psig but below that at 1463 psig will fail and be eliminated. If no failures occur, the absence of flaws in this size range is proven. And, if the pressure in the pipeline is raised to 1625 psig (100% of SMYS), the absence of additional even smaller flaws will be proven. Hence, the benefit of a hydrostatic test can be expressed as the ratio of test pressure to operating pressure. The higher the ratio, the more beneficial the test, because the flaws that can remain will be smaller.

With respect to in-line inspection, most pipeliners are aware that the conventional magnetic-flux leakage tools are reasonably capable of sorting metal-loss anomalies into the following categories: depth of light, moderate, and severe. Lights generally are those with depths of less than 30% of the wall thickness ($d/t < 0.3$). Moderates are those with depths of 30–50% through the wall thickness ($d/t = 0.3–0.5$). Severes have depths greater than 50% of the wall thickness ($d/t > 0.5$). Recalling that the behavior of blunt metal-loss flaws is best represented by the optimum toughness relationships of Figure 2, one can see in Figure 5 (based on Figure 2) how in-line inspection data can be interpreted in terms of the effects on pressure-carrying capacity. The top cross-hatched area corresponds to light, the lower cross-hatched area corresponds to moderate, and the unshaded area corresponds to severe. Note that for long

flaws, the failure pressures corresponding to moderates become low enough to intersect the MOP level. This illustrates the need to excavate and examine moderates in a program of conventional in-line inspection. With the use of the more expensive, high-resolution in-line inspection tools, it is possible to define flaw length and, hence, to avoid many excavations.

Figure 5 provides an important conceptual comparison between in-line inspection and hydrostatic testing. With the aid of this figure, one can see where each has its advantages and disadvantages. The hydrostatic test represented by the 90% SMYS line provides a demonstration of the immediate pressure-carrying capacity up to a pressure level of 1463 psig. The use of conventional magnetic-flux tools cannot provide this kind of assurance unless all severe and moderate anomalies are excavated, leaving only the lights. However, the hydrostatic test does not locate any of the anomalies that do not fail. Hence, very deep pits can survive the test and develop leaks shortly thereafter (as illustrated by the amount of unshaded area—severe anomalies—lying above the horizontal line at 1463 psig). At least with in-line inspection, one can locate and remove or repair the anomalies (especially the deep ones). After having removed or repaired all moderates and severes, one can have a high degree of confidence that the pipeline will not fail or leak for a long time as the result of corrosion-caused metal loss. The same cannot be said after a hydrostatic test. Very short but deep pits could have survived the test and may become leaks (not ruptures) within a short time after the test.

On the other hand, because none of the existing in-line inspection tools is capable of finding cracklike flaws, hydrostatic testing is the only means to assure the integrity of a pipeline that may have sustained cracking in the form of fatigue-crack growth or stress-corrosion cracking.

Determining when to revalidate

Having established the concepts illustrated previously, we can now proceed to the determination of when to revalidate a pipeline. This is done by different means depending on the type of anomaly involved. First, let us consider corrosion-caused metal loss.

Corrosion-caused metal loss. A key parameter for scheduling the revalidation of an externally or internally corroded pipeline is the worst-case corrosion rate. A pipeline operator's corrosion control personnel may be able to estimate such a rate. Alternatively, one can obtain a reasonable estimate if prior corrosion leaks have occurred. If leaks have occurred, the decision to revalidate may be obvious without the use of a model. Nevertheless, having some idea of the corrosion rate is useful for scheduling the "next" revalidation. The rate can be estimated by dividing the nominal wall thickness of the pipeline by the number of years between the time of the original installation and the time of the first leak. This involves assuming a constant rate of corrosion over the life of the pipeline. For any pit that is not a leak, the corrosion rate will be less; it will be proportional to the d/t ratio of the pit. For our example pipeline (0.250-in wall thickness), if we assume that it first developed a leak after 25 years of service, its worst-case corrosion rate is 0.25/25 or 0.010 in per year (10 mils per year). If we postulate that there is a pit on this pipeline that is 80% through the wall, the corrosion rate for that pit is $0.8 \times 0.01 = 0.008$ in per year (8 mils per year). With this kind of a corrosion-rate rationale, we can utilize a figure like Figure 2 to plan the revalidation of a corroded pipeline.

The family of curves represented in Figure 2 is reproduced in Figure 6 for the purpose of examining the effects of time-dependent metal loss. In Figure 6, the pipeline when it was installed is represented by the horizontal line at 1938 psig ($d/t = 0$). If this pipeline is losing metal to corrosion, the effect of the metal loss is to lower the failure pressure.[a] In terms of Figure 6, one can see that as metal loss proceeds, the horizontal line originally at 1938 psig would move downward as indicated by the arrows until it would eventually reach the 1170 psig operating level, and a failure would occur. It is important to note that this occurs at a d/t of 0.75 for a 2-in-long flaw (Arrow 1) and at a d/t of 0.47 for a 10-in-long flaw

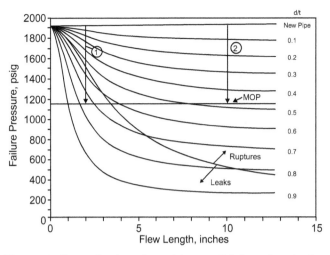

Figure 6. Determination of need for revalidation of a pipeline affected by corrosion (16 in by 0.250 in, X52, blunt flaws).

(Arrow 2). If the company's corrosion specialist has determined that the worst-case corrosion rate is 8 mils per year, then 75% of the wall thickness (250 mils) will be penetrated in less than 23 years, and 47% of the wall thickness will be penetrated in less than 14 years. If the type of metal loss that is occurring is long (10 in or more along the axis of the pipe), this pipeline will require revalidation in less than 14 years. On the other hand, if experience shows that the corrosion is occurring as isolated pits (less than 2 in in axial extent), the time for revalidation would be less than 23 years.

Perhaps a more practical case is that corresponding to a pipeline which has either been hydrostatically tested or has been repaired following in-line inspection. Now the question becomes: When will revalidation be needed again? The question can be answered in terms of Figure 7. Let us consider two alternatives simultaneously. In one case, let us assume that our 16-in O.D. by 0.250-in wall thickness pipeline has been inspected by a conventional magnetic-flux leakage tool and that all anomalies except the lights have been repaired (see the cross-hatched area in Figure 7). In the other case, let us assume that the pipeline has been hydrostatically tested to 1463 psig (90% of SMYS).

With respect to a 2-in-long isolated pit (Arrow 1), the deepest pit remaining after the in-line inspection is represented by Point A ($d/t = 0.3$, pit depth = 75 mils). From the standpoint of the hydrostatic test, the deepest pit that can exist with a 2-in length is represented by Point B ($d/t = 0.6$, pit depth = 150 mils). To cause a leak in service (and it will be a leak rather than a rupture if the length remains less than about 3 in), the pit must grow to Point C ($d/t = 0.76$, pit depth = 190 mils). If the corrosion rate remains at 8 mils per year, the time to grow from A to C is $(190-75)/8 = 14.4$ years, and the time to grow from B to C is $(190-150)/8 = 5$ years.

---

[a] The ASME B31G method for evaluating corroded pipe is based on the same empirical equation that is used to derive the curves of Figure 6. The only difference is that the flaws represented in Figure 6 have been idealized as being rectangular in profile.

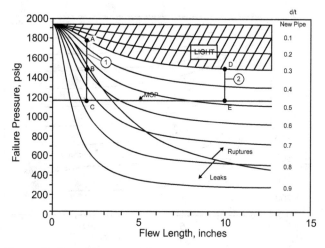

Figure 7. Determination of time to next revalidation after in-line inspection or hydrostatic test (corrosion) (16 in by 0.250 in, X52, blunt flaws).

This shows again the advantage of in-line inspection over hydrostatic testing with respect to dealing with pitting corrosion.

With respect to a long corrosion anomaly (Arrow 2), the deepest possible remaining anomaly is the same whether the present revalidation was by in-line inspection or by hydrostatic test. The anomaly must grow from Point D to Point E to fail in service (and it will fail as a rupture rather than a leak, because it is well above the leak/rupture dividing line). To do so, d/t must change from 0.3 to 0.47. This is a change in wall thickness of 42.5 mils and will take place in 5.3 years at a corrosion rate of 8 mils per year. From this example, it appears that if the corrosion is occurring over long regions as opposed to isolated pits, the advantage of in-line inspection over hydrostatic testing is eroded. This is not quite correct for two reasons. First, the high-resolution tools give reasonably accurate information on the length of the anomaly as well as the depth. Hence, one can focus the repairs on the basis of the remaining pressure-carrying capacity. Second, even with the use of conventional tools, the advantage of locating the corrosion and having the option to examine and repair it provides benefits that do not accrue with hydrostatic testing.

Returning to our analysis, we see that for our example, a revalidation interval of less than 5 years is suggested. An appropriate factor of safety should be applied to this interval, so if this were a real case, the pipeline operator might arbitrarily decide to take action after 2 or 3 years. This probably is an excessively conservative example. Hopefully, with adequate corrosion control, a corrosion rate lower than 8 mils per year can be assured. If so, the revalidation interval would be longer than shown in this example. However, the principles are the same regardless of the corrosion rate.

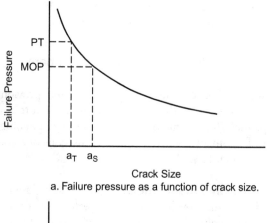

a. Failure pressure as a function of crack size.

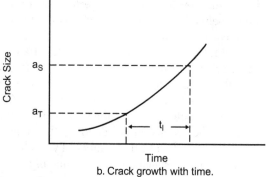

b. Crack growth with time.

Figure 8. Effect of crack growth on pipeline integrity.

Fatigue crack growth

Model for crack growth. As we have seen in terms of Figure 4, if a pipeline is hydrostatically tested to a given pressure, defects of a certain size that are present at the time of the test will fail. Smaller defects would be expected to survive the test but would not be expected to grow further at a static pressure level much below the test pressure level. However, defects of size a_T limited by pressure level PT in Figure 8 could grow during cyclic-pressure service until they reach the size a_s. At that point, failure will occur at the maximum operating pressure (MOP). If a defect as large as a_T remains after a hydrostatic test and if the pipeline is subjected to sufficient pressure fluctuations, the defect will fail after time t_1.

It should be obvious from Figure 8 that flaw enlargement by fatigue-crack growth will erode the effective safety margin provided by a hydrostatic test. This situation may be addressed by analyzing the effects of the service-pressure cycles in terms of the size of the margin between test pressure and operating pressure and assessing when it may be necessary to conduct a subsequent hydrostatic test in the event that flaws that survive the first test become enlarged.

For this purpose, one may use a crack-growth model developed explicitly for pipelines.[2] The model is based upon linear-elastic fracture mechanics. It assumes that crack growth takes place under fluctuating stress levels according to the relationship

$$da/dN = C\Delta K^n \tag{1}$$

where

da/dN is the increment of crack advance per cycle of loading, and

ΔK is the range of the stress intensity factor, which is a function of the change in stress and the initial crack size.

C and n are material properties determined through crack-growth-rate tests. In the use of the model, the initial crack size is fixed by the hydrostatic test. A defect that just barely survives the test is assumed to be the initial crack. The stress range is determined by the maximum and minimum pressure levels. The stress-intensity factor that provides the cracking-driving force is calculated via a Newman-Raju algorithm.[3] The model is capable of handling irregular-size pressure variations. For example, we have analyzed

pressure spectra as complex as that shown in Figure 9. Equation 1 is numerically summed over a number of cycles until the crack becomes large enough to fail in service. The failure criterion for flaw size at failure is that used to generate Figures 1—3.

To demonstrate the importance of a high test-pressure-to-operating-pressure ratio in a situation in which service pressure cycles can cause flaw growth, one can consider an example based on an actual cyclic pressure test. The results of a full-scale pipe experiment involving crack growth are described in Ref. 4 (Figure 47 of Ref. 4) and are shown in Figure 10. In this experiment, a 2-in-long part-through (0.095-in-deep) flaw was placed in the fusion line of the ERW seam of a 12/-3/4-in O.D. by 0.250-in wall thickness X52 material. It was then subjected to constant pressure cycles from 0 to 1020 psig until it grew through the wall and failed (after about 8000 cycles). The actual failure pressure levels are not important, so for purposes of illustration, we can consider the implications of Figure 10 for a material having an ERW seam toughness equivalent to 2 ft-lb of Charpy V-notch impact energy. If this were

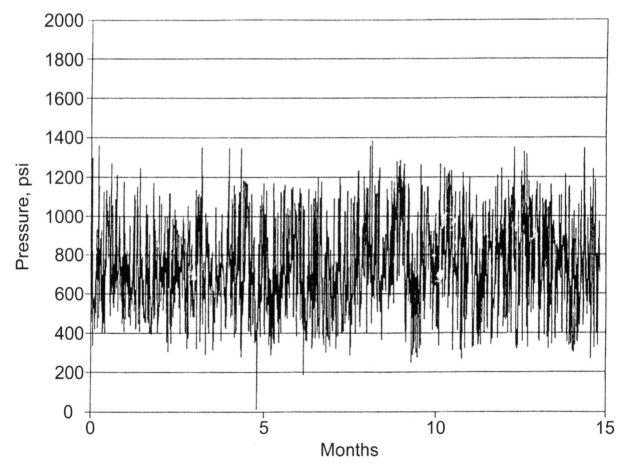

Figure 9. Pressure versus time for a typical petroleum products pipeline.

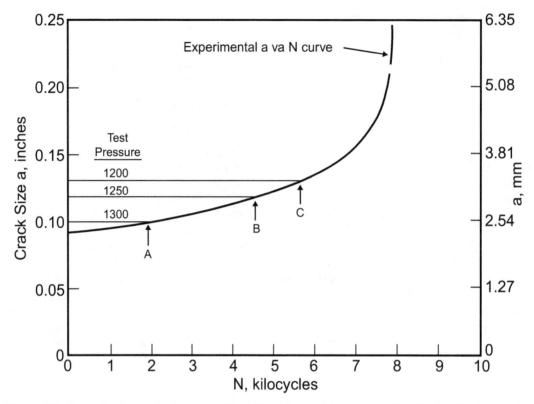

Figure 10. Example of an actual a versus N relationship used to compare the effects of test pressure.

a 2-ft-lb Charpy-energy material, the flaw would have had a failure pressure level of 1300 psig after growing to a depth of 0.1 in as indicated by the 1300 psi "test-pressure" line. Similarly, it would have had failure pressure levels of 1250 and 1200 psig after growing to depths of 0.12 and 0.13 in, respectively. These values were calculated on the basis of the failure-pressure flaw-size relationship of the types shown in Figures 1–3 but for this particular pipe geometry.

Consider the following scenarios based on Figure 10. First, as service cycles are applied and the flaw has grown from 0.095 to 0.1 in (N = 2300 cycles corresponding to Point A), the remaining flaw is exactly the size that would just survive a test to 1300 psig. Since the number of cycles to failure is 8000, the flaw has a remaining life of 5700 cycles. Second, after the flaw has grown to 0.12 in (N = 4800 cycles corresponding to Point B), the remaining flaw is exactly the size that would just survive a test to 1250 psig. It has a remaining life of 3200 cycles. Third, after the flaw has grown to 0.13 in (N = 6000 cycles corresponding to Point C), the remaining flaw is exactly the size that would just survive a test to 1200 psig. It has a remaining life of 2000 cycles. What this shows is the effectiveness of maximizing the test-pressure-to-operating pressure ratio.

In terms of test pressure versus cycles to failure, the results would be as follows:

This exercise illustrates the valuable gain that can be made by small increases in test pressure. For a 4% increase over the base case (1200 psig), the gain in time to failure is 60%. For an 8% increase, the gain is nearly 300%. The above-demonstrated effect of test-pressure-to-operating-pressure ratio on service life provides one of the strongest incentives for high-pressure testing.

Test Pressure (psig)	Cycles to Failure after the Test	Relative Time to Failure
1200	2000	1.0
1250	3200	1.6
1300	5700	2.9

Calculating a retest interval. In the example shown above, let us assume that the pipeline experiences 200 cycles of pressure from 0 to 1020 psig per year. Let us also assume that it was tested to 1200 psig. Such a test probably does not give an adequate margin of safety and, in fact, it does not meet the D.O.T.'s minimum requirement of 1.25 × MOP for a liquid pipeline (Part 195). But it serves as a useful example. As the example shows, the worst-case flaw that might survive the test could be expected to fail in 2000 cycles or 10 years at 200 cycles per year. Prudence dictates that one use a factor of safety as was suggested in our previous discussion on corrosion. The recommended factor of safety is 2, and it is based upon an examination of the "a" versus "N" relationship of Figure 10.

Shown in Figure 11 is a replot of the "a" versus "N" relationship of Figure 10. Early in the life of a crack, the growth is slow. It accelerates, however, as the crack grows deeper, because the stress intensity factor driving the crack is a function of both applied stress and crack size. The implications of this behavior are illustrated as follows. Consider the crack represented by the curve in Figure 11. This crack is initially 0.095 in in depth and will grow to failure in 8000 cycles of the service pressure.

The hydrostatic test and retest level of 1200 psig is represented by the horizontal line at "a" = 0.13. This line is arbitrarily divided into four increments represented by Points 1, 2, 3, 4, and 5. The horizontal line intersects the a versus N relationship at about 5500 cycles. Any time the pipeline is tested after 5500 cycles, the flaw will fail in the test. Suppose one chooses a retest interval equal to the distance from Point 1 to Point 3 (5000 cycles). If the pipeline is tested at Point 3, the flaw will not fail. But it will also not survive until the next test at Point 5 (10,000 cycles). Such a retest interval is too long even though no test failure occurred in the first test.

Now consider what happens if the test interval is cut in half. The retests are then conducted at Points 2, 3, and 4, and the flaw fails during the test at Point 4. Obviously, this is a somewhat contrived example. However, the point is that if the model is reasonably accurate, the times to failure it

predicts should be adjusted by a substantial factor of safety to establish an appropriate retest interval. Because of this, we normally recommend a factor of safety of 2. In our example, this would correspond to a retest interval of only 5 years. For a higher test pressure, say 1300 psig, the interval could be 4700/(2 × 200) = 14 years where the 5700 cycles come from our previous analysis based on Figure 10.

Mechanical damage

Although mechanical damage is frequently inflicted on pipelines and can cause severe degradation of their pressure-carrying capacities, we have not applied any revalidation model to this type of defect for three reasons. First, the state of analysis of mechanical-damage flaws is not as advanced as that for other types of flaws, although both Maxey[5] and British Gas[6] have made attempts to derive algorithms for dents and gouges. Second, the time-dependent nature of mechanical damage is nearly impossible to model. Since its occurrence is often concealed, the "zero" time is seldom known. The rate at which it degrades may depend on creeplike behavior and/or environmental effects. Third, we feel that the best action with respect to mechanical damage is preventative action, patrolling the rights-of-way, participating in one-call systems, conducting public education programs, and providing on-site locating of pipelines at excavation sites. In fact, our primary activity in this respect has been the development of ranking algorithms to calculate relative risk numbers for damage-prevention activities.

It should be mentioned that both in-line inspection and hydrostatic testing are viable remedies for mechanical damage. The problem is that since the existence of concealed damage is difficult to detect short of doing either of these, it is difficult to know when or even if such remedies are needed.

Stress-corrosion cracking

Perhaps the hardest type of defect to deal with is stress-corrosion cracking. With this type of cracking, the main problem is that no reliable model exists to define the crack growth rate. Furthermore, it is suspected that the rate is highly variable with changes in environmental conditions. However, unlike the case for mechanical damage, it is possible to establish a "zero" time based on the date of installation. And if the cracks exist, it is almost certain that they will grow.

A scheme used by several pipeline operators to control stress-corrosion cracking is periodic hydrostatic retesting. In the absence of a crack-growth-rate model, the retest interval has to be set arbitrarily at first and corrected as required by the results of the test. A generalized scheme for setting retest

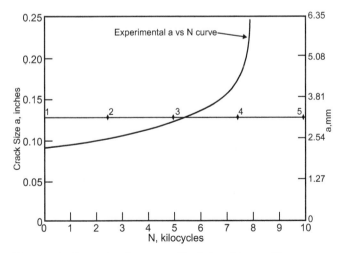

Figure 11. Example of a versus N relationship used to compare the effects of retest interval.

Figure 12. Concept of retesting when defect growth rate is unknown.

intervals to deal with defect growth at unknown rates is illustrated in Figure 12. First, an initial test is performed in which defects may or may not be discovered. But, in either case, the maximum size flaw remaining is fixed by the test pressure, P_f in Figure 12, and the margin of effectiveness is proportional to P_f/P_o, where P_o is the MOP. It is assumed that the defects will grow with the passage of time. Retest 1 is done after some arbitrary period of time short enough to assure that none of the remaining flaws can grow to a size that will fail in service. If this is done correctly, the growing flaws, 1, 2, and 3 in Figure 12, will fail in the test instead of in service at the times suggested by the dashed curves. If no service failures have occurred before Retest 1, then the first interval was adequate (based on the assumption that the crack-growth rate is fairly constant on average). Depending on the outcome of Retest 1, the pipeline operator may decide to increase the interval between retests. If no test failures occur or if they occur only at levels at or near the test pressure, then it makes sense to consider a longer interval. If Retest 2 is then conducted before any flaws grow to failure in service, the new interval is apparently valid and can be lengthened depending again on the failure pressure levels of flaws in Retest 2. In this manner, the pipeline operator is able to control the phenomenon by preventing service failures, and in effect determine the average flaw-growth rates.

It should be emphasized that the scheme portrayed in Figure 12 is valid only if the crack-growth rate is reasonably constant. For example, such a scheme would not be appropriate for fatigue cracks because of the shape of the "a" versus "N" curve as shown in Figures 10 and 11. Where the growth of the flaw accelerates rapidly, it is necessary to determine the

actual growth rate. The setting of retests based upon arbitrarily assumed intervals could easily be defeated by the crack being in a rapid stage of growth, as was illustrated in the discussion of Figure 11.

Summary

We have presented concepts that are frequently used by us and others to attempt to determine when it is time to be concerned about the integrity of a pipeline. The models we use rely on well-established principles and require the use of data that are usually readily available. The models we have discussed can be used to assess revalidation needs for pipelines affected by corrosion, fatigue-crack growth, and stress-corrosion cracking. By means we alluded to but did not discuss, the risks associated with mechanical-damage defects can also be evaluated. While there is never an absolute certainty about the correctness of these types of models, we recommend them to our clients to provide guidance for establishing revalidation policies. We also recommend that they be used in conjunction with sound engineering judgment and generous factors of safety.

References

1. Maxey, W. A., Kiefner, J. E, Eiber, R. J., and Duffy, A. R., "Ductile Fracture Initiation, Propagation, and Arrest in Cylindrical Vessels," Fracture Toughness Proceedings of 1971 National Symposium on Fracture Mechanics, Part II, ASTM STP 514, American Society for Testing and Materials, pp 70–81 (1972).

2. Kiefner, J. F., and Forte, T P., "Model Predicts Hydrostatic Retest Intervals," Oil and Gas Journal (January 7 and January 14, 1985).

3. Newman, J. C., and Raju, I. S., "An Empirical Stress Intensity Factor Equation for Surface Crack," Engineering Fracture Mechanics, Vol. 15, No. 1–2 (1981).

4. Mayfield, M. E., and Maxey, W. A., ERW Weld Zone Characteristics, NG-18 Report No. 130 to American Gas Association, Catalog No. L51427 (June 18, 1982).

5. Maxey, W. A., "Outside Force Defect Behavior," NG-18 Report No. 162, Line Pipe Research Supervisory Committee, Pipeline Research Committee, American Gas Association (August 15, 1986).

6. Hopkins, P., Jones, D. G., and Clyne, A. J., "The Significance of Dents and Defects in Transmission Pipelines," Conference on Pipework Engineering and Operations, I Mech E., London, Paper C376/049 (February 1989).

Modeling for pipeline risk assessment

W. Kent Muhlbauer, PE, WKM Consultancy, LLC

Pipeline risk assessment

The safety of pipelines relative to other modes of transportation is generally not in question: By most measures, moving products by pipeline is safer than by alternate means. Yet even as pipelines claim a superior safety record, catastrophic failures are a real possibility. With an aging infrastructure, ever-encroaching population areas, and increasing economic pressures, the perceived, if not actual, risk of pipeline failure is increasing. Pipeline safety probably can and should be improved.

The ability to predict pipeline failures—where and when the next is to occur—would obviously be a great advantage in reducing risk. Unfortunately, this is not possible except in extreme cases. So, modern risk assessment methodologies provide a surrogate for such predictions. Our risk efforts are *not* attempts to predict how many failures will occur or where the next failure will occur. Rather, the efforts are designed so that we can systematically and objectively capture everything we know and use this information to make better decisions.

Formal versus informal risk management

Although formal pipeline risk management is growing in popularity among pipeline operators, it is important to note that risk management has always been practiced by these pipeline operators. Every time a decision is made to spend resources in a certain way, a risk management decision has been made. This informal approach to risk management has served us well, as evidenced by the very good safety record of pipelines versus other modes of transportation. An informal approach to risk management can have the additional advantages of being simple, being easy to comprehend and to communicate, and being the product of expert engineering consensus built upon solid experience.

However, an informal approach to risk management does not hold up well to close scrutiny, because the process is often poorly documented and not structured to ensure objectivity and consistency of decision making. Expanding public concerns over human safety and environmental protection have contributed significantly to raising the visibility of risk management. Although the pipeline safety record is good, the violent intensity and dramatic consequences of some accidents, an aging pipeline infrastructure, and the continued urbanization of formerly rural areas have increased perceived if not actual risks.

Historical (informal) risk management, therefore, has these pluses and minuses:

Advantages:

- Simple/intuitive
- Consensus is often sought
- Utilizes experience and engineering judgment
- Somewhat successful, based upon pipeline safety record

Reasons to change:

- More at stake from mistakes
- Inefficiencies/subjectivities inherent in informal systems
- Lack of consistency and continuity in a changing workforce unless decision-support systems are in place
- Need to better consider complicated risk factors and their interactions

The beginning of the twenty-first century has witnessed the introduction of more formalized risk programs and studies, as well as an increasing number of private pipeline risk management ventures. Although the general conclusion is typically a recommendation to move the industry toward such formal risk management programs, there is considerably less consensus as to the type and extent of assessment programs to adopt.

Beginning the risk modeling process

Successful risk assessment modeling involves a balance among various issues:

- Identifying an exhaustive list of contributing factors versus choosing the critical few to incorporate in a model (complex versus simple)
- "Hard" data and engineering judgment (how to incorporate widely held beliefs that do not have supporting statistical data)
- Uncertainty versus statistics (how much reliance to place on the predictive power of limited data)
- Flexibility versus situation-specific model (ability to use the same model for a variety of products, geographical locations, facility types, etc.)

It is important that all risk variables be considered, even if only to conclude that certain variables will not be included in the model. In fact, many variables will not be included when such variables do not add significant value but reduce the usability of the model. These decisions are done carefully and with full understanding of the role of the variables in the risk picture.

Note that many simplifying assumptions are often made, especially in complex phenomena such as dispersion modeling, fire and explosion potentials, etc., in order to make the risk model easy to use and still relatively robust.

Both probability variables and consequence variables are examined in most formal risk models. This is consistent with the most widely accepted definition of risk:

$$(\text{event risk}) = (\text{event probability}) \times (\text{event consequence})$$

Several questions to the pipeline operator may better focus a risk management effort and direct the choice of a formal risk assessment technique:

- What data do you have?
- What is your confidence in the predictive value of those data?
- What are the resource demands (and availability) in terms of costs, man hours, and time to set up and maintain a risk model?
- What benefits do you expect to accrue, in terms of cost savings, reduced regulatory burdens, improved public support, and operational efficiency?

Choices in risk-assessment techniques

The goal of any risk-assessment technique is to quantify the risks in either a relative or an absolute sense. The risk-assessment phase is the critical first step in practicing risk management. It is also the most difficult phase. Although we understand engineering concepts about corrosion and fluids flow, predicting failures beyond the laboratory in a complex "real" environment can prove impossible. No one can definitively state where or when an accidental pipeline failure will occur. However, the more likely failure mechanisms, locations, and frequencies can be estimated in order to focus risk efforts.

There is no universally accepted way to assess risks from a pipeline. It is apparent, however, that simply using historical accident counts is not enough. Because of the rare occurrence of pipeline failures, there is a high level of uncertainty associated with pipeline failure prediction and risk assessment. Although this is not always apparent, all risk models, from the simplest to the most complex, make use of probability theory and statistics. In a very simple application, these tools manifest themselves in hard-to-quantify experience factors and engineering judgments, that is, they are the underlying basis of sound judgments. In the more mathematically rigorous models, historical data, statistical theory, and/or mechanistic models (hard data) primarily drive the calculations.

Three general categories of more formal pipeline risk-assessment models can be found in use today:

Simple decision support: Matrix models. One of the simplest risk-assessment approaches is the matrix decision-analysis system. It ranks pipeline risks according to the likelihood and potential consequences of an event by a simple scale, such as high, medium, and low. Each threat to the pipeline is assigned to a cell of the matrix, i.e. the intersection of the perceived likelihood and perceived consequence. Events with both a high likelihood and a high consequence appear higher on the resulting prioritized list. This approach may be as simple as using an expert's opinion or as complicated as using quantitative data to rank risks.

The rigorous approach: Probabilistic/mechanistic models. The more rigorous and complex risk assessment is often called a Probabilistic Risk Assessment (PRA). It usually refers to a technique employed in the nuclear and aerospace industries that generally uses event trees and fault trees to model every aspect of a system. Initiating events are flow-charted forward to all possible concluding events, with probabilities being assigned to each branch along the way. Failures are backward flowcharted to all possible initiating events, again with probabilities assigned to all branches. All possible paths can then be quantified based upon the branch probabilities along the way. This technique usually relies heavily on historical failure rate data. It yields absolute risk assessments for all possible failure events. These more elaborate models are more costly than simpler risk assessments. They are technologically more demanding to develop and integrate via computer applications, they require trained operators, and they benefit from extensive databases.

The hybrid approach: Indexing models. Perhaps the most popular pipeline risk-assessment technique in current use is the "index model". In this approach, a relative weight is assigned to every important condition and activity on the pipeline. This includes both risk-reducing and risk-increasing items. This relative weight reflects the importance of the item in the risk assessment and is based upon statistics where available and upon engineering judgment where data are not available. The risk component scores are subsequently summed for each pipeline section to obtain a relative risk ranking of all pipe sections. The various pipe segments may then be ranked according to their relative risk scores in order to prioritize repairs and inspections. Among pipeline operators today, this technique is widely used and ranges from a very simple 5–20 factor model (where only factors such as leak history and population density are considered) to models with hundreds of factors considering virtually every item that impacts risk. When an indexing model is created from a probabilistic approach—using scenarios, event/fault trees, and all available historical data—this approach is often a solution with the best cost–benefit ratio.

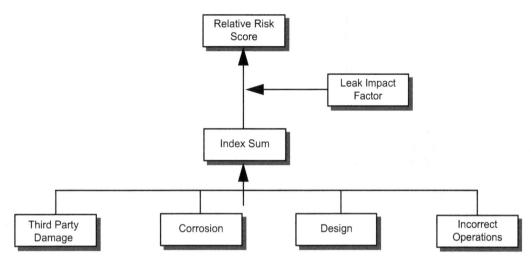

Typical Pipeline Risk Model Structure

Other issues in risk modeling

In comparing risk-assessment approaches, some issues arise that often lead to confusion. The following section discusses some of those issues:

Absolute versus relative risk. Absolute risk implies a universal scale, usually in terms of a measurable consequence such as fatalities or costs per time period by which all risks can be compared. An example would be a risk assessment that produces a value such as 1.25 E-08 fatalities per thousand mile-years of pipeline. Relative risk is designed to prioritize risks by saying that pipeline section B is riskier than section A by an amount equal to X. The former is a predictive measure yielding the probability and consequence of failure, whereas the latter is a comparative tool yielding a ranking of current risks. A criticism of the relative scale is its inability to compare risks of several dissimilar systems (e.g. pipelines versus trucks) and its inability to directly provide failure predictions. The absolute scale fails in that it relies heavily on point estimates, particularly for rare events that are extremely difficult to quantify, and in the unwieldy numbers that often generate a negative reaction from the public.

For purposes of communications and in practical applications, this is not really an important issue. Either scale can be readily converted to the other scale if the circumstances so warrant. A relative risk scale can be converted to an absolute scale by correlating the relative risk scores with known failure rates and ensuring that the relative scales appropriately reflect real-world processes. The absolute scale is converted to more manageable and understandable (nontechnical) relative scales by simple mathematical relationships.

A danger underlying this issue is the possible assumption that a complex number expressed in scientific notation is more accurate than a simple number. In reality, either method should use the same available data pool and be forced to make the same number of assumptions when data are not available.

Quantitative versus qualitative models. These terms are often used to distinguish the amount of historical failure-related data analyzed in the model and the amount of mathematical calculations employed in arriving at a risk answer. A model that exclusively uses hard data is sometimes referred to as quantitative, whereas a model employing relative scales, even if numbers are assigned later, is referred to as qualitative or semiquantitative. The danger in such labels is that they imply a level of accuracy. In reality, the labels often tell more about the level of modeling effort, cost, and data sources. All complete risk models must use both quantitative and qualitative information.

Subjectivity versus objectivity. In theory, a purely objective model will strictly adhere to statistical information and will have no opinion data. A purely subjective model implies complete reliance upon expert opinion. In practice, no pipeline risk model fully adheres to either. Objectivity cannot be purely maintained when dealing with the real-world situation of missing data and variables that are highly confounded. On the other hand, subjective models certainly use "hard" data to form or support judgments.

The use of unquantifiable evidence. In a difficult-to-quantify item, some would argue that nonstatistical analyses are potentially damaging. Although there is always this danger of misunderstanding the role of a factor in a risk assessment, there is similarly the more immediate danger of an incomplete analysis by omission of a factor. For example, public education is seen by most pipeline professionals to be a very important aspect of reducing the number of third-party damages, leak reporting, and emergency response. Quantifying this level of importance and correlating with the many varied approaches to public education is quite difficult. A concerted effort to study these data is needed to best assess the effect on risk, but in the absence of such a study, most would agree that a company that practices public education will be more likely to achieve some level of risk reduction than a company that does not.

Uncertainty. False precision is often assigned to analysis yielding a specific number, especially when such a number appears to be the result of extensive calculations. In reality, uncertainty always exists. Uncertainty can be thought of as a property of the situation. We add to that uncertainty when we measure the situation. The measurement process itself adds uncertainty as measurement errors and inaccuracies accumulate. Separating the noise (the random variations caused in part by measurement uncertainties) from the signal (the true differences in risk) is an arduous task. At best, we hope for a high signal-to-noise ratio. The higher the ratio, the higher our confidence in identifying true signals. All risk-assessment techniques must address uncertainty, and no method can reliably eliminate it or even reduce it significantly.

Choosing a risk-assessment technique

Any or all of the previously described techniques might have a place in risk assessment/management. The choice may well depend on the situation. Understanding the strengths and weaknesses of the different risk assessment methodologies gives the user the basis for choosing one. A case can be made for using each in certain situations. For example, a simple matrix approach crystallizes thinking and is a step above informal risk assessment. However, if the need is to evaluate specific events at any point in time, a narrow focus assessment with well-known event probabilities, then probabilistic risk analysis might be the tool of choice. If the need is to weigh immediate risk tradeoffs or perform inexpensive overall assessments, indexing models might be the best choice. The following shows some tradeoffs associated with each choice:

Pros and cons I (PRA)

+   Rigorous, scientific approach
+   Accepted in other industries
+   Uses all available information

−   Costly
−   Difficult to do resource allocation
−   Difficult to do overall risk management
−   Might create the "illusion of knowledge"
−   Intimidating to nontechnical audience

Pros and cons II (indexing)

+   Uses all available information
+   Intuitive
+   Flexible
−   Possibly more subjective
−   Must be well documented
−   Often requires a subsequent linkage to absolute values

Pros and cons III (matrix)

+   Improvement over informal techniques
+   Forces more logical examination of situation
+   Inexpensive approach
−   Limited number of risk factors (not comprehensive)
−   Subjective

In addition to the three general categories of risk models, supporting tools such as fault trees, event trees, hazard and operability analysis (HAZOPS), what-if scenarios, and others play a role in performing risk assessment. These tools are almost always the foundation and building blocks of the complete risk assessment model.

The following are some ways a particular technique can support a specific need:

Application choices I (PRA, scenarios, event/fault trees)

- Study specific events and/or create risk models
- Perform postincident investigations
- Compare risks of specific failures
- Calculate specific event probabilities

Application choices II (indexing model built from scenarios, event/fault trees, PRA)

- Obtain an inexpensive overall risk model
- Create a resource allocation model
- Model the interaction of many potential failure mechanisms
- Study or create an operating discipline

Application choices III (matrix)

- Better quantify a belief
- Create a simple decision support tool
- Combine several beliefs into a single solution
- Document choices in resource allocation

The ideal risk-assessment model

Regardless of the technique used, some properties of the ideal risk-assessment tool can be identified. A partial list of the characteristics of the "ideal model" includes the following:

Costs. The value or benefits derived from the risk-assessment process should clearly outweigh the costs of setting up, implementing, and maintaining the program.

Learning ability. Because risk is not constant over the length of a pipeline or over a period of time, the model must be able to "learn" as information changes. This means that new data should be easy to incorporate into the model.

Signal-to-noise ratio. Because the model is in effect a measurement tool, it must have a suitable signal-to-noise ratio. This means that the "noise", i.e. the amount of uncertainty in the measurement (due to numerous causes), must be low enough that the "signal", i.e. the risk value of interest, can be read. This is similar to the "accuracy" of the model but involves additional considerations that surround the high level of uncertainty associated with risk management.

In examining a proposed risk-assessment effort, it may be wise to evaluate the risk-assessment model to ensure the following:

- All failure modes are considered
- The model appropriately treats both time-dependent and -independent failure mechanisms
- All risk elements are considered and the most critical are included
- Failure modes are considered, independently as well as in aggregate
- All available information is being appropriately utilized
- Real-world processes are appropriately represented
- Relative risk values are anchored by real failure rates and consequence potential
- Provisions exist for regular updates of information, including new types of data
- Consequence factors are separable from probability factors
- Weightings, or other methods to recognize the relative importance of factors, are established
- The rationale behind weightings is well documented and consistent
- A sensitivity analysis has been performed
- The model reacts appropriately to failures of any type
- Risk elements are combined appropriately ("and" versus "or" combinations)
- Steps are taken to ensure consistency of evaluation
- Risk-assessment results form a reasonable statistical distribution (outliers?)

- There is adequate discrimination in the measured results (signal-to-noise ratio)
- Comparisons can be made against fixed or floating "standards" or benchmarks

Finally, a view to the next step, risk management, should be taken. A good risk-assessment technique will allow a smooth transition into the management of the observed risks. This means that provisions for resource allocation modeling and the evolution of the overall risk model must be made.

Managing risks: The cost connection

In some sense, we have near-complete control of the risk. We can spend nothing on preventing accidents, or we can spend enormous sums of money overdesigning facilities, employing an army of inspectors and routinely shutting down lines for preventive maintenance and replacement. Pragmatically, operators spending too little on preventing accidents will be out of business from regulatory intervention or from the cost of accidents. On the other hand, if an operator spends too much on accident prevention, he can be driven out of business by the competition, even if, perhaps, that competition has more accidents.

Risk management, to a large extent, revolves around the central process of making choices in the design and day-to-day operations of a pipeline system. Many of these choices are mandated by regulations, whereas others are economically (budget) constrained. By assigning a cost to pipeline accidents (a sometimes difficult and controversial thing to do) and including this in the cost of operations, the optimum balance point is the lowest cost of operations.

No operator will ever have all of the relevant information needed to guarantee safe operations. There will always be an element of the unknown. Managers must control the "right" risks with limited resources, as there will always be limits on the amount of time, manpower, or money to apply to a risk situation. Managers must weigh their decisions carefully in light of what is known and unknown. It is, perhaps, this assimilation of complex data and the subsequent integration of competing risk reduction-profit goals that lies at the heart of the debate about how best to manage pipeline risks.

Conclusion

The move to more formal risk techniques is intended to increase operational consistency and credibility, especially when such techniques are offered for public viewing. In a concerted effort to demonstrate its dedication to public and environmental safety while simultaneously seeking to reduce potential liabilities and regulatory requirements, the federal government and the pipeline industry (both gas and liquid) are embracing risk management techniques through integrity management programs.

Even without such broad-based support, risk management has long been recognized as a valuable effort in pipeline operations. Risk-assessment approaches are available to serve many needs, although no single approach is the best at addressing all needs. It is important that those practicing (or planning to practice) risk management understand the strengths and limitations of various approaches and set their expectations accordingly.

References

1. W. Kent Muhlbauer, Pipeline Risk Management Manual, 3rd ed., Gulf Publishing Co., Houston (2004).

Sample risk variable list

Relative Risk = [Index Sum]/[Leak Impact Factor]

Index Sum = [Third Party] + [Corrosion] + [Design]

+ [Incorrect Operation]

Third party damage potential

A	Minimum depth cover	20%	0–20 pts
B	Activity level	20%	0–20 pts
C	Above-ground facilities	10%	0–10 pts
D	One-call system	15%	0–15 pts
E	Public education	15%	0–15 pts
F	Right-of-way condition	5%	0–5 pts
G	Patrol	15%	0–15 pts
		100%	100 pts

Corrosion potential

A	**Atmospheric corrosion**	**0–10 pts**
A1	Atmospheric corrosion	0–5 pts
A2	Atmospheric type	0–2 pts
A3	Atmospheric coating	0–3 pts

B	**Internal corrosion**	**0–20 pts**
B1	Product corrosivity	0–10 pts
B2	Internal protection	0–10 pts

C	**Subsurface corrosion**		**0–70 pts**
C1	Subsurface environment		0–20 pts
	Soil corrosivity	0–15 pts	
	Mechanical corrosion	0–5 pts	
C2	Cathodic protection		0–25 pts
	Effectiveness	0–15 pts	
	Interference potential	0–10 pts	
C3	Coating		0–25 pts
	Fitness	0–10 pts	
	Condition	0–15 pts	
			70 pts

Design risk

A	Safety factor	35%	0–35 pts
B	Fatigue	15%	0–15 pts
C	Surge potential	10%	0–10 pts
D	Integrity verification	25%	0–25 pts
E	Land movements	15%	0–15 pts
		100%	0–100 pts

Incorrect operations

A	Design	30%	0—30 pts
	Hazard identification		0—4 pts
	MOP potential		0—12 pts
	Safety systems		0—10 pts
	Material selection		0—2 pts
	Checks		0—2 pts
B	Construction	20%	0—20 pts
	Inspection		0—10 pts
	Materials		0—2 pts
	Joining		0—2 pts
	Backfill		0—2 pts
	Handling		0—2 pts
	Coating		0—2 pts
C	Operation	35%	0—35 pts
	Procedures		0—7 pts
	SCADA/communications		0—3 pts
	Drug testing		0—2 pts
	Safety programs		0—2 pts
	Surveys/maps/records		0—5 pts
	Training		0—10 pts
	Mechanical error preventers		0—6 pts
D	Maintenance	15%	0—15 pts
	Documentation		0—2 pts
	Schedule		0—3 pts
	Procedures		0—10 pts
		100%	0—100 pts

Leak Impact Factor = (Product Hazard) × (Receptors) × (Spill Size) × (Dispersion)

A Product hazard	
1. (Acute + chronic hazards) *Acute hazards*	
a. Nf	0—4
b. Nr	0—4
c. Nh	0—4
Total (Nh + Nr + Nf)	0—12
2. *Chronic hazard, RQ*	0—1
B Receptors	0—1
Population	
Environment	
High value areas	
C Spill size	0—1
D Dispersion	0—1

Pipeline Risk Assessment

W. Kent Muhlbauer

Risk assessment on any facility is most efficiently done by first dividing the facility into components with unchanging risk characteristics. For a cross-country pipeline, this involves collecting data on all portions of the pipeline and its surroundings and then using these data to "dynamically segment" the pipeline into segments of varying length. Risk algorithms are applied to each of the segments, producing risk estimates that truly reflect changing risks along the pipeline.

The risk estimating algorithms are conceptually very straightforward. However, as with any assessment of a complex mechanical system installed in a varying, natural environment, there are many details to consider. This is illustrated by an example of risk assessment on a hypothetical pipeline.

Varying levels of analytic rigor are available to risk assessors. For example, a resistance estimate might be modeled as simply being related to stress level and pipe characteristics or, for more robust analyses, could include sophisticated finite-element analysis. In this example, a certain amount of detail is omitted in order to better demonstrate the higher-level principles.

To illustrate key concepts, one time-independent failure mechanism (third party damage) and one time-dependent failure mechanism (external corrosion) are assessed. All other failure mechanisms will follow one of these two forms. Estimates from all failure mechanisms can be combined in various ways to meet the needs of the subsequent risk management processes.

Example:

A 120-mile pipeline is to have risk assessment performed. For the assessment, failure is defined as loss of integrity leading to loss of pipeline product. Consequences are measured as potential harm to public health, property, and the environment and are expressed in units of dollars loss—i.e. all consequences are monitized.

Verifiable measurement units for the assessment are as follows:

Measurement	Units
Risk	$/year
Probability of failure (PoF)	Failures/mile-year
Consequence of failure (CoF)	$/failure
Time to failure (TTF)	Years
Exposure	Events/mile-year
Mitigation	%
Resistance	%

Data are collected and include Subject Matter Expert (SME) estimates where actual data are unavailable. The integrated data show changes in risk along the pipeline route—6530 segments are created by the changing data with an average length of 87 ft. This relatively short average length shows that a risk profile with adequate discrimination has been generated.

A level of conservatism is defined as P90 for all inputs that are not based on actual measurements. This is conservative—a bias toward overestimation of actual risks. P90 means that risk is underestimated once out of every 10 inputs, i.e. there will be a negative surprise only 10% of the time. The risk assessors have chosen this level of conservatism to account for plausible (albeit extreme) conditions and to ensure that risks are not underestimated.

For assessing the PoF from time-independent failure mechanisms—those that do not worsen over time, such as third party damage and human error—the summary equation is as follows.

$$PoF\_time\text{-}independent = exposure \times (1 - mitigation)$$
$$\times (1 - resistance).$$

As an example for applying this to the PoF due to time-independent third-party damage, the following inputs are identified (by SME's) for a certain portion of the subject pipeline.

- Exposure (unmitigated "attack") is estimated to be three (3) third-party damage events per mile-year. This means that over this mile of pipeline, excavators will be operating three times per year and, in the absence of mitigation, will cause damage to the pipeline three times per year
- Using a mitigation (defense) effectiveness analysis, SME's estimate that 1 in 50 of these exposures will not be successfully prevented by existing mitigation measures. This results in an overall mitigation effectiveness estimate of 98% mitigated.
- SME's perform a resistance analysis to estimate that, of the exposures that are not mitigated, one in four will cause immediate failure, not just damage. This estimate includes the possible presence of weaknesses due to threat interaction and/or manufacturing and construction issues. So, the pipeline in this area is judged to have a 75% resistance to failure (survivability) from this mechanism, given the failure of mitigations.

Assuming that frequencies and probabilities are practically interchangeable, these inputs result in the following assessment:

PoF_third-party damage = (3 damage events per mile-year)

$$\times \ (1-98\% \ \text{mitigated})$$
$$\times \ (1-75\% \ \text{resistive})$$
$$= 1.5\%(0.015)\text{per mile-year}$$
(a failure every 67 years along this mile of pipeline).

Note that a useful intermediate calculation, i.e. "probability of damage" (but not failure), emerges from this assessment and can be verified by future inspections.

(3 damage events per mile-year) × (1−98% mitigated)

$$= 0.06 \ \text{damage events/mile-year(damage occurring about}$$
once every 17 years).

This same approach is used for other time-independent failure mechanisms and for all portions of the pipeline.

In assessing the PoF due to time-dependent failure mechanisms, e.g. corrosion and cracking, the previous algorithms are slightly modified:

PoF_time-dependent = f(Time-to-Failure, TTF),
TTF = resistance/[exposure × (1 − mitigation)].

To continue the example, SME's have determined that, at certain locations along the 120-mile pipeline, soil corrosivity leads to 5 mpy external corrosion exposure (if left unmitigated). Analyses of coating and CP effectiveness leads SME's to assign a mitigation effectiveness of 90%.

Recent inspections, adjusted for uncertainty and considering possible era-of-manufacture weaknesses, result in an effective pipe wall thickness estimate of 0.220 in (remaining resistance). Use of these inputs in the PoF assessment for the next year is shown below:

$$\text{TTF} = [220 \ \text{mils/5 mpy} \times (1-90\%)] = 440 \ \text{years},$$
$$\text{PoF} = 1/\text{TTF} = [5 \ \text{mpy} \times (1-90\%)]/220 \ \text{mils}$$
$$= 0.11\% \ \text{PoF}.$$

So, the combined PoF from these two threats is estimated to be 0.015 + 0.0011 = 0.016 failures/mile-year. This 1.6% failure probability can now be used with estimates of consequence potential to arrive at overall risk estimates generated by these two threats.

SME's have analyzed potential scenarios and determined the range of possible consequences generated by a failure. After assignment of probabilities to each scenario, a point estimate representing the distribution of all future scenarios yields the value of $18,500 per failure. This can be thought of as a probability-adjusted "average" consequence per failure.

Risk assessors similarly calculate all risk elements for each of the 6530 segments. To estimate the PoF for any portion of the 120-mile pipeline, a probabilistic summation is used to ensure that length effects and the probabilistic nature of the estimates are appropriately considered. To estimate total risk, an expected loss calculation for the full 120 miles yields $252,090 risk exposure from the pipeline per year of operation. The average is $210/mile-year.

Risk Management

The risk estimates generated in this way are extremely useful to decision makers. Such estimates can become part of the budget setting and valuation processes. In this example, the company first uses these values to compare to, among other benchmarks, a U.S. national average for similar pipelines $350/mile-year. The comparison needs to consider the P90 level of conservatism employed. Often, a P90 or higher level of conservatism is appropriate for determining risk management on specific pipeline segments, but it will not compare favorably to historical incident data since those generally reflect P50 estimates.

Understanding how each pipeline segment contributes to the overall risk sets the stage for efficient risk management.

For risk management at specific locations, cost/benefits of various risk mitigation measures can be compared by running "what if" scenarios using the same equations with anticipated mitigation effectiveness arising from the proposed actions(s).

These estimates can also be used to establish "safe enough" limits by following predetermined risk acceptability criteria such as those proposed in CSA Z662 Annex 0.

Used with permission, W. Kent Muhlbauer—Austin, Texas.

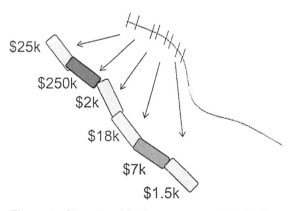

Figure 1. Changing risk along a segmented pipeline.

22: Conversion Factors

Units of measurement convert from one system to another

To Convert	To	Multiply by
A		
abampere	ampere	1.00 E+01
abcoulomb	Statcoulombs	2.998 E+10
acceleration, standard	m per sec squared	9.80556
acceleration, standard	ft/sec/sec (ft/s²)	32
acre	Sq. chain (Gunters)	10
acre	sq. rods	160
acre	Square links (Gunters)	1 E+05
acre	Hectare or sq. hectometer	0.4047
acres	sq feet	43,560.00
acres	sq meters	4,046.87
acres	sq miles	1.562 E-03
acres	sq yards	4,840.00
acre-feet	cu feet	43,560.00
acre-feet	gallons	3.259 E+05
acre-feet	cu. meters	1.233489 E+03
amperes/sq cm	amps/sq in.	6.452
amperes/sq cm	amps/sq meter	1 E-04
amperes/sq in.	amps/sq cm	0.155
amperes/sq in.	amps/sq meter	1,550.00
amperes/sq meter	amps/sq cm	E-04
amperes/sq meter	amps/sq in.	6.452 E-04
ampere-hours	coulombs	3,600.00
ampere-hours	faradays	0.03731
ampere-turns	gilberts	1.257
ampere-turns/cm	amp-turns/in.	2.54
ampere-turns/cm	amp-turns/meter	100
ampere-turns/cm	gilberts/cms	1.257
ampere-turns/in.	amp-turns/cm	0.3937
ampere-turns/in.	amp-turtis/meter	39.37
ampere-turns/in.	gilberts/cms	0.495
ampere-turns/ meter	amp/turns/cm	0.01
ampere-turns/ meter	amp-turns/in.	0.0254
ampere-turns/ meter	gilberts/cms	0.01257
Angstrom unit	Inch	3937 E-09
Angstrom unit	Meter	1 E-10
Angstrom unit	Micron or (Mu)	1 E-04
Are	Acre (US)	0.02471
Ares	sq. yards	119.6
ares	acres	0.02471

To Convert	To	Multiply by
ares	sq meters	100
Astronomical Unit	Kilometers	1.495 E+08
atmospheres	Ton/sq.inch	0.007348
atmospheres	bar	1.01325
atmospheres	cms of mercury	76
atmospheres	ft of water (at 4°C)	33.9
atmospheres	in. of mercury (at 0°C)	29.92
atmospheres	kgs/sq cm	1.0333
atmospheres	kgs/sq meter	10,332.00
atmospheres	kiloPascal	101.33
atmospheres	pounds/sq in.	14.7
atmospheres	lbs/square foot	2116.2
atmospheres	tons/sq ft	2.058
atmospheres	kPa	101.33
B		
Barrels (U.S., dry)	cu.inches	7056
Barrels (U.S., drY)	quarts (dry)	105
Barrels (U.S., liquid)	gallons	31.5
barrels (oil)	gallons (oil)	42
barrels	cu. inches	9702
barrels	cu. Meters	0.15899
barrels/hr	cu. meters/hr	1.589873 E-01
barrels/hr	gallons/min	1.4286
barrels/day	cu. meters/day	1.589873 E-01
barrels/day	cu. meters/hr	6.624471 E-03
barrels/day	cu. meters/year	5.803036 E+01
barrels/day	gallons/min	0.02917
barrels/ft	cu. meters/meter	5.216119 E-01
barrels/mi	cu. meters/km	9.879013 E-02
barrels/US ton	cu. Meters/ton	0.17525
bars	atmospheres	0.9869
bars	dynes/sq cm	1 E+06
bars	kgs/sq meter	1.020 E+04
bars	kiloPascal	100
bars	pounds/sq ft	2,089.00
bars	pounds/sq in.	14.5
Baryl	Dyne/sq. cm	1
board feet	cu. inches	144
board feet	cu. Meters	0.00236
Bolt (US Cloth)	Meters	36.576
Btu	Liter-Atmosphere	10.409
Btu	ergs	1.0550 E+10

To Convert	To	Multiply by
Btu	foot-lbs	778.3
Btu	gram-calories	252
Btu	horsepower-hrs	3.931 E-04
Btu	joules	1,054.80
Btu	kilograin-calories	0.252
Btu	kilograin-meters	107.5
Btu	kilowatt-hrs	2.928 E-04
Btu/cu. ft.	megajoules/cu. meter	3.725895 E-02
Btu/hr	foot-pounds/sec	0.2162
Btu / hr	gram-cal/sec	0.07
Btu/hr	horsepower-hrs	3.929 E-04
Btu/hr	watts	0.2931
Btu/hr-foot 2 oF/foot	Watt/meter-Kelvin	1.73074
Btu/lb	Joules/gram	2.326
Btu/lb	watt hour/kilogram	0.64611
Btu/lb mol	Joules/mol	2.326
Btu/lb	megajoules/kg	2.326000 E-03
Btu/min	foot-lbs/sec	12.96
Btu/min	ft-lb/min	778.2
Btu/min	gram-calorie/sec	4.203
Btu/min	gram/sec	4.203
Btu/min	metric hp	0.0239
Btu/min	ft-lb/sec	778.,2
Btu/min	horsepower	0.02356
Btu/min	kilowatts	0.01757
Btu/min	watts	17.57
Btu/sq ft/min	watts/sq in.	0.1221
Btu/UK gal	megajoules/cu. meter	2.320800 E-01
Btu/US gal	mega joules/cu. meter	2.787163 E+04
Bucket (Br. dry)	Cubic Cm.	1.818 E+04
bushels	cu ft	1.2445
bushels	cu in.	2,150.40
bushels (UK	cu meters	0.36368720
bushels (US)	cu meters	0.03523907
bushels	liters	35.24
bushels	pecks	4
bushels	pints (dry)	64
bushels	quarts (dry)	32
	C	
cable	mile (nautical)	0.1
cable	meters	185.3
caliber	inches	0.01
caliber	millimeters	0.254
calories	Joules	4.1868
calories/gram	absolute Joule/ gram	4.184

To Convert	To	Multiply by
calories/gram	Int. Joule/gram	4.1833
calories/gram	Int. cal/gram	0.99935
calories/gram	Btu/lb	1.7988
calorie (int) per hour	watt	1.163+E-03
Calories/pound	Joules/kilogram	9.22414
Calories, gram (mean)	B.T.U. (mean)	3.9685 E-03
Candle/sq- cm	Lamberts	3.1420
Candle/sq. inch	Lamberts	0.487
centares (centiares)	sq meters	1
Centigrade	Fahrenheit	(oCx9/5)+32
centigrams	grams	0.01
Centiliter	Ounce fluid (US)	0.3382
Centiliter	Cubic inch	0.6103
Centiliter	drams	2,705
centiliters	liters	0.01
centimeters	feet	3.281 E-02
centimeters	inches	0.3937
centimeters	kilometers	1 E-05
centimeters	meters	0.01
centimeters	miles	6.214 E-06
centimeters	millimeters	10
centimeters	mils	393.7
centimeters	yards	1.094 E-02
centimeter-dynes	cm-grams	1.020 E-03
centimeter-dynes	meter-kgs	1.020 E-08
centimeter-dynes	pound-feet	7.376 E-08
centimeter-grams	cm-dynes	980.7
centimeter-grams	meter-kgs	1 E-05
centimeter-grams	pound-feet	7.233 E-05
centimeters of mercury	atmospheres	0.01316
centimeters of mercury	feet of water	0.4461
centimeters of mercury	kgs/sq meter	100
centimeters of mercury	pascal	1333.2239
centimeters of mercury	pounds/sq ft	27.85
centimeters of mercury	pounds/sq in.	0.1934
centimeters of water (4°C)	pascal	98.0638
centeimers of water (conventional)	pascal	98.0665
centimeters/sec	feet/min	1.1969
centimeters/sec	feet/sec	0.03281

(*continued on next page*)

To Convert	To	Multiply by
centimeters/sec	kilometers/hr	0.036
centimeters/sec	knots	0.1943
centimeters/sec	meters/min	0.6
centimeters/sec	miles/hr	0.02237
centimeters/sec	miles/min	3.728 E-04
centimeters/sec/sec	feet/sec/sec	0.03281
centimeters/sec/sec	kms/hr/sec	0.036
centimeters/sec/sec	meters/sec/sec	0.01
centimeters/sec/sec	miles/hr/sec	0.02237
centipoise	newton-second/meter^2	0.001
centipoise	pascal-second	0.001
centistoke	mm^2/second	1
Chain	Inches	792
Chain	meters	20.12
Chains (surveyors'or Gunter's)	yards	22
circular mils	sq cms	5.067 E-06
circular mils	sq mils	0.7854
Circumference	Radians	6.283
circular mils	sqinches	7.854 E-07
Cords	cord feet	8
Cord feet	cu. feet	16
Coulomb	Statcoulombs	2.998 E+09
coulombs	faradays	1.036 E-05
coulombs/sq cm	coulombs/sq in.	64.52
coulombs/sq cm	coulombs/sq meter	1 E+04
coulombs/sq in.	coulombs/sq cm	0.155
coulombs/sq in.	coulombs/sq meter	1,550.00
coulombs/sq meter	coulombs/sq cm	1 E-04
coulombs/sq meter	coulombs/sq in.	6.452 E-04
cubic centimeters	cu feet	3.531 E-05
cubic centimeters	cu inches	0.06102
cubic centimeters	cu meters	1 E-06
cubic centimeters	cu yards	1.308 E-06
cubic centimeters	gallons (U. S. liq.)	2.642 E-04
cubic centimeters	liters	0.001
cubic centimeters	pints (U.S. liq.)	2.113 E-03
cubic centimeters	quarts (U.S. liq.)	1.057 E-03
cubic feet	bushels (dry)	0.8036
cubic feet	cu cms	28,320.00
cubic feet	cu inches	1,728.00
cubic feet	cu meters	0.02832
cubic feet	cu yards	0.03704
cubic feet	gallons (U.S. liq.)	7.48052
cubic feet	gallons (Imperial)	6.229
cubic feet	liters	28.32
cubic feet	pints (U.S. liq.)	59.84
cubic feet	quarts (U.S. liq.)	29.92

To Convert	To	Multiply by
cubic feet/min	cu cms/sec	472
cubic feet/min	gallons/sec	0.1247
cubic feet/min	liters/sec	0.4719474
cubic feet/min	pounds of water/min	62.43
cubic feet/sec	million gals/day	0.646317
cubic feet/sec	gallons/min	448.831
cubic inches	cu cms	16.39
cubic inches	cu feet	5.787 E-04
cubic inches	cu meters	1.639 E-05
cubic inches	cu yards	2.143 E-05
cubic inches	gallons	4.329 E-03
cubic inches	liters	0.01639
cubic inches	mil-feet	1.061 E+05
cubic inches	pints (U.S. liq.)	0.03463
cubic inches	quarts (U.S. liq.)	0.01732
cubic meters	bushels (dry)	28.38
cubic meters	cu cms	1 E+06
cubic meters	cu feet	35.31
cubic meters	cu inches	61,023.00
cubic meters	cu yards	1.308
cubic meters	gallons (U.S. liq.)	264.2
cubic meters	gallons (Imperial)	220
cubic meters	liters	1,000.00
cubic meters	pints (U.S. liq.)	2,113.00
cubic meters	quarts (U.S. liq.)	1,057.00
cubic yards	cu feet	27
cubic yards	cu inches	46,656.00
cubic yards	cu meters	0.7646
cubic yards	gallons (U.S. liq.)	202
cubic yards	gallons (Imperial)	168.2
cubic yards	liters	764.6
cubic yards	pints (U.S. liq.)	1,615.90
cubic yards	quarts (U.S. liq.)	807.9
cubic yards/min	cubic ft/sec	0.45
cubic yards/min	gallons/sec	3.367
cubic yards/min	liters/sec	12.74
	D	
Dalton	Gram	1.650 E-24
days	seconds	86,400.00
decigrams	grams	0.1
deciliters	liters	0.1
decimeters	meters	0.1
degrees (angle)	quadrants	0.01111
degrees (angle)	radians	0.01745
degrees (angle)	seconds	3,600.00
degrees F	(F-32)x0.55555	degrees C
degrees C	(C+17.78) x 1.8	degrees F

To Convert	To	Multiply by
degrees/sec	radians/sec	0.01745
degrees/sec	revolutions/min	0.1667
degrees/sec	revolutions/sec	2.778 E-3
dekagrams	grams	10
dekaliters	liters	10
dekameters	meters	10
Drams (apothecaries' or troy)	ounces (avoidupois)	0.1371429
Drams (apothecaries' or troy)	ounces (troy)	0.125
Drams (U.S., fluid or apoth.)	cubic cm.	3.6967
drams	grams	1.7718
drams	grains	27.3437
drams	ounces	0.0625
Dyne/cm	Erg/sq. millimeter	0.01
Dyne/sq. cm.	Atmospheres	9.869 E-07
Dyne/sq. cm.	Inch of Mercury at 0°C	2.953 E-05
Dyne/sq. cm.	Inch of Water at 4°C	4.015 E-04
dynes	grams	1.020 E-03
dynes	joules/cm	1 E-07
dynes	joules/meter (newtons)	1 E-05
dynes	kilograms	1.020 E-06
dynes	poundals	7.233 E-05
dynes	pounds	2.248 E-06
dynes/sq cm	bars	1 E-06

To Convert	To	Multiply by
	E	
Ell	Cm.	114.3
Ell	Inches	45
Em, Pica	Inch	0.167
Em, Pica	Cm.	0.4233
Erg/sec	Dyne - cm/sec	1
ergs	Btu	9.480 E-11
ergs	dyne-centimeters	1
ergs	foot-pounds	7.367 E-08
ergs	gram-calories	0.2389 E-07
ergs	gram-cms	1.020 E-03
ergs	horsepower-hrs	3.7250 E-14
ergs	joules	1 E-07
ergs	kg-calories	2.389 E-11
ergs	kg-meters	1.020 E-08
ergs	ki lowatt-hrs	0.2778 E-13
ergs	watt-hours	0.2778 E-10
ergs/sec	Btu/min	5,688 E-09

To Convert	To	Multiply by
ergs/sec	ft-lbs/min	4.427 E-06
ergs/sec	ft-lbs/sec	7.3756 E-08
ergs/sec	horsepower	1.341 E-10
ergs/sec	kg-calories/min	1.433 E-09
ergs/sec	kilowatts	1 E-10
	F	
farads	microfarads	1 E+06
Faraday/sec	Ampere (absolute)	9.6500 E+04
faradays	ampere-hours	26.8
faradays	coulombs	9.649 E+04
Fathom	Meter	1.828804
fathoms	feet	6
feet	centimeters	30.48
feet	kilometers	3.048 E-04
feet	meters	0.3048
feet	miles (naut.)	1.645 E-04
feet	miles (stat.)	1.894 E-04
feet	millimeters	304.8
feet	mils	1.2 E+04
feet of water	atmospheres	0.0295
feet of water	in. of mercury	0.8826
feet of water	kgs/sq cm	0.03048
feet of water	kgs/sq meter	304.8
feet of water	pounds/sq ft	62.43
feet of water	pounds/sqin.	0.4335
feet of water	mm Hg (32°F)	22.418
feet of water	in Hg (32°F)	0.8828
feet of water	kPa	2.989
feet/min	cms/sec	0.508
feet/min	feet/sec	0.01667
feet/min	kms/hr	0.01829
feet/min	meters/min	0.3048
feet/min	miles/hr	0.01136
feet/sec	cms/sec	30.48
feet/sec	kms/hr	1.097
feet/sec	knots	0.5921
feet/sec	meters/min	18.29
feet/sec	miles/hr	0.6818
feet/sec	miles/min	0.01136
feet/sec/sec	cms/sec/sec	30.48
feet/sec/sec	kms/hr/sec	1.097
feet/sec/sec	meters/sec/sec	0.3048
feet/sec/sec	miles/hr/sec	0.6818
feet/100 feet	percent grade	1
Foot -candle	Lumen/sq. meter	10.764

(*continued on next page*)

To Convert	To	Multiply by
foot-pounds	Btu	1.286 E-03
foot-pounds	ergs	1.356 E+07
foot-pounds	grain-calories	0.3238
foot-pounds	hp-hrs	5.050 E-07
focit-pounds	joules	1.356
foot-pounds	kg-calories	3.24 E-04
foot-pounds	kg-meters	0.1383
foot-pounds	kilowatt-hrs	3.766 E-07
foot-pounds/min	Btu/min	1.286 E-03
foot-pounds/min	foot-pounds/sec	0.01667
foot-pounds/min	horsepower	3.030 E-05
foot-pounds/min	kg-calories/min	3.24 E-04
foot-pounds/min	kilowatts	2.260 E-05
foot-pounds/sec	Btu/hr	4.6263
foot-pounds/sec	Btu/min	0.07717
foot-pounds/sec	horsepower	1.818 E-03
foot-pounds/sec	kg-calories/min	0.01945
foot-pounds/sec	kilowatts	1.356 E-03
Furlongs	miles (U.S.)	0.125
furlongs	rods	40
furlongs	feet	660
	G	
gallons	cu cms	3,785.00
gallons	cu feet	0.1337
gallons	pints (liq.)	8
gallons	quarts	4
gallons	cu inches	231
gallons (US)	cu meters	3.785412 E-03
gallons (UK)	cu. meters	4.546092 E-03
gallons	cu yards	4.951 E-03
gallons	liters	3.785
gallons (liq. Br. Imp.)	gallons (U.S.liq.)	1.20095
gallons (U.S.)	gallons (imp.)	0.83267
gallons (Imp)	cubic inches	277.4
gallons (Imp)	cubic feet	0.1605
gallons (Imp)	cubic centimeters	4546.1
gallons (Imp)	liters	4.546
gallons of water	pounds of water	8.3453
gallons/min	cu ft/sec	2.228 E-03
gallons/min	liters/sec	0.06308
gallons/min	cu ft/hr	8.0208
gallons/min	barrels/hour	0.7
gausses	lines/sq in.	6.452
gausses	webers/sq cm	1 E-08
gausses	webers/sq in.	6.452 E-08
gausses	webers/sqmeter	1 E-04

To Convert	To	Multiply by
gilberts	ampere-turns	0.7958
gilberts/cms	amp-turns/cm	0.7958
gilberts/cms	amp-turns/in	2.021
gilberts/cms	amp-turns/meter	79.58
Gills (British)	cubic cm.	142.07
gills	ltters	0.1183
gills	pints (liq.)	0.25
Grade	Radian	0.01571
Grains	drams (avoirdupois)	0.03657143
grains (troy)	grains (avdp)	1
grains (troy)	grams	0.0648
grains (troy)	ounces (avdp)	2.0833 E-03
grains (troy)	pennyweight (troy)	0.04167
grains/U.S. gal	parts/million	17.118
grains/U.S. gal	pounds/million gal	142.86
grains/Imp. gal	parts/million	14.286
grams	dynes	980.7
grams	grains	15.43
grams	joules/cm	9.807 E-05
grams	joules/meter (newtons)	9.807 E-03
grams	kilograms	0.001
grams	milligrams	1,000.00
grams	ounces (avdp)	0.03527
grams	ounces (troy)	0.03215
grams	poundals	0.07093
grams	pounds	2.205 E-03
grams/cm	pounds/inch	5.600 E-03
grams/cu cm	pounds/cu ft	62.43
grams/cu cm	pounds/cuin	0.03613
grams/cubic centimeter	kg/m^3	1000
grams/cu cm	pounds/mil-foot	3.405 E-07
grams/liter	grains/gal	58.417
grams/liter	pounds/ 1,000 gal	8.345
grams/ liter	pounds/cu ft	0.062427
grams/liter	parts/million	1,000.00
grams/sq cm	pounds/sq ft	2.0481
gram-calories	Btu	3.9683 E-03
gram-calories	ergs	4.1868 E+07
gram-calories	foot-pounds	3.088
gram-calories	horsepower-hrs	1.5596 E-06
gram-calories	joules	4.1833
gram-calories	kilowatt-hrs	1.1630 E-06
gram-calories	watt-hrs	1.1630 E-06
gram-calories/sec	Btu/hr	14.286
gram-centimeters	Btu	9.297 E-08
gram-centimeters	ergs	980.7

To Convert	To	Multiply by
gram-centimeters	joules	9.807 E-05
gram-centimeters	kg-cal	2.343 E-08
gram-centimeters	kg-meters	1 E-05
H		
Hand	Cm.	10.16
hectares	acres	2.471
hectares	sq feet	1.076 E+05
hectares	square meters	10000
hectograms	granis	100
hectoliters	liters	100
hectometers	meters	100
hectowatts	watts	100
henries	millihenries	1,000.00
Hogsheads (British)	cubic ft.	10.114
Hogsheads (U.S.)	cubic ft.	8.42184
Hogsheads (U.S.)	gallons (U.S.)	63
horsepower	Btu/min	42.44
horsepower	foot-lbs/min	33,000.00
horsepower	foot-lbs/sec	550
horsepower (metric) (542.5 ft lb/sec)	horsepower (550 ft lb/sec)	0.9863
horsepower (550 ft lb/sec)	horsepower (metric) (542.5 ft lb/sec)	1.014
horsepower	kg-calories/min	10.68
horsepower	kilowatts	0.7457
horsepower	watts	745.7
horsepower	Btu/minute	42.41
horsepower	Btu/hour	2544.5
horsepower	ft-lb/sec	550
horsepower	ft-lb/min	33000
horsepower	gm-calorie/sec	178.2
horsepower	metric horsepower	1.104
metric horsepower	horsepower	0.9863
metric horsepower	watts	735.5
metric horsepower	Btu/min	41.83
metric horsepower	Btu/hour	2509.6
metric horsepower	ft-lb/sec	542.5
metric horsepower	ft-lb/minute	32.55
metric horsepower	gram-calories/sec	175.7
horsepower (boiler)	Btu/hr	33479
horsepower (boiler)	kilowatts	9.810554
horsepower-hrs	Btu	2,547.00
horsepower-hrs	Pegs	2.6845 E+13
horsepower-hrs	foot-lbs	1.98 E+06
horsepower-hrs	gram-calories	641,190.00

To Convert	To	Multiply by
horsepower-hrs	joules	2.684 E+06
horsepower-hrs	kg-calories	641.1
horsepower-hrs	kg-meters	2.737 E+05
horsepower-hrs	kilowatt-hrs	0.7457
hours	days	4.167 E-02
hours	weeks	5.952 E-03
Hundredweights (long)	pounds	112
Hundredweights (long)	tons (long)	0.05
Hundredweights (short)	ounces (avoirdupois)	1600
Hundredweights (short)	pounds	100
Hundredweights (short)	tons (metric)	0.0453592
Hundredweights (short)	tons (long)	0.0446429
hydraulic horsepower	kilowatts	7.460430 E-01
I		
inches	centimeters	2.54
inches	feet	0.0833
inches	meters	0.0254
inches	miles	1.578 E-05
inches	millimeters	25.4
inches	mils	1,000.00
inches	yards	2.778 E-02
inches of mercury	atmospheres	0.03342
inches of mercury	feet of water	1.133
inches of mercury	kgs/sq cm	0.03453
inches of mercury	kgs/sq meter	345.3
inches of mercury	pounds/sq ft	70.73
inches of mercury	mm Hg (32°F)	25.4
inches of mercury	kPa	3.387
inches of mercury	pounds/sqin.	0.4912
inches of mercury @ 60°F	kiloPascal	3.376850 E+00
inches of mercury @ 32°F	kiloPascal	3.386380 E+00
inches of water @ 4°C	atmospheres	2.458 E-03
inches of water @ 4°C	inches of mercury	0.07355
inches of water @ 4°C	kgs/sq cm	2.540 E-03
inches of water @ 4°C	ounces/sq in.	0.5781
inches of water @ 4°C	pounds/sq ft	5,204

(*continued on next page*)

To Convert	To	Multiply by
inches of water @ 4°C	pounds/sq in.	0.03613
inches of water @ 39.2°F	kiloPascal	2.490820 E-01
inches of water @ 60°F	kiloPascal	2.488400 E-01
international Ampere	Ampere (absolute)	0.9998
international Volt	Volts (absolute)	1.0003
International volt	Joules (absolute)	1.593 E-19
International volt	Joules	9.654 E+04

	J	
joules	Btu	9.480 E-04
joules	ergs	1 E+07
joules	foot-pounds	0.7376
joules	kg-calories	2.389 E-04
joules	kg-meters	0.102
joules	watt-hrs	2.778 E-04
Absolute joule/gram	intl joule/gram	0.99984
Absolute joule/gram	calorie/gram	0.23901
Absolute joule/gram	intl calorie/gram	0.23885
Absolute joule/gram	Btu/pound	0.42993
intl Joule/gram	absolute Joule/gram	1.000165
intl Joule/gram	calorie/gram	0.23904
intl Joule/gram	intl calorie/gram	0.23892
intl Joule/gram	Btu/pound	0.43
joules/cm	grams	1.020 E+04
joules/cm	dynes	1 E+07
joules/cm	joules/meter (newtons)	100
joules/cm	poundals	723.3
,joules/cm	pounds	22.48

	K	
kilograms	dynes	980,665.00
kilograms	grams	1,000.00
kilograms	joules/cm	0.09807
kilograms	joules/meter (newtons)	9.807
kilograms	poundals	70.93
kilograms	pounds	2.20462
kilograms	ounces	35.27000
kilograms	tons (long)	9.842 E-04
kilograms	tons (short)	1.102 E-03
kilograms/cu meter	grams/cu cm	0.001
kilograms/cu meter	pounds/cu ft	0.06243
kilograms/cu meter	pounds/cuin.	3.613 E-05
kilograms/cu meter	pounds/mil-foot	3.405 E-10
kilograms cal-sec	BTU/sec	3.968

To Convert	To	Multiply by
kilograms cal-sec	ft-lb/sec	3.086
kilograms cal-sec	HP	5.6145
kilograms cal-sec	watts	4186.7
kilograms cal-min	ft-lb/min	3085.9
kilograms cal-min	HP	0.9351
kilograms cal-min	watts	69.733
kilograms/meter	pounds/ft	0.672
Kilogram/sq. cm.	Dynes	980,665
kilograms/sq. cm	atmospheres	0.9678
kilograms/sq. cm	feet of water	32.81
kilograms/sq cm	inches of mercury	28.96
kilograms/sq cm	pounds/sq ft	2,048.00
kilograms /sq cm	pounds/sq in.	14.22
kilograms/sq meter	atmospheres	9.678 E-05
kilograms/sq meter	bars	98.0 E-06
kilograms/sq meter	feet of water	3.281 E-03
kilograms/sq meter	inches of mercury	2.896 E-03
kilograms/sq meter	pounds/sq ft	0.2048
kilograms/sq meter	pounds/sqin.	1.422 E-03
kilograms/sq mm	kgs/sq meter	1 E+06
kilograin-calories	Btu	3.968
kilograin-calories	foot-pounds	3,088.00
kilograin-calories	hp-hrs	1.560 E-03
kilograin-calories	joules	4.186.
kilograin-calories	kg-meters	426.9
kilograin-calories	kilojoules	4,186
kilograin-calories	kilowatt-hrs	1.163 E-03
kilogram meters	Btu	9.294 E-03
kilogram meters	ergs	9.804 E+07
kilogramm meters	foot-pounds	7.233
kilogram meters	joules	9.804
kilogram meters	kg-calories	2.342 E-03
kilogram meters	kilowatt-hrs	2.723 E-06
kilolines	maxwells	1,000.00
kiloliters	liters	1,000.00
kilometers	centimeters	1 E+05
kilometers	feet	3,281.00
kilometers	inches	3.937 E+04
kilometers	meters	1,000.00
kilometers	miles	0.6214
kilometers	millimeters	1 E+06
kilorneters	yards	1,094.00
kilometers/hr	cms/sec	27.78
kilometers/hr	feet/min	54.68
kilometers/hr	feet/sec	0.9113
kilometers/hr	knots	0.5396
kilometers/hr	meters/min	16.67
kilometers/hr	miles/hr	0.6214

To Convert	To	Multiply by
kilometers/hr/sec	cms/sec/sec	27.78
kilometers/hr/sec	ft/sec/sec	0.9113
kilometers/hr/sec	meters/sec/sec	0.2778
kilometers/hr/sec	miles/hr/sec	0.6214
kilowatts	Btu/min	56.869
kilowatts	Btu/hour	3412.8
kilowatts	ft-lb/sec	737.1
kilowatts	ft-lb/min	44.254
kilowatts	gram-calorie/sec	239
kilowatts	metric horsepower	1.369
kilowatts	foot-lbs/min	4.426 E+04
kilowatts	foot-lbs/sec	737.6
kilowatts	horsepower	1.341
kilowatts	kg-calories/min	14.34
kilowatts	watts	1,000.00
kilowatt-hrs	Btu	3,413.00
kilowatt-hrs	ergs	3.600 E+13
kilowatt-hrs	foot-lbs	2.655 E+06
kilowatt-hrs	gram-calories	859,850.00
kilowatt-hrs	horsepower-hrs	1.341
kilowatt-hrs	joules	3.6 E+06
kilowatt-hrs	kg-calories	860.5
kilowatt-hrs	kg-meters	3.671 E+05
kilowatt-hrs	pounds of water evaporated from and at 212° F.	3.53
kilowatt-hrs	pounds of water raised from 62° to 212° F.	22.75
kiloPascal	pounds/sq. ft.	20.898
kiloPascal	atmospheres	0.009869
kiloPascal	kilogram/sq. cm.	0.010197
kiloPascal	mm Hg (32°F)	7.5146
kiloPascal	in Hg (32°F)	0.29526
kiloPascal	feet of water	0.33456
knots	feet/hr	6,080.00
knots	kilometers/hr	1.8532
knots	nautical miles/hr	1
knots	statute miles/hr	1.151
knots	yards/hr	2,027.00
knots	feet/sec	1.689
knots	meters per second	0.5144444

L		
league	miles (approx.)	3
Light year	Miles	5.9 E+12
Light year	Kilometers	9.46E+12
lines/sq cm	gausses	1
lines/sq in.	gausses	0.155

To Convert	To	Multiply by
lines/sq in.	webers/sq cm	1.550 E-09
lines/sq in.	webers/sq in.	1 E-08
lines/sq in.	webers/sq meter	1.550 E-05
links (engineer's)	inches	12
links (surveyor's)	inches	7.92
liters	bushels (U.S. dry)	0.02838
liters	cu cm	1,000.00
liters	cu feet	0.03531
liters	cu inches	61.023
liters	cu meters	0.001
liters	cu yards	1.308 E-03
liters	gallons (U.S. liq.)	0.2642
liters	gallons (imp.)	0.22
liters	pints (U.S. liq.)	2.113
liters	quarts (U.S. liq.)	1.057
liters/min	cu ft/sec	5.886 E-04
liters/min	gals/sec	4.403 E-03
lumens/sq ft	foot-candles	1
Lumen	Spherical candle power	0.07958
Lumen	Watt	0.001496
Lumen/sq. ft.	Lumen/sq. meter	10.76
lux	foot-candies	0.0929

M		
maxwells	kilolines	0.001
maxwells	webers	1 E-08
megalines	maxwells	1 E+06
megohms	microhms	1 E+12
megohms	ohms	1 E+06
meters	centimeters	100
meters	feet	3.281
meters	inches	39.37
meters	kilometers	0.001
meters	miles (naut.)	5.396 E-04
meters	miles (stat.)	6.214 E-04
meters	millimeters	1,000.00
meters	yards	1.094
meters	varas	1.179
meters/min	cms/sec	1,667
meters/min	feet/min	3.281
meters/min	feet/sec	0.05468
meters/min	kms/hr	0.06
meters/min	knots	0.03238
meters/min	miles/hr	0.03728
meters/sec	feet/min	196.8

(*continued on next page*)

To Convert	To	Multiply by
meters/sec	feet/sec	3,281
meters/sec	kilometers/hr	3.6
meters/sec	kilometers/min	0.06
meters/sec	miles/hr	2.237
meters/sec	miles/min	0.03728
meters/sec/sec	cms/sec/sec	100
meters/sec/sec	ft/sec/sec	3.281
meters/sec/sec	kms/hr/sec	3.6
meters/sec/sec	miles/hr/sec	2.237
meter-kilograms	cm-dynes	9.807 E+07
meter-kilograms	cm-grams	1 E+05
meter-kilograms	pound-feet	7.233
microfarad	farads	1 E-06
micrograms	grams	1 E-06
microhms	megohms	1 E-12
microhms	ohms	1 E-06
microliters	liters	1 E-06
Microns	meters	1 E-06
miles (naut.)	feet	6,080.27
miles (naut.)	kilometers	1.853
miles (naut.)	meters	1,853.00
miles (naut.)	miles (statute)	1.1516
miles (naut.)	yards	2,027.00
miles (statute)	centimeters	1.609 E+05
miles (statute)	feet	5,280.00
miles (statute)	inches	6.336 E+04
miles (statute)	kilometers	1.609
miles (statute)	meters	1,609.00
miles (statute)	miles (naut.)	0.8684
miles (statute)	yards	1,760.00
miles/hr	cms/sec	44.7
miles/hr	feet/min	88
miles/hr	feet/sec	1.467
miles/hr	kms/hr	1.609
miles/hr	kms/min	0.02682
miles/hr	knots	0.8684
miles/hr	meters/min	26.82
miles/hr	miles/min	0.1667
miles/hr/sec	cms/sec/sec	44.7
miles/hr/sec	feet/sec/sec	1.467
miles/hr/sec	kms/hr/sec	1.609
miles/hr/sec	meters/sec/sec	0.447
miles/min	cms/sec	2,682.00
miles/min	feet/sec	88
miles/min	kms/min	1.609
miles/min	knots/min	0.8684
miles/min	miles/hr	60

To Convert	To	Multiply by
mil-feet	cuinches	9.425 E-06
milliers	kilograms	1,000.00
Millimicrons	meters	1 E-09
Milligrams	grains	0.01543236
milligrams	grams	0.001
milligrams/liter	parts/million	1
millihenries	henries	0.001
milliliters	]iters	0.001
millimeters	centimeters	0.1
millimeters	feet	3.281 E-03
millimeters	inches	0.03937
millimeters	kilometers	1 E-06
millimeters	meters	0.001
millimeters	miles	6.214 E-07
millimeters	mils	39.37
millimeters	yards	1.094 E-03
mm Hg	pounds/sq in	0.0193
mm Hg	pounds/sq. ft.	2.785
mm Hg	inches Hg	0.0394
mm Hg	feet of water	0.0446
mm Hg	kPa	0.13307
million gals/day	cu ft/sec	1.54723
mils	centimeters	2.540 E-03
mils	feet	8.333 E-05
mils	inches	0.001
mils	kilometers	2.540 E-08
mils	yards	2.778 E-05
miner's inches	cu ft/min	1.5
Minims (British)	cubic cm.	0.059192
Minims (U.S., fluid)	cubic cm.	0.061612
minutes (angles)	degrees	0.01667
minutes (angles)	quadrants	1.852 E-04
minutes (angles)	radians	2.909 E-04
minutes (angles)	seconds	60
myriagrams	kilograms	10
myriameters	kilometers	10
myriawatts	kilowatts	10
	N	
natuical mile (international)	meter	1852
nepers	decibels	8.686
Newton	Dynes	1 E+05
	O	
OHM (International)	OHM (absolute)	1.0005

To Convert	To	Multiply by
ohms	megohms	1 E-06
ohms	microhms	1 E+06
ounces	drams	16
ounces	grains	437.5
ounces	grams	28.349527
ounces	pounds	0.0625
ounces	ounces (troy)	0.9115
ounces	tons (long)	2.790 E-05
ounces	kilograms	28.35
ounces	tons (metric)	2.835 E-05
ounces (fluid)	cu inches	1.805
ounces (fluid)	liters	0.02957
ounces (troy)	grains	480
ounces (troy)	grams	31.103481
ounces (troy)	ounces (avdp.)	1.09714
ounces (troy)	pennyweights (troy)	20
ounces (troy)	pounds (troy)	0.08333
Ounce/sq. inch	Dynes/sq. cm.	4309
ounces/sq. in.	pounds/sqin.	0.0625

P		
Parsec	Miles	19 E+12
Parsec	Kilometers	3.084 E+13
parts/million	grains/U.S. gal	0.0584
Pascal	Psi	0.000145037738
parts/million	grains/Imp. gal	0.07016
parts/million	pounds/million gal	8.345
Pecks (British)	cubic inches	554.6
Pecks (British)	liters	9.091901
Pecks (U.S.)	bushels	0.25
Pecks (U.S.)	cubic inches	537.605
Pecks (U.S.)	liters	8.809582
Pecks (U.S.)	quarts (dry)	8
pennyweights (troy)	grains	24
pennyweights (troy)	ounces (troy)	0.05
pennyweights (troy)	grams	1.55517
pennyweights (troy)	pounds (troy)	4.1667 E-03
pints (dry)	cu inches	33.6
pints (ljq.)	cu cms.	473.2
pints (liq.)	cu feet	0.01671
pints (liq.)	cu inches	28.87
pints (liq.)	cu meters	4.732 E-04
pints (ljq.)	cu. yards	6.189 E-04
pints (liq.)	gallons	0.125

To Convert	To	Multiply by
pints (liq.)	liters	0.4732
pints (liq.)	quarts (liq.)	0.5
Planck's quantum	Erg -second	6.624 E-27
Poise	Gram/cm. sec.	1
Pounds (avoirdupois)	ounces (troy)	14.5833
poundals	dynes	13,826.00
poundals	grams	14.1
poundals	joules/cm	1.383 E-03
poundals	joules/meter (newtons)	0.1383
poundals	kilograms	0.0141
poundals	pounds	0.03108
pounds	drams	256
pounds	dynes	44.4823 E+04
pounds	grains	7,000.00
pounds	grams	453.5924
pounds	joules/cm	0.04448
pounds	joules/meter (newtons)	4.448
pounds	kilograms	0.4536
pounds	ounces	16
pounds	ounces (troy)	14.5833
Pounds	poundals	32.17
pounds	pounds (troy)	1.21528
pounds/sq. ft.	kPa	0.04788
pounds/sq. inch	kg/square cm	0.0703
pounds/sq. inch	mm Hg (32°F)	51.713
pounds/sq. ft	mm Hg (32°F)	0.3591
pounds/aq. In.	kPa	6.895
pounds	tons (short)	0.0005
pounds/sq. in./ft	kiloPascal/meter	2.262059 E+01
pounds/sq. in/mile	kiloPascal/ kilometer	4.284203 E+00
pounds (troy)	grains	5,760.00
pounds (troy)	grams	373.24177
pounds (troy)	ounces (avdp.)	13.1657
pounds (troy)	ounces (troy)	12
pounds (troy)	pennyweights (troy)	240
pounds (troy)	pounds (avdp.)	0.822857
pounds (troy)	tons (long)	3.6735 E-04
pounds (troy)	tons (metric)	3.7324 E-04
pounds (troy)	tons (short)	4.1143 E-04
pounds/US gal (gas)	kilograms/cu. meter	1.601846 E+01
pounds/US gal (liquid)	kilograms/cu. meter	1.198264 E+02

(*continued on next page*)

To Convert	To	Multiply by
pounds/UK gal (liquid)	kilograms/cu. meter	9.977633 E+01
pounds/cu. foot	kilograms/cu. meter	1.601846 E+01
pounds/barrel	kilograms/cu. meter	2.853010 E+00
pounds of water	cu. feet	0.01602
pounds of water	cu. inches	27.68
pounds of water	gallons	0.1198
pounds of water/min	cu. ft/sec	2.670 E-04
pound-feet	cm-dynes	1.356 E+07
pound-feet	cm-grams	13,825.00
pound-feet	meter-kgs	0.1383
pounds/cu ft	grams/cu. cm	0.01602
pounds/cu ft	kgs/cu. meter	16.02
pounds/cu ft	pounds/cu. in.	5.787 E-04
pounds/cu ft	pounds/mil-foot	5.456 E-09
pounds/cu in.	gms/cu. cm	27.68
pounds/cu. in.	kgs/cu. meter	2.768 E+04
pounds/cu. in.	pounds/cu. ft	1,728.00
pounds/cu. in.	pounds/mil-foot	9.425 E-06
pounds/ft	kgs/meter	1.488
pounds/in.	gms/cm	178.6
pounds/mil-foot	gms/cu cm	2.306 E+06
pounds/sq ft	atmospheres	4.725 E-04
pounds/sq ft	feet of water	0.01602
pounds/sq ft	inches of mercury	0.01414
pounds/sq ft	kgs/sq meter	4.882
pounds/sq ft	pounds/sqin.	6.944 E-03
pounds/sq in.	atmospheres	0.06804
pounds/sq in.	feet of water	2.307
pounds/sq in.	inches of mercury	2.036
pounds/sq i n.	kgs/sq meter	703.1
pounds/sq in.	pounds/sq ft	144

Q		
quadrants (angle)	degrees	90
quadrants (angle)	minutes	5,400.00
quadrants (angle)	radians	1.571
quadrants (angle)	seconds	3.24 E+05
quarts (dry)	cu inches	67.2
quarts (liq.)	cu cms	946.4
quarts (liq.)	cu feet	0.03342
quarts (liq.)	cu inches	57.75
quarts (liq.)	cu meters	9.464 E-04
quarts (liq.)	cu yards	1.238 E-03
quarts (liq.)	gallons	0.25

To Convert	To	Multiply by
quarts (liq.)	liters	0.9463

R		
radians	degrees	57.3
radians	minutes	3,438.00
radians	quadrants	0.6366
radians	seconds	2.063 E+05
radians/sec	degrees/sec	57.3
radians/sec	revolutions/min	9.549
radians/sec	revolutions/sec	0.1592
radians/sec/sec	revs/min/min	573
radians/sec/sec	revs/min/sec	9.549
radians/sec/sec	revs/sec/sec	0.1592
revolutions	degrees	360
revolutions	quadrants	4
revolutions	radians	6.283
revolutions/min	degrees/sec	6
revolutions/min	radians/sec	0.1047
revolutions/min	revs/sec	0.01667
revolutions/min/min	radians/sec/sec	1.745 E-03
revolutions/min/min	revs/min/sec	0.01667
revolutions/min/min	revs/sec/sec	2.778 E-04
revolutions/sec	degrees/sec	360
revolutions/sec	radians/sec	6.283
revolutions/sec	revs/min	60
revolutions/sec/sec	radians/sec/sec	6.283
revolutions/sec/sec	revs/min/min	3,600.00
revolutions/sec/sec	revs/min/sec	60
Rod	Chain (Gunters)	0.25
Rod	Meters	5.029
Rods (Surveyors' meas.)	yards	5.5
rods	feet	16.5

S		
Scruples	grains	20
seconds (angle)	degrees	2.778 E-04
seconds (angle)	minutes	0.01667
seconds (angle)	quadrants	3.087 E-06
seconds (angle)	radians	4.848 E-06
Slug	Kilogram	14.59
Slug	Pounds	32.17
Sphere	Steradians	12.57
square centimeters	circular mils	1.973 E+05
square centimeters	sq feet	1.076 E-03

To Convert	To	Multiply by
square centimeters	sq inches	0.155
square centimeters	sq meters	0.0001
square centimeters	sq miles	3.861 E-11
square centimeters	sq millimeters	100
square centimeters	sq yards	1.196 E-04
square feet	acres	2.296 E-05
square feet	circular mils	1.833 E+08
square feet	sq cms	929
square feet	sq inches	144
square feet	sq meters	0.0929
square feet	sq miles	3.587 E-08
square feet	sq millimeters	9.290 E+04
square feet	sq yards	0.1111
square inches	circular mils	1.273 E+06
square inches	sq cms	6.452
square inches	sq feet	6.944 E-03
square inches	sq millimeters	645.2
square inches	sq mils	1 E+06
square inches	sq yards	7.716 E-04
square kilometers	acres	247.1
square kilometers	sq cms	1 E+10
square kilometers	sq ft	10.76 E+06
square kilometers	sq inches	1.550 E+09
square kilometers	sq meters	1 E+06
square kilometers	sq miles	0.3861
square kilometers	sq yards	1.196 E+06
square meters	acres	2.471 E-04
square meters	sq cms	1 E+04
square meters	sq feet	10.76
square meters	sq inches	1,550.00
square meters	sq miles	3.861 E-07
square meters	sq millimeters	1 E+06
square meters	sq yards	1.196
square miles	acres	640
square miles	sq feet	27.88 E+06
square miles	sq kms	2.50
square miles	sq meters	2.590 E+06
square miles	hectares	259
square miles	sq yards	3.098 E+06
square millimeters	circular mils	1,973.00
square millimeters	sq cms	0.01
square millimeters	sq feet	1.076 E-05
square millimeters	sq inches	1.550 E-03
square mils	circular mils	1.273
square mils	sq cms	6.452 E-06
square mils	sq inches	1 E-06

To Convert	To	Multiply by
square yards	acres	2.066 E-04
square yards	sq cms	8,361.00
square yards	sq feet	9
square yards	sq inches	1,296.00
square yards	sq meters	0.8361
square yards	sq miles	3.228 E-07
square yards	sq millimeters	8.361 E+05
T		
temperature ($^{\circ}$C) + 273	absolute temperature ($^{\circ}$C)	1
temperature ($^{\circ}$C) + 17.78	temperature ($^{\circ}$F)	1.8
temperature ($^{\circ}$F) + 459.67	absolute temperature($^{\circ}$F)	1
temperature ($^{\circ}$F) −32	temperature ($^{\circ}$C)	($^{\circ}$F-32)*0.5556
temperature ($^{\circ}$R)	temperature ($^{\circ}$K)	$^{\circ}$R/1.8
therm/US gal.	megajoules/cu. meter	2.787163 E+04
therm/US	joule	1.05E+08
tons (long)	kilograms	1,016.00
tons (long)	pounds	2,240.00
tons (long)	tons (short)	1.12
tons (metric)	kilograms	1,000.00
tons (metric)	pounds	2,205.00
tons (metric)	tons (short)	1.10
tons (short)	kilograms	907.1848
tons (short)	ounces	32,000.00
tons (short)	ounces (troy)	29,166.66
tons (short)	pounds	2,000.00
tons (short)	pounds (troy)	2,430.56
tons (short)	tons (long)	0.89287
tons (short)	tons (metric)	0.9078
tons (UK)	kilograms/sec	2.822353 E-01
tons (US)	kilograms/sec	2.519958 E-01
tons (short)/sq ft	kgs/sq meter	9,765.00
tons (short)/sq ft	pounds/sqin.	2,000.00
tons of water/24 hrs	pounds of water/hr	83.333
tons of water/24 hrs	gallons/-min	0.16643
tons of water/24 hrs	cu ft/hr	1.3349
tons of refigeration	BTU/hr	12000
tons of refigeration	watts	3517.2
V		
Volt/inch	Volt/cm.	0.39370
Volt (absolute)	Statvolts	0.003336

(continued on next page)

To Convert	To	Multiply by
W		
watts	Btu/hr	3.4129
watts	Btu/min	0.05688
watts	ergs/sec	107
watts	foot-lbs/min	44.27
watts	foot-lbs/sec	0.7378
watts	horsepower	1.341 E-03
watts	horsepower (metric)	1.360 E-03
watts	kg-calories/min	0.01433
watts	kilowatts	0.001
watts	Btu/hour	3.413
watts	ft-lb/sec	0.7376
watts	ft-lb/min	44.25
watts	gram-calories/sec	0.239
watts	hp (metric)	0.00136
watts (Abs.)	B.T.U. (mean)/min.	0.056884
watts (Abs.)	joules/sec.	1
watt-hours	Btu	3.413
watt-hours	ergs	3.60 E+10
watt-hours	foot-pounds	2,656.00
watt-hours	gram-calories	859.85
watt-hours	horsepower-hrs	1.341 E-03
watt-hours	kilogram-calories	0.8605
watt-hours	kilogram-meters	367.2
watt-hours	kilowatt-hrs	0,001

To Convert	To	Multiply by
Watt (International)	Watt (absolute)	1.0002
webers	maxwells	1 E+08
webers	kilolines	1 E+05
webers/sq in.	gausses	1.550 E+07
webers/sq in.	lines/sq in.	1 E+08
webers/sq in.	webers/sq cm	0.155
webers/sq in.	webers/sq meter	1,550.00
webers/sq meter	gausses	1 E+04
webers/sq meter	lines/sq in.	6.452 E+04
webers/sq meter	webers/sq cm	1 E-04
webers/sq meter	webers/sq in.	6.452 E-04
	Y	
yards	centimeters	91.44
yards	kilometers	9.144 E-04
yards	meters	0.9144
yards	miles (naut.)	4.934 E-04
yards	feet	3
yards	inches	36
yards	miles (stat.)	5.682 E-04
yards	millimeters	914.4
year (common)	days	365
year (common)	hours	8760
year (leap)	days	366
year (leap)	hours	8784

Viscosity—equivalents of absolute viscosity

Absolute or Dynamic Viscosity		Centipoise (μ)	Poise $\dfrac{\text{Gram}}{\text{Cm Sec}}$ $\dfrac{\text{Dyne Sec}}{\text{Cm}^2}$ $(100\ \mu)$	$\dfrac{\text{Slugs}}{\text{Ft Sec}}$ $\dfrac{\text{*Pound}_f\ \text{Sec}}{\text{Ft}^2}$ (μ'_e)	$\dfrac{\text{†Pound}_m}{\text{Ft Sec}}$ $\dfrac{\text{Poundal Sec}}{\text{Ft}^2}$ (μ_e)
Centipoise	(μ)	1	0.01	$2.09\ (10^{-5})$	$6.72\ (10^{-4})$
Poise $\dfrac{\text{Gram}}{\text{Cm Sec}}$ $\dfrac{\text{Dyne Sec}}{\text{Cm}^2}$	$(100\ \mu)$	100	1	$2.09\ (10^{-3})$	0.0672
$\dfrac{\text{Slugs}}{\text{Ft Sec}}$ $\dfrac{\text{*Pound}_f\ \text{Sec}}{\text{Ft}^2}$	(μ'_e)	47 900	479	1	g or 32.2
$\dfrac{\text{†Pound}_m}{\text{Ft Sec}}$ $\dfrac{\text{Poundal Sec}}{\text{Ft}^2}$	(μ_e)	1487	14.87	$\dfrac{1}{g}$ or .0311	1

*Pound$_f$ = Pound of Force   †Pound$_m$ = Pound of Mass

To convert absolute or dynamic viscosity from one set of units to another, locate the given set of units in the left-hand column and multiply the numerical value by the factor shown horizontally to the right under the set of units desired.

As an example, suppose a given absolute viscosity of 2 P is to be converted to slugs/foot second. By referring to the table, we find the conversion factor to be $2.09\ (10^{-3})$. Then, 2 (poise) times $2.09\ (10^{-3}) = 4.18\ (10^{-3}) = 0.00418$ slugs/foot second.

Courtesy Crane Co. (Technical Paper #410).

General liquid density nomograph

X and Y Values for Density Nomograph

Compound	X	Y	Compound	X	Y	Compound	X	Y
Acetic acid	40.6	93.5	Ethyl chloride	42.7	62.4	Methyl sulfide	31.9	57.4
Acetone	26.1	47.8	Ethylene	17.0	3.5	n-Nonane	16.2	36.5
Acetonitrile	21.8	44.9	Ethyl ether	22.6	35.8	n-Octadecane	16.2	46.5
Acetylene	20.8	10.1	Ethyl formate	37.6	68.4	n-Octane	12.7	32.5
Ammonia	22.4	24.6	Ethyl propionate	32.1	63.9	n-Pentadecane	15.8	44.2
Isoamyl alcohol	20.5	52.0	Ethyl propyl ether	20.0	37.0	n-Pentane	12.6	22.6
Ammine	33.5	92.5	Ethyl auindo	25.7	55.3	n-Nonadecane	14.9	47.0
Benzene	32.7	63.0	Fluorobenzene	41.9	87.6	Isopentane	13.5	22.5
n-Butyric acid	31.3	78.7	n-Heptadecane	15.6	45.7	Phenol	36.7	103.8
Isobutane	13.7	16.5	n-Heptane	12.6	29.8	Phosphine	28.0	22.1
Isobutyric acid	31.5	75.9	n-Hexadecane	15.8	45.0	Propane	14.2	12.2
Carbon dioxide	78.6	45.4	n-Hexane	13.5	27.0	Propionic acid	35.0	83.5
Chlorobenzene	41.7	105.0	Methanethiol	37.3	59.5	Piperidene	27.5	60.0
Cyclohexane	19.6	44.0	Methyl acetate	40.1	70.3	Propionitrile	20.1	44.6
n-Decane	16.0	38.2	Methyl alcohol	25.8	49.1	Propyl acetate	33.0	65.5
n-Dedecane	14.3	41.4	Methyl n-butyrate	31.5	65.5	Propyl alcohol	23.8	50.8
Diethylamine	17.8	33.5	Methyl isobutyrate	33.0	64.1	Propyl formate	33.8	66.7
n-Elconane	14.8	47.5	Methyl chloride	52.3	62.9	n-Tetradecane	15.8	43.3
Ethane	10.8	4.4	Methyl ether	27.2	30.1	n-Tridecane	15.3	42.4
Ethanethiol	32.0	55.5	Methyl ethyl ether	25.0	34.4	Triethylamine	17.9	37.0
Ethyl acetate	35.0	95.0	Methyl formate	46.4	74.6	n-Undecane	14.4	39.2
Ethyl alcohol	24.2	48.6	Methyl propionate	36.5	68.3			z

Reprinted by permission of Tubular Exchanger Manufacturers Association (TEMA).

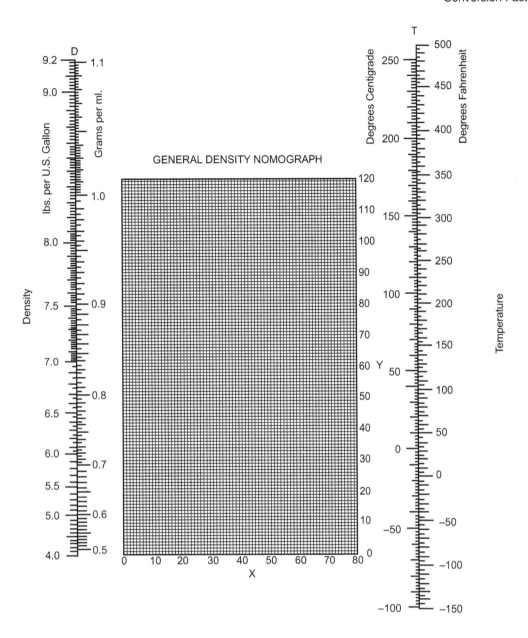

GENERAL DENSITY NOMOGRAPH

Density Conversions

To convert the numerical value of a property expressed in one of the units in the left-hand column of the table to the numerical value expressed in one of the units in the top row of the table, multiply the former value by the factor in the block common to both units.

Units→ ↓	$\frac{g}{cu\ cm}$	$\frac{g}{ml}$	$\frac{lb}{cu\ in.}$	$\frac{lb}{cu\ ft}$	$\frac{lb}{gal\ (U.S.)}$
$\frac{g}{cu\ cm}$	1	1.000028	0.036128	62.428	8.3455
$\frac{g}{ml}$	0.99997	1	0.036126	62.427	8.3452
$\frac{lb}{cu\ in.}$	27.680	27.681	1	1,728	231
$\frac{lb}{cu\ ft}$	0.016018	0.016019	5.7870×10^{-4}	1	0.13368
$\frac{lb}{gal\ (U.S.)}$	0.11983	0.11983	4.3290×10^{-3}	7.4805	1

Chart gives specific gravity/temperature relationship for petroleum oils

C₂H₆ = Ethane
C₃H₈ = Propane iC₄H₁₀ = Isobutane
C₄H₁₀ = Butane iC₅H₁₂ = Isopentane

Example: The specific gravity of an oil at 60 °F is 0.85. The specific gravity at 100 °F = 0.83.

To find the weight density of a petroleum oil at its flowing temperature when the specific gravity at 60 °F/60 °F is known, multiply the specific gravity of the oil at flowing temperature (see chart above) by 62.4, the density of water at 60 °F.

Courtesy Crane Co. (Technical Paper #410).

Weight density and specific gravity of various liquids

Liquid	Temp.	Weight Density	Specific Gravity	Liquid	Temp.	Weight Density	Specific Gravity	
	t	*ρ*	*S*		*t*	*ρ*	*S*	*Liquid at 60 F referred to water at 60 F.
	Deg. Fahr.	Lbs. per Cu. Ft.			Deg. Fahr.	Lbs. per Cu. Ft.		
Acetone	60	49.4	0.792	Mercury	20	849.74	...	†Milk has a weight density of 64.2 to 64.6.
Ammonia, Saturated	10	40.9	...	Mercury	40	848.03	...	
Benzene	32	56.1	...	Mercury	60	846.32	13.570	‡100 Viscosity Index.
Brine, 10% Ca Cl	32	68.05	...	Mercury	80	844.62	...	
Brine, 10% Na Cl	32	67.24	...	Mercury	100	842.93	...	
Bunkers C Fuel Max.	60	63.25	1.014	Milk	...	†	...	
Carbon Disulphide	32	80.6	...	Olive Oil	59	57.3	0.919	
Distillate	60	52.99	0.850	Pentane	59	38.9	0.624	
Fuel 3 Max.	60	56.02	0.898	SAE 10 Lube‡	60	54.64	0.876	
Fuel 5 Min.	60	60.23	0.966	SAE 30 Lube‡	60	56.02	0.898	
Fuel 5 Max.	60	61.92	0.993	SAE 70 Lube‡	60	57.12	0.916	
Fuel 6 Min.	60	61.92	0.993	Salt Creek Crude	60	52.56	0.843	
Gasoline	60	46.81	0.751	32.6° API Crude	60	53.77	0.862	
Gasoline, Natural	60	42.42	0.680	35.6° API Crude	60	52.81	0.847	
Kerosene	60	50.85	0.815	40° API Crude	60	51.45	0.825	
M. C. Residuum	60	58.32	0.935	48° API Crude	60	49.16	0.788	

Courtesy Crane Co. (Technical Paper #410).

True vapor pressure of crude oil stocks with a Reid vapor pressure of 2–15 psi

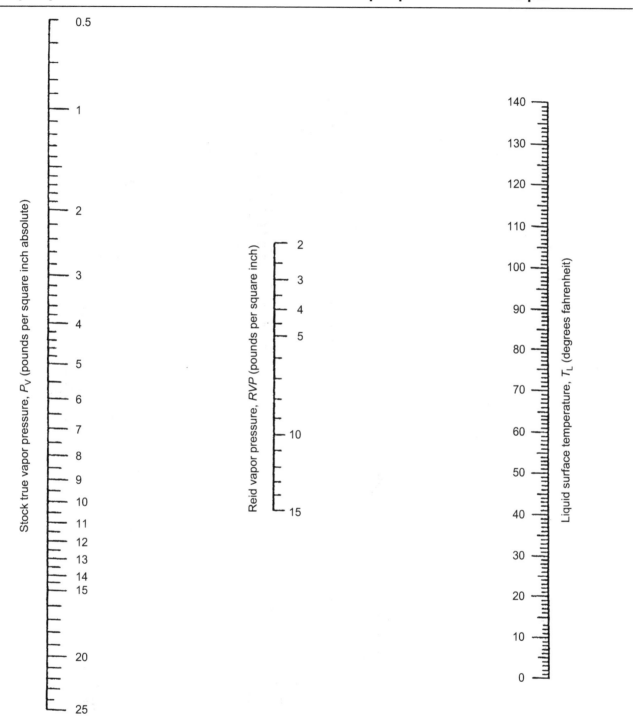

True vapor pressure (P_v) of crude oil stocks with a Reid vapor pressure of 2–15 psi. Reprinted courtesy of the American Petroleum Institute.

Low temperature vapor pressure of light hydrocarbons

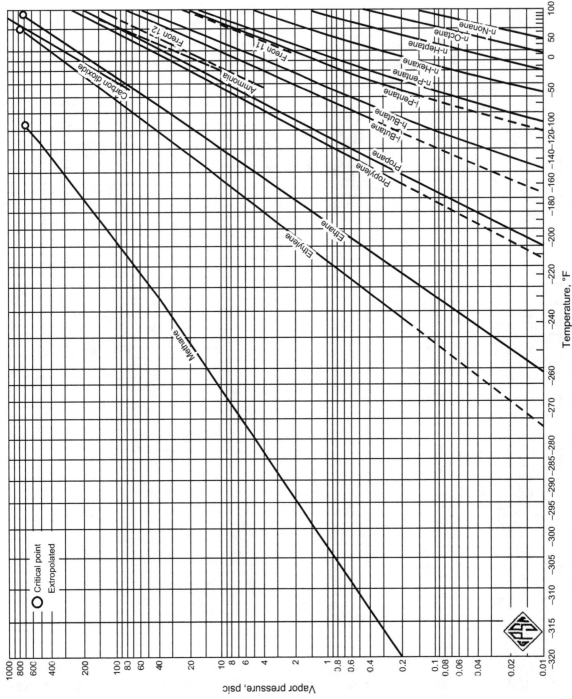

Low-temperature vapor pressures for light hydrocarbons. Reprinted with permission—Gas Processors Suppliers Association.

High temperature vapor pressures for light hydrocarbons

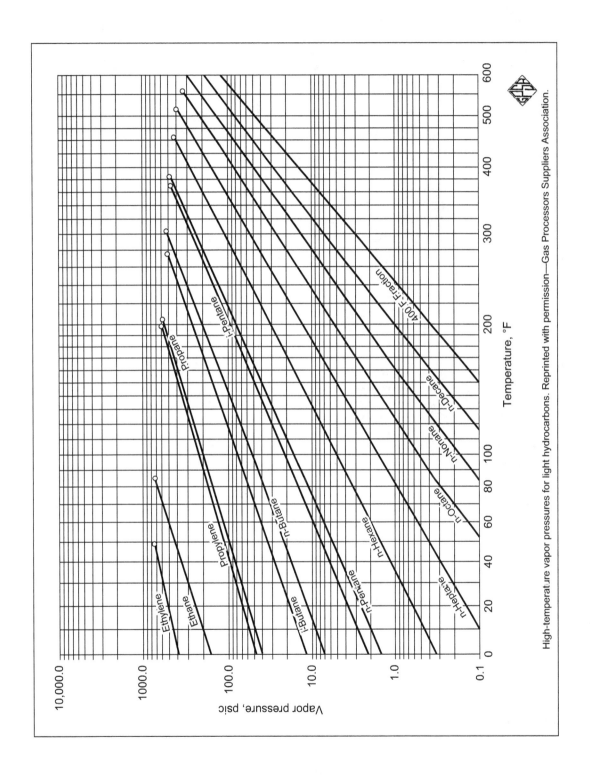

High-temperature vapor pressures for light hydrocarbons. Reprinted with permission—Gas Processors Suppliers Association.

Hydrocarbon gas viscosity

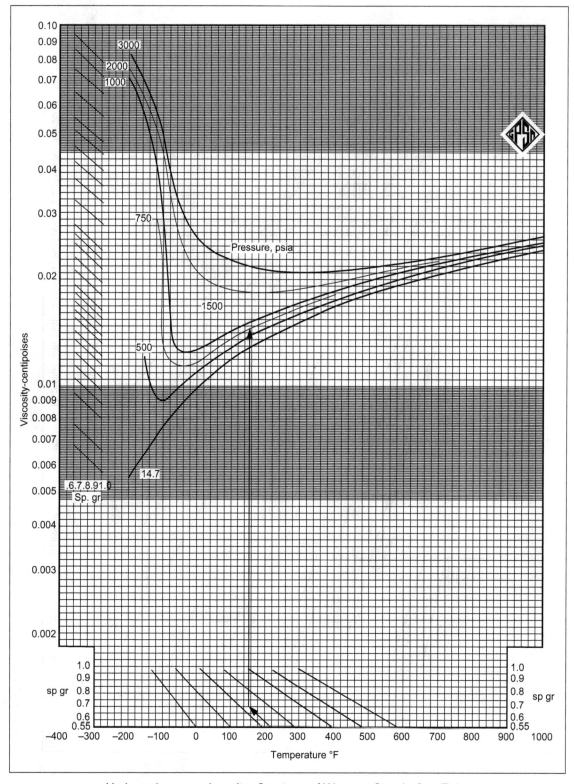

Hydrocarbon gas viscosity. Courtesy of Western Supply Co., Tulsa.

Metric conversions—metric to English, English to metric

To convert from	To	Multiply by	To convert from	To	Multiply by
Length			**Length**		
mm	inches	.03937	inches	mm	25.40
cm	inches	.3937	inches	cm	2.540
meters	inches	39.37	inches	meters	.0254
meters	feet	3.281	feet	meters	.3048
meters	yards	1.0936	feet	km	.0003048
km	feet	3280.8	yards	meters	.9144
km	yards	1093.6	yards	km	.0009144
km	miles	.6214	miles	km	1.609
Area			**Area**		
sq mm	sq inches	.00155	sq inches	sq mm	645.2
sq cm	sq inches	.155	sq inches	sq cm	6.452
sq meters	sq feet	10.764	sq feet	sq meters	.09290
sq meters	sq yards	1.196	sq yards	sq meters	.8361
sq km	sq miles	.3861	sq miles	sq km	2.590
hectares	acres	2.471	acres	hectares	.4047
Volume			**Volume**		
cu cm	cu inches	.06102	cu inches	cu cm	16.387
cu cm	fl ounces	.03381	cu inches	liters	.01639
cu meters	cu feet	35.314	cu feet	cu meters	.02832
cu meters	cu yards	1.308	cu feet	liters	28.317
cu meters	US gal	264.2	cu yards	cu meters	.7646
liters	cu inches	61.023	fl ounces	cu cm	29.57
liters	cu feet	.03531	US gal	cu meters	.003785
liters	US gal	.2642	US gal	liters	3.785
Weight			**Weight**		
grams	grains	15.432	grains	grams	.0648
grams	ounces†	.0353	ounces†	grams	28.350
kg	ounces†	35.27	ounces†	kg	.02835
kg	pounds†	2.2046	pounds†	kg	.4536
kg	US tons	.001102	pounds†	tonnes	.000454
kg	long tons	.000984	US tons	kg	907.2
tonnes	pounds†	2204.6	US tons	tonnes	.9072
tonnes	US tons	1.1023	long tons	kg	1016
tonnes	long tons	.9842	long tons	tonnes	1.0160

To convert from	To	Multiply by	To convert from	To	Multiply by
Unit weight			**Unit weight**		
gr/sq cm	lb/sq in	.01422	lb/ft	kg/m	1.4881
gr/cu cm	lb/cu in	.0361	lb/sq in	gr/sq cm	70.31
kg/sq cm	lb/sq in	14.22	lb/sq in	kg/sq cm	.07031
kg/cu m	lb/cu ft	.0624	lb/cu in	gr/cu cm	27.68
kg/m	lb/ft	.6720	lb/cu ft	kg/cu m	16.018
Unit volume			**Unit volume**		
liters/min	US gpm	.2642	US gpm	liters/min	3.785
liters/min	cfm	.03531	US gpm	liters/hr	227.1
liters/hr	US gpm	.0044	US gpm	cu m/hr	.2271
cu m/min	cfm	35.314	cfm	liters/min	28.317
cu m/hr	cfm	.5886	cfm	cu m/min	.02832
cu m/hr	US gpm	4.4028	cfm	cu m/hr	1.6992
Power			**Power**		
watts	ft-lb/sec	.7376	ft-lb/sec	watts	1.356
watts	hp	.00134	hp	watts	745.7
kw	hp	1.3410	hp	kw	.7457
cheval-vap	hp	.9863	hp	cheval-vap	1.0139
Heat			**Heat**		
gr-cal	Btu	.003969	Btu	gr-cal	252
kg-cal	Btu	3.9693	Btu	kg-cal	.252
kg-cal/kg	Btu/lb	1.800	Btu/lb	kg-cal/kg	.5556
gr-cal/sq cm	Btu/sq ft	3.687	Btu/sq ft	gr-cal/sq cm	.2713
kg-cal/cu m	Btu/cu ft	.1124	Btu/cu ft	kg-cal/cu m	8.899
Work or energy			**Work or energy**		
joule	ft-lb	.7376	ft-lb	joule	1.356
meter-kg	ft-lb	7.2330	ft-lb	meter-kg	.1383
gr-cal	ft-lb	3.087	ft-lb	gr-cal	.3239
kg-cal	ft-lb	3087	ft-lb	kg-cal	.000324
hp-hr	ft-lb	1,980,000	ft-lb	hp-hr	5.05×10^{-7}
kwhr	ft-lb	2,655,000	ft-lb	kwhr	3.766×10^{-7}
Btu	ft-lb	778.0	ft-lb	Btu	.001285

†avoirdupois pounds and ounces.

Courtesy Cameron Hydraulic Data Book, *Ingersoll-Rand Company.*

Temperature conversion—centigrade to Fahrenheit or Fahrenheit to centigrade

NOTE: The numbers in boldface refer to the temperature in degrees, either Centigrade or Fahrenheit, which it is desired to convert into the other scale. If converting from Fahrenheit to Centigrade degrees, the equivalent temperature will be found in the left column; while if converting from degrees Centigrade to degrees Fahrenheit, the answer will be found in the column on the right.

Centigrade		Fahrenheit	Centigrade		Fahrenheit	Centigrade		Fahrenheit	Centigrade		Fahrenheit
−73.3	**−100**	−148.0	2.8	**37**	98.6	33.3	**92**	197.6	293	**560**	1040
−67.8	**−90**	−130.0	3.3	**38**	100.4	33.9	**93**	199.4	299	**570**	1058
−62.2	**−80**	−112.0	3.9	**39**	102.2	34.4	**94**	201.2	304	**580**	1076
−59.4	**−75**	−103.0	4.4	**40**	104.0	35.0	**95**	203.0	310	**590**	1094
−56.7	**−70**	−94.0	5.0	**41**	105.8	35.6	**96**	204.8	316	**600**	1112
−53.9	**−65**	−85.0	5.6	**42**	107.6	36.1	**97**	206.6	321	**610**	1130
−51.1	**−60**	−76.0	6.1	**43**	109.4	36.7	**98**	208.4	327	**620**	1148
−48.3	**−55**	−67.0	6.7	**44**	111.2	37.2	**99**	210.2	332	**630**	1166
−45.6	**−50**	−58.0	7.2	**45**	113.0	37.8	**100**	212.0	338	**640**	1184
−42.8	**−45**	−49.0	7.8	**46**	114.8	43	**110**	230	343	**650**	1202
−40.0	**−40**	−40.0	8.3	**47**	116.6	49	**120**	248	349	**660**	1220
−37.2	**−35**	−31.0	8.9	**48**	118.4	54	**130**	266	354	**670**	1238
−34.4	**−30**	−22.0	9.4	**49**	120.2	60	**140**	284	360	**680**	1256
−31.7	**−25**	−13.0	10.0	**50**	122.0	66	**150**	302	366	**690**	1274
−28.9	**−20**	−4.0	10.6	**51**	123.8	71	**160**	320	371	**700**	1292
−26.1	**−15**	5.0	11.1	**52**	125.6	77	**170**	338	377	**710**	1310
−23.3	**−10**	14.0	11.7	**53**	127.4	82	**180**	356	382	**720**	1328
−20.6	**−5**	23.0	12.2	**54**	129.2	88	**190**	374	388	**730**	1346
−17.8	**0**	32.0	12.8	**55**	131.0	93	**200**	392	393	**740**	1364
−17.2	**1**	33.8	13.3	**56**	132.8	99	**210**	410	399	**750**	1382
−16.7	**2**	35.6	13.9	**57**	134.6	100	**212**	414	404	**760**	1400
−16.1	**3**	37.4	14.4	**58**	136.4	104	**220**	428	410	**770**	1418
−15.6	**4**	39.2	15.0	**59**	138.2	110	**230**	446	416	**780**	1436
−15.0	**5**	41.0	15.6	**60**	140.0	116	**240**	464	421	**790**	1454
−14.4	**6**	42.8	16.1	**61**	141.8	121	**250**	482	427	**800**	1472
−13.9	**7**	44.6	16.7	**62**	143.6	127	**260**	500	432	**810**	1490
−13.3	**8**	46.4	17.2	**63**	145.4	132	**270**	518	438	**820**	1508
−12.8	**9**	48.2	17.8	**64**	147.2	138	**280**	536	443	**830**	1526
−12.2	**10**	50.0	18.3	**65**	149.0	143	**290**	554	449	**840**	1544
−11.7	**11**	51.8	18.9	**66**	150.8	149	**300**	572	454	**850**	1562
−11.1	**12**	53.6	19.4	**67**	152.6	154	**310**	590	460	**860**	1580
−10.6	**13**	55.4	20.0	**68**	154.4	160	**320**	608	466	**870**	1598
−10.0	**14**	57.2	20.6	**69**	156.2	166	**330**	626	471	**880**	1616
−9.4	**15**	59.0	21.1	**70**	158.0	171	**340**	644	477	**890**	1634
−8.9	**16**	60.8	21.7	**71**	159.8	177	**350**	662	482	**900**	1652
−8.3	**17**	62.6	22.2	**72**	161.6	182	**360**	680	488	**910**	1670
−7.8	**18**	64.4	22.8	**73**	163.4	188	**370**	698	493	**920**	1688
−7.2	**19**	66.2	23.3	**74**	165.2	193	**380**	716	499	**930**	1706
−6.7	**20**	68.0	23.9	**75**	167.0	199	**390**	734	504	**940**	1724
−6.1	**21**	69.8	24.4	**76**	168.8	204	**400**	752	510	**950**	1742
−5.6	**22**	71.6	25.0	**77**	170.6	210	**410**	770	516	**960**	1760
−5.0	**23**	73.4	25.6	**78**	172.4	216	**420**	788	521	**970**	1778
−4.4	**24**	75.2	26.1	**79**	174.2	221	**430**	806	527	**980**	1796
−3.9	**25**	77.0	26.7	**80**	176.0	227	**440**	824	532	**990**	1814
−3.3	**26**	78.8	27.2	**81**	177.8	232	**450**	842	538	**1000**	1832
−2.8	**27**	80.6	27.8	**82**	179.6	238	**460**	860	566	**1050**	1922
−2.2	**28**	82.4	28.3	**83**	181.4	243	**470**	878	593	**1100**	2012
−1.7	**29**	84.2	28.9	**84**	183.2	249	**480**	896	621	**1150**	2102
−1.1	**30**	86.0	29.4	**85**	185.0	254	**490**	914	649	**1200**	2192
−0.6	**31**	87.8	30.0	**86**	186.8	260	**500**	932	677	**1250**	2282
0.0	**32**	89.6	30.6	**87**	188.6	266	**510**	950	704	**1300**	2372
0.6	**33**	91.4	31.1	**88**	190.4	271	**520**	968	732	**1350**	2462
1.1	**34**	93.2	31.7	**89**	192.2	277	**530**	986	760	**1400**	2552
1.7	**35**	95.0	32.2	**90**	194.0	282	**540**	1004	788	**1450**	2642
2.2	**36**	96.8	32.8	**91**	195.8	288	**550**	1022	816	**1500**	2732

The formulas at the right may also be used for converting Centigrade or Fahrenheit degrees into the other scale.

Degrees Cent., $C° = \frac{5}{9}(F° + 40) - 40$ Degrees Fahr., $F° = \frac{9}{5}(C° + 40) - 40$

Courtesy Natural Gas Processors Suppliers Association.

Viscosity—equivalents of kinematic viscosity

Kinematic Viscosity		Centistokes (v)	Stokes $\dfrac{Cm^2}{Sec}(100\,v)$	$\dfrac{Ft^2}{Sec}(v')$
Centistokes	(v)	1	0.01	1.076 (10^{-5})
Stokes $\dfrac{Cm^2}{Sec}$	($100\,v$)	100	1	1.076 (10^{-3})
$\dfrac{Ft^2}{Sec}$	(v')	92,900	929	1

To convert kinematic viscosity from one set of units to another, locate the given set of units in the left-hand column and multiply the numerical value by the factor shown horizontally to the right, under the set of units desired.

As an example, suppose a given kinematic viscosity of 0.5 sq ft/s is to be converted to centistokes. By referring to the table, we find the conversion factor to be 92,900. Then, 0.5 (sq ft/s) times 92,900 = 46,450 cSt.

Courtesy Crane Co. and American Society for Testing and Materials.

Viscosity—equivalents of kinematic and Saybolt Universal Viscosity

Kinematics Viscosity, Centistokes v	Equivalent Saybolt Universal Viscosity, s	
	At 100 °F Basic Values	At 210 °F
1.83	32.01	32.23
2.0	32.62	32.85
4.0	39.14	39.41
6.0	45.56	45.88
8.0	52.09	52.45
10.0	58.91	59.32
15.0	77.39	77.93
20.0	97.77	98.45
25.0	119.3	120.1
30.0	141.3	142.3
35.0	163.7	164.9
40.0	186.3	187.6
45.0	209.1	210.5
50.0	232.1	233.8
55.0	255.2	257.0
60.0	278.3	280.2
65.0	301.4	303.5
70.0	324.4	326.7
75.0	347.6	350.0
80.0	370.8	373.4
85.0	393.9	396.7
90.0	417.1	420.0
95.0	440.3	443.4
100.0	463.5	466.7
120.0	556.2	560.1
140.0	648.9	653.4
160.0	741.6	
180.0	834.2	

Kinematics Viscosity, Centistokes v	Equivalent Saybolt Universal Viscosity, s	
	At 100 °F Basic Values	At 210 °F
200.0	926.9	
220.0	1019.6	
240.0	1112.3	
260.0	1205.0	
280.0	1297.7	
300.0	1390.4	
320.0	1483.1	
340.0	1575.8	Saybolt seconds equal centistokes times 4.6347
360.0	1668.5	
380.0	1761.2	
400.0	1853.9	
420.0	1946.6	
440.0	2039.3	
460.0	2132.0	
480.0	2224.7	
500.0	2317.4	
Over 500	Saybolt seconds equal centistokes times 4.6347	

Note: To obtain the Saybolt Universal viscosity equivalent to a kinematic viscosity determined at t, multiply the equivalent Saybolt Universal viscosity at 100 °F by $1 + (t - 100)\,0.000064$. For example, $10v$ at 210 °F are equivalent to 58.91 multiplied by 1.0070 or 59.32 see Saybolt Universal at 210 °F.

Courtesy Crane Co. (Technical Paper #410).

Viscosity—equivalents of kinematic and Saybolt Furol Viscosity at 122 °F

Kinematics Viscosity, Centistokes ν	Equivalent Saybolt Universal Viscosity, s	
	At 122 °F Basic Values	At 210 °F
48	25.3	
50	26.1	25.2
60	30.6	29.8
70	35.1	34.4
80	39.6	39.0
90	44.1	43.7
100	48.6	48.3
125	60.1	60.1
150	71.7	71.8
175	83.8	83.7
200	95.0	95.6
225	106.7	107.5
250	118.4	119.4
275	130.1	131.4
300	141.8	143.5
325	153.6	155.5
350	165.3	167.6
375	177.0	179.7
400	188.8	191.8
425	200.6	204.0
450	212.4	216.1
475	224.1	228.3
500	235.9	240.5
525	247.7	252.8
550	259.5	265.0
575	271.3	277.2
600	283.1	289.5
625	294.9	301.8
650	306.7	314.1
675	318.4	326.4
700	330.2	338.7
725	342.0	351.0

Kinematics Viscosity, Centistokes ν	Equivalent Saybolt Universal Viscosity, s	
	At 122 °F Basic Values	At 210 °F
750	353.8	363.4
775	365.5	375.7
800	377.4	388.1
825	389.2	400.5
850	400.9	412.9
875	412.7	425.3
900	424.5	437.7
925	436.3	450.1
950	448.1	462.5
975	459.9	474.9
1000	471.7	487.4
1025	483.5	499.8
1050	495.2	512.3
1075	507.0	524.8
1100	518.8	537.2
1125	530.6	549.7
1150	542.4	562.2
1175	554.2	574.7
1200	566.0	587.2
1225	577.8	599.7
1250	589.5	612.2
1275	601.3	624.8
1300	613.1	637.3
Over 1300	*	†

*Over 1300 cSt at 122 °F:
Saybolt Fluid Sec = Centistokes × 0.4717.
†Over 1300 cSt at 210 °F:
Log (Saybolt Furol Sec − 2.87) = 1.0276 [Log (Centistokes)] − 0.3975.

Courtesy Crane Co. and American Society for Testing and Materials.

Viscosity—general conversions

Approximate Conversion Table

Seconds Saybolt Universal	Seconds Saybolt Furol	Engler Degrees	Engler Time	Barbey	Redwood Standard	Redwood Admiralty	Kinematic Viscosity Stokes
35		1.18	60	2800	32		0.026
50		1.60	82	880	44		0.074
75		2.30	102	460	65		0.141
100	15	3.00	153	320	88		0.202
150	19	4.40	230	205	128		0.318
200	23	5.90	305	148	170	18	0.431
250	28	7.60	375	118	212	23	0.543
300	33	8.90	450	98	254	27	0.651
400	42	11.80	550	72	338	36	0.876
500	52	14.50	750	59	423	45	1.10
600	61	17.50	900	48	518	53	1.32
700	71	20.6	1050	41	592	62	1.54
800	81	23	1200	36.5	677	71	1.76
900	91	27	1300	32.0	762	78	1.98
1000	100	29	1500	29.5	846	87	2.20
1500	150	42	2300	19.5	1270	135	3.30
2000	200	59	3000	14.5	1695	175	4.40
2500	250	73	3750	11.5	2120	230	5.50
3000	300	87	4500	9.6	2540	260	6.60
4000	400	117	6000	7.4	3380	350	8.80
5000	500	145	7500	6.0	4230	435	11.00
6000	600	175	9000	5.2	5080	530	13.20
7000	700	205	10,500	4.1	5925	610	15.40
8000	800	230	12,000	3.7	6770	700	17.60
9000	900	260	13,500	3.2	7620	780	19.8
10,000	1000	290	15,000	2.9	8460	870	22
20,000	2000	590	30,000	1.4	16,920	1760	44
40,000	4000	1170	60,000		33,850	3600	88
60,000	6000	1750	90,000		50,800	5300	132
80,000	8000	2300	120,000		67,700	7000	176
100,000	10,000	2900	150,000		84,600	8700	220
200,000	20,000	5900	300,000		169,200	17,600	440
400,000	40,000	11,700	600,000		338,500	36,000	880
600,000	60,000	17,500	900,000		508,000	53,000	1320
800,000	80,000	23,000	1,200,000		677,000	70,000	1760
1,000,000	100,000	29,000	1,500,000		846,000	87,000	2200

Absolute Viscosity is the force required to move a unit plane surface over another plane surface at a unit velocity when surfaces are separated by a layer of fluid of unit thickness.

Kinematic Viscosity is the absolute viscosity divided by the density of the fluid.

SAE Viscosity Numbers are arbitrary numbers that have been assigned to classify lubricating oils and do not have any functional relationship with other systems.

Viscosity Index is an arbitrary system relating the effect of temperature change to viscosity. It is the slope of a plot of viscosity versus temperature for a given fluid.

1 St = 1 P/Density in Grams

1 cSt = 1076×10^{-5} ft^2/s

1 P = 1 g/(cm)(s)

1 cP = 1/100 P

1 cSt = 1/100 St

1 cP = 0.000672 lbs/(ft)(s)

1 cP = 0.000672 (poundals)(s)/ft^2

1 cP = 0.0000209 (lbs)(s)/ft^2

1 cP = 0.0000209 slugs/(ft)(s)

1 cP = 2.42 lbs/(ft)(h)

1 slug/(ft)(s) = 1/0.002089 P

1 ft^2/s = 92,903 cSt

A.S.T.M. standard viscosity temperature chart

Pressure conversion chart

This line chart provides an easy method for converting units of pressure. Draw a line perpendicular to the scale lines through a known value of flow and read the equivalent value on any of the other scales. For values off the scale, known values can be read in multiples of 10 or 100.

278 ft of water
1,920 oz per in^2
17,400 lb/ft^2
3,300 in of water
242 in of mercury
6,180 mm of mercury

Example. Known value—120 psi is equivalent to:

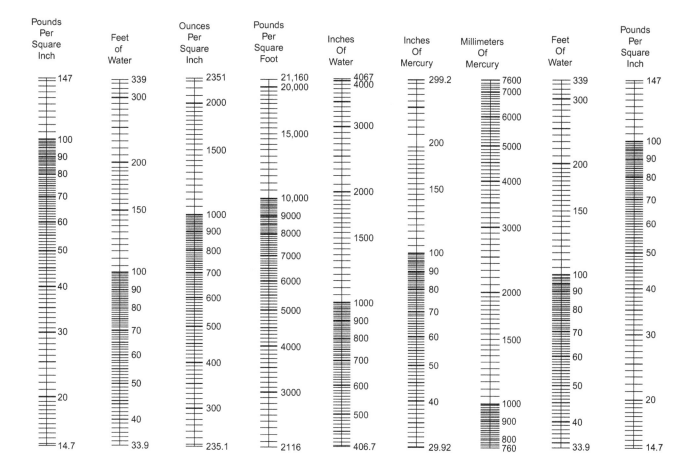

A simple method to determine square root

Problem. Determine the square root of the number 6280.

Solution.

1. *Guess the square root to an even number. For the example, use 80 since* $(80)^2$ *equals 6400.*

2. *Divide the number used into the number of which the square is desired and carry the quotient to only two figures. Thus:*

$$6{,}280 \div 80 = 78$$

3. *Average this quotient with the divisor.*

$80 + 78 = 158$, and $158 \div 2 = 79$

4. *Divide this figure into the number* and carry quotient to four or five figures, if desired.

$6,280 \div 79 = 79.5$

5. *Average this quotient with #3 for the square root.*

$79 + 79.5 = 158.5$, and $158.5 \div 2 = 79.25$

6. *To check answer*, multiply answer by itself.

$79.25 \times 79.25 = 6,280.5$

SI data[a]

Examples of SI-Derived Units

Quantity	SI Unit	
	Name	Symbol
Area	square meter	m^2
Volume	cubic meter	m^3
Speed, velocity	meter per second	m/s
Acceleration	meter per second squared	m/s^2
Density	kilogram per cubic meter	kg/m^3
Concentration (of amount of substance)	mole per cubic meter	mol/m^3
Specific volume	cubic meter per kilogram	m^3/kg
Luminance	candela per square meter	cd/m^2

SI Prefixes

Factor	Prefix	Symbol
10^{18}	exa	E
10^{15}	peta	P
10^{12}	tera	T
10^{9}	giga	G
10^{6}	mega	M
10^{3}	kilo	k
10^{2}	hecto	h
10^{1}	deka	da
10^{-1}	deci	d
10^{-2}	centi	c
10^{-3}	milli	m
10^{-6}	micro	μ
10^{-9}	nano	n
10^{-12}	pico	p
10^{-15}	femto	f
10^{-18}	atto	a

Examples of SI-Derived Units with Special Names

Quantity	SI Unit			Expression in Terms of Other Units	Expression in Terms of SI Unit
	Name	Symbol			
Frequency	hertz	Hz			s^{-1}
Force	newton	N			$m\,kg\,s^{-2}$
Pressure	Pascal	Pa		N/m^2	$m^{-1}\,kg\,s^{-2}$
Energy, work, quantity of heat	joule	J		Nm	$m^2\,kg\,s^{-3}$
Power, radiant flux	watt	W		J/s	$m^2\,kg\,s^{-3}$
Electric potential, potential differences, electromotive force	volt	V		W/A	$m^2\,kg\,s^{-3}\,A^{-1}$
Electric resistance	ohm	Ω		V/A	$m^2\,kg\,s^{-3}\,A^{-2}$
Conductance	siemens	S		AV	$m^{-2}\,kg\,s^{-1}\,A^2$

[a]Reprinted from API *Manual of Petroleum Measurement Standards*, Chapter 15, "Guidelines for the Use of the International System of Units (SI) in the Petroleum and Allied Industries," with permission of American Petroleum Institute.

Energy conversion chart

This nomograph provides an easy method for determining the fuel requirements for various types of fuel. Since the Btu content per unit volume of natural gas and crude oil vary over a limited but significant range of values, nomographs for the energy value of several gases and oils are included rather than using an average value.

Example. 30 cu ft of a gas with 1100 Btu per cu ft yields 33,000 Btu. Conversion of 33,000 Btu to the other units of energy is obtained by drawing a straight line connecting the values 33,000 on two end Btu scales. Using the right-hand nomograph, an oil with 11,000 Btu per gallon (about 4.6 million Btu per barrel) will produce 33,000 Btu from 0.3 gallon (about 0.0072 barrel).

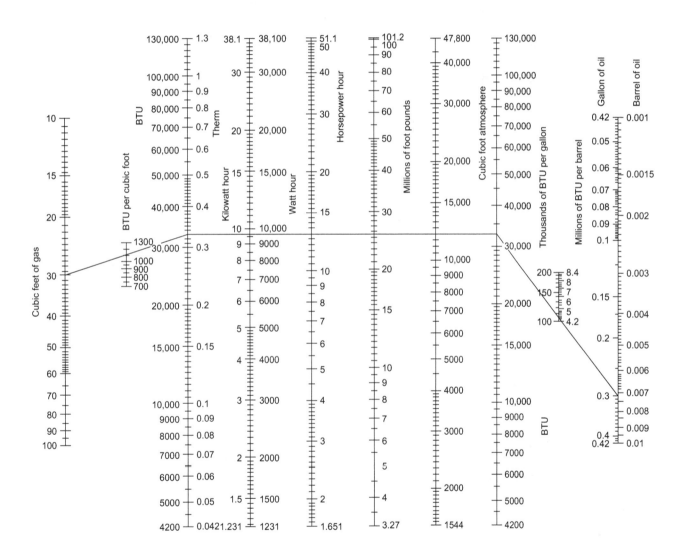

Flow conversion chart

This line chart provides an easy method for converting units of volume flow. Simply draw a line perpendicular to the scale lines through a known value of flow and read the equivalent value on any of the other scales.

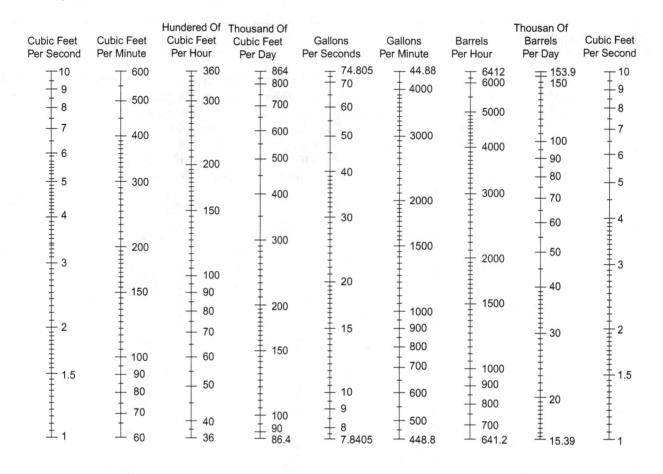

Conversions involving different types of fuel

Natural gas

The calorific value of natural gas varies according to its composition, and virtually no two fields produce gas of identical quality; the composition of gas from any one field can vary over its producing life. Thus, there is no such thing as a "standard" natural gas, nor a corresponding "standard" calorific value. However, when referring to natural gas in hypothetical cases it is quite common to indicate an "average" gross calorific value of 1000 Btu/ft^3 (the equivalent of which is taken, for purposes here, to be 9400 kcal/Nm^3 gross; see note on Table 3). Although in the United States most natural gas transmitted is at or slightly above this heat value, this does not generally apply in other countries. As an important example, Groningen (Netherlands) gas, because of a high nitrogen content, has a gross calorific value no greater than 898 Btu/ft^3 measured at 60 °F, 30 in Hg, dry (7980 kcal/mg^3_s or 8420 kcal/Nm^3).

When precise values are required, reference should be made to the proprietors of the gas field concerned or to published data.

Manufactured gases

These are derived from primary energy sources by processes involving chemical reaction, e.g. gases produced from coal and naphtha, also propane, butane, etc. Gas as piped to consumers can comprise both manufactured gases and natural gas used for enrichment. Calorific values of the component gases can thus vary widely depending on origin, but the resulting blend of what is known as town gas has a fairly constant heating value that, depending on locality, is usually about half that of natural gas. Exact figures can only come from the distribution company concerned.

Solid fuels

The calorific value of solid fuels varies widely between countries, grades, qualities, and even from seam to seam. It is particularly important, therefore, to use local data wherever possible. Measurements in Btu/lb or kcal/kg will normally include moisture and ash content as received.

N.B. Btu/lb × 0.5556 = kcal/kg

kcal/kg × 1.8 = Btu/lb

Where actual values are not known, an average gross calorific value of 12,400 Btu/lb (6890 kcal/kg) is considered reasonable for hard coal grades of good quality. This also serves as the basis of a common unit, "coal equivalent," to which the many grades of solid fuel and other forms of energy may be reduced for the purposes of aggregation. Standard fuel for the purposes of European Economic Community statistics is considered to have a heat value of 7000 kcal/kg.

Table 1
Petroleum fuels volume/weight/calorific value factors

Hydrocarbons	Specific Gravity	Weight/ Volume bbl/ton	Gross Calorific Value Btu/ Imp.gal	Btu/ lb	kcal/ kg	10^6 Btu/ bbl	10^6 kcal/ bbl	b.o.e./ bbl
Propane	0.501	12.790	...	21,500	11,943	3.795	0.956	0.654
Butane	0.579	11.060	...	21,140	11,742	4.315	1.088	0.744
Motor gasoline	0.733	8.740	148,000	20,160	11,200	5.176	1.304	0.892
Kerosene	0.780	8.213	156,000	20,000	11,110	5.456	1.377	0.941
Vaporizing oil	0.823	7.783	163,000	19,800	11,000	5.701	1.442	0.983
Gas oil	0.840	7.625	164,000	19,500	10,835	5.736	1.445	0.988
Fuel oils	0.850	7.536	165,660	19,490	10,830	5.794	1.460	0.999
	0.860	7.448	166,840	19,400	10,780	5.835	1.470	1.006
	0.870	7.362	168,080	19,320	10,730	5.878	1.481	1.013
	0.880	7.278	169,310	19,240	10,690	5.921	1.492	1.021
	0.890	7.196	170,440	19,150	10,640	5.961	1.502	1.028
	0.900	7.116	171,640	19,070	10,590	6.002	1.513	1.035
	0.910	7.038	172,730	18,980	10,540	6.041	1.522	1.042
	0.920	6.961	173,800	18,890	10,490	6.078	1.532	1.048
	0.930	6.887	174,860	18,800	10,440	6.115	1.541	1.054
	0.940	6.813	175,890	18,710	10,390	6.151	1.550	1.061
	0.950	6.741	176,910	18,620	10,340	6.187	1.559	1.067
	0.960	6.671	177,910	18,530	10,290	6.222	1.568	1.073
	0.970	6.602	178,890	18,440	10,240	6.256	1.577	1.079
	0.980	6.535	179,860	18,350	10,190	6.290	1.585	1.084
	0.990	6.469	180,800	18,260	10,140	6.323	1.593	1.090
	1.000	6.404	181,730	18,170	10,090	6.356	1.602	1.096

Sources:
W.P.C. Publication "Technical Data on Fuel" (Editor H. M. Spiers) 6th Edition 1961; all products except fuel oil. Latter from H. Hyams, "Petroleum Cargoes, Measurement and Sampling, 1966."

Notes:
1. Btu/gal × 34.9726 = Btu/bbl.
2. Btu/bbl × 0.252 = kcal/bbl.
3. b.o.e./bbl = $\dfrac{10^6 \text{ Btu/bbl}}{5.8}$.
4. kcal/kg × 1.8 = Btu/lb.
5. To convert 10^6 Btu/bbl to ft^3 natural gas (1000 Btu), multiply by 10^3.
6. To convert 10^6 kcal/bbl to m^3 natural gas (9400 kcal), multiply by 106.4.

Conversion factors for calorific values of gases under different conditions of measurement

Table 2

B A	Btu/ft³ @60°F 30 inHg dry	Btu/ft³ @60°F 30 inHg saturated	Btu/ft³ @60°F 760 mmHg (60°F) dry	kcal/m³ @0°C 760 mmHg dry	kcal/m³ @ 0°C 760 mmHg saturated	kcal/ m³ @ 15°C 760 mmHg dry	kcal/m³ @ 15°C 760 mmHg saturated
Btu/ft³ @ 60°F 30 inHg dry	1	0.9826	0.9946	9.381	9.325	8.893	8.743
Btu/ft³ @ 60°F 30 inHg saturated	1.0177	1	1.0122	9.547	9.490	9.050	8.898
Btu/ft³ @ 60°F 760 mmHg (60°F) dry	1.0054	0.9879	1	9.431	9.375	8.940	8.790
kcal/m³ @ 0°C 760 mmHg dry	0.10659	0.10474	0.10602	1	0.9940	0.9480	0.9320
kcal/m³ @ 0°C 760 mmHg saturated	0.10724	0.10537	0.10666	1.0060	1	0.9537	0.9376
kcal/m³ @ 15°C 760 mmHg dry	0.11245	0.11049	0.11184	1.0548	1.0486	1	0.9832
kcal/m³ @ 15°C 760 mmHg saturated	0.11437	0.11238	0.11375	1.0729	1.0665	1.0171	1

Notes

1. For equal conditions of temperature, pressure and water vapor content, the following factors apply:

   1 Btu/ft³ = 8.8992 kcal/m³

   1 kcal/m³ = 0.11,237 Btu/ft³

2. To convert one unit of A to its equivalent heat value in terms of alternative units/conditions B, multiply by the appropriate factor.

Electricity

The basic unit of electric power is the *watt* (W), with the *kilowatt-hour* (kWh) as the derivative of most importance.

$$1 \text{ kilowatt-hour (kWh)} = 3,412.14 \text{ Btu}$$
$$= 859.845 \text{ kcal}$$
$$= 3,600 \text{ kJ}$$

See note to Table 4.

Liquid fuels

In the case of liquid fuels, the actual calorific value of the particular fuel under consideration should be used where possible for conversions to other units, for example when calculating the natural gas equivalent of some specific application. For more general conversions, refer to Table 1.

Standard reference fuels

To convert different types and grades of fuel to a common basis for the purposes of presentation or comparison, it is the practice to express them in terms of some standard unit of a defined calorific value, for example:

Barrel of Oil Equivalent (b.o.e.), with a heat content of 5.8×10^6 Btu gross.

Tons of Oil Equivalent (t.o.e.), simply a hypothetical "ton of oil" with an average heating value of 43×10^6 Btu gross.

Oil equivalents may, if required, also be expressed in terms of standard fuel oil equivalent with a calorific value of 41.4×10^6 Btu per ton gross, i.e. 18,500 Btu/lb × 2240.

Metric Ton Coal Equivalent (t.c.e.), with a heat content of 27.337×10^6 Btu gross, i.e. 12,400 Btu/lb × 2204.62.

The heat values of these different units and conversions between them are given in Table 3, together with natural gas equivalents.

Heat value conversions and natural gas equivalents of various fuel units

Table 3

	10^6 Btu gross	10^6 kcal gross	bbl oil equiv.	tons oil equiv.	metric t.c.e.	ft³ natural gas	m³ natural gas	10^6 kWh
1 barrel oil equivalent	5.8	1.462	1	0.1349	0.21217	5800	155.50	0.0017
1 metric ton coal equivalent	27.337	6.888	4.713	0.6357	1	27,337	732.90	0.0080
1 ton oil equivalent	43	10.836	7.414	1	1.5730	43,000	1152.82	0.0126
1 ton fuel oil equivalent	41.4	10.433	7.138	0.9628	1.5144	41,400	1109.92	0.0121

Notes
The natural gas equivalents above and in Tables 1 (notes) and 4 are based on gas with a gross heat content of 1000 Btu/ft³ measured dry at 60 °F and 30 in Hg, which within 0.2% is equal to 9400 kcal/m³ gross measured dry at 0 °C and 760 mmHg. On this basis, 37.3 cu ft = 1 cu m.

The abbreviation m.t.c.e. is often used for metric tons of coal equivalent or million tons coal equivalent. Neither practice is sound, as small m is the symbol for meter and not for metric or million. Instead, the ton unit used should be stated in full and the quantity expressed by the use of exponents; e.g. 36×10^6 metric t.o.e.

Conversions above and in Table 4 are based on gross calorific values in line with normal statistical practice both inside and outside the gas industry. Net calorific values may, however, be more appropriate in situations where it is desired to compare or convert fuels in terms of useful heat content, due allowance having been made for changes in end-use efficiency.

Conversion for daily/annual rates of energy consumption (gross heat basis)

Table 4

	1×10^6 metric t.c.e. per year	1×10^6 ft³ nat. gas per day	1×10^9 m³ nat. gas per year	bbl daily oil equiv.	1×10^{12} (US trillion) ft³ nat. gas per year	1×10^6 kWh per year	1×10^6 thermies per year
1×10^6 metric t.c.e. per year =	1	74.896	0.7329	12,913	0.027337	8012	6891
100×10^6 ft³ nat. gas per day =	1.3352	100	0.9785	17,241	0.0365	10,697	9201
1×10^9 m³ nat. gas per year =	1.3645	102.2	1	17,620	0.0373	10,932	9403
1×10^4 bbl daily oil equiv. =	0.7744	58	0.5675	1×10^4	0.0212	6204	5336
1×10^{12} (US trillion) ft³ nat. gas per year =	36,580	2,740	26,808	472,367	1	293,071	252,074
1×10^6 kWh per year =	0.1248×10^{-3}	0.9348×10^{-2}	0.9147×10^4	1.6118	0.341×10^{-5}	1	0.8601
1×10^6 thermies per year =	0.1451×10^{-3}	0.1087×10^{-1}	0.1064×10^{-3}	1.8739	0.397×10^{-5}	1.1626	1

Notes

The table is good for natural gas conversions on the basis of a gas of 1000 Btu/ft³ gross (9400 kcal/Nm³). See note to Table 3 for conditions of measurement. Corrections may be applied as follows.

Gases with other calorific values: apply a correction based on the ratio of the calorific values; e.g. to use for Groningen gas, multiply the horizontal figures shown against 1×10^9 m³ per year by 8420 ÷ 9400, i.e. 0.896 and the corresponding column by 1.116.

Gases measured at other conditions of temperature and pressure: apply a correction based on the appropriate factor shown in Table 2; e.g. for gas measured saturated at 60 °F and 30₆ inHg, multiply the horizontal figures shown against 100×10^6 ft³ per day by 0.9826 and the corresponding column by 1.0177.

Calorific value and conditions of measurement different: apply a combined correction; e.g. to use for Gronigen gas measured dry at 15 °C and 760 mmHg (m³ₛ), multiply the horizontal figures shown against 1×10^9 m³ by 0.948 × 8420 = 9400, i.e. 0.849 and the corresponding column by the reciprocal of this number, i.e. 1.178.

To use the table for 1×10^6 *therms per year*, multiply US trillion row by 10^{-4} and the corresponding column by 10^4.

To use the table for 1×10^6 *tons oil equivalent*, multiply the horizontal figures shown against 1×10^6 metric t.c.e. per year by 1.573 and the corresponding column by 0.6357.

The factors relating to kilowatt-hours are based on the heat content of the electricity when used and not to the corresponding power station intakes of coal, oil, or natural gas, which would need to reflect station-generating efficiencies and load factor.

Weight of water per cubic foot at various temperatures

Temperature °F	Weight per Cubic Foot, lbs	Temperature °F	Weight per Cubic Foot, lbs	Temperature °F	Weight per Cubic Foot, lbs	Temperature °F	Weight per Cubic Foot, lbs	Temperature °F	Weight per Cubic Foot, lbs
32	62.42	150	61.18	260	58.55	380	54.36	500	48.7
40	62.42	160	60.98	270	58.26	390	53.94	510	48.1
50	62.41	170	60.77	280	57.96	400	53.5	520	47.6
60	62.37	180	60.55	290	57.65	410	53.0	530	47.0
70	62.31	190	60.32	300	57.33	420	52.6	540	46.3
80	62.23	200	60.12	310	57.00	430	52.2	550	45.6
90	62.13	210	59.88	320	56.66	440	51.7	560	44.9
100	62.02	212	59.83	330	56.30	450	51.2	570	44.1
110	61.89	220	59.63	340	55.94	460	50.7	580	43.3
120	61.74	230	59.37	350	55.57	470	50.2	590	42.6
130	61.56	240	59.11	360	55.18	480	49.7	600	41.8
140	61.37	250	58.83	370	54.78	490	49.2	...	...

From "National Pipe Standards," The National Tube Co.

Engineering constants

Atmosphere:

Theoret. air to burn 1 lb carbon = 11.521 bs
Vol. of 1 lb air at 32 °F and 14.7 lbs per sq in = 12.387 cu ft
Vol. of 1 lb air at 62 °F and 14.7 lb per sq in = 13.144 cu ft
14.223 lb per sq in = 1 kg per sq cm. = 1 metric atm.
28.8 = equivalent mol. wgt of air
30 in Hg at 62° = atm. press
0.07608 lb = wgt 1 cu ft air at 62 °F and 14.7 lb per sq in
0.08073 lb = wgt 1 cu ft air at 32 °F and 14.7 lb per sq in

Heat:

8 lb oxygen required to burn 1 lb hydrogen (H)
777.5 ft lb = 1 Btu
62,000 Btu = cal. val. hydrogen (H)
14,600 Btu per lb cal. val. of carbon (C)
1.8 °F = 1 °C
2547 Btu per hour = 1 h.p.
−270 °C = absolute zero
3145 Btu = 1 kWh
34.56 lb = wt air to burn 1 lb hydrogen (H)
3.9683 Btu = 1 kg calorie
4050 Btu = cal. val. of sulfur (S)
4.32 lb = wgt air to burn 1 lb sulfur (S)

Water:

Freezing point = 0 °C, 32 °F
Boiling point, atm. press = 100 °C, 212 °F
Water power available from 1 cu ft-s falling 1 ft = 0.1134 h.p.
1.134 ft at 62 °F = 1 in Hg at 62 °F
2.309 ft at 62 °F = 1 lb per sq in
27.71 in at 62 °F = 1 lb per sq in
33.947 ft at 62 °F = atm. press.
5.196 lb per sq ft = 1 in at 62 °F
59.76 lb = wgt 1 cu ft at 212 °F

62.355 lb = wgt 1 cu ft at 62 °F
1 in at 62 °F = 0.0735 in Hg at 62 °F
8.3356 lb = wgt 1 gal at 62 °F
970.4 Btu = latent heat of evap. at 212 °F

Steam:

Total heat sat. at atm. press = 1150.4 Btu 0.47 Btu
per lb °F = approx. specific heat of superheated steam at atm. press.

Surfaces:

1,273,239 circular mils = 1 sq in
1.4142 = square root of 2
144 sq in = 1 sq ft
1728 cu in = 1 cu ft
1.7321 = square root of 3
2.54 cm = 1 in
57.296° = 1 rad (angle)
0.0624281 b per cu ft = 1 kg per cu meter
Area of circle = 0.7854 (= 3.1416 − 4) × diameter squared

Miscellaneous:

2116.3 lb per sq ft = atm. press
2.3026 × log 10a log ea
2.7183 = c = base hyperbolic logs
288,000 Btu per 24 h = 1 ton of refrigeration
3.1416 = π = ratio of circumference of circle to diameter = ratio area of circle to square of radius
550 ft-lb per sec = 1 h.p.
58.349 grains per gal = 1 g per liter
61.023 cu in = 1 l
745.7 W = 1 h.p.
8760 h = 1 year of 365 days
88 ft per sec (min) = 1 mile per min (h)
2.666 lb = wgt oxygen required to burn 1 lb carbon

Mensuration units

Diameter of a circle × 3.1416 = Circumference
Radius of a circle × 6.283185 = Circumference

Square of the radius of a circle × 3.1416 = Area
Square of the diameter of a circle × 0.7854 = Area
Square of the circumference of a circle × 0.07985 = Area
Half the circumference of a circle × half its diameter = Area

Circumference of a circle × 0.159155 = Radius
Square root of the area of a circle × 0.56419 = Radius

Circumference of a circle × 0.31831 = Diameter
Square root of the area of a circle × 1.12838 = Diameter

Diameter of a circle × 0.86 = Side of an inscribed equilateral triangle
Diameter of a circle × 0.7071 = Side of an inscribed square
Circumference of a circle × 0.225 = Side of an inscribed square
Circumference of a circle × 0.282 = Side of an equal square
Diameter of a circle × 0.8862 = Side of an equal square

Base of a triangle × ½ the altitude = Area
Multiplying both diameter and 0.7854 together = Area of an ellipse
Surface of sphere × 1/6 of its diameter = Solidity

Circumference of a sphere × its diameter = Surface
Square of the diameter of a sphere × 3.1416 = Surface
Square of the circumference of a sphere × 0.3183 = Surface

Cube of the diameter of a sphere × 0.5236 = Solidity
Cube of the radius of a sphere × 4.1888 = Solidity
Cube of the circumference of a sphere × 0.016887 = Solidity

Square root of the surface of a sphere × 0.56419 = Diameter
Square root of the surface of a sphere × 1.772454 = Circumference

Cube root of the solidity of a sphere × 1.2407 = Diameter
Cube root of the solidity of a sphere × 3.8978 = Circumference

Radius of a sphere × 1.1547 = Side of inscribed cube
Square root of ($\frac{1}{3}$ of the square of) the diameter of a sphere = Side of inscribed cube

Area of its base × $\frac{1}{3}$ of its altitude = Solidity of a cone or pyramid, whether round, square, or triangular

Area of one of its sides × 6 = Surface of cube
Altitude of trapezoid × ½ the sum of its parallel sides = Area

Minutes to decimal hours conversion table

Minutes	Hours	Minutes	Hours
1	0.017	31	0.517
2	0.034	32	0.534
3	0.050	33	0.550
4	0.067	34	0.567
5	0.084	35	0.584
6	0.100	36	0.600
7	0.117	37	0.617
8	0.135	38	0.634
9	0.150	39	0.650
10	0.167	40	0.667
11	0.184	41	0.684
12	0.200	42	0.700
13	0.217	43	0.717
14	0.232	44	0.734
15	0.250	45	0.750
16	0.267	46	0.767
17	0.284	47	0.784
18	0.300	48	0.800
19	0.317	49	0.817
20	0.334	50	0.834
21	0.350	51	0.850
22	0.368	52	0.867
23	0.384	53	0.884
24	0.400	54	0.900
25	0.417	55	0.917
26	0.434	56	0.934
27	0.450	57	0.950
28	0.467	58	0.967
29	0.484	59	0.984
30	0.500	60	1.000

How to compare costs of gas and alternate fuels

Growing shortage of natural gas prompts consideration of practical replacements, including coal, heavy fuel oil, middle distillates, and LPG

Conversion from natural gas to other fuels has progressed from possibility to reality for many companies and will become necessary for many others in months and years ahead.

Fuels considered practical replacements for gas include coal, heavy fuel oils, middle distillates (such as kerosene-type turbo fuel and burner fuel oils) and liquefied petroleum gas (LPG).

To compare relative costs of fuels, the quoted prices normally are converted to the price per million British thermal units (Btu). With HHV being the higher (gross) heating value of the fuel, the following equations may be used.

Coal:

$$\text{Cents/million Btu} = \frac{\text{dollars/ton} \times 100 \times 10^6}{\text{lb/ton} \times \text{HHV}}$$

Heavy fuel oils:

$$\text{Cents/million Btu} = \frac{\text{dollars/barrel} \times 100 \times 10^6}{\text{gallons/barrel} \times \text{HHV}}$$

Middle distillates and LPG:

$$\text{Cents/million Btu} = \frac{\text{cents/gallon} \times 10^6}{\text{HHV}}$$

Natural gas:

$$\text{Cents/million Btu} = \frac{\text{cents/Mcf} \times 106}{\text{HHV} \times 1,000}$$

The approximate heat contents (HHV) of the fuels are as follows:

Coal:

Anthracite	13,900 Btu/lb
Bituminous	14,000 Btu/lb
Sub-bituminous	12,600 Btu/lb
Lignite	11,000 Btu/lb

Heavy fuel oils and middle distillates:

Kerosene	134,000 Btu/gal
No. 2 burner fuel oil	140,000 Btu/gal
No. 4 heavy fuel oil	144,000 Btu/gal
No. 5 heavy fuel oil	150,000 Btu/gal
No. 6 heavy fuel oil, 2.7% sulfur	152,000 Btu/gal
No. 6 heavy fuel oil, 0.3% sulfur	143,800 Btu/gal

Gas:

Natural	1000 Btu/cu ft
Liquefied butane	103,300 Btu/gal
Liquefied propane	91,600 Btu/gal

Typical characteristics of fuel oils

	Kerosene	No. 2 Burner Fuel
Gravity, °API	41.4	36.3
Pounds per gallon	6.814	7.022
Flash point, °F	133	165
Btu/gal	134,000	140,000
Viscosity, SSU at 100 °F	31.5	35.0
Pour point, °F	−30	
Sulfur, wt%	0.04	0.18
Carbon residue, wt%	0.01	0.05
BS&W, %	—	1.30[1]

[1] mg/100 ml.

LPG

Vapor pressure, psig at 100 °F	188
Distillation, 95% point, °F	−40
Residue, vol. %	0.05
Oil-ring test	none

Index

CPSIA information can be obtained
at www.ICGtesting.com
Printed in the USA
FFHW010329270319
51242054-56745FF

9 780123 876935